T5-BCI-288

Lecture Notes in Biomathematics

Managing Editor: S. Levin

68

The Dynamics of Physiologically Structured Populations

Edited by J. A. J. Metz and O. Diekmann

Springer-Verlag

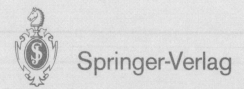

Lecture Notes in Biomathematics

Vol. 1: P. Waltman, Deterministic Threshold Models in the Theory of Epidemics. V, 101 pages. 1974.

Vol. 2: Mathematical Problems in Biology, Victoria Conference 1973. Edited by P. van den Driessche. VI, 280 pages. 1974.

Vol. 3: D. Ludwig, Stochastic Population Theories. VI, 108 pages. 1974.

Vol. 4: Physics and Mathematics of the Nervous System. Edited by M. Conrad, W. Güttinger, and M. Dal Cin. XI, 584 pages. 1974.

Vol. 5: Mathematical Analysis of Decision Problems in Ecology. Proceedings 1973. Edited by A. Charnes and W.R. Lynn. VIII, 421 pages. 1975.

Vol. 6: H.T. Banks, Modeling and Control in the Biomedical Sciences. V. 114 pages. 1975.

Vol. 7: M.C. Mackey, Ion Transport through Biological Membranes, An Integrated Theoretical Approach. IX, 240 pages. 1975.

Vol. 8: C. DeLisi, Antigen Antibody Interactions. IV, 142 pages. 1976.

Vol. 9: N. Dubin, A Stochastic Model for Immunological Feedback in Carcinogenesis: Analysis and Approximations. XIII, 163 pages. 1976.

Vol. 10: J.J. Tyson, The Belousov-Zhabotinskii Reaktion. IX, 128 pages. 1976.

Vol. 11: Mathematical Models in Medicine. Workshop 1976. Edited by J. Berger, W. Bühler, R. Repges, and P. Tautu. XII, 281 pages. 1976.

Vol. 12: A.V. Holden, Models of the Stochastic Activity of Neurones. VII, 368 pages. 1976.

Vol. 13: Mathematical Models in Biological Discovery. Edited by D.L. Solomon and C. Walter. VI, 240 pages. 1977.

Vol. 14: L.M. Ricciardi, Diffusion Processes and Related Topics in Biology. VI, 200 pages. 1977.

Vol. 15: Th. Nagylaki, Selection in One- and Two-Locus Systems. VIII, 208 pages. 1977.

Vol. 16: G. Sampath, S.K. Srinivasan, Stochastic Models for Spike Trains of Single Neurons. VIII, 188 pages. 1977.

Vol. 17: T. Maruyama, Stochastic Problems in Population Genetics. VIII, 245 pages. 1977.

Vol. 18: Mathematics and the Life Sciences. Proceedings 1975. Edited by D.E. Matthews. VII, 385 pages. 1977.

Vol. 19: Measuring Selection in Natural Populations. Edited by F.B. Christiansen and T.M. Fenchel. XXXI, 564 pages. 1977.

Vol. 20: J.M. Cushing, Integrodifferential Equations and Delay Models in Population Dynamics. VI, 196 pages. 1977.

Vol. 21: Theoretical Approaches to Complex Systems. Proceedings 1977. Edited by R. Heim and G. Palm. VI, 244 pages. 1978.

Vol. 22: F.M. Scudo and J.R. Ziegler, The Golden Age of Theoretical Ecology: 1923–1940. XII, 490 pages. 1978.

Vol. 23: Geometrical Probability and Biological Structures: Buffon's 200th Anniversary. Proceedings 1977. Edited by R.E. Miles and J. Serra. XII, 338 pages. 1978.

Vol. 24: F.L. Bookstein, The Measurement of Biological Shape and Shape Change. VIII, 191 pages. 1978.

Vol. 25: P. Yodzis, Competition for Space and the Structure of Ecological Communities. VI, 191 pages. 1978.

Vol. 26: M.B Katz, Questions of Uniqueness and Resolution in Reconstruction from Projections. IX, 175 pages. 1978.

Vol. 27: N. MacDonald, Time Lags in Biological Models. VII, 112 pages. 1978.

Vol. 28: P.C. Fife, Mathematical Aspects of Reacting and Diffusing Systems. IV, 185 pages. 1979.

Vol. 29: Kinetic Logic – A Boolean Approach to the Analysis of Complex Regulatory Systems. Proceedings, 1977. Edited by R. Thomas. XIII, 507 pages. 1979.

Vol. 30: M. Eisen, Mathematical Models in Cell Biology and Cancer Chemotherapy. IX, 431 pages. 1979.

Vol. 31: E. Akin, The Geometry of Population Genetics. IV, 205 pages. 1979.

Vol. 32: Systems Theory in Immunology. Proceedings, 1978. Edited by G. Bruni et al. XI, 273 pages. 1979.

Vol. 33: Mathematical Modelling in Biology and Ecology. Proceedings, 1979. Edited by W.M. Getz. VIII, 355 pages. 1980.

Vol. 34: R. Collins, T.J. van der Werff, Mathematical Models of the Dynamics of the Human Eye. VII, 99 pages. 1980.

Vol. 35: U. an der Heiden, Analysis of Neural Networks. X, 159 pages. 1980.

Vol. 36: A. Wörz-Busekros, Algebras in Genetics. VI, 237 pages. 1980.

Vol. 37: T. Ohta, Evolution and Variation of Multigene Families. VIII, 131 pages. 1980.

Vol. 38: Biological Growth and Spread: Mathematical Theories and Applications. Proceedings, 1979. Edited by W. Jäger, H. Rost and P. Tautu. XI, 511 pages. 1980.

Vol. 39: Vito Volterra Symposium on Mathematical Models in Biology. Proceedings, 1979. Edited by C. Barigozzi. VI, 417 pages. 1980.

Vol. 40: Renewable Resource Management. Proceedings, 1980. Edited by T. Vincent and J. Skowronski. XII, 236 pages. 1981.

Vol. 41: Modèles Mathématiques en Biologie. Proceedings, 1978. Edited by C. Chevalet and A. Micali. XIV, 219 pages. 1981.

Vol. 42: G.W. Swan, Optimization of Human Cancer Radiotherapy. VIII, 282 pages. 1981.

Vol. 43: Mathematical Modeling on the Hearing Process. Proceedings, 1980. Edited by M.H. Holmes and L.A. Rubenfeld. V, 104 pages. 1981.

Vol. 44: Recognition of Pattern and Form. Proceedings, 1979. Edited by D.G. Albrecht. III, 226 pages. 1982.

Vol. 45: Competition and Cooperation in Neutral Nets. Proceedings, 1982. Edited by S. Amari and M.A. Arbib. XIV, 441 pages. 1982.

Vol. 46: E. Walter, Identifiability of State Space Models with applications to transformation systems. VIII, 202 pages. 1982.

Vol. 47: E. Frehland, Stochastic Transport Processes in Discrete Biological Systems. VIII, 169 pages. 1982.

Vol. 48: Tracer Kinetics and Physiologic Modeling. Proceedings, 1983. Edited by R.M. Lambrecht and A. Rescigno. VIII, 509 pages. 1983.

Vol. 49: Rhythms in Biology and Other Fields of Application. Proceedings, 1981. Edited by M. Cosnard, J. Demongeot and A. Le Breton. VII, 400 pages. 1983.

Vol. 50: D.H. Anderson, Compartmental Modeling and Tracer Kinetics. VII, 302 pages. 1983.

Vol. 51: Oscillations in Mathematical Biology. Proceedings, 1982. Edited by J.P.E. Hodgson. VI, 196 pages. 1983.

Vol. 52: Population Biology. Proceedings, 1982. Edited by H.I. Freedman and C. Strobeck. XVII, 440 pages. 1983.

Vol. 53: Evolutionary Dynamics of Genetic Diversity. Proceedings, 1983. Edited by G.S. Mani. VII, 312 pages. 1984.

Vol. 54: Mathematical Ecology. Proceedings, 1982. Edited by S.A. Levin and T.G. Hallam. XII, 513 pages. 1984.

Vol. 55: Modelling of Patterns in Space and Time. Proceedings, 1983. Edited by W. Jäger and J.D. Murray. VIII, 405 pages. 1984.

Vol. 56: H.W. Hethcote, J.A. Yorke, Gonorrhea Transmission Dynamics and Control. IX, 105 pages. 1984.

Vol. 57: Mathematics in Biology and Medicine. Proceedings, 1983. Edited by V. Capasso, E. Grosso and S.L. Paveri-Fontana. XVIII, 524 pages. 1985.

ctd. on inside back cover

Lecture Notes in Biomathematics

Managing Editor: S. Levin

68

T.M

The Dynamics of Physiologically Structured Populations

Edited by J. A. J. Metz and O. Diekmann

Springer-Verlag
Berlin Heidelberg New York London Paris Tokyo

WILLIAM MADISON RANDALL LIBRARY UNC AT WILMINGTON

Editorial Board

M. Arbib H.J. Bremermann J.D. Cowan W. Hirsch S. Karlin
J.B. Keller M. Kimura S. Levin (Managing Editor) R.C. Lewontin R. May J. Murray
G.F. Oster A.S. Perelson T. Poggio L.A. Segel

Editors

Johan A.J. Metz
Odo Diekmann
Institute of Theoretical Biology
Groenhovenstraat 5, 2311 BT Leiden, The Netherlands
and
Centre for Mathematics and Computer Science
Kruislaan 413, 1098 SJ Amsterdam, The Netherlands

Mathematics Subject Classification (1980): 92A15, 92A17; 45K05, 47D05

ISBN 3-540-16786-2 Springer-Verlag Berlin Heidelberg New York
ISBN 0-387-16786-2 Springer-Verlag New York Berlin Heidelberg

This work is subject to copyright. All rights are reserved, whether the whole or part of the material is concerned, specifically those of translation, reprinting, re-use of illustrations, broadcasting, reproduction by photocopying machine or similar means, and storage in data banks. Under § 54 of the German Copyright Law where copies are made for other than private use, a fee is payable to "Verwertungsgesellschaft Wort", Munich.

© Springer-Verlag Berlin Heidelberg 1986
Printed in Germany

Printing and binding: Beltz Offsetdruck, Hemsbach/Bergstr.
2146/3140-543210

QH352
.D964
1986

Preface

Physiological processes within individuals and behaviour patterns displayed by individuals rank high among the subject matters of biology. The catch, handling and digestion of a prey by a predator, the uptake of nutrients by and the growth and fission of a unicellular organism, and the succession of larval stages of an insect, all are biological phenomena attracting great interest.

When trying to describe and predict the development of real world or laboratory populations by means of mathematical models one usually concentrates on numbers of individuals only, while ignoring those aspects of the life-cycle which would necessitate a further classification of individuals.

Structured population models bridge the gap between the individual and the population level. They allow us to derive information about the dynamics of populations from information about the dynamical properties of individuals or vice versa.

These lecture notes arose out of a colloquium on the Dynamics of Structured Populations, held at the Centre for Mathematics and Computer Science in Amsterdam during the first half of 1983. The colloquium consisted of seven one-day meetings; the number of participants, mainly biologists, ranged from 57 to 21 with an average of 34. At the start of the colloquium the following ideals and aims were formulated:

Wishful thinking

1. To use knowledge about the behaviour and the physiology of individual animals or plants to describe, understand and predict the dynamics of real world and laboratory populations.
2. To build a qualitative theory of (possibly nonlinear) first order partial differential equations (balance laws) in which *non-local* terms or side conditions occur.

Practical objectives

1. To take stock of existing knowledge, literature, models, analytical techniques,....
2. To collect examples of biological mechanisms which force us to distinguish individuals from each other on the basis of certain traits.
3. To train the participants in formulating structured models as functional-partial differential equations. (One has to get accustomed to the mathematical formalism before it becomes a convenient language for expressing ones thoughts.)
4. To explore the connections between structured models and models formulated as ordinary differential equations in $\mathbb{R}^k$, through limit arguments, the taking of appropriate projections, etc.
5. To practice using both biological relevance and mathematical tractability as criteria for choosing problems and making simplifications.
6. To train the participants in interpreting mathematical results in biological terms.
7. To learn analytical (and also numerical) techniques which are useful when studying functional partial differential equations.
8. To elaborate simple prototype problems in detail in order to enhance the intuitive understanding of the relations between the dynamics of individuals and the dynamics of populations.
9. To show how qualitative information obtained from simplified models complements quantitative information obtained from simulation studies.
10. To let the participants become conversant with each other and to provide them with a common basis, and by doing so, to stimulate cooperation.

The present notes have grown out of the seeds sown at that colloquium. The text has grown and ripened during a period of intense interdisciplinary interaction of the core participants after the colloquium itself had finished. Part A consists of a systematic exposition of the present state of the mathematical theory. Part B is a collection of papers on special topics. The connections with the main theory are manifest from the many forward references in part A. Each of the topical sections of part B carries a short preface placing the contributions in perspective. The remaining part of this preface refers mainly to part A.

The first step in building structured population models consists of finding a suitable explicit parametrization of the state of the individuals (satiation, size, age, ...). The state of the population is then given by a density function describing their distribution in the individual state space. In the course of time the state of each individual changes smoothly (owing to digestion, growth, aging, ...). Moreover individuals are born, die or may make jumps in the individual state space (for instance, the satiation of a predator will change abruptly upon the swallowing of a prey). In the second step one draws up the balance of these processes. At this point two roads depart.

The first road leads to the computer. The biological description of the various smooth and jump transitions is translated directly into a computer program and subsequently biological reality is *simulated* by calculating the result of all transitions during a certain period of time. This approach is touched upon in the contributions by Sabelis and coworkers and by Goudriaan in part B, but it is certainly not the main theme of these notes. Whenever the individual state space is high-dimensional and/or the constants and functions specifying the model can be determined quantitatively in advance, this is an appropriate road. The reward at the end of it is a *quantitative* understanding of the relation between the behaviour of the individuals and the dynamics of the population.

The second road leads to (first order functional partial) differential equations describing the infinitesimal changes in population state. The coefficients in the equation describe the functioning and the behaviour of individuals, but their solution pertains to the population as a whole. Subsequently we look for typical properties of solutions, such as convergence towards relatively simple time independent or periodic solutions. (Finding such properties can be difficult, and often a mixture of numerical and analytical techniques will be used.) Whenever we use modelling as a *strategic* tool for understanding biological reality, this second road is the appropriate one. The reward at the end is insight into *qualitative* aspects of the relation between the behaviour of individuals and the dynamics of the population.

The aim of part A is to map out this second road. We want to describe its present state, and to stimulate our readers to use it, to improve it, to strengthen its foundations, and to enlarge its number of lanes. The required mathematical theory has only just started and a wealth of biological phenomena is awaiting a mathematical description which fits into this framework.

In these notes we deliberately address biologists and mathematicians alike. We try to present mathematical concepts, techniques and results together with their biological interpretation in the context of specific models in order to introduce biological readers to the mathematics and we try to present biological phenomena and their verbal models together with their mathematical counterparts (exotic equations with transformed arguments) in order to introduce mathematical readers to the underlying biology, as painlesly as possible. Since one needs to practice a language to learn it, we have included numerous exercises.

Perhaps at this point it should be made clear that we don't want to join issue with advocates of simulation methods. Simulation models undoubtedly can come closer to reality. This is a clear advantage. But it can be a disadvantage as well, since the inherent complexity can make it difficult to interpret the results. Between present day realistic simulation models and the usual idealized total population models formulated as ordinary differential equations lies quite a gap. Models of the type discussed in these notes are specifically designed to fill this gap. A true understanding of natural phenomena quite often requires a continuous spectrum of supplementary models and not just one. Thus we strive for symbiosis, not competition.

Functional partial differential equations yield a new and interesting class of infinite-dimensional dynamical systems which holds great appeal for mathematicians. In recent years many papers were written on this topic, mainly dealing with age-structured populations. One of the basic objectives of structured population dynamics is the use of biological knowledge to describe the interaction of a population and its environment, including other populations, by specifying in detail how birth, death and growth processes depend on environmental quantities. In this respect "age" usually fails to be a sufficient characteristic for distinguishing individuals, as it provides but scant information about their physiological condition, except in the special case of constant environmental conditions (and then only indirectly). Indeed, in most nonlinear age-structured models density dependence is incorporated ad hoc, without any physiological foundation. Although this may lead to interesting mathematical problems, from a biological viewpoint it hardly improves upon the usual unstructured models. We hope that these notes will stimulate applied mathematicians to spend more time and energy on the equally (or rather more, we feel) interesting infinite-dimensional dynamical systems which derive from choosing some physiological variable such as size as distinguishing characteristic. (This also makes it necessary to pay more attention to the modelling aspect. Chapters I and III provide some clues).

The mathematical theory of the class of *linear* functional partial differential equations corresponding to structured population models is basically in good shape (see chapters II, IV and V), though a great deal of amendments and extensions are certainly needed. This in sharp contrast with the nonlinear case where hardly any systematic theory exists (except for the case of purely age-structured models, but even there the work is far from finished; in chapter VI we present some general thoughts about the directions we feel that such a theory ought to take). Yet another aim we have is to stimulate our mathematical readers to bring about drastic changes in this situation. So, in a sense, we hope that these notes will soon be out of date!

Amsterdam and Leiden
december 1985

Odo Diekmann and Hans Metz

Hints for reading part A

dependence between the chapters:

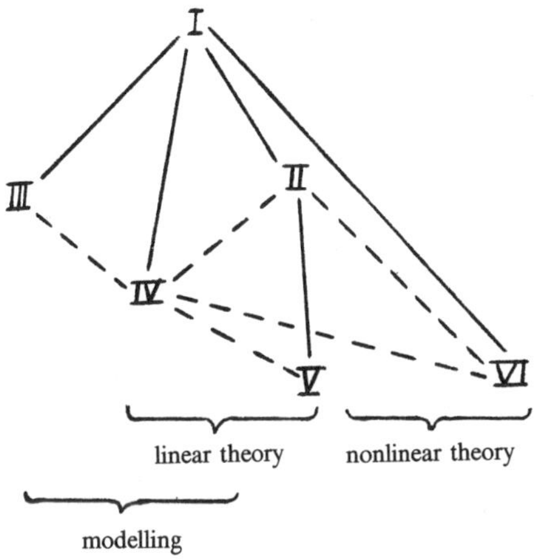

primary subject:

The starred parts of chapters I, III and IV are of a more mathematical nature than the rest of these chapters.

Acknowledgements

The editors wish to thank Henk Heijmans for all his help in the organization of the colloquium and the editing of these notes, Henk Heijmans, Mats Gyllenberg, Bas Kooijman, Roger Nisbet, Mous Sabelis, Horst Thieme and John Tyson for illuminating discussions on various topics from part A, Steve Blythe, Frank van den Bosch, Bas Kooijman, André de Roos and especially Henk Heijmans and Sean Nee for constructive comments on various drafts of part A, Hans Heesterbeek for preparing the table of contents and the index of examples, Frank van den Bosch, Kees Lankester and André de Roos for drawing the figures, and Mrs G. Verloop for her skill in dealing with the text processor.

Hans Metz wishes to thank the Netherlands organization for the advancement of pure research ZWO for a stipendium and the Department of Applied Physics of the University of Strathclyde, the School of Biological Sciences of the University of Sussex, and the Edward Grey Institute of Field Ornithology, Oxford for their hospitality during the academic year 1984/85.

Odo Diekmann wishes to thank the University of Amsterdam for giving him the opportunity to lecture on structured populations during the fall of 1983.

And all of us wish to thank the Centre for Mathematics and Computer Science, Amsterdam, for providing facilities for the 1983 Colloquium on Structured Populations and for the assistance in the preparation of these notes.

List of Contributors

T. ALDENBERG, RIVM (National Institute for Public Health and Environmental Hygiene), Postbus 1, 3720 BA Bilthoven, the Netherlands.

S.P. BLYTHE, Department of Applied Physics, University of Strathclyde, 107 Rottenrow, Glasgow G4 ONG, Scotland, U.K.

F. VAN DEN BOSCH, Center for Mathematics and Computer Science, Kruislaan 413, 1098 SJ Amsterdam, the Netherlands (Present address: Institute of Theoretical Biology, University of Leiden, Groenhovenstraat 5, 2311 BT Leiden, the Netherlands).

N. DAAN, RIVO (Netherlands Institute for Fishery Investigations),Postbus 68, 1970 AB IJmuiden, the Netherlands

O. DIEKMANN, Center for Mathematics and Computer Science, Kruislaan 413, 1098 SJ Amsterdam, the Netherlands & Institute of Theoretical Biology, University of Leiden, Groenhovenstraat 5, 2311 BT Leiden, the Netherlands

J. GOUDRIAAN, Department of Theoretical Production Ecology, Agricultural University, Bornsesteeg 65, 6708 PD Wageningen, the Netherlands

W.S.C. GURNEY, Department of Applied Physics, University of Strathclyde, 107 Rottenrow, Glasgow G4 ONG, Scotland, U.K.

H.J.A.M. HEIJMANS, Center for Mathematics and Computer Science, Kruislaan 413, 1098 SJ Amsterdam, the Netherlands

A.L. KOCH, Department of Biology, Indiana University, Bloomington, Indiana 47405, U.S.A.

S.A.L.M. KOOIJMAN, Biologisch Laboratorium, Vrije Universiteit, Postbus 7161, 1007 MC Amsterdam, the Netherlands & Division of Technology for Society TNO, Postbus 217, 2600 AE Delft, the Netherlands

W.E.M. Laane, Department of Animal Ecology, Agricultural University, Postbus 61, 6700 AB Wageningen, the Netherlands

J. VAN DER MEER, Department of Population Biology, University of Leiden, Kaiserstraat 63, 2300 RA Leiden, the Netherlands

J.A.J. METZ, Institute of Theoretical Biology, University of Leiden, Groenhovenstraat 5, 2311 BT Leiden, the Netherlands

R.M. NISBET, Department of Applied Physics, University of Strathclyde, 107 Rottenrow, Glasgow G4 ONG, Scotland, U.K.

P.A.C. RAATS, Institute for Soil Fertility, Postbus 30003, 9750 RA Haren (Gr.), the Netherlands

M.W. SABELIS, Department of Population Biology, University of Leiden, Kaiserstraat 63, 2300 RA Leiden, the Netherlands.

N.M. VAN STRAALEN, Biologisch Laboratorium, Vrije Universiteit, Postbus 7161, 1007 MC Amsterdam, the Netherlands

H.R. THIEME, Sonderforschungsbereich 123, Universität Heidelberg, Im Neuenheimer Feld 294, 6900 Heidelberg, Bundesrepublik Deutschland

W.J. VOORN, Department of Electron Microscopy and Molecular Cytology, University of Amsterdam (Present address: Department of Medical Microbiology, University of Amsterdam, Meibergdreef 15, 1105 AZ Amsterdam, the Netherlands)

TABLE OF CONTENTS

Part A. Mathematical Models for Physiologically Structured Populations: a Systematic Exposition 1

I. A Gentle Introduction to Structured Population Models: Three Worked Examples 3

J.A.J. Metz & O. Diekmann

1. Introduction 3

2. The invertebrate functional response 5

2.1. Introduction 5
2.2. Holling's disk equation 1: the underlying time scale argument, as exemplified by the finite state predator 6
2.3. Holling's disk equation 2: general handling times 8
2.4. The general invertebrate predator 1: the basic biology 12
2.5. The general invertebrate predator 2: the population equation 14
2.6. The general invertebrate predator 3: calculating the functional response 16
2.7. The general invertebrate predator 4: the Rashevsky limit 18
2.8. Concluding remarks and summary 20

3. Size dependent reproduction in ectothermic animals 21

3.1. The dynamics of individuals 21
3.2. Formulation of the population equations 23
3.3. Constant environments 24
3.4. Constant environments: reduction to an age-dependent problem 26
3.5. Variable environments 29
3.6. Summary 31

4. The cell size distribution 31

4.1. The dynamics of individuals 32
4.2. Formulation of the p equations 34
4.3. The stable cell size distribution 35
4.4. The inverse problem 38
4.5. Limits to growth 39
4.6. Summary 42

5. On semigroups and generators 43

II. The Cell Size Distribution and Semigroups of Linear Operators 46

O. Diekmann.

1. Formulation of the problem 46

2. Strongly continuous semigroups of bounded linear operators 49

3. Do growth, death and division generate a semigroup? 53

4. The spectrum of A 56

5. The characteristic equation 58

6. Decomposition of the population state space X 60

7. Relations between the spectra of A and of $T(t)$ 61

8. Exponential estimates 62

9. The stable size distribution 65

10. Interlude: integration along characteristics 68

11. The merry-go-round 71

12. The merry-go-round with an absorbing exit 74

13. Remarks about positivity 76

14. A somewhat special nonlinear problem 76

III. Formulating Models for Structured Populations. 78

J.A.J. Metz & O. Diekmann.

1. Introduction: six examples for later use 78

2. Some modelling philosophy 83

2.1. The state concept 83
2.2. Obtaining an i-state representation 86
2.3. From the individual to the population level 88

3. Mass balance 90

3.1. Mass transport due to continuous i-state movement 92
3.1.1. A biologist's shortcut 92
3.1.2. The mathematician's derivation 94
3.2. The (local) loss of p-mass 96
3.2.1. The local disappearance of p-mass from the interior of Ω 96
3.2.2. The disappearance of p-mass across the boundary of Ω 98
3.3. The (re)appearance of p-mass 99
3.3.1. The (re)appearance of p-mass in the interior of Ω 100
3.3.2. The (re)appearance of p-mass across the boundary of Ω 101
3.4. Boundaries and side conditions: picking up the strands 101
3.5. Summary and concluding remarks 103

4. Integration along characteristics, transformation of variables, and the following of cohorts through time 104

4.1. Integration along characteristics 104
4.2. Transformation of variables 107
4.3. Following cohorts through time 111

5. About delta-functions and related topics 112

5.1. The delta-function formalism 113
5.2. Delta-functions and transition conditions 114
5.3. Delta-functions in initial conditions 115
5.4. Delta-functions in more dimenions 116

6. Limiting processes and model simplification 117

6.1. Introduction: the role of limit arguments 117
6.2. Time scale arguments 118
6.2A. An explicit expression for F from the nursery competition model from example 6.2.4 122
6.3. Laws of large numbers on the individual level: the step from particulate to nonparticulate 123
* 6A. A justification for the limit arguments: the Trotter-Kato Theorem 124

A. Calculus in $\mathbb{R}^n$, a short refresher 125

A1. Differentiation 125
A2. Integration 127
A3. Some useful relations from linear algebra: the differentiation of determinants 129

B. Stochastic continuous *i*-state movements 130

C. The *p*-equations for the examples from section 1 133

IV. Age Dependence 136

J.A.J. Metz & O. Diekmann

1. Age as a substitute for comprehension 136

1.1. Why this special attention 136
1.2. Which problems allow an age representation? 136
1.3. Integral equations as a natural modelling tool 138
1.4. The calculation of some birth kernels 140

2. Linear theory 143

2.1. An explicit expression for the population birth rate 143
2.2. Renewal theorems 144
2.3. Semigroup approaches 148
2.3.1. The age distribution 148
2.3.2. Two semi-groups derived directly from the renewal equation itself 149
* 2.3.3. Finite representability 151
2A. Moments, cumulants and some approximations for r 153
2B. The dependence of the long run population size on the initial age distribution 155

3. Extensions of the linear theory 156

3.0. Introduction 156
3.1. Scar distributions in yeast 157
3.1.1. Budding yeasts 157
3.1.2. Fission yeasts 160
3.2. Colony size in the diatom *Asterionella* 163

4. Some nonlinear extensions of the linear theory 169

4.1. Kermack's and McKendrick's (1927) general epidemic and the nonlinear renewal theorem 169
4.2. Population decline in ectotherms 174

5. Models allowing a reduction to a differential equation on $\mathbb{R}^k$ 176

5.1. Models with a separable death rate 176

5.2. Linear chain trickery 178

A. The Laplace transformation 182

V. The Dynamical Behaviour of the Age-Size-Distribution of a Cell Population 185

H.J.A.M. Heijmans

0. Introduction 185

1. The model 186

2. Reduction to an abstract renewal equation 188

3. Existence and uniqueness of solutions 189

4. Laplace transformation 190

5. Positive operators 191

6. Location of the singular points 192

7. Computation of the residue in λ_d 195

8. The inverse Laplace transform 196

9. Interpretation, conclusions and final remarks 199

A. Appendix 201

VI. Nonlinear Dynamical Systems: Worked Examples, Perspectives and Open Problems 203

O. Diekmann & H.J.A.M. Heijmans (with contributions by F. van den Bosch).

1. Basic terminology and an outline of the program 203

1.1. Fundamental concepts of dynamical systems theory 203
1.2. Linearized stability and bifurcation theory in the context of ordinary differential equations 207
1.3. An impressionistic sketch of some global aspects 211

2. An example of the construction of a dynamical system: an epidemic model with temporary immunity 214

2.1. The model 214
2.2. Existence and uniqueness 216
2.3. The stability of the steady states 218

3. Hopf bifurcation in scalar nonlinear renewal equations and nursery competition 220

3.1. Introduction to the theory 220
3.2. A first application 222
3.3. Nursery competition 224

4. Lyapunov functions and monotone methods: the G-M model in cell kinetics 227

4.1. The model 227
4.2. Existence and uniqueness 229

4.3. Boundedness of solutions 231
4.4. Extinction of the population 231
4.5. Existence of a nontrivial equilibrium and monotonicity on an invariant subset 232
4.6. Global stability of the nontrivial equilibrium 235
4.7. Final remarks 236

5. Reduction to an ODE-system: a chemostat model for a cell population reproducing by unequal fission 237

5.1. The model 237
5.2. The linear equation 239
5.3. An ODE system related to the nonlinear problem 240
5.4. The nonlinear problem 240

6. Interaction through the environment: some open problems 241

Bibliography 244

Index of examples 261

Part B. From Physiological Ecology to Population Dynamics: a Collection of Papers 263

Topic I. Individuals and laboratory populations. 265

S.A.L.M. Kooijman, Population dynamics on basis of budgets. 266

M.W. Sabelis, The functional response of predatory mites to the density of two-spotted spider mites. 298

M.W. Sabelis & J. van der Meer, Local dynamics of the interaction between predatory mites and two-spotted spider mites 322

M.W. Sabelis & W.E.M. Laane, Regional dynamics of spider-mite populations that become extinct locally because of food source depletion and predation by phytoseiid mites *(Acarina: Tetranychidae, Phytoseiidae).* 345

Topic II. Field populations. 376

N. Daan, Age structured models for exploited fish populations 377

N.M. van Straalen, The "inverse problem" in demographic analysis of stage-structured populations 393

T. Aldenberg, Structured population models and methods of calculating secondary production 409

Topic III. Cell populations. 429

W.J. Voorn & A.L. Koch, Characterization of the stable size distribution of cultured cells by moments 430

P.A.C. Raats, The kinematics of growing tissues 441

Topic IV. Numerical approaches. 452

J. Goudriaan, Boxcartrain methods for modelling of ageing, development, delays and dispersion. 453

W.S.C. Gurney, R.M. Nisbet & S.P. Blythe, The systematic formulation of models of stage-structured populations. 474

Topic V. Analytical approaches and a novel type of *i*-state. 495

H.R. Thieme, A differential-integral equation modelling the dynamics of populations with a rank structure 496

Part A

Mathematical Models for Physiologically Structured Populations
a Systematic Exposition

I. A Gentle Introduction to Structured Population Models: Three Worked Examples

J.A.J. Metz & O. Diekmann

1. Introduction

In population models the basic unit is the individual. Therefore it is the task of the model builder to translate his/her knowledge about mechanisms on the individual level into models for the change in the number of such individuals. Generally, if one talks to an experimental ecologist (s)he has all kinds of alluring stories to tell about such mechanisms. However, as soon as it comes to writing down equations usually all that remains is but a handwaving reference when some mathematically convenient relationship between, say, death rate and population size is pulled out of the hat. The main reason for this unsatisfactory state of affairs probably is that applied mathematics seems to revolve around differential equation models which in the simplest case of ordinary differential equations necessarily start at a rather high phenomenological level. And biologists cannot but comply (but see *e.g.* McKendrick (1926) for an early exception!). What clearly is needed, therefore, is a modelling methodology which in principle can accomodate any necessary amount of biological detail and yet is sufficiently near to the mainstream of applied mathematics that its tools can be brought to bear. This is the background to our efforts as set forward in these notes.

In this first section we shall give a very brief sketch of our modelling philosophy. If you find this sketch overly abstract we ask you to bear with us for a while: the examples in the next sections almost certainly will clarify the issue. A more detailed, and more technical, exposition of the basics of the modelling process may be found in chapter III. Chapter II proceeds with the general development of the mathematical toolkit that is needed to put the models thus derived to good use. This program is pushed on in chapters V and VI, while chapter IV concentrates on an alternative, slightly more phenomenological, modelling approach which can be very effective but applies only to a relatively special class of problems.

The basic concept in structured population modelling is that of an individual's state (*i*-state; below the prefixes *i*- and *p*- will be used to distinguish comparable concepts at the individual and population levels). Roughly speaking this amounts to a collection of variables which 1) at any one time fully determine the population dynamical properties of that individual, like rate of resource consumption or probability of dying or giving birth, and 2) such that their future values are fully determined by their present values and the intervening environmental history as encountered by that individual. Examples of potential *i*-state variables are satiation (of a predator), size, age or energy reserves.

Given a state representation of our individuals the population is conceived as a frequency distribution n: $\Omega \to \mathbb{R}^+$, where Ω denotes the set of possible *i*-states, the so-called *i*-state space. This distribution evolves over time due to physiological processes within the individuals, (local) removal of individuals from Ω due to deaths (or jumps to elsewhere in Ω), and addition of individuals due to births (and returns from jumps). In this conception the frequency distribution n fulfills the same role at the population level as the *i*-state fulfills at the individual level. Therefore we shall refer to it as the *p*-state. The question then is how to write down equations which do the right kind of bookkeeping, and given such equations how to extract information from them.

Before going on we should make clear that even if it is perfectly conceivable to construct stochastic population models of the structured kind, we shall confine ourselves in these notes to deterministic models only. This is not to say that an individual's behaviour is conceived as being completely fixed by its environment (in fact in almost all of our models stochastic effects at the individual level play an essential role), but only that there are sufficiently many individuals that any chance fluctuations deriving from individual stochasticity are completely ironed out.

Generally bookkeeping is easier for infinitesimally short time intervals as then the various processes contributing to population change tend to operate independently so that we can simply add their contributions. (On longer time scales we have to account for example for the fact that individuals that were eaten cannot give birth any longer). This explains the preponderance of differential equations as a modelling tool. But in the case of structured population models these cannot be ordinary differential equations as we need infinitely many numbers to characterize the p-state. In fact it turns out that they generally are (systems of) first order partial differential equations where the partial derivatives come from the smooth shift of the p-state due to continuous physiological processes. Apart from this the equations contain source and sink terms representing births and deaths respectively. An interesting point for mathematicians is that in general at least some of the source and sink terms are non-local, *e.g.* due to children being born from parents present elsewhere in the i-state space. To express this fact we shall frequently refer to the p-equations as functional partial differential equations.

It is when writing down expressions for the various terms in our bookkeeping operation that the second fundamental modelling concept comes into play, that of the "law of mass action". Adoption of this law amounts to saying that all the various individual contributions to for example source and sink terms, like births from different individuals or individual deaths can simply be added up. This summation then leads to the appearance of linear expressions like $\alpha_0(x)n(x)$ for a (local) disappearance rate due to random deaths (where $\alpha_0(x)$ is the probability per unit of time that any one individual in i-state x dies), $\int \lambda(x,y)n(y)dy$ for a (local) birth rate (where $\lambda(x,y)dx$ denotes the (mean) rate at which one parent in i-state y gives birth to children with i-states in an interval of size dx around x), but also to bilinear expressions like $\int \alpha_1(x,y)m(y)dy \cdot n(x)$ for a (local) disappearance rate due to predation (where $\alpha_1(x,y)$ is the rate at which one predator in i-state y searches for prey in i-state x and $m(y)dy$ is the density of predators in an interval of i-states of size dy around y), and also, through appropriate limit arguments, to the appearance of those partial derivatives referred to earlier.

Our first example below will indicate how starting from nothing but law of mass action considerations more complicated functional relationships may be derived by the use of time scale arguments. We want to stress, however, that it is only the basic derivation from law of mass action considerations which allows us to interpret the resulting relations in terms of individual behavioural characteristics.

Having derived population equations we can start studying them mathematically. One of the things we might do for example is simplify them by various kinds of limiting arguments. This may even lead to ordinary differential equations. But now the expressions occurring in these equations are not just phenomenological but can be interpreted in real, mechanistic terms. We may also go through the usual mathematical procedure of studying the large time behaviour of solutions, possibly with some numerical help in the later stages of the calculations, or we may immediately solve the equations numerically to obtain insight in the transient behaviour. The examples below and elsewhere in these notes will surely give you a taste of the various possible models and approaches subsumed under our general framework.

REMARK 1.1: Concerning the law of mass action: (*i*) The probabilistic nature of our models on the individual level intimately links the concepts of law of mass action and that of i-state. (A quantity like $\alpha_0(x)$ encountered above is nothing but a conditional probability per unit of time, where the condition is that the individual be in state x.) As far as we are concerned the detailed nature of this link still needs further clarification. Our present understanding of the matter is set forth in chapter III.
(*ii*) In our discussion of the p-state and the p-equation we were (and will remain) rather sloppy in that we constantly referred to population *numbers* whereas we meant *densities* i.e. *numbers per unit of* (spatial) *area or volume.* The reasons why densities and not numbers are paramount are a) our deterministic models dealing with smoothly varying quantities should rightly be interpreted as limits of stochastic models dealing with integers, the limit being taken by letting the area/volume under consideration grow large, and with it the number of individuals it contains, and b) only for p-equations dealing with densities are the coefficients in any bilinear expressions properties of the interacting individuals concerned: for example $\alpha_1(x,y)$ encountered above is the area/volume effectively searched per predator per unit of time; if we rephrase the p-equation in terms of numbers, by multiplying with the total area under consideration, at the place of α_1 we get the fraction of the total area searched per predator per unit of time.

REMARK 1.2: Concerning the role of simulation methods. With the introduction of advanced simulation methods at the end of the sixties there also seemed to come a promise of bridging the gap between the mathematical models as studied by the theoretical community and the detailed mechanistic considerations as dealt with verbally by experimental ecologists. However, what actually happened is that a third breed of scientists emerged with their own preoccupations and linguistic pecularities. Naturally it is our hope that at least the language of structured population models will provide the necessary links. In this vision there are two important uses for the simulation approach: (i) at the level of the individual, where discrete stochastic simulation models closer represent an experimental biologists way of thinking; given such a simulation model it is usually fairly easy to re-express it in terms of an i-state and corresponding i-equations. And (ii) at the level of the population where simulation often is nothing but numerically solving the p-equations.

2. The invertebrate functional response

2.1. Introduction

The very first models for predator prey interaction by VOLTERRA (1926) and LOTKA (1925) were based on the law of mass action applied to whole predator and prey populations:

$$\text{prey:} \quad \frac{dx}{dt} = rx - apx \qquad\qquad \text{predators:} \quad \frac{dp}{dt} = -cp + haxp \tag{2.1.1}$$

where h is the conversion efficiency of prey into predator biomass. Population experiments by GAUSE (1934) soon led to a refutation: instead of the neutrally stable oscillations predicted, Gause found consistently that the oscillations immediately grew out of bounds, followed by extinction of the prey and subsequently the predator. His explanation was that there necessarily is a maximum to the rate at which a single predator can reproduce and/or ingest prey. Therefore at higher prey densities the amount of prey eaten and/or the number of offspring cannot be assumed to be simply proportional to prey density. A modification of the equations using various empirical formulae to take account of the 'satiation effect' indeed could generate the observed population developments (if also the assumption is made that a population went extinct when its density dropped below say one individual per culture vessel). The function relating prey density to number of prey eaten by a predator nowadays is called the functional response. Throughout this section we shall denote it as F.

EXERCISE 2.1.1: Write down the analog of equation (2.1.1) for a general functional response.

REMARK 2.1.2: The name functional response originates in the work of SOLOMON (1949) who rather loosely coined the phrase to describe the additional mortality at higher prey densities effected by predators, as opposed to the subsequent numerical response, *i.e.* population increase due to good feeding conditions, of those same predators.

The first detailed quantative experiments on predator feeding behaviour were done by IVLEV (1955). He found that for fish feeding on *Daphnia* the empirical formula

$$F(x) = c(1 - e^{-(a/c)x}) \tag{2.1.2}$$

gave an exceedingly good fit to the number of prey eaten in his experiments. (Incidentally, the same formula was already used earlier by Gause to fit data on parasitoid reproduction). (2.1.2) is still commonly used by ecologists under the name of Ivlev functional response even if there really is not much to recommend it as the discussion below will make clear.

The first mechanistic models for the functional response were published in 1959 by HOLLING and by RASHEVSKY.

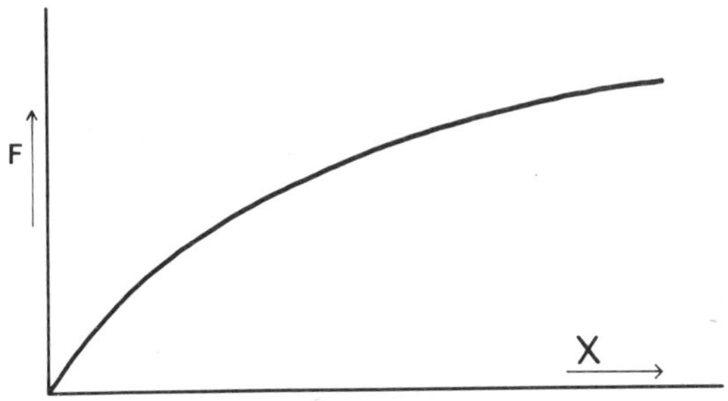

Fig. 2.1.1. Hollings "disk equation" functional response.

Holling started from a partitioning of the time budget of an individual predator. First of all he assumed that the number of prey caught per predator is proportional to prey density and the time spent in actual search. Then he noted that the time spent searching is less than the total amount of time allocated to foodgathering activities by the time needed to handle individual prey items. So, if N denotes the number of prey caught during a foodgathering period, T the duration of that period, x the prey density, a the search rate and b handling time

$$N = ax(T - bN)\,, \tag{2.1.3}$$

and therefore

$$F(x) = \frac{N}{T} = \frac{ax}{1+abx}. \qquad (2.1.4)$$

(See figure 2.1.1). In the ecological literature (2.1.4) is known as Holling's disk equation, after an experiment in which a blindfolded assistant searched for sandpaper disks laid out randomly on a table. The overwhelming advantage of (2.1.4) over purely empirical formulae like (2.1.2) is that all parameters in (2.1.4) have a direct empirical interpretation. This i) makes it possible to extend the model to other situations, like predators searching for a variety of prey items, and ii) makes the model and its extensions amenable to direct empirical testing. By now the model has spawned a whole industry of theoretical as well as empirical research.

Rashevsky's main inspiration were the experiments done by Ivlev. Therefore he took as his starting point a large predator snapping up small prey. He conjectured that the rate of search would be regulated mainly by the filling of the stomach. This he translated, apparently without any direct experimental evidence, into the assumptions that i) search rate decreases linearly with satiation, and ii) the rate at which satiation decreases due to digestion is precisely proportional to satiation. Combined with the usual law of mass action assumption this gave him for the satiation of the individual predator

$$\frac{ds}{dt} = -\mu s + xa(1-s/c). \qquad (2.1.5)$$

From this model he was able to explain Ivlev's formula (2.1.2) as the result of transient behaviour when very hungry fish were brought in contact with high food densities for a short time, in accordance with Ivlev's experimental protocol. The long term catch rate at constant food densities, however, is given by Holling's formula (2.1.4) with $b = (\mu c)^{-1}$. (See exercises 2.7.3 and 2.7.2).

In the ecological literature Rashevsky's work was largely ignored, and with it the role of satiation as an underlying *i*-state variable, until the publication in 1966 of another remarkable paper by Holling. In that paper Holling conceptually decomposed predatory behaviour into its component processes, measured all these processes in detail for a particular predator, the praying mantid *Hierodula crassa,* and finally reassembled the components in a relatively complicated computer simulation program. Independent experiments to test this model showed as good an agreement of observed and predicted catch as ever can be had in a model for a real organism.

Holling's mantid work was later refined, and extended to a number of other species, by various authors among whom FRANSZ (1974) and SABELIS (1981, this volume). Yet the interest in satiation as an ecological variable remained mainly confined to the simulation literature. This probably is due to the seeming cumbersomeness of the resulting simulation models and certainly not to satiation being only a minor factor in determining functional responses. For as far as the data go generally just the opposite seems to be the case!

Below we shall derive a structured population version of Holling's "mantid" model due to METZ & VAN BATENBURG (1984, 1985 a,b). But before doing so we shall discuss how Holling's disk equation fits into the general framework laid out in section 1, as this allows us to introduce the necessary technical tools one at a time.

2.2. *Holling's disk equation 1: the underlying time scale argument, as exemplified by the finite state predator*

We shall start with a version of the model which can be formulated in terms of ordinary differential equations in order to ease the transition to the partial differential equations that will be needed eventually. To this end we assume that the predator can exist in only two states, 0: searching and 1: busy handling prey, leading to the population equations (*p*-equations)

$$\frac{dx}{dt} = rx - ap_0x \qquad (2.2.1a)$$

$$\frac{dp_0}{dt} = -axp_0 + \beta p_1 + [\text{births}] - [\text{deaths}] \qquad (2.2.1b)$$

$$\frac{dp_1}{dt} = axp_0 - \beta p_1 + [\text{births}] - [\text{deaths}]$$

Here we have intentionally left the birth and death terms unspecified as their specification becomes much easier after we have simplified our system of equations a bit.

To bring about this simplification we observe that generally prey densities are much higher than predator densities, and that moreover the rate of return from handling a prey, β, generally is of the same order of magnitude as the catch rate ax. This means that the exchange rates between states 0 and 1 is much faster than any changes in the overall predator density $p = p_0+p_1$ due to births or deaths. Moreover, the relative rate of decrease of prey density due to predation, ap_0, is small. We shall assume that r, the relative rate of change of x due to births and (other) deaths, is of the same order of magnitude as ap_0. The consequence is that the relative proportions of predators in states 0 and 1 equilibrates much faster than either x or p changes.

We shall first deal with the short time scale. To this end we neglect the birth and death terms in (2.2.1b) and concentrate on the predation submodel per se:

$$\frac{dp_0}{dt} = -axp_0 + \beta p_1 \qquad \frac{dp_1}{dt} = axp_0 - \beta p_1 \tag{2.2.2}$$

These equations necessarily satisfy the conservation condition $d(p_0 + p_1)/dt = 0$. Setting either dp_0/dt or dp_1/dt equal to zero and using the additional condition $p_0 + p_1 = p$ then gives the (pseudo) equilibrium values

$$\tilde{p}_0 = \frac{\beta p}{\beta + ax}. \tag{2.2.3}$$

This equilibrium can easily be shown to be globally attractive in the linear manifold $p_0 + p_1 = p$. Using (2.2.3) we can calculate the functional response as

$$F(x) = \frac{ax\tilde{p}_0}{p} = \frac{ax}{1+abx}, \tag{2.2.4}$$

with $b = \beta^{-1}$.

Given the functional response it is easy to specify the birth and death terms for the predator population. First of all we have to account for biomass loss due to deaths and respiration, and secondly we have a biomass gain due to food intake. Together this gives (compare 2.1.1)

$$\frac{dx}{dt} = rx - \frac{axp}{1+abx}, \tag{2.2.5a}$$

$$\frac{dp}{dt} = -cp + h\frac{axp}{1+abx}. \tag{2.2.5b}$$

(Note that we have artfully muddled the distinction between predator numbers and biomass. We simply have to do this if we want to neglect the size structure of the predator population!).

EXERCISE 2.2.1: Draw a picture of the linear manifold $p_0 + p_1 = p$ in the (p_0, p_1)-plane and sketch the vector field defined by (2.2.2). Also show analytically that the equilibrium (2.2.2) is stable within the manifold $p_0 + p_1 = p$. Hint: Eliminate p_1.

EXERCISE 2.2.2: Let $\pi_i \overset{def}{=} p_i/p$ so that $\pi_0 + \pi_1 = 1$. Then π_i, $i=0,1$, also satisfies (2.2.2). And $F(x) = ax\tilde{\pi}_0$ where $\tilde{\pi}_0$ is the equilibrium value of π_0. Here π_0 and π_1 can be interpreted as the probabilities that an individual predator is in state 0 or 1 respectively. Seen from this vantage point the state process of the individual predator is a continuous time Markov chain with "differential generator" *

$$B = \begin{bmatrix} -ax & \beta \\ ax & -\beta \end{bmatrix}$$

where b_{ij} is the transition rate from state j to state i and b_{ii} is the rate at which state i is left for any other state. Figure 2.2.1 shows a predator centered diagrammatic representation. What is the probability distribution of the handling time? What is the mean handling time? Hint: Consider a 'Gedanken-Experiment' in which at $t = 0$ first the prey density is made infinite for a very short time and then all prey is removed. The probability that the predator still is busy handling prey at t then is equal to the probability $\mathcal{H}(t)$ that the handling time is larger than t. The probability density of the handling time equals $-d\mathcal{H}/dt$.

---

* In contradistinction to the usual practice in probability theory we shall treat probability distributions as column vectors, as this fits better in the rest of our notation.

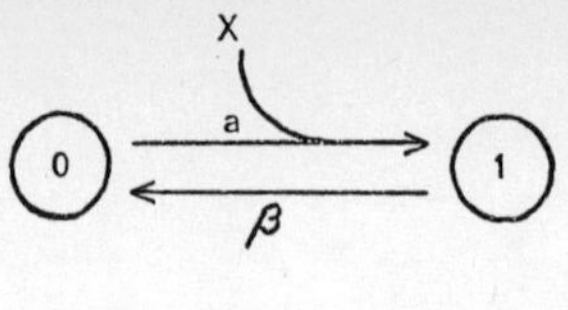

Fig. 2.2.1: Diagrammatic representation of the two state predator.

EXERCISE 2.2.3: Generalize the previous calculations to a predator eating two types of prey; to a predator eating n types of prey.

EXERCISE 2.2.4: There is quite a lot of evidence that at least vertebrate predators start searching more efficiently for a prey when they encounter it sufficiently frequently, resulting in an overrepresentation of the commoner prey in the diet. The simplest finite state model that incorporates this effect is represented in fig. 2.2.2. Calculate the functional response for this model.

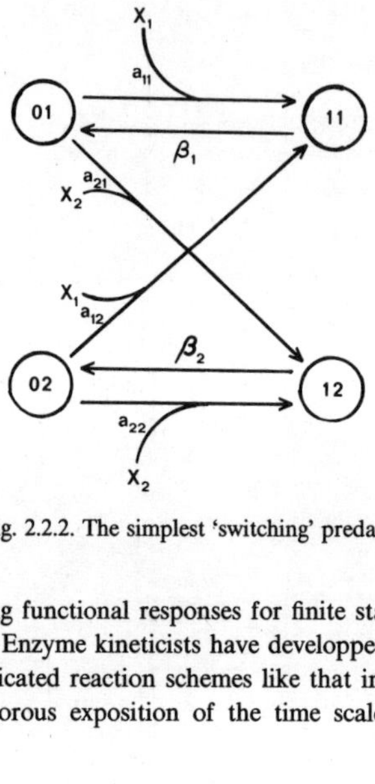

Fig. 2.2.2. The simplest 'switching' predator

REMARK 2.2.5: The problem of calculating functional responses for finite state predators is the same as that of calculating rate equations in enzyme kinetics. Enzyme kineticists have developped various efficient graphical techniques for deriving rate equations from more complicated reaction schemes like that in figure 2.2.2. See *e.g.* CORNISH-BOWDEN (1979) and WONG (1975). The first rigorous exposition of the time scale argument may be found in HEINEKEN, TSUCHIYA & ARIS (1967).

2.3. *Holling's disk equation 2: general handling times*

One not particularly biological feature implicit in the ordinary differential equation model is that prey handling time is assumed to be terminated randomly, *i.e.* the probability distribution of handling time is exponential (compare exercise 2.2.2), whereas in reality handling time distributions are usually strongly peaked around some fixed value. In itself the assumption of exponentially distributed handling times is no more artificial than the assumption that prey are born and die at random which underlies the ordinary differential equation form of the long term equation (2.2.5). However, there is the more serious question how β can be estimated from observations made on the individual level. And the quantitative testing of any predictions derived from the model, and thereby the value of the model as a piece of scientific theory, hinges on our ability to make such estimates!

Below we shall indicate how in principle a structured population framework can be used to cope with general handling time distributions. In the end it turns out that it is only the mean handling time which counts.

REMARK 2.3.1: This result could also be arrived at in other ways, *e.g.* by using standard results from (Markov) renewal theory (Cox, 1962). The method chosen here, however, has the advantage that it can readily be extended to much more complicated models, *e.g.* incorporating satiation as well as handling time.

For didactical reasons we shall describe the basic ideas while confining ourselves to the case of a fixed handling time. General handling time distributions will be dealt with in the exercises at the end.

Let at any time t the number of searching predators be $P_0(t)$ and the number of predators which are handling prey $P_1(t)$. (The reason for the capital letters will become clear below). However, predators busy handling prey differ in how long ago they encountered their prey, and this difference affects their future encounter rate by determining when they will restart searching. Therefore we assign to each busy predator the time τ, $0 \leq \tau < b$, elapsed since its last prey encounter (imagine a stopwatch which starts running at the moment our predator meets a prey), and introduce a function $p_1(t,\tau)$ such that the number of predators with internal clock time between α and β is given by $\int_\alpha^\beta p_1(t,\tau)d\tau$. Consequently

$$P_1(t) = \int_0^b p_1(t,\tau)d\tau \,. \tag{2.3.1}$$

We shall refer to p_1 as the τ-distribution, or, when we feel fussy, the τ-density.
Clearly

$$p_1(t,\tau) = p_1(t-\Delta,\ \tau-\Delta) \tag{2.3.2}$$

whenever $\Delta < \tau < b$, and in particular

$$p_1(t,b) = p_1(t-b,0) \tag{2.3.3}$$

A good way to visualize these relations is to imagine a heap of sand being transported by a conveyor belt (see fig. 2.3.1). In this image (2.3.3), for example, relates the amount of sand leaving the belt per unit of time to the amount put on it at the other end some while ago.

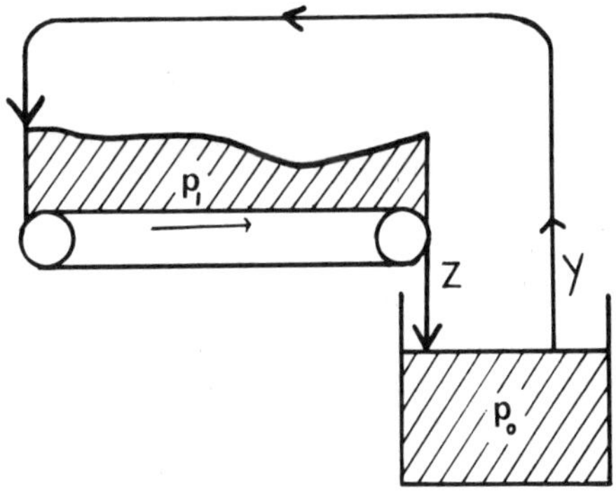

Fig. 2.3.1: The conveyor belt analogy for the structured population type formulation of Holling's disk model with fixed handling time.

The next step is to use (2.3.2) to derive a differential equation for p_1. Taking the limit $\Delta \downarrow 0$ in

$$\Delta^{-1}[p_1(t,\tau)-p_1(t-\Delta,\tau)] = \Delta^{-1}[p_1(t-\Delta,\ \tau-\Delta)-p_1(t-\Delta,\tau)]$$

we obtain the partial differential equation

$$\frac{\partial p_1}{\partial t}(t,\tau) = -\frac{\partial p_1}{\partial \tau}(t,\tau) \tag{2.3.4a}$$

as a description of the way in which p_1 changes due to the movement of the belt (the running of the stopwatch) during infinitesimally short time intervals, for all positions $\tau > 0$. To complete the description we have to specify how many predators step on the belt per unit of time. As during one time unit exactly one unit of belt length becomes available at its left hand end, $p_1(t,0)$ precisely equals this "arrival rate". To derive this more formally let the arrival rate be denoted by y. Then

$$\int_0^\Delta p_1(t,\tau)d\tau = \int_{t-\Delta}^{t} y(s)ds$$

or after dividing by Δ and letting Δ go to zero

$$p_1(t,0) = y(t)\,. \tag{2.3.4b}$$

The usual law of mass action consideration lets us calculate y as

$$y(t) = axP_0(t)\,. \tag{2.3.4c}$$

A relation like (2.3.4b) is called *a boundary condition* as it tells us something about the behaviour of p_1 at the boundary of its domain. Together (2.3.4b) and (2.3.4c) are also known as a *side condition* as in combination they amount to a restriction on the class of possible p-states (P_0,p_1) at any one time.

The companion equation for P_0 is found from law of mass action considerations to be

$$\frac{dP_0}{dt} = -axP_0 + z \tag{2.3.4d}$$

where z is the rate at which predators leave the belt at $\tau = b$. By repeating the reasoning which led to (2.3.4b) we get

$$z(t) = p_1(t,b)\,. \tag{2.3.4e}$$

Equation (2.3.4) forms a dynamical mathematical formulation of our simple model. This description moreover is such that if we provide an initial condition $P_0(0) = \Psi_0$, $p_1(0,\tau) = \psi_1(\tau)$ such that $\Psi_0 + \int_0^b \psi_1(\tau)d\tau = P$ then $P_0(t) + \int_0^b p_1(t,\tau)d\tau = P$ for all t.

EXERCISE 2.3.2: Verify the last statement. Hint: Differentiate relation (2.3.1) for t, and substitute (2.3.4a) to arrive at $dP_1 / dt = y - z$.

At equilibrium all derivatives with respect to time vanish. From (2.3.4a) we then find that the equilibrium value of p_1, $\hat{p}_1$, is constant, *i.e.* independent of τ. Both (2.3.4b) with (2.3.4c), and (2.3.4d) with (2.3.4e) then lead to the conclusion that $\hat{p}_1 = ax\hat{P}_0$. In combination with the conservation condition $b\hat{p}_1 + \hat{P}_0 = P$ this gives

$$\hat{P}_0 = \frac{P}{1+abx} \tag{2.3.5}$$

(Compare (2.2.3)), and therefore

$$F(x) = \frac{ax\hat{P}_0}{P} = \frac{ax}{1+abx} \tag{2.3.6}$$

as before.

REMARK 2.3.3: The fact that we could either use (2.3.4b) and (2.3.4c) or (2.3.4d) and (2.3.4e) to arrive at $\hat{p}_1 = ax\hat{P}_0$ shows that the various components of the equilibrium equation, derived from setting the time derivatives equal to zero, are not independent. It is therefore that we have to add the condition $b\hat{p}_1 + \hat{P}_0 = P$. All this is a direct consequence of the conservation of predator number inherent in (2.3.4). We already encountered this phenomenon in a more obvious fashion in the finite state predator. Exercise 2.2.1 provides a geometrical illustration.

The mathematical justification for using the equilibrium catch rate (2.3.6) as a component of our model for the long term development of predator and prey populations hinges on the stability of the equilibrium (2.3.5). This we will not prove here. But after you have read the following chapters you should be quite able to repair this omission yourself.

The exercises below put our predation (sub)model in some different, possibly illuminating, perspectives. They also extend the previous calculations to general handling time distributions.

EXERCISE 2.3.4: In a cooking timer you set the time needed to bring the cooking of a dish to completion, and then the timer starts to run backwards. We can just as well use this analogy instead of that of the stopwatch as a basis for our i-state description. Let $\sigma = b-\tau$ be the state of the cooking timer and let q be the σ-density, then

$$\frac{\partial q}{\partial t}(t,\sigma) = \frac{\partial q}{\partial \sigma}(t,\sigma)\,. \tag{2.3.6a}$$

Derive this equation from first principles and also from (2.3.4) using the transformation rule

$$q(t,\sigma) = p_1(t,b-\sigma) \tag{2.3.7}$$

Also draw the analogue of figure 2.3.1. What is the analogue of equation (2.3.4b)?

EXERCISE 2.3.5: Complete the partial p-equations derived in the previous exercise. What is the equilibrium? Calculate the functional response.

EXERCISE 2.3.6: To give (2.3.6a) a direct biological interpretation consider a predator which only searches on an empty gut. Gobbling a prey brings its gut content, or satiation s, to w. Prey is digested at a constant rate v. So $b = w/v$. Write down population equations on the basis of the i-state variable s. Call the s-density u. As a check of your results also calculate the functional response from these equations. Finally rescale satiation by setting $\sigma = s/v$, and rederive the p-equation from the previous two exercises, using the substitution

$$q(t,\sigma) = vu(t,v\sigma)\,, \tag{2.3.8}$$

where the factor v derives from the fact that the σ-density q has to accomodate as many predators on a length of conveyor belt $b = w/v$, as the s-density u has to on a length w.

EXERCISE 2.3.7: Now consider the situation in which some prey are bigger than others, leading to a probability distribution of handling times with density f. Try to convince yourself that instead of (2.3.6a) we now get

$$\frac{\partial q}{\partial t}(t,\sigma) = \frac{\partial q}{\partial \sigma}(t,\sigma) + y(t)f(\sigma)\,. \tag{2.3.9}$$

First assume that there is no maximum to the possible handling times so that q lives on the whole positive half axis. Complete the p-equations on this assumption and calculate the functional response.
Hint: Let the so called survivor function of the handling times be

$$\mathcal{F}(\sigma) = \int_\sigma^\infty f(\rho)d\rho\,. \tag{2.3.10}$$

Use partial integration to show that

$$\int_0^\infty \mathcal{F}(\sigma)d\sigma = \int_0^\infty \sigma f(\sigma)d\sigma\,, \tag{2.3.11}$$

the mean handling time.

EXERCISE 2.3.8: Consider the same setting as the previous exercise, but now assume that there is a maximum σ_m to the possible handling times, *i.e.* $f(\sigma) = 0$ for $\sigma > \sigma_m$. It then seems expedient to restrict the domain of q to the reachable set of σ values, just as was done in exercise 2.3.4. In that case we have to add the boundary condition

$$q(t,\sigma_m) = 0\,. \tag{2.3.12}$$

Why? Check that at equilibrium (2.3.12) is automatically satisfied. Can you think of a reason why this should be the case? Hint: Look at remark 2.3.3.

EXERCISE 2.3.9: The variable handling time problem can also be set in a stopwatch as opposed to a cooking timer perspective. Use the relation

$$p_1(t,\tau) = \frac{\mathcal{F}(\tau)}{\mathcal{F}(\tau-\Delta)}\,p_1(t-\Delta,\,\tau-\Delta) \tag{2.3.13}$$

where $\mathcal{F}(\tau)/\mathcal{F}(\tau-\Delta)$ is the probability that a handling time will be longer than τ conditionally on it being longer than $\tau-\Delta$, to derive

$$\frac{\partial p_1}{\partial t}(t,\tau) = -\frac{\partial p_1}{\partial \tau}(t,\tau) - \mu(\tau)p_1(t,\tau) \tag{2.3.14}$$

with

$$\mu(\tau) = -d\log\mathcal{F}(\tau)/d\tau = f(\tau)/\mathcal{F}(\tau) \tag{2.3.15}$$

Complete the p-equation and calculate the functional response.

* EXERCISE 2.3.10 (for delay equation fans): Use (2.3.4e), (2.3.4b) with (2.3.4c) and (2.3.3) to reformulate (2.3.4d) as a linear autonomous retarded functional differential equation

$$\frac{dP_0}{dt} = -axP_0(t) + axP_0(t-b). \tag{2.3.16}$$

Show that $\lambda = 0$ is a root of the characteristic equation

$$\lambda = ax(e^{-b\lambda} - 1) \tag{2.3.17}$$

and that all the other roots lie in the left half plane. Use this to prove that the solution to (2.3.4) converges exponentially to its equilibrium.

2.4. *The general invertebrate predator 1: the basic biology*

After the excursion into handling time based models we are now ready to consider the effect of satiation. Figure 2.4.1 shows a representation of the prey catching process broken down into its main components according to HOLLING (1966). The rectangular boxes correspond to the various directly observable activities of a generalized invertebrate predator, with between parentheses a reference to the form this activity takes in a particular predator, the praying mantid, which Holling studied in great detail. The duration and/or success of each of these activities may be influenced by the predator's satiation.

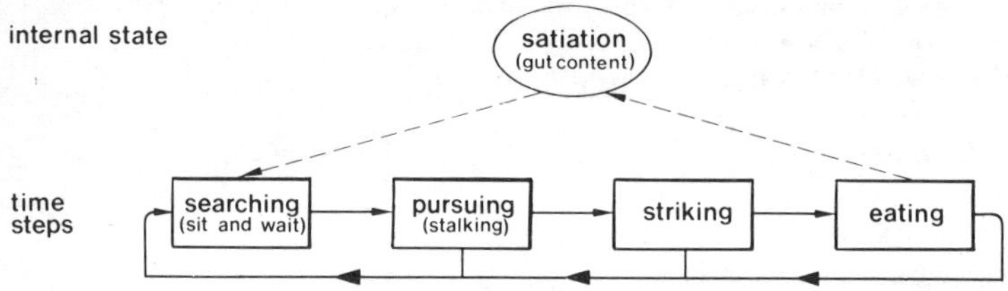

Fig. 2.4.1: Decomposition of the prey catching process according to Holling.

We shall defer to chapter III (sections 1, 5.1, appendix B_1, and exercises) the presentation of a complex model which takes into account both satiation and the durations of pursuit and eating. Often, however, these durations are so short that they can safely be neglected. In that case satiation remains as the only component of the *i*-state. Experiments (HOLLING (1966); another reference is SABELIS (1981)) suggest that, for invertebrates at least, satiation can effectively be considered a one-dimensional quantity, to be equated here naively with gut content, decreasing exponentially between meals and increasing in the course of a meal. If we neglect eating duration, we have to assume that the capture of a prey leads to an instantaneous upward jump of the satiation (compare exercise 2.3.6). The size of this jump will equal the prey weight w if a prey is eaten completely (provided we measure satiation and prey weight in the same units) but it may be smaller when satiation is very high (see below). In between the meals the satiation dynamics is given by

$$\frac{ds}{dt} = f(s) = -as, \tag{2.4.1}$$

where a is the rate constant of digestion.

REMARK 2.4.1: The implicit assumption that prey weight has but one value can easily be relaxed. See METZ & VAN BATENBURG (1985a).

The essential reason for introducing the *i*-state variable satiation is that the search rate is satiation dependent, or, in the terminology of the law of mass action, the rate constant of prey capture depends on it. We shall denote this rate constant as g. The rate at which an individual predator captures prey then is $xg(s)$

INTERLUDE 2.4.2: *Determining g from experimental data, the mantid case.*

The rate constant of prey encounter during search is determined by the form and the size of the search field and the velocity distributions of predator and prey. In the case of a 'sit and wait' predator like the mantid, the predator's velocity is zero. Therefore the rate constant of prey encounter equals the mean prey velocity times the mean width of the search field (compare fig. 2.4.2). Holling found that this width decreases linearly with satiation until it becomes zero. To get the rate constant of prey capture we still have to multiply with pursuit and strike success. Strike success in the mantid is constant. Pursuit success can be calculated from the observation that prey escape during pursuit at a constant rate, that pursuit velocity is constant and that the distance that has to be bridged for a successful pursuit equals the distance at first sighting minus the distance bridged by the strike. The resulting formula for the satiation dependent rate constant of prey capture is

$$g(s) = g_0(s)g_1(s)g_2(s) \tag{2.4.2a}$$

where $g_0(s)$ is the rate constant of prey encounter, $g_1(s)$ pursuit success and $g_2(s)$ strike success, and

$$g_0(s) = b(1-s/c)^+, \quad g_1(s) = \exp[-\beta(1-s/\gamma)^+], \quad g_2(s) = q \tag{2.4.2b}$$

where the $^+$ indicates that a negative quantity should be replaced by zero, and where

- c is the satiation threshold, *i.e.* the value of s at which the width of the search field becomes zero,
- b is the (mean) width of the search field times the (mean) prey velocity,
- $\gamma = c(R_m - R_s)/R_m$, with R_m the search radius at zero satiation and R_s the distance bridged by a strike,
- $\beta = \mu(R_m - R_s)/v$, with μ the escape rate of the prey and v the pursuit velocity, and
- q is strike success.

(In the formula for g_1 we made the simplifying assumption that the search field is circular. See METZ & VAN BATENBURG (1985a) for the general case).

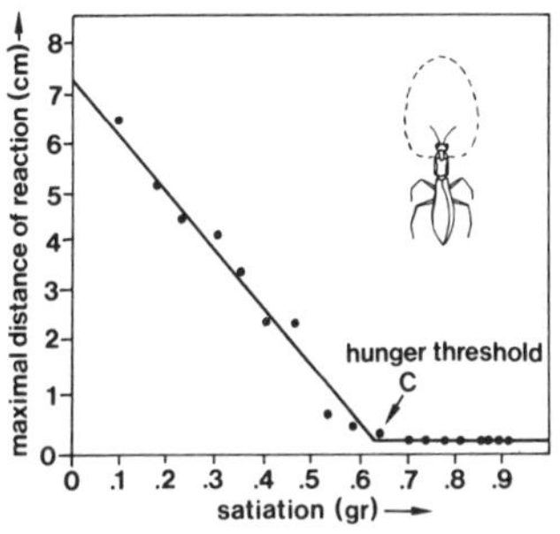

Fig. 2.4.2: Maximum distance of prey recognition as a function of satiation, and shape of the search field for the mantid *Hierodula crassa.* From HOLLING (1966).

REMARK 2.4.3: A complication in Holling's mantid experiments was that the speed of the prey, house flies, appeared to decrease with prey density. This can be accounted for by interpreting x as the effective prey density, *i.e.* the prey density multiplied by the speed reduction relative to the speed at zero density.

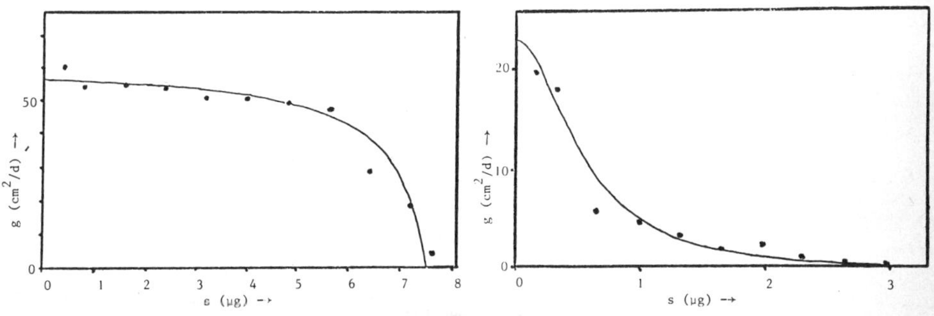

Fig. 2.4.3: Rate constant of prey capture as a function of satiation for the predatory mites *Mesaseiulus occidentalis* (left) and *Phytoseiulus persimilis* (right) feeding on eggs of *Tetranychus urticae* at 27°C. Adapted from Sabelis (1981).

We take g to be a continuous, decreasing function of s which vanishes for $s \geq c$ for some $c>0$. The previous interlude shows this to be the case for the mantid. As additional support for our assumption we refer to figure 2.4.3, adapted from SABELIS (1981) which shows g for two different species of predatory mites feeding on spider mite eggs (see also the contributions of Sabelis to part B).

As a final point in the description of our predation (sub)model we still have to consider the biological detail which we rather brushed under the carpet at the start. Both Holling's mantid and Sabelis' mites still caught prey at satiations at which they were not able to eat all of its weight. When this happened they simply filled their gut to its maximum capacity s_m, and discarded the prey remains. (In Sabelis' case this effect in fact dominates the form of the functional response. In most of Holling's mantid experiments, including the final population experiments, prey size w was in fact so small that $c + w < s_m$).

The functions f and g, and the constants w and s_m totally specify our predation model. There remains the task of calculating the functional response from these ingredients. In contrast to the disk equation case there now is no straightforward simple argument which immediately yields the right answer, and we indeed have to put to work the whole machinery of deriving p-equations, showing that the dynamical system so defined allows a unique (globally) stable equilibrium state which can be calculated at least numerically, finally to obtain the functional response as the corresponding mean catch rate per predator.

2.5. *The general invertebrate predator 2: the population equation*

The p-state for our population of individually satiating predators corresponds to a density over the satiation axis, again denoted by p, so that $P(t,(u,v)) := \int_u^v p(t,s)ds$ equals the number of predators with satiation between u and v at time t, and (with some slight abuse of notation), for $\Omega = (0,s_m]$ the i-state space,

$$\int_\Omega p(t,s)ds = P(t) \tag{2.5.1}$$

where $P(t) := P(t,\Omega)$ is the total predator population size.

Deducing the p-equation is slightly more complicated than it was in our treatment of the disk equation as now the velocity of the i-state f depends on s. Our conveyor belt from subsection 2.3 has become *elastic:* at some places it moves faster and at other places slower, and the differences in local velocities have to be accomodated by shrinking or stretching. In chapter III.3 we shall present a potpourri of systematic approaches to the general problem of setting up p-equations. Here we shall give but one, heuristic, derivation. For the devotees exercise 2.5.1 adds some more rigour.

To make things simple at the start we begin assuming that there are no prey present so that we may concentrate on the part of the p-equation representing the deterministic movement of individuals through Ω due to digestion. As we humans are better at reasoning with numbers or masses we shall first concentrate on $P(t,(s,s+\Delta_s))$, the number of predators in a small satiation interval of length Δ_s beginning at s (see fig. 2.5.1).

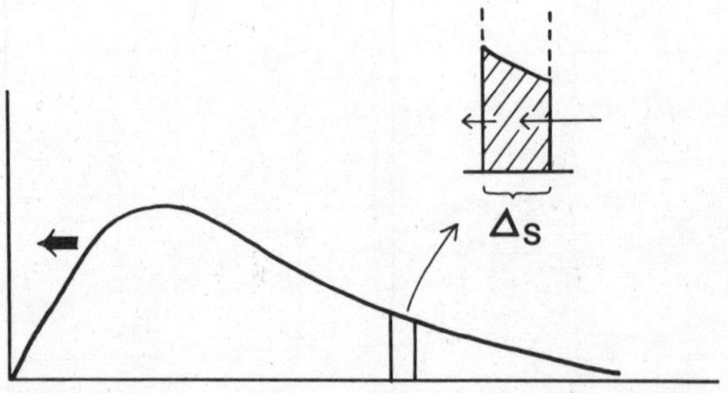

Fig. 2.5.1: The p-state of the predation (sub)model, and the subset of the total predator population in the satiation bracket $(s,s+\Delta_s)$. The horizontal arrows indicate movement of individuals due to digestion.

This number changes due to predators entering the satiation interval under consideration at its right hand boundary and predators leaving the interval at its left hand boundary. The two local flows equal the local speeds of the conveyor belt times the local predator densities. So, remembering that f is negative,

$$\frac{dP(t,(s,s+\Delta_s))}{dt} = f(s)p(t,s) - f(s+\Delta_s)p(t,s+\Delta_s) . \tag{2.5.2}$$

Assuming sufficient smoothness of f and p we may write*

$$f(s+\Delta_s)p(t,s+\Delta_s) = f(s)p(t,s) + \frac{\partial f(s)p(t,s)}{\partial s}\Delta_s + o(\Delta_s)$$

and also

$$P(t,(s,s+\Delta_s)) = p(t,s)\Delta_s + o(\Delta_s) .$$

Substituting this in (2.5.2), dividing by Δ_s and letting $\Delta_s \downarrow 0$ results in

$$\frac{\partial}{\partial t}p(t,s) = -\frac{\partial}{\partial s}(f(s)p(t,s)) \tag{2.5.3}$$

which is our sought equation.

EXERCISE 2.5.1: Derive (2.5.2) by the following argument. For Δ_t sufficiently small the predators that by time $t+\Delta_t$ have left our satiation interval had at time t satiation between s and $\underline{s}$ where $\underline{s}$ is defined by $\sigma(t) = \underline{s}$ with σ satisfying $d\sigma/dt = f(\sigma)$ with $\sigma(t+\Delta_t) = s$. So for small Δ_t we have $\underline{s} = s - f(s)\Delta_t + o(\Delta_t)$. Also calculate the number of predators that have entered. Use these quantities to calculate $\overline{P}(t+\Delta_t,(s,s+\Delta_s))$ and apply the usual limiting arguments.

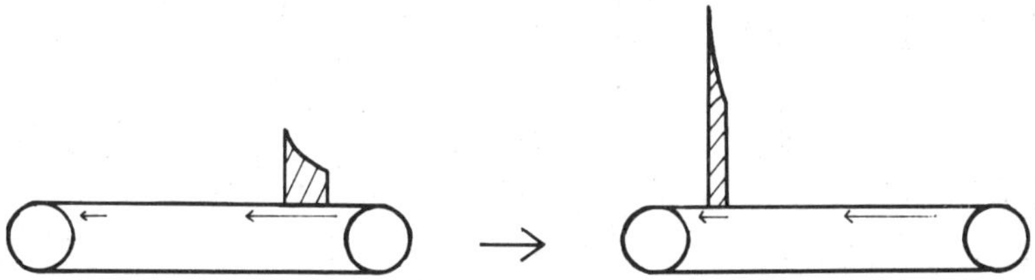

Fig. 2.5.2: The elastic conveyor belt at two subsequent times to illustrate how its contraction during its movement results in a higher stacking of the mass it carries.

Equation (2.5.3) can be rewritten as

$$\frac{\partial}{\partial t}p(t,s) = -f(s)\frac{\partial}{\partial s}p(t,s) - p(t,s)\frac{d}{ds}f(s) . \tag{2.5.4}$$

When f is a constant only the first term remains (compare exercise (2.3.6). What then is the meaning of the second term? To understand this we return to our image of an elastic conveyor belt. (see fig. 2.5.2). At the right the belt moves faster than at the left since digestion goes faster at higher satiation. Therefore the conveyor belt has to contract while moving to the left. As a result the mass on it becomes stacked higher and higher as the length of conveyor belt available to it becomes shorter and shorter. It is this change in available length combined with the conservation of predator number which leads to the additional term $p(t,s)df(s)/ds$.

EXERCISE 2.5.2: Assume that at $t = 0$ a magician comes along and colours our predators in different shades of red, smoothly depending on their current satiation. After this the magician vanishes into thin air in the customary manner so that our predators are doomed to stay the same colour for the rest of their ever more hungry lives. Let $y(t,s)$ be the colourshade of a predator which at time t has satiation s. Derive a partial differential equation for y. (This problem also has some more useful counterparts in the modelling of our present magicwise rather impoverished world!) Hint: Use the approach from subsection 2.3.

* The notation $y = o(x)$ means that $y/x \to 0$ for either $x \to 0$ or $x \to \infty$, which of the two being clear from the context. In the present context $o(\Delta_s)$ means a term which is relatively negligable compared to $\frac{\partial f(s)p(t,s)}{\partial s}\Delta_s$.

Now assume that we add highly toxic prey (at density x) so that the predators die immediately on catching a prey item. This changes (2.5.3) into

$$\frac{\partial}{\partial t}p(t,s) = -\frac{\partial}{\partial s}(f(s)p(t,s)) - xg(s)p(t,s)\,. \tag{2.5.4}$$

If the prey are not toxic a predator with satiation s on catching a prey item still disappears at s, but only to "resurrect" higher up on the satiation axis, viz. at $s+w$. This means that to complete the bookkeeping at s we have to add an arrival term representing the predators coming in from $s-w$. The final result is the functional partial differential equation

$$\frac{\partial}{\partial t}p(t,s) = -\frac{\partial}{\partial s}(f(s)p(t,s)) - xg(s)p(t,s) + xg(s-w)p(t,s-w)\,. \tag{2.5.5.a}$$

(Functional, since the last term contains a so-called *transformed argument* making it *non-local:* the change of p at s depends on the value of p at $s-w$.) Of course the meaning of the last term still has to be defined for $0<s\leqslant w$ as then $s-w \notin \Omega = (0,s_m]$. This will be taken care of in the following

CONVENTION 2.5.3: If a transformed argument falls outside Ω we shall assume that the term in which it occurs equals zero.

Equation (2.5.5a) only tells us what happens inside Ω. To this we have to add a boundary condition at s_m, the point of departure of the conveyor belt. In a small time interval of length Δ_t the number of predators whose eyes were bigger than their belly, and which therefore arrive at s_m, are

$$\int_{s_m-w}^{s_m} xg(s)p(t,s)ds\Delta_t + o(\Delta_t)\,.$$

These arrivals have to be accomodated on a stretch of conveyor belt of length $-f(s_m)\Delta_t + o(\Delta_t)$. The resulting side condition is

$$-f(s_m)p(t,s_m) = \int_{s_m-w}^{s_m} xg(s)p(t,s)ds\,. \tag{2.5.5b}$$

REMARK 2.5.4: When $s_m>c+w$ (2.5.5b) reduces to

$$p(t,s_m) = 0\,. \tag{2.5.6}$$

In that case after some initial period there will be no predators left with a satiation in $(c+w,s_m]$, *i.e.* $p(t,s) = 0$ for $s>c+w$, and t sufficiently large and therefore also, by continuity,

$$p(t,c+w) = 0\,. \tag{2.5.5b'}$$

If we also start without any predators in the upper satiation ranges we can just as well choose $\Omega = (0,c+w]$, using (2.5.5b′) as the boundary condition.

EXERCISE 2.5.5: Prove that the total number of predators

$$P(t) = \int_0^{s_m} p(t,s)ds$$

is constant by integrating (2.5.5a) from zero to s_m and using (2.5.5b) to mop up the remains.
Hint: Use that p should be integrable and that $f(s) = -as$ as an indication that $\lim_{s\downarrow 0} p(t,s) = 0$ (see also exercise 2.6.2). (Do not forget convention 2.5.3).

2.6. *The general invertebrate predator 3: calculating the functional response*

The equilibrium p-state of our predation model can be calculated from

$$0 = -\frac{d}{ds}(f(s)\hat{p}(s)) - xg(s)\hat{p}(s) + xg(s-w)\hat{p}(s-w) \tag{2.6.1a}$$

together with the conservation condition

$$\int_{\Omega} \hat{p}(s)ds = P . \tag{2.6.1b}$$

(The side condition (2.5.5b) is satisfied automatically! See exercise 2.6.3.) A proof, using methods similar to those developed in chapter II, that the equilibrium is globally stable can be found in HEIJMANS (1984).

To solve (2.6.1) one can start by taking $\tilde{p}(w) = \theta$, some arbitrary positive constant, and (numerically) integrating the ordinary differential equation (remember convention 2.5.3!) from w down to 0. Thereafter the known values of $\tilde{p}$ on $(0,w]$ can be used to integrate from w to $2w$ and so on. Next one defines the auxiliary function

$$\bar{p}(s) = \frac{\tilde{p}(s)}{\int_{\Omega} \tilde{p}(\sigma)d\sigma} . \tag{2.6.2}$$

Since (2.6.1a) is linear it is also satisfied by $\bar{p}$. Finally we obtain $\hat{p}$ as $\hat{p} = P\bar{p}$.

Once $\bar{p}$ is calculated for various values of x the functional response can be obtained from

$$F(x) = x\int_0^c g(s)\bar{p}(s)ds . \tag{2.6.3}$$

Figures 2.6.1 and 2.6.2 show some examples of $\bar{p}$ and F calculated in this manner.

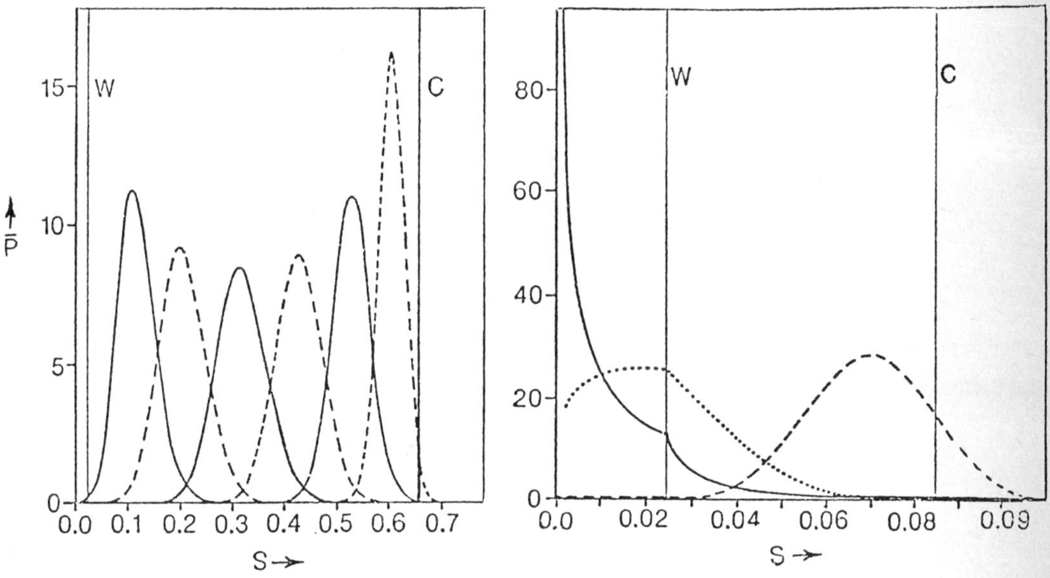

Fig. 2.6.1: The stationary satiation distribution for the mantids *Hierodula crassa* (left) and *Mantis religiosa* (right) for various values of the prey density.

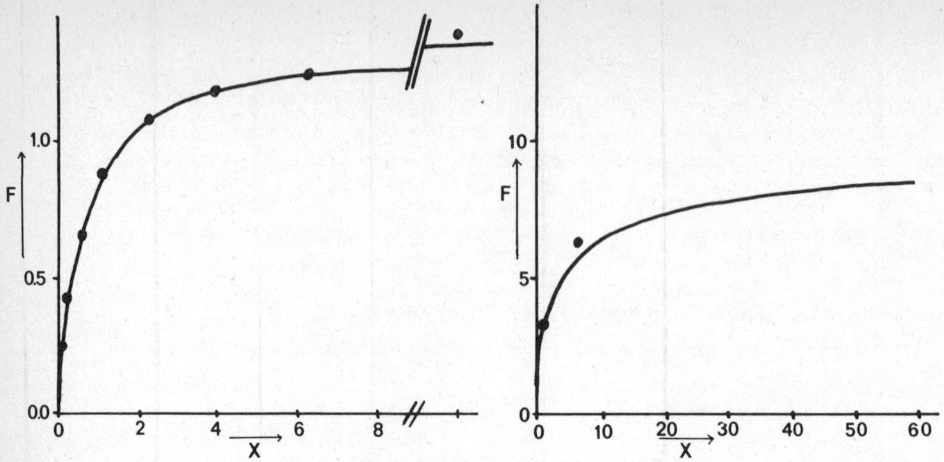

Fig. 2.6.2: The functional responses of the mantid *Hierodula crassa* and the mite *Metaseiulus occidentalis:* ● exact calculation, — from approximation formula (2.7.5).

REMARK 2.6.1: Unless one takes great care numerical schemes for solving (2.6.1a) tend to be very unstable. (Good checks are whether positivity is conserved near $s=0$ and whether the side condition (2.5.5b) is approximately satisfied.) However, the straightforward scheme

$$0 \approx f(n\Delta)\tilde{p}(n\Delta) - f((n+1)\Delta)\tilde{p}((n+1)\Delta) - g(n\Delta)\tilde{p}(n\Delta) + g((n-k)\Delta)\tilde{p}((n-k)\Delta) \tag{2.6.4}$$

with $k = w / \Delta$ behaves very nicely. It not only guarantees the conservation of positivity but also ensures that an appropriately discretized version of the side condition (2.5.5b) is satisfied automatically. In SABELIS (1981, and part B) (2.6.4) is derived from a finite state predator model approximating our continuous one.

EXERCISE 2.6.2: Replace $g(s)$ by $g(0)$ to derive an approximation for $\tilde{p}$ on $(0,\epsilon]$, $\epsilon << 1$, assuming $\tilde{p}(\epsilon)$ to be known. What does $\hat{p}$ look like near $s = 0$? And $\widehat{fp}$?

EXERCISE 2.6.3: Prove that $\hat{p}$ automatically satisfies (2.5.5b). Also explain why this should be the case.
Hint: Integrate (2.6.1a) from 0 to s_m and use the result from the previous exercise. A look at remark 2.3.3 also pays.

2.7. *The general invertebrate predator 4: The Rashevsky limit*

Although (2.6.1), (2.6.3) allow the numerical calculation of F in specific cases, it is difficult to derive general properties from these equations. Moreover, we would like to have some relatively simple approximation formulae for F for use in strategic models for the long term dynamics of prey and predator populations. We shall derive one such approximation here, assuming that the prey are very small but also very numerous. More extensive approximations may be found in METZ & VAN BATENBURG (1985a; a proof of the essential correctness of the limit argument used below can be found in HEIJMANS (1984).

For small but numerous prey it is the biomass that counts. Therefore we introduce as new variables the biomass density.

$$\xi := xw \, , \tag{2.7.1}$$

and the biomass functional response

$$\Phi(\xi) := wF \, . \tag{2.7.2}$$

For small w and sufficiently smooth p

$$xg(s-w)p(t,s-w) = xg(s)p(t,s) - xw\frac{\partial}{\partial s}(g(s)p(t,s)) + x\cdot o(w)$$

and in the limit for $w\downarrow 0$, $x\uparrow\infty$, ξ constant, equation (2.5.5a) reduces to

$$\frac{\partial}{\partial t}p(t,s) = -\frac{\partial}{\partial s}((f(s) + \xi g(s))p(t,s)) , \tag{2.7.3}$$

In (2.7.3) the jump contribution has been replaced by a contribution $\xi g(s)$ to the smooth movement of the satiation. In other words in the limit the originally particulate prey have been turned into some continuous broth which the predator just sucks in.

The i-state dynamics corresponding to (2.7.3) is described by the ordinary differential equation

$$\frac{ds}{dt} = f(s) + \xi g(s) = -as + \xi g(s) , \tag{2.7.4}$$

where f accounts for digestion and ξg for ingestion (see figure 2.7.1). (2.7.4) is nothing but a slight generalization of Rashevsky's model described in subsection 2.1. A graphical argument immediately shows that for each $\xi>0$ (2.7.4) allows a unique globally attracting equilibrium $\hat{s}(\xi)$. For (2.7.3) this means that from everywhere our "conveyor belt" shrinks towards $\hat{s}$. So any initial distribution will contract in the course of time towards an infinitely peaked distribution, or delta "function", at $\hat{s}$.

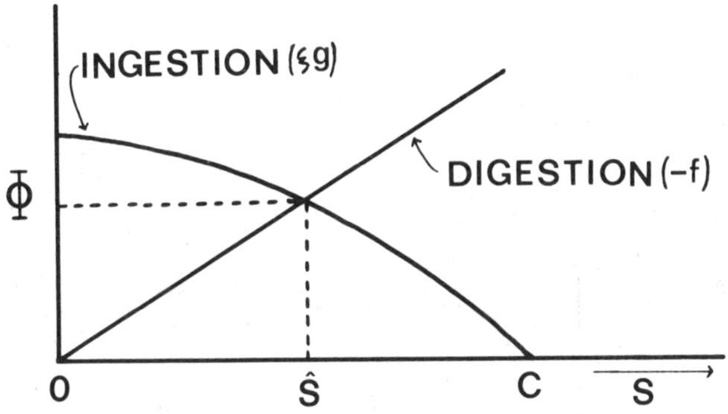

Fig. 2.7.1: The processes contributing to the satiation dynamics in the Rashevsky limit. Also indicated are the equilibrium satiation $\hat{s}$ and the value of the biomass functional response Φ.

The biomass functional response for our limit model can be calculated as

$$\Phi(\xi) = a\hat{s}(\xi) \tag{2.7.5}$$

since

$$\Phi(\xi) = \xi\int_0^c g(s)\delta(s-\hat{s}(\xi))ds = \xi g(\hat{s}(\xi))$$

and from the definition of $\hat{s}$

$$\xi g(\hat{s}(\xi)) = -f(\hat{s}(\xi)) = a\hat{s}(\xi) . \tag{2.7.6}$$

Figure 2.6.2 shows how (2.7.5) performs in two practical examples. (The large discrepancy between the numerically calculated functional response and the result of approximation (2.7.5) in the right graph is mainly due to the phenomenon of predators discarding half eaten prey at higher satiation. See SABELIS (1981) or METZ & VAN BATENBURG (1985a)). The exercises provide some examples of g for which Φ can be calculated explicitly. However, in practice a numerical solution of (2.7.6) is about as fast as using a complicated explicit formula. Any qualitative information, if needed, can relatively easily be extracted from (2.7.6). For example it is easily shown that Φ always increases with ξ and that if g is concave Φ is concave as well.

EXERCISE 2.7.1: Prove the last statements. Hint: Differentiate (2.7.6) for ξ.

EXERCISE 2.7.2: Calculate Φ for $g(s) = \alpha(1-s/c)^+$ (Rashevsky's choice for g) and for

$$g(s) = \alpha\left[\frac{1-\beta s}{1-\gamma s}\right]^+ , \ \beta>\gamma$$

(the g from figure 2.4.3 right).

EXERCISE 2.7.3: Rashevsky's explanation of Ivlev's findings. (In this exercise it pays to revert to the notation of subsection 2.1: $f(s) = -\mu s$ and $g(s) = a(1-s/c)^+$.) Ivlev's experiments i) were of short duration, ii) were done at high Daphnia densities, and iii) were done with (initially) starved fish. i) and ii) make that the digestion term in (2.7.4), or equivalently (2.1.5), can be neglected. Therefore the total amount of prey eaten equals the difference between the satiation at the start and at the end of the experiment. iii) means that the initial satiation was zero. Calculate the amount of prey eaten on the assumption that the duration of the experiment was one time unit. Compare the result with (2.1.2).

* EXERCISE 2.7.4: Let $r(t,s)$ be the solution to $dr/dt = h(r)$ with $h(r) = f(r) + \xi g(r)$ and $r(0,s) = s$. Then $r(t_2,r(t_1,s)) = r(t_1 + t_2,s)$. Therefore $r(t,s) = r(t-\Delta, r(\Delta,s)) = r(t-\Delta, s+\Delta h(s))+o(\Delta)$. Use this relation to show that

$$\frac{\partial}{\partial t}r(t,s) = h(s)\frac{\partial}{\partial s}r(t,s) = h(r(t,s)) . \tag{2.7.7}$$

Use this result to prove that the solution to (2.7.3) with $p(0,s) = \psi(s)$ equals

$$p(t,s) = \psi(r(-t,s))\frac{\partial r(-t,s)}{\partial s} . \tag{2.7.8}$$

How should the two terms in the product be interpreted in the light of the discussion of the elastic conveyor belt in subsection 2.5?

2.8. *Concluding remarks and summary*

For pedagogical reasons we argued in this section mainly through worked examples. As a finish we shall go once more over the main steps, rephrasing them in general terms.

We started arguing that generally the time scales of the interaction of predator and prey populations on the one hand, and the predation process of individual predators on the other hand, differ to such an extent that we can model the latter while assuming that predator and prey densities (and age and/or size structures, if present) stay constant, and the former while assuming that the distribution of predators with respect to the relevant traits is in a pseudo equilibrium pertinent to the current prey density. This allowed us to concentrate on models for the predation process per se.

As a next step we gave a detailed analysis, following HOLLING (1966), of the various components of the predation process, culminating in the writing down of a linear functional partial differential equation describing how the p-state of the resulting predation model changes over time on an infinitesimal basis, as set forth in METZ & VAN BATENBURG (1985a, b). This equation generates a dynamical system which can be analyzed using the techniques from chapters II, IV and V. From this analysis it appears that in the linear manifold defined by the constancy of the total population size there exists a unique, globally attractive p-state, or stable i-state distribution. For various simple limiting cases this distribution can be calculated explicitly, but in general it will be necessary to revert to numerical calculations. After normalizing this distribution, so that its integral equals unity, it can be used to calculate the long term per capita predation rate as the average of the corresponding i-state specific rates.

In the following sections you will see various parts of this pattern of arguments repeated for total population models, in particular the extensive biological discussion, the deriving of functional partial differential equations, the calculation of stationary i-state distributions, and the use of these distributions to arrive at some biologically relevant end result.

** EXERCISE 2.8.1: The sketch of the argument above is slightly idealized, for we certainly did not prove everything we say. In fact there are still two open problems, both having to do with the structure of the time scale argument. In essence this is a singular perturbation argument using the predator prey ratio as the small parameter. Sketches of a more mathematical version, in an enzyme kinetical context, may be found in RUBINOV (1975) or MURRAY (1977).

i) In many predator- prey models the predator and prey densities may go through huge fluctuations. For example, if in (2.2.5a) we add a term $-dx^2$ to represent self-limiting effects in the prey population, the system may show limit cycles which pass close to the axes. How does this affect the argument?
ii) As yet the general extension of the usual singular perturbation theory for ordinary differential equations to the class of functional partial differential equations describing the dynamics of structured populations still has to be made.

3. Size dependent reproduction in ectothermic animals

In our second example we turn to real population dynamics, using individual size as the main *i*-state variable.

Models for the growth of size structured populations were first derived in the late sixties by SINKO & STREIFER (1967). Yet these models did not catch on as well as they should have done (otherwise there would have been no need to write these notes!) One probable reason is that the early workers did not fall back on specific physiological mechanisms but just left the equations in a fairly general form. This is very well if one wishes to prove general results. But for the interactive dealing with empirical data sufficiently specified parameter scarce families of models are almost a *sine qua non.*

Below we shall consider a specific model introduced by KOOIJMAN & METZ (1984), and reformulated as a structured population model by DIEKMANN, METZ, KOOIJMAN & HEIJMANS (1984). All functions involved will be specified immediately from first principles. The model was designed with simple ectothermic animals in mind. In our presentation we shall concentrate on the waterflea *Daphnia magna* as the experimental animal. More detail on the individual level as well as empirical evidence substantiating our assumptions can be found in Kooijman & Metz (op. cit.). See the contribution by Kooijman in part B for a follow up.

3.1. The dynamics of individuals

We shall derive our *i* equations by drawing up a balance of energy flows in combination with some dimensional considerations and a specific assumption about energy chanelling. We start our specification with a look at food intake, next we consider growth, reproduction and finally deaths. In the following a will denote age, l length and x food density. We shall assume that during growth the animals just scale up in a selfsimilar way, and interpret $w := l^3$ as weight (expressed in a suitable unit to avoid a constant of proportionality).

Our first assumption is that the ingestion of food by an individual of length l at food density x, equals

$$[\text{food intake}] = f(x)l^2, \tag{3.1.1a}$$

where explicitly

$$f(x) = \frac{\nu\xi x}{1 + \xi x}. \tag{3.1.1b}$$

The basis for the assumption that ingestion is proportional to l^2 is that *i*) the maximum rate of food intake should equal the maximum digestion rate which scales with the surface area of the digestive apparatus, and ii) for filter feeders (and many other feeding types as well) the food intake at low food densities is proportional to the surface area of the feeding apparatus times the food density. The hyperbolic relation (3.1.1b) between food density and intake results from many micro models of the food catching process *(e.g.* HOLLING (1959) and RASHEVSKY (1959); see subsection 2.1, and especially formula (2.1.4)). In the ecological literature it is known as the Holling (type II) functional response, in the microbiological literature as the Monod curve, and in the biochemical literature as Michaelis Menten uptake kinetics. Figure 3.1.1 shows some experimental evidence substantiating our choice of (3.1.1).

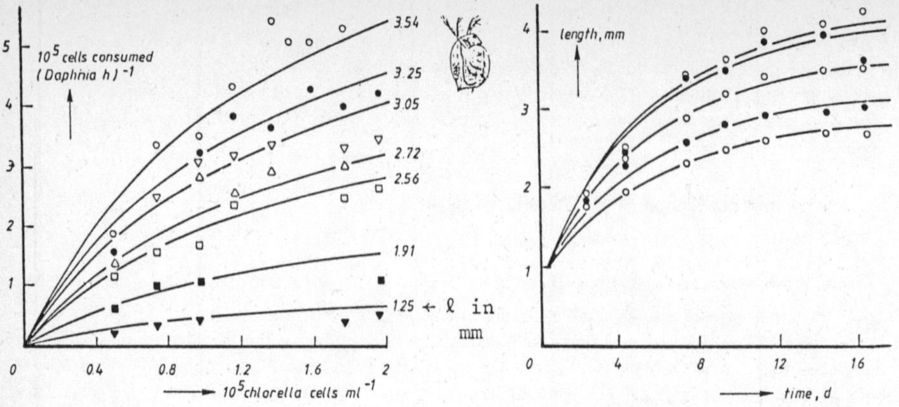

Fig. 3.1.1. Left: Feeding rate at 20°C of the waterflea *Daphnia magna* on the alga *Chlorella* as a function of food density x and body length l. The fitted curves are given by $y = l^2 f(x)$, with $f(x) = \nu\xi x/(1+\xi x)$ with $\nu = 0.75\times 10^5 cells/h.mm^2$, $\xi = 0.7\times 10^{-5} ml/cell$. From KOOIJMAN & METZ (1984). Data from MCMACON & RIGLER (1963), reproduced by WULFF (1980).
Right: length l of *Daphnia magna* as a function of age a for various food densities x. The fitted curves are given by $l = \exp(-\gamma a)(l_b - \underline{l}(x)) + \underline{l}(x)$, with $\underline{l}(x) = 2.89, 3.24, 3.72, 4.17, 4.31 mm$, $l_b = 0.8mm$ and $\gamma = 0.17d^{-1}$.

To arrive at the individual growth equation we assume that a fraction κ of the ingested food is chanelled to maintenance and growth, and a fraction $1-\kappa$ to reproduction. The rate at which maintenance needs energy (expressed in food units) is assumed to be ζw, where w is the metabolizing tissue weight. What remains for growth is $\kappa f(x) w^{2/3} - \zeta w$, provided this quantity is positive. We assume that when it is negative animals stay of constant size (sea-anemones and flatworms do shrink during food scarcity, but more highly organized animals like Daphnids generally don't), the necessary maintenance energy being rechanelled from reproduction. Thus, remembering that $l^2 = w^{2/3}$, we obtain a growth equation of so-called von Bertalanffy type (VON BERTALANFFY, 1934)

$$\frac{dw}{da} = \eta^{-1}(\kappa f(x) w^{2/3} - \zeta w)^+ , \tag{3.1.2}$$

where η is a conversion factor relating weight units to food units and $z^+ := \max(0,z)$.

Below we shall use length as our *i*-state variable in preference to weight, as it both leads to simpler equations and, at least for small aquatic animals, is easier to measure non- destructively. Transforming (3.1.2) to length (using $dl/da = (dl/dw)(dw/da))$ and defining

$$g(x,l) = (3\eta)^{-1}(\kappa f(x) - \zeta l)^+ \tag{3.1.3a}$$

we find

$$\frac{dl}{da} = g(x,l). \tag{3.1.3b}$$

For Daphnids, as for all animals that go through a regular reproductive cycle, birth length is constant to a very good approximation. Therefore we supplement (3.1.3a, b) with the initial condition

$$l(0) = l_b . \tag{3.1.3c}$$

Figure 3.1.1 also shows some laboratory observations on the growth of *Daphnia magna* at constant food densities together with the fitted solutions to (3.1.3) as a further substantiation of our assumptions.

To calculate the individual reproductive rate we recall that a fraction $1-\kappa$ of the ingested food was assumed to be chanelled to reproduction. However, Daphnids (as most other invertebrates) only start reproducing when they have grown sufficiently large. (Presumably the "reproductive" energy surplus is invested in building up the reproductive apparatus and other pieces of body equipment without adding to overall growth; we certainly do not see an abrupt change in growth rate at the abrupt start of reproduction.) The length at which reproduction starts will be called l_J. Moreover we had introduced a(nother) modification of the energy chanelling rule as maintenance was assumed to take priority over reproduction when food densities drop so far that the animal stops growing. Taking stock we assume that the individual size specific birth rate λ is given by

$$\lambda(x,l) = \begin{cases} 0 & \text{for } l_b \leqslant l < l_J, \\ (\omega w_b)^{-1}(1-\kappa)f(x)l^2 & \text{for } l_J \leqslant l \leqslant \underline{l}(x), \\ (\omega w_b)^{-1}(f(x)l^2 - \zeta l^3) & \text{for } \max(l_J, \underline{l}(x)) \leqslant l \leqslant \overline{l}(x), \end{cases} \tag{3.1.4}$$

where $w_b = l_b^3$ and ωw_b is the amount of food needed to produce one young,

$$\underline{l}(x) := \zeta^{-1}\kappa f(x) \tag{3.1.5}$$

is the size at which growth just stops at the current food density, and

$$\overline{l}(x) := \zeta^{-1} f(x) \tag{3.1.6}$$

is the size for which at the current food density precisely all ingested food is needed for maintenance. We assume that animals for which $l > \overline{l}(x)$ die instantaneously, in order to be consistent with our earlier assumptions about energy chanelling and maintenance. (In part B Kooijman develops a model which also accounts for energy reserves).

Finally we have to specify the death rate. It appears that, apart from deaths from starvation when $\overline{l}(x)$ crosses l, the death rate is mainly determined by age and not by size, at least under laboratory conditions. This observation would entail the introduction of age as a second state variable (and we shall do so in III.1), but here we shall make the simplifying assumption of a constant death rate, denoted as μ, while easing our conscience with the observation that under natural conditions deaths from old age are usually negligible compared with deaths due to other hazards such as predation.

Our description of the *i*-dynamics being completed we conclude this subsection with stressing once more that both the strong size dependence of reproduction observed in most ectothermic animals and the size dependence of resource use force us to take account of size as an *i*-state variable. Other aspects of the *Daphnia* life cycle, like the production of males and subsequently winter eggs (see fig. 11 in Kooijman in part B) as a reaction to deteriorating circumstances, we deliberately leave out of the model. (Likewise we do not consider the effect of temperature fluctuations.) This leaves us with l as the only *i*-state variable with associated *i*-state space the interval

$$\Omega = [l_b, l_m)\,, \tag{3.1.7a}$$

where

$$l_m = \zeta^{-1}\kappa\nu \tag{3.1.7b}$$

is the absolute upper bound to the individual length.

EXERCISE 3.1.1.: Calculate l as a function of a at constant food density.

EXERCISE 3.1.2.: Calculate the total number of offspring produced per individual during its lifetime at constant food density.

3.2. Formulation of the population equations

The *p*-state corresponds to a density over Ω, to be denoted as n. So $\int_a^b n(t,l)dl$ is the number of animals with lengths between a and b at time t. (Or more precisely the number of animals per unit of volume; in the following we shall omit this qualification.) Using exactly the same reasoning as in section 2.5 we find that

$$\frac{\partial}{\partial t}n(t,l) = -\frac{\partial}{\partial l}(g(x,l)n(t,l)) - \mu n(t,l) \quad \text{for } l_b < l \leqslant \overline{l}(x)\,, \tag{3.2.1a}$$

while, of course,

$$n(t,l) = 0 \quad \text{for } l > \overline{l}(x)\,. \tag{3.2.1b}$$

At l_b we have an influx of newborn individuals equal to the population birth rate $b(t)$. This influx should match the flux of individuals away from l_b, which equals $g(x, l_b)n(t,l_b)$, the product of the local individual "velocity" (*i.e.* the *i*-growth rate) and the local density (compare the derivation of (2.3.4b) and (2.5.5b)). The total number of births per unit of time equals the integral of λ times n. Therefore n should satisfy the side condition

$$g(x,l_b)n(t,l_b) = b(t) \tag{3.2.1c}$$

with

$$b(t) = \int_{\Omega} \lambda(x,l)n(t,l)dl \,. \tag{3.2.1d}$$

REMARK 3.2.1: Our model formulation contains one hidden assumption which is slightly embarassing from a biological point of view: the production of young is assumed to depend instantaneously on energy intake. In other words, an individual needs not accumulate the necessary energy ωw_b. Thus we implicitly assume that all individuals at each time add some infinitesimal amount of young tissue to a communal pool from which, by some miracle, individual young are created. This assumption is commonly made in the literature of mathematical biology, but always implicitly. However in III.6.3 it will be shown that (3.2.1) can be justified as an approximation for the case when very many young of a very small weight are produced, each of which has only a small chance of surviving its early youth. Aldenberg's contribution in part B is bent on clarifying the confusion in the theory of biological production created by this hidden assumption.

3.3. Constant environments

If we wish to use (3.2.1) to arrive at conclusions about the possible behaviour of *Daphnia* populations we shall have to specify the dynamics of the food density x. The simplest assumption that we can make is that x is constant. This will *e.g.* be approximately the case when the total population density is still very small and the food population (algae in the case of *Daphnia)* is at its own dynamical equilibrium. Anyhow, throughout this and the next subsection we assume x to be constant and sufficiently large for $\underline{l}(x)$ to be larger than l_J (for otherwise no *Daphnia* would ever reproduce).

Generalizing from other models for density independent population growth one guesses that asymptotically (for $t \to \infty$) the population will grow exponentially and that the length distribution will stabilize. The theory developed in chapters II, IV and V shows that this is indeed the case. Here we shall confine ourselves to calculating the asymptotic growth rate and the stable size distribution.

Mathematically our guess amounts to*

$$n(t,l) = ce^{rt}(\psi(l) + o(1)) \quad \text{for } t \to \infty \,, \tag{3.3.1}$$

where r is the (asymptotic) specific growth rate of the population, ψ the stable length distribution and c a constant depending on the initial condition. Neglecting the $o(1)$ term we insert (3.3.1) into the differential equation (3.2.1a) to find

$$\frac{d}{dl}(g(x,l)\psi(l)) = -(\mu + r)\psi(l) \,. \tag{3.3.2}$$

This equation should hold for $l_b \leq l \leq \bar{l}(x)$, as ψ has to vanish for $\bar{l}(x) < l < l_m$. In order to simplify the notation we write

$$g(x,l) = \gamma(\underline{l}(x) - l)^+ , \quad \gamma := \zeta / (3\eta) \,. \tag{3.3.3}$$

Solving (3.3.2) gives us the stable size distribution as

$$\psi(l) = \begin{cases} \psi(l_b) \left[\dfrac{\underline{l}(x) - l}{\underline{l}(x) - l_b} \right]^{\frac{r+\mu}{\gamma} - 1} & \text{for } l_b \leq l < \underline{l}(x) \\ 0 & \text{for } \underline{l}(x) < l < l_m, \end{cases} \tag{3.3.4a}$$

where $\psi(l_b)$ is still at our disposal as a free parameter and can be used to satisfy a normalization condition, such as the natural one

* $o(1)$ refers to a term which becomes arbitrarily small for sufficiently large t, *i.e.* $o(1) \to 0$ for $t \to \infty$.

$$\int_{\Omega} \psi(l)dl = 1\,, \tag{3.3.4b}$$

(see also exercise 3.3.4).

In (3.3.4) we have found an *explicit* expression for the stable size distribution, involving the still unknown population growth rate r. To calculate r we substitute (3.3.1), neglecting the $o(1)$ term, into the side condition (3.2.1c,d) to arrive at the *characteristic equation*

$$1 = \pi_x(r)\,, \tag{3.3.5}$$

where by definition

$$\pi_x(s) = \frac{1}{g(x,l_b)} \int_{l_b}^{\underline{l}(x)} \lambda(x,l) \left[\frac{\underline{l}(x)-l}{\underline{l}(x)-l_b}\right]^{\frac{s+\mu}{\gamma}-1} dl\,, \tag{3.3.6}$$

and hence

$$\pi_x(s) = \frac{(1-\kappa)f(x)}{\omega w_b \gamma(\underline{l}(x)-l_b)} \int_{l_j}^{\underline{l}(x)} l^2 \left[\frac{\underline{l}(x)-l}{\underline{l}(x)-l_b}\right]^{\frac{s+\mu}{\gamma}-1} dl \tag{3.3.7}$$

(see exercise 3.3.3 below for an evaluation of the integral). The integral in (3.3.7) diverges for $s \leqslant -\mu$ and consequently π_x is defined for $s > -\mu$ only (we shall only consider real arguments in accordance with the biological interpretation). Since $(\underline{l}(x)-l)/(\underline{l}(x)-l_b) < 1$ for $l_j \leqslant l < \underline{l}(x)$ the integrand in (3.3.7) strictly decreases with s. Therefore $\pi_x(s)$ strictly decreases too for $s > -\mu$. Moreover $\lim_{s \downarrow -\mu} \pi_x(s) = \infty$ and $\lim_{s \uparrow \infty} \pi_x(s) = 0$. We conclude that (3.3.5) has a unique simple* real solution r.

r corresponds to what in the classical age-dependent theory is called the *intrinsic rate of natural increase* (compare subsection 3.4). The advantage of the mechanistic approach followed here is that it allows r to be calculated as a function of the food density x, thereby making our theoretical developments amenable to experimental tests.

For practical applications equation (3.3.5) has to be solved numerically. However, many qualitative properties of r can be derived without actually solving it. An example is provided by the extremely useful equivalence

$$\pi_x(0) \gtrless 1 \Leftrightarrow r \gtrless 0\,. \tag{3.3.8}$$

(In the next subsection we shall derive a biological interpretation of the quantity $\pi_x(0)$.) It is also relatively easy to show that r strictly increases with x, as is to be expected. Without loss of generality we may assume that $r>0$ for sufficiently large x (biological generality: animals whose populations go extinct at any food density do not exist). Moreover $r \downarrow -\mu$ for $x \downarrow \underline{x}$ defined by $\underline{l}(x) = l_j$ (see exercise 3.3.2 below). We conclude that there exists a unique food density x_c defined by

$$\pi_{x_c}(0) = 1\,, \tag{3.3.9}$$

such that $-\mu < r < 0$ for $x < x_c$ and $0 < r < r_m$ for $x > x_c$, where r_m is the solution to $\pi_\infty(r) = 1$. So x_c is the critical food density which the population needs to survive. For given values of the model parameters $\mu, \nu, \xi, \zeta, \eta, \omega, l_b$ and l_j one can easily calculate x_c numerically from the defining relation (3.3.9). So altogether we are able to give a rather neat and intuitively appealing characterisation of the dependence of the intrinsic rate of natural increase on the food density x.

EXERCISE 3.3.1: Prove (3.3.8).

EXERCISE 3.3.2: Prove that r strictly increases with x, and that $r \downarrow -\mu$ for $x \downarrow \underline{x}$. Hint: use that $\pi_x(s)$ decreases with x, and that $\pi_x(s) \downarrow 0$ for $x \downarrow \underline{x}$, for all $s > -\mu$.

EXERCISE 3.3.3: Show that

$$\pi_x(s) = \frac{(1-\kappa)f(x)}{\omega w_b} \left\{ \frac{\underline{l}^2(x)}{\mu+s} A^{\frac{\mu+s}{\gamma}} - \frac{2\underline{l}(x)(\underline{l}(x)-l_b)}{\mu+s+\gamma} A^{\frac{\mu+s+\gamma}{\gamma}} + \frac{(\underline{l}(x)-l_b)^2}{\mu+s+2\gamma} A^{\frac{\mu+s+2\gamma}{\gamma}} \right\} \tag{3.3.10a}$$

with

---

* A solution y to $h(x)=0$ is called simple if $\frac{dh}{dx}(y) \neq 0$.

$$A = \frac{l(x)-l_J}{l(x)-l_b}\,. \tag{3.3.10b}$$

Hint: use A as the integration variable in (3.3.7).

EXERCISE 3.3.4: Calculate $\psi(l_b)$ from condition (3.3.4b).

EXERCISE 3.3.5: Devise a strategy for finding a starting value such that a Newton iteration applied to (3.3.5) is guaranteed to converge. Hint: use the graphical interpretation of the Newton iteration and exploit the fact that $\pi_x(s)$ is a convex function of s.

EXERCISE 3.3.6: What can you tell about the dependence of r on the other parameters of the model?

EXERCISE 3.3.7: What can you tell about the dependence of x_c on the parameters of the model?

3.4. *Constant environments: reduction to an age-dependent problem*

The assumptions on the individual level imply that for constant food density there will be a fixed relation between size and age given by the solution to the equation for individual growth (3.1.3). Therefore we can also approach the problem of density independent *Daphnia* population growth from the angle of classical age dependent theory. By far the easiest way to do that is to start right from scratch, or more specifically from the considerations presented in subsection 3.1; and you are invited to do so yourself (possibly with some reference to IV.1, where we describe how to approach age dependent problems in general). In this subsection, however, we shall, for pedagogical reasons, show how the age dependent formalism can be derived from the more basic equation (3.2.1), as this allows us to introduce in a relatively painless manner the extremely useful mathematical concept of *characteristic,* which in this particular case is nothing but the trajectory of an individual in $\mathbb{R}^+\times\Omega$, where $\mathbb{R}^+$ is our time set. Moreover, the actual calculations performed to effect the transformation are in many ways exemplary of the bookkeeping encountered in dealing with structured population models in general. The reward at the end of the road will be a straightforward biological interpretation of the crucial quantity $\pi_x(0)$. The greater gain, however, will be the acquired facility with some types of calculations which initially may seem forbidding, but really are quite easy provided you have had some practice.

The mathematical reasoning underlying the introduction of characteristics generally, starts by rewriting (3.2.1a) as

$$\left[\frac{\partial}{\partial t} + g(x,l)\frac{\partial}{\partial l}\right]n(t,l) = -(g_l(x,l) + \mu)n(t,l)\,, \tag{3.4.1}$$

where

$$g_l(x,l) := \frac{\partial}{\partial l}g(x,l)\,. \tag{3.4.2}$$

This then suggests considering t and l as functions of an independent variable a such that precisely

$$\frac{d}{da} = \frac{dt}{da}\frac{\partial}{\partial t} + \frac{dl}{da}\frac{\partial}{\partial l} = \frac{\partial}{\partial t} + g(x,l)\frac{\partial}{\partial l} \tag{3.4.3}$$

(the first equality being nothing but the chain rule for functions of more than one variable), as this allows us to write (3.4.1), and perforce (3.2.1a), as an *ordinary* differential equation

$$\frac{d}{da}n(t(a),l(a)) = -(g_l(x,l(a)) + \mu)n(t(a),l(a))\,, \tag{3.4.4}$$

which can easily be solved explicitly (always assuming that we can find ways to provide an initial condition). The final step is to equate coefficients of like differential operators in (3.4.3) to arrive at

$$\frac{dt}{da} = 1\,, \tag{3.4.5a}$$

$$\frac{dl}{da} = g(x,l)\,. \tag{3.4.5b}$$

Solving (3.4.5) for various initial conditions (t,l) defines a family of curves in $\mathbb{R}^+\times\Omega$, such that through each point $(t,l)\in\mathbb{R}^+\times\Omega$ passes one and only one curve. These are the characteristics of (3.4.1) or equivalently (3.2.1a).

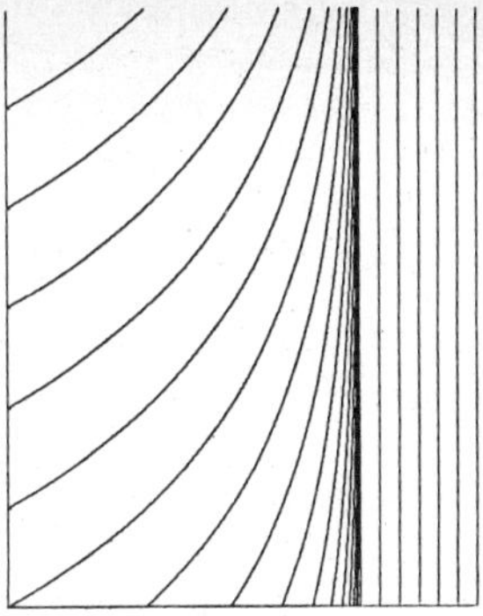

Figure 3.4.1. The characteristics of the *Daphnia* model

We shall now clarify the sense in which the characteristics can be interpreted as trajectories of individuals. (3.4.5a) tells us that a and clock time proceed at the same speed, suggesting an interpretation as age. This suggestion is supported by the observation that (3.4.5b) is nothing but the differential equation for individual growth (3.1.3b). The only problem is choosing the appropriate initial conditions. At age zero we should have $l(0) = l_b$. So for the characteristics starting from the left hand boundary $\{(t_0,l_b)|t_0>0\}$ in figure 3.4.1 we can conventionally set $t(0) = t_0$ and $l(0) = l_b$, and interpret the characteristic starting from (t_0,l_b) as the trajectory of an individual born at t_0. For the characteristics that start on the lower boundary $\{(0,l_0)|l_0\in\Omega\}$, we shall just set $t(0) = 0$ and $l(0) = l_0$, and interpret a as time since the start of the experiment, and the characteristic itself as the trajectory of an individual already present in the starting population, and having length l_0 at $t = 0$. (Of course we could formally extend the characteristics into the region of negative time in order to keep the age interpretation (but only formally, individuals in the starting population may have experienced different feeding histories!), and we shall do so in III.4.2, but for our present purposes such a ploy would be counterproductive.) With these conventions the characteristics are given by

$$t = t_0+a, \qquad l = e^{-\gamma a}(l_b-\underline{l}(x)) + \underline{l}(x)\,, \tag{3.4.6a}$$

for the characteristics which pass through the left hand boundary of $\mathbb{R}^+\times\Omega$, and

$$t = a, \qquad l = \begin{cases} e^{-\gamma a}(l_0-\underline{l}(x)) + \underline{l}(x) & \text{if } l_0<\underline{l}(x), \\ l_0 & \text{if } l_0\geqslant\underline{l}(x), \end{cases} \tag{3.4.6b}$$

for the characteristics passing through the lower boundary.

The whole reason for introducing the concept of characteristic is that it allows us to get rid of the partial differential equation (3.4.1), or equivalently (3.2.1a), by *integrating* it *along the characteristics, i.e.* by solving the ordinary differential equation (3.4.4), the initial condition for (3.4.4) being provided by either the initial size distribution $n(0,l)$ for the characteristics defined by (3.4.6b) or the boundary condition (3.2.1c), here rewritten as

$$g(x,l_b)n(t,l_b) = b(t)\,, \tag{3.4.7}$$

where b is the population birth rate for the characteristics defined by (3.4.6a). Substituting the explicit expression for $g_l(x,l)$ turns (3.4.4) into

$$\frac{dn(t(a),l(a))}{da} = \begin{cases} (\gamma-\mu)n(t(a),l(a)) & \text{for } l(a)<\underline{l}(x), \\ -\mu n(t(a),l(a)) & \text{for } \underline{l}(x)\leqslant l(a) \end{cases} \tag{3.4.8}$$

so that

$$n(t,l) = \begin{cases} e^{(\gamma-\mu)a}n(t_0,l_b) & \text{for } l<L(t) \\ e^{(\gamma-\mu)a}n(0,l_0) & \text{for } L(t)\leqslant l<\underline{l}(x) \\ e^{-\mu a}n(0,l_0) & \text{for } \underline{l}(x)\leqslant l \end{cases} \tag{3.4.9}$$

where

$$L(a) := e^{-\gamma a}(l_b-\underline{l}(x)) + \underline{l}(x) \tag{3.4.10}$$

is the individual growth curve. (3.4.9) relates the density, *i.e.* number of individuals per length unit, now at l to either the density of newborns or the density of individuals in the starting population from which they have grown. The factor $e^{-\mu a}$ represents survival. The factor $e^{\gamma a}$ derives from the fact that we are dealing with a contracting conveyor belt. The more we approach $\underline{l}(x)$ the more the conveyor belt contracts and so the more individuals have to be stowed away on the same amount of length.

To put (3.4.9) to work we still need to express a and either t_0 or l_0 as functions of t and l. To this end we introduce the function

$$A(l):=L^{(-1)}(l)=\gamma^{-1}\ln\left[\frac{\underline{l}(x)-l_b}{\underline{l}(x)-l}\right], \tag{3.4.11}$$

the age at which an individual first reaches length l. Then

$$t_0=t-A(l),\quad a=t-t_0\quad \text{for } l<L(t), \tag{3.4.12a}$$

$$a=t,\quad l_0=\begin{cases}L(A(l)-t) & \text{for } L(t)\leqslant l<\underline{l}(x),\\ l & \text{for } \underline{l}(x)\leqslant l\,.\end{cases} \tag{3.4.12b}$$

EXERCISE 3.4.1: Express (3.4.12) verbally, in biological terms.

EXERCISE 3.4.2: Reexpress (3.4.12b) more explicitly.

By combining (3.4.9) and (3.4.12) with (3.4.7) we have found an explicit expression for $n(t,l)$, but in terms of the as yet unknown birth rate b. However, we also have available (3.2.1d) expressing the birth rate in terms of n. By combining these two expressions we end up with an equation for the unknown b:

$$b(t)=\frac{(1-\kappa)f(x)}{\omega w_b}\int_{l_J}^{\max(l_J,L(t))} l^2 e^{(\gamma-\mu)a}\,\frac{b(t-a)}{\gamma(\underline{l}(x)-l_b)}\,dl+h(t)\,, \tag{3.4.13}$$

where h represents all births from parents which were already present at $t=0$:

$$h(t)=\frac{(1-\kappa)f(x)}{\omega w_b}\int_{L(t)}^{\underline{l}(x)} l^2 e^{(\gamma-\mu)t} n(0,l_0)dl+\frac{1}{\omega w_b}\int_{\underline{l}(x)}^{\bar{l}(x)} (f(x)l^2-\zeta l^3)e^{-\mu t} n(0,l)dl\,. \tag{3.4.14}$$

The integral in (3.4.13) can be rendered in a more intuitively appealing way by using $a=A(l)$ as the integration variable, using $dl\,/\,da=-\gamma(l-\underline{l}(x))=\gamma e^{-\gamma a}(\underline{l}(x)-l_b)$:

$$b(t)=\frac{(1-\kappa)f(x)}{\omega w_b}\int_{a_J}^{\max(a_J,t)} L^2(a)e^{-\mu(a)}\,b(t-a)da+h(t) \tag{3.4.15}$$

where

$$a_J:=A(l_J)\,, \tag{3.4.16}$$

the age at which an individual becomes reproductive. In words: the birth rate at t equals the cumulation of the births from all individuals born at different times between time zero and now, which are still among the living, plus the births produced by the remains of the initial population.

Equation (3.4.15) is a so-called *renewal equation* (or linear Volterra integral equation of the convolution type). In IV.2 we shall discuss a theorem telling that under fairly general conditions, one of which is that h does not grow too fast, the solution to equations like (3.4.15) will grow exponentially for large t. The speed of exponential growth can be determined by substituting

$$b(t)=ce^{rt}(1+o(t)) \tag{3.4.17}$$

in (3.4.15), dividing both sides by ce^{rt} and formally taking the limit for $t\to\infty$ (IV.2 will give a slightly more rigorous but considerably less appealing rendering), to arrive at the characteristic equation

$$1=\frac{(1-\kappa)f(x)}{\omega w_b}\int_{a_J}^{\infty} L^2(a)e^{-\mu a}\,e^{-ra}da\,. \tag{3.4.18}$$

Which brings us full circle.

EXERCISE 3.4.3: Combine (3.4.17) with (3.4.9), (3.4.12a) and (3.4.7) to rederive the stationary length distribution ψ.

From its derivation it should be intuitively clear that (3.4.18) can be nothing but the characteristic equation (3.3.5) in a different disguise. Therefore its right hand side provides another way of calculating π_x. This can also be checked algebraically. However, the form in which π_x is written in (3.4.18) gives a much better clue to the biological interpretation of the condition $\pi_x(0) \gtrless 1$. Since μ is the (constant) death rate, $e^{-\mu a}$ is precisely the probability that an individual survives to age a. The factor $(\omega w_b)^{-1}(1-\kappa)f(x)L^2(a)$ is the fertility of an individual aged a. So $\pi_x(0)$ equals the number of offspring a newborn individual is expected to bear. When this number is greater than one the population grows, otherwise it goes extinct.

A small sleight of hand and concentrating on the borderline case $\pi_x(0) = 1$ also makes it possible to interpret the characteristic equation: $\pi_x(s)$ equals the expected number of offspring per individual if we harvest a population by randomly removing individuals at a rate s, *i.e.* replacing μ by $\mu + s$. If the harvesting rate precisely equals the natural population growth rate the population size should remain constant, which in turn is equivalent to the expected number of offspring per individual in the harvested population being precisely equal to one.

EXERCISE 3.4.4: Check algebraically that indeed (3.4.18) is just (3.3.5) written in a different form.

EXERCISE 3.4.5: It is also possible to attack age dependent problems by interpreting age as an *i*-state variable. Write down the *p*-equations, direct from *a priori* considerations. * Also rederive (3.2.1) from the age dependent equations and the function L which expresses l in terms of a.

3.5. *Variable environments*

The simple hypothesis of constant food density will seldomly be fulfilled in practice except for very short periods. Food densities fluctuate due to external causes and as a result of the feeding activity of the *Daphnia*. In simple laboratory systems the external causes are kept constant, so only the feeding activity of the population has to be taken into account. The simplest assumption we can make about the food dynamics is

$$\frac{dx}{dt} = k(x) - f(x) \int_{\Omega} l^2 n(t,l)dl \,, \tag{3.5.1}$$

where k describes the dynamics of the food population in the absence of *Daphnia,* and the second term is precisely the amount of food ingested by all individuals together. For k we may *e.g.* choose

$$k(x) = \alpha - \beta x \,, \tag{3.5.2}$$

$$k(x) = \alpha x - \beta x^2 \,. \tag{3.5.3}$$

The first assumption corresponds to a constant inflow of fresh, nonreproducing, food particles combined with a constant food deterioration, *e.g.* through sinking (or dilution in a flow culture). The second assumption corresponds to logistic growth of an unstructured food population. More generally k may be assumed to be positive for $0<x<x_e$ and negative for $x>x_e$ for some parameter x_e.

If we consider the combined food-*Daphnia* system the first step will be the determination of any possible equilibria $\hat{n}$, $\hat{x}$. Clearly $\hat{n}(l) = 0$, $\hat{x} = x_e$ is an equilibrium: no *Daphnia* present and the food density at its unexploited equilibrium. This is usually called the trivial equilibrium or also the boundary equilibrium. The results of the previous sections imply that in any non-trivial (or internal) equilibrium the food density $\hat{x}$ must be such that

$$\pi_{\hat{x}}(0) = 1 \,, \tag{3.5.4}$$

since, given a constant food density $\hat{x}$, the *Daphnia* numbers should remain constant. In subsection 3.3 we already found that this equation has a unique solution

$$\hat{x} = x_c \tag{3.5.5}$$

provided $\pi_\infty(0)>1$. We conclude that at the nontrivial equilibrium the food density is solely determined by the *Daphnia* dynamics, independent of the details of the food dynamics as expressed in the function k. The corresponding equilibrium *Daphnia* density is necessarily of the form

$$\hat{n}(l) = c\psi_{x_c}(l) \,, \tag{3.5.6}$$

where ψ_{x_c} is the (normalized) stable size distribution (at the critical density x_c given by (3.3.4)), and c is a constant which has to be chosen such that the right hand side of (3.5.1) is zero for $x = \hat{x} = x_c$ and $n(t,l) = \hat{n}(l)$.

$$c = \frac{k(x_c)}{f(x_c) \int_{l_b}^{l(x_c)} l^2 \psi_{x_c}(l)dl} \,. \tag{3.5.7}$$

But of course we also require that $c>0$, and therefore a necessary and sufficient condition for the existence of a nontrivial equilibrium is that

$$x_c<x_e\,. \tag{3.5.8}$$

Condition (3.5.8) also emerges from a consideration of the stability of the boundary equilibrium. If we add a very small number of *Daphnia* to a food population which is at its own dynamical equilibrium x_e, the dynamics of the *Daphnia* will, initially at least, conform approximately to the assumption of a constant food density. The results from section 3.3 then tell us that the *Daphnia* population will go extinct whenever $x_e<x_c$ and will start to grow when $x_c<x_e$. In IV.4.2 this argument is extended to include the borderline case $x_e = x_c$: there it is proved that the boundary equilibrium is globally attractive if $x_0 \leqslant x_c$. Our results till now therefore can be summarized by saying that there exists a unique internal equilibrium whenever the autonomous food equilibrium is invadable, and when this is not the case no *Daphnia* population can maintain itself.

The stability of the internal equilibrium is a more complicated matter. The general results from bifurcation theory, discussed in chapter VI, suggest that the internal equilibrium will be asymptotically stable for parameter values which are such that x_e is but slightly larger than x_c.

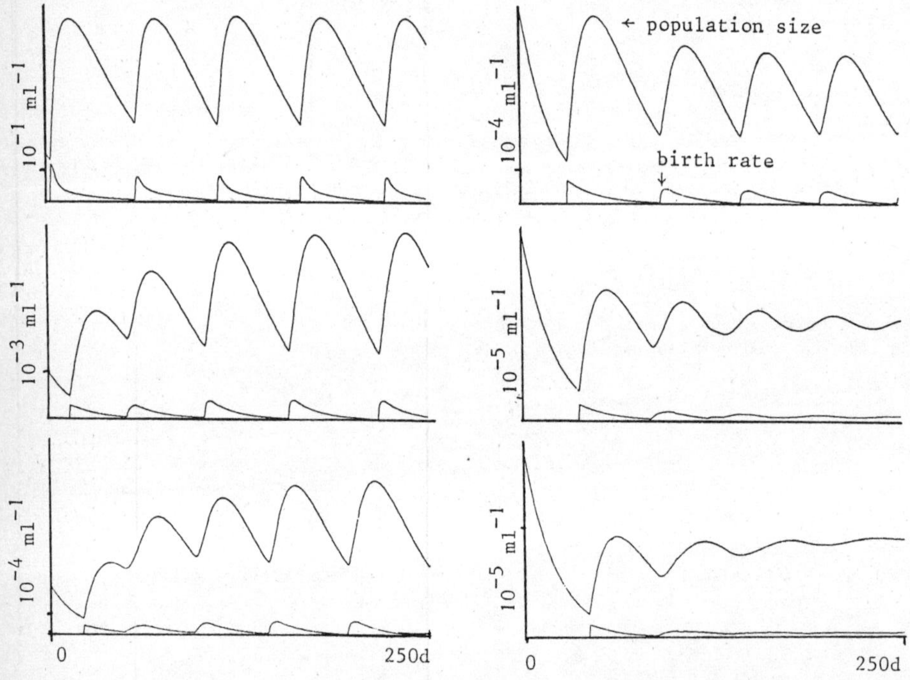

Fig. 3.5.1: Food regulated *Daphnia* dynamics: numerical results for a flow culture with continuous feeding (food dynamics (3.5.2)) for various values of the feeding rate. Upper trace: total population size. Lower trace: birth rate. Parameters $l_b = .8mm$, $l_j = 2.5mm$, $l_m = \kappa\nu / \zeta = 6.6mm$, $\gamma = \zeta / 3\eta = .17d^{-1}$, $\lambda_m = (1-\kappa)l_m\nu / \omega w_b = 10d^{-1}$, $\kappa = .3333$, $\xi = .7\times10^{-6}ml / cell$, $\nu = 1.8\times10^{6} cell / mm^2$, $\beta = .5d^{-1}$, $\alpha = 10^6, 5\times10^4, 4.5\times10^4, 4.37\times10^4, 4.36\times10^4 cell / ml.d$ (from top to bottom, left to right).

The numerical results shown in figure 3.5.1 indicate that for other parameter values it may become unstable and that asymptotically stable oscillations result. The suggestion is that the transition comes about through a so-called Hopf bifurcation, but before such a result can be proven rigorously, the theory of structured population models still needs to be developed a great deal further.

EXERCISE 3.5.1: Show that two species of *Daphnia* cannot coexist in stable equilibrium unless they have equal values of x_c. If this is not the case which species would you tip as winner?

* EXERCISE 3.5.2: Even if there were no numerical results one would suspect that any transition to instability would be

through a Hopf bifurcation. Why? Hint: Refer to the results from bifurcation theory sketched in VI.1.

3.6. Summary

In this section we approached the problem of modelling the population dynamics of ectothermic animals on the basis of simple physiological considerations, concentrating on "size" as the major *i*-state variable.

We started with some simple energetic and dimensional considerations on the individual level along the lines laid out by KOOIJMAN & METZ (1984). This way we arrived at a simple model for individual behaviour, taking account of feeding, growth, basal metabolism, reproduction (after reaching adult size), and starvation, all in dependence on individual length (or equivalently weight as we assumed selfsimilar growth), and food density. From this *i*-model we then derived a set of *p*-equations which formed the basis of our further considerations.

At a constant food density the population after some time will grow exponentially, with rate constant r, while the size distribution stabilizes. An explicit expression for the stable size distribution can be given once r is known, but r itself has to be calculated numerically from the characteristic equation. r increases with the food density x, being positive only for x above some critical value x_c which again can easily be calculated numerically.

At a constant food density, size is uniquely related to age, implying that then there should exist an equivalent age based model, and we have gone to some effort to derive this model from the equations of the size based model, mainly in order to demonstrate some useful mathematical concepts and techniques, in particular integration along characteristics.

In the final subsection we considered a dynamic food source, the consumption by the animals being an essential component of its dynamics. When the autonomous food equilibrium x_e is larger than x_c there exists a unique non-trivial equilibrium. Otherwise the consumers go extinct. At equilibrium the food density necessarily equals x_c and the size distribution of the consumers is given by the stable size distribution of the linear problem studied previously, for $r = 0$. Although the non-trivial equilibrium may be stable we found numerically that often it is replaced by a limit cycle as the stable attractor.

4. The cell size distribution

In the final example of this chapter we shall turn to unicellular organisms. From the very beginning of their science microbiologists have been interested in population growth curves *i.e.* the number of individuals in a culture plotted as a function of time. Microorganisms also provided the conventional testing ground for the first theoretical developments in population modelling. These early models were all phrased in terms of ordinary differential equations, as are most models today, without taking into account any aspect of the population structure, even if it was recognized that population growth characteristics, at least initially, tend to depend on the origin (and therefore the composition) of the inoculum.

Clearly individual cells do not divide purely at random as is assumed in the conventional differential equation models. Such traits as age, size, weight and biochemical composition all contribute to determining when the next cell division will occur. A natural approach, and the one advocated here, is to seek to describe the cell state as completely as is feasible by a limited number of numerical variables, and to derive equations which enable us to deduce relations between, on the one hand, the statistical and dynamical behaviour of populations of cells, and, on the other hand, the physiological processes which take place in the individual cells. Moreover, nowadays sophisticated equipment enables us automatically to gather statistical data on various aspects of population composition, thus putting those models on a level beyond that of mere figments of imagination.

At the end of the sixties at least three groups of theoreticians have, independently, derived such equations (BELL & ANDERSON, 1967; FREDRICKSON, RAMKRISHNA & TSUCHIYA, 1967; SINKO & STREIFER, 1967). They philosophised about the structure of mathematical models, did computer simulations and fitted experimental data. Yet after this strong start the development stagnated again, probably due to a combination of causes. Routine automated data gathering as well as routine computing were much less well developed than they are today. But the main factor probably was that the equations were so far removed from the available mathematical tools of the time, that there was no incentive for any concerted effort from the mathematical and theoretical biological research communities. All this has changed considerably by now, and the theory of structured cell population models is showing distinct signs of a renewed bloom.

Below we shall present a relatively simple model, in which cell size is the only *i*-state variable. This model will also be the main example around which the next chapter, dealing with the mathematical toolkit, revolves. In chapter V a more complicated model taking account of age as well will be developed, and chapter VI also contains two applications to cell population dynamics. In part B Voorn & Koch consider various statistical aspects of size structured cell populations and Kooijman explores the consequences of extending the *i*-state description to account for reserves.

4.1. The dynamics of individuals

Clearly the step from no population structure to a realistic structure, in which, say, all available knowledge about the biochemical processes within the cell is taken into account, is much too large. We therefore take an intermediate position and assume that the state of a cell is completely characterized by one quantity x which moreover obeys a physical conservation law, in the sense that at cell division it is divided up among the daughters. One can think of x as meaning: the length of oblong cells, the volume of a cell, dry mass, nitrogen content etc. But obviously not age. We shall call x size.

Given our choice of the *i*-state a cell has only three behavioural possibilities: growing, dividing and dying (including wash-out in a chemostat). Any other processes like the merging of cells as part of some sexual form of reproduction we simply leave out of the account.

As in *Daphnia* we shall assume growth to be completely deterministic, occurring at a rate V:

$$\frac{dx}{dt} = V(x), \quad V>0 .$$

Of course V will also depend on environmental variables such as temperature and nutrient concentration, but we shall only express this in our notation when the variation of these variables actually will be accounted for in the model. Otherwise the environment will simply be assumed constant.

Fission we shall treat as a stochastic process. In fact our description of the state by only one variable is by necessity rather poor. The best we may hope for is that it is the main variable determining the occurrence of division. The observed variability in the size at division we account for by building a stochastic element into the model. Interestingly enough it turns out that this non-mechanistic artifice needs further specification when the growth rate is subject to environmental variation (DIEKMANN, LAUWERIER, ALDENBERG & METZ, 1983). Several possibilities present themselves. We shall only consider what in a sense are the two extreme cases:

(i) We postulate that the rate, or *probability per unit of time,* at which cells of size x undergo fission is given by a function

$$b(x)\geqslant 0$$

which we assume to be known.

(ii) We postulate that each cell has a *stochastically predetermined size* at which fission has to occur, independent of any characteristics of its forebears, provided the cell does not die before that time. This may also be expressed more graphically by saying that division occurs when cell size reaches a stochastic threshold (cf fig. 4.1.1). The probability density of this hypothetical "size at division precluding death" will be denoted as ϕ_b, *i.e.* if we consider a large population of cells all starting life at so small a size that they cannot possibly divide yet, a fraction

$$\int_{x_1}^{x_2} \phi_b(x)dx$$

will divide between sizes x_1 and x_2, provided we have eliminated all possible causes of death. (In the terminology of PAINTER & MARR (1968), ϕ_b is called the probability density for the size at division in a sample of newborn cells; the supposition behind this term clearly being that a) even the largest newborn cells are so small that any smaller cell cannot divide, and b) no cells ever die (see exercise 4.1.5).)

In order to investigate the relation between these two possible paradigms we imagine a cohort of N_0 immortal cells all having size a at time $t = 0$, where a is the minimal size at which division can possibly occur. As time increases their size changes according to

$$\frac{dx}{dt} = V(x), \quad x(0) = a , \tag{4.1.1}$$

and their number according to

$$\frac{dN}{dt} = -b(x(t))N, \quad N(0) = N_0 . \tag{4.1.2}$$

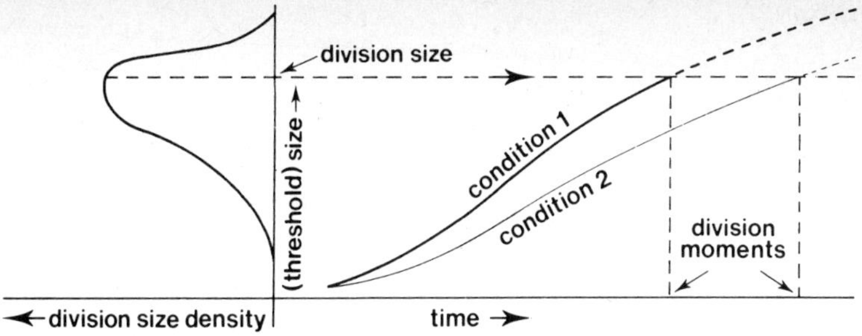

Fig. 4.1.1 The division on crossing a stochastic size threshold paradigm. Under condition 2 the ages at cell division are larger than under condition 1 but the distribution of the size at division remains the same.

Hence

$$N(t) = N_0 \exp\{-\int_0^t b(x(\tau))d\tau\} . \tag{4.1.3}$$

Using $\xi = x(\tau)$ as a new integration variable we see that

$$N_0 \exp\{-\int_a^x \frac{b(\xi)}{V(\xi)}\, d\xi\} \tag{4.1.4a}$$

cells reach size x. But this number should equal

$$N_0(1-\int_a^x \phi_b(\xi)d\xi) . \tag{4.1.4b}$$

And by differentiating these two expressions we deduce that

$$\phi_b(x) = \frac{b(x)}{V(x)} \exp\{-\int_a^x \frac{b(\xi)}{V(\xi)} d\xi\}, \tag{4.1.5}$$

and, conversely, by differentiating their logarithms,

$$b(x) = V(x) \frac{\phi_b(x)}{1-\int_a^x \phi_b(\xi)d\xi} . \tag{4.1.6}$$

So, if V does not depend either explicitly or implicitly on time, (i) and (ii) are just two different ways of doing the bookkeeping. But if V does vary with time, then at least one of the two functions b or ϕ_b has to be nonconstant as well. Fig. 4.1.2 depicts the essential dynamical difference between the two mechanisms. In that case we have to make a choice: what is the *intrinsic* characteristic of the cell, b, ϕ_b or still something else? This need not bother us for the time being as in most of this section and chapter II we shall assume that V is just a fixed function of x only. However, near the end of both the distinction will become a crucial modelling ingredient.

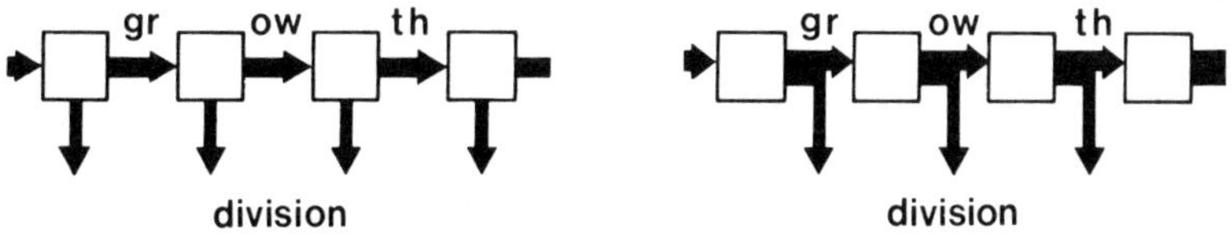

Fig. 4.1.2 The essential dynamical difference between division mechanisms (i) and (ii).

We complete the description of the division process by assuming that division always is into two exactly equal daughters, and our description of individual behaviour by assuming that deaths occur randomly at a possibly size dependent rate

$$\mu(x) \geq 0 .$$

EXERCISE 4.1.1: What is the probability that a cell having birth size a, in which division is inhibited but that is otherwise the same, survives till age τ? That it dies before τ? What is the probability density of its age at dying?

EXERCISE 4.1.2: What is the probability that the cell of the previous exercise survives till size x? The probability density of its size at dying?

EXERCISE 4.1.3: What is the probability that a cell that may both divide and die ever reaches age τ, size x? Assume again that it has birth size a.

EXERCISE 4.1.4: What is the probability that the cell from the previous exercise divided between ages τ and $\tau + d\tau$? What is the probability that it dies before reaching division?

EXERCISE 4.1.5: What is for the cells of the previous exercise the probability density of the size at division (counting only cells that reach division)?

EXERCISE 4.1.6: Assume that there is a lower limit a to the size at division, and that a is larger than the size of any newborn cell. What is the expected number of daughters reaching size a, of a cell that itself at this moment passes size a?

4.2. Formulation of the p equations

Just as for *Daphnia* the size density will fulfill the role of p-state. We shall denote it as n, *i.e.* $\int_y^z n(t,x)dx$ is the number of cells with sizes between y and z at time t. Drawing up the balance of growth, death and division during an arbitrarily small time interval, like we did in the previous sections, we find

$$\frac{\partial}{\partial t} n(t,x) = -\frac{\partial}{\partial x} (V(x)n(t,x)) - \mu(x)n(t,x) - b(x)n(t,x) + 4b(2x)n(t,2x) . \tag{4.2.1a}$$

The factor 4 in the birth term may come as a bit of a surprise. The first of the two 2's from which it is made up comes from the cells dividing into two. The second 2 derives from the fact that cells born into the interval $(x, x + dx)$ come from parents in the twice as long interval $(2x, 2(x + dx))$. So when the mass is lifted from the conveyor belt, doubled and deposited again it also is compressed sideways by a factor one half resulting once again in a doubling of its height.

The next step, as usual, is to see whether any side conditions have to be imposed to arrive at a well-defined mathematical problem. This also means that we have to specify more carefully what the individual state space Ω looks like. Let us assume that there exists a smallest size a at which fission can occur, *i.e.*

$$a = \inf\{x \mid b(x) > 0\} \tag{4.2.2}$$

is positive. Then $\frac{1}{2}a$ is the smallest size a cell can possibly have. Integration of (4.2.1a) with respect to x yields

$$\frac{d}{dt} \int_{\frac{1}{2}a}^{\infty} n(t,x)dx = V(\tfrac{1}{2}a)n(t,\tfrac{1}{2}a) - \int_{\frac{1}{2}a}^{\infty} \mu(x)n(t,x)dx + \int_{a}^{\infty} b(x)n(t,x)dx . \tag{4.2.3}$$

In words: the number of cells changes due to resp. (i) cells growing beyond $\frac{1}{2}a$, (ii) cells that die, and (iii) cells that divide (at each division cell number increases by one). We want the first term to be zero (indeed, where would those cells come from?) Since $V(\frac{1}{2}a) > 0$ we *have* to impose the boundary condition

$$n(t,\tfrac{1}{2}a) = 0 . \tag{4.2.1b}$$

We shall also assume that there exists a maximal size x_{max}, in the sense that cells will divide with probability one before reaching x_{max}, in accordance with accepted biological wisdom. Without loss of generality we may set $x_{max} = 1$.

(We can always adjust the unit of size to suit our purpose.) How should this biological assumption be expressed in our mathematical formulation? In terms of the probability density ϕ_b our assumption is equivalent to

$$\int_a^1 \phi_b(x)dx = 1\,,$$

which, in view of the equality of (4.1.4a, b), in turn is equivalent to b having a non-integrable singularity at $x = 1$.

Our two assumptions of a minimal size a at division and a maximal size $x_{max} = 1$, imply that $\Omega = (\frac{1}{2}a, 1)$. As n only "lives" on Ω the term $4b(2x)n(t,2x)$ in (4.2.1a) is undefined for $x > \frac{1}{2}$ but for convention 2.5.3, which we repeat here as

CONVENTION 4.2.1: If a transformed argument falls outside Ω we shall assume that the term in which it occurs equals zero.

REMARK 4.2.2: Also (4.2.1b) should be interpreted as $\lim_{x \downarrow \frac{1}{2}a} n(t,x) = 0$.

Our assumptions on the i-level also, somewhat unexpectedly, preclude us from prescribing just any size distribution on $(\frac{1}{2}a, 1)$ as initial condition for (4.2.1)! Cells with larger sizes become increasingly scarce relatively, and therefore any size density compatible with our basic mechanism will have to taper off near $x = 1$. In the next chapter it will be shown that the collection of size densities $\nu(x)$ such that $\nu(x)/(1-\int_{\frac{1}{2}a}^{x} \phi_b(\xi)d\xi)$ is bounded for $x \uparrow 1$ is invariant under (4.2.1). Confining the attention to compatible size densities also prevents the birth term $4b(2x)n(t,2x)$ growing out of bounds for $x \uparrow \frac{1}{2}$ (provided ϕ_b is bounded). It precisely is this sort of subtlety which makes the theory of structured populations so complicated mathematically, but also so enticing.

EXERCISE 4.2.3: Can you also derive (4.2.1b) from a heuristic conveyor belt type argument?

EXERCISE 4.2.4: Multiply (4.2.1) with x and integrate over x to derive a differential equation for the "total biomass", and interpret the result in biological terms. What happens to the last two terms, and why?

EXERCISE 4.2.5: Assume in the previous exercise that μ is constant and that $V(x) = \alpha x$. How does the total biomass develop over time? (Unfortunately the special case $V(x) = cx$ has rather unpleasant properties in other respects; see below.)

EXERCISE 4.2.6. Another way to guarantee that the cells divide before reaching size $x_{max} = 1$, is to assume that $V(x) \downarrow 0$ for $x \uparrow 1$. Does this assumption have any consequences for b? How can one distinguish experimentally between the two assumptions using the (population) distribution of the time until division. (This quantity is relatively easy to obtain experimentally, such in stark contrast to the individual growth curve)?

4.3. The stable cell size distribution

Equation (4.2.1) is dealt with at great length in chapter II. Therefore we shall confine ourselves here to giving a short heuristic introduction to the main results, stressing the biological interpretation. In this subsection we shall concentrate on the large time behaviour for constant environments (*i.e.* we assume V to depend on x only), in subsection 4.4 we shall discuss how the results from the present subsection can be used to make inferences about individual behaviour from observations on the population level, and in subsection 4.5 we shall consider a specific case of population regulation through resource limitation.

Proceeding by analogy with the previous two examples we substitute a trial solution $n(t,x) = e^{kt}\lambda(x)$ in (4.2.1) (again we use the notation of PAINTER & MARR (1968)), and find

$$\frac{dV(x)\lambda(x)}{dx} = -(k+\mu(x))\lambda(x) - b(x)\lambda(x) + 4b(2x)\lambda(2x)\,. \tag{4.3.1a}$$

For $x > \frac{1}{2}$ the last term drops out, and (4.3.1) becomes an ordinary differential equation which we can solve provided we prescribe some arbitrary initial condition, at $x = \frac{1}{2}$. Given λ on $[\frac{1}{2}, 1)$ we can then work our way backwards and solve λ on $(\max\{\frac{1}{2}a, \frac{1}{4}\}, \frac{1}{2})$ and so on till we have filled Ω. For the ease of presentation we shall from now on assume

$$a \geq \frac{1}{2}, \tag{4.3.2}$$

so that we only have to make one retrograde step. (The general case is treated in great detail in HEIJMANS (1985b)). From the requirement that λ satisfies the boundary condition

$$\lambda(\tfrac{1}{2}a) = 0 \tag{4.3.1b}$$

it then is found that k should satisfy the *characteristic equation*

$$\pi(k) = 1 \tag{4.3.3a}$$

with

$$\pi(s) = 2\int_a^1 \frac{b(\xi)}{V(\xi)} \exp\left[-\int_{\xi/2}^{\xi} \frac{s+\mu(\eta)+b(\eta)}{V(\eta)}\,d\eta\right] d\xi\,, \tag{4.3.3b}$$

and (up to multiplication with an arbitrary constant)

$$\lambda(x) = \frac{\rho(x)}{V(x)} \exp\left[-\int_{a/2}^{x} \frac{k+\mu(\xi)+b(\xi)}{V(\xi)}\,d\xi\right] \tag{4.3.4a}$$

with

$$\rho(x) = \begin{cases} 1 & \text{for } \frac{1}{2} \leq x \leq 1 \\ 2\displaystyle\int_a^{2x} \frac{b(\xi)}{V(\xi)} \exp\left[-\int_{\xi/2}^{\xi} \frac{k+\mu(\eta)+b(\eta)}{V(\eta)}\,d\eta\right] d\xi & \text{for } \frac{1}{2}a \leq x \leq \frac{1}{2}. \end{cases} \tag{4.3.4b}$$

(Note that the relation $\pi(k) = 1$ ensures that ρ is continuous at $x = \frac{1}{2}$.)

REMARK 4.3.1: Instead of integrating backwards from $x = \frac{1}{2}$ we could also have integrated forwards from $x = \frac{1}{2}a$ using (4.3.1b) as initial condition. (4.3.3) then results from the requirement that λ be continuous at $x = \frac{1}{2}$.

Since the functions V, μ and b are nonnegative the function π is strictly decreasing on the real axis. Moreover $\pi(-\infty) = +\infty$ and $\pi(+\infty) = 0$. We conclude that there indeed exists precisely one simple real solution which we have chosen to call k. Given expressions for V, b and μ it is relatively easy to determine k numerically *(e.g.* by Newton iteration). As in the *Daphnia* case the sign of k can easily be determined from the relation

$$\pi(0) \gtreqless 1 \Leftrightarrow k \gtreqless 0\,. \tag{4.3.6}$$

INTERLUDE 4.3.2: *A biological interpretation of the quantities* $\pi(0)$ *and* $\pi(s)$.
Since $a \geq \frac{1}{2}$ and size at birth $x_b \leq \frac{1}{2}$ each cell needs to pass size $x = a$ before it can possibly divide. Hence the average contribution of an arbitrary cell passing size $x = a$ to the next generation of cells can be measured by counting how many of its daughters on the average reach size $x = a$ themselves. This quantity is made up of the following components:

(i) he probability that a potential mother reaches size ξ is $\exp(-\int_a^{\xi} \frac{\mu(\eta)+b(\eta)}{V(\eta)}d\eta)$

(ii) The probability per unit of size that it divides at size ξ, given that it has reached that size is $b(\xi)/V(\xi)$ (the factor $1/V(\xi)$ converts probability per unit of time to probability per unit of size). The number of daughters is exactly 2 and each has size $\frac{1}{2}\xi$.

(iii) The probability that a cell born with size $\frac{1}{2}\xi$ safely grows up to size a is $\exp(-\int_{\xi/2}^{a} \frac{\mu(\eta)}{V(\eta)}d\eta)$.

By integrating over all possible values of the size at division ξ we arrive at the interpretation: $\pi(0)$ is the average number of offspring of cells sampled at size a growing up to at least size a again.

To interpret $\pi(s)$ for general s we observe that in (4.3.3) s occupies exactly the same position as μ, allowing us to interpret it as an additional death rate which can be imposed on the population in order to induce exact replenishment: $\pi(s) = 1$. (Compare the argument at the end of section 3.4.)

EXERCISE 4.3.3: Calculate $\pi(0)$ for the special case $\mu(x) = 0$ and interpret the answer. Hint: Use that $b(x) = 0$ for $x<a$.

Now that we have found a stable size distribution we should ask the convergence question: will the population state, after suitable normalization, converge to the stable size distribution, and will the population size settle on asymptotic exponential growth, independent of the initial condition (the composition of the inoculum)? Thus far we have ducked this question and for very good reasons. The answer turns out to be : it depends ; we know an (important) special case where it doesn't! More precisely:

(i) If $V(2x) \neq 2V(x)$ for some $x \in (\frac{1}{2}a, \frac{1}{2})$ then for $t \to \infty$

$$e^{-kt}n(t,x) \to C\lambda(x) \tag{4.3.7}$$

where C depends on the initial condition (in chapter II we show how C can be computed; an explicit expression for C may be found in DIEKMANN (1985)).

(ii) If, however, $V(2x) = 2V(x)$ for all $x \in (\frac{1}{2}a, \frac{1}{2})$ then

$$e^{-kt}n(t,x) \to p(t - \int_{a/2}^{x} \frac{d\xi}{V(\xi)})\lambda(x) \tag{4.3.8}$$

where p is a periodic function, which depends on the initial condition, with period

$$\int_{a/2}^{a} \frac{d\xi}{V(\xi)}$$

(in chapter II we show how p can be computed).

And the exceptional case (ii) is not just a bizarre counterexample living only in the feverish imagination of mathematicians: the expression for V most often encountered in the microbiological literature, $V(x) = \alpha x$, precisely belongs to this class! (Exercise 4.2.5 shows though that for this special case total biomass behaves quite nicely, even if cell number doesn't.)

In hindsight there is a simple biological argument, due to BELL & ANDERSON (1967), who were the first to notice the non-convergence problem, which makes the special case $V(2x) = 2V(x)$ easy to understand. Consider two cells A and B which have equal sizes at $t = 0$, and assume that A splits immediately into two daughters a_1 and a_2 and B only splits at $t = t_1 > 0$ into b_1 and b_2. During the time interval $[0,t_1)$, B grows twice as fast as each of the a's. So at t_1 the a's and b's have exactly equal size! The relation "equal size" therefore is hereditary, extending over the generations, and therefore many properties of the initial condition will stay manifest over time.

The Bell & Anderson explanation also immediately points to a "biological resolution" of the non-convergence problem. The anomaly apparently hinges on a combination of the two contingencies that (i) $V(2x) = 2V(x)$ for all x in $(\frac{1}{2}a, \frac{1}{2}]$, and (ii) cells split into two exactly equal parts. Therefore one expects that as soon as either (i) or (ii) is replaced by a different assumption the trouble will disappear. As the case $V(x) = \alpha x$ clearly is an important special case the next thing to do is to see whether modifying (ii) is of help. And indeed, HEIJMANS (1984b) has proved that when a mother of size x can split into two daughters of sizes qx and $(1-q)x$, where q is a random variable with density $d(q)$ independent of x, convergence to a stable size distribution occurs even when $V(2x) = 2V(x)$. (See also exercise 4.3.6 and exercise II.11.3) So clearly our quandary depended on the unlucky combination of two simplifying assumptions. Yet both assumptions seem to be met to a good approximation in many practical cell populations. In such cases the present model predicts slow convergence towards a stable distribution (in the limiting case there is no convergence at all). A rough estimate of the relaxation time gives for a standard deviation of q equal to 0.01 a relaxation time of 60 generations versus 2½ generations for s d q = 0.05.

REMARK 4.3.4: If the probability density d is highly peaked near $q = \frac{1}{2}$, formulae (4.3.3) and (4.3.4) stil will give a good approximation to the population growth rate k and the stable size distribution λ, even for the case $V(x) = \alpha x$. But we have to interpret λ given by (4.3.4) as the limit of the stable size of the more complicated model for the variance in q going to zero, and not as the limiting size distribution of the simplified model for $t \to \infty$.

The common observation in cell cultures, where environmental conditions are kept constant by regularly changing the culture medium, is that after a certain initial time the normalized size distribution hardly changes, while numbers and other relevant quantities (total biomass, amount of DNA) grow exponentially. Cell biologists then say that the culture is in *steady state* or in a state of *balanced exponential growth* (see PAINTER & MARR (1968) for a precise description of the terminology). So here clearly is a case indeed where our results (4.3.3) and (4.3.4) are highly relevant.

* EXERCISE 4.3.5: Equation (2.5.5a) (the equation for the p-state of the invertebrate predator), with boundary condition (2.5.5b′), also allows an exceptional class of digestion rates f for which there is no convergence to a stable stationary solution. (Clearly $f(s) = -as$ does not belong to this category). Could you guess which ones? Hint: In the predator model jumps were additive in the i-state variable, in the cell growth model multiplicative, and the anomaly apparently has to do with a relation between the nature of the continuous drift (f or V) and that of the jumps.

EXERCISE 4.3.6: Show that the unequal division model referred to above leads to the equation

$$\frac{\partial}{\partial t}n(t,x) = -\frac{\partial}{\partial x}(V(x)n(t,x)) - \mu(x)n(t,x) - b(x)n(t,x) + 2\int_x^1 \frac{d(q)}{q}\, b(\frac{x}{q})n(t,\frac{x}{q})dq\,. \tag{4.3.9}$$

Check that necessarily $d(q) = d(1-q)$, *i.e.* d is symmetric about $q = \frac{1}{2}$. Moreover, check that the equation (4.2.1a) is recovered by formally putting $d(q) = \delta(q-\frac{1}{2})$ where δ is the Dirac delta function(al). Exercise II.11.3 contains further information about (4.3.9).

EXERCISE 4.3.7: Show by integration of equation (4.3.1a) that $k \geqslant 0$ when $b(x) \geqslant \mu(x)$ for $x \in \Omega$ (which presupposes that $\mu(x) = 0$ for $x \in (\frac{1}{2}a, a]$, *i.e.* preparing for division may be risky, but deaths of pre-division cells can be neglected) and $k \leqslant 0$ when $b(x) \leqslant \mu(x)$ for $x \in \Omega$ (which requires that μ too has a singularity at $x = 1$). Use $\lambda(\frac{1}{2}a) = \lambda(1) = 0$, and $\lambda \geqslant 0$. Interpret the result.

EXERCISE 4.3.8: Multiply (4.3.1a) with x and integrate. Show that $k > 0$ when $V(x) \geqslant x\mu(x)$ for $x \in \Omega$ while $k \leqslant 0$ when $V(x) \leqslant x\mu(x)$ for $x \in \Omega$. Interpret the result.

4.4. *The inverse problem*

Thus far we have assumed that V, b and μ are known, but in practice this is hardly ever the case. In fact in cell- and microbiological applications often the main reason for the whole modelling exercise is that we want to use the experimentally obtained (stable) size distribution λ, and the specific population growth rate k to obtain information about V, b and μ, since it is too difficult (or downright impossible) to determine these functions direct from experimental observations on individual cells. This is called the *inverse problem:* measurements at the population level are used to get information about physiological processes at the individual level (see also the contributions by Van Straalen and Voorn & Koch in part B).

The inverse problem comes in many guises, depending on what we may assume to be known. Moreover, there are the practical problems of measurement errors and all other kinds of noise. However, as these notes are first and foremost about modelling we shall start by ignoring these problems, notwithstanding their great practical importance, and concentrate on the identifyability problem per se. Moreover we shall throughout make the simplifying assumption that $\mu = 0$ (cell deaths indeed usually are negligible).

Putting $\mu = 0$ and integrating once in (4.3.1) we find the identity

$$V(x) = \frac{1}{\lambda(x)} \int_{a/2}^{x} [4b(2\xi)\lambda(2\xi) - b(\xi)\lambda(\xi) - k\lambda(\xi)]d\xi \tag{4.4.1}$$

which cell biologists call the equation of Collins & Richmond. So if b were known we can determine V from k and $\lambda(x)$. Unfortunately b usually is equally unknown. An ingenious experimental way to overcome this difficulty is provided by the observation that $b(x)\lambda(x)$ is proportional to the size distribution ϕ of cells which are in the process of division with the constant of proportionality equal to the population growth rate k (in reality division is not instantaneous so that ϕ is indeed observable), and $2b(2x)\lambda(2x)$ is proportional to the size distribution ψ of newborn cells (which often are recognizable for a short time after division occurred). So if one can measure either ϕ or ψ experimentally one can determine V from (4.4.1).

If V is known but b is not, we can determine $b(x)$ from either

$$b(x) = -\frac{1}{\lambda(x)} \frac{d(V(x)\lambda(x))}{dx} - k, \quad a \leqslant x < 1 \tag{4.4.2a}$$

or

$$b(2x) = \frac{1}{4\lambda(2x)} \{\frac{d(V(x)\lambda(x))}{dx} + k\}, \quad \tfrac{1}{2}a \leqslant x < \tfrac{1}{2}\,. \tag{4.4.2b}$$

(Recall the assumption $a \geq \frac{1}{2}$.) The requirement that these two formulae should yield a similar appearance of b can be used as a check for an assumed form of V. Unfortunately the fact that measurements usually yield information about λ in the form of a histogram somewhat spoils the easy elegance of these formulae.

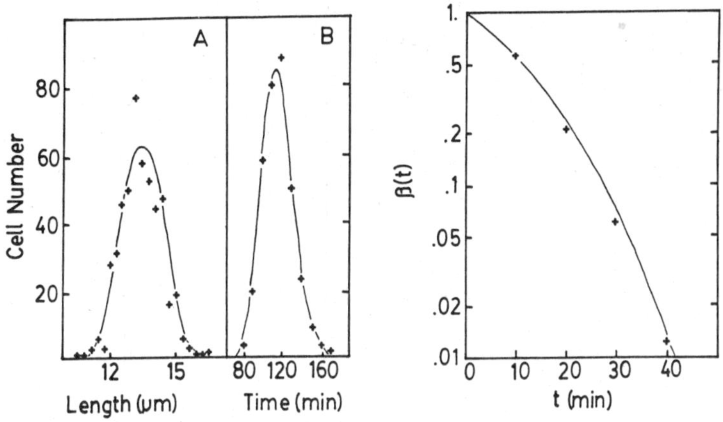

Fig. 4.4.1 Probability density of division size (A) and generation time (B) and probability β that two sister cells differ in generation time by $t(C)$, for an exponentially growing population of fission yeast *(Schizosaccharomyces pombe)*. Data from Miyata *et al* (1978). The continuous curves are predictions from the model with $V(x) = kx$ and $b(x) = \hat{b}(x-a)^2/(1-x)$ for $x>a, a = 0.6$ and $\hat{b}/k = 100$. From Tyson & Diekmann (1986).

When data are few and noisy, as they almost invariably are, the only way to proceed is to make plausible guesses for the functional forms of b and V, still depending on a limited number of parameters. Subsequently the values of the parameters are estimated by juggling them around to get as good a match as possible between λ calculated from the explicit expression (4.3.4) and the experimentally observed size distribution. The closeness of the fit obtained provides a first check of the model. However, the really convincing checks are provided by the ability of the model to predict other experimentally observable quantities. An example is provided by the work of Tyson & Diekmann (1986), who, using data on fission yeast reported by Miyata, Miyata & Ito (1978) fitted the "sloppy size control" model of this section in the way indicated, and subsequently found remarkably good agreement between observed and predicted generation time distributions (see fig. 4.4.1).

We conclude our too brief sketch with stressing once more that the inverse problem has many variants and is frought with many subtle difficulties, which are worthy of a great deal of attention. But this is all outside the scope of this chapter. The paper by Voorn & Koch in part B deals with the moment relations between the various distributions involved, using (a generalized version of) the Collins-Richmond equation as its main tool. Some interesting early references which also contain concrete applications, are Bell & Anderson (1967) and Anderson, Bell, Petersen & Tobey (1969).

4.5. Limits to growth

In the previous sections we found that the number of cells will grow exponentially when $k>0$. In reality this will be prevented by exhaustion of space, nutrients etc. In our model conception the causal chain leading to a stagnation of population growth starts with a stagnation in individual growth which in turn slows down the rate of cell fission, and thereby the increase in cell numbers. In this section we shall assume that it is the nutrients which are in short supply. In that case the obvious way to model the feedback loop between population size and nutrient availability is to introduce an additional finite dimensional variable R representing the availability of the various necessary nutrients and to specify (i) how the individual growth rate depends on R, (ii) how nutrient consumption depends on both R and cell size, and (iii) what other contributions there are to the change in R. Since in this set up the growth rate depends implicitly on time the distinction made in subsection 4.1 between various conceptions about the detailed probabilistic nature of the fission process becomes very relevant. In this section we shall opt for the stochastically predetermined division size, or stochastic threshold, assumption, mainly since this choice allows us to exploit the linear machinery developed previously (and more fully in chapter II) to analyse an inherently nonlinear situation. We refer to chapter

VI for a discussion of some other approaches.

EXERCISE 4.5.1: Discuss ways to distinguish experimentally between the two possible assumptions about the division process from subsection 4.1 by manipulating the food availability. Which of the two possibilities has a greater plausibility in your opinion?

For definiteness we shall assume that the food dynamics is that of a chemostat. This is a vessel into which a nutrient broth is pumped at a constant rate, while excess medium is removed at the same rate so that the total living volume remains the same. (see figure 4.5.1.) The vessel is continually stirred to keep the cells in suspension and to guarantee spatial homogeneity of the culture conditions. We shall assume moreover that all nutrients except one so-called *limiting nutrient* (also called *limiting substate)* are available in excess, making R effectively one dimensional.

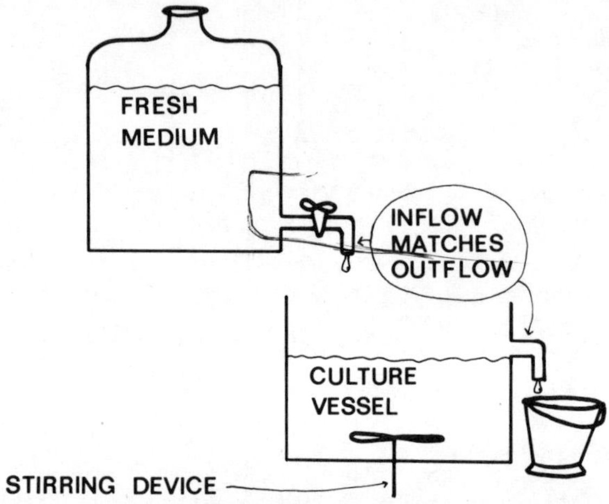

Fig. 4.5.1. Schematic representation of a chemostat

Moreover we assume that the growth rate factorizes in the sense that (with some abuse of notation)

$$V = \beta(R)V(x) \tag{4.5.1}$$

In DIEKMANN, LAUWERIER, ALDENBERG & METZ (1984) this is called the *structural nutrient hypothesis,* since it seems a reasonable assumption when the growth rate is limited by the uptake of nutrients like nitrate or phosphate which are all transformed into cell biomass, as opposed to energy which also is consumed in basal metabolism (compare the *Daphnia* model of the previous section).

REMARK 4.5.2: The assumption commonly made in the microbiological literature is that β is the Monod (or Michaelis-Menten) function

$$\beta(R) = \frac{k_1 R}{1+k_2 R} \tag{4.5.2}$$

(compare section 3.1), which is approximately proportional to R for small R and saturates for large R.

In combination with the stochastic threshold assumption (4.5.1) gives us for the division rate

$$b(R,x) = \beta(R)V(x)\delta(x)\,, \tag{4.5.3}$$

where

$$\delta(x) = \frac{\phi_b(x)}{1-\int_a^x \phi_b(\xi)d\xi}\,. \tag{4.5.4}$$

Finally, keeping to the structural nutrient interpretation also forces us to assume that the consumption of limiting nutrient by the cell population as a whole is just proportional to the total biomass accretion of the cell population

$$\int_{a/2}^{1} \beta(R)\, V(x) n(t,x) dx\ .$$

If we also assume that cell death can be neglected, so that the only "death" term is due to washout of cells in the effluent medium we get

$$\frac{\partial}{\partial t} n(t,x)) = \beta(R(t))\{-\frac{\partial}{\partial x}(V(x)n(t,x)) - V(x)\delta(x)n(t,x) + 4V(2x)\delta(2x)n(t,2x)\} - Dn(t,x) \tag{4.5.5}$$

$$\frac{dR}{dt} = D(R_i - R(t)) - \alpha\beta(R(t)) \int_{a/2}^{1} V(\xi)n(t,\xi)d\xi \tag{4.5.6}$$

where

$$D \equiv \text{dilution rate} \equiv \frac{\text{flow rate of medium}}{\text{volume of culture vessel}}$$

$R_i \equiv$ concentration of limiting substrate in the fresh medium

$\alpha \equiv$ conversion factor relating biomass units substrate units

EXERCISE 4.5.3: Check that total biomass $W(t) = \int_{a/2}^{1} xn(t,x)dx$ satisfies

$$\frac{dW(t)}{dt} = \beta(R(t)) \int_{a/2}^{1} V(\xi)n(t,\xi)d\xi - DW \tag{4.5.7}$$

and that $Z(t) = \alpha W(t) + R(t)$ satisfies

$$\frac{dZ}{dt} = D(R_i - Z(t)) \tag{4.5.8}$$

and that consequently $Z(t) \to R_i$ for $t \to \infty$. Also argue that this result is immediately obvious from the (chemical) interpretation of Z.

From the way equation (4.5.5) is written it appears that the change in the size distribution is made up of two components: (i) the usual growth cum division, and (ii) washout. The nutrient availability only influences the first process by changing its overall speed. As the washout is not size selective it only influences population numbers but not the relative frequencies of the cell sizes. In other words, the process by which the size distribution equilibrated in the linear case is still there except that it now proceeds at a variable speed. This idea can be captured in the following theorem which is proved in chapter II:

THEOREM 4.5.4: Suppose that $V(2x) \neq 2V(x)$ for some $x \in (\frac{a}{2}, \frac{1}{2}]$. Then for large t the solution to (4.5.4) behaves like

$$n(t,x) = \rho(t)\{\lambda(x) + o(1)\} \tag{4.5.9}$$

where ρ satisfies

$$\frac{d\rho}{dt} = (k\beta(R) - D)\rho \tag{4.5.10}$$

and where λ and k are defined by (4.3.4) and (4.3.3) resp., with $\mu(x) = 0$ and $b(x) = V(x)\delta(x)$.

Substituting (4.5.9) in (4.5.6) gives, neglecting the o(1) term,

$$\frac{dR}{dt} = D(R_i - R) - \alpha_0 \beta(R)\rho \tag{4.5.11}$$

with

$$\alpha_0 = \alpha \int_{a/2}^{1} V(\xi)\lambda(\xi)d\xi\ .$$

(4.5.10) and (4.5.11) together govern the asymptotic behaviour of ρ and R, and by (4.5.9) also the asymptotic behaviour of the solution to (4.5.5) and (4.5.6).

Equations (4.5.10) and (4.5.11) are just the traditional chemostat equations of unstructured population dynamics, the asymptotic behaviour of which is well known (see the exercises below), implying that here at least we have a structured population model which can be solved in toto; but at the cost of some rather special, even if not unreasonable, assumptions.

An indirect check of our special assumptions is provided by the prediction that the equilibrium size distribution (which can easily be observed noninvasively in the effluent medium) should be independent of the chemostat parameters D and R_i. One set of possibly relevant data can be found in WILLIAMS (1971). These certainly appear to be at variance with our prediction (but unfortunately the experimental conditions are indicated rather loosely, so that this cannot be considered a definite check even in this special case). We would be very grateful for any future data of this type. Only by considering a large number of special cases it is possible to evaluate the range of our approach.

Even without any experimental evidence it can be seen that there should be a wide range of cases in which our simplifying assumptions cannot possibly hold good. Therefore there is every reason to study other special cases in which either (i) the function $V = V(R,x)$ is not the product of two factors $\beta(R)$ and $V(x)$, or (ii) b is not a product of $V(R,x)$ and some function $\delta(x)$ indepent of R. Unfortunately such cases are essentially more complicated since it will no longer be possible to reduce the problem to a linear one through a simple time scaling trick. We are inevitably led to enter the domain of infinite dimensional nonlinear dynamical systems; but this is postponed to chapter VI.

EXERCISE 4.5.5: Show that

$$\int_{a/2}^{1} V(\xi)\lambda(\xi)d\xi = k \int_{a/2}^{1} \xi\lambda(\xi)d\xi \,. \tag{4.5.12}$$

What is the interpretation of this relation? Hint: Use (4.3.1) with $\mu = 0$.

EXERCISE 4.5.6: Let β be a strictly increasing function with $\beta(0) = 0$ and $\beta(\infty)<\infty$. We denote by β^{-1} the inverse function of β. Show that the system (4.5.10) and (4.5.11) has equilibrium points $(0,R_i)$ and $(\hat{\rho}, \hat{R})$ where

$$\hat{\rho} = \frac{k}{\alpha_0}(R_i - \beta^{-1}(D/k)) \qquad \hat{R} = \beta^{-1}(D/k) \tag{4.5.13}$$

provided that $D/k<\beta(R_i)$. If $D/k\geqslant\beta(R_i)$ only the trivial equilibrium remains. Interpret this difference between the two cases in biological terms.

EXERCISE 4.5.7: Use the result of exercise (4.5.3) together with theorem 4.5.4 to show that

$$R(t) + \frac{\alpha_0}{k}\rho(t) \rightarrow R_i \text{ for } t\rightarrow\infty.$$

Conclude that the asymptotic behaviour of ρ is described by the scalar differential equation

$$\frac{d\rho}{dt} = (k\beta(R_i - \frac{\alpha_0}{k}\rho) - D)\rho \,, \tag{4.5.14}$$

and from this that all solutions converge towards $\rho = 0$ when $\beta(R_i)\leqslant D/k$ and to $\rho=\hat{\rho}$ when $\beta(R_i)>D/k$.

EXERCISE 4.5.8: Discuss how β can be estimated by observing the composition of the effluent medium while (slowly) changing the flow rate. Hint: How would you determine α_0, k?

EXERCISE 4.5.9: Assume that two species of microorganisms satisfying all the assumptions of this subsection and both dependent on the same limiting substrate are grown together in a chemostat. Analyse the outcome of the competition on the assumption that β_1 and β_2 both strictly increase with R. Hint: analyse the ordinary differential equations describing the large time behaviour, using the asymptotic conservation condition for total nutrient concentration to reduce the problem to a two-dimensional one.

EXERCISE 4.5.10: Consider a microorganism which satisfies all the assumptions of this subsection except for the structural nutrient hypothesis. Assume that $\partial V(x,R)/\partial R>0$. Analyse the existence and uniqueness of any nontrivial equilibrium. Hint: Follow the pattern laid out in the *Daphnia* story of the previous section.

4.6. Summary

Following BELL & ANDERSON (1967), FREDRICKSON, RAMKRISHNA & TSUCHIYA (1967) and SINKO & STREIFER (1967) we have derived a balance law for the size distribution of unicellular organisms reproducing by fission. Basic ingredients for this equation were a deterministic (mechanistic, physiological) conception of individual growth and a statistical description of fission and death. In doing so we hit upon some unexpected, mechanistic ambiguities inherent in the statistical description which only become overt when individual growth is under environmental control: is the probability of division related to the crossing of a certain size range or does it depend on the available time?

By means of the semigroup theory developed in chapter II it can be shown that generally under constant environmental conditions the population will grow exponentially with rate parameter k while the size distribution stabilizes, the exception being provided by the case of individual growth rates satisfying $V(2x) = 2V(x)$ (which includes the biologically non-trivial case of exponential *i*-growth) and equal division. We have moreover given an equation from which k can be calculated, as well as an explicit expression for the asymptotic size distribution, and we have briefly indicated how these results may help to solve the inverse problem of determining the individual characteristics from data on the population growth rate and size distribution.

Finally we discussed substrate limited growth in a chemostat for the special case that (i) the probability of fission depends on the size range crossed, and (ii) the individual growth rate can be written as a product of a size dependent and a nutrient dependent term (the structural nutrient hypothesis). These two assumptions were found to guarantee that the size distribution converges to a stable distribution which is independent of the chemostat parameters, and that total population size and substrate concentration asymptotically satisfy the usual chemostat equations of unstructured population dynamics.

We ended with a plea for the study of nonlinear structured cell population models under less restricted conditions.

5. On semigroups and generators

Underlying the three examples we have just treated is one abstract, mathematical pattern. In this final section we shall bring this pattern out in the open, both for its own intrinsic interest and as introduction to the ensuing mathematical chapters. For here we have one of those pleasing cases where in fact the biological imagery and the mathematical toolkit have converged to a considerable extent.

The basic ingredient in our approach was the concept of population state, in the guise of a density function

$$x \mapsto n(t,x) .$$

Here "state" is interpreted as "an amount of information about the population sufficient to completely fix the future development of that population". In other words, if we know at t_0 the function $n(t_0,\cdot)$ and the time course of the environment, then at least in principle we can calculate $n(t,\cdot)$ for all $t > t_0$.

In all the cases of variable environments we have dealt with, the environment itself was subject to dynamical equations, just like the population was, with the state of the population as the only input, making the total, coupled system autonomous, *i.e.* independent of outside influences (mathematically this means that all our equations are invariant under a translation of the time axis). Below we shall concentrate on autonomous systems only. (A rough, heuristic sketch of the framework needed for dealing with general non-autonomous systems is given in chapter III.) Moreover we shall assume for definiteness that the only changes are in the population as this simplifies the notation, and, less harmlesly, we shall restrict ourselves mainly to linear systems, *i.e.* systems in which there also is no direct interaction between the individuals of our population.

For an autonomous system an initial condition

$$n(0,x) = n_0(x) , \tag{5.1}$$

(with n_0 assumed to be known) completely and uniquely determines $n(t,\cdot)$ for all $t \geq 0$. Sometimes we write $n(t,x,n_0)$ to emphasize this dependence on the initial condition. Another way of expressing that n_0 for any given t completely determines $n(t,\cdot)$ is to write

$$n(t,x,n_0) = (S(t)n_0)(x) , \tag{5.2}$$

where for each $t \geq 0$ the symbol $S(t)$ stands for an *operator,* which maps functions of x into functions of x. (See fig. 5.1).

The definition immediately implies

$$S(0) = I \tag{5.3a}$$

$$S(t)S(\tau) = S(t + \tau) \qquad t,\tau \geq 0 \tag{5.3b}$$

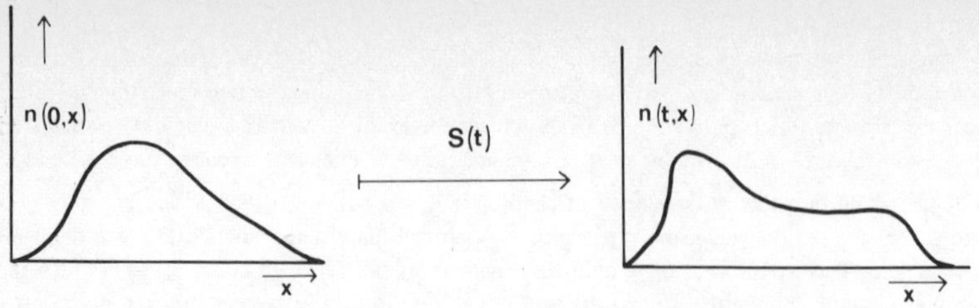

Fig. 5.1. The action of the next state operator S on the population state

One calls (5.3b) the *semigroup* property, where the prefix "semi" reflects the restriction to positive t.

EXERCISE 5.1: Use uniqueness and translation invariance to verify (5.3b). Hint: Refer to figure 5.2.

$$n(0,\cdot) \xmapsto{\ S(t)\ } n(t,\cdot) \xmapsto{\ S(\tau)\ } n(t+\tau,\cdot)$$
$$S(t+\tau)$$

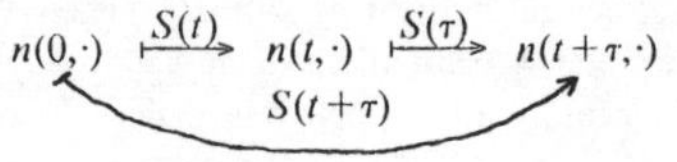

Fig. 5.2. The causal chain connecting the population state at $t + \tau$ to the initial state through the state at the intermediate time t.

There exists an extensive mathematical theory of (one-parameter) *semigroups of linear operators* (Pazy, 1983a).
Here one considers a family $\{S(t)\}_{t\geqslant 0}$ of linear operators with properties (5.3a, b) and in addition some continuity properties. The interesting thing is that one can arrive at such semigroups direct from the biological interpretation.
This should motivate a study of the corresponding mathematical theory. Chapter II gives an introduction especially geared towards applications in structured population models.

From a fundamental point of view the operators $S(t)$ are precisely the objects of one's inquiry. But in practice they are hard to obtain direct in explicit form. Indeed, how can we express them in terms of for example the functions V, b and μ characterizing the behaviour of individual cells, as introduced in the previous section? In finite time intervals the intrinsic entanglement of the effects of growth, births and deaths almost invariably complicates the bookkeeping to an unmanageable extent. The solution to this problem is as old as the invention of differential calculus: in small time intervals there will hardly be any "compound" events, and in the limit they can be neglected altogether. In other words, in infinitesimal time intervals the contributions of the different processes to the change in the population state are uncoupled and therefore easily calculated. So instead of directly specifying $S(t)$ we derive differential equations.

Abstractly we can write

$$\frac{dn}{dt} = An \qquad n(0) = n_0 \tag{5.4}$$

where the operator A (acting on functions of x) is defined by

$$Am = \lim_{t\downarrow 0}\frac{1}{t}(S(t)m - m) \tag{5.5}$$

for those functions m for which this limit exists indeed. One calls A the infinitesimal generator of the semigroup $S(t)$.
For example in the last example (the cell population) A is defined by

$$(Am)(x) = -\frac{dV(x)m(x)}{dx} - b(x)m(x) + 4b(2x)m(2x) - \mu(x)m(x), \tag{5.6a}$$

$m:[\frac{1}{2}a, 1)\to\mathbb{R}$ a representative *i*-state distribution, together with the condition that m is such that the expression on the right hand side makes sense and

$$m(\tfrac{1}{2}a) = 0 \tag{5.6b}$$

(Operators, like all maps, are defined by telling what their domain is and by giving a recipe to calculate the result of their action).

Equation (5.4) with (5.6) is just an abstract reexpression of our old equation (4.2.1). The difference is that (5.4) is an (abstract) ordinary differential equation for the function n of one variable t which assumes values in a space of functions of x (so that we can write $n(t)(x)$), whereas (4.2.1) is a (functional) partial differential equation. It should be clear that these are just two ways of looking at the same object. However the first way certainly is the more profitable one.

In chapter II we shall sketch how for linear structured population models one can construct the unique solution to (5.4) by means of a so-called *generation expansion, i.e.* a series representation in which the subsequent terms are just the i-state densities of the individuals in the subsequent biological generations (in the predation example we shall consider a jump in satiation as a combined death-birth event). Unfortunately no such general technique is available in the nonlinear case. In chapter VI some examples are given how one may proceed to prove existence and uniqueness of solutions in some special cases.

The generation expansion construction is quite useful for calculating $S(t)$ for small t, but clearly it is unsuited to the calculation of the large time behaviour. For the latter purpose we have to take recourse to *spectral theory,* much in the same way as the asymptotic behaviour of the solution of the ordinary differential equation $dX/dt = BX$, $X(t) \in \mathbb{R}^n$ and B and $n \times n$ matrix, can be determined with the aid of the eigenvalues and eigenvectors of B. The extension of this theory from the finite to the infinite dimensional case is beset with subtle and mathematically interesting difficulties, keeping it an area of ever active mathematical research. Luckily in the case of structured population models there is a helpful constraint deriving direct from the interpretation: population densities are nonnegative. So the semigroup $S(t)$ should leave the so-called *cone* of nonnegative functions invariant. By exploiting this positivity property it can often be shown (in chapters II and V it will be told how) that the eigenvalue equation

$$A\nu = \lambda\nu$$

allows precisely one nonnegative *eigenfunction* ν and a corresponding real *eigenvalue* λ (compare equation (4.3.1), equations (3.3.2), (3.3.5) and (3.3.6), and equation (2.6.1a); in the latter equation the conservation of total predator number guarantees that $\lambda = 0$). Moreover the space of possible population states can be decomposed into two components, a one-dimensional space spanned by ν and a "remainder" component, such that

$$n(t,x) = c(n_0)e^{\lambda t}\nu(x) + R(t,x,n_0)$$

($c(n_0)$ a real number), where the remainder R is such that

$$e^{-\lambda t}R(t,x) \to 0$$

in some appropriate sense.

The eigenfunction ν is the stable i-state distribution which kept recurring in our discussions, and the corresponding eigenvalue λ is the intrinsic rate of natural increase. In the first example of this chapter apparently $\lambda = 0$, and in the other two examples it was possible to derive a relatively simple characteristic equation from which λ can easily be calculated numerically. Moreover in the last two examples we were able to give explicit expressions for ν, and in the first example we had available an easy recipe from which ν could be calculated numerically. And in all three cases we could reasonably argue that the essential biological information provided by the model was contained precisely in that real number λ and that eigenfunction ν.

EXERCISE 5.1: Give another explanation, phrased in terms of eigenvalues and characteristic equations, why the solution to (2.6.1) automatically satisfies the side condition (2.5.5b).

II. The Cell Size Distribution and Semigroups of Linear Operators

O. Diekmann

In this chapter we present relevant parts of the theory of strongly continuous semigroups of linear operators in the context of a concrete example: the time evolution of the size distribution of a proliferating cell population. Our aim is to give a motivated introduction to the general mathematical theory of linear semigroups and to demonstrate its usefulness for the study of density independent structured population models. We want to show that abstract arguments and concrete calculations may be combined to arrive at strong conclusions.

In section 1 we perform a preliminary transformation of the cell size density equation introduced in section I.4, and we specify the population state space in which we want to work. In section 2 we collect some basic definitions and results from semigroup theory and we illustrate these by means of some elementary but enlightening examples. Subsequently we show in section 3 that one can indeed associate a semigroup with the size structured cell model. Notions from spectral theory are introduced in section 4 and subsequently it is shown that the spectrum of the infinitesimal generator of the cell semigroup can be found by solving a so-called characteristic equation. This characteristic equation is studied in section 5, and in section 6 we show that the state space can be decomposed according to a subdivision of the spectrum. Sections 7 and 8 are again concerned with the general theory. First we discuss various relations between the spectrum of the generator and the spectrum of the semigroup operators, and then we deal with exponential estimates. In section 9 the results of the preceding sections are applied to the cell model in the special case that $V(2x)<2V(x)$, where V is the individual cell growth rate as a function of size x. The conclusion is that asymptotically for large time the population grows exponentially, while its size distribution converges towards a (stable) distribution which is independent of the initial condition (the dynamics is asymptotically one-dimensional). Section 10 continues the introduction in section I.3.4 of integration along characteristics, a very important auxiliary technique. The results are used in section 11 to study the exceptional case that $V(2x) = 2V(x)$ for all relevant x. In section 12 the case that $V(2x) = 2V(x)$ in some interval and $V(2x)<2V(x)$ in another interval is treated, and the conclusions for the general case are summarized. Some clarifying remarks about the role of positivity are made in section 13. Finally, we show in section 14 that the linear theory can be used in a rather special case of a general model for substrate limited growth in a chemostat.

1. Formulation of the problem.

In chapter I, section 4, we showed that the size-density of a population of unicellular organisms reproducing by binary fission is governed (under constant environmental conditions) by the linear functional-partial differential equation

$$\frac{\partial}{\partial t}n(t,x) = -\frac{\partial}{\partial x}(V(x)n(t,x)) - \mu(x)n(t,x) - b(x)n(t,x) + 4b(2x)n(t,2x) \tag{1.1}$$

$$n(t,\tfrac{1}{2}a) = 0$$

where V, μ and b denote, respectively, the growth, death and fission rate of individual cells and a = min support of b = minimal size at which fission can possibly occur. In order to describe that cells will divide with certainty before reaching a maximal size, which we take without loss of generality to be $x = 1$, we shall assume that b has a non-integrable singularity in $x = 1$ and we shall restrict the domain of the size variable x to the interval $[\tfrac{1}{2}a, 1]$. We repeat convention I.4.2.1 as:

CONVENTION 1.1. If the x-argument of a function exceeds one, the value of the function is by definition zero.

This convention serves to assign to the term $4b(2x)n(t,2x)$ the value zero for $x>\frac{1}{2}$, in accordance with the fact that no baby cell will have size greater than one half. Throughout this chapter we assume that $0<a<1$, but in many parts we make the much stronger assumption $\frac{1}{2} \leq a < 1$ which entails that the smallest mother is still larger than the biggest daughter.

For each fixed t we conceive of $n(t,\cdot)$ as a function defined on $[\frac{1}{2}a, 1]$.

INTERLUDE 1.2. *About Banach spaces and linear operators: a quick introduction for readers with little mathematical background.* It is convenient to consider such a function of x as an *element of (point in) a space* in much the same way as an N-tuple of real numbers is considered as a point in $\mathbb{R}^N$, the N-dimensional Euclidean space. A space of functions on $[\frac{1}{2}a, 1]$ will necessarily be infinite-dimensional. Quite naturally we want it to be a *linear* space: if we multiply an element by a *scalar* (a real or complex number) we want to obtain another element and likewise if we add two elements.

It turns out that in an infinite dimensional space one can introduce several non-equivalent notions of *convergence*. So we have to specify explicitly our ideas about nearness. A convenient way to do this is to define the *distance* between two elements, in such a way that certain natural (intuitive) geometrical properties hold. In a *normed* linear space X there is associated with each element f of X (notation: $f \in X$) a nonnegative real number $\|f\|$, called the norm of f, such that

(i) $\|f+g\| \leq \|f\| + \|g\|$,

(ii) $\|\alpha f\| = |\alpha| \, \|f\|$ for any scalar α,

(iii) $\|f\| = 0$ implies $f = 0$.

In such a space we can identify the distance between f and g with $\|f-g\|$. Note that by (i) the triangle inequality

$$\|f-g\| \leq \|f-h\| + \|h-g\|$$

holds. Finally we want to include in our axioms the technical condition that sequences for which it is reasonable to expect that they converge, do indeed converge to an element of the space. An infinite sequence $\{f_n\}$ of elements of X is called a *Cauchy sequence* if for any $\epsilon > 0$ there exists an integer n_0 such that the relations $k \geq n_0$ and $l \geq n_0$ imply that $\|f_k - f_l\| < \epsilon$. A *Banach space* X is a normed linear space in which every Cauchy sequence $\{f_n\}$ converges to some $f \in X$ (i.e. $\|f_n - f\| \to 0$ for $n \to \infty$).

The defining properties of a Banach space are such that:

(i) they allow the construction of a rich and powerful mathematical theory,

(ii) many concrete function spaces, which arise in practical applications, have these properties.

Function spaces which we shall meet in the following are:

(i) $C[\alpha,\beta]$, the space of continuous functions defined on the interval $[\alpha,\beta]$ with values in $\mathbb{R}$ provided with the *supremum norm* $\|f\| = sup\{|f(x)| \mid \alpha \leq x \leq \beta\}$

(ii) $C_0[\alpha,\beta]$, the subspace of $C[\alpha,\beta]$ of functions which are zero for $x = \alpha$, provided with the supremum norm;

(iii) $L_p(\alpha,\beta)$, the space of integrable functions for which $\|f\| = (\int_\alpha^\beta |f(x)|^p dx)^{\frac{1}{p}}$ is finite (strictly speaking this is a space of equivalence classes of functions; see RUDIN, 1974);

(iv) $L_\infty(\alpha,\beta)$, the space of (essentially) bounded measurable functions provided with the supremum norm.

Let X and Y be Banach spaces with norms $\|\|_X$ and $\|\|_Y$ and let L be a linear operator from X into Y (that is, to every $f \in X$ there is associated $Lf \in Y$ and $L(\alpha f) = \alpha Lf$ and $L(f+g) = Lf + Lg$). To say that L is *continuous* means that Lf_n converges to Lf in Y, whenever f_n converges to f in X, but one can prove that this property is equivalent to L being *bounded* in the sense that for some positive constant K

$$\|Lf\|_Y \leq K\|f\|_X \text{ for all } f \in X. \tag{1.2}$$

Hence we use for linear operators the words "continuous" and "bounded" interchangeably. The set of all bounded linear operators from X into Y is itself a linear space $\mathcal{L}(X,Y)$ which we can norm by

$$\|L\|_{L(X,Y)} := \sup\{\|Lf\|_Y \mid \|f\|_X \leq 1\}. \tag{1.3}$$

One can prove that equipped with this so-called *operator norm* $L(X,Y)$ is a Banach space. An alternative equivalent definition of the operator norm is

$$\|L\|_{L(X,Y)} = \inf\{K \mid \text{relation (1.2) holds}\}. \tag{1.4}$$

Frequently we will omit the indices X, Y and $L(X,Y)$, which distinguish the various norms from each other, simply because the context unambiguously stipulates which norm is meant.

It may happen that one can define a linear operator L on a subspace of X without being able to extend L to a bounded operator defined on all of X (we will meet several examples below). In such a case L is said to be *unbounded* and one has to specify carefully both the subspace on which L is defined, to be denoted by $\mathcal{D}(L)$ and to be called the *domain* of L, and the action of L. If the domains of two unbounded operators are not identical they have to be considered as different objects even if their action is identical. Frequently $\mathcal{D}(L)$ will be *dense* in X (i.e. for any $f \in X$ and any $\epsilon > 0$ one can find $g \in \mathcal{D}(L)$ such that $\|f-g\|<\epsilon$ or, equivalently, the closure $\overline{\mathcal{D}(L)}$ of $\mathcal{D}(L)$ is all of X).

Occasionaly we shall have to integrate Banach space valued functions (see, for instance, the variation-of- constants formula (3.6) below). For our purposes the Riemann integral of a continuous function is adequate. We refer to LADAS & LAKSHMIKANTHAM (1972) for a quick introduction to the theory of integration in Banach space.

It remains to specify the function space Y in which $t \mapsto n(t,\cdot)$ is supposed to take its values. At this point many roads depart. From the point of view of probability theory one may wish to interpret $n(t,\cdot)$ as a measure. Our "definition" of n in section I.4 suggests to take $L_1(\frac{1}{2}a, 1)$ as the state space Y. But continuous functions may be easier to work with.

Likewise we have to make assumptions about the measurability, integrability and/or continuity of V, μ and b and we can either strive for the utmost generality by assuming as little as possible about these ingredients of our model or, on the contrary, we may assume everything which seems reasonable and makes life easy. Noting that it is largely a matter of taste we adopt the second attitude. We make the

ASSUMPTION 1.3:

H_V: V is a strictly positive continuous function on $[\frac{1}{2}a, 1]$;

H_μ: μ is a non-negative continuous function on $[\frac{1}{2}a, 1]$;

H_b: b is non-negative and continuous on $[\frac{1}{2}a, 1)$; $b(x) = 0$ for $x \in [\frac{1}{2}a, a]$ and $b(x) > 0$ for $x \in (a, 1)$ while

$$\lim_{x \uparrow 1} \int_a^x b(\xi) d\xi = +\infty .$$

Note that one of the assumptions concerning b expresses that every cell with size greater than a has a positive probability per unit of time of division.

Before we define Y we make a transformation. Motivated by the cohort calculations in subsection I.4.1 we define

$$E(x) = \exp\left[-\int_{a/2}^{x} \frac{b(\xi) + \mu(\xi)}{V(\xi)} d\xi\right] . \tag{1.5}$$

The transformation

$$m(t,x) = \frac{V(x)}{E(x)}\, n(t,x) \tag{1.6}$$

leads to

$$\begin{cases} \frac{\partial}{\partial t} m(t,x) = -V(x)\frac{\partial}{\partial x} m(t,x) + k(x)m(t,2x) \\ m(t, \frac{1}{2}a) = 0 \end{cases} \tag{1.7}$$

where

$$k(x) = 4\, \frac{V(x)}{E(x)}\, \frac{E(2x)}{V(2x)}\, b(2x) \ , \ \tfrac{1}{2}a \leqslant x \leqslant \tfrac{1}{2} . \tag{1.8}$$

EXERCISE 1.4: Check that (1.7) is correct.

We emphasize that Convention 1.1 implies that $k(x)m(t,2x) = 0$ for $x > \frac{1}{2}$ and for convenience later on we define

$$k(x) = 0 \ , \ x > \tfrac{1}{2} . \tag{1.9}$$

The profit of transformation (1.6) is technical in character: it turns out that (1.7) is easier to handle than (1.1). We can interpret m as the, with a survival-factor corrected, flux of cells (recall that the flux at x is by definition the number of cells which pass x per unit of time and therefore equals $V(x)n(t,x)$, the velocity times the density).

We are going to look for a "solution" $t \mapsto m(t,\cdot)$ of (1.7) which takes values in the Banach space X of continuous functions defined on $[\frac{1}{2}a, 1]$ which are zero in $x = \frac{1}{2}a$, provided with the supremum norm:

$$X = \{\psi \in C[\tfrac{1}{2}a, 1] \mid \psi(\tfrac{1}{2}a) = 0\} = C_0[\tfrac{1}{2}a, 1]. \tag{1.10}$$

This amounts to considering $n(t,\cdot)$ as an element of a weighted C-space Y with tailor-made norm

$$\|\phi\| = \sup\{\frac{V(x)}{E(x)} \mid \phi(x)\mid \ \mid \ \tfrac{1}{2}a \leq x \leq 1\}. \tag{1.11}$$

In particular it implies that $n(t,x)$ has to go to zero for $x \uparrow 1$ at least as fast as $E(x)$ and that the behaviour near $x = 1$ carries extra weight (note that $E(1) = 0!$). In view of the cohort calculations these are natural properties. Moreover, as an a posteriori justification we shall demonstrate in subsequent sections that these properties are preserved in the course of time (so working in Y is a restriction on the initial condition $n(0,x)$ only).

We are now ready to give a precise mathematical formulation of the problem and to outline the program of this chapter:

(i) we want to interpret (1.7) as an abstract equation

$$\frac{dm}{dt} = \tilde{A}m$$

in X, where $\tilde{A}$ is the generator of a semigroup $\tilde{T}(t)$ acting on X

(ii) we want to study the large time behaviour of $\tilde{T}(t)$

(iii) as an important tool in (ii) we shall use a detailed study of the spectral properties of $\tilde{A}$.

REMARK 1.5. The tildes serve here to emphasize the distinction with the generator A and the semigroup $T(t)$ corresponding to $n(t,\cdot)$ and introduced in section I.5. In the following sections we will omit the tildes.

EXERCISE 1.6. Define $H: Y \to X$ by (cf. (1.6))

$$(H\phi)(x) = \frac{V(x)}{E(x)}\phi(x). \tag{1.12}$$

Convince yourself that H is continuous and that its inverse H^{-1} is continuous as well (H is an isomorphism). Verify that

$$T(t) = H^{-1}\tilde{T}(t)H$$

REMARK 1.7. (About the notation.) In the following we shall use the symbols λ, ϕ, ψ and k to denote mathematical objects which are different from the various distributions and the population growth rate which were denoted by these symbols in section I.4. We trust that this will not lead to confusion.

2. Strongly continuous semigroups of bounded linear operators

In the first part of this section and in sections 7 and 8 we present without proof some basic mathematical material concerning linear semigroups of operators. For proofs and additional results we refer to PAZY (1983a) and DAVIES (1980). The bible remains HILLE & PHILLIPS (1957). Other convenient sources are BALAKRISHNAN (1976), BELLENI-MORANTE (1979), BUTZER & BERENS (1967), DUNFORD & SCHWARTZ (1958), KATO (1976), LADAS & LAKSHMIKANTHAM (1972), SCHAPPACHER (1983), WEBB (1985a) and YOSIDA (1980). The second part of this section is devoted to some examples (in particular translation semigroups) which serve to pave the way for our treatment of the fission equation in section 3.

Let X be a Banach space with norm $\|\cdot\|$. Let $\{T(t)\}_{t \geq 0}$ be a family of bounded linear operators from X into itself. We will call $\{T(t)\}$ a *strongly continuous (one-parameter) semigroup* iff

(i) $T(0) = I$

(ii) $T(t)T(\tau) = T(t+\tau)$, $t,\tau \geq 0$,

(iii) $\lim_{t \downarrow 0} \|T(t)\phi - \phi\| = 0$, $\forall \phi \in X$.

(The adverb "strongly" indicates that (iii) amounts to convergence in the strong operator topology). Assumption (iii) is based on the wish to be assured of continuous orbits $t \mapsto T(t)\phi$ and so it may seem strange that (iii) only requires continuity in $t = 0$. Therefore we propose

EXERCISE 2.1. Show that (i) - (iii) guarantee that for each fixed $\phi \in X$ the mapping $t \mapsto T(t)\phi$ is continuous from $\mathbb{R}_+$ to X.

The operator A defined by

$$A\phi = \lim_{t\downarrow 0} \frac{1}{t}(T(t)\phi - \phi)$$

with domain of definition $\mathcal{D}(A)$ consisting of those $\phi \in X$ for which $\frac{1}{t}(T(t)\phi - \phi)$ converges in X to a limit as $t\downarrow 0$, is called the *infinitesimal generator* of $\{T(t)\}$. We will frequently omit the adjective "infinitesimal".

THEOREM 2.2. (Uniqueness of the correspondence). *A linear operator can be the generator of at most one semigroup.*

DEFINITION 2.3. A linear operator $L:X \to Y$ (where X and Y are Banach spaces) is called *closed* iff for each sequence $\{x_n\}$ in X such that (i) $x_n \in \mathcal{D}(L)$; (ii) $\exists x \in X$ such that $x_n \to x$ for $n \to \infty$; (iii) $\exists y \in Y$ such that $Lx_n \to y$; necessarily $x \in \mathcal{D}(L)$ and $Lx = y$.

REMARKS (i) One can show that L is closed iff the *graph* $\{(x, Lx) \mid x \in \mathcal{D}(L)\}$ is a closed subset of $X \times Y$.
(ii) Among unbounded operators some behave better than others. We will see later that closed operators are good guys in the sense that for these a satisfactory spectral theory exists.
(iii) Differential operators are, as a rule, closed.

In principle the generator A of a semigroup $\{T(t)\}$ could be a "bad" operator with a rather small domain of definition. The next result states that this is actually impossible.

THEOREM 2.4. *A is a closed operator and $\mathcal{D}(A)$ is dense in X.*

Intuitively A is the derivative of $T(t)$ at $t = 0$. Indeed we have

THEOREM 2.5. *If $\phi \in \mathcal{D}(A)$ then $T(t)\phi \in \mathcal{D}(A)$ for all $t \geqslant 0$. On $\mathcal{D}(A)$ the operators $T(t)$ and A commute and*

$$\frac{d}{dt}T(t)\phi = AT(t)\phi = T(t)A\phi, \text{ if } \phi \in D(A).$$

One of the high-lights of semigroup theory gives a precise characterization of those operators which generate semigroups:

THEOREM 2.6. (Hille-Yosida). *A closed operator A with dense domain $\mathcal{D}(A)$ is the infinitesimal generator of a strongly continuous semigroup iff real numbers M and ω exist such that for all $\lambda > \omega$ the operator $\lambda I - A$ has a bounded inverse and*

$$\|((\lambda I - A)^{-1})^n\| \leqslant \frac{M}{(\lambda - \omega)^n}, \quad n = 1, 2, \cdots .$$

REMARKS (i) Of course the symbol $\|\cdot\|$ here indicates the operator norm (see Interlude 1.2).
(ii) The most important way to establish that some given operator generates a semigroup is to verify the Hille-Yosida conditions (see PAZY, 1983a). However, for the kind of problems we are dealing with in this chapter there exist, as we will show in detail, easier ways. As a consequence we are not going to use Theorem 2.6 (it is stated here only for the sake of completeness).

EXAMPLE 2.7. Let $X = BUC(\mathbb{R})$, the space of bounded, uniformly continuous functions from $\mathbb{R}$ into $\mathbb{R}$ provided with the supremum norm $\|\phi\| = \sup\{|\phi(x)| \mid -\infty < x < +\infty\}$. Defining

$$(T(t)\phi)(x) = \phi(x - t),$$

we clearly obtain a strongly continuous semigroup (note that the strong continuity of translation is guaranteed by our restriction to *uniformly* continuous functions).
Claim: $A\phi = -\phi'$, with $\mathcal{D}(A) = \{\phi \mid \phi$ is continuously differentiable and ϕ' belongs to $BUC(\mathbb{R})\}$.

PROOF. Suppose that $\phi \in \mathcal{D}(A)$ and $A\phi = \psi$ then by definition

$$\lim_{t\downarrow 0} \sup\{|\frac{\phi(x-t) - \phi(x)}{t} - \psi(x)| \mid -\infty < x < +\infty\} = 0,$$

from which we conclude that ψ is the left-derivative of $-\phi$. Since ψ is (uniformly) continuous the inequality

$$\left|\frac{\phi(x+t)-\phi(x)}{-t}-\psi(x)\right| \leq \left|\frac{\phi(y-t)-\phi(y)}{t}-\psi(y)\right| + |\psi(y)-\psi(y-t)|,$$

where $y = x + t$, shows that ϕ is differentiable with derivative $-\psi$. If, on the other hand, $\phi' \in BUC(\mathbb{R})$ then

$$\left|\frac{\phi(x-t)-\phi(x)}{t} + \phi'(x)\right| = \left|\int_0^t (\phi'(x)-\phi'(x-\tau))d\tau\right| \overset{t\downarrow 0}{\rightarrow} 0$$

uniformly for $x \in \mathbb{R}$ since

$$\forall \epsilon > 0 \ \exists \ \delta > 0 \text{ such that } |\phi'(x)-\phi'(x-\tau)| \leq \epsilon \text{ provided } |\tau| \leq \delta \ \square$$

REMARKS (i) A is unbounded. Indeed, consider $\phi(x) = \sin mx$ then $\|\phi\| = 1$ and $\|A\phi\| = m$ and consequently for no $K \in \mathbb{R}$ the inequality $\|A\phi\| \leq K\|\phi\|$ can hold for all $\phi \in X$.

(ii) The fact that $\mathfrak{D}(A)$ is dense in X is well-known in this special example.

(iii)One can easily verify directly that A is closed by taking limits in the relation $\phi_n(x) = \phi_n(0) + \int_0^x \phi_n'(\xi)d\xi$.

(iv)If ϕ is differentiable then so is its translate and the order of translation and differentiation may be changed without changing the result. Moreover, $\frac{d}{dt}\phi(x-t) = -\phi'(x-t)$. Compare this with Theorem 2.5.

(v) Defining $n(t,x) = (T(t)\phi)(x)$ we may, if $\phi \in \mathfrak{D}(A)$, rewrite the abstract equation $\frac{d}{dt}T(t)\phi = AT(t)\phi$ as $\frac{\partial n}{\partial t} + \frac{\partial n}{\partial x} = 0$, a first order partial differential equation with initial condition $n(0,x) = \phi(x)$. We observe that both partial derivatives exist iff $\phi \in \mathfrak{D}(A)$, but that the semigroup yields a very natural extension of the solution concept. This can be made more explicit by interpreting $\frac{\partial}{\partial t} + \frac{\partial}{\partial x}$ as the directional derivative D in the (1,1) - direction:

$$Df(t,x) := \lim_{\epsilon \to 0} \frac{1}{\epsilon}\{f(t+\epsilon, x+\epsilon) - f(t,x)\}$$

(see Appendix IIIA for a general definition). Indeed, the directional derivative Dn exists, even though n is not C^1, and $Dn = 0$. Thus $n(t,x) = (T(t)\phi)(x)$ is a solution of $Dn = 0$, $n(0,x) = \phi(x)$.

The example above illustrates a general principle: the generator of translation is differentiation. A technical elaboration of this principle in various function spaces amounts to a precise description of the domain of definition of the operator of differentiation. For the next result we refer to RUDIN, 1974, Chapter 8.

DEFINITION 2.8. A function $\phi:[a,b] \to \mathbb{R}$ is called *absolutely continuous* if $\forall \epsilon > 0 \ \exists \ \delta(\epsilon) > 0$ such that for each finite collection of disjunct open intervals $(x_1,y_1), \cdots, (x_N,y_N)$ the condition $\sum_{i=1}^N (y_i - x_i) < \delta$ implies $\sum_{i=1}^N |\phi(y_i) - \phi(x_i)| < \epsilon$.

THEOREM 2.9. *(i) Suppose* $\phi(x) = \phi(a) + \int_a^x \psi(\xi)d\xi$ *with* $\psi \in L_1[a,b]$. *Then* ϕ *is absolutely continuous.*

(ii) Let ϕ *be absolutely continuous. Then* ϕ *is differentiable for almost all* x *and* $\phi' \in L_1[a,b]$ *while* $\phi(x) = \phi(a) + \int_a^x \phi'(\xi)d\xi$.

EXAMPLE 2.10. Let $X = L_1(\mathbb{R})$ and define

$$(T(t)\phi)(x) = \phi(x-t). \tag{2.1}$$

Then $\{T(t)\}$ is a strongly continuous semigroup (more general the translation operator is continuous in L_p-spaces with $1 \leq p < \infty$; note however that translation is not continuous in L_∞). In view of the theorem above it comes as no surprise that $A\phi = -\phi'$ with $\mathfrak{D}(A) = \{\phi | \phi$ is absolutely continuous and $\phi' \in L_1(\mathbb{R})\}$.

EXAMPLE 2.11. In bounded domains we have to incorporate a boundary condition. Let $X = L_1[0,1]$ and define

$$(T(t)\phi)(x) = \begin{cases} \phi(x-t) & \text{for } x \geq t \\ 0 & \text{for } x < t \end{cases} \tag{2.2}$$

Then $A\phi = -\phi'$ with $\mathcal{D}(A) = \{\phi | \phi$ is absolutely continuous and $\phi(0) = 0\}$.

EXAMPLE 2.12. The semigroup of the last example cannot be defined on $C[0,1]$ since, when $\phi(0) \neq 0$, $T(t)\phi$ would have a jump discontinuity. But with $X = C_0[0,1]$, the space of continuous functions on [0,1] which vanish for $x = 0$, provided with the supremum norm, everything is fine and $A\phi = -\phi'$ with $\mathcal{D}(A) = \{\phi | \phi \in C^1[0,1]$ and $\phi'(0) = 0\}$.

EXERCISE 2.13. Verify the last assertion.

EXAMPLE 2.14. The last two examples above correspond to the first order partial differential equation

$$\frac{\partial n}{\partial t} + \frac{\partial n}{\partial x} = 0$$

with boundary condition $n(t,0) = 0$ and initial condition $n(0,x) = \phi(x)$. As a next step towards the fission problem we change the equation into

$$\frac{\partial n}{\partial t} + V(x)\frac{\partial n}{\partial x} = 0 , \tag{2.3}$$

where V is a continuous and strictly positive function. In order to show that essentially (in a sense specified below) nothing has changed we try to find a transformation of variables

$$y = G(x)$$

which reduces the new problem to the old. Since

$$\frac{\partial}{\partial y} = \frac{\partial x}{\partial y}\frac{\partial}{\partial x} = (G^{-1})'(G(x))\,\frac{\partial}{\partial x}$$

we want to choose G such that $(G^{-1})'(G(x)) = V(x)$. Differentiating the identity $G^{-1}(G(x)) = x$ we obtain $(G^{-1})'(G(x))G'(x) = 1$ and therefore we choose

$$G(x) = \int_{a/2}^{x} \frac{d\xi}{V(\xi)} . \tag{2.4}$$

Let $X = C_0[\frac{a}{2},1]$, $Y = C_0[0,G(1)]$ and define $L: X \to Y$ by

$$(L\phi)(y) = \phi(G^{-1}(y)) ,$$

and $L^{-1}: Y \to X$ by

$$(L^{-1}\psi)(x) = \psi(G(x)) .$$

Let the semigroup $T(t)$ acting on Y be defined by (2.2). Then

$$\tilde{T}(t) = L^{-1}T(t)L \tag{2.5}$$

defines a semigroup on X with generator

$$\tilde{A} = L^{-1}AL \ , \quad \mathcal{D}(\tilde{A}) = L^{-1}\mathcal{D}(A) . \tag{2.6}$$

Explicitly we obtain

$$(\tilde{T}(t)\phi)(x) = \phi(G^{-1}(G(x)-t)) \tag{2.7}$$

(where we define $G^{-1}(y) = 0$ for $y \leq 0$) and

$$\tilde{A}\phi = -V\phi' , \quad \mathcal{D}(\tilde{A}) = \{\phi \mid \phi \in C^1[\tfrac{a}{2},1] \ \& \ \phi(\tfrac{a}{2}) = \phi'(\tfrac{a}{2}) = 0\} . \tag{2.8}$$

EXERCISE 2.15. Verify the preceding calculations. Verify that (2.7) defines a solution of (2.3).

Semigroups $\{\tilde{T}(t)\}$ and $\{T(t)\}$ which are related as in (2.5) are called intertwined or conjugated. In the present example the representation (2.5) shows how the action of $\tilde{T}(t)$ can be decomposed into translation and deformation.

EXERCISE 2.16. If V goes to zero sufficiently fast in the left endpoint of the interval (the "stream in" point) we don't need to prescribe a boundary condition (nor can we). First solve formally

$$\begin{cases} \dfrac{\partial n}{\partial t} + x\dfrac{\partial n}{\partial x} = 0 \ , \ 0 < x < +\infty, \\ n(0,x) = \phi(x) \qquad , \ 0 \leq x < +\infty \end{cases}$$

and subsequently define the corresponding semigroup on $X = C[0,+\infty)$ and calculate the generator.

EXERCISE 2.17. In continuation of Exercise 1.6 show that

$$A = H^{-1}\tilde{A}H \quad \text{with } \mathfrak{D}(A) = H^{-1}\mathfrak{D}(\tilde{A}) .$$

REMARK 2.18. The examples 2.11 and 2.12 show that in an L_1-space (and more generally in L_p-spaces) a boundary condition may show up only in the domain of the generator, whereas in the C-context one is forced to include it in the definition of the space. But then one of the conditions characterizing the domain of the generator will be a boundary condition for the derivative!

EXERCISE 2.19. Let A be a *bounded* linear operator from X into X. Define for each $t \in \mathbb{R}$ the bounded linear operator e^{At} by the Taylor series

$$e^{At} = \sum_{l=0}^{\infty} \frac{t^l A^l}{l!} ,$$

which converges in the operator norm. Convince yourself that $\{e^{At}\}$ is a semigroup with generator A.

EXERCISE 2.20. If $X = \mathbb{R}^N$ then any bounded linear operator A is, for a given basis, represented by a matrix. One way to compute the matrix e^{At} is to bring A in Jordan canonical form (HIRSCH & SMALE, 1974) and to use
LEMMA. If $B = PAP^{-1}$ then $e^B = Pe^A P^{-1}$, and
LEMMA. If $A_1 A_2 = A_2 A_1$ (i.e. A_1 and A_2 commute) then $e^{A_1}e^{A_2} = e^{A_1 + A_2}$.
Show in this manner that

$$(i) \quad e^{\begin{bmatrix} -3 & 1 \\ 1 & -3 \end{bmatrix} t} = \frac{1}{2}\begin{bmatrix} e^{-2t} + e^{-4t} & e^{-2t} - e^{-4t} \\ e^{-2t} - e^{-4t} & e^{-2t} + e^{-4t} \end{bmatrix}$$

$$(ii) \quad e^{\begin{bmatrix} \alpha & -\beta \\ \beta & \alpha \end{bmatrix} t} = e^{\alpha t}\begin{bmatrix} \cos\beta t & -\sin\beta t \\ \sin\beta t & \cos\beta t \end{bmatrix} .$$

3. Do growth, death and division generate a semigroup?

Our aim is to associate a semigroup with the problem

$$\begin{cases} \frac{\partial}{\partial t} m(t,x) = -V(x)\frac{\partial}{\partial x} m(t,x) + k(x)m(t,2x) \\ m(t,\frac{1}{2}a) = 0 . \end{cases} \tag{3.1}$$

Instead of verifying the Hille-Yosida conditions we will first study the rather easy problem obtained by dropping the term $k(x)m(t,2x)$ and subsequently re-introduce this term as a relatively innocent perturbation. So we define an unbounded operator B on X by

$$\begin{cases} (B\phi)(x) = -V(x)\phi'(x) \\ \mathfrak{D}(B) = \{\phi \in X \mid \phi \in C^1[\frac{1}{2},1] \ \&\ \phi'(\frac{1}{2}a) = 0\} \end{cases} \tag{3.2}$$

As in Example 2.14 it follows that B generates the strongly continuous semigroup

$$(U_0(t)\phi)(x) = \phi(G^{-1}(G(x)-t)) \tag{3.3}$$

with

$$G(x) = \int_{a/2}^{x} \frac{d\xi}{V(\xi)} . \tag{3.4}$$

EXERCISE 3.1. Let $F(t)$ be the unique solution of $\frac{dF}{dt} = V(F)$, $F(0) = \frac{1}{2}a$. Interpret $F(t)$ as the size of a cell at time

t given that the cell had size $\frac{1}{2}a$ at time zero. Verify that $F(t) = G^{-1}(t)$, i.e., $G(F(t)) = t$ and $F(G(x)) = x$.

EXERCISE 3.2. Interpret $G(x)$ as the time which a cell needs to grow from size $\frac{1}{2}a$ to size x.

EXERCISE 3.3. Let $X(t,y)$ denote the solution of the initial value problem $\frac{dx}{dt} = V(x)$, $x(0) = y$. Show that $X(t,y) = G^{-1}(G(y)+t)$ and that, consequently, (3.3) can be rewritten as $(U_0(t)\phi)(x) = \phi(X(-t,x))$. Interpret this representation in biological terms.

EXERCISE 3.4. (a technical point) Convince yourself that the strict positivity of V guarantees that the solution of the initial value problem $\frac{dx}{dt} = V(x)$, $x(0) = y$ is *unique* (without any Lipschitz condition on V).

Next we state a perturbation result:

THEOREM 3.5. *Let B be the generator of a semigroup $U_0(t)$ and let C be a bounded linear operator. Then $B+C$ generates a semigroup $T(t)$.*

We shall sketch some instructive aspects of the proof. The differential equation

$$\frac{dm}{dt} = Bm + Cm\ , \quad m(0) = \phi \tag{3.5}$$

and the *variation-of-constants formula* (see HIRSCH & SMALE, 1974, section V.5) suggest to look for solutions of the integral equation

$$m(t) = U_0(t)\phi + \int_0^t U_0(t-\tau)Cm(\tau)d\tau\ . \tag{3.6}$$

Using a certain straightforward estimate (PAZY, 1983a, section 3.1, proposition 1.2) one can show that the *successive approximations*

$$m_k(t;\phi) = \sum_{j=0}^{k} U_j(t)\phi \tag{3.7}$$

where

$$U_{j+1}(t)\phi = \int_0^t U_0(t-\tau)CU_j(\tau)\phi d\tau \tag{3.8}$$

converge to a solution $m = m(t;\phi)$ and, moreover, that there is at most one solution. Finally one can verify that $T(t)\phi = m(t;\phi)$ defines a strongly continuous semigroup of bounded linear operators with infinitesimal generator $A = B + C$, $\mathcal{D}(A) = \mathcal{D}(B)$.

Returning to the cell population we formally write

$$(C\phi)(x) = k(x)\phi(2x) \tag{3.9}$$

and ask ourselves whether or not this defines a bounded operator on X.

EXERCISE 3.6. Check that (3.9) defines a bounded operator if and only if $k(\frac{1}{2}) = 0$ or, in other words, $b(x)E(x) \to 0$ for $x \uparrow 1$. (Hint: recall Convention 1.1).

Although $xb(x)E(x)$ is integrable on $[\frac{1}{2}a, 1]$, the behaviour of this function in $x = 1$ may be such that C is unbounded, and consequently Theorem 3.5 is not strong enough for our purposes. There do exist many generalizations (KATO, 1976). In the present problem one can easily generalize the constructive procedure (3.7) - (3.8) to solve (3.6), by exploiting the embedding of X into $L_1[\frac{1}{2}a, 1]$. Using that (3.3) defines a semigroup on $L_1[\frac{1}{2}a, 1]$ as well, that C is always continuous from X to $L_1[\frac{1}{2}a, 1]$ and that, as one can verify, the integration with respect to τ produces a continuous function of x one can prove (see DIEKMANN, HEIJMANS & THIEME (1984) for the details):

THEOREM 3.7. *The operator*

$$(A\phi)(x) = -V(x)\phi'(x) + k(x)\phi(2x) \tag{3.10}$$

$\mathfrak{D}(A) = \{\phi \in X \mid \phi$ is continuously differentiable on$[\frac{1}{2}a, \frac{1}{2}) \cup (\frac{1}{2}, 1]$ and the limits

$\lim_{x \uparrow \frac{1}{2}} -V(x)\phi'(x) + k(x)\phi(2x)$ and $\lim_{x \downarrow \frac{1}{2}} -V(x)\phi'(x)$ exist and equal each other, and $\phi'(\frac{1}{2}a) = 0\}$

is the generator of a semigroup $T(t)$ on X.

REMARKS (i) Note that $\mathfrak{D}(A) = \mathfrak{D}(B)$ if and only if $k(\frac{1}{2}) = 0$ and that, in general, the domains of B and C "interact" to produce $\mathfrak{D}(A)$.
(ii) The terms in the series $T(t)\phi = \sum_{j=0}^{\infty} U_j(t)\phi$ have a straightforward biological interpretation. $U_0(t)\phi$ is the contribution to the density of those cells which were present at $t = 0$ and have not yet divided. We call it the *zero'th generation.* Inductively we find that $U_j(t)\phi$, the *j-th generation,* gives the contribution of those cells which arose from fission of cells of the $(j-1)$th generation and which have not yet divided themselves. Thus we speak about a *generation expansion.*
(iii) Note the monotone convergence of the successive approximations (3.7) if ϕ is a non-negative function.
(iv) The zero'th generation becomes extinct at $t = G(1)$. One can prove inductively that numbers e_j exist such that the jth generation becomes extinct at $t = e_j$. Mathematically this amounts to the jth term being identically zero for $t \geq e_j$.
(v) Assume that $a > \frac{1}{2}$ (i.e., the smallest mother is still larger than the biggest daughter or, in other words, a cell which is just created cannot divide). Since a newly created daughter has a size less than or equal to $\frac{1}{2}$, it needs a certain time to grow up to size a at which it can divide. Consequently there will exist integers $J(\sigma)$ such that for $t \leq \sigma$ the generations with $j \geq J(\sigma)$ cannot yet exist. Mathematically this amounts to all terms with index $j \geq J(\sigma)$ being zero for $t \leq \sigma$, i.e., for each finite t the generation expansion has only finitely many non-zero terms. The assumption $a > \frac{1}{2}$ makes the exposition easy but is not needed for the result. One can prove (DIEKMANN, HEIJMANS & THIEME, 1984):

LEMMA 3.8. *Choose $\sigma > 0$. Let the bounded linear operator H from $C([0,\sigma];X)$ into itself be defined by*

$$(Hm)(t) = \int_0^t U_0(t-\tau)Cm(\tau)d\tau \tag{3.11}$$

Then $H^j = 0$ for $j \geq \frac{2\sigma}{a} \|V\|_\infty + m$, where m is such that $2^{-m} \leq \frac{a}{2} < 2^{-m+1}$ and $\|V\|_\infty = \sup\{|V(x)| \mid \frac{1}{2}a \leq x \leq 1\}$. (In other words: H is nilpotent.)

EXERCISE 3.9. (not difficult but time consuming) Prove this lemma for the special case that $a > \frac{1}{2}$ and $g(x) = 1$ for all x.

In conclusion of this section we address the problem of specifying the sense in which $T(t)\phi$ satisfies the partial differential equation in (3.1). Combining Remark (v) about and below Example 2.7 and the representation (2.5) we now define

$$(Df)(t,x) = \lim_{\epsilon \to 0} \frac{1}{\epsilon}\{f(t+\epsilon, G^{-1}(G(x)+\epsilon)) - f(t,x)\} \tag{3.12}$$

EXERCISE 3.10. Verify that $m(t,x) = (T(t)\phi)(x)$ satisfies

$$(Dm)(t,x) = k(x)m(t,2x) \tag{3.13}$$

where, to be precise, one should take for $x = \frac{1}{2}$ the limits $\epsilon \uparrow 0$ and $\epsilon \downarrow 0$ in (3.12) separately if $k(\frac{1}{2}) \neq 0$.

EXERCISE 3.11. Use the identity $G^{-1}(G(x)) = x$ and the chain rule to show that $(G^{-1})'(G(x)) = V(x)$.

REMARK. Since $G^{-1}(G(x)+\epsilon) = x + \epsilon V(x) + o(\epsilon)$ we have

$$(Df)(t,x) = \lim_{\epsilon \to 0} \frac{1}{\epsilon}\{f(t+\epsilon,\ x+\epsilon V(x)) - f(t,x)\}$$

if f is smooth. However, the derivative in the direction $(1, V(x))$ at the right hand side may not exist, while still the limit in (3.12) is well defined. In section 10 and Chapter III, section 4 we shall interpret D as a derivative along *characteristics.*

The first step in our analysis of the fission model is now completed: we showed existence and uniqueness of a solution which depends continuously on the initial data ϕ. The solution is constructed as a series in which every subsequent term is easily calculated from the preceding one. However, this representation of the solution is not of much help for an analysis of the large time behaviour. We need spectral theory.

EXERCISE 3.12. Apply the constructive proof of Theorem 3.5 to the case where B is the generator of translation in $BUC(\mathbb{R})$ (i.e., the operator called A in Example 2.7) and $(C\phi)(x) = k\phi(x)$. Find an explicit expression for $S(t)$ by evaluating the series.

EXERCISE 3.13. (a little tricky) Repeat the foregoing exercise with $(C\phi)(x) = k(x)\phi(x)$ where $k \in BUC(\mathbb{R})$. Note that the explicit expression for $T(t)$ defines a semigroup for a much wider class of functions k.

4. The spectrum of A.

Let X be a Banach space over the complex numbers $\mathbb{C}$ and let L denote a closed linear operator with domain $\mathcal{D}(L) \subset X$ and range $\mathcal{R}(L) \subset X$. Let λ be a complex number.

DEFINITION 4.1. λ is an element of the *resolvent set* $\rho(L)$ iff the *resolvent* $(\lambda I - L)^{-1}$ exists and is bounded, i.e.,

(*i*) $\lambda I - L$ is one–to–one (injective)

(*ii*) $\mathcal{R}(\lambda I - L)$ is dense in X

(*iii*) $(\lambda I - L)^{-1}$ is bounded*

The so-called *spectrum* $\sigma(L)$ is by definition the complement of $\rho(L)$. The *point spectrum* $P\sigma(L)$ is the set of those $\lambda \in \mathbb{C}$ for which $\lambda I - L$ is not one-to-one, i.e., $L\phi = \lambda\phi$ for some $\phi \neq 0$. One then calls λ an *eigenvalue* and ϕ an *eigenvector* corresponding to λ. The null space $\mathcal{N}(\lambda I - L)$ is called the *eigenspace* and its dimension the *geometric multiplicity* of λ. The *generalized eigenspace* $\mathcal{M}(\lambda I - L)$ is the smallest closed linear subspace that contains $\mathcal{N}((\lambda I - L)^j)$ for $j = 1,2,3,...$ and its dimension is called the *algebraic multiplicity* of λ.

Although we will not need the following definitions, we present them for completeness. The *continuous spectrum* $C\sigma(L)$ is the set of those $\lambda \in \mathbb{C}$ for which $\lambda I - L$ is one-to-one and $\mathcal{R}(\lambda I - L)$ is dense in X but $(\lambda I - L)^{-1}$ is unbounded. The *residual spectrum* $R\sigma(L)$ is the set of those $\lambda \in \mathbb{C}$ for which $\lambda I - L$ is one-to-one but $\mathcal{R}(\lambda I - L)$ is not dense in X.

Throughout the rest of this chapter (!) we make

ASSUMPTION 4.2. $a \geq \frac{1}{2}$

(see section II.1 for the biological interpretation). It will turn out that the analysis of the spectrum of the operator A defined in (3.10), with k defined in (1.8) and (1.9), is rather simple under this biologically reasonable restriction. Moreover, the results are representative for the general case, which is elaborated in detail in HEIJMANS (1985b).

According to the rules we first try to construct $(\lambda I - A)^{-1}$ for as many $\lambda \in \mathbb{C}$ as possible. Note that we now work in the complexification of X (again denoted by X) which means that all functions take values in $\mathbb{C}$. The abstract inhomogeneous equation $(\lambda I - A)\psi = f$ implies

$$\begin{cases} V(x)\psi'(x) + \lambda\psi(x) = f(x) \ , & \frac{1}{2} \leq x \leq 1 \\ V(x)\psi'(x) + \lambda\psi(x) = f(x) + k(x)\psi(2x) \ , & \frac{1}{2}a \leq x \leq \frac{1}{2} \\ \psi(\frac{1}{2}a) = 0 \, . \end{cases} \tag{4.1}$$

* This assumption means that $(\lambda I - L)^{-1}$, which at first is only defined on $\mathcal{R}(\lambda I - L)$, can be extended to a *bounded* operator defined on the whole space X. For *closed* operators the conditions actually *imply* that $\mathcal{R}(\lambda I - L) = X$; so the formulation above is unnecessarily cumbersome in the present context, and was chosen solely because it is the standard formulation.

Using the variation-of-constants formula for ordinary differential equations we can solve the first equation to obtain

$$\psi(x) = Ce^{\lambda(G(\frac{1}{2})-G(x))} + \int_{\frac{1}{2}}^{x} e^{\lambda(G(\xi)-G(x))}\frac{f(\xi)}{V(\xi)}d\xi \tag{4.2}$$

for $\frac{1}{2}\leqslant x\leqslant 1$ with the constant C still to be determined. The right hand side of the second equation in (4.1) can now be expressed in known functions and the unknown constant C and we find, taking account of the third equation,

$$\psi(x) = \int_{a/2}^{x} e^{\lambda(G(\xi)-G(x))}\,\{Ce^{\lambda(G(\frac{1}{2})-G(2\xi))}k(\xi) + f(\xi) + k(\xi)\int_{\frac{1}{2}}^{2\xi} e^{\lambda(G(\eta)-G(2\xi))}\frac{f(\eta)}{V(\eta)}d\eta\}\,\frac{d\xi}{V(\xi)} \tag{4.3}$$

for $\frac{1}{2}a\leqslant x\leqslant\frac{1}{2}$. The first question is whether or not (4.2) and (4.3) define an element ψ of X. The only thing that could possibly be wrong is the continuity in $x = \frac{1}{2}$. Now $\lim_{x\downarrow\frac{1}{2}}\psi(x) = C$ and $\lim_{x\uparrow\frac{1}{2}}\psi(x) = \pi(\lambda)C + \zeta(\lambda,f)$, where by definition

$$\pi(\lambda) = \int_{\frac{1}{2}a}^{\frac{1}{2}} e^{\lambda(G(\xi)-G(2\xi))}\frac{k(\xi)}{V(\xi)}d\xi \tag{4.4}$$

and

$$\zeta(\lambda,f) = \int_{\frac{1}{2}a}^{\frac{1}{2}} e^{\lambda(G(\xi)-G(\frac{1}{2}))}\{f(\xi) + k(\xi)\int_{\frac{1}{2}}^{2\xi} e^{\lambda(G(\eta)-G(2\xi))}\frac{f(\eta)}{V(\eta)}d\eta\}\frac{d\xi}{V(\xi)}\,. \tag{4.5}$$

In order that ψ is continuous we have to choose C such that

$$(1-\pi(\lambda))C = \zeta(\lambda,f)\,. \tag{4.6}$$

If $\pi(\lambda)\neq 1$ we can, for arbitrary $f\in X$, indeed satisfy this compatibility condition. The explicit expressions (4.2) - (4.3) with

$$C = (1-\pi(\lambda))^{-1}\zeta(\lambda,f) \tag{4.7}$$

define a *continuous* mapping $f\mapsto\psi$ from X into X and our construction guarantees that $\psi\in\mathfrak{D}(A)$ and $(\lambda I-A)\psi = f$. Thus we proved

THEOREM 4.3. $\pi(\lambda)\neq 1$ *implies* $\lambda\in\rho(A)$.

Next, consider those λ for which $\pi(\lambda) = 1$. If we choose $f(x)\equiv 0$ the function ψ defined by (4.2) - (4.3) is, for arbitrary $C\in\mathbb{C}$, an element of $\mathfrak{D}(A)$ such that $(\lambda I-A)\psi = 0$. Hence we have

THEOREM 4.4. $\pi(\lambda) = 1$ *implies* $\lambda\in P\sigma(A)$.

So the spectrum of A is precisely the set $\{\lambda|\pi(\lambda) = 1\}$ and A has only a point spectrum. By analogy with the well-known situation for ordinary differential equations we will call the equation

$$\pi(\lambda) = 1 \tag{4.8}$$

the *characteristic equation.* In subsection I.4.3 we discussed its biological interpretation and in the next section we discuss the position of its roots in the complex plane.

EXERCISE 4.5. The assumption $a\geqslant\frac{1}{2}$ made that we had to consider only two subintervals when constructing the resolvent. Use three subintervals to derive the corresponding characteristic equation for the case $a\geqslant\frac{1}{4}$.

The construction above implies that elements of $\mathfrak{N}(\lambda I-A)$ are unique modulo the constant C. So the dimension of $\mathfrak{N}(\lambda I-A)$ (i.e., the geometric multiplicity) is at most one (here the restriction $a\geqslant½$ is essential). In conclusion of this section we prove

THEOREM 4.6. *Suppose* $\pi(\lambda) = 1$. *The eigenvalue* λ *is simple (i.e. has algebraic multiplicity one) iff* $\pi'(\lambda)\neq 0$.

PROOF. Any element ϕ of $\mathfrak{N}((\lambda I-A)^2)$ that does not belong to $\mathfrak{N}(\lambda I-A)$ necessarily satisfies $(\lambda I-A)\phi = \theta\psi$, for some non-zero constant θ, where ψ is defined by (4.2) - (4.3) with $f(x)\equiv 0$ and, say, $C = 1$. Just as before we find the solvability condition $(1-\pi(\lambda))\phi(\frac{1}{2}) = \theta\zeta(\lambda,\psi)$. Since $\pi(\lambda) = 1$ this condition can be satisfied iff $\zeta(\lambda,\psi) = 0$. Now $\zeta(\lambda,\psi) = (1) + (2)$ where

$$(1) = \int_{\frac{1}{2}a}^{\frac{1}{2}} e^{\lambda(G(\xi)-G(\frac{1}{2}))}\frac{\psi(\xi)}{V(\xi)}\,d\xi = \int_{\frac{1}{2}a}^{\frac{1}{2}}\int_{\frac{1}{2}a}^{\xi} e^{\lambda(G(\eta)-G(2\eta))}\,k(\eta)\frac{d\eta}{V(\eta)}\,\frac{d\xi}{V(\xi)} = \int_{\frac{1}{2}a}^{\frac{1}{2}}(G(\tfrac{1}{2})-G(\xi))e^{\lambda(G(\xi)-G(2\xi))}k(\xi)\frac{d\xi}{V(\xi)}$$

$$(2) = \int_{\frac{1}{2}a}^{\frac{1}{2}} e^{\lambda(G(\xi)-G(\frac{1}{2}))}\,k(\xi)\int_{\frac{1}{2}}^{2\xi} e^{\lambda(G(\eta)-G(2\xi))}\,\frac{\psi(\eta)}{V(\eta)}\,d\eta\,\frac{d\xi}{V(\xi)} = \int_{\frac{1}{2}a}^{\frac{1}{2}} e^{\lambda(G(\xi)-G(2\xi))}\,k(\xi)(G(2\xi)-G(\tfrac{1}{2}))\,\frac{d\xi}{V(\xi)}\,.$$

Hence $\zeta(\lambda,\psi) = \int_{\frac{1}{2}a}^{\frac{1}{2}} e^{\lambda(G(\xi)-G(2\xi))}k(\xi)(G(2\xi)-G(\xi))\,\frac{d\xi}{V(\xi)}$. On the other hand we obtain the same expression for $-\pi'(\lambda)$ by differentiation of (4.4). □

EXERCISE 4.7. Determine $\sigma(A)$ for A from Example 2.7.

EXERCISE 4.8. Do the same for A from Example 2.12 and for B from Example 2.14.

5. The characteristic equation

Using the definition (1.5), (1.8), (3.4) and (4.4) we can rewrite $\pi(\lambda)$ as

$$\pi(\lambda) = 2\int_a^1 \frac{b(\xi)}{V(\xi)}\exp\left[-\int_{\xi/2}^{\xi}\frac{\lambda+\mu(\eta)+b(\eta)}{V(\eta)}d\eta\right]d\xi\,. \tag{5.1}$$

Since, V,μ and b are nonnegative π is *strictly decreasing* on the real axis. Clearly $\pi(-\infty) = +\infty$ and $\pi(+\infty) = 0$. We conclude that there exists *precisely one real root* which we call λ_d (d means *dominant,* a terminology explained in Remark 5.6 (i) below; note that in section I.4 we called the real root k; so $k = \lambda_d$). From the explicit expression for π' obtained by differentiation of (5.1) it readily follows that $\pi'(\lambda_d)<0$. We conclude from Theorem 4.6:

THEOREM 5.1. *A has precisely one real eigenvalue λ_d and this is a simple eigenvalue.*

Since V,μ and b are real valued $\overline{\pi(\lambda)} = \pi(\bar{\lambda})$, where the bar denotes complex conjugation. Hence non-real roots of $\pi(\lambda) = 1$ occur in complex conjugate pairs. From the definition (5.1) it follows almost directly that π is an analytic function. Hence the roots are isolated points, without any finite point of accumulation. In order to obtain further information about the position of the roots we try to write π as a Laplace transform. Formula (4.4) suggests the transformation

$$\tau = G(2\xi)-G(\xi) \tag{5.2}$$

Since

$$\frac{d\tau}{d\xi} = \frac{2}{V(2\xi)} - \frac{1}{V(\xi)}\,, \tag{5.3}$$

we need a condition on V for (5.2) to define a one-to-one relationship. In the following we shall treat three cases:

DEFINITION 5.2.

Case I : $V(2x)<2V(x)\quad ,\ \frac{1}{2}a\leqslant x\leqslant\frac{1}{2}\,,$

Case II : $V(2x) = 2V(x)\quad ,\ \frac{1}{2}a\leqslant x\leqslant\frac{1}{2}\,,$

Case III : $\left\{\begin{array}{ll} V(2x) = 2V(x)\ , & \beta<x\leqslant\frac{1}{2}, \\ V(2x)<2V(x)\ , & \frac{1}{2}a\leqslant x\leqslant\beta, \end{array}\right\}$ for some $\beta\in(\frac{1}{2}a,\frac{1}{2})$.

Exercise 5.3. Verify that the general solution of $V(2x) = 2V(x)$, $\frac{1}{2}a \leq x \leq \frac{1}{2}$, is given by $V(x) = xp(\ln x/\ln 2)$ with p an arbitrary one-periodic function.

Remarks 5.4. (i) Mathematically there is no difference between $V(2x)<2V(x)$ and $V(2x)>2V(x)$, for all x, but biologically the first seems reasonable and the second absurd.
(ii) The list of cases is far from exhaustive. Occasionaly we shall state remarks, exercises and results for still other cases. See Theorem 5.10 and the end of section 12, in particular Theorem 12.3.

Theorem 5.5. *In Case I there exists* $\epsilon>0$ *such that every root* $\lambda \neq \lambda_d$ *of the characteristic equation* $\pi(\lambda) = 1$ *satisfies* $\mathrm{Re}\lambda \leq \lambda_d - \epsilon$.

Proof. Let $\xi(\tau)$ be the inverse function of $\tau(\xi)$ defined in (5.2). Then

$$\pi(\lambda) = \int_{G(a)}^{G(1)-G(\frac{1}{2})} e^{-\lambda\tau} \frac{k(\xi(\tau))}{V(\xi(\tau))} \frac{d\xi}{d\tau}(\tau)d\tau . \tag{5.4}$$

Hence π is of the form

$$\pi(\lambda) = \int_{c_1}^{c_2} e^{-\lambda\tau} K(\tau)d\tau \tag{5.5}$$

with $c_2>c_1>0$ and $K(\tau)\geq 0$ and not identically zero. Putting $\lambda = \mu + i\nu$ we obtain

$$\pi(\lambda) = \int_{c_1}^{c_2} \cos\nu\tau \, e^{-\mu\tau} K(\tau)d\tau - i \int_{c_1}^{c_2} \sin\nu\tau \, e^{-\mu\tau} K(\tau)d\tau ,$$

from which we infer that for $\nu \neq 0$

$$|\mathrm{Re}\pi(\lambda)| < \pi(\mu) = \pi(Re\lambda) .$$

Since $0<\pi(\mu)\leq 1$ for $\mu \geq \lambda_d$ necessarily $\mathrm{Re}\lambda<\lambda_d$ if λ is a root. According to the Lemma of Riemann-Lebesgue (Rudin, 1974, 5.14)

$$\lim_{|\nu|\to\infty} \int_{c_1}^{c_2} \cos\nu\tau \, e^{-\mu\tau} K(\tau)d\tau = 0$$

uniformly for μ in compact sets. So in each vertical strip $\mu_1 \leq \mathrm{Re}\lambda \leq \mu_2$ there can be at most finitely many roots. The conclusion of the theorem is now obvious. □

Remarks 5.6. (i) Since λ_d is the eigenvalue with largest real part we call it the dominant eigenvalue.
(ii) One can use Hadamard's Factorization Theorem for entire functions of order one to show that there exist infinitely many roots. See chapter VIII of Titchmarsh (1979).

Theorem 5.7. *In Case II the roots of* $\pi(\lambda) = 1$ *are given explicitly by*

$$\lambda_l = \frac{1}{G(a)}\{\ln \int_{a/2}^{1/2} \frac{k(\xi)}{V(\xi)} d\xi + 2l\pi i\} , \ l \in \mathbb{Z} .$$

Proof. In this case $G(2x) - G(x) = \text{constant} = G(a)$ and consequently

$$\pi(\lambda) = e^{-\lambda G(a)} \int_{a/2}^{1/2} \frac{k(\xi)}{V(\xi)} d\xi .$$

Taking logarithms in the equation $\pi(\lambda) = 1$ yields the result (recall that the "complex" logarithm is multi valued or see Exercise 7.5 below). □

Theorem 5.8. *In Case III the conclusion of Theorem 5.5 holds.*

Proof.

$$\pi(\lambda) = e^{-\lambda G(a)} \int_{\frac{1}{2}a}^{\beta} \frac{k(\xi)}{V(\xi)} d\xi + \int_{G(2\beta)-G(\beta)}^{G(1)-G(\frac{1}{2})} e^{-\lambda\tau} \frac{k(\xi(\tau))}{V(\xi(\tau))} \frac{d\xi}{d\tau}(\tau)d\tau$$

with $\xi(\tau)$ implicitly defined by (5.2). The proof of Theorem 5.5 carries over almost verbatim. □

EXERCISE 5.9. Assume that $V(2x)>2V(x)$ for $\frac{1}{2}a\leq x<\beta$ and $V(2x)<2V(x)$ for $\beta<x\leq\frac{1}{2}$ for some $\beta\in(\frac{1}{2}a,\frac{1}{2})$. Prove that the conclusion of Theorem 5.5 holds.

By now it should be clear that Case II is really exceptional and we summarize and extend our conclusions in:

THEOREM 5.10. *If $V(2x)\neq 2V(x)$ for some $x\in[\frac{1}{2}a,\frac{1}{2}]$ the real eigenvalue λ_d is strictly dominant. If $V(2x)=2V(x)$ for all $x\in[\frac{1}{2}a,\frac{1}{2}]$ then, on the contrary, there exist countably many eigenvalues on the line $\mathrm{Re}\lambda=\lambda_d$, which form an additive subgroup of this line. In all cases the eigenvector corresponding to λ_d is positive.*

In section 13 we shall put these findings in the right perspective and there we also present several important references.

6. Decomposition of the population state space X

Let ψ_d denote the eigenvector corresponding to the real eigenvalue λ_d. Then

$$T(t)\psi_d = e^{\lambda_d t}\psi_d\,. \tag{6.1}$$

EXERCISE 6.1. Prove this identity. Hint: compute $\frac{d}{dt}T(t)\psi_d$ using Theorem 2.5.

An obvious conjecture is now that for arbitrary $\phi\in X$ the X-valued function $T(t)\phi$ will have for $t\to\infty$ its fastest growth "in the direction" of ψ_d when λ_d strictly leads the field of real parts of eigenvalues of A. To begin with we have to give a precise meaning to "in the direction".

The eigenvector ψ_d spans the linear subspace $\mathfrak{N}(\lambda_d I-A)$ which clearly is invariant under $\{T(t)\}$. Our plan is to decompose X into $\mathfrak{N}(\lambda_d I-A)$ and another *invariant* subspace and to prove subsequently that the restriction of $\{T(t)\}$ to that second subspace obeys an exponential estimate with exponent $\lambda_d-\epsilon$ when $\mathrm{Re}\lambda<\lambda_d-\epsilon$ for all eigenvalues λ other than λ_d. In this section we carry out the first and easiest half of the plan only, postponing the second half to sections 8 and 9.

DEFINITION 6.2. Let X be a Banach space. A bounded linear operator $P:X\to X$ is called a *projection* if and only if $P^2=P$.

DEFINITION 6.3. X is the *direct sum* of two linear subspaces Y and Z if and only if for each $x\in X$ there exist a *unique* $y\in Y$ and $z\in Z$ such that $x=y+z$. NOTATION: $X=Y\oplus Z$.

EXERCISE 6.4. Let $P:X\to X$ be a projection. Show that $X=\mathfrak{R}(P)\oplus\mathfrak{N}(P)$ and that $\mathfrak{N}(P)=\mathfrak{R}(I-P)$. Show that $\mathfrak{N}(P)$ and $\mathfrak{R}(P)$ are closed.

EXERCISE 6.5. Let $X=Y\oplus Z$ with Y and Z *closed* linear subspaces. Define linear operators $P,Q:X\to X$ by $Px=y$ and $Qx=z$ where $y\in Y$ and $z\in Z$ are defined by $x=y+z$. Show that P and Q are projections and that $Q=I-P$. (Hint: In order to show that P and Q are *bounded* one can use the *closed graph theorem:* If the linear operator $L:X_1\to X_2$ is closed then L is continuous; see Definition 2.3 for the notion of a closed operator).

The next theorem is one of the key-stones of our approach. It shows that one can associate with spectral values which are poles of the resolvent a natural direct sum decomposition of the underlying space. For the proof we refer to YOSIDA, 1980, VIII.8.

THEOREM 6.6. *Let L be a closed linear operator on the complex Banach space X and let λ_0 be an isolated point of $\sigma(L)$. Then $\lambda\mapsto(\lambda I-L)^{-1}$ is a holomorphic mapping (in a punctured neighbourhood $\Omega\setminus\{\lambda_0\}$ of λ_0) admitting the Laurent expansion*

$$(\lambda I-L)^{-1}=\sum_{-\infty}^{+\infty}(\lambda-\lambda_0)^k A_k \tag{6.2}$$

where for each $k \in \mathbb{Z}$

$$A_k = \frac{1}{2\pi i} \int_\Gamma (\lambda-\lambda_0)^{-k-1}(\lambda I - L)^{-1} d\lambda \tag{6.3}$$

with Γ *a (counter-clockwise oriented) circumference* $|\lambda-\lambda_0| = \eta$, *where* η *is so small that the circle* $|\lambda-\lambda_0| \leqslant \eta$ *does not contain singularities of* $(\lambda I - L)^{-1}$ *other than* λ_0 *itself. The operator* A_{-1} *is a projection on X.*

If λ_0 *is a pole of* $(\lambda I - L)^{-1}$ *of order m (i.e.,* $A_{-m} \neq 0$ *and* $A_k = 0$ *for* $k < -m$*) then* λ_0 *is an eigenvalue of L and for* $k \geqslant m$

$$\mathcal{R}(A_{-1}) = \mathcal{N}((\lambda_o I - L)^k)$$

$$\mathcal{R}(I - A_{-1}) = \mathcal{R}((\lambda_0 I - L)^k)$$

so that, in particular,

$$X = \mathcal{N}((\lambda_0 I - L)^m) \oplus \mathcal{R}((\lambda_0 I - L)^m) \tag{6.4}$$

REMARK 6.7. Calculations involving integrals of complex variable operator-valued functions can be performed in precisely the same way as calculations involving ordinary complex functions. Thus the expression (6.3) for A_k is obtained from the Cauchy formula. To determine A_{-1} in practice one simply calculates the "coefficient" of $(\lambda-\lambda_0)^{-1}$ in the expansion of $(\lambda I - L)^{-1}$ in powers of $\lambda - \lambda_0$ (see below for a concrete example).

We are now going to apply Theorem 6.6 to the operator A. The calculations in section 4, notably formulas (4.2) - (4.7) imply that

$$((\lambda I - A)^{-1} f)(x) = (1-\pi(\lambda))^{-1}\, \zeta(\lambda,f)\Psi(\lambda,x) + R(\lambda,f,x) \tag{6.5}$$

where

$$\Psi(\lambda,x) = \begin{cases} e^{\lambda(G(\frac{1}{2}) - G(x))} , & \frac{1}{2} \leqslant x \leqslant 1, \\ \int\limits_{a/2}^{x} e^{\lambda\{G(\xi) - G(2\xi) - G(x) + G(\frac{1}{2})\}} \frac{k(\xi)}{V(\xi)} d\xi , & \frac{1}{2}a \leqslant x \leqslant \frac{1}{2}, \end{cases} \tag{6.6}$$

and where $\lambda \mapsto R(\lambda,f,x)$ is analytic. Hence the singularities of $\lambda \mapsto (\lambda I - A)^{-1}$ are precisely the zeros of $1-\pi(\lambda)$ and these are poles. The order of the pole equals the order of the zero. In the case of λ_d we have $\pi'(\lambda_d) \neq 0$ and the order is one (cf. Theorem 5.1). The residue of $\lambda \mapsto ((\lambda I - A)^{-1} f)(x)$ in $\lambda = \lambda_d$ is given by

$$\frac{\zeta(\lambda_d,f)}{-\pi'(\lambda_d)} \Psi(\lambda_d,x) = \frac{\zeta(\lambda_d,f)}{-\pi'(\lambda_d)} \psi_d(x)$$

so the projection A_{-1}, here denoted by P, is given by

$$Pf = \frac{\zeta(\lambda_d,f)}{-\pi'(\lambda_d)} \psi_d \,. \tag{6.7}$$

EXERCISE 6.8. Verify that P is a projection using the calculations in the proof of Theorem 4.6.

COROLLARY 6.9. $X = \mathcal{N}(\lambda_d I - A) \oplus \mathcal{R}(\lambda_d I - A)$ *and the corresponding projection onto* $\mathcal{N}(\lambda_d I - A)$ *is P defined in (6.7).*

EXERCISE 6.10. Verify that $\mathcal{R}(\lambda_d I - A)$ is invariant under $\{T(t)\}$ and show that P commutes with $T(t)$.

7. Relations between the spectra of A and of $T(t)$

In this and the following section X denotes a Banach space, $\{T(t)\}$ a strongly continuous semigroup of bounded linear operators on X and A the infinitesimal generator of $\{T(t)\}$. We intend to explore the conclusions about the behaviour of $\{T(t)\}$ (especially for large t), that can be drawn from our knowledge of the spectrum of A. In this section we concentrate on the spectrum of $T(t)$ and in the next one we shall consider exponential estimates.

NOTATION: If W is a subset of $\mathbb{C}$ then $e^{tW} = \{e^{tw} \mid w \in W\}$.

THEOREM 7.1. $e^{t\sigma(A)} \subset \sigma(T(t))$, $t \geq 0$.

The proof will be delegated to the next two exercises.

EXERCISE 7.2. Define $Z(t)\phi = \int_0^t e^{\lambda(t-\tau)} T(\tau)\phi d\tau$. Show that

(i) $\mathcal{R}(Z(t)) \subset \mathcal{D}(A)$;

(ii) $AZ(t)\phi = Z(t)A\phi$, $\forall \phi \in \mathcal{D}(A)$;

(iii) $(\lambda I - A)Z(t) = e^{\lambda t}I - T(t)$.

EXERCISE 7.3. Prove Theorem 7.1 by showing that for $e^{\lambda t} \in \rho(T(t))$ the operator $Z(t)(e^{\lambda t}I - T(t))^{-1}$ is the inverse of $\lambda I - A$.

EXERCISE 7.4. Assume that $e^{-\lambda t}\|T(t)\| \to 0$ for $t \to \infty$ and $\mathrm{Re}\lambda > \omega_0$ (in the next section we prove that one can always find ω_0 such that this holds). Define $R(\lambda) = \int_0^\infty e^{-\lambda\tau} T(\tau) d\tau$ for $\mathrm{Re}\lambda > \omega_0$. Show that $\mathcal{R}(R(\lambda)) \subset \mathcal{D}(A)$ and that $(\lambda I - A)R(\lambda) = I$. The observation that A and $R(\lambda)$ commute on $\mathcal{D}(A)$ implies that $R(\lambda)$ is the resolvent of A for $\mathrm{Re}\lambda > \omega_0$. Note the analogy with the identity

$$\int_0^\infty e^{-\lambda\tau} e^{\alpha\tau} d\tau = (\lambda - \alpha)^{-1} .$$

EXERCISE 7.5. (For those who have little experience with complex exponentials and logarithms). For fixed $t > 0$, analyse the mapping $\lambda \mapsto e^{\lambda t}$ (from $\mathbb{C}$ into $\mathbb{C}$) and its multivalued inverse. Concentrate in particular on vertical and horizontal lines and their images. (Draw each line in some colour and draw its image in the same colour .) Use the periodicity with respect to the imaginary part of λ as a motivation to divide the plane in horizontal strips

$$\{\lambda \mid (2l-1)\frac{\pi}{t} < \mathrm{Im}\lambda \leq (2l+1)\frac{\pi}{t}\}, \; l \in \mathbb{Z} .$$

One can show by means of examples (see, for instance, Example 8.6) that the converse of Theorem 7.1 does not hold. The situation is more surveyable if we restrict our attention to the point spectrum.

THEOREM 7.6. $e^{tP\sigma(A)} \subset P\sigma(T(t)) \subset \left[e^{tP\sigma(A)} \cup \{0\}\right]$; *more precisely we have that* $e^{\lambda t} \in P\sigma(T(t))$ *if* $\lambda \in P\sigma(A)$ *and that for at least one* $l \in \mathbb{Z}$, $\lambda + \dfrac{2\pi i l}{T} \in P\sigma(A)$ *if* $e^{\lambda t} \in P\sigma(T(t))$.

EXERCISE 7.7. Prove the first inclusion by means of the identity (on $\mathcal{D}(A)$) $e^{\lambda t}I - T(t) = (\lambda I - A)Z(t) = Z(t)(\lambda I - A)$.

THEOREM 7.8. $\mathcal{N}(e^{\lambda t}I - T(t))$ *is the closed linear subspace spanned by the linear subspaces* $\mathcal{N}(\lambda_l I - A)$ *with* $\lambda_l \in P\sigma(A)$ *such that* $e^{\lambda_l t} = e^{\lambda t}$.

EXERCISE 7.9. Let $\phi \in \mathcal{N}((\lambda I - A)^2)$. Show that

$$T(t)\phi = e^{\lambda t}\phi + te^{\lambda t}(A - \lambda I)\phi$$

and deduce from this identity that $\phi \in \mathcal{N}((e^{\lambda t}I - S(t))^2)$ while

$$\phi \in \mathcal{N}(e^{\lambda t}I - T(t)) \text{ iff } \phi \in \mathcal{N}(\lambda I - A) .$$

EXERCISE 7.10. Prove by induction that $\mathcal{N}((\lambda I - A)^k) \subset \mathcal{N}((e^{\lambda t}I - S(t))^k)$.

REMARK 7.11. NUSSBAUM (1984) proves the analogue of Theorem 7.8 for the k-times iterated operators. A spectral mapping theorem similar to Theorem 7.6 holds for the residual spectrum, but the continuous spectrum of the semigroup need not be faithful to that of the generator.

8. Exponential estimates

Aiming at exponential estimates we shall study the large time behaviour of $\frac{1}{t}\log\|T(t)\|$. It will appear that this function has a limit for $t \to \infty$. The next and far more difficult problem is to characterize the limit in terms of (known) properties of the generator A and we will find that certain technical (so-called compactness) conditions are very helpful if not essential. To begin with we introduce an important concept.

Let L be a bounded linear operator on X. The *spectral radius* $r_\sigma(L)$ can be defined by

$$r_\sigma(L) = \inf_{k \geq 1} \|L^k\|^{\frac{1}{k}} \tag{8.1}$$

although it clearly deserves its name because an alternative definition is provided by point (ii) of:

LEMMA 8.1.

(i) $\lim_{k \to \infty} \|L^k\|^{\frac{1}{k}}$ *exists and equals* $r_\sigma(L)$

(ii) $\sup_{\lambda \in \sigma(L)} |\lambda| = r_\sigma(L)$.

The proof of (i) is based on a discrete version of the following auxiliary result which we need below.

LEMMA 8.2. *Let* $p:[0,\infty) \to \mathbb{R}$ *be bounded on each finite subinterval and subadditive (i.e.* $p(t_1 + t_2) \leq p(t_1) + p(t_2)$, $\forall t_1, t_2 \geq 0$). *Define* $\nu = \inf_{t>0} \frac{p(t)}{t}$. *Then* $\lim_{t \to \infty} \frac{p(t)}{t}$ *exists and equals* ν.

We want to apply this lemma with $p(t) = \log\|T(t)\|$ and therefore we need the following:

LEMMA 8.3. $\|T(t)\|$ *is bounded on bounded intervals.*

This lemma is a straightforward consequence of the uniform boundedness principle (the Banach-Steinhaus theorem, see RUDIN, 1974, 5.8).

We now define

$$\omega_0 = \omega_0(T(t)) = \inf_{t>0} \frac{1}{t} \log\|T(t)\| \tag{8.2}$$

and find

THEOREM 8.4.

i) $\lim_{t \to \infty} \frac{1}{t} \log\|T(t)\|$ *exists and equals* ω_0.

ii) $\forall \omega > \omega_o \ \exists M(\omega)$ *such that* $\|T(t)\| \leq M(\omega) e^{\omega t}$, $t \geq 0$.

iii) $r_\sigma(T(t)) = e^{\omega_0 t}$ for $t \geq 0$.

EXERCISE 8.5. Prove points (ii) and (iii).

It remains to characterize the so-called *growth bound* ω_0. The following example (due to GREINER, VOIGT & WOLFF, 1981) shows that the obvious conjecture $\omega_0 = s(A)$, where by definition

$$s(A) = \sup\{\mathrm{Re}\lambda | \lambda \in \sigma(A)\},$$

is false.

EXAMPLE 8.6. Define $\|f\|_1 = \int_0^\infty e^\tau |f(\tau)| d\tau$ and let for some $p \in (1,\infty)$

$$X = \{f \in L_p(0,\infty) \mid \|f\|_1 < \infty\}$$

provided with the norm $\|f\| = \|f\|_{L_p} + \|f\|_1$. Then X is a Banach space and the translation semigroup $(T(t)f)(x) = f(x+t)$, $t \geq 0$, is strongly continous. Clearly $\|T(t)\| \leq 1$. In order to show that $\|T(t)\| = 1$ we introduce

$$f_\epsilon(\tau) = \begin{cases} 1 & \text{for } t \leq \tau \leq t + \epsilon^p, \\ 0 & \text{elsewhere} \end{cases}$$

A simple calculation shows that

$$\|f_\epsilon\| = \epsilon + \epsilon^t(e^{t^{\ell}} - 1) \text{ and } \|T(t)f_\epsilon\| = \epsilon + e^{\ell} - 1$$

so that

$$\|T(t)f_\epsilon\| = \|f_\epsilon\|(1 - o(1)) \text{ for } \epsilon\downarrow 0\,.$$

We conclude that $\|T(t)\| = 1$ and $\omega_0 = 0$.

The infinitesimal generator is $Au = u'$ with $\mathcal{D}(A) = \{u \mid u$ is absolutely continuous and $u' \in X\}$. We shall demonstrate that $\{\lambda \mid \mathrm{Re}\lambda > -1\} \subset \rho(A)$. The abstract equation $(\lambda I - A)u = f$ leads to $\lambda u - u' = f$ and hence to

$$u(t) = \int_0^\infty e^{-\lambda\tau} f(t+\tau)d\tau = e^{\lambda t}\int_t^\infty e^{-\lambda\tau} f(\tau)d\tau$$

(in principle we could add a term $ce^{\lambda t}$ but such a term does not belongs to X for $\mathrm{Re}\lambda > -1$). Straightforward estimates show that u thus defined belongs to X for $\mathrm{Re}\lambda > -1$ and consequently $u' = \lambda u - f \in X$. We conclude that $u = (\lambda I - A)^{-1} f$ and that $\lambda \in \rho(A)$. Finally, $t \mapsto e^{\lambda t}$ is an eigenvector corresponding to the eigenvalue λ if $\mathrm{Re}\lambda < -1$. Hence $\sup_{\lambda \in \sigma(A)} \mathrm{Re}\lambda = -1 \neq 0 = \omega_0$.

In view of Theorem 7.6 we know that $\omega_0 = \sup\{\mathrm{Re}\lambda \mid \lambda \in \sigma(A)\}$ whenever $T(t)$ has, for some $t > 0$, an eigenvalue on the circumference $|z| = r_\sigma(T(t)) = e^{\omega_0 t}$. Such is certainly the case if *all* spectral values, except possibly $z = 0$, are eigenvalues. An important class of operators, viz. *compact* operators, has this property. We need some terminology.

A subset W of X is called *compact* if every cover of X by open sets contains a finite subcover. An equivalent but more imaginitive condition for compactness is that every sequence in W has a subsequence that converges to an element of W (N.B. The equivalence holds for metric spaces and therefore certainly for Banach spaces; see HUTSON & PYM, 1980). W is called *precompact* (or, *relatively compact)* if the closure $\overline{W}$ of W is compact.

The compact subsets of $\mathbb{R}^N$ are precisely the bounded, closed sets. If X is some space of functions defined on a domain $\Omega \subset \mathbb{R}^N$ one can sometimes find a reasonable simple criterion for the (pre)compactness of subsets of X (see KUFNER, JOHN & FUCIK, 1977). We present one well-known example which we need later on.

Let Ω be a compact subset of $\mathbb{R}^N$. A set W of continuous functions on Ω is called *uniformly bounded* if it is a bounded subset of $C(\Omega)$ (i.e., there exists a constant K such that $\|f\| = \sup_{x\in\Omega}|f(x)| \leq K$ for all $f \in W$). The set W is called *equicontinuous* if $\forall \epsilon > 0\ \exists \delta = \delta(\epsilon) > 0$ such that $\forall f \in W$ and $\forall x,y \in \Omega$ with $|x-y| < \delta$ the estimate $|f(x) - f(y)| < \epsilon$ holds.

THEOREM 8.7. (Arzela-Ascoli) *A subset W of $C(\Omega)$ is precompact if and only if W is uniformly bounded and equicontinuous.*

A linear operator is called *compact* (or *completely continuous)* if bounded sets are mapped onto precompact sets. Again there is an equivalent condition: the image of any bounded sequence should contain a convergent subsequence. It is not difficult to prove that compact linear operators are necessarily bounded and that the compact linear operators form a *closed* linear subspace of the space of bounded linear operators (provided with the operator norm). Moreover, the product of a compact and a bounded operator is compact.

THEOREM 8.8. *Suppose* $\dim X = \infty$ *and let $L: X \to X$ be a compact linear operator. Then $0 \in \sigma(L)$. (0 can belong to $P\sigma(L)$, $R\sigma(L)$ or $C\sigma(L)$) and $\sigma(L) \setminus \{0\}$ consists of either a finite number of eigenvalues or an infinite sequence of eigenvalues that converges to zero.*

COROLLARY 8.9. *Suppose that $T(t_0)$ is compact for some $t_0 > 0$. Then $\omega_0 = s(A) = \sup\{\mathrm{Re}\lambda \mid \lambda \in P\sigma(A)\}$.*

REMARK. The semigroup property implies that $T(t)$ is compact for $t \geq t_0$ whenever $T(t_0)$ is compact. Indeed, $T(t) = T(t_0)T(t-t_0)$ and the product of a compact and a bounded operator is compact.

Although one can use Corollary 8.9 in many applications it is not strong enough to cover many others. We need some refinements.

The (Kuratowski) *measure of noncompactness* $\alpha(W)$ of a bounded set $W \subset X$ is the infimum of the positive numbers d for which W can be covered by finitely many sets of diameter less than or equal to d (recall that the diameter of a set U is the supremum of $\{\|x-y\| \mid x,y \in U\}$; note that W is compact iff $\alpha(W) = 0$). The *measure of noncompactness* $|L|_\alpha$ of a bounded linear operator L is the infimum of the positive numbers θ for which $\alpha(L(W)) \leq \theta\alpha(W)$ for all bounded sets $W \subset X$ (hence L is compact iff $|L|_\alpha = 0$; $|\cdot|_\alpha$ defines a seminorm on X, cf. NUSSBAUM, (1970).

The (Browder) *essential spectrum* $\sigma_e(L)$ of a closed operator L is defined as the set of those $\lambda \in \sigma(L)$ for which at least one of the following conditions is satisfied

(i) $\mathcal{R}(\lambda I - L)$ is not closed;

(ii) λ is an accumulation point of $\sigma(L)$;

(iii) the generalized eigenspace corresponding to λ is infinite-dimensional.

It is known that the complement $\sigma(L)\setminus\sigma_e(L)$ consists of isolated poles of finite order of the resolvent. The elements of $\sigma(L)\setminus\sigma_e(L)$ are called *normal eigenvalues.* Finally, we define the *essential spectral radius*

$$r_e(L) = \sup\{\,|\lambda|\;|\lambda\in\sigma_e(L)\} \tag{8.3}$$

and quote the following result of NUSSBAUM (1970):

LEMMA 8.10. $r_e(L) = \lim_{k\to\infty} |L^k|_\alpha^{\frac{1}{k}} = \inf_{k\geq 1} |L^k|_\alpha^{\frac{1}{k}}$

The idea to use these concepts and results in the context of linear semigroups seems to be due to PRÜSS (1981). Detailed proofs of the following theorems can be found in WEBB (1985a). In analogy with the definition of ω_0 in (8.2) we introduce

$$\omega_e = \omega_e(T(t)) = \inf_{t>0} \frac{1}{t} \log |T(t)|_\alpha \tag{8.4}$$

(with the convention that $\log 0 = -\infty$).

THEOREM 8.11.

i) $\lim_{t\to\infty} \frac{1}{t}\log|T(t)|_\alpha$ *exists and equals* ω_e

ii) $r_e(T(t)) = e^{\omega_e t}$, $t>0$ *(with the convention* $e^{-\infty} = 0$).

THEOREM 8.12.

$$\omega_0 = \max\{\omega_e, \omega_n\} \textit{ where } \omega_n = \sup\{\mathrm{Re}\lambda \;|\lambda \text{ is a normal eigenvalue of } A\} \tag{8.5}$$

So, provided we can show (using Theorem 8.11) that $\omega_e<\omega_n$, we have obtained a characterization of ω_0 in terms of the spectrum of A.

REMARK 8.13. Using Theorem 8.12 one can show that $\omega_0 = \max\{\omega_e, s(A)\}$.

9. The stable size distribution

Let us return to our concrete example, the semigroup $\{T(t)\}$ and the generator A that go with the model for cell proliferation. The first thing we try is to prove that the semigroup is compact after finite time. Since the semigroup is constructively defined by the generation expansion we shall scrutinize the terms in this expansion.

The zero'th generation

$$(U_0(t)\phi)(x) = \phi(G^{-1}(G(x)-t)) \tag{9.1}$$

transforms and translates the initial function ϕ without changing the smoothness, and a glance at the Arzela-Ascoli Theorem 8.7 should suffice to conclude that $U_0(t)$ is not compact for $t<G(1)$. For $t\geq G(1)$, however, $U_0(t)$ is the zero-operator and thus certainly compact.

The first generation corresponds to the operator

$$U_1(t)\phi = \int_0^t U_0(t-\tau)CU_0(\tau)\phi d\tau\,. \tag{9.2}$$

EXERCISE 9.1. Show that explicitly

$$(U_1(t)\phi)(x) = \int_0^t k(X(-t+\tau,x))\phi(X(-\tau,2X(-t+\tau,x)))d\tau \tag{9.3}$$

where (cf. Exercise 3.3)

$$X(-t,x) = G^{-1}(G(x)-t) \tag{9.4}$$

and where, for convenience, $G(x) = G(1)$ for $x \geqslant 1$ and $k(x) = 0$ for $x > \frac{1}{2}$.

EXERCISE 9.2. Let $\xi(\tau,t,x) = G(X(-\tau,2X(-t+\tau,x))) = G(2X(-t+\tau,x))-\tau = G(2G^{-1}(G(x)-t+\tau))$. Show that

$$\frac{\partial \xi}{\partial \tau} = \frac{2V(X(-t+\tau,x))}{V(2X(-t+\tau,x))} - 1 . \tag{9.5}$$

In order to prove compactness of $U_1(t)$ we try to rewrite (9.3) such that the argument of ϕ does not contain the variable x anymore (indeed, we need equicontinuity without knowing more about ϕ than some sup-norm bound). Formula (9.5) implies that we can do so provided we restrict V to Case I of Definition 5.2:

ASSUMPTION 9.3: For the rest of this section we assume that $V(2x) < 2V(x)$, for $\frac{1}{2}a \leqslant x \leqslant \frac{1}{2}$.

Combining the results of Exercises 9.1 and 9.2 we have

$$(U_1(t)\phi)(x) = \int_{G(2X(-t,x))}^{G(2x)-t} k(X(-t+\tau(\xi,t,x)))\phi(G^{-1}(\xi))\frac{\partial \tau}{\partial \xi}(\xi,t,x)d\xi \tag{9.6}$$

where $\tau(\xi,t,x)$ is the inverse function of $\xi(\tau,t,x)$ defined in Exercise 9.2. From this representation one can prove that $U_1(t)$ is compact, but the proof is rather technical. The essential ideas can be conveniently demonstrated in the simple special case $V(x) \equiv 1$.

EXERCISE 9.4. Show that (9.6) reduces to

$$(U_1(t)\phi)(x) = \int_{2x-2t}^{2x-t} k(t-x+\xi)\phi(\xi)d\xi , \tag{9.7}$$

with the convention $k(x) = 0$ for $x \leqslant \frac{1}{2}a$ and $x > \frac{1}{2}$, when $V(x) \equiv 1$.

THEOREM 9.5. *Under Assumption 9.3 the first generation operator $U_1(t)$ is compact for all $t \geqslant 0$.*

PROOF (for $V(x) \equiv 1$). For any $y > x$ we obtain from (9.7) that

$$|(U_1(t)\phi)(x) - (U_1(t)\phi)(y)| \leqslant ((1) + (2) + (3))\|\phi\|$$

where

$$(1) = \int_{2x-2t}^{2x-t} |k(t-x+\xi) - k(t-y+\xi)|\, d\xi \leqslant \int_{-\infty}^{\infty} |k(\tau) - k(\tau+x-y)| d\tau$$

$$(2) = \int_{2x-2t}^{2y-2t} k(t-y+\xi)d\xi = \int_{2x-y-t}^{y-t} k(\tau)d\tau$$

$$(3) = \int_{2x-t}^{2y-t} k(t-y+\xi)d\xi = \int_{2x-y}^{y} k(\tau)d\tau$$

As $y-x \downarrow 0$, (1) goes to zero since translation is continuous in $L_1(\mathbb{R})$ (which is easy to prove using the fact that C^∞-functions with compact support are dense in $L_1(\mathbb{R})$) and (2) and (3) go to zero since the indefinite integral of an L_1-function is continuous (even absolutely continuous, see Definition 2.8 and Theorem 2.9). We conclude that $\{U_1(t)\phi \mid \|\phi\| \leqslant K\}$ is equicontinuous. Since $|(U_1(t)\phi)(x)| \leqslant \int_{\frac{1}{2}a}^{\frac{1}{2}} k(\tau)d\tau \, \|\phi\|$ this set is also uniformly bounded. □

We next define inductively the lth generation operator

$$U_l(t) = \int_0^t U_0(t-\tau)C\, U_{l-1}(\tau)d\tau . \tag{9.8}$$

Exactly the same type of arguments (note that in the proof of Theorem 9.5 the equicontinuity is uniform for $t \in [0,s]$ for any $s < \infty$) yield

THEOREM 9.6. *Under Assumption 9.3 the lth generation operator $U_l(t)$ is compact for all $t \geqslant 0$ and all $l \in \mathbb{N}$.*

As an aside we remark that a short proof applies when C is bounded (i.e., $k(\frac{1}{2}) = 0$): the (Riemann) integral is a limit of finite sums and the set of compact linear operators is closed.

COROLLARY 9.7. *Under Assumption 9.3 the semigroup*

$$T(t) = \sum_{l=0}^{\infty} U_l(t) \tag{9.9}$$

is compact for $t \geqslant G(1)$.

PROOF. Only finitely many terms of the series are non-zero (see Lemma 3.8) so we don't have to worry about the sense of convergence. □

EXERCISE 9.8. If $V(2x) = 2V(x)$ for $\frac{1}{2}a \leqslant x \leqslant \frac{1}{2}$ then $\frac{\partial \xi}{\partial \tau} = 0$ and consequently $\xi(\tau,t,x) = \xi(t,t,x) = G(X(-t,2x)) = G(2x)-t$. Show that now

$$(U_1(t)\phi)(x) = \phi(G^{-1}(G(2x)-t)) \int_0^t k(G^{-1}(G(x)-t+\tau))d\tau \tag{9.10}$$

and convince yourself that $U_1(t)$ will be compact only when the first generation has become extinct, i.e. for $t \geqslant G(2)$. Inductively one shows that $U_l(t)$ is not compact before $t = G(l+1)$ and therefore $T(t)$ will *never* be compact.

EXERCISE 9.9. Compute $G^{-1}(G(2x)-t)$ for the special case $V(x) = x$.

EXERCISE 9.10. Compute $U_2(t)$ for the special case $V(x) = x$.

EXERCISE 9.11. Show that $G(2x)-G(x) = \text{constant} = G(a)$ when $V(2x) = 2V(x)$ and give a biological interpretation of this identity.

We are now ready to reap the fruits of our efforts in sections 5 - 8. In particular we are going to apply Corollary 8.9 to the restriction of $T(t)$ to the invariant subspace $\mathcal{R}(\lambda_d I - A)$ (see Corollary 6.9 and Exercise 6.10). Denoting this restriction by $T_R(t)$ we observe that its generator is A_R, the restriction of A to $\mathcal{R}(\lambda_d I - A)$. Since $\sigma(A_R) = \sigma(A) \setminus \{\lambda_d\}$ we conclude from Theorem 5.5 that $\sup\{\text{Re}\lambda \mid \lambda \in \sigma(A_R)\} < \lambda_d - \epsilon$ for some $\epsilon > 0$ and subsequently from Corollary 8.9 that $\omega_0(T_R(t)) < \lambda_d - \epsilon$ where $\omega_0(T_R(t))$ denotes the growth bound of $T_R(t)$, and, finally, from Theorem 8.4 (ii) that

$$\|T_R(t)\| \leqslant Me^{(\lambda_d - \epsilon)t}, \quad t \geqslant 0. \tag{9.11}$$

Recalling that P, defined by (6.7), denotes the projection onto $\mathcal{N}(\lambda_d I - A)$ along $\mathcal{R}(\lambda_d I - A)$, we write

$$T(t)\phi = T(t)(P\phi + (I-P)\phi) = e^{\lambda_d t}P\phi + T_R(t)(I-P)\phi = e^{\lambda_d t}(P\phi + O(e^{-\epsilon t})),\ t \to \infty,$$

and summarize our conclusions as one of the main results of this chapter.

THEOREM 9.12. (The stable size distribution). *Under Assumption 9.3*

$$T(t)\phi = e^{\lambda_d t}\left(-\frac{\zeta(\lambda_d,\phi)}{\pi'(\lambda_d)}\psi_d + O(e^{-\epsilon t})\right),\ t \to \infty, \tag{9.12}$$

or, in words: the dominant term in the asymptotic expansion of $T(t)\phi$ for $t \to \infty$ is the product of three factors

i) $e^{\lambda_d t}$: *an exponential function of time with exponent λ_d, the dominant eigenvalue*

ii) ψ_d: *a fixed element of X, the eigenvector of A corresponding to λ_d*

iii) $-\frac{\zeta(\lambda_d,\phi)}{\pi'(\lambda_d)}$: *a time-independent scalar which is the only factor that depends on the specific initial condition ϕ*

and the remainder terms are relatively exponentially smaller with an exponent that is determined by the distance along the real axis of λ_d and the other eigenvalues of A.

EXERCISE 9.13. The eigenfunction ψ_d corresponding to λ_d is defined by (4.2) - (4.3) with $\lambda = \lambda_d$, $f(x) \equiv 0$ and, say, $C = 1$ as a normalization. Thus ψ_d is a positive function. Show that $\zeta(\lambda_d,\phi) > 0$ whenever $\phi \geqslant 0$ and $\phi \not\equiv 0$.

We conclude that the cell population will grow exponentially when $\lambda_d>0$ (respectively, die out exponentially when $\lambda_d<0$ or approach a constant level when $\lambda_d = 0$) while at the same time the (normalized, transformed) size distribution converges to a fixed distribution ψ_d that does not depend on the initial condition. For that reason one calls ψ_d the (transformed) *stable size distribution.* We refer back to section I.4 for a discussion of the biological information that is contained in this result.

The techniques of this section required that we restricted our attention to the (biologically most important) Case I of Definition 5.5 (see Assumption 9.3). Our next objective will be to analyse the asymptotic behaviour of $T(t)$ in Case II: $V(2x) = 2V(x)$, $\frac{1}{2}a\leqslant x\leqslant 1$. In order to understand what happens in that case we need an auxiliary result (the difference equation (10.18) below) which can be obtained by a very important and useful technique called *integration along characteristics.*

10. Interlude: integration along characteristics

In this section we shall introduce and illustrate the technique of integration along characteristics by analysing in detail its application to the cell proliferation model. In Chapter III, section 4 we return to it in a more general context and there we explain some of its features in a more geometric language. Here we concentrate on those aspects that involve straightforward systematic calculations, repeating to some extent our treatment in I.3.4.

The basic idea is to consider the independent variables t and x temporarily as functions of one variable s such that

$$\frac{d}{ds} = \frac{dt}{ds}\frac{\partial}{\partial t} + \frac{dx}{ds}\frac{\partial}{\partial x} = \frac{\partial}{\partial t} + V(x)\frac{\partial}{\partial x}\,. \tag{10.1}$$

Thus it appears that we have to choose

$$\frac{dt}{ds} = 1 \Rightarrow t = s + c_1\,, \tag{10.2}$$

$$\frac{dx}{ds} = V(x) \Rightarrow \frac{dx}{V(x)} = ds \Rightarrow G(x) = s + G(c_2) \Rightarrow x = G^{-1}(s + G(c_2))\,, \tag{10.3}$$

where c_1 and c_2 are arbitrary constants still at our disposal.

First consider a function $m(t,x)$ satisfying

$$\frac{\partial m}{\partial t} + V(x)\frac{\partial m}{\partial x} = 0\,. \tag{10.4}$$

Then $\dfrac{d\overline{m}}{ds} = 0$ where $\overline{m}(s) := m(t(s),x(s))$ and consequently $\overline{m}(s) = \text{constant} = \overline{m}(0)$ or

$$m(s + c_1,\, G^{-1}(s + G((c_2))) = m(c_1,c_2)\,. \tag{10.5}$$

The mapping $s\mapsto(s+c_1,G^{-1}(s + G(c_2)))$ corresponds to an orbit in the (t,x)-plane with starting point (c_1, c_2). Such orbits are called *characteristics.* In our biological model they correspond to the orbits that individual cells follow as a result of their growth.

Next we have to realize that (t,x) points are restricted to the strip $\{(t,x) \mid t\geqslant 0, \frac{1}{2}a\leqslant x\leqslant 1\}$ and that the homogeneous equation (10.4) only holds in $\{(t,x) \mid t\geqslant 0, \frac{1}{2}\leqslant x\leqslant 1\}$. The *boundary* of the latter domain consists of the lines $x = \frac{1}{2}$ and $x = 1$ and the segment $t = 0$, $\frac{1}{2}\leqslant x\leqslant 1$. At the line $x = \frac{1}{2}$ and the segment the characteristics enter this domain in the sense that $(s + c_1, G^{-1}(s + G(c_2))$ belongs to the inside for $s>0$ when either $c_2 = \frac{1}{2}$ or $c_1 = 0$. At the line $x = 1$ they leave the domain.

So we can use (10.5) to express $m(t,x)$ for $t\geqslant 0$ and $\frac{1}{2}\leqslant x\leqslant 1$ in terms of $m(0,x) = \phi(x)$ (the initial condition) and $m(t,\frac{1}{2})$, which we, pretending that it is a known function, baptize $y(t)$. The procedure is as follows:

i) Take $c_1 = 0$ and $c_2\geqslant\frac{1}{2}$ then

$$m(s,G^{-1}(s + G(c_2)) = \phi(c_2)\,. \tag{10.6}$$

In order to transform back from the variables (s,c_2) to the variables (t,x) we put $(t,x) = (s,G^{-1}(s + G(c_2)))$ and obtain $s = t$, $c_2 = G^{-1}(G(x)-t)$ and finally

$$m(t,x) = \phi(G^{-1}(G(x)-t)),\ t\leqslant G(x)-G(\tfrac{1}{2})\,. \tag{10.7}$$

ii) Take $c_2 = \frac{1}{2}$ and $c_1 > 0$ then

$$m(s + c_1, G^{-1}(s + G(\tfrac{1}{2}))) = y(c_1). \tag{10.8}$$

If $(t,x) = (s + c_1, G^{-1}(s + G(\frac{1}{2})))$ then $s = G(x) - G(\frac{1}{2})$ and $c_1 = t - G(x) + G(\frac{1}{2})$ so that

$$m(t,x) = y(t - G(x) + G(\tfrac{1}{2})),\ t > G(x) - G(\tfrac{1}{2}) \tag{10.9}$$

This ends our calculations for $\frac{1}{2} < x \leqslant 1$.

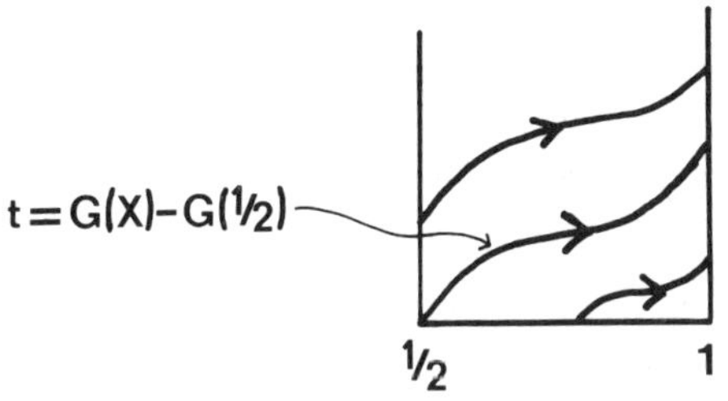

For $\frac{1}{2}a \leqslant x \leqslant \frac{1}{2}$ we have to deal with the inhomogeneous equation

$$\frac{\partial m}{\partial t} + V(x)\frac{\partial m}{\partial x} = h(t,x), \tag{10.10}$$

and the boundary condition

$$m(t, \tfrac{1}{2}a) = 0. \tag{10.11}$$

For the time being h is considered as a known function but later we will substitute $h(t,x) = k(x)m(t,2x)$. The characteristics enter the strip $\{(t,x) \mid t \geqslant 0, \frac{1}{2}a \leqslant x \leqslant \frac{1}{2}\}$ at the segment $t = 0$, $\frac{1}{2}a \leqslant x \leqslant \frac{1}{2}$ and at the line $x = \frac{1}{2}a$, so again we have to distinguish between two choices of coordinates.

i) Take $t = s$ and $x = G^{-1}(s + G(c_2))$ with $\frac{1}{2}a \leqslant c_2 \leqslant \frac{1}{2}$ then

$$\begin{cases} \dfrac{d\overline{m}}{ds} = h(s, G^{-1}(s + G(c_2)) \\ \overline{m}(0) = m(0,c_2) = \phi(c_2) \end{cases}$$

$$\Rightarrow \overline{m}(s) = \phi(c_2) + \int_0^s h(\sigma, G^{-1}(\sigma + G(c_2)))d\sigma \Rightarrow$$

$$m(t,x) = \phi(G^{-1}(G(x) - t)) + \int_0^t h(\sigma, G^{-1}(\sigma + G(x) - t))d\sigma, \text{ for } t \leqslant G(x). \tag{10.12}$$

ii) Take $t = s + c_1$ and $x = G^{-1}(s)$ with $c_1 \geqslant 0$ then

$$\begin{cases} \dfrac{d\overline{m}}{ds} = h(s + c_1, G^{-1}(s)) \\ \overline{m}(0) = m(c_1, \frac{1}{2}a) = 0 \end{cases}$$

$$\Rightarrow \overline{m}(s) = \int_0^s h(\sigma + c_1, G^{-1}(\sigma))d\sigma \Rightarrow$$

$$m(t,x) = \int_0^{G(x)} h(\sigma + t - G(x), G^{-1}(\sigma))d\sigma = \int_{a/2}^{x} h(G(\tau) + t - G(x), \tau)\frac{d\tau}{g(\tau)}, \text{ for } t > G(x). \tag{10.13}$$

To round off our calculations we have to substitute $h(t,x) = k(x)m(t,2x)$, with m given by (10.7) or (10.8), into (10.12) and (10.13), thus expressing m for all $t \geqslant 0$, $\frac{1}{2}a \leqslant x \leqslant 1$ in terms of the (known) initial condition ϕ and the (unknown) function y. If our approach is to be consistent, taking $x = \frac{1}{2}$ in (10.12) and (10.13) should produce $y(t)$!

Thus we find an *equation* for y which, assuming $2V(x) \geq V(2x)$, takes for $t \geq G(1)-G(\frac{1}{2})$ the form

$$y(t) = \int_{a/2}^{\frac{1}{2}} \frac{k(\tau)}{V(\tau)}\, y(t + G(\tau)-G(2\tau))d\tau\,. \tag{10.14}$$

EXERCISE 10.1. Assume that $2V(x)>2V(x)$. Show that

$$y(t) = \phi(G^{-1}(G(\tfrac{1}{2})-t)) + \int_0^t k(G^{-1}(\sigma + G(\tfrac{1}{2})-t))\, \phi(G^{-1}(G(2G^{-1}(\sigma + G(\tfrac{1}{2})-t))-\sigma))d\sigma, \quad t \leq G(\tfrac{1}{2})\,, \tag{10.15}$$

$$y(t) = \int_{a/2}^{\frac{1}{2}} \frac{k(\tau)}{V(\tau)} y(t + G(\tau)-G(2\tau))d\tau + \int_{\rho(t)}^{\frac{1}{2}} \frac{k(\tau)}{V(\tau)}\, \phi(G^{-1}(G(2\tau)-G(\tau)-t + G(\tfrac{1}{2})))d\tau \tag{10.16}$$

for $G(\frac{1}{2}) \leq t \leq G(1)-G(\frac{1}{2})$, where $\rho(t)$ is the unique solution of $G(2\rho)-G(\rho) = t$.

If $2V(x)>V(2x)$ we can use the transformation $\sigma = G(2\tau)-G(\tau)$ and its inverse $\tau = \rho(\sigma)$ to rewrite (10.14) as the *Volterra convolution equation*

$$y(t) = \int_{G(a)}^{G(1)-G(\frac{1}{2})} \frac{k(\rho(\sigma))}{V(\rho(\sigma))} \frac{d\rho}{d\sigma}(\sigma) y(t-\sigma)d\sigma\,. \tag{10.17}$$

If, on the other hand, $2V(x) = V(2x)$, then equation (10.14) is a *difference equation* (cf. Exercise 9.11)

$$y(t) = \int_{a/2}^{\frac{1}{2}} \frac{k(\tau)}{V(\tau)} d\tau\, y(t-G(a))\,. \tag{10.18}$$

EXERCISE 10.2. Consider Case II: $V(2x) = 2V(x)$. Show that

$$G(2G^{-1}(y)) = y + G(a)\,, \tag{10.19}$$

and use this result to derive the following relations from (10.7), (10.9), (10.12) and (10.13)

$$m(t,x) = \begin{cases} \phi(G^{-1}(G(x)-t)) & ,t \leq G(x)-G(\frac{1}{2}) \\ y(t-G(x) + G(\frac{1}{2})) & ,t>G(x)-G(\frac{1}{2}) \end{cases} \quad x>\frac{1}{2} \tag{10.20}$$

$$m(t,x) = \left.\begin{cases} \phi(G^{-1}(G(x)-t)) + \displaystyle\int_{G^{-1}(G(x)-t)}^{x} \frac{k(\sigma)}{V(\sigma)} d\sigma\, \phi(G^{-1}(G(x) + G(a)-t))\,, & t \leq G(x)\,. \\ \displaystyle\int_{a/2}^{x} \frac{k(\sigma)}{V(\sigma)} d\sigma\, \phi(G^{-1}(G(a) + G(x)-t))\,, & G(x)<t \leq G(x) + G(a)-G(\frac{1}{2})\,. \\ \displaystyle\int_{a/2}^{x} \frac{k(\sigma)}{V(\sigma)} d\sigma\, y(t-G(a) + G(\frac{1}{2})-G(x))\,, & t>G(x) + G(a)-G(\frac{1}{2})\,. \end{cases}\right\} \text{for } x \leq \frac{1}{2}\,. \tag{10.21}$$

EXERCISE 10.3. Use the relations above to deduce that

$$y(t) = \phi(G^{-1}(G(\tfrac{1}{2})-t)) + \int_{G^{-1}(G(\frac{1}{2})-t)}^{\frac{1}{2}} \frac{k(\sigma)}{V(\sigma)}\, d\sigma\, \phi(G^{-1}(G(\tfrac{1}{2}) + G(a)-t)), \quad t \leq G(\tfrac{1}{2}) \tag{10.22}$$

$$y(t) = \int_{a/2}^{\frac{1}{2}} \frac{k(\sigma)}{V(\sigma)}\, d\sigma\, \phi(G^{-1}(G(a) + G(\tfrac{1}{2})-t)), \quad G(\tfrac{1}{2})<t \leq G(a)\,. \tag{10.23}$$

EXERCISE 10.4. Use formulas (10.20) - (10.23) to show that

$$m(G(a),x) = \begin{cases} \int_{a/2}^{x} \frac{k(\sigma)}{V(\sigma)} d\sigma\{\phi(x) + \int_{x}^{\frac{1}{2}} \frac{k(\sigma)}{V(\sigma)}\, d\sigma\, \phi(G^{-1}(G(x) + G(a))) & , \frac{1}{2}a \leqslant x \leqslant \frac{1}{2}, \\ \int_{a/2}^{\frac{1}{2}} \frac{k(\sigma)}{V(\sigma)}\, d\sigma\, \phi(x) & , \frac{1}{2} \leqslant x \leqslant a, \\ \phi(G^{-1}(G(x) - G(a)) + \int_{G^{-1}(G(x)-G(a))}^{\frac{1}{2}} \frac{k(\sigma)}{V(\sigma)} d\sigma\, \phi(x) & , a \leqslant x \leqslant 1. \end{cases} \tag{10.24}$$

REMARKS 10.5. (i) Again we found that relevant features depend crucially on the function $2V(x) - V(2x)$.
(ii) The results of this section constitute a first step towards an alternative proof of the existence and uniqueness of solutions. Indeed, one can use (10.15) - (10.17) to give a straightforward constructive proof of the existence of a unique solution y and subsequently (10.7), (10.9), (10.12) and (10.13) to define $m(t,x)$ for $x \neq \frac{1}{2}$. Moreover, the existence of a stable size distribution in case $2V(x) > V(2x)$ can be deduced from (10.17) as well (we refer to Chapter IV section 2 for an exposition of the relevant material). In Case II ($V(2x) = 2V(x)$) formulas (10.18) - (10.23) in fact provide us with an *explicit* representation of the solution.
(iii) We were able to derive a scalar equation for $y(t) = m(t,\frac{1}{2})$ since every potential mother cell necessarily passes size $\frac{1}{2}$ during her life time (recall the reflections about the interpretation of $\pi(0)$ in Interlude 4.3.2 in Chapter I. Here the Assumption 4.2: $a \geqslant \frac{1}{2}$ is essential. If we relax this assumption to $a \geqslant 2^{-k}$ we need k equations for the variables $m(t,2^{-l})$, $l = 1,...,k$.

EXERCISE 10.6. Assume $a \geqslant \frac{1}{4}$ and $2V(x) \geqslant V(2x)$. Derive the analogue of equation (10.14).

11. The merry-go-round

In this section we concentrate on Case II of Definition 5.2.: $V(2x) = 2V(x)$. Firstly, we summarize the knowledge obtained so far. The characteristic equation has countably many simple roots

$$\lambda_l = \delta + \frac{2l\pi i}{\tau}, \quad l \in \mathbb{Z}, \tag{11.1}$$

where

$$\delta = \frac{1}{\tau} \ln \int_{a/2}^{1/2} \frac{k(\xi)}{V(\xi)}\, d\xi . \tag{11.2}$$

and by definition

$$\tau = G(a),$$

the size doubling time (see Exercise 9.11). The corresponding eigenvectors are

$$\psi_l(x) = \theta(x) e^{\lambda_l (G(\frac{1}{2}) - G(x))} \tag{11.3}$$

where

$$\theta(x) = \frac{\int_{a/2}^{x} \frac{k(\xi)}{V(\xi)} d\xi}{\int_{a/2}^{\frac{1}{2}} \frac{k(\xi)}{V(\xi)}\, d\xi} \tag{11.4}$$

(with the usual convention that $k(x) = 0$ for $x > \frac{1}{2}$).

From (10.20) and (10.21) it follows that for $t > \tau$

$$m(t,x) = \theta(x) y(t - G(x) + G(\tfrac{1}{2})) \tag{11.5}$$

where y is a solution of

$$y(t) = \int_{a/2}^{\frac{1}{2}} \frac{k(\sigma)}{V(\sigma)} d\sigma \, y(t-\tau) \, , \; t>\tau \, , \qquad (11.6)$$

(with $y(t)$ for $0 \leq t \leq \tau$ determined by the initial function ϕ as described in (10.22), (10.23)).

Thus an obvious conjecture is that the action of the semigroup is some kind of combination of multiplication and periodic continuation, and our objective is to verify this conjecture while simultaneously making it more precise. In order to separate the multiplication from the periodic continuation we introduce

$$z(t) = e^{-\delta t} y(t) \qquad (11.7)$$

and find upon substitution into (11.6) that

$$z(t) = z(t-\tau) \quad , \; t \geq \tau \, , \qquad (11.8)$$

which implies that z is a τ periodic function. Hence

$$m(t,x) = e^{\delta t} \, \theta(x) e^{\delta(G(\frac{1}{2}) - G(x))} z(t - G(x) + G(\tfrac{1}{2})) \qquad (11.9)$$

or, in words: for $t \geq \tau$, $T(t)\phi$ is the product of three factors

i) $e^{\delta t}$: an exponential function of time with exponent δ

ii) $\theta(x) e^{\delta(G(\frac{1}{2}) - G(x))}$: a fixed element of X

iii) $z(t - G(x) + G(\frac{1}{2}))$: a τ-periodic function, with argument $t - G(x) + G(\frac{1}{2})$, which is the only factor that depends on the specific initial condition ϕ.

Note that (11.9) shows that $\omega_0 = \delta = s(A)$ and in addition (since $|\kappa L|_\alpha = \kappa |L|_\alpha$ for $\kappa > 0$) that $\omega_e = \delta$ as well. An alternative proof of the latter identity is given by the observation that $\{e^{\lambda_l t} \mid l \in \mathbb{Z}\}$ lies dense on the circumference $\{\lambda \mid |\lambda| = e^{\delta t}\}$ when t/τ is irrational (indeed, use the definition of essential spectrum, Theorem 8.11 (ii) and $\omega_e \leq \omega_0 = \delta$).

Recalling Theorem 7.8 we observe that $\mathfrak{N}(e^{\delta\tau} I - T(\tau))$ is the infinite dimensional subspace which is obtained by taking the closure of the set spanned by

$$\psi_l(x) = \theta(x) e^{\delta(G(\frac{1}{2}) - G(x))} e^{\frac{2l\pi i}{\tau}(G(\frac{1}{2}) - G(x))} \, , \; l \in \mathbb{Z} \, . \qquad (11.10)$$

Well known results from Fourier analysis imply that this is precisely the subspace of functions of the form

$$\theta(x) e^{-\delta G(x)} q(G(x)) \, ,$$

with q a continuous τ-periodic function. Clearly (11.9) implies that $T(t)\phi$ belongs to this subspace for $t \geq \tau$.

It follows likewise from Fourier theory that the sequence ψ_l does not constitute a basis for $\mathfrak{N}(e^{\delta\tau} I - T(\tau))$, i.e. the expansion in a series $\sum c_l \psi_l$ does not necessarily converge in our supremum norm topology!

Apart from some transient phenomena during a time interval of length τ, the dynamics takes place in the dominant subspace $\mathfrak{N}(e^{\delta\tau} I - T(\tau))$. Since this is an infinite dimensional subspace, infinitely many characteristics of the initial condition (like zeros, extrema etc.) remain manifest for all time. This is in sharp contrast with the one-dimensional asymptotic dynamics in the case $V(2x) < 2V(x)$.

A priori the existence of a canonical (i.e. commuting with the semigroup) projection P onto $\mathfrak{N}(e^{\delta\tau} I - T(\tau))$ is not guaranteed. Calculations involving the residues in the isolated poles λ_l are not directly applicable since the resulting series might be divergent. (A remedy for this deficiency is provided by Cesaro summation (this was pointed out to the author by Prof. R. Nagel; see SCHAEFER, 19 1974, III.7).) Happily, however, we can in the present case easily obtain an explicit representation of P from a detour: given ϕ we can calculate $T(\tau)\phi$ from (10.24) and subsequently determine $P\phi \in \mathfrak{N}(e^{\delta\tau} I - T(\tau))$ from the condition that $T(\tau)P\phi = T(\tau)\phi$, by exploiting the fact that (11.9) gives an explicit expression for the (group) action of $T(t)$ on $\mathfrak{N}(e^{\delta\tau} I - T(\tau))$. Thus we obtain*

---

* Side-remark: For readers interested in functional differential equations we point out that for these the corresponding construction cannot be carried out because one has neither the precise characterization of the relevant subspace nor the explicit representation of the dynamics on it. See VERDUYN-LUNEL (1984) for a recent treatment of retarded functional differential equations admitting solutions that vanish after finite time.

$$(P\phi)(x) = \begin{cases} \theta(x)\{\phi(x) + \int_x^{\frac{1}{2}} \frac{k(\sigma)}{V(\sigma)}\, d\sigma \cdot \phi(G^{-1}(G(x) + G(a)))\} & , \frac{1}{2}a \leqslant x \leqslant \frac{1}{2}, \\ \phi(x) & , \frac{1}{2} \leqslant x \leqslant a, \\ \dfrac{1}{\int_{a/2}^{\frac{1}{2}} \frac{k(\sigma)}{V(\sigma)} d\sigma} \{\phi(G^{-1}(G(x) - G(a))) + \int_{G^{-1}(G(x)-G(a))}^{\frac{1}{2}} \frac{k(\sigma)}{V(\sigma)}\, d\sigma \cdot \phi(x)\} & , a \leqslant x \leqslant 1 . \end{cases} \tag{11.11}$$

EXERCISE 11.1. Verify that $P^2 = P$, that $\mathcal{R}(P) \subset \mathcal{N}(e^{\delta\tau}I - T(\tau))$ and that the restriction of P to $\mathcal{N}(e^{\delta\tau}I - T(\tau))$ reduces to the identity.

REMARK 11.2. Put a tape in a loop and slowly but surely turn the loudspeakers on. What you hear resembles to some extent the meaning of (11.9). Another convenient mental picture is the spiral staircase. Recall from Exercise 9.11 that the phase period $\tau = G(a)$ equals the time which individual cells need to double their size.

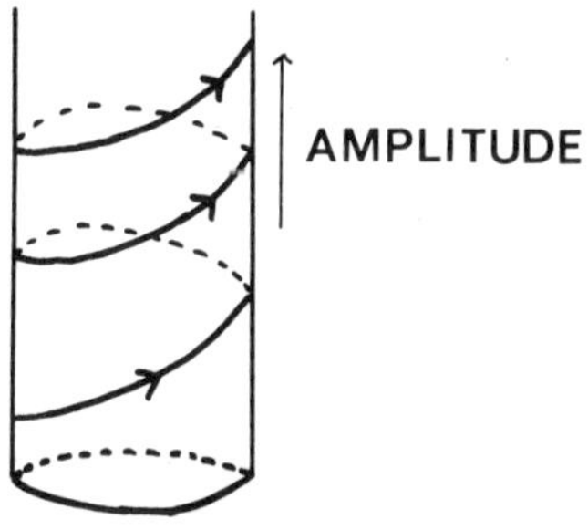

EXERCISE 11.3. In Exercise I.4.3.6 the equation

$$\frac{\partial}{\partial t} n(t,x) = -\frac{\partial}{\partial x}(V(x)n(t,x)) - \mu(x)n(t,x) - b(x)n(t,x) + 2\int_x^1 \frac{d(p)}{p} b(\frac{x}{p}) n(t,\frac{x}{p}) dp \tag{11.12}$$

was derived as a description of cell proliferation, when the probability that a mother cell of size x splits into one daughter of size px and one of size $(1-p)x$ is given by the x-independent probability density $d(p)$. Assume that $d(p) = 0$ for $p \notin (\frac{1}{2} - \Delta, \frac{1}{2} + \Delta)$ for some $\Delta \in (0, \frac{1}{2})$ and that $a > \frac{1}{2} + \Delta$, where, as before, $b(x) = 0$ for $x \leqslant a$ and $b(x) > 0$ for $x > a$. Derive the characteristic equation and analyse it. Put special emphasis on the case $V(x) = x$. Cautionary note: the minimal size is now $(\frac{1}{2} - \Delta)a$.

EXERCISE 11.4. In the special case that fission occurs exactly when reaching size $x = 1$ the analogue of (11.12) takes the form

$$\frac{\partial n}{\partial t}(t,x) = -\frac{\partial}{\partial x}(V(x)n(t,x)) - \mu(x)n(t,x) + 2d(x)V(1)n(t,1) , \tag{11.13}$$

supplemented by the boundary condition

$$n(t, \frac{1}{2} - \Delta) = 0 \tag{11.14}$$

Analyse this problem.

REMARK 11.5. An alternative way to derive (11.9) goes as follows. Put

$$m(t,x) = e^{\delta t} \theta(x) e^{\delta(G(\frac{1}{2}) - G(x))} p(t,x) \tag{11.15}$$

then

$$\frac{\partial p}{\partial t}(t,x) + V(x)\frac{\partial p}{\partial x}(t,x) = \frac{k(x)e^{-\delta\tau}}{\theta(x)}[p(t,2x) - p(t,x)] \tag{11.16}$$

Since $\theta(\frac{1}{2}a) = 0$ we are led to require $p(t,\frac{1}{2}a) = p(t,a)$ in order to keep the right hand side bounded. If $p(t,2x) = p(t,x)$ the right hand side reduces to zero and therefore $p(t,x) = p(t-G(x),\frac{1}{2}a) = p(t-G(x),a)$. Hence $p(t,a) = p(t-\tau,a)$, i.e. $p(t,a)$ is a τ-periodic function. Consistency now requires that $p(t,2x) = p(t-G(2x),a) = p(t-G(x),a), = p(t,x)$ but this requirement is fulfilled since $G(2x)-G(x) = G(a)$. Thus we can solve (11.16) explicitly after a transient period of length $\tau = G(a)$.

12. The merry-go-round with an absorbing exit

In Case III of Definition 5.2 V satisfies the relation $V(2x) = 2V(x)$ on the interval $\frac{1}{2}a \leqslant x \leqslant \beta$ while $V(2x) < 2V(x)$ for $\beta < x \leqslant \frac{1}{2}$. Here β is some number between $\frac{1}{2}a$ and $\frac{1}{2}$. Because of Theorem 5.8 (the existence of a strictly dominant real eigenvalue λ_d) we expect that in this case the normalized size distribution will converge towards a stable distribution. But calculations as in Exercise 9.8 indicate that $T(t)$ will not be compact after finite time. Therefore we need more subtle arguments involving the measure of non-compactness and in particular the result $\omega_0 = \max\{\omega_e, \omega_n\}$ (Theorem 8.12). Remarkably our mathematical procedure can to a large extent be described in biological terms and this we will do first.

One can conceive of the population as the union of two subpopulations. The first of these consists of those cells which were either present at $t = 0$ or arose from divisions of ancestors which at the moment of division had a size smaller than or equal to 2β (in other words, none of the ancestors has divided after $t = 0$ with size greater than 2β). The second subpopulation is the complement and consists of those cells for which at least one of the ancestors has undergone a division after $t = 0$, while having a size greater than 2β.

The dynamics of the first subpopulation is of the merry-go-round type. The members of the second subpopulation, on the other hand, are obtained by successive application of generation operators of which one at least is compact, and hence the product is compact. Moreover, the traffic between the two subpopulations goes one way only since offspring of the first subpopulation can be a member of the second but not vice-versa. Therefore the second will grow faster than the first (indeed, the chance that an arbitrary newborn cell that is determined to divide will do so with size less than or equal to 2β is less than one and $q^l \to 0$ when $l \to \infty$ and $|q| < 1$!). Hence there is hope that the compact part of the semigroup operator is asymptotically dominant over the non-compact part in the sense that $\omega_e < \omega_0$.

NOTATION: we shall denote functions that describe the first subpopulation by a dash above the letter, e.g. $\bar{m}$, and functions that describe the second subpopulation by a hat, e.g. $\hat{m}$.

Define

$$\bar{k}(x) = \begin{cases} k(x) & , \frac{1}{2}a \leqslant x \leqslant \beta, \\ 0 & , \beta < x \leqslant 1, \end{cases} \tag{12.1}$$

and

$$\hat{k}(x) = \begin{cases} k(x) & , \beta < x \leqslant \frac{1}{2}, \\ 0 & , \frac{1}{2} < x \leqslant 1 . \end{cases} \tag{12.2}$$

Let m satisfy

$$\begin{cases} \dfrac{\partial m}{\partial t} + V(x)\dfrac{\partial m}{\partial x} = k(x)m(t,2x) \\ m(0,x) = \phi(x) \end{cases} \tag{12.3}$$

then we may write $m = \bar{m} + \hat{m}$ where

$$\begin{cases} \dfrac{\partial \bar{m}}{\partial t} + V(x)\dfrac{\partial \bar{m}}{\partial x} = \bar{k}(x)\bar{m}(t,2x) \\ \bar{m}(0,x) = \phi(x) \end{cases} \tag{12.4}$$

$$\begin{cases} \dfrac{\partial \hat{m}}{\partial t} + V(x)\dfrac{\partial \hat{m}}{\partial x} = k(x)\hat{m}(t,2x) + \hat{k}(x)\bar{m}(t,2x) \\ \hat{m}(0,x) = 0 \end{cases} \tag{12.5}$$

The definition

$$\bar{T}(t)\phi = \bar{m}(t,\cdot;\phi) \tag{12.6}$$

produces a merry-go-round semigroup and consequently

$$\omega_0(\bar{T}(t)) = \frac{1}{G(a)}\ln\int_{a/2}^{1/2}\frac{\bar{k}(\xi)}{V(\xi)}\,d\xi = \frac{1}{G(a)}\ln\int_{a/2}^{\beta}\frac{k(\xi)}{V(\xi)}\,d\xi\,. \tag{12.7}$$

Next we define

$$\hat{U}_1(t)\phi = \int_0^t U_0(t-\tau)\hat{C}U_0(\tau)\phi d\tau \tag{12.8}$$

where

$$(\hat{C}\phi)(x) = \hat{k}(x)\phi(2x)\,. \tag{12.9}$$

Explicitly we have (cf. Exercise 9.1)

$$(\hat{U}_1(t)\phi)(x) = \int_0^t \hat{k}(X(-t+\tau,x))\phi(X(-\tau,2X(-t+\tau,x)))d\tau\,. \tag{12.10}$$

Now the point is that we can perform the transformation of Exercise 9.2 for precisely the τ-domain for which $\hat{k}(X(-t+\tau,x))\neq 0$ and consequently the compactness of $\hat{U}_1(t)$ follows in precisely the same way as the compactness of the operator $U_1(t)$ of section 9. Let

$$\hat{T}(t)\phi = \hat{m}(t,\cdot;\phi) \tag{12.11}$$

(nota bene that $\{\hat{T}(t)\}$ is *not* a semigroup) then repetition of the argument yields that $\hat{T}(t)$ is compact for all t. From

$$T(t) = \bar{T}(t) + \hat{T}(t) \tag{12.12}$$

we infer that $|T(t)|_\alpha = |\bar{T}(t)|_\alpha$ and so

$$\omega_e(T(t)) = \lim_{t\to\infty}\frac{\log|T(t)|_\alpha}{t} = \lim_{t\to\infty}\frac{\log|\bar{T}(t)|_\alpha}{t} \leqslant \omega_0(\bar{T}(t))\,.$$

EXERCISE 12.1. (i) Use the definition (4.4) of $\pi(\lambda)$ and the definition (12.1) of $\bar{k}$ to show that $\lambda_d(k)>\lambda_d(\bar{k})$. (ii) Deduce from (i) that (recall the definition $s(A) = \sup\{\mathrm{Re}\lambda \mid \lambda\in\sigma(A)\}$) $s(A)>s(\bar{A})$. (iii) Use $s(\bar{A}) = \omega_0(\bar{T}(t))$ and (ii) to obtain $s(A)>\omega_e(\bar{T}(t)) = \omega_e(T(t))$ and conclude that $\omega_0(T(t)) = s(A) = \lambda_d(k)$ and, moreover, $\omega_e(T(t))\leqslant\lambda_d(k)-\epsilon$ for some $\epsilon>0$.

Let as before $T_R(t)$ denote the restriction of $T(t)$ to the invariant subspace $\mathcal{R}(\lambda_d I-A)$ then $\omega_0(T_R(t)) = \max\{\omega_e(T_R(t)),\, s(A_R)\}\leqslant\max\{\omega_e(T(t)),\, s(A_R)\}\leqslant\lambda_d-\epsilon$ for some $\epsilon>0$. It follows that $T_R(t)$ satisfies an exponential estimate with exponent $\lambda_d-\epsilon$, where ϵ is the minimum of the distances along the real axis to, on the one hand, the other eigenvalues and, on the other hand, the essential spectrum. We summarize the result in

THEOREM 12.2. *With V as in Case III of Definition 5.2 the conclusion of Theorem 9.12 about the stable size distribution remains valid, but the characterization of the exponent in the remainder term has to be modified as indicated above.*

The assumption on V in Case III is such that the set $\{x \mid V(2x) = 2V(x),\ \frac{1}{2}a\leqslant x\leqslant\frac{1}{2}\}$ is just one interval and, moreover, such that the complement in $[\frac{1}{2}a,\frac{1}{2}]$ is just one interval as well. In the general case in which $V(2x)\neq 2V(x)$ for some $x\in[\frac{1}{2}a,\frac{1}{2}]$, these sets might consist of many intervals and in addition $V(2x)-2V(x)$ might assume both positive and negative values. This complicates the notation and the presentation of the arguments, but apart from this one can use essentially the ideas of this section to prove

THEOREM 12.3. *A necessary and sufficient condition for the existence of a stable size distribution is that $V(2x)\neq 2V(x)$ for at least one $x\in[\frac{1}{2}a,1]$.*

A somewhat different proof of this result is presented in DIEKMANN, HEIJMANS & THIEME (1984, part II), a paper which mainly deals with extensions of the above results to the case of time-periodic rates b,μ and V.

13. Remarks about positivity

Performing explicit calculations we derived the characteristic equation $\pi(\lambda) = 1$ for the spectrum of the generator A and subsequently we found from an analysis of π that two possibilities exist:

(i) either there exists a real eigenvalue λ_d which is strictly dominant in the sense that $\mathrm{Re}\lambda \leq \lambda_d - \epsilon$ for all spectral values $\lambda \neq \lambda_d$ and some $\epsilon > 0$, or

(ii) there exists a vertical line $\mathrm{Re}\lambda = \delta$ on which lie countably many eigenvalues which constitute an additive subgroup of $\mathbb{R}$ in the sense that $\delta + il\eta$, $l \in \mathbb{Z}$, is an eigenvalue whenever $\delta + i\eta$ is; and all other eigenvalues (if any; there were none in our case) satisfy $\mathrm{Re}\lambda \leq \delta - \epsilon$ for some $\epsilon > 0$.

The aim of this section is to draw attention to the fact that it is frequently possible, notably in population problems, to obtain such conclusions even when explicit calculations are impossible or just cumbersome. The mathematical theory which deals with such matters goes under the heading of "spectral theory of positive operators and positive semigroups". Note that clearly $T(t)$ maps positive functions onto positive functions as required by our interpretation of $T(t)\phi$ as a population density. The set of all positive functions in X is an example of a *cone* and an operator is called *positive* if it maps some cone into itself. As a generalization of the famous Perron-Frobenius theorem on the eigenvalues of a matrix with positive entries, there exists a collection of results which describes to some extent the structure of the spectrum of a positive operator (SCHAEFER, 1974). Moreover, analogues of such results for positive semigroups and their generators are known (GREINER, 1981; also see NAGEL, 1984, GREINER, 1984, and the references given there). The fact that we did not need these results in the present chapter (simply because we had other means to analyse the spectrum) detracts nothing from the merits of positive operator theory in the context of structured population models. Indeed, in Chapter V positivity arguments will play a major role and we refer to that chapter for an outline of the relevant theory. In addition we refer to the paper "Structured populations, linear semigroups and positivity" (HEIJMANS, 1984a) for a systematic exposition and a wealth of examples.

14. A somewhat special nonlinear problem

In this section we show how the conclusions about the model for *substrate limited growth in the chemostat* presented in subsection I.4.5 can be derived from the linear theory developed so far. At the risk of causing confusion we shall now use again the symbols A and $T(t)$ in the context of the original variables (recall the transformation (1.6)).

Abstractly we can write the balance equation (I.4.5.5) in the form (with S instead of R for the substrate concentration)

$$\frac{dn}{dt} = \beta(S)An - Dn \tag{14.1}$$

where

$$(A\phi)(x) = -(V(x)\phi(x))' - V(x)\delta(x)\phi(x) + 4V(2x)\delta(2x)\phi(2x) \tag{14.2}$$

with the appropriate domain of definition (so note that this A corresponds to the untransformed problem and is different from, but intertwined with, the A of the foregoing sections). We know that A generates a linear semigroup, let us call it $T(t)$, on a space of continuous functions with a tailor-made norm (1.11). Pretending that the substrate concentration S is a *known* function of time we can solve the equation for n quasi-explicitly:

$$n(t,\cdot;\phi) = e^{-Dt}\, T\Big(\int_0^t \beta(S(\tau))d\tau\Big)\phi\,, \tag{14.3}$$

where ϕ denotes the initial condition for n. As a side remark we mention that (14.3) delivers us from the obligation to define the notion of "a solution" for the nonlinear problem: we simply refer back to the end of section 3.

Since $\beta(S(t)) \geq 0$ the integral $\int_0^t \beta(S(\tau))d\tau$ approaches a limit as $t \to \infty$. If this limit would be finite then necessarily we would have that $S(t) \to 0$ for $t \to \infty$. But then (14.3) implies that the biomass $W(t) \to 0$ as well and hence $Z(t) = \alpha W(t) + S(t) \to 0$ which is in contradiction with the result of Exercise I.4.5.3. We conclude that $\int_0^t \beta(S(\tau))d\tau \to \infty$ for $t \to \infty$. Assuming that for some $x \in [\frac{1}{2}a, \frac{1}{2}]$, $V(2x) \neq 2V(x)$, this implies that

$$n(t,\cdot;\phi) = \rho(t)\{\phi_d + o(1)\}\ ,\quad t\to\infty\ , \tag{14.4}$$

where ϕ_d is the stable size distribution and $\rho(t)$ a real valued function which needs further investigation. Note that for constant death rates μ the stable distribution does *not* depend on the precise value of the death rate although, of course, the dominant eigenvalue does; if below we write λ_d we mean the dominant eigenvalue corresponding to $\mu = 0$.

Substituting (14.4) into the differential equation for n we find

$$\frac{d\rho}{dt} = (\lambda_d\beta(S) - D)\rho\ . \tag{14..5}$$

EXERCISE 14.1. Verify that the o(1) term is rightly left out of (14.5) since the o(1) term in (14.4) lies in $\mathcal{R}(\lambda_d I - A)$ while $\mathcal{N}(\lambda_d I - A) \cap \mathcal{R}(\lambda_d I - A) = \{0\}$.

Substitution into the differential equation (I.4.5.6) for S leads to

$$\frac{dS}{dt} = D(S^i - S) - \alpha_o\beta(S)\rho + \beta(S)\rho\cdot o(1)\ , \tag{14.6}$$

where

$$\alpha_0 = \alpha \int_{a/2}^{1} V(\xi)\phi_d(\xi)d\xi\ . \tag{14.7}$$

Because $Z(t) = \alpha W(t) + S(t)$ remains bounded for $t\to\infty$ and both W and S are positive, each of them remains bounded as well. The boundedness of W implies the boundedness of ρ. So as far as the asymptotic behaviour is concerned we may forget about the $\beta(S)\rho o(1)$ term in the equation for $\frac{dS}{dt}$. We now refer back to subsection I.4.5, and in particular to the Exercises 4.5.6 and 4.5.7, for a formulation of the conclusions which can be drawn.

III. Formulating Models for Structured Populations

J.A.J. Metz & O. Diekmann

1. Introduction: six examples for later use

By now the examples from chapter I and/or the subsequent theoretical elaboration in chapter II should have given you a taste for structured population models. In this chapter we shall develop a do-it-yourself kit enabling you to build models incorporating various amounts of biological detail.

We start in this section by introducing six examples that will be used throughout the chapter to illustrate various complications of the modelling process. In the next section we shall expose some of our modelling philosophy and in Section 3 the mathematical formalism for dealing with "mass" balance at the population level will be explicated in detail. Section 4 deals with some alternative parametrizations of the individual and population state spaces and sections 5 and 6 treat various kinds of limit arguments which may be used to simplify the model formulation. Finally appendix A contains a short refresher of vector notation and calculus in $\mathbb{R}^n$ written to assist readers with a mainly biological background, appendix B considers the extension of the formalism necessary to deal with stochastic components in the continuous i-state movement, a subject otherwise outside the scope of these notes, and appendix C gives the complete p-equations of the six examples introduced in this section (and with that the answers to most of the exercises in this chapter).

EXAMPLE 1.1. *The invertebrate functional response continued*

This example has already been introduced in I.2. There we made the simplifying assumption that pursuit and eating durations could safely be neglected. If we wish to take account of these durations we need a more complicated i-state space than the one spanned by satiation alone. Here we shall give a verbal account of how such a state space may be constructed. Later in this chapter we shall derive the corresponding p-equations. A more detailed discussion of the biological rationale for our assumptions as well as various extensions may be found in Metz & van Batenburg (1985a,b).

We start asssuming that for a searching predator the rate of search, g_0, depends only on its satiation, and that the rate of decrease of satiation, $-f_0$, depends only on that satiation itself as $f_0(s) = -as$. Accordingly for a searching predator its satiation still is a sufficient state description.

When a prey is sighted our predator starts pursuing. We shall assume here that the future course of the pursuit depends only on the distance which still remains between predator and prey. Moreover we assume that 1) this distance shortens at a constant speed, 2) prey escape, by flying away, at a constant rate μ, and 3) the distance at which pursuit is started depends only on the predator's satiation at that time (remember: the width of the search field depended on satiation). Accordingly for a pursuing predator the pair satiation *cum* distance-to-prey suffices as a state description.

When the predator arrives within a sufficient distance of its prey it strikes. We shall assume that the probability that this strike is successful is a constant q. A successful strike ends with a transition to eating. Otherwise the prey escapes and the predator starts searching again.

For the sake of the exposition we shall assume here that our predator eats at a constant speed, that a prey item is eaten *in toto* and that prey size also is constant. Accordingly remaining meal size and satiation together suffice as a state description for an eating predator.

To complete our model description we still have to prescribe the dynamics of the satiation during pursuit and eating. For pursuing predators we shall assume that satiation decreases at the same rate as during search. For eating predators we shall assume that satiation rises as $f_1 = f_0 + u$, where u denotes the rate of ingestion.

A point in favour of the state space described above is that the state variables all have an immediate physical interpretation allowing a direct simultaneous measurement, at least in principle. Such a type of model formulation should generally be preferred as it makes it relatively easy to extend the model to different experimental circumstances. However, the state space itself has a rather odd appearance. It consists of three unconnected pieces, the predator jumping from one piece to the other in the course of the predation process. In general this is exactly how it is. However, the simplifying assumptions made above were made with the specific purpose in mind of arriving at a state space allowing an easier visualisation. For, using these assumptions, we can transform the dimensions of both additional state variables to time by dividing the first one by the speed of pursuit after subtracting the strike distance and by dividing the second one by the speed of eating. The resulting two state variables then can be combined by simple addition into one new state variable: maximum time still to be spent handling a prey. This new state variable we shall call τ. For searching predators we shall put $\tau = 0$. The predator's i-state space then is contained in the product of the s- and τ-axes.

For practical calculations it is important that we delimit ourselves to the reachable states only. The resulting i-state space Ω for Holling's mantid is depicted in figure 1.1, together with some segments of possible trajectories representing various behaviour sequences that may occur after a prey has entered the predator's search field. Boundary segment (2) is given by $\tau = \tau_m(s)$, *i.e.* the maximum time that handling a prey sighted at satiation s may take. Boundary segments (1) and (4) both satisfy $ds \,/\, dt = -as + u$.

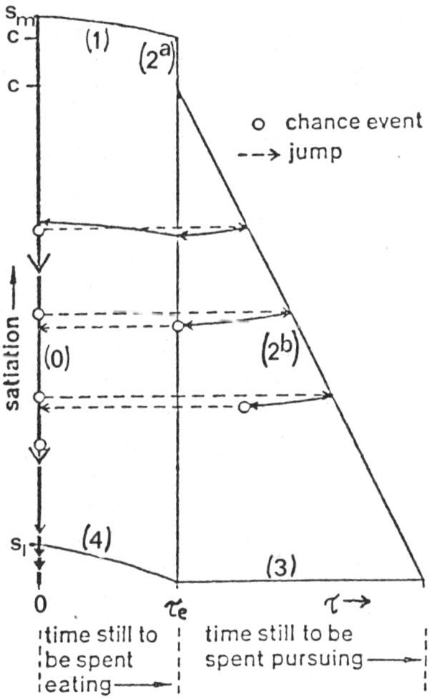

Figure 1.1: The individual state space, Ω, for Holling's mantid. In the state space some segments of trajectories are depicted showing the various possible events that may happen after the sighting of a prey. The starting points of the segments on the s-axis are arbitrary. The upper segment corresponds to a successful prey capturing sequence. In the second segment the strike is unsuccessful and in the third segment the prey escapes during pursuit. The segments of the boundary of Ω are numbered for later reference. From Metz & van Batenburg (1985a).

EXAMPLE 1.2: *Size dependent reproduction in ectothermic animals, continued*

This example has already been introduced in I.3. The only addition which we shall make in this chapter is that we shall relax the assumption that the death rate is constant.

Under laboratory conditions the death rate of Daphnids does not seem to depend on size except that large animals may die from starvation if feeding levels are suddenly and strongly reduced. However there is a pronounced dependence on age. The maximum age *Daphnia magna* may reach is about 70 days. Moreover there has sometimes been observed an increased death rate already considerably prior to this age. In Diekmann et al (1984) the following death rates have been proposed

$$\mu(l,a) = \begin{cases} \mu_0 & a < a_{\max} \\ \infty & a \geqslant a_{\max} \end{cases} \tag{1.1}$$

or

$$\mu(l,a) = \mu_0 + \psi(a) \text{ with } \psi(a) = (a_{max} - a)^{-1} \tag{1.2}$$

The reachable i-state space Ω for this example is depicted in figure 1.2 (see also exercise 1.3).

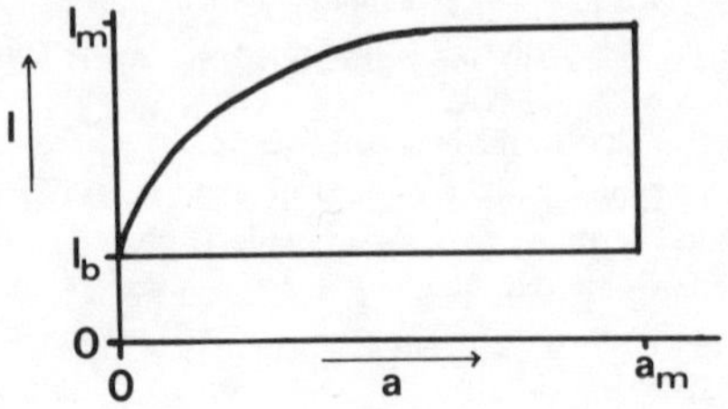

Figure 1.2. The reachable i-state space of the *Daphnia* model

EXERCISE 1.3: Prove that also for death rate (1.2) animals never reach an age higher than a_{max}. What happens if we replace $\psi(a)$ by $\epsilon / (a_{max} - a)$ and let ϵ go to zero? Hint: The answer to the last question is given at the end of section 5.2.

EXAMPLE 1.4: *Reproduction by binary fission, continued*

Just as in the previous example we may also have to introduce age as an additional state variable in the binary fission model from I.4. One reason may be that cells can only divide after sufficient time has elapsed to duplicate their DNA, independent of their size. The state space then again is a subset of the age size plane.

EXERCISE 1.5: Assuming that cells only can divide after they have reached a certain age a_0 construct the subset of the age size plane that is reachable by newborn cells starting life at all possible sizes $x>0$. Assume that there are no further restrictions on division age or division size and that the division rate $b(a,x) \leq b_{max} < \infty$, and assume that cell growth rate V depends only on cell size and that

$$0 < V_{min} \leq V(x) \leq V_{max} < \infty.$$

Which region of the age size plane can be reached by their first generation descendants, their second generation descendants etc? What, therefore, does the reachable state space Ω for naturally occurring cells look like?
Hints: First assume that $V(x) = V_0$ to get the feel of the problem. The answers to this exercise can be found in chapter V.

EXAMPLE 1.6: *Colony size distribution in the diatom* Asterionella.

Asterionella is a planctonic diatom which occurs in small star-shaped colonies. These colonies generally have sizes which are powers of two. The probable mechanism behind this phenomenon is that cells within one colony divide synchronously, two daughter cells staying linked by an (inanimate) bond, thereby doubling colony size, and that colonies break into equal parts as a result of the progressive weakening of the bonds over time.
In natural populations colony size distribution varies considerably. This probably is related to the growth rate of the population: when interdivision times are short *e.g.* as a result of better nutritional conditions, colony size will tend to be larger. The question therefore is how colony size distribution relates to the distribution of the interdivision times and hence to population growth rate.
The previous discussion makes clear that at least the ages of the various bonds should be among the state variables characterizing a colony. (Of course we could use other quantities, like bond strength, which are monotonically related to bond age, but doing so would force us to make additional assumptions, unbacked by any data, without affecting our predictions at the population level.) For instance in a colony of size 2^3 there is one oldest bond of age a_3, two bonds of age $a_2 < a_3$ and four bonds of age $a_1 < a_2$. To keep things as simple as possible we shall assume that cell division is also (cell) age dependent. We do not need a separate state variable representing cell age, however, as this equals the age of the youngest bond a_1, except in colonies of size 2^0, *i.e.* single cells. Our i-state space therefore corresponds to a (subset of) $\mathbb{R}^+ \cup \bigcup_{k=1}^{\infty} (\mathbb{R}^+)^k$, where $\mathbb{R}^+$ denotes the non-negative half axis. If bonds break at exactly age A and there is no lower bound to the age at division the reachable i-state space Ω equals

$$\{a_0 | A \leq a_0\} \cup \bigcup_{n=1}^{\infty} \{(a_1,..,a_n) \mid 0 \leq a_1 < \cdots < a_n < A\}.$$

EXERCISE 1.7: What does Ω look like when bonds may break at any age but always all bonds older than a certain age break together? How does it change Ω if we assume that there is a minimal age D before division can occur?

REMARK 1.8: To ensure convergence to a stable colony size distribution we shall have to assume some spread in the ages at which cell division occurs (see exercise 1.9 and the treatment of the *Asterionella* problem in IV.3.2). However, from a strict mechanistic point of view this assumption is incompatible with the assumption of perfect synchrony of division: the assumption that cell age is the sole determinant of division implies the independence of the separate cells within a colony. The explanation for this seeming inconsistency is that age is only a convenient but rather artificial state variable, which need not even be related in a one to one manner to the "real" *i*-state process. One may think for example that in reality division is contingent upon the accumulation of sufficient energy or nutrients, like in the size based model from I.4, except that the division threshold should be asssumed to be a fixed quantity to ensure the independence of mothers and daughters. If colonies are swept through regions with different light intensities or nutrient availabilities by turbulent water movement then the microscopic differences in environmental histories of the various colonies will give rise to a stochastic appearance of the *i*-state processes of representative cells of separate colonies but cells within one colony develop in exactly the same manner. (A prediction from our mechanistic picture would be that in those cases where colonies have distinct "outer" cells (as in colonies of size four) these outer cells should start dividing slightly earlier. This indeed has been observed). In the age type description we simply take account of the apparent distribution of the ages at which division occurs, without bothering about the detailed generating mechanism. (A further discussion of the strengths and weaknesses of age based models may be found in chapter IV.)

EXERCISE 1.9: Give an explicit expression for the age distribution $n(t,a)$ for a cell population in which all cells exactly on reaching age one divide into two daughters of age zero. Assume that there are no cell deaths. Call the initial age distribution $n(0,a) = n_0(a)$. Does $n(t,a)$ converge to a stable age distribution? Also given an expression for the rate $b(t)$ at which new cells are born into the population.

Example 1.10 *The prey-predator-patch (PPP) problem*

In the first three examples the individuals were indeed individuals in the biological sense. In the last example we no longer dealt with individuals in the strict sense, but the state parametrization still was in terms of individual properties. In this example we shall go still one step further: our individuals will be patches and the *i*-state variables will be population sizes.

Often prey occur in local patches, e.g. spider mite colonies on single or a few leaves (see the contribution by Sabelis in part B of this volume). Such prey patches are started by a single inseminated foundress. Pure prey patches end their existence as a result of the local exhaustion of resources and the concomitant emigration of the prey, or the arrival of a predator. Till that time the local prey population grows about exponentially to a first approximation. Within a prey patch a predator almost need not search for prey so its predation rate may to a first approximation be assumed to be constant. By the same argument a local predator population will grow approximately exponentially till either all prey are consumed or the prey patch ceases its existence due to resource exhaustion, followed by emigration of the remaining prey and the predators.

If we temporarily forget about the resource availability, our previous arguments imply that a (potential) patch is characterized by two state variables taking values in $\mathbb{R}^+$: size of prey population x and size of predator population y. The deterministic *i*-state movement is given by

$$\frac{dx}{dt} = \alpha x - \beta y \, , \quad \frac{dy}{dt} = \gamma y \, . \tag{1.3}$$

These equations hold for empty patches ($x = y = 0$), pure prey patches ($x>0$, $y = 0$) as well as for predator patches ($x>0$, $y>0$).

The previous description still leaves open the question how empty patches transform into prey patches and prey patches into predator patches: (1.3) leaves the origin as well as the x-axis invariant. We shall assume that these transitions are brought about by the random arrival of immigrants recruted from the prey leaving exhausted patches and the predators leaving empty prey patches. However, we are still in for some trouble. When a prey foundress arrives in an empty patch, into what does this patch transform? Our deterministic differential equation models for the local populations in essence are but limits of stochastic models dealing with integer numbers of individuals, the limit being taken by concentrating on densities and letting both the numbers of individuals and the area involved go to infinity. So within the deterministic framework we cannot properly account for the arrival of a single individual having noticeable effects in a finite time. Apparently we run into an incompatibility of our simplifying assumptions! In a biologically consistent model we have to assume that we are dealing with finite numbers of individuals all the time and we have to forsake (1.3) for a much more complicated stochastic model for the behaviour of the *i*-state. Such a model will certainly be intractable. Therefore we shall make a compromise and assume that the arrival of a prey foundress in an empty patch simply sets x equal to one (or ϵ if you like) and also that the arrival of a predator in a prey patch sets y equal to one, but that from then on (1.3) applies again.

In our discussion of the patch state we still left the resources out of the consideration. If each patch starts with a fixed amount of resources which are consumed at a rate proportional to the prey density we have

$$\frac{dr}{dt} = -\delta x \tag{1.4}$$

So if no predator arrives a prey patch is exhausted when it has reached age a_m defined by

$$r_0 = \delta \int_0^{a_m} x(a)da = \delta \int_0^{a_m} e^{\alpha a} da = \delta\alpha^{-1}(e^{\alpha a_m} - 1),$$

where r_0 is the amount of resource available in as yet unexploited patches. But the age of a pure prey patch and the prey density in it are monotonically related as $a = \alpha^{-1}\log(x)$. Therefore a patch is exhausted when x reaches the value

$$x_m = \alpha r_0 / \delta + 1 \tag{1.5}$$

In the same manner we can calculate a boundary line in the (x,y) plane corresponding to the resource exhaustion of the predator patches. For each point in the (x,y) plane corresponds to a unique previous history of prey population size within that patch. So we do not need r as an additional *state variable:* all the necessary information about the dynamics of r can be expressed in terms of x and y.

REMARK 1.11: An easier way to derive (1.5) derives from the observation that

$$\frac{dr}{da} + \frac{\delta}{\alpha}\frac{dx}{da} = 0, \quad r(0) = r_0\,, \quad x(0) = 1 \tag{1.6}$$

and therefore for all $a \leqslant a_m$,

$$r(a) + (\delta / \alpha)x(a) = r_0 + (\delta / \alpha\,), \tag{1.7}$$

which combined with the fact that $r(a_m) = 0$ gives (1.5) again.

EXERCISE 1.12: Calculate the resource exhaustion boundary and draw Ω. (Assume that a predator patch is left by the predator as soon as the prey population size drops below one, but not earlier.)
Hint: Extend the method used in the previous remark.

EXAMPLE 1.13: *Deterministic binary fission combined with stochastic individual growth.*

This example is analogous to the cell kinetics example treated in I.4, except that now we assume that (i) cells divide into two exactly equal parts as soon as they reach size x, and that (iia) cell growth is stochastic and, of course, (iib) continuous. (Some motivation for these assumptions has already been provided in remark 1.8). Assumptions (iia) and (iib) together imply that cell size follows a stochastic process of the diffusion type (see e.g. Goel & Richter-Dyn (1974) or Karlin & Taylor (1981)), which also allows cell size to decrease. Therefore we need some assumption to prevent cells from becoming too small. The assumption which we shall make here is that cells die on reaching size x_0 ($x_0 < \frac{1}{2}x_1$).

In these notes we intend to concentrate on models in which the continuous i-state movements are wholy deterministic. The main role of this example, now and in the future, is to complete the list of possible structural elements and to remind us of the implied unexplored possibilities.

The previous discussion concentrated on the i-state process and the reachable i-state space Ω. Before going on to the next section a few words are needed still about the corresponding p-states. In all cases this is a distribution over Ω, but for the purpose of writing down the p-equation it is often necessary to decompose this distribution into a number of separate components corresponding to the components of Ω.

In examples 1.4 and 1.13, reproduction by binary fission, Ω is a simply connected subset of $(\mathbb{R}^+)^k$ so there is no need to decompose the p-state. In example 1.10, the *PPP* problem, Ω consists of three separate sets corresponding to respectively the empty patches, the prey patches and the predator patches. Since these sets have different dimensions the same applies to the distribution living on them. If $n = (n_0, n_1, n_2)$ where n_0 corresponds to the empty patches, n_1 to the prey patches and n_2 to the predator patches, then n_0 has dimension number of patches (per unit of area, but we shall omit this qualification from now on) n_1 number of patches per unit of prey density, and n_2 number of patches per unit of prey density per unit of predator density.

In example 1.1, the invertebrate functional response, the p-state also has to be decomposed on dimensional grounds into a component n_0, corresponding to the searching predators and a component n_1 corresponding to the predators pursuing or eating prey. Moreover n_1 makes a sharp jump at the line $\tau = \tau_e$ due to the missed strikes. So the p-state process can satisfy a partial differential equation only away from this line.

In example 1.6, colony size distribution in *Asterionella*, Ω consists of infinitely many parts with different dimensions. So the *p*-state has to be decomposed into infinitely many components $n = (n_i)_{i=0}^{\infty}$ corresponding to colony sizes $2^i, i = 0, 1, \cdots$.

Finally in example 1.2, size dependent growth in ectotherms, Ω is a simply connected subset of $(\mathbb{R}^+)^2$ as in example 1.4 (reproduction by binary fission). Yet this example has one peculiarity which sets it apart from all other examples discussed in this section: There is no possible mechanism of dispersion of one cohort. All animals are born at the same size and all animals of the same age share the same feeding history, so their size has remained the same. As a result the *p*-state is no longer a frequency distribution over the full *i*-state space. Instead it is concentrated on some, continually changing, curve in the age-length plane. This fact that the *p*-state necessarily has only a one-dimensional support thwarts our attempt to describe the *p*-state process by a partial differential equation in the usual manner. In section 4.3 we shall therefore develop an alternative formalism in which the *p*-state is described in terms of the age-distribution together with the prevalent age-length relation.

2. Some modelling philosophy

In the previous examples the key concept was that of state, of individuals as well as of populations. The reason is that in these notes we wish to stress the mechanistic approach to modelling as opposed to the facile introduction of equations which may look attractive but lack a detailed biological underpinning. Formulating one's model in state space form, usually is a healthy way to bring out one's conception of the physical or biological mechanism one is trying to represent.

In this section we shall give a heuristic introduction to the mathematics of the state concept. We start in the first subsection with, heuristically phrased, definitions of the three allied concepts of state, next state transformation and output map, which are sufficiently general to encompass both varying environments and stochastic behaviour. The terminology in this subsection will be mainly that of the individual but with appropriate modifications everything we say applies to the population level as well. In the next subsection we consider the problem of obtaining state space models capturing the relevant aspects of i-behaviour. In the final subsection we shall argue why structured population models as defined in these notes, *i.e.* models in which the *p*-state is a distribution over an *i*-state space, are the inevitable outcome of a program in which one whishes to explain population dynamics in terms of mechanisms on the individual level.

2.1. The state concept

Throughout this book we concentrate on short term causal, as opposed to evolutionary, questions. The main purpose of our modelling effort then is to find such a description of an empirical *system*, be it an individual or a population, that its future behaviour can be predicted in terms of its initial "preparation" and the intervening environmental conditions. Here *behaviour* is everything that is of interest to us and that can be measured at least in principle. In the case of populations this may be e.g. numbers or total biomass, in the case of individuals reactions to a prey item, the acts of giving birth or dying, or weight, depending on the needs of the encompassing population model.

We shall start our discussion on the assumption that exact prediction is possible, at least in principle. Later on we shall have a look at stochastic models.

A *state* is such a collection of (hypothetical or empirically measurable) quantities X, that
(i) given $X(t)$ and the environmental history *(input)* between t and $t+\tau$, $u_{[t,t+\tau)}$, the state at $t+\tau$ is determined by

$$X(t+\tau) = T_{u[t,t+\tau)}X(t)\,, \tag{2.1.1}$$

where the transformations $T_{u_{[t,t+\tau)}}$ satisfy the *semigroup property* (see figure 2.1.1; we assume that the set of possible input functions u is such that the u-segments form a semigroup under the operation "gluing head to tail" (a more formal definition can be found in Metz (1981)), and
(ii) the behaviour of our system at t (the *output* at t), is completely determined by the value of X at t and possibly the condition of the environment at the same time. (The relation between state *cum* environment and corresponding behaviour is called the *output map).*

SYSTEM IN STATE SPACE FORM

$X \equiv$ state, $U \equiv$ input

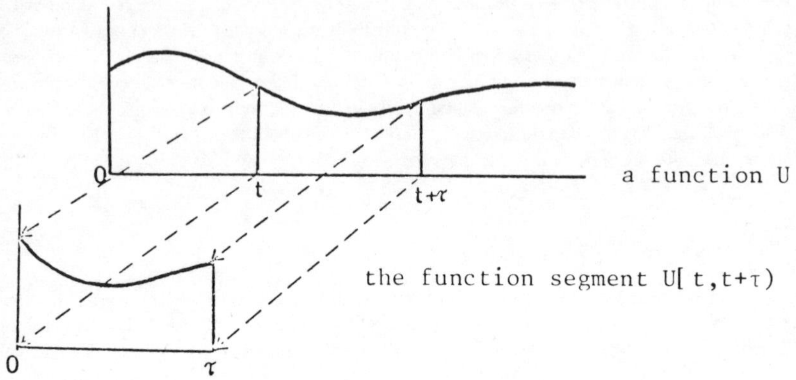

Next state map:

$$X(t+\tau) = T_{U[t,t+\tau)} X(t)$$

Semigroup property:

$$T_{U[t_2,t_3)} T_{U[t_1,t_2)} = T_{U[t_1,t_3)}, \quad T_{U[t,t)} = Id$$

Often with additional requirements:

(1) $\lim_{\tau \downarrow 0} T_{U[t,t+\tau)} X = X$

(2) $T_{U[t,t+\tau)} X$ continuous in X

Fig. 2.1.1. The semigroup property. *Id* denotes the identity mapping.

Examples are (*a*) individual size in *Daphnia,* with food density as the input and individual reproduction, dying from starvation, and possibly feeding rate, as the output (random deaths we still have to leave out of the picture till we are ready to deal with stochastic models as well), and (*b*) satiation *cum* maximum time still to be spent handling prey in Holling's mantid, with the (relative) positions of as yet unnoticed prey items as the input, and the catching of a prey item as the output.

The reason to stress a formalism which is sufficiently rich to account for a possibly varying input, is that this enables us to construct models by combining separately constructed building blocks. On the level of the individual one individual's behaviour may be made a component of its own or another individual's input, on the level of the population the population output like total reproduction, total feeding rate or total search rate, may act as a component of the input of the same or another population, like that of its prey or its predators. (A detailed discussion of the interface between the individual and population levels will be deferred to subsection 2.3.) When the input is constant our formalism reduces to that of the one-parameter semigroups of next state transformations dealt with in the previous two chapters (with the parameter t the length of the input segment). Some further specialized terminology for autonomous systems may be found in chapter VI. As it is usually next to impossible to reach any interesting conclusions about the behaviour of non-autonomous population systems we shall generally strive to end up with population models in which the subsystems are coupled in such a way that there are not any "loose ends" at the input side.

In general it is very difficult to calculate the operators $T_{u_{[t,t+\tau)}}$ explicitly as all kinds of processes interact in determining the movement of the state X. However, usually these interactions occur only through the changing of the state itself. In that case the contributions of the various processes can be considered separately if τ is infinitesimally small (and if condition (1) from figure 2.1.1 is fulfilled). Moreover, often for infinitesimally small τ the total change in X is proportional to τ. (Here we assume implicitly that the state space has a vector (Banach) space structure at least locally; see chapter II.) In that case we can write

$$\frac{d}{dt}X = A_v X\,, \tag{2.1.2}$$

where A_v is some operator still depending on the current value of the input $v = u(t)$. The family of operators A_v is called the differential generator of the transformation semigroup $\{T_{u[t,t+\tau)}\}$. It is the differential generators which are our main modelling tool: coupling of subsystems generally is done in terms of differential generators, as is the step from the individual to the population level.

EXAMPLE 2.1.1: In our original *Daphnia* example from I.3 we assumed that the weight w of an individual changed as a result of two processes, ingestion and basal metabolism. This eventually led to the i-differential equation*

$$\frac{dw}{dt} = \eta^{-1}(k\nu f(x)w^{2/3} - \zeta w)^{+}\,. \tag{2.1.3}$$

In two animals of the same age the one which has been kept under richer circumstances will have a higher ingestion rate as well as basal metabolism, but this is only due to it being larger; the feeding history is assumed not to have any other direct or indirect effects on present ingestion rate or metabolism.
For the *Daphnia* population we obtained in I.3.2 the p-differential equation

$$\frac{\partial n}{\partial t} = -\frac{\partial g(x)n}{\partial l} - \mu n \quad \text{for } l \leqslant \bar{l}(x)\,, \tag{2.1.4a}$$

$$n = 0 \quad \text{for } l > \bar{l}(x)\,,$$

$$g(x,l_b)n(t,l_b) = \int_{l_b}^{l_m} \lambda(x,l)n(t,l)dl\,, \tag{2.1.4b}$$

showing that in our model formulation x as a function of t may act as an input both on the individual and on the population level. At a later modelling stage in I.3.5 x was treated as being coupled dynamically to the *Daphnia* population again.

REMARKS 2.1.2: (i) Equation (2.1.4) contains a side condition relating the value of the i-state distribution at the boundary of Ω to its values in the interior (equation 2.1.4b)). An operator, like any map, is defined by (a) telling what its domain is and (b) giving a recipe telling how it acts on the elements of its domain. Side conditions form part of the definition of the domain. Their occurrence in the definition of a differential generator A_v is typical for infinite dimensional state spaces.
(ii) The p-differential generators of the elementary structured population models are always linear. Any nonlinearities are always due to feedbacks through the inanimate or animate environment, *i.e.* to the coupling of the population input to its or another population's output. Making this explicit is another advantage of our modular approach to model building.
(iii) Generally the basis for arriving at some prospective differential generator A_v is entirely heuristic. Therefore it is not *a priori* clear whether indeed a unique semigroup corresponding to A_v exists. We may for example have added an incompatible, or forgotten an essential side condition. Proving existence and uniqueness can be a deep mathematical problem, especially in the nonlinear case (see chapter VI for an example and some references). However, in this chapter we shall put ourselves in the shoes of the biologist and proceed on the assumption that generator and semigroup are well-defined and uniquely related.

* Throughout this chapter we shall follow the notational convention, that arguments of functions will be suppressed when not explicitly needed.

The definition of the state concept for the stochastic case is exactly analogous to that for the deterministic one except that all the maps involved may be stochastic. As there are some technical difficulties in treating the problem of how to define output maps in full generality, we shall first concentrate on the next state maps. The counterpart of (2.1.1) now is the assumption that for every given present state $X(t)$ and input segment $u_{[t,t+\tau)}$ we are given a so-called *transition probability* distribution for the resulting state $X(t+\tau)$. In this view the probability distribution for $X(t)$ plays a role which is exactly analogous to the p-state, and the transitions probabilities form a semigroup allowing a differential generator in the same manner as we are already used to.

EXAMPLE 2.1.3: If in our mantid model from I.2 or example 1.1 we assume that prey arrive at a rate proportional to their density and proportional to the width of the mantid's search field, we arrive at a stochastic model for an individual predator's behaviour which has prey density instead of prey positions as an input. If we normalize p from I.2 such that its integral over s equals one it may be interpreted as probability density of an individual predator's satiation, and formula (I.2.5.5) shows us the differential generator of the transition probabilities (also compare exercise I.2.2.2). The deterministic p-equation as studied in these notes is in this case a direct shadow of the equation applying to the state-transition probability density of individuals, obtained by applying the law of large numbers to a population of independently operating predators.

Generalizing the concept of output map to the stochastic case is a bit more complicated than generalizing that of next state map. First of all there are good reasons to allow the present output to depend not only on the present state and environmental condition as in the deterministic case, but also on the state transition that is presently occurring. For example, in the mantid model from I.2, with handling times neglected, catches were coincident with state *jumps*.

REMARK 2.1.4: For models in which all state transitions are deterministic it does not make a difference whether we allow the output to depend on the transitions as opposed to on the state only. For models in which all stochastic transitions are of the jump type the only additional sorts of outputs allowed are coupled to the jumps. The additional possibilities for models with stochastic continuous i-state movements are pretty scary. However, in a population dynamical context we only need to consider pure state dependent outputs and outputs which are coupled to jump transitions.

The second point is that we have to allow the output map to be stochastic. Why will become clear in a few lines when we have discussed how this statement should be interpreted. Consider for the time being a hypothetical situation in which both environment and state stay constant for a while. If we allow at each time point one of a number of possible output values to be chosen randomly, our animal would switch its behaviour infinitely often in any small time interval, unless we assume that the output map is such that one particular output value is chosen with probability one. Yet, as each time interval infinitely many choices are made, this assumption does not preclude the occurrence of other possible output values as long as these occurrences are point events having zero duration. These point events necessarily occur totally at random, or, in the language of probability theory, in independent Poisson processes, the concept of probability of occurrence of an output value (or range of output values) being taken over by that of the rate of the corresponding Poisson process. (where rate in this case may be defined as the *mean* number of occurrences per unit of time). Now assume that state and environment vary over time. This means that the rates of the Poisson processes start to vary as well, in a manner which deterministically depends on the state *cum* environment. This then is the concept of "stochastic output map" we need. An example is provided by the idea of an age dependent birth rate, the births themselves being modelled as randomly occurring point events. (Ask yourself what state space we would need in this case if we only allowed deterministic output maps!)

Having indicated the two main extensions of the concept of output map to the case of stochastic systems we leave it as an exercise to work out the various possible, and often mindboggling, combinations.

2.2. *Obtaining an i-state representation*

We shall confine ourselves here to i-state spaces which consist of subsets of $\mathbb{R}^k, k = 0,1,2, \cdots$ as these are the only ones playing a role in these notes. We shall indicate the i-state as X.

The easiest case, experimentally as well as conceptually, occurs when X wholly consists of physiological or physical quantities allowing direct measurement at least in principle. Examples are weight, amount of a toxic chemical present in the individual, fat reserves, gut content, *etc.* X is then uniquely determined, at least in principle and it is usually rather simple to write down a differential generator for it on the basis of its interpretation. Similarly it is not difficult to specify the value of X at the moment of birth or after a state jump. For practical reasons we should choose as simple an i-state representation as possible, *i.e.* we should identify states which do lead to the same behaviour now and in the future (minimal dimension of the i-state space) and we should confine the attention to those states that can be reached during the population situations that have our interest (the reachable i-state space Ω), but that is all there is to it. Here behaviour again means everything that is population dynamically relevant like giving birth, dying, capturing prey, *etc.*

Life becomes more complicated if on the individual level we have only input-output data at our disposal like the observed moments of molting or egglaying as a function of the temperature regime, or the observed reactions to prey items as a function of the feeding history. In that case we have to dream up an i-state representation satisfying our requirements *(i.e.* (i) and (ii) from the previous subsection). Examples of state variables of an input-output provenance are physiological age, degree of developmental commitment, satiation (as opposed to the physiological variable gut content with which it is frequently and uncritically identified; see *e.g.* Doucet & van Straalen (1980)) etc.. If our individuals are totally deterministic and if we can do all input-output experiments that we want, then systems theorists have proved that under very general conditions it is possible to construct a state space representation generating those input-output data and that moreover all minimal state space representations that do the same job are equal up to a choice of parametrization (see *e.g.* Metz (1977, 1981) for a review). Minimal here means again that there are no behaviourally indistinguishable states, and that all states can be reached given suitable inputs. However in practice the number of experiments necessary to reach uniqueness is forbidding except in the simplest cases. Therefore we have to satisfy ourselves with educated guesses. Moreover, there is the problem of extending our model to environmental circumstances not subsumed under the original experiments.

It should be remarked that "deterministic" refers here to the "internal" determinism of the individual machinery. The total i-state process may well be stochastic due to stochastic but i-experimentally observable inputs. Holling's mantid again provides an example. The mantid itself is totally deterministic; all stochastic events can be ascribed to prey behaviour, which can be considered an input quantity as far as the construction of the i-model is concerned.

The uniqueness theorem no longer holds when the i-process is stochastic "internally". This means that it is possible to construct two models which assume a totally different physiology but still predict exactly the same behaviour. Therefore in the stochastic case input-output based state space representations do not have the same epistemological status as those based on exactly reproducible input-output experiments. An example of such a "weak" state space representation is the use of age to model the change over time of an individual's birth rate or probability of dying.

EXAMPLE 2.2.1: Under constant circumstances both the stochastic cell growth model from the previous section (example 1.13) and the simple *Daphnia* model from I.3 each allow equivalent age representations. However in the *Daphnia* example there is a one to one correspondence between age and size whereas no such relation exists in the stochastic individual growth case. Both size dependent model formulations allow rather simple extensions to varying nutritional conditions, whereas this is not the case for the age dependent models.

REMARK 2.2.2: The stochastic fine structure of individual behaviour is often largely irrelevant to the gross behaviour of a deterministic structured population model derived from it. This sometimes makes it possible to replace an i-model by a population dynamical equivalent i-model of a simpler kind. This phenomenon still bears closer study. At present only scattered examples can be found in the literature, mainly pertaining to age-dependent models. The following example of our own making may illustrate the idea. In chapter IV, on age dependence, we shall consider some other examples when discussing the relationships between various kinds of equations describing population behaviour.
If we change the stochastic cell growth model from example 1.13 a bit by assuming that an individual divides into unequal parts the sizes of which bear a constant relation to each other, we may speak of an individual giving birth each time it hits the division threshold and at the same time falling back in size by a fixed amount. The two population dynamically relevant output quantities are being alive (or dying) and giving birth. We shall also assume that the environment as perceived by the individuals is constant so that their behaviour does not depend on absolute time or on each other.
Now assume that we do not have our physiologically based model. We then can construct an i-state space representation on a purely behavioural basis by defining the state of an individual to be the best possible summary of previous behaviour as far as the prediction of future behaviour is concerned. The state space of this representation is $\mathbb{R}^+ \times \{0,1\}$, where the second component tells us whether the individual is nulliparous or that it already has given birth at least once, and the first component tells us the time since the last event in its life, whether it be being born or giving birth. Such a model is behaviourally equivalent to our original physiological model. By its very construction knowing the frequencies of the individuals in the various states of our behaviourally based representation gives us a prediction of the future population dynamical constributions by these individuals which is the best possible one if we have nothing else to go by than their externally observable behaviour. (Knowing the frequencies of their sizes would probably give a, slightly, better prediction.)

The reason why we did not end up with age, *i.e.* time since birth, as the only state variable, is that the events of an individual dying and giving birth are dependent: when an individual has just given birth we know it is not near to the lower, death, boundary of its physiological state space. Yet, a model in which we use age as the only descriptor of an individual, while not a genuine model of its behaviour, can do almost as good a population dynamical job as our more complicated model, given that we match both the probability distribution of the age at death, and the mean rate of giving birth (conditional on the individual being alive) to those of our other two *i*-models. The only difference between the deterministic population models based on the three *i*-models is that the more detailed models will be better in predicting births and deaths occurring in the remainder of the founder population. But once this ancestral population is all gone all three models behave exactly the same as far as births, deaths and total population number are concerned.

In practice most *i*-state space representations will be based on a mixture of physiological and input-output considerations, the exact procedural details being dependent on the experimental possibilities particular to the case. Good examples are provided by the work of Kooijman and of Sabelis reported in part B of these notes.

2.3. *From the individual to the population level*

The pervading theme in these notes (with the possible exception of chapter IV) is that (i) from a reductionist/mechanistic point of view the sensible choice for the *p*-state should take the form of an *i*-state distribution, and (ii) we wish to formulate our models in terms of *p*-equations which (iii) should be based, either directly or indirectly, on the law of mass action. Yet in our presentation up till now the status of (i) to (iii) was to a large extent that of articles of faith, as the argument presented in I.1 was at best rather sketchy. In this subsection we intend to fill in some of the gaps.

The p-state

From a mechanistic viewpoint the purpose of our modelling effort should be the explanation of the behaviour of populations in terms of the doings of the individuals constituting them. Therefore our models should keep track of those doings at least to such an extent as is necessary for calculating the population properties in which we are interested. These properties can for example be numbers of individuals, (or rather densities, *i.e.* number per unit of area or volume, see remark I.1.1 (ii)) total biomass *etc.* To simplify the argument we shall from now on assume that the only properties in which we are interested are total numbers. Individuals therefore only have to be distinguished insofar they have different prospects of contributing to changes in the numbers of their own kind or of any other population in which we are interested, either directly, *e.g.* through catching prey, giving birth, being caught by a predator or migrating, or indirectly by influencing resource availability etc. This shows that our *p*-state should at least contain complete information about the frequency distribution of the *i*-states. The question is under what conditions this information is also sufficient.

For the *i*-state distribution to be sufficient information clearly a first (1) necessary condition is that individuals do not differ in influences on their behaviour other than through their *i*-states. Therefore we have to assume that they all experience the same inputs. This leads us to *define* a population as a collection of individuals all experiencing the same circumstances, be it "physical" circumstances like temperature or resource availability, or "biotic" circumstances like prevailing prey or predator densities. (The mantid example makes clear how we should (re)phrase our *i*-models such that they have densities (in the spatial sense) of other individuals as their input. That we have to phrase our *i*-models this way is a consequence of condition (2) below.) This means that by definition the environment of the population and that of the individuals is the same thing, and also that the environmental circumstances are considered as extraneous to the population. They should be generated by some other part of our model or be given *a priori, e.g.* by telling that the prey are just the smaller and the predators are just the larger members of the same population, or by giving the densities of prey and predators as functions of time.

REMARKS 2.3.1: (i) To conform to the usual biological interpretation of population we shall also wish to add to our definition that the individuals belong to the same "species", *i.e.* that any two individuals are equivalent in the sense that at least one member of the pair is equal in all relevant aspects to a potential descendant of the other member of the pair.
(ii) If the environmental circumstances cannot be considered the same for all individuals we wish to deal with, then the total or *meta*-population should be split into subpopulations, *e.g.* on the basis of location in space, such that within each subpopulation the circumstances are comparable. If we characterize individuals with an extended "type"-label consisting of a "location" label plus an *i*-state label, we can write down *p*-equations for the meta-population by going through exactly the same kind of routines as when writing down an ordinary *p*-equation.

(iii) The linearity of our elementary p-equations is due precisely to our equating of population and individual inputs.

A second (2) necessary condition for the i-state distribution to qualify as a p-state is that the population output can be calculated from it. This is by definition the case for any population properties we set out to explain. However, it should also apply to those outputs which we need as inputs for other parts of our model. Therefore the impact of individuals on their environment, be it physical or biotic, should only be through their "mass action", *i.e.* it should be possible to express this impact in terms of the spatial densities of the individuals in various i-states, without taking account of the behaviour of the separate individuals as individuals.

Together conditions (1) and (2) are also sufficient for the i-state distribution to qualify as p-state.

REMARK 2.3.2: (i) Biologically (1) and (2) mean that the individuals either do not interact at all, or if they do so, "mix homogeneously". This assumption will often apply as a first approximation to animals and also to plants interacting through a fast mixing chemical environment, but not for example to plants interacting with a small number of fixed neighbours through mutual shading.
(ii) When individuals sometimes temporarily form closely knitted groups but otherwise mix randomly we can still conform to (1) and (2) by a suitable redefinition of what we call an individual. An example is provided by the *Asterionella* colonies from example 1.6.

Before going on to the next topic, determinism, we still have to voice a warning. Even though a framework in which the elements are populations which are only coupled through their inputs and outputs in the sense of (1) and (2) is essentially sound, a too rigorous adherence to it may lead to the impractical consequence that we have to duplicate parts of our p-equations when interactions between individuals cannot be considered point events: For example, what we called pursuing predators in the invertebrate predator example 1.1 are in fact predator-prey pairs. So the part of the predator i-state-space that corresponds to pursuit (see fig. 1.1) also appears as part of the prey i-state-space, and the component of the predator p-state representing the pursuing predators exactly equals the component of the prey p-state representing pursued prey. Clearly it is more practical to treat the predator-prey pairs locked in pursuit as one, new, population in addition to the populations of single prey and predators.

p-equations and determinism

The use of a differential equation as a means of generating population behaviour is independent upon the assumption that our populations behave in a deterministic fashion. Biologically this means that (3) the number of individuals in the population is sufficiently large that any chance processes on the individual level are ironed out, and (4) all essential environmental conditions which are not outputs of the p-systems comprising our model are given as deterministic functions of time.

REMARK 2.3.3: (i) In checking (3) for spatially distributed populations as introduced in remark 2.3.1 (ii) we have to distinguish two cases. When the location label is discrete all subpopulations should be sufficiently large. However, in the case of a continuous location label (like positional coordinates in $\mathbb{R}^2$) it is only necessary that large numbers of individuals are present in the largest areas over which the environmental circumstances do not change overmuch. For direct biotic interactions depending on relative distances this condition is equivalent to: for all individuals the number of individuals potentially capable of interacting with them within a short time interval is large.
(ii) In our phrasing of assumptions (1) and (2) we already anticipated on the assumption of determinism, or, more particularly, on assumption (3). When numbers of individuals are not infinite the behaviour of one population can also show a dependence on the behaviour of another population, which cannot be captured in terms of a dependence on the state of that other population only: we have to allow for a dependence on state transitions as well. For example, in a model for predator-prey interaction dealing with finite numbers of individuals we have to account for the fact that the capturing of a prey by a predator is *exactly* coincident with a drop in the prey number. This is a stronger coupling of the two population processes than just a dependence of the dynamics of the prey population on the satiation distribution of the predators combined with a dependence of the dynamics of that satiation distribution on the prey density. But in large populations we soon loose sight of this kind of coupling at the level of the microscopic probabilistic structure.

The law of mass action

The "law of mass action" states not only that the rate per unit of area/volume at which a reaction occurs depends on no more than the spatial densities of the two reactants, but also that this rate is linear in the densities of each of them (with the understanding that if the two reactants are of the same kind the reaction rate is quadratic in the reactant density).

The law of mass action is a direct consequence of the assumptions that (i) two identical reactant individuals have the same probability per unit of time to undergo a reaction, where (ii) this probability depends only on the density of the other individuals, if any, involved in the reaction (and possibly on the physical circumstances), and moreover (iii) the maximum number of individuals involved in any reaction is two. Adding up over all identical individuals in a unit of volume gives us the expected number of reactions per unit of time, the reaction rate.

A more careful analysis of the previous argument shows that there is an intimate connection between the law of mass action and the concept of *i*-state. (i) follows from the fact that two individuals belonging to the same population and being in the same *i*-state should certainly have the same probability of reacting. After all the concept of *i*-state was introduced precisely to guarantee this kind of behavioural equivalence. The reasoning showing that the probability of reacting may only depend on the density of the individuals with which it reacts and *not* on the densities of other sorts of individuals is more complicated. We start with the observation that whitin our framework reactions cannot have a duration, for if an interaction between individuals or a behavioural transition costs time, this time has to be accounted for in terms of the *i*-state space (compare our discussion of the invertebrate predator in example 1.1) Next we throw in the assumption (A) that within a finite time interval an individual can only interact with finitely many other individuals (see also remark 2.3.4 (iii) below). Therefore the only reactions in which (a) other individuals are involved and which (b) have a noticeable effect, are births, deaths and *i*-state jumps. Only the third type of events results in a protracted effect on an individual's behaviour, *but only through a change in its i-state.* The direct involvement of any other individual in the generation of a reaction automatically classifies that individual as a reactant in the sense of the law of mass action, thus precluding any direct effect of non-reactants in the probability per unit of time that an individual in a given state reacts. Finally (iii) follows from a stronger version of our earlier assumption (A): (A ′) for any individual in any condition the probability per unit of time of interacting with an other individual is finite.

Yet there exist apparently reaction rates which do not conform to the law of mass action. Our discussion of the invertebrate functional response in I.2 indicates how such reaction rates may be fitted into the general framework by paying attention to faster time scales and correspondingly more complicated *i*-state spaces. Note, however, that that time scale argument was based essentially on the prey population being much larger than the predator population. It is only in the limit when the ratio between prey and predator densities has gone to infinity that we obtain nothing more complicated than a nonlinear relation between overall predation rate and (rescaled) prey density. But in taking this limit we gave up treating the predator and prey populations on equal footing.

REMARK 2.3.4: (i) The reason to stress in the present context the per unit of area/volume aspects is that an individual cannot "see" the total number of individuals present with which it can react, but only the local density of individuals surrounding it (compare remark I.1.1 (ii)).
(ii) For the initial argument leading up to the law of mass action it is already sufficient that the *expected* number of reactions undergone by individuals are equal. This is the basis for the construction of population dynamically equivalent simplified *i*-models as discussed in remark 2.2.3.
(iii) Our discussion of the invertebrate functional response in I.2 also shows how an apparently continuous *i*-movement may result from very many interactions each having a very small effect. Again this result was reached by a procedure which precludes treating the prey as a population of individuals au par with that of the predators.

3. Mass balance

In this section we shall discuss the various building blocks of the differential generator for the *p*-state transitions. The whole problem of specifying this differential generator boils down to an appropriate accounting of *p*-mass: individuals may be assumed to (dis)appear only through known (or at least named) causes like births, deaths or transport to or from some other place in the reachable *i*-state-space Ω. Moreover, during infinitesimally small time intervals these processes act independently. Therefore we can specify the *p*-differential generator by simply adding the various separate contributions to the local change of the *p*-mass. Depending on whether we concentrate on *i*-states inside or at a boundary of Ω our accounting gives us components of the defining formula of the *p*-differential generator A or components of a side- condition at that boundary. Since our discussion will be in terms of the biological processes involved we shall be intermittently dealing with one or the other.

First we introduce some notation. We start with repeating that throughout this chapter we shall adhere to the

CONVENTION: we shall suppress any arguments of functions that are not explicitly operated on or referred to.

The i-state will be denoted as $x = (x_1, \cdots, x_k)^T$, k the dimension of the component of Ω under consideration.* The density of individuals at x we shall indicate as $n(x)$, the p-state as n, *i.e.*

$$n(x)\Pi dx_i \equiv \text{number of individuals in the infinitesimal rectangle } \Pi(x_i, x_i + dx_i)$$

or, if your inclination is mathematical instead of scientific, for any $\Omega_0 \subset \Omega$:

$$\int_{\Omega_0} n(x)dx \equiv \text{number of individuals in } \Omega_0 \,.$$

(where dx denotes Lebesgue measure†).

In the interior of Ω the differential generator A can be written as

$$\frac{\partial n}{\partial t} = \delta_t n + \delta_{s+} n - \delta_{s-} n \qquad (3.1)$$

where

$\delta_t n \equiv$ contribution to the local change due to continuous $i-$state movement (the t stands for transport),

$\delta_{s+} n \equiv$ contribution to the local change due to births or jumps from elsewhere in Ω (the s stands for source)

$-\delta_{s-} n \equiv$ contribution to the local change due to deaths or jumps to elsewhere in Ω (the s stands for sink)

In principle

$$\delta_t n = \delta_{td} n + \delta_{ts} n \qquad (3.2)$$

where

$\delta_{td} \equiv$ contribution of the deterministic (local average) component of the continuous $i-$state movement

$\delta_{ts} n \equiv$ contribution of the purely stochastic component of the continuous $i-$state movement

However, we shall conventionally assume that $\delta_{ts} n = 0$, at least in the main body of the chapter, deferring to appendix B the discussion of the complications that arise when $\delta_{ts} n \neq 0$. *Also, when concentrating on one particular kind of contribution we shall often without further notice assume that all other contributions are zero.*

Side conditions can be decomposed in essentially the same manner as the defining formula for A and the same conventions will apply.

The various standard notational conventions of vector analysis and calculus in $\mathbb{R}^k$, as well as a discussion of the underlying concepts geared towards a biological audience, can be found in appendix A.

---

* We shall think of x as a column vector. ^{T} denotes transposition.

† If you are not familiar with the concept of Lebesgue integration you can just consider this a piece of notation. Appendix A tries to convey the idea behind the concept in a manner geared to the minimal use made of it in this chapter.

3.1. *Mass transport due to continuous i-state movement*

This subsection is divided into two parts. In the first part we follow the biologists line of approach, with the stress on the heuristics and minimizing the algebra. In the second part we essentially repeat the argument in a manner to suit the taste of a mathematician.

3.1.1. *A biologist's shortcut*

In chapter I we derived that the one dimensional conveyor belt is described by

$$\delta_t n = -\frac{\partial}{\partial x}(vn) = -(v\frac{\partial n}{\partial x} + n\frac{\partial v}{\partial x})$$

where v denotes its local speed (the speed of the i-movement). For $k>1$ we can just add the contributions of the transport in the various directions, *i.e.*

$$\delta_t n = -\sum_i \frac{\partial}{\partial x_i}(vn) \tag{3.1.1}$$

where v_i denotes the speed of movement in the direction of the positive x_i axis. Using the notation from vector analysis this can also be written as

$$\delta_t n = -\nabla\cdot(vn) \tag{3.1.2}$$

where v denotes the vector of local speeds.

The previous line of argument concentrated on one locality in Ω. Some more insight in the nature of (3.1.2) can be gained by rederiving it in a different manner. To this end we imagine that we move along through Ω with the flow generated by the vector field v, or more biologically phrased, that we stick to one individual and observe what happens to the content of a little box around it, currently of the form $\Pi(x_i,x_i + dx_i)$, and having a volume Πdx_i. At $t + dt$ this box has been transferred to a position slightly more downstream and has changed in shape and therefore also in volume (see fig. 3.1.1). The new position is given by $x \mapsto x+vdt$, the new value of the i'th state variable being $(x + vdt)_i = x_i + v_i dt$. To calculate the volume of our little box after its transfer downstream we start by noting that the map transforming neighbourhoods of x into neighbourhoods of $x + vdt$ may to first order in the dx_i be approximated by a linear map. This linear map transforms the original square box into a parallelepipid. Therefore any curviness seen in fig. 3.1.1 can safely be neglected as being due to higher order terms in the dx_i only. Next we concentrate on one i-state variable at a time, say the i-th, *i.e.* we consider the dilation of the box in the direction of the i-th unit vector $e^{(i)}$, while temporarily neglecting any changes in the other sides of the box. To this end we look what happens to the two adjacent corners x and $x+e^{(i)}dx_i$. The first corner transforms into $x+v(x)dt$, and the second one into

$$x+e^{(i)}dx_i + v(x+e^{(i)}dx_i)dt = x+e^{(i)}dx_i + (v(x)+\frac{\partial v}{\partial x_i}dx_i)dt\ .$$

Apparently under our linear map $e^{(i)}dx_i \mapsto (e^{(i)}+\frac{\partial v}{\partial x_i}dt)dx_i$ only the change in the $e^{(i)}$ direction contributes to the volume change (see fig. 3.1.2A (or B)).

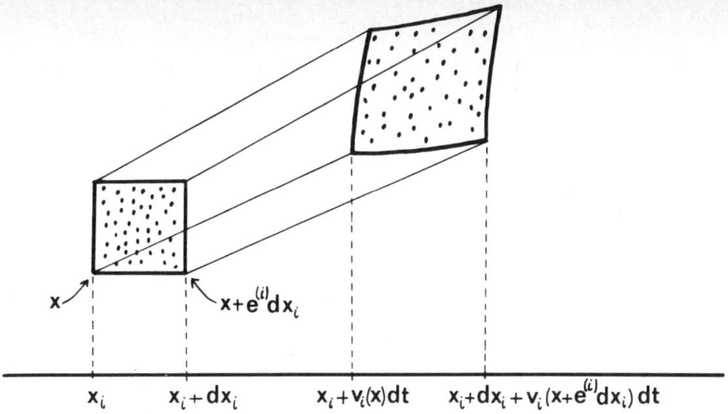

Fig. 3.1.1. The change of an infinitesimal volume when it moves along with the stream, and the contribution to this change due to changes in the i-th state variable.

Singling out the i-th coordinate we find $(e^{(i)}+\frac{\partial v}{\partial x_i}dt)_i dx_i = (1+\frac{\partial v_i}{\partial x_i}dt)dx_i$. Therefore the dilation of our box in the direction of the i-th coordinate adds a fractional increase of $\frac{\partial v_i}{\partial x_i}dt$ to its volume. If, as before, we neglect higher order terms we can simply add up the contributions of the volume changes in the directions of the various i-state variables (see fig. 3.1.2C, and/or exercise 3.1.1) to arrive at a total fractional volume change of $\sum_i \frac{\partial v_i}{\partial x_i} dt$. Equating the numbers of individuals in the box before and after its infinitesimal transformation then gives

$$n(t+dt,x+vdt)\Pi dx_i(1 + \sum\frac{\partial v_i}{\partial x_i}dt) = n(t,x)\Pi dx_i$$

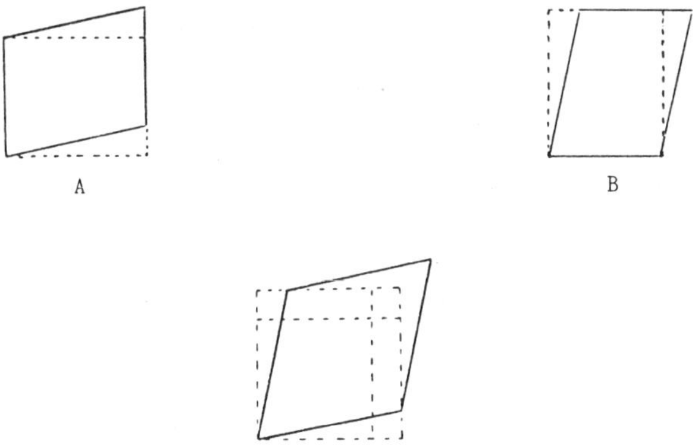

Fig. 3.1.2. Decomposing the infinitesimal volume change of the box from fig. 3.1.1 into the component changes along the axes.

Hence

$$[n(t+dt,x+vdt)-n(t,x+vdt)] + [n(t,x+vdt)-n(t,x)] + n(t+dt,x+vdt)\sum\frac{\partial v_i}{\partial x_i}dt = 0 .$$

Dividing by dt finally gives

$$\frac{\partial n}{\partial t} + \sum \frac{\partial n}{\partial x_i} v_i + n \sum \frac{\partial v_i}{\partial x_i} = 0 , \tag{3.1.3}$$

which is equivalent to (3.1.1).

The various terms in (3.1.3) bear special names referring to their interpretation. The quantity

$$n \sum \frac{\partial v_i}{\partial x_i} = n \nabla . v \tag{3.1.4}$$

is called the *dilation* (or dilatation) since it accounts for the contraction or expansion of our volume element $\Pi(x_i, x_i + dx_i)$. The remaining quantity

$$\sum \frac{\partial n}{\partial x_i} v_i = \nabla n . v \tag{3.1.5}$$

accounts for the pure transportation of individuals with the *i*-flow and therefore is called the *convection term.* (compare the discussion after (I.2.5.4)). Finally the quantity

$$\frac{Dn}{Dt} \stackrel{def}{=} \frac{\partial n}{\partial t} + \nabla n . v \tag{3.1.6}$$

is called the *material* or *substantial derivative* as it describes what happens around one individual which we follow through its life (see also section 4 on integration along characteristics).

As a final point it should be noticed that v in (3.1.1) to (3.1.3) may depend explicitly or implicitly on time, *e.g.* as a result of a dependence on the state of other or the same populations. This does not influence at all our derivation of these formulae.

EXERCISE 3.1.1: Show that the linear map transforming a neighbourhood of x into a neighbourhood of $x + vdt$ allows the matrix representation $I + Dvdt$, where the derivative D is taken with respect to x. Therefore the volume of our little box after transformation equals $\Pi dx_i \det(I + Dvdt)$. Show that this is equal to $\Pi dx_i(1 + \sum \frac{\partial v_i}{\partial x_i} dt + 0(dt^2))$.

EXERCISE 3.1.2: Write down the transport part of the p-differential generators of examples 1.1, 1.2, 1.4, 1.6, and 1.10 from section 1. (Mind the decomposition of Ω!)

3.1.2. The mathematician's derivation

The more rigorous derivation of the transport component of the p-differential generator to be presented in this subsection uses the physically inspired concept of flux, which will not be introduced in intuitive terms untill well into subsection 3.2.2. So if you are not familiar with this concept already, you are advised first to read on and only afterwards to return to this subsection.

Just as with the "derivation" in subsection 3.1.1. there are two possible approaches depending on whether we imagine ourselves as staying put at a fixed point in Ω or as moving along with the stream. These two approaches are called after the mathematicians Euler and Lagrange respectively. The Euler approach has the advantage that it still applies in situations where the flux vector ϕ depends in a more complicated manner on the local density than just $\phi = vn$ (e.g. due to random components in the continuous i-state movement), but the Lagrange approach adds more to our insight into the nature of the transport equation and its solutions.

The ideas presented below arose in the context of continuum mechanics. We refer to Lin & Segel (1974), Segel (1977 a, b) for an introduction to this subject. Also see Aris (1978).

The Euler approach

Let Ω_0 be a domain in Ω with a piecewise smooth boundary $\partial\Omega_0$. If ϕ denotes the flux vector and ν the outward normal at a point x on $\partial\Omega_0$, the outward flux at x, be it positive or negative, is given by $\nu{\cdot}\phi$. Conservation of mass implies that

$$\frac{d}{dt} \int_{\Omega_0} n dx + \int_{\partial\Omega_0} \nu{\cdot}\phi d\sigma = 0 , \tag{3.1.7}$$

where σ denotes Lebesgue measure on $\partial\Omega_0$. Applying the divergence theorem we find

$$\frac{d}{dt} \int_{\Omega_0} n dx + \int_{\Omega_0} \nabla{\cdot}\phi dx = 0$$

or

$$\int_{\Omega_0} \left[\frac{\partial n}{\partial t} + \nabla \cdot \phi \right] dx = 0 .$$

Since Ω_0 is arbitrary this is equivalent to (by the so-called Dubois-Reymond lemma)

$$\frac{\partial n}{\partial t} + \nabla \cdot \phi = 0 , \tag{3.1.8}$$

and in particular when $\phi = vn$, as is the case when the continuous *i*-state movements are all deterministic (see subsection 3.2.2),

$$\frac{\partial n}{\partial t} + \nabla \cdot vn = 0 . \tag{3.1.9}$$

Langrange approach

The calculations for the Lagrange approach are considerably more complicated. However, the technicalities involved also have some interest in themselves. For example, in part B both Aldenberg and Roots use calculations of this kind in the context of concrete biological examples.

As said before, the Lagrange approach is based on following an individual and looking at what happens in its immediate vicinity. Let such an individual be characterized by its *i*-state x_0 at $t = 0$, x_0 in the interior of Ω. Its *i*-state at t can be determined from the *i*-equation

$$\frac{dx}{dt} = v(t,x) \quad x(0) = x_0 . \tag{3.1.10}$$

We assume that (3.1.10) allows a unique solution which can be written as

$$x(t,x_0) = T_{[0,t)} x_0 \tag{3.1.11}$$

where $T_{[0,t)}$ is the next *i*-state operator acting over the interval $[0,t)$. The uniqueness of $x(t,x_0)$ guarantees that $T_{[0,t)}$ allows an inverse $T_{[0,t)}^{-1}$ at least in a neighbourhood of $t = 0$, $x = x_0$.

Now consider an arbitrary domain $\Omega_0(0)$ in the interior of Ω, and let

$$\Omega_0(t) := \{x(t,x_0) \mid x_0 \in \Omega_0(0)\} .$$

With this notation the conservation of individuals expresses itself as

$$\int_{\Omega_0(t)} n(t,x)dx = \int_{\Omega_0(0)} n(0,x)dx \tag{3.1.12}$$

This relation forms the basis of the Lagrange derivation of the transport equation. But before we proceed to the derivation we introduce some auxiliary notation

$$\rho(t,x_0) := n(t,x(t,x_0)), \quad J(t,x_0) := \det D_{x_0} x(t,x_0) = \det DT_{[0,t)} x_0 . \tag{3.1.13}$$

(NB The function ρ is not a density! If one wishes to transform variables and keep the interpretation of densities one should include the Jacobian of the transformation, in this case J.) To calculate J we define the matrix valued function M by

$$M(t,x_0) := D_{x_0} x(t,x_0) = DT_{[0,t)} x_0 . \tag{3.1.14}$$

By differentiating (3.1.10) for x_0 we find

$$\frac{dM}{dt} = D_x v(t,x(t,x_0))M, \quad M(0) = I . \tag{3.1.15}$$

This is a linear time varying o.d.e. Elementary linear algebra (see appendix A) shows that $\det M$ satisfies

$$\frac{dJ}{dt} = (\mathrm{trace} D_x v)J = (\nabla \cdot v)J, \quad J(0,x_0) = 1 \tag{3.1.16}$$

and therefore

$$J(t,x_0) = \exp\left[\int_0^t \nabla \cdot v(\tau,x(\tau,x_0))d\tau\right] > 0 . \tag{3.1.17}$$

(The last inequality also shows that for v differentiable in x the inverse map $T_{[0,t)}^{-1}$ exists at least in a neighbourhood of

$t = 0,\ x = x_0$.)

Having introduced all necessary instruments we are ready to derive the transport equation. To this end we shall calculate $\frac{d}{dt}\int_{\Omega_0(t)} n(t,x)dx$ by first applying the transformation $x = T_{[0,t)}x_0$ and then use the inverse map $T_{[0,t)}^{-1}$ to transform back again. Owing to the positivity of J we can drop the absolute value signs.

$$\int_{\Omega_0(t)} n(t,x)dx = \int_{\Omega_0(0)} n(t,x(t,x_0))J(t,x_0)dx_0 = \int_{\Omega_0(0)} \rho(t,x_0)J(t,x_0)dx_0\ . \tag{3.1.18}$$

Therefore

$$0 = \frac{d}{dt}\int_{\Omega_0(t)} n(t,x)dx = \int_{\Omega_0(0)} \{\frac{\partial\rho}{\partial t}(t,x_0)J(t,x_0) + \rho(t,x_0)\frac{\partial J}{\partial t}(t,x_0)\}dx_0$$

$$= \int_{\Omega_0(0)} \{\left[\frac{\partial}{\partial t}n(t,x(t,x_0)) + \nabla n(t,x(t,x_0))\cdot v(t,x(t,x_0))\right]J(t,x_0) + n(t,x(t,x_0))\nabla\cdot v(t,x(t,x_0))J(t,x_0)\}dx_0$$

$$= \int_{\Omega_0(0)} \left[\frac{\partial}{\partial t}n(t,x(t,x_0)) + \nabla n(t,x(t,x_0))\cdot v(t,x(t,x_0)) + n(t,x(t,x_0))\nabla\cdot v(t,x(t,x_0))\right]J(t,x_0)dx_0$$

$$= \int_{\Omega_0(t)} \{\frac{\partial n}{\partial t} + \nabla n\cdot v + n\nabla\cdot v\}dx = \int_{\Omega_0(t)} (\frac{\partial n}{\partial t} + \nabla\cdot(nv))dx$$

which again implies (3.1.9).

The next two exercises provide some links with Aldenberg's use in part B of the flux in deriving the balance law.

EXERCISE 3.1.3: Consider the special case that Ω is 1-dimensional and v time independent. Show that $J(t,x_0) = v(x(t,x_0))\,/\,v(x_0)$.
Hint: Use that $\frac{d}{dt}v(x(t,x_0)) = v(x(t,x_0))\frac{\partial v}{\partial x}(x(t,x_0))$, *i.e.* $v(x(t,x_0))$ satisfies the same differential equation as J but with a different initial condition.

EXERCISE 3.1.4.: Let

$$\frac{\partial n}{\partial t} + \nabla\cdot(vn) = -\mu n\ ,$$

$\mu(t,x_0)$ a death rate. And let

$$\zeta(t,x_0) \overset{def}{=} \rho(t,x_0)J(t,x_0)\ ,$$

i.e. if there were no deaths we would have $\zeta(t,x_0) = n(0,x_0)$. Show that

$$\frac{d\zeta}{dt} = -\mu\zeta\ .$$

Interpret the result in intuitive terms.

EXERCISE 3.1.5: What are the dimensions of ρ and J, of ζ?

3.2. *The (local) loss of p-mass*

P-mass can disappear locally from the interior of Ω through deaths (or emigration) and through jumping to some other place in Ω. Of course in the latter case the mass returns elsewhere in Ω. On the other hand *p*-mass can disappear through transport across the boundary $\partial\Omega$ of Ω. We shall discuss these two forms of disappearance of *p*-mass separately. In subsection 5.2 the connection will be made again.

3.2.1. *The local disappearance of p-mass from the interior of* Ω.

From law of mass-action considerations the local disappearance of *p*-mass at x can always be expressed as

$$-\delta_{s-}n(x) = -\alpha(x)n(x)\,. \tag{3.2.1}$$

α itself may be a sum of various terms like natural mortality rate, predation intensity, rate of jumping to elsewhere in Ω *etc.* Each of these terms can depend in its own way on time or on the state of the p-system as a whole. For example, if x denotes size and α_P denotes a predation component of α and m denotes the distribution of predator size y, then for randomly searching predators (*i.e.* predators for which satiation effects for example play no role)

$$\alpha_P(x) = \int_{\Omega_y} \beta(x,y)m(y)dy\,, \tag{3.2.2}$$

where $\beta(x,y)dx$ is the searching intensity for a predator of size y for prey in the size bracket $(x, x + dx)$ (we have to consider a prey size interval here as predators search for prey, not for prey per unit of prey size!). (In part B Daan discusses *i.a.* the possible functional forms of β and the problems of determining it for commercially relevant marine fishes).

The reasoning for jumps goes as follows: let $\gamma(x,y)\Pi dy_i$ be the probability per unit of time that an individual at x will jump to $\Pi(y_i, y_i + dy_i)$ then the contribution α_J to α due to this kind of jump is

$$\alpha_J(x) = \int_{\Omega} \gamma(x,y)\Pi dy_i\,. \tag{3.2.3}$$

This can always be rewritten as

$$\gamma(x,y) = \alpha_J(x)\,p(x,y)\,, \tag{3.2.4}$$

where $p(x,\cdot)$ is the probability density of the "touch down" position for jumps starting at x.

REMARK 3.2.1: γ and p may contain delta-"functions" with respect to y. See section 5 for a discussion centering around the problem of the occurrence of delta-functions in the definition of the p-differential generator.

Sometimes we have two or more touch down positions due to the individual splitting into two or more separate individuals. In that case it is still possible to write $\gamma(x,y) = \alpha_J(x)\,p(x,y)$ but the probability density $p(x,\cdot)$ now is a mixture of two or more probability densities, those of the 1st, 2nd, 3rd *etc.* division product. Even more generally the number of division products may be a random variable with distribution $i \mapsto q_i(x)$. In that case

$$p(x,y) = \frac{1}{r(x)} \sum_i q_i(x) \sum_{j=1}^{i} p_{ij}(x,y) \tag{3.2.5a}$$

with

$$r(x) = \sum_i i q_i(x) \tag{3.2.5b}$$

where $p_{ij}(x,\cdot)$ is the probability density of the touch down position of the j'th division product of an individual splitting into i products, and $r(x)$ is the mean number of division products derived from an individual which divided while it was in state x.

Quantities like α_J and $p(x,y)$ may depend again on time or on the state of the system as a whole. They should be calculated from submodels on the individual level. We shall not go into the various possibilities here, with one exception: the generalization to higher dimensional i-state spaces of the "divide on crossing a stochastic threshold" mechanism from I.4.1. Such a generalization will be needed for example if we wish to extend the model from I.4 to include cell age as an i-state variable as in example 1.4. from section 1. As another example you may think of a modification of the PPP-problem from example 1.10, in which it is assumed that the amount of resources in as yet unexploited patches is not constant but varies randomly over the patches. A final example comes from environmental toxicology: one of the i-state variables is concentration of poison in the individual and death occurs when the poison level passes some random threshold (which is assumed to have been set at birth).

In order to be specific we shall proceed as if crossing the stochastic threshold results in death. Let $\mathcal{F}(t)$ denote the probability that an individual born alive at $t = 0$ is still alive at t. The probability that it dies between t and $t + dt$, provided it survived till t, is precisely what we called $\alpha_D(x(t),t)dt$. Therefore $\mathcal{F}$ can be calculated from

$$\frac{d\mathcal{F}}{dt} = -\alpha_D(x(t),t)\,\mathcal{F}(t)\,, \quad \mathcal{F}(0) = 1$$

or

$$\mathcal{F}(t) = \exp\left(-\int_0^t \alpha_D(x(\tau), \tau)d\tau\right).$$

Let $\mathcal{G}(x)$ denote the probability that x is below threshold. Then

$$\mathcal{F}(t) = \mathcal{G}(x(t)) \,/\, \mathcal{G}(x(0))$$

Therefore

$$\alpha_D(x,t) = -\frac{d\log\mathcal{F}(t)}{dt} = -D(log(\mathcal{G}(x))\frac{dx}{dt} = \Sigma\mu_i(x)v_i(x,t) \tag{3.2.6a}$$

with

$$\mu_i = -\frac{\partial}{\partial x_i}\log(\mathcal{G}(x))\,. \tag{3.2.6b}$$

(NB. We assume here that the i-state movement is such that $\mathcal{G}(x(t))$ is guaranteed to decrease with time!)

REMARK 3.2.2: The definition of the μ_i implies that

$$\frac{\partial\mu_i}{\partial x_j} = \frac{\partial\mu_j}{\partial x_i}\,. \tag{3.2.7}$$

Conversely condition (3.2.7) is also sufficient (provided Ω is simply connected) for a set of μ_i to define a function $\mathcal{G}$, unique up to a multiplicative constant, through

$$\mathcal{G}(x) = \exp(-\int_{x_0}^{x}\mu(x)\cdot\frac{dx}{ds}\,ds) \tag{3.2.8}$$

for some path Σ joining x_0 and x and parametrized by s. Formulae (3.2.8) and (3.2.6) are the direct generalization to higher dimensional state spaces of (I.4.1.5) and (I.4.1.6) with (for Ω 1-dimensional)

$$\mathcal{G}(x) = 1-\int_0^x\phi_b(\xi)d\xi\,. \tag{3.2.9}$$

(Working with $\mathcal{G}$ instead of ϕ_b allows us to bypass detailing the relation between a single threshold variable and the k state variables, $k>1$.)

EXERCISE 3.2.3: Write down the local parts of the differential generators of example 1.1 (the invertebrate predator) and 1.10 (the PPP-problem with the amount of resources in newly discovered patches fixed). Call the densities of prey and predator in search of a patch P and Q respectively, and the corresponding rate constants ζ and $\eta(x)$.

EXERCISE 3.2.4: Write down the local mass loss part of the differential generator of the binary fission model from example 1.4 from section 1 on the assumptions that (i) division rate depends only on size, not on age, (ii) growth may be time dependent, (iii) the same stochastic division threshold model applies as has been introduced in I.4, and (iv) cell death rate is only age dependent. Where does the lost mass reappear?

EXERCISE 3.2.5: In the *Asterionella* example 1.6 from section 1 assume that breakage of colonies is due to randomly occurring randomly sized shocks that break all bonds older than a certain age. In that case there exists an increasing function $b:\mathbb{R}^+\to\mathbb{R}^+$ such that $b(a_i)dt$ equals the probability of the breaking of the ith and older bonds, and $[b(a_{i+1})-b(a_i)]dt$ equals the probability that the $(i+1)$th and older bonds break but not the ith and any younger bonds. Write down the local mass loss terms of the p-differential generator. (Assume that division rate d only depends on cell age and that there is no death.) Where does the lost mass reappear in Ω? (NB colonies of size 2 are a bit special as far as reappearance of mass from division is concerned.)

EXERCISE 3.2.6: Write down the local mass loss term of the modified PPP-problem, with the initial amount of resources in a patch a random variable. Hint: Let $\mathcal{H}$ be the survivor function of the initial amount of resources, *i.e.* $P\{r_0>\rho\} = \mathcal{H}(\rho)$. Use the calculations of exercise 1.12 to express $\mathcal{G}$ in $\mathcal{H}$.

3.2.2. The disappearance of p-mass across the boundary of Ω

P-mass can get lost across the boundary $\partial\Omega$ of Ω *e.g.* when animals die, or cells divide, when their i-state reaches a fixed barrier. Such a loss of p-mass does not show up in the defining formula for calculating the differential generator A but in the definition of its domain. The fact that A is assumed to act on functions on Ω is part of its definition. In the case that we are dealing with loss of p-mass and deterministic flow only, no side conditions are needed.

Another point is that p-mass lost by flowing across the boundary of Ω may reappear elsewhere. Therefore we have to be able to calculate how much mass flows out of Ω and where along $\partial\Omega$. As a start we notice that, except for some singularities which may occur when v is no longer smooth at the boundary of Ω and which shall be dealt with in subsection 3.4, p-mass can only leave Ω when the local speed vector v points out of Ω. The part of the boundary where this is the case we shall call $\partial_-\Omega$. To calculate the i-state "volume" that flows out of Ω per unit of time we have to decompose the local flow rate into a component orthogonal to the boundary which contributes to the outflow of p-mass, and a component parallel to the boundary which does not contribute. This can be done by means of the outward normal vector ν (see fig. 3.2.1, or appendix A). The volume that flows across a small unit of area $d\sigma$ of $\partial\Omega$ surrounding x in a small time interval dt equals $v(x)dt\cdot\nu(x)d\sigma$.

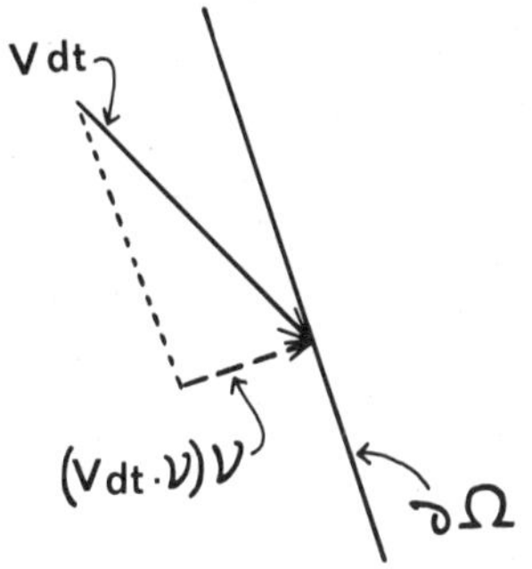

Figure 3.2.1

Therefore the local mass flow rate per unit of area, or, as it is usually called, the flux, out of Ω equals

$$[\text{local flux out of } \Omega \text{ at } x\in\partial_-\Omega] = v(x)\cdot\nu(x)n(x) \tag{3.2.10}$$

and

$$[\text{total mass leaving } \Omega \text{ across } \partial_-\Omega] = \int_{\partial_-\Omega} v\cdot\nu n d\sigma\,. \tag{3.2.11}$$

For further calculations it is sometimes helpful to introduce a new quantity, the flux vector $\phi = (\phi_1, \cdots, \phi_k)^T$. ϕ is defined as the amount of mass that flows per unit of time and per unit of area along a point in the positive x_1 to x_k directions. For a deterministic movement through i-state space we therefore have

$$\phi = vn\,. \tag{3.2.11}$$

This way we can write the local flux across a $(k-1)$ dimensional surface Σ as

$$[\textit{local flux across } \Sigma \text{ at } x] = \phi(x)\cdot\nu(x)\,. \tag{3.2.12}$$

The importance of this formula is that it still applies when ϕ has to be calculated in a different manner, *e.g.* when there are also stochastic components to the continuous i-state movements. It also forms the basis for the Euler derivation of the transport equation set forth in subsection 3.1.2.

EXCERCISE 3.2.7. What is the local rate of mass flow out of the interior of Ω across boundary segment (0), (see fig. 1.1), and therefore the local reappearance rate on the s-axis in example 1.1 from section 1 (the invertebrate predator). What is the arrival rate locally at the line $\tau = \tau_e$? Where does this mass reappear?

EXERCISE 3.2.8: Assume in the PPP-problem with fixed initial resources in a partch (example 1.10) that prey "take the wing" when a patch gets exhausted. What is the total production of prey emigrants per unit of time? Assume that predators leave empty prey patches as well. What is the production rate of predator emigrants?

EXERCISE 3.2.9: Assume that *Asterionella* colonies break only at the exact moment that the oldest bond reaches age A. What is the local rate of mass loss at this boundary? Where in Ω (and how) does this mass reappear? (NB colonies of size 1 are a bit special as far as the reappearance of mass is concerned.)

3.3. The (re)appearance of p-mass

New p-mass can appear in two manners: locally in the interior of Ω and at the boundary of Ω. We shall discuss these two possibilities separately. In subsection 5.2 the connection will be made again.

3.3.1. The (re)appearance of p-mass in the interior of Ω

If per unit of time an amount of p-mass $b\Pi dx_i$ is added to a volume element $\Pi(x_i, x_i + dx_i)$ this results in a component

$$\delta_{s+}n = b \tag{3.3.1}$$

in the defining formula for the p-differential generator A. There remains the question of how b can be calculated. In principle b is the sum of two terms: the results of new births and of jumps of the i-state process. Of course, if an individual divides into two or more parts it depends on our conventions with respect to the concept of "parent" whether we speak of a jump and birth(s) or of births only. To simplify the presentation we shall call every "touch down" in Ω a birth, independent of whether this is a real birth or just a reappearance after a jump.

To calculate b we split up the various births with respect to their origin, calculate the corresponding birth rates through a law of mass action argument and cumulate over origins. That is, if the separate components of Ω are denoted as Ω_i,

$$b(x) = \sum_i\{\int_{\Omega_i} r(y)p(y,x)\alpha_J(y)n(y)dY + \int_{\partial_-\Omega_i} r(y)p(y,x)\phi(y)\cdot\nu(y)d\sigma\}\ , \tag{3.3.2}$$

(assuming that one can only be born out of individuals of the same population and that there is no immigration), where r,p and α_J in the first integral have been defined in (3.2.3) to (3.2.5), r and p in the second integral have the same interpretation as in the first integral, ϕ and ν and $d\sigma$ have been introduced in subsection 3.2.2 (a more extended discussion of surface integrals may be found in appendix A).

There is still one snag in calculating the various terms in (3.3.2) which has to do with the way in which p is calculated. We shall show the problems involved by means of an example.

EXAMPLE 3.3.1: *Cell division into unequal parts.* Consider a cell division model in which the size of the two daughter cells are two random fractions, θ and $1-\theta$, of the parents size. Let f denote the probability density of θ (with $f(\theta) = f(1-\theta)$ so that $f(\theta) + f(1-\theta) = 2f(\theta)$), Then, if α_d denotes the division rate as a function of size,

$$b(x) = \int_0^1 2f(\theta)\theta^{-1}\alpha_d(\frac{x}{\theta})n(\frac{x}{\theta})d\theta\ . \tag{3.3.3}$$

The reason for the factor θ^{-1} is that for a fixed value of θ births in the range $(x, x + dx)$ come from parents in the range $(\frac{x}{\theta}, \frac{x}{\theta} + \frac{dx}{\theta})$ and we have to discount a compression factor θ of the interval during the jump. More formally, if $p(y,x)dx$ equals the probability of touchdown in $(x, x + dx)$ of a randomly chosen division product of a parent cell of size y, like in (3.2.5), then (using integration rule (ii) from appendix A with $\theta = \frac{x}{y}$)

$$\int_{\theta_1}^{\theta_2} f(\theta)d\theta = \int_{x_1=\theta_1 y}^{x_2=\theta_2 y} p(y,x)dx = \int_{x_1}^{x_2} f(\frac{x}{y})y^{-1}dx\ .$$

So we find

$$p(y,x) = f(\frac{x}{y})y^{-1}\ .$$

According to (3.3.2) we have to integrate $2\alpha_d(y)p(y,x)n(y)$ with respect to y to get $b(x)$. This way we find again (applying integration rule (ii) with $y = \frac{x}{\theta}$)

$$\int_\Omega 2p(y,x)\alpha_d(y)n(y)dy = \int_1^0 2f(\theta)\alpha_d(\frac{x}{\theta})n(\frac{x}{\theta})\frac{\theta}{x}\frac{d}{d\theta}(\frac{x}{\theta})d\theta = \int_0^1 2f(\theta)\alpha_d(\frac{x}{\theta})n(\frac{x}{\theta})\theta^{-1}d\theta$$

as in (3.3.3.).

Numbers are conserved, not densities. Therefore we have to be very careful in ascertaining how the volume elements at the places of take off and touch down are connected, as the various volume conversion factors enter into the calculation in a nontrivial way. An analogous remark applies to the connection of surface and volume elements for individuals taking off across $\partial_- \Omega$. An example of the latter type of calculation will be shown in the next subsection where the instream of p-mass across $\partial\Omega$ is considered.

EXERCISE 3.3.2: Calculate the local birth terms for the *Asterionella* models from exercises 3.2.5 and 3.2.9.

3.3.2. The (re)appearance of p-mass across the boundary of Ω

P-mass can only stream into Ω across $\partial\Omega$ at those points where v is directed inwards, or more formally, where $v.\nu$ is negative. We shall call that part of the boundary $\partial_+ \Omega$. On $\partial_+ \Omega$ the arrival rate per unit of surface area b_∂ should equal the inward flux at the boundary, giving rise to the boundary condition

$$-\nu(x)\cdot\phi(x) = -\nu(x)\cdot v(x)n(x) = b_\partial(x) \tag{3.3.4}$$

There remains the problem of calculating $b_\partial(x)$. This can be done along the same lines as the calculation of the interior local arrival rate (3.3.2). Only now p_∂ has the dimension (surface area) $^{-1}$ instead of (volume) $^{-1}$, and we have to keep careful track how volume or surface elements at take off transform into surface elements at touch down. We shall but give an example.

EXAMPLE 3.3.3: *The invertebrate functional response continued.* In the excercises (3.1.2, 3.2.3 and 3.2.7) we have already encountered various parts of the differential generators for the invertebrate predator of example 1.1. The full p-differential generator is displayed in subsection 4.1. Here we shall concentrate on $\partial_+ \Omega$. $\partial_+ \Omega$ consists of two parts. The maximum satiation of searching predators s_m bounding $\Omega_0 \subset \mathbb{R}$ from above and the boundary segment (2) bounding $\Omega_1 \subset \mathbb{R}^2$ to the right. (see figure 1.1). At s_m we just have to make sure that there is no instream of maximally satiated searching predators by setting

$$\phi(s_m) = -f(s_m)n_0(s_m) = 0\,, \tag{3.3.4}$$

(where f was the digestion rate), or equivalently

$$n_0(s_m) = 0\,. \tag{3.3.5}$$

At boundary segment (2) there is an instream of predators having just started search. The predators arriving at $(s,\tau_m(s))$ come from $(s,0)$. Predators leave there at a rate $xg_0(s)n_0(s)$. However, we need the arrival rate per unit of boundary length. A line element of length ds on the s-axis connects with a line element of length $d\sigma = ((ds)^2 + (\frac{d\tau_m}{ds}ds)^2)^{\frac{1}{2}} = (1 + (\frac{d\tau_m}{ds})^2)^{\frac{1}{2}}ds$. (Compare formula (A2.6); we parametrize the touch down boundary with the position at take off and match number of predators leaving per unit of time, an integral over a segment of the s-axis, with number of predators arriving, an integral over a corresponding segment of the touch down boundary.) The arrival rate per unit of length equals

$$b_\partial(s,\tau_m(s)) = xg_0(s)n(s)(1 + (\frac{d\tau_m(s)}{ds})^2)^{-\frac{1}{2}} \tag{3.3.6}$$

This should equal

$$\phi(s,\tau_m(s))\cdot -\nu(s,\tau_m(s)) = n_1(s)\begin{bmatrix}1\\ -as\end{bmatrix}\cdot\begin{bmatrix}1\\ \frac{d\tau_m}{ds}(s)\end{bmatrix}(1 + (\frac{d\tau_m(s)}{ds})^2)^{-\frac{1}{2}}\,. \tag{3.3.7}$$

Therefore on boundary segment (2) we should have the side condition

$$(1 - as\frac{d\tau_m}{ds}(s))n_1(s,\tau_m(s)) = xg_0(s)n_0(s)\,. \tag{3.3.8}$$

EXERCISE 3.3.4: Derive appropriate side conditions for example 1.4 (binary fission), 1.6 *(Asterionella)* and 1.10 (the PPP-problem) of section 1. In the *Asterionella* example consider both the cases covered by the assumptions made about the breakage mechanism in exercise 3.2.5 and in exercise 3.2.9.

3.4. Boundaries and side conditions: picking up the strands

In subsection 3.2.2 we encountered the boundary $\partial_- \Omega$ defined by $v\cdot\nu>0$, and in subsection 3.3.2 the boundary $\partial_+ \Omega$

defined by $v \cdot \nu < 0$. The definition of the differential generator A will have to include side conditions on $\partial_+ \Omega$ for otherwise A will certainly not define a unique semigroup: if we do not prescribe what is to be put onto a conveyor belt at its start we cannot foretell its contents. On $\partial_- \Omega$ no side conditions are needed. In fact if we would prescribe side conditions on $\partial_- \Omega$ there would probably nó longer exist a semigroup associated with the would be differential generator so defined: once we have prescribed what happens at the start of as well as on a conveyor belt we are no longer free to prescribe what is going to leave it at its end.

Apart from $\partial_- \Omega$ and $\partial_+ \Omega$ we may have one piece of boundary left: $\partial_0 \Omega$ for which $v \cdot \nu = 0$. If the vector field v is smooth on $\bar{\Omega}$ (the closure of Ω) then it is impossible for individual trajectories to reach or leave $\partial_0 \Omega$ so no boundary condition has to be prescribed nor should it be. The following example shows that this need not hold if our smoothness assumption is relaxed.

EXAMPLE 3.4.1: *Von Bertalanffy growth starting from zero initial size.* In the model for size dependent population growth in ectotherms from I.3 we encountered the equation

$$\frac{\partial n}{\partial t} = -\frac{\partial g_1 n}{\partial l} - \mu n \tag{3.4.1a}$$

with

$$g_1(l) = (3\eta)^{-1}(\kappa \nu f - \zeta l)^+ \tag{3.4.1b}$$

and

$$\phi(l_b) = g_1(l_b)n(l_b) = b \tag{3.4.1c}$$

with

$$b = \int_\Omega \lambda_1(l)n(l)dl \,. \tag{3.4.1d}$$

Biological considerations imply that birth length l_b should be positive. However, from a mathematical point of view nothing keeps us from considering the limiting case $l_b = 0$. (In fact, this can be a useful step in dealing with models in which evolution drives the population towards smaller and smaller l_b).
So far there is no problem. $g_1(0) > 0$ and (3.2.1c) provides the necessary boundary condition. But now consider the same model but with weight as the *i*-state. Derived from first principles the *p*-equation now looks like

$$\frac{\partial n}{\partial t} = -\frac{\partial g_3 n}{\partial w} - \mu n \tag{3.4.2a}$$

$$g_3(w) = \eta^{-1}(\kappa \nu f w^{2/3} - \zeta w)^+ \tag{3.4.2b}$$

$$\phi(w_b) = g_3(w_b)n(w_b) = b \tag{3.4.2c}$$

$$b = \int_\Omega \lambda_3(w)n(w)dw \tag{3.4.2d}$$

In the limiting case for $w_b \downarrow 0$ the left hand boundary at w_b belongs to $\partial_0 \Omega$. Still, since (3.4.1) and (3.4.2) are equivalent, (3.4.2) should be a genuine model also for $w_b = 0$, provided that the boundary condition (3.4.2c) is replaced by

$$\lim_{w \downarrow 0} \phi(w) = \lim_{w \downarrow 0} g_3(w)n(w) = b \tag{3.4.2c'}$$

What happens in this example is that indeed for genuine weight distributions, *i.e.* functions n: $\Omega \rightarrow \mathbb{R}^+$ such that

$$\int_\Omega n dw < \infty \tag{3.4.3}$$

(3.4.2c′) is a perfectly valid boundary condition as it is equivalent to

$$n(w) = \frac{\eta b}{k \nu w^{2/3}} + o(w^{-2/3}) \text{ for } w \downarrow 0$$

which is compatible with (3.4.3). However, would $g_3(w)$ have gone to zero smoothly for $w \downarrow 0$ then an expression like (3.4.2.c′) could never be fulfilled by a function satisfying (3.4.3).
On the individual level the anomaly has to do with the fact that for the initial condition $w(0) = 0$ the von Bertalanffy growth equation $dw / da = g_3(w)$ does not have a unique solution: a solution may either stay zero for ever (as we expect from $g_3(0) = 0$) or may leave zero at any nonnegative time. It is this possibility of growing away from $0 = \partial_0 \Omega$ that makes that we have to prescribe a boundary condition at $\partial_0 \Omega$. For growth laws which are smooth also in 0 like g_1 no such anomalous behaviour is possible: either 0 belongs to $\partial_+ \Omega$ and we have to prescribe a boundary

condition, or it belongs to $\partial_0\Omega$ or $\partial_-\Omega$ and we should not do so. (Of course there remains the biological anomaly that $g_1(0)>0$, making individuals grow away from zero size. The explanation here is that g_3 and g_1 were based on the assumption that individuals of all sizes are structuraly similar, which can only apply from a sufficient size onwards.)

In the same manner as we may have inflow of mass across a subset $\partial_{0+}\Omega$ of $\partial_0\Omega$, we may have outflow across a subset $\partial_{0-}\Omega \subset \partial_0\Omega$, but biological examples do not seem to abound.

If we have taken care of prescribing the right side conditions on $\partial_+\Omega$ and $\partial_{0+}\Omega$, then it is generally possible, in the linear case at least, to prove existence and uniqueness of solutions of our p-differential equation and therefore the existence and the uniqueness of an associated semigroup of next p-state transformations, by means of a generation expansion, *i.e.* a series expansion in which the subsequent terms are just the i-state densities of individuals in subsequent biological "generations" (where generation should be interpreted in a generalized sense in that i-state jumps are treated as combined death-birth events). Therefore the side conditions which have passed in revue are the only ones that are needed.

REMARK 3.4.2: It is not uncommon for $\partial\Omega$ to consist of a finite number of smooth surfaces connected by edges (see *e.g.* figure 1.1). In an edgepoint there does not exist a well defined normal ν. Usually it is clear from the interpretation to which of its adjoining surfaces we should assign such a point and therefore what we should choose for ν in our calculations. However, in general we can just leave unspecified what happens as the edges only form a subset of $\partial\Omega$ of measure zero.

EXERCISE 3.4.3: Discuss the possibilities of calculating an optimal birth weight under r-selection in our *Daphnia* model from I.3. Here r-selection refers to selection for the maximum possible asymptotic population growth rate r as introduced in I.3.3. Assume an energetic trade off between birth rate and birth weight as expressed by the factor $(\omega w_b)^{-1}$ in formula (I.3.1.4).

3.5. *Summary and concluding remarks*

We now have come at the end of the model building sections. In the following sections we shall discuss how our basic p-equations may be used to derive other types of equations by means of various transformation rules and/or limiting arguments. However, before going on it may be useful once more to put the variety of arguments that have passed in revue into perspective.

The specification of a structured population model embraces

- the choice of an individual state space Ω, a domain in $\mathbb{R}^k$ (or the union of a number of such domains, possibly for different k)
- a recipe for calculating the velocity v
- a recipe for the calculation of the per capita disappearance rate α
- a recipe for the calculation of the (re)appearance rates b and b_∂.

In subsections 2.2 and 2.3 we have discussed respectively the conceptual intricacies of constructing models for individual behaviour and the coupling of individuals into populations. Together these two subsections covered all four ingredients mentioned above, albeit in a rather abstract, philosophical fashion. The examples from section 1 serve to add some concreteness. In the present section we discussed how the abstract considerations from section 2 can be translated into our main working tool: p-equations. In 3.1, 3.2.1 and 3.3.1 we discussed the bits and pieces corresponding respectively to v, α and b, of the defining recipe of the p-differential generator, telling how it acts on the elements of its domain. In subsections 3.2.2 and 3.3.2 and 3.4 we discussed the other necessary component of its definition, the description of its domain, with special reference to respectively Ω and b_∂. In 3.1 and 3.2.2 we dealt with the process of deterministic *change:* in 3.1 we considered the continuous shifting of the i-state distribution in the interior of Ω resulting from v, in 3.2.2 we saw how this shifting may lead to deaths or i-state jumps by individuals being pushed inexorably over the boundary $\partial_-\Omega$. In 3.2.1 and 3.3 we considered *chance* in the form of random deaths, births and i-state jumps: in 3.2.1 deaths and take-offs represented by α, in 3.3 births and touch downs represented by b and b_∂.

There is still one possible mode of i-state behaviour left which we refrained from discussing: stochastic continuous i-state movements. These are covered in appendix B.

EXERCISE 3.5.1 The stochastic threshold models from 3.2.1 hold a kind of intermediate position with respect to the change versus chance dichotomy: in these models we can make the occurrence of deaths or i-state jumps deterministic by shoving in an extra i-state variable representing the threshold, but at the price of a higher dimensional i-state space and additional randomness in the state at birth. Rephrase the "division on reaching a stochastic threshold" model from I.4 in the manner indicated. Also integrate the resulting p-equation over the threshold variable to rederive the p-equation from I.4.

4. Integration along characteristics, transformation of variables, and the following of cohorts through time.

In our treatment of size dependent reproduction in ectothermic animals in I.3 we achieved for the case of constant food density a transformation into an equivalent renewal equation through a technique called integration along characteristics. In this section we shall consider this technique and various related techniques for transforming p-equations from a more general perspective. Of course the usefulness of these techniques in any specific instance is contingent upon the resulting equation being relatively more tractable than the original one. Whether this is the case depends on the presence of some exploitable special properties which are brought to the fore by the transformation.

In the following we shall proceed as if Ω consists of only one k-dimensional component. If Ω has more than one component then each component can be treated separately along the lines discussed.

4.1. Integration along characteristics

Characteristics are curves in $(k+1)$-space $\Omega \times \mathbb{R}^+$, spanned by the i-state coordinates and time, along which the material derivative

$$\frac{Dn}{Dt} = \frac{\partial n}{\partial t} + \nabla n \cdot v$$

reduces to an ordinary (time) derivative. These curves correspond to the trajectories of individuals conditional on the non-occurrence of i-state jumps. Sometimes we shall be sloppy and identify these curves with their projections on Ω which will then be called characteristics as well. When the time dependence of the individual speed v through Ω is explicitly given and the same applies to the relative jump/death rate α, the birth/reappearance rate b and the (re)appearance rate b_∂ of p-mass at $\partial_+\Omega$, then it is possible to obtain explicit expressions for n along these curves. If instead one or more of the quantities v, α, b and b_∂ are generated by the model we can still proceed as if they were known. The result is a formula expressing n in terms of some as yet unknown functions. This expression for n is equivalent to the partial differential equation from which it was derived. If we substitute it in the remaining part of the p-equation we end up with an equation of a different type containing the same information. Obviously we may only expect this new equation to be more manageable if sufficiently many of the quantities v,α,b and b_∂ are specified explicitly. The calculations from I.3.4 provide a good example.

REMARK 4.1.1.: In the general theory of partial differential equations one sometimes refers to the characteristics as curves in $(k+2)$-space, spanned by the i-state coordinates, time and n. Our characteristics are projections of these curves on $(k+1)$ -space.

Let θ be a parameter which is equivalent to time, except for a temporary arbitrariness of its origin, then the characteristics are defined by

$$\frac{dt}{d\theta} = 1, \quad \frac{dx}{d\theta} = v(t,x)\,. \tag{4.1.1}$$

If the p-equation is given by

$$\frac{\partial n}{\partial t} + \nabla n \cdot v = -n \nabla \cdot v - \alpha n + b \tag{4.1.2}$$

then along the characteristics

$$\frac{dn}{d\theta} = \frac{\partial n}{\partial t}\frac{dt}{d\theta} + \nabla n \cdot \frac{dx}{d\theta} = \frac{\partial n}{\partial t} + \nabla n \cdot v$$

and therefore

$$\frac{dn}{d\theta} = -n(\nabla \cdot v + \alpha) + b\,. \tag{4.1.3}$$

This equation is linear, and allows the explicit solution

$$n(\theta) = n(0)\exp(-\int_0^\theta (\nabla\cdot v + \alpha)d\theta_1) + \int_0^\theta b\exp(-\int_{\theta_2}^\theta (\nabla\cdot v + \alpha)d\theta_1)d\theta_2 \,. \tag{4.1.4}$$

If for example we choose points of $\partial_+\Omega$ as initial conditions for (4.1.1) we can calculate $n(0)$ from the corresponding side condition. The dilation factor $\exp\,(-\int_0^\theta \nabla\cdot v d\theta_1)$ describes the result of the contraction or expansion of the conveyor belt in the course of its movement, the factor $\exp\,(-\int_0^\theta \alpha d\theta_1)$ describes the cumulated local relative loss of p-mass through jumps or deaths and the term after the plus sign consists of the cumulation of the various additions to the p-mass after accounting for the changes undergone during transport.

Of course points on $\partial_+\Omega$ provide only one possible starting point for the integration of (4.1.2) and (4.1.3). The calculations in II.10 for the cell proliferation model from I.4 (see exercise 4.1.5 below) show for example that it can also pay to consider starting points inside Ω.

In I.3.4 we substituted the result from our integration along characteristics into the side conditions to arrive at a renewal equation. The end result of the calculations in II.10 also was a renewal equation. The example below shows that renewal equations are by no means the only possible end result. (In chapter IV we shall discuss in detail the i-features that allow p-behaviour to be captured by a renewal equation.)

EXAMPLE 4.1.2: *The invertebrate functional response.* The p-equation for example 1.1 from section 1 is, in obvious notation

$$\frac{\partial n_0(t,s)}{\partial t} = \underbrace{\frac{-\partial f_0(s)n_0(t,s)}{\partial s}}_{\text{digestion}} \underbrace{- xg_0(s)n_0(t,s)}_{\text{starting pursuit}} + \tag{4.1.5a}$$

$$+ \underbrace{\int_{\tau_e}^{\tau_m(s)} \mu n_1(t,s,\tau)d\tau}_{\text{from abortive pursuits}} \underbrace{+ (1-q)n_1(t,s,\tau_e+)}_{\text{from failed strikes}} \underbrace{+ n_1(t,s,0+)}_{\text{from finished meals}}$$

$$\frac{\partial n_1(t,s,\tau)}{\partial t} = \underbrace{\frac{-\partial f_1(s,\tau)n_1(t,s,\tau)}{\partial s}}_{\text{digestion / ingestion}} + \underbrace{\frac{\partial n_1}{\partial \tau}(t,s,\tau)}_{\text{proceeding of time}} \underbrace{-\lambda(\tau)n_1(t,s,\tau)}_{\text{escaping prey}} , \tag{4.1.5b}$$

with the "transition condition" *

$$n_1(t,s,\tau_e-) = qn_1(t,s,\tau_e+) \qquad \text{[from successful strikes]} \tag{4.1.5c}$$

and the boundary conditions

$$n_0(t,s_m) = 0 \,, \tag{4.1.5d}$$

$$(1-as\frac{d\tau_m(s)}{ds})n_1(t,s,\tau_m(s)), = xg_0(s)n_0(t,s) \text{ [from started pursuits]}, \tag{4.1.5e}$$

where

$$f_0(s) = -as \,, \quad s_m = e^{-a\tau_m(c)}c + (1-e^{-a\tau_e})u/a \,,$$

$$f_1(s) = \begin{cases} -as + u \\ -as \end{cases} , \quad \lambda(\tau) = \begin{cases} 0 \\ \mu \end{cases} \text{for} \begin{cases} \tau<\tau_e \\ \tau>\tau_e \end{cases} ,$$

and

$$\tau_m(s) = \tau_e + \tau_p(s) \,,$$

τ_e the total meal duration, τ_p the duration of a succesful pursuit, and c the search threshold, *i.e.* $g_0(s)=0 \Leftrightarrow s\geqslant c$. These are two partial differential equations, in t and one and two i-state variables. However, the second equation, that

---

* If $f:\mathbf{R}\rightarrow\mathbf{R}$, then $f(x+) := \lim_{y\downarrow x} f(y)$ and $f(x-) := \lim_{y\uparrow x} f(y)$, provided these limits exist.

for n_1, can be eliminated by integrating it along characteristics, leaving us with one partial differential equation with delayed arguments. As we are especially interested in the calculation of equilibria (remember I.2), and since at equilibrium time delays drop out, this presents a considerable simplification.
The characteristics of (4.1.5b) are given by

$$t = t_0 + \theta \,, \tag{4.1.6a}$$

$$\tau = \tau_0 - \theta = \tau_m(s_0) - \theta = \tau_e + \tau_p(s_0) - \theta \,, \tag{4.1.6b}$$

$$s = e^{-a\theta} s_0 + (u/a)[1 - e^{-a(\theta - \tau_p(s_0))^+}] \,. \tag{4.1.6c}$$

with s_0 and τ_0 the intersection of the characteristic with $\partial_+\Omega_1$ ((2) in figure 1.1). Given a value of (s,τ) we can solve (4.1.6b) and (4.1.6c) for s_0 and θ. Using these values we can express (s_0,τ_0) in (s,τ). For s_0

$$\tau>\tau_e : \log(s_0) - \tau_p(s_0) = \tau_e - a\tau + \log(s) \,, \tag{4.1.7a}$$

$$\tau<\tau_e : \log(s_0) - \tau_p(s_0) = \tau_e - a\tau + \log\{s - (u/a)[1 - e^{-a(\tau_e - \tau)}]\} \,. \tag{4.1.7b}$$

Our assumption that τ_p is a decreasing function of s guarantees that these equations have a unique solution. For later use we introduce the notation

$$s_1(s) := s_0(s,\tau_e) \,, \quad s_2(s) := s_0(s,0) \,, \quad \tau_i := \tau_m(s_i) \,.$$

(Note that for the parameter values of a realistic predator $s_2(s)<s$, whereas $s_1(s)>s$, and $s_0(s,\tau)>s$ for $\tau>\tau_e$.) For n_1 we find

$$\tau>\tau_e : n_1(t,s,\tau) = xg_0(s_0)n_0(t-\tau_0+\tau,s_0)(1 - as_0\frac{d\tau_m(s_0)}{ds_0})^{-1}\, e^{-(a-\mu)(\tau_0-\tau)} \,,$$

$$\tau<\tau_e : n_1(t,s,\tau) = xg_0(s_0)n_0(t-\tau_0+\tau,s_0)(1 - as_0\frac{d\tau_m(s_0)}{ds_0})^{-1}\, e^{a(\tau_0-\tau)-\mu(\tau_0-\tau_e)}(1-q).$$

Substituting this in (4.1.5a) finally gives

$$\frac{\partial n_0(t,s)}{\partial t} = \frac{\partial asn_0(t,s)}{\partial s} - xg_0(s)n_0(t,s) + \tag{4.1.8a}$$

$$\int_{\tau_e}^{\tau_m(s)} \mu xg_0(s_0(s,\tau))n_0(t-\tau_0(s,\tau)+\tau,s_0(s,\tau))(1 - as_0(s,\tau)\frac{d\tau_m(s_0(s,\tau))}{ds_0})^{-1} e^{-(a-\mu)(\tau_0(s,\tau)-\tau)} d\tau \; +$$

$$(1-q)xg_0(s_1(s))n_0(t-\tau_1(s)+\tau_e,s_1(s))(1 - as_1(s)\frac{d\tau_m(s_1(s))}{ds_1})^{-1} e^{(a-\mu)(\tau_1(s)-\tau_e)} \; +$$

$$qxg_0(s_2(s))n_0(t-\tau_2(s),s_2(s))(1 - as_2(s)\frac{d\tau_m(s_2(s))}{ds_2})^{-1} e^{(a-\mu)\tau_2(s)-\mu\tau_e} \,,$$

$$n_0(t,s_m) = 0 \,, \tag{4.1.8b}$$

which is the sought result.

Setting $\frac{\partial \hat{n}_0}{\partial t} = 0$ gives us an equation for the equilibrium $\hat{n}_0$. This has to be combined with the normalization condition

$$\int_{\Omega_0} \hat{n}_0 ds + \int_{\Omega_1} \hat{n}_1 ds d\tau = 1 \tag{4.1.9b}$$

Given $\hat{n}_0$ we can calculate $\hat{n}_1$, and from this the functional response F as

$$F(x) = \int q\hat{n}_1(s,\tau_e+)ds \,. \tag{4.1.9c}$$

An easy way to calculate the integrals will be derived in example 4.2.4.

EXERCISE 4.1.3: Write the equation (4.1.9a) for $\hat{n}_0$ out in full.

EXERCISE 4.1.4: Calculate the characteristics of the PPP-model from example 1.10. Also calculate the densities n_1 and n_2 on Ω_x and Ω_{xy} resp. on the assumption that the time course of n_0, the number of empty patches, and of the numbers of the free searching, patchless prey and predators are known.

EXERCISE 4.1.5: Under constant nutrient conditions the reproduction by binary fission model from I.4 can be transformed into a renewal equation. Assume to this end that the division rate b is positive only on $(a,1)$ with $a>\frac{1}{2}$. Integrate the p-equation along its characteristic from a to 1, assuming $n(a)$ to be known. Next do the same on $(\frac{1}{2}a,a)$, using the boundary condition $n(\frac{1}{2}a) = 0$ and the known n on $(\frac{1}{2}a,a)$. Use the results to derive a renewal equation for $n(t,a)$. (The answer to this exercise can be found in II.10.)

4.2. *Transformation of variables*

Integration along characteristics provides one way of transforming our p-equations. Another way is provided by a change in i-state variables. An example of two p-equations which are thus connected can be found in excample 3.4.1, Von Bertalanffy growth starting from zero initial size. Below we shall first discuss the reparametrization of the i-state in general. At the end of this subsection the two ideas of characteristic and i-state reparametrization will be combined.

Let y be the new i-state parametrization,

$$y = h(t,x), \quad x = g(t,y) \tag{4.2.1.}$$

where h and g are each others inverses with respect to their second argument, *i.e.* for all $x,y \in \mathbb{R}^k$ and all $t \geq 0$

$$x = g(t,h(t,x)), \quad y = h(t,g(t,y)) \tag{4.2.2}$$

and let m be the corresponding p-state, then (by integration rule (ii) from appendix A)

$$m(t,y) = n(t,g(t,y))\det|D_y\, g(t,y)|\,. \tag{4.2.3a}$$

Now let n satisfy

$$\frac{\partial n(t,x)}{\partial t} + \nabla_x \cdot (v(t,x)n(t,x)) = -\alpha(t,x)n(t,x) + b(t,x)\,. \tag{4.2.4}$$

For m we have the corresponding equation

$$\frac{\partial m(t,y)}{\partial t} + \nabla_y \cdot (w(t,y)m(t,y)) = -\beta(t,y)m(t,y) + c(t,y) \tag{4.2.3b}$$

The question is: what is the relationship between the various terms in these two equations?

We start with v and w. On the individual level we have

$$w(t,y) = \frac{dy}{dt} = \frac{d}{dt}h(t,x) = D_x h(t,x)\frac{dx}{dt} + D_t h(t,x)$$

$$= D_x h(t,g(t,y))v(t,g(t,y)) + D_t h(t,g(t,y)), \tag{4.2.3c}$$

or equivalently

$$= [D_y g(t,y)]^{-1}[v(t,g(t,y)) - D_t g(t,y)]\,. \tag{4.2.3c'}$$

Here we used

$$0 = D_t h(t,g(t,y)) + D_x h(t,g(t,y))D_t g(t,y)\,, \tag{4.2.5}$$

(note that the differentiation for t in the first term only applies to the argument of h!) and

$$I = D_x h(t,g(t,y))D_y g(t,y), \tag{4.2.6}$$

obtained from differentiating (4.2.2) for t and y respectively.

To find the transformation for α and b we use dimensional considerations. The dimensions of α and β are both time^{-1}. Therefore

$$\beta(t,y) = \alpha(t,g(t,y))\,. \tag{4.2.3d}$$

The dimension of b is (number of individuals) / (i-state volume $\times$ time). The factor transforming i-state volumes was $|\det D_y g(t,y)|^{-1}$ (compare (4.2.3a)). Therefore

$$c(t,y) = b(t,g(t,y))|\det D_y g(t,y)| \tag{4.2.3e}$$

(Obviously, if b has to be calculated from some integral over x, we also have to make the appropriate changes to transform this into an integral over y.) Boundary conditions transform in an analogous way: To calculate c_∂ we have to multiply b_∂ with a factor expressing how a length element of $\partial_+\Omega_x$ relates to a length element of $\partial_+\Omega_y$.

* INTERLUDE 4.2.1: *An algebraic check of the transformation rules.*

We start with the case of time independent g and h. Multiplying (4.2.3b) with $|\det D_x h| = |\det D_y g|^{-1}$ gives

$$\frac{\partial n}{\partial t} + |\det D_x h| \nabla_y \cdot (wm) = -\alpha n + b$$

So all we have to show is that

$$\nabla_x \cdot (vn) = |\det D_x h| \nabla_y \cdot (wm) \,. \tag{4.2.7}$$

To simplify the notation we shall assume from now on that $\det D_x h > 0$; the case $\det D_x h < 0$ only necessitates some selfevident changes in the calculations below. We start by observing that $n(x) = m(h(x))\det D_x h(x)$. Therefore, by the chain rule

$$\nabla_x \cdot (vn) = \sum_i \frac{\partial}{\partial x_i}(v_i(x)n(x)) = \sum_{i,j} \frac{\partial}{\partial y_j}[v_i(g(y))m(y)\det D_x h(g(y))]\frac{\partial h_j}{\partial x_i} \,.$$

This should be equal to

$$\det D_x h \sum_{i,j} \frac{\partial}{\partial y_j}[(D_x h(g(y))_{ji} v_i(g(y))m(y)] \,.$$

To show that this is the case it is sufficient to show that

$$\sum_j \frac{\partial}{\partial y_j}(\det D_x h(g(y)))\frac{\partial h_j}{\partial x_i} = \det D_x h \sum_j \frac{\partial}{\partial y_j}(D_x h(g(y)))_{ji} \tag{4.2.8}$$

(just differentiate the products inside the square brackets and concentrate on one value of i only). Applying the chain rule once again we find that (4.2.8) is equivalent to

$$\sum_j \Big(\sum_k \frac{\partial}{\partial x_k}(\det D_x h)\frac{\partial g_k}{\partial y_j}\Big)\frac{\partial h_j}{\partial x_i} = \det D_x h \sum_{j,k} \frac{\partial^2 h_j}{\partial x_i \partial x_k} \frac{\partial g_k}{\partial y_j} \,. \tag{4.2.9}$$

Since, from (4.2.6), $\sum_j \frac{\partial g_k}{\partial y_j}\frac{\partial h_j}{\partial x_i}$ equals 1 or 0 depending on whether $i = k$ or not, the left hand side of (4.2.9) reduces to $\frac{\partial}{\partial x_i}(\det D_x h)$, which equals (see appendix A)

$$\sum_{j,k} \frac{\partial^2 h_j}{\partial x_k \partial x_i} \,[\text{cofactor} \frac{\partial h_j}{\partial x_k}] \,.$$

But since $D_y g$ and $D_x h$ are each others inverses

$$(\det D_x h)^{-1} \,[\text{cofactor } \frac{\partial h_j}{\partial x_k}] = [D_y g]_{kj} = \frac{\partial g_k}{\partial y_j} \,.$$

In the time dependent case we get on multiplying (4.2.3.b) with $\det D_x h$

$$\det D_x h\Big(\frac{\partial m}{\partial t} + \nabla_y \cdot (D_x h v m) + \nabla_y \cdot (D_t h m)\Big) = -\alpha n + b \,.$$

By the same calculations as before we find that

$$\nabla_y \cdot (D_x h v m) = \nabla_x v n \,.$$

So it remains to show that

$$\det D_x h\Big(\frac{\partial m}{\partial t} + \nabla_y \cdot D_t h m\Big) = \frac{\partial n}{\partial t} \,. \tag{4.2.10}$$

Differentiating $n(t,x) = m(t,h(t,x))\det D_x h(t,x)$ for t gives

$$\frac{\partial n}{\partial t} = \Big(\frac{\partial m}{\partial t} + \nabla_y m \cdot D_t h\Big)\det D_x h + m\frac{\partial \det D_x h}{\partial t}$$

Next we observe that (see appendix A)

$$\frac{\partial}{\partial t}\det D_x h = \sum_{j,k} \frac{\partial^2 h_j}{\partial x_k \partial t} \,[\text{cofactor } \frac{\partial h_j}{\partial x_k}] = \det D_x h \sum_{j,k} \frac{\partial^2 h_j}{\partial x_k \partial x_t} \frac{\partial g_k}{d y_j} =$$

$$\det D_x h \sum_j \frac{\partial}{\partial y_j} (\frac{\partial h_j}{\partial t}(t,g(t,y))) = \det D_x h \nabla_y \cdot D_t h(t,g(t,y))$$

Hence

$$\frac{\partial n}{\partial t} = \det D_x h(\frac{\partial m}{\partial t} + D_t h \cdot \nabla_y m + m \nabla_y \cdot D_t h)$$

which is equivalent to (4.2.10).

We have considered here the most general case, where the transformation relating x and y may be time dependent. One reason to use a time dependent i-state parametrization may be that this way we can sometimes bring equations which are time dependent into time independent form. However, time independent i-state transformations are more commonly called to help.

When v is time independent an often useful (re)parametrization is given by the parameter θ indicating position along a characteristic as in subsection 4.1, in conjunction with some $(k-1)$-dimensional variable ξ indicating where this characteristic enters Ω through $\partial_+\Omega$ (see fig. 4.2.1). If there are no i-state jumps and all newborns necessarily start at $\partial_+\Omega, \theta$ is just what we call age. The map $g:(\theta,\xi) \mapsto x$ from (4.3.1) for this parametrization obviously has to be defined by

$$\frac{dg}{d\theta} = v(g), \quad \text{with } g(0,\xi) = x_0(\xi)\,. \tag{4.2.11}$$

The great advantage of this choice of parametrization is that $\Omega_{\theta,\xi}$ is filled with nothing but parallel ordinary conveyor belts roling at unit speed, and distinguished by the *parameter* ξ. This is not only an easy image to deal with, it also makes for easy algebra and easy numerics. (In the next subsection we shall consider the generalization of this idea to the case of time dependent v. There we shall also check algebraically that the transformation defined by (4.2.11) indeed has the desired properties.)

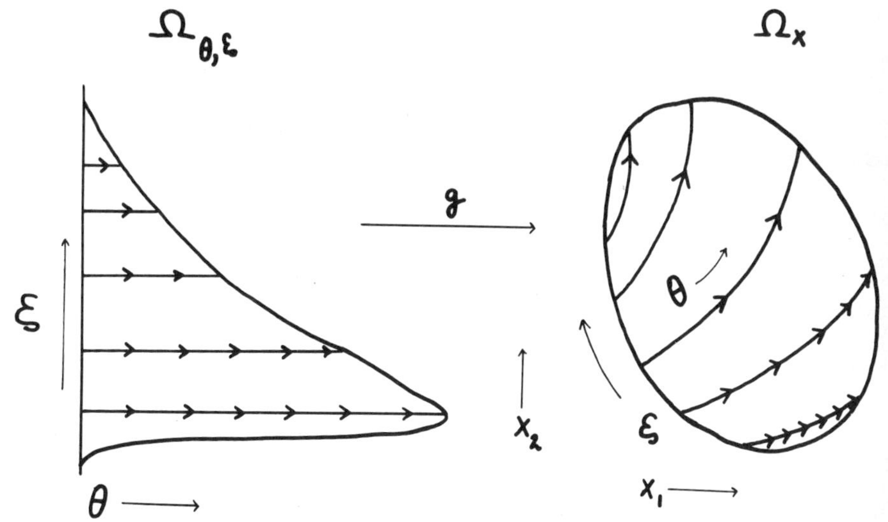

Figure 4.2.1 Reparametrizing the i-state x with the time θ needed to move to x along a characteristic starting at $x_0(\xi) \in \partial_+\Omega_x$, and a parameter ξ indicating position along $\partial_+\Omega_x$. (The distance between the arrow points on the characteristics is supposed to correspond to one time unit.)

EXAMPLE 4.2.2: *Size dependent reproduction in ectothermic animals.* In I.3 we have already seen how for this example under constant feeding conditions integration along characteristics could be used to arrive at a renewal equation for the population birth rate. The parameter θ characterizing position along the characteristic in this case was called a for age. For animals born after $t = 0$ the age-length relation is given by the function L defined in (I.3.4.10). The inverse of this function was called A. (see I.3.4.11). If we define $a = A(l)$ also for characteristics which do not start at

$\partial_+\Omega\times\mathbb{R}^+$ (but at $\Omega\times\{0\}$), we can write an equation for the age distribution as

$$\frac{\partial m}{\partial t} = -\frac{\partial m}{\partial a} - \mu m \qquad (4.2.12)$$

$$m(0) = \int_0^\infty \lambda(L(a))m(a)da$$

EXERCISE 4.2.3: Derive (4.2.12) from (I.3.2.1) and vice versa.

EXAMPLE 4.2.4: *The invertebrate functional response.* If we apply the procedure indicated above to the predators busy handling prey from example 1.1 we get

$$\frac{\partial m_1}{\partial t} = -\frac{\partial m_1}{\partial \theta} - \lambda(\tau(\theta))m_1$$

$$m_1(s,\tau_p(s)+) = qm_1(s,\tau_p(s)-) \qquad (4.2.13)$$

$$m_1(s,0) = xg_0(s)n_0(s)$$

At equilibrium

$$\hat{m}_1(s,\theta) = \begin{cases} xg_0(s)\hat{n}_0(s)e^{-\mu\theta} & \text{for } \theta<\tau_p(s) \\ xg_0(s)\hat{n}_0(s)qe^{-\mu\tau_p(s)} & \text{for } \theta>\tau_p(s) \end{cases} \qquad (4.2.14)$$

Therefore

$$\int_{\Omega_{s,\tau}} \hat{n}_1\, dsd\tau = \int_{\Omega_{s,\theta}} \hat{m}_1(s,\theta)dsd\theta = \int_0^{S_m} \hat{n}_0(s)xg_0(s)\,[\mu^{-1}(1-e^{-\mu\tau_p(s)}) + \tau_e qe^{-\mu\tau_p(s)}]ds\,,$$

making (4.1.9b) reduce to

$$\int_0^{S_m} \hat{n}_0(s)[1 + xg_0(s)\,(\mu^{-1}(1-e^{-\mu\tau_p(s)}) + \tau_e qe^{-\mu\tau_p(s)})]ds = 1\,. \qquad (4.2.15)$$

The factor following $xg_0(s)$ can be interpreted as the mean time a predator spends on sighted prey. $\mu^{-1}(1-e^{-\mu\tau_p(s)})$ represents the mean time spent on pursuits, both the successful and the unsuccessful ones. $qe^{-\mu\tau_p(s)}$ is the probability that a prey that is sighted is also caught.
The functional response can be calculated from

$$\int q\hat{m}_1(s,\tau_p(s))ds = \int_0^c xg_0(s)qe^{-\mu\tau_p(s)}\hat{n}_0(s)ds\,. \qquad (4.2.16)$$

In the following exercises we shall not only effect a transformation of the *i*-state of the PPP-problem in the manner indicated above, but we shall also replace the partial differential equation in three variables, t, x and y, for the predator patches by an equivalent pair of partial differential equations in two variables only. This transformed system forms the starting point of the calculations by Sabelis in part B.

EXERCISE 4.2.5: Transform the p-equation, including boundary conditions, for the predator patches by replacing the variables x and y by the variables θ_2, "age" of the predator patch, and ξ, the value of x at which the first predator arrived. Give also a formula for the rates R and S at which free searching, patchless prey and predators are produced.

EXERCISE 4.2.6: Transform the p-equations for both the predator and prey patches by replacing for the prey patches the variable x by θ_1, the "age" of the prey patch, and for the predator patches ξ by θ_1, age of prey patch at which the first predator arrived. Give also formulae for R and S.

EXERCISE 4.2.7: Replace θ_2 from the previous exercises by the time τ till the predator patch ceases its existence, due either to resource exhaustion or extinction of the prey. Again give formulae for R and S.

EXERCISE 4.2.8: Let R and S in the previous exercise be calculated as

$$R = \int \rho(\theta_1)m_2(\theta_1,0)d\theta_1 + m_1(a_m)x_m\,,\quad S = \int \sigma(\theta_1)m_2(\theta_1,0)d\theta_1\,, \qquad (4.2.17)$$

where m_1 denotes the density function over Ω_{θ_1} and m_2 the density function over $\Omega_{\theta_1,\tau}$. Now define

$$r(\tau) = \int \rho(\theta_1)\, m_2(\theta_1,\tau)d\theta_1\ , \quad s(\tau) = \int \sigma(\theta_1) m_2(\theta_1,\tau)d\theta_1\ . \tag{4.2.18}$$

Derive partial differential equations and boundary conditions for r and s.
Observe that the equations for n_0, the number of empty patches, m_1,r and s together form a complete state space model for the within patch "black box" as far as the calculation of R and S is concerned. If we couple these equations with p-equations for the free searching prey and predators the end result is a complete, autonomous model. The course of the state of this reduced model over time is sufficient information to calculate the time course of the state of the original PPP-model from an integration along characteristics.

4.3. Following cohorts through time

When v does depend on t, implicitly or explicitly, the characteristics in Ω are no longer constant curves but change over time. Yet, if newborns only start at $\partial_+\Omega$, it may still be useful to characterize an individual by its age a and its state at birth, again parametrized by ξ. The reparametrization connecting the i-state x of an individual to the characterization of that same individual by a pair (a,ξ) is necessarily time dependent *i.e.* the map $g:(a,\xi)\mapsto x$ depends on t. As g plays an essential role in the determination by a and ξ of any x-dependent quantities like individual birth rate, the mechanism by which the frequency distribution m over the space $\Omega_{a,\xi}$ of reachable (a,ξ)-pairs changes is time dependent as well. If we want a description of our p-system in which any time dependence is as usual accounted for by a dependence on the p-input we have to include g in our population description as a necessary adjunct to the frequency distribution m. Together the pair (m,g) indeed qualifies as the p-state of yet another alternative representation of our population. The differential generator of the corresponding p-system then consists of a differential equation which updates m plus a differential equation doing the same job for g. We propose to call this idea after its originatrix the Murphy-trick (Murphy, 1983a).

The differential equation for m, given g, can easily be written down from first principles. The differential equation for g can immediately be written down by observing that a may be considered a position along a conveyor belt moving with speed one. Following a point on the surface of the conveyor belt corresponds to following one individual through its life. If we follow a representative individual born at t_0 in state $x_0(\xi)$ and having state x at time $t = t_0 + a$

$$g(t_0+a,a,\xi) = x(a,x_0,t_0) \text{ and } \frac{dx}{da} = \frac{dg}{da} = \frac{\partial g}{\partial t} + \frac{\partial g}{\partial a}\ .$$

Since also

$$\frac{dx}{da} = v(t_0+a,x)\ ,$$

we have

$$\frac{\partial g}{\partial t}+\frac{\partial g}{\partial a} = v(t,g)\ , \quad g(0,\xi) = x_0(\xi)\ . \tag{4.3.1}$$

(NB (i) when $dim\Omega>1$,g is vector-valued, (ii) the combined equation for g and m will in general be nonlinear)

REMARK 4.3.1: In our above exposition of the Murphy trick we restricted ourselves to individuals born after $t = 0$. The individuals present at $t = 0$ can for example be described by their state at $t = 0$ plus (an analogue of) the function g telling their state at t in dependence on both t and their state at $t = 0$. The two sets of coordinates, ξ and a, and state at $t = 0$ and t, have in common that they are especially adapted to integration along characteristics.

EXAMPLE 4.3.2: *Size dependent reproduction in ectothermic animals.* If food density x is not constant the analogue of (4.2.11) - (4.2.12) becomes (NB g in (4.3.2a) corresponds to v from (4.3.1))

$$\frac{\partial l}{\partial t} + \frac{\partial l}{\partial a} = g(x,l), \qquad l(0) = l_b\ , \tag{4.3.2a}$$

$$\frac{\partial m}{\partial t} + \frac{\partial m}{\partial a} = -\mu m, \qquad \text{for } l(a)\leqslant\bar{l}(x)\ , \tag{4.3.2b}$$

$$m(a) = 0, \qquad \text{for } l(a)>\bar{l}(x)\ ,$$

$$m(0) = \int_0^\infty \lambda(x,l(a))m(a)da \tag{4.3.2c}$$

One reason to call attention to the Murphy trick is that it seems to hold considerable potential for the development of numerical procedures for coping in a uniform way with a large class of structured population models. A second reason is that Murphyzed equations often provide a good starting point for further transformation procedures like "linear chain trickery" (see *e.g.* Nisbet, Gurney, Blythe & Metz (1985); IV 5.2 and the contribution of Gurney, Nisbet & Blythe in these notes introduce the concept for age resp. physiological age dependent models.) The final, and most important reason is that it enables us to cope in an easy manner with situations in which the p-distribution is concentrated on some lower dimensional manifold in Ω, as discussed in section 5.

* INTERLUDE 4.3.3: *The algebra of transforming to age and i-state at birth*

We follow the notation of subsection 4.2, with some minor adaptations. So $h(t,x)$ denotes the inverse of $g(t,a,\xi)$ with respect to the last two variables, *i.e.*

$$\binom{a}{\xi} = h(t,g(t,a,\xi))\,.$$

From (4.2.3a)

$$m(t,a,\xi) = n(t,g(t,a,\xi))\det D_{(a,\xi)}g(t,a,\xi)$$

and from (4.2.3c) the speed in the (a,ξ) plane equals

$$w(t,a,\xi) = D_x h(v - \frac{\partial g}{\partial t})\,.$$

But

$$D_x h = (D_a g,\ D_\xi g)^{-1} = (v - \frac{\partial g}{\partial t},\ D_\xi g)^{-1}\,,$$

where the last equality follows from (4.3.1). Therefore

$$\sum_j (D_x h)_{ij}(v_j - \frac{\partial g_j}{\partial t}) = \begin{cases} 1 & \text{for } i=1 \\ 0 & \text{for } i\neq 1 \end{cases}$$

EXERCISE 4.3.4: Derive (I.3.2.1) from (4.3.2) and *vice versa.*

EXERCISE 4.3.5: Murphyze the cell kinetics model from example 1.4. Also derive the ordinary p-equations from the Murphy-type equations. Assume that the division rate b is not time dependent.

* EXERCISE 4.3.6: Formulate possible converses of interlude 4.3.3.

EXERCISE 4.3.7: How should the concept of Murphy equations be extended to cover the cell kinetics model from I.4? Write down the Murphy equations both for the probability per unit of time and the stochastically predetermined size assumptions about the division process.
Hint: There are two possible approaches. One is to keep the characterization of individuals by age and i-state at birth. In that case one should follow the lead of exercise 4.3.5 and as the last step integrate over the spurious variable age. The other possibility is to use a quantity analogous to θ from the previous section, as ones i-description. $\theta = h(x)$ can for example be defined as the age a hypothetical individual that started at $\partial_+\Omega$ would have if it now has size x.

5. About delta-functions and related topics

In section 3 we already mentioned a few times that some of the formulae there were not completely general, unless we allowed the expressions involved to contain delta- "functions" to represent sharp localized effects in a formalism that otherwise only allows smoothly distributed ones. In such a formalism delta-functions are the natural representations of point masses versus mass densities, instantaneous death probabilities versus death rates, *etc.* In this section we shall subject this suite of ideas to a closer scrutiny.

The main advantage of the delta-function formalism is its conciseness and unifying generality. This also makes the delta-function formalism an ideal tool for expressing our biological assumptions. The penalty which we have to pay is that in each specific instance we have to find out what the formalism exactly means for our equations, *i.e.* we have to translate our formal expressions containing delta-functions into less slim looking but concretely manageable ones. In this section we shall give some of the rules for achieving this. We shall mainly concentrate on the one-dimensional case. Only in the last subsection we shall demonstrate by means of examples the extension to higher dimensional *i*-state spaces.

5.1. The delta function formalism

In these notes we shall take the line that delta-functions are terms that we put into formal integral expressions in order to have one formalism for quantities which in conceptually related cases can be calculated as classical integrals, but not in the case under consideration. For such a line to make sense the two kind of cases should at least be related through some kind of limit argument.

EXAMPLE 5.1.1: If we define for $c \in C[-1,1]$,

$$f(c) = \int c(x)w(x)dx\ , \tag{5.1.1}$$

where w is some integrable weight function, then f defines a continuous linear map from $C([-1,1])$ to $\mathbb{R}$. Such a map is called a continuous linear functional on $C([-1,1])$. Unfortunately not every such functional can be represented by (5.1.1) for some w. The linear functional $g_0(c) = c(0)$ provides an immediate counterexample. However, $g_0(c)$ and similar linear functionals can be obtained as limits of expressions of the form (5.1.1). For example, let

$$w_i(x) = \begin{cases} \frac{1}{2}i & \text{for } x \in (-i^{-1}, i^{-1}) \\ 0 & \text{elsewhere} \end{cases}\ , \tag{5.1.2a}$$

then

$$g_0(c) = \lim_{i\to\infty} \int c(x)w_i(x)dx\ . \tag{5.1.2b}$$

The formal limit* of the sequence w_i, and of any other similar sequence, like

$$w_i(x) = \begin{cases} 2i^2(i^{-1}-x) & \text{for } x \in (0,i^{-1}) \\ 0 & \text{elsewhere} \end{cases}\ , \tag{5.1.2a$'$}$$

such that (5.1.2b) holds for any $c \in C[-1,1]$, is denoted as δ, so that we may write

$$g_0(c) = \int c(x)\delta(x)dx\ . \tag{5.1.2c}$$

As an indication of the power of this formalism we write for $a \in (-1,1)$

$$\delta_a(x) := \delta(x-a)\ , \tag{5.1.3}$$

$$g_a(c) := \int c(x)\delta_a(x)dx\ , \tag{5.1.4}$$

which implies

$$g_a(c) = c(a)\ . \tag{5.1.5}$$

For $a = \pm 1$ (5.1.3) is no longer well defined (to see this consider the two sequences (5.1.2a) and (5.1.2a$'$): for the first sequence (5.1.2b) would give us $g_a(c) = \frac{1}{2}c(a)$, and for the second sequence $g_a(c) = c(a)$). In order to include these cases as well we shall modify our definition of δ to include the requirement that in any formula in which δ occurs we shall always confine the attention to weighting functions which have their support contained in $[-1,1]$. With this modification (5.1.3) and (5.1.4) imply (5.1.5) also for $a = \pm 1$.

REMARK 5.1.2: Delta *functions* are very much a scientist's tool. The real mathematical objects are not the delta-functions δ_a but the delta *functionals* g_a. The latter objects are the subject of the theory of distributions as developed by Schwartz (1950, 1951) (see also Mikusińsky (1959), Gelfand & Shilov (1964), Rudin (1973), Trèves (1967) and Yosida (1980)) which has the specific purpose to cope in a mathematically consistent way with delta-functions and the like. However, we do not intend to make the connection with this theory here in any but the most handwaving manner.

---

* For mathematicians only: Formal limits can be defined as equivalence classes of sequences, under the equivalence relation "leading to the same limit result in the formulae under consideration".

The simplest use of delta-functions in the specification of the p-differential generator is precisely in the manner of the following example: the distribution of the "touch down" positions after an i-state jump is assumed to contain one or more (multiples of) delta-functions.

EXAMPLE 5.1.3: *Multiplication by division.* The cell proliferation model from I.4, where division was assumed to be into two equal parts is but a special case of the more general cell division model expounded in section 3.3, formula (3.3.3), in which $f(\theta) = \delta(\theta - \frac{1}{2})$, leading to a birth rate at x of

$$\int_0^1 2f(\theta)\theta^{-1}\alpha_d(\frac{x}{\theta})n(\frac{x}{\theta})d\theta = 4\alpha_d(2x)n(2x)\,, \tag{5.1.6}$$

in accordance with (I.4.2.1).

Some slightly more subtle uses of delta-functions will be explicated in the next subsection.

5.2. *Delta-functions and transition conditions*

Life becomes more complicated when we write delta functions not explicitly under an integral sign but only implicitly, like in the right hand side of a differential equation.

We shall start considering the case where a disappearance term αn contains a delta function. Suppose we write

$$\alpha = \alpha_0 + \beta\,\delta(x-a) \tag{5.2.1}$$

with α_0 bounded, *i.e.* we assume that apart from a bounded death rate we also have a concentrated occurrence of deaths at $x = a$. The question then is how many deaths occur at $x = a$ and what does (5.2.1) mean for the p-differential generator? To answer this question we consider again the sequence w_i from (5.1.2a) that converges to a delta-function. To pass the interval $(a - \frac{1}{2}\epsilon,\ a + \frac{1}{2}\epsilon)$ the individual needs a time $\epsilon\,/\,|v(a)| + o(\epsilon)^*$. During that time the disappearance probability per unit of time is $\beta\epsilon^{-1} + O(1)$. So the probability of still being present after having passed the critical interval is $\exp(-\beta\,/\,|v(a)\,| + o(1))$, or in the limit for $\epsilon\downarrow 0$, for $v(a)>0$ (remember that ϕ denoted the flux)

$$\phi(a+) = \exp(-\beta\,/\,v(a))\phi(a-)\,, \tag{5.2.2}$$

or, since $\phi = vn$,

$$n(a+) = \exp(-\beta\,/\,v(a))\; n(a-)\,. \tag{5.2.3}$$

Some thought shows that (5.2.3) holds for $v(a)<0$ as well.

EXERCISE 5.2.1: Show that the second result of the previous calculations is independent of the specific choice of the sequence of weighting functions by which we represent δ.
Hint: Consider only positive weighting functions. Calculate an explicit expression for the probability of surviving through an interval of length ϵ around a, ϵ fixed. Take the delta-function limit, and only then let $\epsilon\downarrow 0$. (The answer to this exercise can be found in example 6.2.3.
In the limit n is no longer differentiable or even continuous at a. So n no longer satisfies a partial differential equation in the whole interior of Ω but only in the interior of the two subsets $x<a$ and $x>a$. Conditions like (5.2.2) and (5.2.3) that relate two subsets of Ω over a boundary at which ϕ or n makes a jump are called *transition conditions.* We have already met an example of a transition condition in the invertebrate predator model where a strike led to an instantaneous probability of escape as opposed to the escape probability per unit of time during pursuit (formula 4.1.5c).

Another possibility is that the (re)appearance term b contains a delta function at a like

$$b(x) = b_0 + c\delta(x-a)\,. \tag{5.2.4}$$

Applying the same reasoning as before we find the transition condition

$$\phi(a+) = \phi(a-) + c\,, \tag{5.2.5}$$

---

* The notation $y = o(x)$ ($y = O(x)$) means that $y\,/\,x \to 0$ ($y\,/\,x$ stays bounded).

or equivalently

$$n(a+) = n(a-) + c\,/\,v(a)\,. \tag{5.2.6}$$

In the special case that a lies exactly at the boundary of Ω, say the left hand boundary and $v>0$, and where there is no influx of mass across the boundary, (5.2.5) reduces to

$$v(a)n(a+) = v(a)n(a-) + c = c\,, \tag{5.2.7}$$

where we rewrote the usual boundary condition in the absence of influx as

$$n(a-) = 0\,.$$

So the delta-function formalism unifies the two hitherto separate ways in which mass may (re)appear.

It may come as a surprise that taking limits in the defining formula of a p-differential generator may lead to side conditions. However what really matters is the convergence of the associated semigroup (which we did not prove!) and not the subtleties of the connections between semigroups and generators. A more detailed account of this problem may be found in appendix 6A. A second manner in which what effectively amounts to a side condition was generated through a limiting argument pertaining to the interior of Ω, can be found in exercise 1.3: A death rate $\epsilon\,/\,(a_{max}-a)$ naturally restricts age a to $[0,a_{max})$. If we let ϵ go to zero the restriction of a remains, but in the interior of Ω there are no visible remnants of the mechanism by which this restriction came about.

EXERCISE 5.2.2: In the age as well as size dependent cell proliferation model from example 1.4 assume that a stochastic size threshold for division model applies like expounded in I.4.1, except that the division is only allowed to occur in cells older than age a_0. What does the p-equation look like? (Assume that the distribution of the size threshold and the growth pattern are such that some cells already reach the threshold before age a_0.)

* EXERCISE 5.2.3: One would like to have a good distributional interpretation of partial differential equations with distributional source and sink terms. However, this is actually rather complicated in the sink case. In the source case the following calculations are illustrative.
Let n satisfy

$$\frac{\partial n}{\partial t} + \frac{\partial}{\partial x}(vn) = f(t,x)$$

with v independent of time. Derive a differential equation for the flux $\phi = vn$. Solve this differential equation by the method of characteristics on the assumption that $\phi(x_0) = b(t)$. Now let $f(t,x) \to c(t)\delta(x-a)$. What does the formula for ϕ become? In this case there is no problem with a distributional interpretation as ϕ is just the sum of a Heaviside function and a smooth function, and the δ-"function" is precisely the distributional derivative of the Heaviside component. (This also was the reason why we stated our results first in terms of ϕ and only afterwards in terms of n.)

5.3 *Delta-functions in initial conditions*

Often we wish to allow delta-functions as initial conditions. (For example, we may wish to start a predation experiment with a population of totally satiated predators.) The natural way to do this would be to work with a p-state space rich enough to contain such objects. (Of course such objects can never belong to the domain of the differential generator as derived in section 3. Chapter II discusses how such a differential generator can still uniquely define a semigroup on the total p-state space.)

* REMARK 5.3.1: The best move in many cases probably is to work in the space of Borel measures on Ω, provided with the weak * topology derived from its interpretation as the dual of $C(\Omega)$. This space is just sufficiently rich to contain all objects corresponding to even our wildest thought experiments, and nearness in the weak * topology is an immediate counterpart of being close as far as observations go. An example of this approach may be found in Heijmans' treatment of our invertebrate predator example (Heijmans, 1984d).

However, often it is easier for technical reasons to restrict the p-state space to some space of better behaved functions, like the Lebesgue integrable or continuous ones. One way to smuggle in delta-functions in such a framework exploits the observation that usually the state jumps have a sufficient smoothing action for the delta-function to have disappeared among the "first generation" descendants of the delta-function wise concentrated "zero'th generation". In that case we can simply treat the zero'th generation as a separate entity.

EXAMPLE 5.3.2: Assume that in the predator of I.2.5, *i.e.* the predator with negligible handling time, we start our experiment with N_0 predators at satiation s_m, the maximum of the gut capacity. Than we can formulate the p-problem as

$$\begin{cases} \dfrac{dS}{dt} = -f(S), \quad S(0) = s_m \\ \dfrac{dN}{dt} = -xg(S), \quad N(0) = N_0 \end{cases} \tag{5.3.1a}$$

$$\begin{cases} \dfrac{\partial n}{\partial t} = -\dfrac{\partial fn}{\partial s} - xg(s)n + xg(s-w)n(s-w) + xg(S)N(S)\delta(s-S-w) \\ -f(s_m)n(s_m) = \displaystyle\int_{s_m-w}^{c} xg(s)n(s)ds + \chi(S>s_m-w)xg(S)N \\ n(s,0) = 0\,. \end{cases} \tag{5.3.1b}$$

where $\chi(A)$ denotes the indicator function of the event between the brackets (*i.e.* $\chi(S>s_m-w) = 1$ if $S>s_m-w$ and zero otherwise).

In example 5.3.2 we obtain a piecewise continuous first generation. If we wish continuity we should also keep the first generation separate, *etc.* (provided g is sufficiently smooth).

Formula (5.3.1a) describes the fate of a delta function of initial size N_0 when we follow it along a characteristic.

EXERCISE 5.3.3: Replace (5.3.1b) by a separate (formal) partial differential equation for the first generation distribution and a further partial differential equation for the second plus higher generations distribution. Next replace the equation for the first generation by two partial differential equations on contiguous, moving, domains, coupled by a transition condition. Solve these equations by integration along their characteristics. This solution may be used to account for the first generation in case we only wish to confine our dealings with p-states to continuous i-state distributions.

EXERCISE 5.3.4: Equation (5.3.1a) does not contain a dilation term. Such in contrast to the equations which we get when following the local behaviour of an ordinary p-distribution along a characteristic. Explain this phenomenon, in connection to our definition of a delta function as a formal limit of a sequence of progressively more concentrated p-distributions.
Hint: Ask yourself what happens to the sequence of weight functions from (5.1.2a) when they move along with the conveyor belt.

EXERCISE 5.3.5: How should the equations of the example from I.3 (size dependent reproduction in ectothermic animals) be modified if we start our population with N_0 neonates?

5.4. *Delta-functions in more dimensions*

In the example we have already encountered a number of applications of the delta-function formalism in higher dimensional state spaces, like the transition condition at the passage from pursuit to eating in our predator model. All these examples had in common that the delta-functions were essentially one dimensional, *i.e.* they could be written as a product of a delta-function in one direction times an ordinary function in the other directions (and the deterministic flow v was transversal to those "other" directions). Higher dimensional delta-functions which are integrated over a variable of the same dimension can be treated exactly as in the one-dimensional case. As a rule essentially higher dimensional delta-functions do not appear in the sink term αn. Our concern therefore should be higher dimensional delta-functions in the source term b: these do not disappear but move on with the stream in the same manner as delta-functions in initial conditions. We shall not try to give a general survey of all the possibilities but restrict ourselves to two examples. (In both examples the delta-functions appear in the boundary condition at $\partial_+\Omega$. The analysis from subsection 5.2 makes clear that such boundary conditions can be interpreted as delta-function valued source terms contiguous to the boundary.)

EXAMPLE 5.4.1: *Size dependent reproduction in ectothermic animals, finale.* Using the delta-function formalism the p-equations for example 1.2 from the start of this chapter are

$$\frac{\partial}{\partial t}n(t,l,a) + \frac{\partial}{\partial a}n(t,l,a) + \frac{\partial}{\partial l}(g(x,l)n(t,l,a)) = -\mu(l,a)n(t,l,a)\,, \tag{5.4.1a}$$

$$n(t,l,0) = \delta(l-l_b) \int_{\Omega} \lambda(x,l)n(t,l,a)dlda \ , \tag{5.4.1b}$$

where (5.4.1b) expresses the fact that newborns all have length l_b. The action of the two dimensional conveyor belt symbolized by (5.4.1a) moves the delta function appearing in the side condition (5.4.1b) bodily into the interior of Ω. The result is a one dimensional delta-function which lives on a line in two-dimensional space. This is nice as an image goes but it does not help in analyzing our population model. It is here that the Murphy trick from subsection 4.3 brings relief: we project all p-mass on the age axis and introduce a separate equation to calculate the instantaneous age-length relation. The resulting p-equation we encountered already as (4.3.2).

EXAMPLE 5.4.2: *Colony size distribution in the diatom Asterionella.* Till now in our discussion of this example we have concentrated on the main sequence of colony sizes 2^k, $k = 0,1,2, \cdots$. However, other colony sizes do occur even if relatively rarely. Since synchronous division is fairly well documented this must be due to a breakage mechanism in which the bonds are less than completely dependent. This is also in accordance with intuition. If we wish to relax the assumption of complete dependence we are left with a staggering choice of possible mechanisms. As the observed colony size distributions are the only data available on which we can decide between the various alternatives it is of no use to explore even a moderate range of possibilities. We shall therefore only consider the extreme opposite of complete dependence as a possibly useful null-model and assume that the bonds break completely independently.
If the bonds are not completely dependent there is no longer a one to one correspondence between the age of the youngest bond and cell age. Therefore we introduce a different parametrization of the i-state space. To this end we observe that in a colony the bonds are ordered in a linear sequence. So we can choose as an i-state the cell age a_0 plus the ages a_1 to a_r of the various bonds counted from left to right. Here we may think of "left" and "right" as being determined by which of the two observationally indistinguishable flat sides of a colony happens to be up when it is put onto the microscope slide; the arbitrariness of this procedure does not matter as the breakage and division mechanisms are necessarily invariant under the operation of flipping over a colony and all its descendants. The corresponding p-state is denoted as $n_{r+1}(a_0,a_1, \cdots ,a_r)$, $r = 0,1, \cdots$ ($r+1$ is the colony size), with the convention that $n_{r+1}(a_0,a_1, \cdots ,a_r) = n_{r+1}(a_0, a_r, \cdots ,a_1)$.
If we denote the age specific division rate as d and the bond-age specific breaking rate as b we get as our p-equations

$$\frac{\partial n_{r+1}}{\partial t} + \sum_{i=0}^{r} \frac{\partial n_{r+1}}{\partial a_i} = -d(a_0)n_{r+1} - \sum_{i=1}^{r} b(a_i)n_{r+1} \tag{5.4.2a}$$

$$+ \sum_{j=r+1}^{\infty} \int_0^{\infty} \cdot\cdot \int_0^{\infty} b(a_{r+1}) \, [n_{j+1}(a_0,a_1, \cdots ,a_r,a_{r+1}, \cdots ,a_j)$$

$$+ \, n_{j+1}(a_0,a_r, \cdots ,a_1,a_{r+1}, \cdots a_j)] \, da_{r+1} \cdots da_j$$

with the side conditions

$$n_{r+1}(0,a_1, \cdots a_r) = \begin{cases} \delta(a_1)\delta(a_3) \cdots \delta(a_r) \int_0^{\infty} d(a_0) n_{(r+1)/2}\,(a_0,a_2, \cdots ,a_{r-1})da_0 & \text{for } r+1 \text{ even} \\ 0 & \text{for } r+1 \text{ odd .} \end{cases} \tag{5.4.2b}$$

Just as in the previous example the delta-functions are moved bodily into the interior of Ω along the lines $a_1 = a_2 = .. = a_r$. While moving they decrease in size due to bonds breaking, at the same time spawning lower dimensional delta-functions corresponding to the fragmentation products. (This picture may seem hopelessly complicated. Yet, in chapter IV we shall extract some information from this model albeit by totally different methods.)

EXERCISE 5.4.3: Try to rewrite this model without using delta-functions. Introduce to this end a special notation for the p-masses on the various diagonal planes supporting delta-functions, and write equations for the movement of all these separate components of the p-state.
(Hint: we didn't dare try doing this exercise ourselves!)

6. Limiting processes and model simplification

6.1. *Introduction: the role of limit arguments*

Structured population models are considerably more resistant to mathematical analysis than their nonstructured counterparts. A mathematical toolbox for dealing with these models is only on the verge of being developed. Chapters II, IV, V and VI give some indication of the state of the art. Numerical work also tends to be extremely time consuming except in certain special cases (see *e.g.* the contributions by Goudriaan and by Gurney, Nisbet and Blythe in part B). Finally these models generally are relatively parameter rich which poses a problem to the experimental and theoretical biologist alike who respectively have to supply the parameter values for particular applications or should explore the parameter space to get a general idea about the possible dynamics. Therefore, when the initial modelling stage is over it is of the utmost importance to simplify one's model as far as possible. How well a simplification performs should be judged mainly on the extent to which the connection between population phenomena and individual mechanisms is preserved, *i.e.* a good simplified model should match the original model's prediction to an extent determined by the application in mind while at the same time keeping to the essentials of the mechanistic assumptions of the original model.

The main simplification technique is through limiting procedures. That is, we observe that some parameters are small or large relatively and then consider a limit in which these parameters are set equal to zero or infinity. In this final section we shall give an indication of some of the possibilities.

Some examples of limiting procedures we have already encountered in section 5: our approach to delta-functions was essentially through appropriate limit arguments.

EXAMPLE 6.1.1: *Multiplication by division.* In this example we let the density f of the relative size after division θ approach $\delta(\theta - \frac{1}{2})$. The resulting limit model may be a good approximation if the variance of f is small as it usually is. It both has fewer parameters and allows an easier calculation of the stationary size distribution.
There is one snag. In the practically important special case of exponential individual growth the limit model allows a unique stationary size distribution, but the population does not converge to it in the course of time, (see II.11), whereas for nonzero variance of f such convergence does occur, (see VI.5 and Heijmans (1984b)). Still the stationary size distribution for the limit model can be shown to be a good approximation to the stationary size distribution for small variances. (But the lack of a smoothing action in the limit model may well play havoc with some numerical procedures for dealing with nonlinear variants of the model.)

The example shows that there may be various subtleties involved where it comes to the convergence of the associated semigroups. In the case of exponential individual growth the semigroup certainly cannot converge uniformly for all positive time. Appendix 6A gives a short introduction to the mathematical aspects of the convergence problems. In the main text we shall confine ourselves to formal calculations, while stressing the biological side of the argument. Nowhere shall we deal generally and in depth with the kind of pathologies sketched above. But we trust that our intuition will tell us how to deal with them in any specific instance.

6.2. *Time scale arguments*

Time scale arguments provide one of the most, if not *the* most, important ways of simplifying models.

EXAMPLE 6.2.1: *Size dependent reproduction in ectotherms.* In the model of this name from chapter I we assumed eventually that individual food supply x was coupled dynamically to the p-state. For model parameters based on laboratory experiments with *Daphnia magna* it turned out that food dynamics was extremely fast compared to the population reaction. Therefore in the numerical simulations we assumed that the (constant) food supply to the population α was always in equilibrium with the feeding rate of the *Daphnia* population and the (linear) rate of food deterioration βx.

This use of a time scale argument is completely analogous to that in ordinary differential equation population models. Some population state variables are assumed to have relaxation times grossly different from those of the remaining state variables so that we can treat the slow variables as constants on the time scale of the fast variables, and the fast variables as being permanently in equilibrium on the time scale of the slow variables. In the example the fast subspace was one dimensional. An example with an infinite dimensional fast subspace is provided by our model for the invertebrate functional response from I.2, where we assumed the time scale on which the satiation distribution equilibrates to be fast relative to the prey dynamics (and to individual growth; compare the use of the functional response in I.3, size dependent reproduction in ectothermic animals!).

In structured population models it is also possible to use time scale arguments on the individual level, as a means of lowering the dimensionality of the i-state space Ω. An example is provided by the initial neglecting of prey handling times in our invertebrate predator model.

EXERCISE 6.2.2: Derive formally equation (I.2.5.5) from equation (4.1.8).

Below we shall consider three examples in which individual and population time scales are combined in a non-trivial manner. The important message to be derived from these examples is that in such more complicated cases one should carry out the simplification in detail rather than try to specify the simple model direct on an *ad hoc* basis.

EXAMPLE 6.2.3: *Egg eating predators.*

Consider an age structured prey population obeying the balance law

$$\frac{\partial n}{\partial t} + \frac{\partial n}{\partial a} = -\alpha n \ , \quad n(0) = b \ , \tag{6.2.1}$$

where α denotes the relative death rate and b the birth rate. For the sake of the exposition we assume temporarily that the only cause of death is predation. So, assuming as a start a simple law of mass action interaction we may put

$$\alpha(t,a) = \beta(a)\, p(t) \ , \tag{6.2.2}$$

where p denotes the density of predators. β is an age specific vulnerability index (the variation in β with age may be due for example to the differences in escape probabilities of differently aged prey).

Some species do only predate on individuals of a prey population which are in an early stage of their life cycle *(e.g.* eggs or larvae or molluscs with a shell sufficiently thin to drill through). In that case β will differ from zero only for $0 \leqslant a < \epsilon$, where ϵ is small. As an idealization one may want to put $\epsilon = 0$, *i.e.,* only individuals of age zero are vulnerable to predation so that, effectively, predation only affects the rate of recruitment to the prey population. This idea was first introduced by Gurtin & Levine (1979), who just made an *ad hoc* assumption about the predation effect on recruitment, not based on any more detailed model for the predation process. This assumption was later criticized by Thompson, DiBiasio & Mendes (1982). However, the alternative which these authors proposed was equally lacking a mechanistic basis (they argued as if they were concerned with numbers whereas they were dealing with rates). The right formulation from a mechanistic viewpoint is contained already in the general arguments from section 5. In order for the predation effect not to become negligible when ϵ goes to zero we have to assume that the integral of β stays equal to some constant, say θ. Then β itself converges to θ times a delta-function at zero, and by (5.2.3) we arrive at

$$n(0+) = b\exp(-\theta p) \ , \tag{6.2.3}$$

which is the correct recruitment rate as modified by "egg predation". The factor $e^{-\theta p}$ is the probability that an egg survives the predation window, the risk of succumbing to predation being $1-e^{-\theta p}$. The quantity θ measures the intensity of the predator prey interaction. It can be interpreted as the effective "deadly" area/volume surrounding one predator (remember p was the number of predators per unit of area/volume).

Before going on to the next step, the incorporation of a functional response, we shall derive (6.2.3) again from first principles, as we need the intermediate steps at a later stage. We start by setting

$$\beta(a) = \epsilon^{-1}\zeta(a\,/\,\epsilon) \ ,$$

where ζ is some nonnegative function with support* in [0,1] and

$$\int_0^1 \zeta(\tau)d\tau = \theta \ ,$$

to express the fact that β may only differ from zero on $[0,\epsilon]$, and that $\int\beta(a)da = \theta$. Integrating (6.2.1) along the characteristics (see section 4) gives

$$n(t,a) = b(t-a)\exp(-\int_0^a \epsilon^{-1}\zeta(\alpha\,/\,\epsilon)p(t-a+\alpha)d\alpha = b(t-a)\exp(-\int_0^{a/\epsilon} \zeta(\tau)p(t-a+\epsilon\tau)d\tau) \ ,$$

which on taking formal limits for $\epsilon\downarrow 0$ reduces to

$$n(t,a) = b(t-a)\exp(-\theta p(t-a))$$

for all $a>0$. Taking the limit for $a\downarrow 0$ final gives (6.2.3) again.

The first who tried to modify Gurtin's & Levine's model by the incorporation of a functional response was Frauenthal (1983). Again no mechanistic route was taken. Following Diekmann *et al* (1986) (who concentrate on cannibalism) we shall now show how the previous derivation should be extended in this case.

We start assuming that

* The support of a function f is defined as closure $\{x | f(x) \neq 0\}$

$$\alpha(t,a) = \beta(a)p(t)\Phi(c)\,, \tag{6.2.4a}$$

where $0 \leqslant \Phi(c) \leqslant 1$ is a reduction factor *e.g.* due to prey handling times and

$$c(t) = \int_0^\infty h(a)\beta(a)n(a)da\,, \tag{6.2.4b}$$

where h is an age specific weight function *e.g.* describing how handling time depends on prey age (compare exercise I.2.2.3). The well known Holling disk equation gives $\Phi(c) = 1/(1+c)$ but other functions may be inserted. As before we find for $\epsilon \downarrow 0$ the recruitment rate

$$n(0+) = be^{-\theta p \Phi(c)}\,. \tag{6.2.5a}$$

However, the calculation of the limiting c for $\epsilon \downarrow 0$ (remember that β in (6.2.4b) was ϵ-dependent) is a more subtle problem. From

$$c(t) = \int_0^1 h(\epsilon\sigma)\zeta(\sigma)b(t-\epsilon\sigma)e^{-\int_0^\sigma \zeta(\tau)p(t-\epsilon\sigma-\epsilon\tau)\Phi(c(t-\epsilon\sigma-\epsilon\tau))d\tau}\,d\sigma$$

we obtain for $\epsilon \downarrow 0$

$$c(t) = h(0)b(t)\int_0^1 \zeta(\sigma)e^{-p(t)\Phi(c(t))\int_0^\sigma \zeta(\tau)d\tau}\,d\sigma = \frac{h(0)b(t)}{p(t)\Phi(c(t))}\,(1-e^{-\theta p(t)\Phi(c(t))})\,,$$

which we can rewrite as

$$\frac{c\Phi(c)}{h(0)} = \frac{b}{p}\,(1-e^{-\theta p \Phi(c)})\,. \tag{6.2.5b}$$

(6.2.5b) is a consistency relation: The left hand side equals the limit for $\epsilon \downarrow 0$ of $\int_0^\infty \beta(a)n(a)da\Phi(c)$ which is the number of prey eaten per predator per unit of time *(i.e.* the functional response), and the right hand side equals the total number of prey eaten per unit of time divided by the number of predators.
The question remains whether for given b and p (6.2.5.b) defines a unique c. The biological interpretation requires the left hand side to be an increasing function of c which is zero for $c = 0$. If, as in Holling's disk equation, Φ is a decreasing function of c with $\Phi(\infty) = 0$ then a unique solution $c = c(b,p)$ exists (see figure 6.2.1) (It seems possible in principle that several solutions coexist if the functional response is *s*-shaped but we did not pursue this complication).

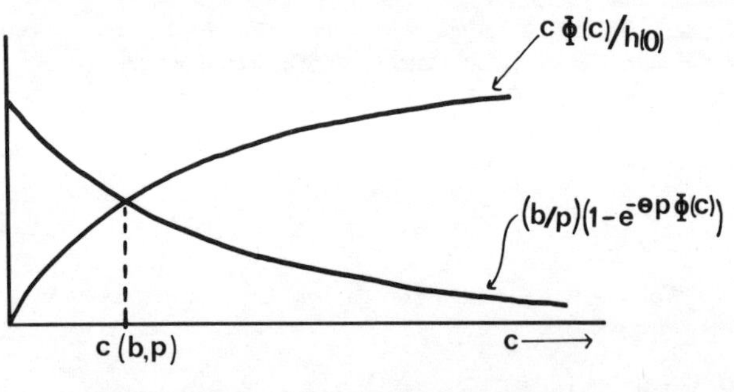

Fig. 6.2.1: The determination of the quantity $c(b,p)$.

In the case of cannibalism p corresponds to a certain part of the prey population itself. Thus one might take

$$p = \int_0^\infty k(a)n(a)da\,, \tag{6.2.6}$$

where k is an age specific index of cannibalistic tendency. (See Diekmann *et al* (1985) for a detailed analysis of this situation.) If cannibalism acts indiscriminately among juveniles only, as *e.g.* in *Tribolium* (see Fujii (1978) and the references therein) then β and k have the same support and moreover are constant on this support (or, if we make the more general assumption that the juvenile period has a stochastic duration, β and k are both proportional to the probability that an individual of age a is still in the juvenile stage). In that case in the limit for $\epsilon \downarrow 0$, p equals some fixed constant times c. This special case is elaborated further in VI.3. Exercise 6.2.8 discusses, in the context of the next example, how the model should be modified when the coincidence between the vulnerable and cannibalistic classes is less than exact.

EXAMPLE 6.2.4: *Nursery competition.* Technically this example is closely related to the previous one but the biological process is completely different. We consider a larval biotope which contains a limited number of "safe places" *(e.g.* spots to hide for predators or spots with abundant food). Larvae have to compete for these places. Those larvae which are temporarily excluded from the safe places have an increased risk of dying. Using a time scale argument we model this by letting the death rate depend on the fraction of the time that the individual is not in a safe place. The latter quantity of course depends on the number of competitors m. So instead of (6.2.2) we put

$$\alpha(t,a) = \beta(a)F(m) \tag{6.2.7a}$$

and identify (compare the discussion at the end of the previous example)

$$m = \int_0^\infty \beta(a)n(a)da\,. \tag{6.2.7b}$$

Thus β describes the age dependence of both the activity in the competition process and the need for a safe place (in principle these two could also be different, safe places being occupied by individuals who do not really need them), and F describes how the death rate depends on the effective number of competitors m. (In the appendix to this subsection we shall derive an explicit expression for F from a mechanistic submodel on a fast time scale.)
With F at our disposal we may normalize β to have integral one. Setting again $\beta(a) = \epsilon^{-1}\zeta(a/\epsilon)$ and letting $\epsilon\downarrow 0$ we arrive at

$$n(0+) = be^{-F(m)}\,, \tag{6.2.8a}$$

$$m = (b/F(m))(1-e^{-F(m)})\,. \tag{6.2.8b}$$

The consistency relation (6.2.8b) again allows a simple interpretation. To this end we rewrite it as

$$mF(m) = b(1-e^{-F(m)})\,.$$

Both the left and the right hand sides now equal the number of larvae dying per unit of time.
The biological interpretation requires that F is an increasing function of m. This implies (see exercise 6.2.6 below) that (6.2.8b) has a unique solution $m = m(b)$.
Combining (6.2.8) with the usual balance laws for births and deaths gives the population model

$$\frac{\partial n}{\partial t} + \frac{\partial n}{\partial a} = -\nu n\,,\quad n(0+) = be^{-F(m)}\,, \tag{6.2.9}$$

$$b = \int_0^\infty B(a)n(a)da\,,\quad m = (b/F(m))(1-e^{-F(m)})\,,$$

where B and ν are the age specific birth and death rates.
In IV.1.3 we shall reformulate this model as a nonlinear renewal equation in b (see also exercise 6.2.7). In VI.3 this renewal equation will be analyzed for its dynamic properties.

EXERCISE 6.2.5: Derive (6.2.8b)

EXERCISE 6.2.6: Prove that for increasing F (6.2.8b) allows a unique solution.
Hint: First show that $f(x) = (1-e^{-x})/x$ decreases for $x>0$, and then appeal to a graphical argument.

EXERCISE 6.2.7: Convert (6.2.9) into an integral equation for b.
Hint: Introduce the function h defined by $h(b) = be^{-F(m)}$, m the solution of (6.2.8b). The answer to this exercise can be found in IV example 1.3.1.

EXERCISE 6.2.8: An essential point in the argument in example 6.2.4 was that the vulnerability to competition and the involvement in competition depend in exactly the same manner on age, as expressed by the appearance of the same function β in both (6.2.7a) and (6.2.7b). To see this set $\alpha(t,a) = \epsilon^{-1}\zeta(a/\epsilon)F(m)$ and $m = \epsilon^{-1}\int_0^\infty \eta(a/\epsilon)da$. For $\epsilon\downarrow 0$ we get again $n(0+) = e^{-F(m)}$ in combination with an equation for m of the form $m = bG(m)$. Find an expression for G. This expression will still contain a number of integrals. In order to arrive at an algebraic expression assume that ζ is a step function which equals 1 for $a<1$ and zero after, and that η is the same sort of step function except that the step is made at $1+\delta$, $\delta>-1$. Calculate an explicit expression for G.

EXAMPLE 6.2.9: *The PPP-problem: the air plancton.* Till now we have refrained from discussing the searching, patchless prey and predators. In the hints to the various exercises we have simply said that the overall densities of searching prey and predators were to be denoted as P and Q respectively.
P and Q have to be calculated from an equation for the "air plancton", to coin a phrase from Sabelis (these notes), *i.e.* those prey and predator individuals that have taken the wing after their patch got exhausted. It seems reasonable to describe these populations by an age dependent model, to describe the effects of exhaustion, desiccation *etc.* Assuming that exhaustion only influences death rates, but not searching efficiency, we may write

$$P = \int_0^\infty p(a)da\,, \qquad \frac{\partial p}{\partial t} = -\frac{\partial p}{\partial a} - \mu p - \zeta p n_0, \qquad p(0) = R\,, \tag{6.2.10}$$

$$Q = \int_0^\infty q(a)da, \qquad \frac{\partial q}{\partial t} = -\frac{\partial q}{\partial a} - \nu q - q\int_0^{x_m} \eta(x) n_1(x)dx, \qquad q(0) = S\,, \tag{6.2.11}$$

(remember n_0 and n_1 were the densities of the empty and the prey patches respectively and x was the within patch prey density), where R and S denote respectively the production of searching prey and predators due to patch exhaustion. ζ is the rate constant of empty patch encounter by searching prey and η is the rate constant of prey patch encounter by searching predators.
When μ and ζ are both large (and continuous in their arguments) (6.2.10) can be simplified to (with some selfevident abuse of notation)

$$\zeta P = \frac{\zeta R}{\mu(0) + \zeta n_0}\,, \tag{6.2.12}$$

and when ν and η are both large (6.2.11) can be simplified to

$$\eta Q = \frac{\eta S'}{\nu(0) + \int_1^{x_m} \eta(\xi) n_1(\xi) d\xi}\,. \tag{6.2.13}$$

(NB ζP and ηQ are precisely the quantities that appear in the p-equations for the patches.)
EXERCISE 6.2.10: Write the time scale arguments by which (6.2.12) and (6.2.13) are derived out in full.

6.2A. *An explicit expression for F from the nursery competition model from example 6.2.4.*

During our derivation the number of competitors is assumed to be fixed. We assume that the total number of safe places is fixed at s. These places can be divided into occupied, s_1, and empty ones, s_0. The number of competitors temporarily deprived of a safe place equals $m - s_1$, (*i.e.* we assume that a safe place can be occupied by at most one individual). A law of mass action assumption now leads to

$$\frac{ds_1}{dt} = -\lambda s_1 + \mu\, s_0(m - s_1) = -\lambda s_1 + \mu(s - s_1)(m - s_1)\,,$$

which has the stable equilibrium

$$\hat{s}_1 = \frac{m + s + (\lambda/\mu) - \sqrt{(m - s - \lambda/\mu)^2 + 4(\lambda/\mu)m}}{2}$$

The number of competitors which has to do temporarily without a safe place is then given by

$$H(m) = m - \hat{s}_1 = \tfrac{1}{2}\{m - \xi + \sqrt{(m-\xi)^2 + \rho m}\} \tag{6.2.A1a}$$

where by definition

$$\xi = s + \lambda/\mu\,, \quad \rho = 4\lambda/\mu\,. \tag{6.2A.1b}$$

Finally, we observe that the fraction of the time that an arbitrary competitor is unsafe equals the probability that it is unsafe at any particular time, which equals $H(m)/m$. Therefore we put

$$F(m) = \theta H(m)/m\,, \tag{6.2A.1c}$$

where θ is the factor introduced by normalizing β to have integral one.

In the following exercises some properties of F are derived which will be needed in Chapter VI.3.

EXERCISE 6.2A.1: Show that $H(m) = \frac{\rho}{4\xi}\{m + \frac{1}{\xi}(1-\frac{\rho}{4\xi})m^2 + O(m^3)\}$ for $m\downarrow 0$. Note that (6.2A.1b) implies that $1-\frac{\rho}{4\xi}>0$!

EXERCISE 6.2A.2: Show that $dH/dm = \frac{H(m)+\rho/4}{\sqrt{(m-\xi)^2+\rho m}}>0$.
Hint: differentiate the quadratic equation for H.

EXERCISE 6.2A.3: Show that $dH/dm<1$ by first observing that $dH(0)/dm<1$ and then deriving a contradiction from the assumption $dH/dm = 1$ for some m.

EXERCISE 6.2A.4: Show that $d^2H/dm^2 = 2dH/dm\frac{1-dH/dm}{\sqrt{(m-\xi)^2+\rho m}}>0$.

EXERCISE 6.2A.5: Show that $\lim_{m\downarrow 0}\frac{mdH/dm-H(m)}{m^2} = \frac{1}{4}\frac{\rho}{\xi^2}(1-\frac{\rho}{4\xi})>0$.

EXERCISE 6.2A.6: Show that $mdH/dm-H>0$.

EXERCISE 6.2A.7: Show that $\frac{dF}{dm} = \theta\frac{mdH/dm-H}{m^2}>0$.

6.3. *Laws of large numbers on the individual level: the step from particulate to nonparticulate*

A second type of limit argument that often comes in useful is related to the law of large numbers from probability theory or the continuum approximations from physics. We have encountered this type of limit already in the first example from chapter I, the invertebrate predator. There we introduced the assumption that the prey were very numerous as well as very small. The result was an equation in which the jump terms were replaced by an additional contribution to the continuous *i*-state movement, which we called the Rashevsky limit. Below we shall give one more example to illustrate the idea.

EXAMPLE 6.3.1: *Size dependent reproduction in ectotherms, 2nd finale.* When treating this model in chapter I we remarked that in our birth term

$$b = \int_0^{l_m}\lambda(l)n(l)dl\,,\quad g(l_b)n(l_b) = b\,, \tag{6.3.1}$$

we made the implicit and very unbiological assumption that all individuals in the population contributed infinitesimal shares to a common pool from which the individual young are produced. Here we shall derive this, or rather a slightly modified, birth term from a limit argument, showing that at least (6.3.1) can be given a mechanistic underpinning. (Aldenberg's contribution in part B shows, in the context of biological production calculations, that such hair-splitting indeed has its uses!).

Since we wish to argue in energetic terms we shall in our arguments replace length l by weight w. Now consider an animal that reproduces by forming buds of size w_0. The *p*-differential generator for such an animal may take the form

$$\frac{\partial n}{\partial t} = -\frac{\partial gn}{\partial w}-\alpha n + \alpha(w+w_0)n(w+w_0) - \mu n\,, \tag{6.3.2}$$

$$g(w_0)n(w_0) = \int_{w_0}^{w_m}\alpha(w)n(w)dw\,.$$

We now derive an approximating *p*-equation by letting w_0 decrease to zero. As a first step we observe that for the same amount of energy spent one may make many small or one large young. Therefore we set $\alpha(w) = w_0^{-1}\beta(w)$. As a next step we consider the jump terms in the differential equation,

$$-w_0^{-1}\beta(w)n(w) + w_0^{-1}\beta(w+w_0)n(w+w_0)\,.$$

Taylor expanding the second term gives for $w_0\downarrow 0$

$$-w_0^{-1}\beta(w)n(w) + w_0^{-1}(\beta(w)n(w) + \frac{\partial}{\partial w}(\beta(w)n(w)w_0 + o(w_0)) = \frac{\partial}{\partial w}\beta(w)n(w) + o(1)\,.$$

Therefore in the limit for $w_0 \downarrow 0$ the p-differential equation becomes

$$\frac{\partial n}{\partial t} = -\frac{\partial}{\partial w}((g-\beta)n) - \mu n\,, \tag{6.3.3a}$$

which corresponds to the i-differential equation

$$\frac{dw}{dt} = g(w) - \beta(w)\,,$$

i.e. the energy chanelled to reproduction is subtracted from growth.

At the boundary we get into trouble. First we have to assume that individuals can grow away from zero. (Our discussion of the boundary $\partial_{0+}\Omega$ for this example in 3.4 shows how this is naturally the case for Von Bertalanffy growth). But also the number of young born per unit of time explodes for $w_0 \downarrow 0$. This can be repaired by the biologically reasonable assumption that very small young have a much larger chance of dying in the early stages of their life than large young have. To make things simple we shall assume that these deaths occur immediately after birth. The probability of "egg" survival we put equal to γw_0. In that case for $w_0 \downarrow 0$

$$g(0)n(0) = \int \gamma\beta(w)n(w)dw\,. \tag{6.3.3b}$$

REMARKS 6.3.2: (i) (6.3.3b) differs from (6.3.1) in the location of the boundary, l_b in (6.3.1), 0 in (6.3.3b). In practice this will not make a great difference if g is relatively large on $(0,l_b)$ as is the case in the Von Bertalanffy growth equation used in our model formulation.

(ii) For many invertebrates the chanelling of energy to reproduction happens in a manner different from that assumed in (6.3.2): the weight of egg masses or their precursors cannot be lumped with individual weight. This means that we have to introduce egg mass as an additional i-state variable. In the spirit of our previous setup we may for example assume that this variable increases at a rate β, and that the animal reproduces as soon as the egg mass reaches a given value which equals the clutch size times the mass of one egg. However, there is a certain risk in such an approach. Once the course of the food density as a function of time is given, the course of the state of an individual is completely determined as well. This means that there is no spread in the size-egg mass plane among individuals in one cohort. So we get into the same kind of trouble as we sketched for the age-size dependent model from example 1.2. More important, however, is that the birth events of a cohort become completely predictable: the mean birth rate of a representative individual becomes a sequence of delta-functions. Example 6.1.1 shows that such a non-smooth production of offspring may entail a lack of convergence of the p-state to a stable i-state distribution. This is exactly the kind of effects we have to be on our guard for when analyzing structured population models. In our simplified *Daphnia* model from I.3 the smoothing action of the implicit randomness of the reproduction process kept any such trouble at bay.

EXERCISE 6.3.3: Instead of pulling an egg survival factor out of the hat we can introduce an age dependent death rate μ such that $\mu(a)\uparrow\infty$ for $a\downarrow 0$, which we rig in such a way that the increase in the production of young when we let $w_0\downarrow 0$, is precisely compensated by a decreasing survival through the initial age interval. What should this μ look like?

It is also possible to combine time scale arguments and "law of large number limits" in a more intricate manner than was done above. For example Metz & van Batenburg (1985a) describe a limit for the invertebrate predator model of the form, for the individual,

$$\frac{ds}{dt} = -as + xg_0(s)\Phi(xg_0(s),s) \tag{6.3.4}$$

where s is again individual satiation and Φ is a handling time correction factor as in (6.2.4a). And this is but one more of the variety of possible limit arguments: we cannot describe them all, but we hope that you have got the taste from our examples.

* 6A. *A justification for the limit arguments: the Trotter-Kato Theorem*

In order to justify the limit transitions discussed in this and the previous section mathematically, we can sometimes use a so-called Trotter-Kato theorem, provided we are dealing with linear equations all the way. Such a theorem relates the convergence of differential generators (or, rather, resolvents; see chapter II) to the convergence of the associated semigroups. Here we shall state a version which seems useful in the context of linear structured population models.

Let X be a Banach space. Let for all $n \in \mathbb{N}$ the closed operator A_n with (densely defined) domain $\mathfrak{D}(A_n)$ generate a strongly continuous semigroup $T_n(t)$. Let X_0 be a closed subspace of X. Let $T(t)$ be a strongly continuous semigroup on X_0 with infinitesimal generator A. Finally let $D \subseteq \mathfrak{D}(A)$ satisfy: for all $\psi \in \mathfrak{D}(A)$ there exists a sequence $\psi_n \in D$ such that $\psi_n \to \psi$ for $n \to \infty$ (D is sometimes called a *core* for A).

THEOREM *(Trotter-Kato). If for all $\psi \in D$, there is a $\psi_n \in \mathfrak{D}(A_n)$, $n \in \mathbb{N}$, such that $\psi_n \to \psi$, $A_n \psi_n \to A\psi$ for $n \to \infty$, then for all $t_0 > 0$ and $\psi \in X_0$ one has*

$$\lim_{n\to\infty} \sup_{0 \leq t \leq t_0} \|T_n(t)\psi - T(t)\psi\| = 0 .$$

We refer to Davies (1980) and Pazy (1983a) for alternative versions of the Trotter-Kato theorem.

As an example of the application of this theorem we refer to Heijmans (1985), where the Trotter-Kato theorem is used to justify the Rashevsky limit from I.2.7.

A. Calculus in $\mathbb{R}^n$, a short refresher

A1. Differentiation

Let $f: X = \mathbb{R}^n \to Y = \mathbb{R}^m$ (or more generally let X and Y be Banach spaces, see II interlude 1.2). We call f *(Fréchet) differentiable* at x_0 if there exists a linear map denoted as $Df(x_0)$ such that

$$\|f(x_0 + h) - f(x_0) - Df(x_0)h\| = o(\|h\|) \text{ for } \|h\| \downarrow 0 , \tag{A1.1}$$

where $\|z\| := (\Sigma z_i^2)^{\frac{1}{2}}$ is the Euclidian norm in $\mathbb{R}^n$ (or the norm of the Banach space of which z is an element), and

$$g(h) = o(\|h\|) :\Leftrightarrow \|g(h)\| / \|h\| \downarrow 0 \text{ for } \|h\| \downarrow 0 .$$

The function $Df: X \to \mathfrak{L}(X, Y): x \mapsto Df(x)$ (where $\mathfrak{L}(X, Y)$ denotes the space of continuous linear maps from X to Y), if it exists, is called the *derivative* of f.

Some rules applying to Df are*

(i) $D(g(f(x)) = Dg(f(x)) \cdot Df(x)$ *(chain rule)*

(ii) If $Y = \prod_{i=1}^{k} Y_i$ then $Df = \begin{bmatrix} Df_1 \\ \cdot \\ \cdot \\ Df_k \end{bmatrix}$ *(componentwise differentiation)*

(iii) If $X = \prod_{i=1}^{k} X_i$ then $Df(x) = \sum_{i=1}^{k} D_i f(x) h_i = (D_1 f(x), \cdots, D_k f(x))h$ *(partial differentiation)*

Applying (ii) and (iii) to the one dimensional components of x and y gives

$$Df(x) = \begin{bmatrix} \frac{\partial f_1}{\partial x_1} & \cdots & \frac{\partial f_1}{\partial x_n} \\ \cdot & & \cdot \\ \frac{\partial f_m}{\partial x_1} & \cdots & \frac{\partial f_m}{\partial x_n} \end{bmatrix} \tag{A1.2}$$

This matrix representation of $Df(x)$ is called the *Jacobian matrix.*

(A 1.2) allows us to identify Df with the map $Df: \mathbb{R}^n \to \mathbb{R}^{m \times n}$, and define the second derivative of f as $D^2 f := DDf$, and so on. (When X and Y are more general Banach spaces $\mathfrak{L}(X, Y)$ is supplied with the usual operator norm, see interlude II.1.2, to arrive at the same result.) The notation $C(\Omega, Y)$ or $C^0(\Omega, Y)$, $\Omega \subseteq X$, is used to denote the space of all bounded continuous functions $f: \Omega \to Y$ provided with the norm†

* We shall think of $\mathbb{R}^n$ as a space of "column vectors". ^{T} will denote transpose.

† The concept of norm in essence is a generalization of the concept of length of a vector. A formal definition may be

$$\|f\|_0 := \sup_{x\in\Omega} \|f(x)\| \,, \tag{A1.3}$$

and more generally $C^k(\Omega, Y)$ is used to indicate the space of all k times differentiable functions $f:\Omega\to Y$ with continuous derivatives up to the k'th, provided with the norm

$$\|f\|_k := \sum_{i=0}^{k} \sup_{x\in\Omega} \|D^i f(x)\| \,. \tag{A1.4}$$

When $Y = \mathbb{R}$ one generally writes $C^k(\Omega)$ instead of $C^k(\Omega, Y)$. (The spaces $C^k(\Omega, Y)$ can all be shown to be Banach spaces.)

For $h\in X$ the *directional derivative* of f in the direction of h at x_0 is defined as

$$\partial_h f(x_0) = \lim_{\epsilon\downarrow 0} \epsilon^{-1}(f(x_0 + \epsilon h) - f(x_0)) \,. \tag{A1.5}$$

When f is differentiable in x_0 we have by the chain rule

$$\partial_h f(x_0) = Df(x_0)h \,. \tag{A1.6}$$

A class of functions of special interest to us, as the p-states belong to this class, are the functions $f:\mathbb{R}^n\to\mathbb{R}$. In $\mathbb{R}^n$ we define the usual inner product

$$x\cdot y := x^T y = \sum_{i=1}^{n} x_i y_i \,.$$

So $\|x\|^2 = x\cdot x$. When a function $u:\mathbb{R}^n\to\mathbb{R}$ is differentiable the transpose of its derivative Du is called the *gradient* of u, and denoted as ∇u. Thus the gradient is a column vector of partial derivatives

$$\nabla u = \begin{bmatrix} \dfrac{\partial u}{\partial x_1} \\ \cdot \\ \cdot \\ \dfrac{\partial u}{\partial x_n} \end{bmatrix} \tag{A1.7}$$

and $Duh = \nabla u\cdot h$ and also $\partial_h u = \nabla u \cdot h$ (provided ∇u is well defined).

A second class of functions of special interest as these comprise both the vector fields of i-state speeds in $\mathbb{R}^n$ as well as the reparametrizations of i-state space, are the functions $w:\mathbb{R}^n\to\mathbb{R}^n$. For such a function w the trace of its Jacobian matrix $Dw = \left[\dfrac{\partial w_i}{\partial x_j}\right]_{\substack{i=1,\ldots,n\\ j=1,\ldots,n}}$ is called the *divergence* of w and denoted as $\nabla\cdot w$. So

$$\nabla\cdot w = \sum_{i=1}^{n} \frac{\partial w_i}{\partial x_i} \,. \tag{A1.8}$$

(Note that (A1.8) represents the recipe by which we calculate $\nabla\cdot w$; the definition $\nabla\cdot w :=$ trace (Dw) is just convenient for analytical purposes.) The determinant of the Jacobian matrix is known as the Jacobian determinant, or *Jacobian* for short. It is also denoted as

$$\frac{\partial(w_1, \cdots, w_n)}{\partial(x_1, \cdots, x_n)} := \det(Dw) \tag{A1.9}$$

(Note that the Jacobian matrix, the divergence and the Jacobian are all functions of x!) The so-called inverse function theorem tells us that w is invertible on a neighbourhood of x whenever $\det(Dw)(x)\neq 0$.

REMARKS A1: (i) In the often used Einstein convention one drops the summation sign, while it is understood that summation takes place over any repeated indices (like i in the last two summations.)
(ii) Often one writes $\vec{x}$ for vectors and $\vec{w}$ for vector valued functions.
(iii) The so-called Laplacian, occurring in diffusion equations, is the divergence of the gradient

$$\Delta u = \nabla\cdot\nabla u = \sum_{i=1}^{n} \frac{\partial^2 u}{\partial x_i^2} \tag{A1.10}$$

---

found in interlude II.1.2.

A2. Integration

For functions defined on $\mathbb{R}^n$ differentiation and integration are no longer inverse operations. Since integration of vector-valued functions can be carried out componentwise we concentrate on functions $f:\mathbb{R}^n\to\mathbb{R}$. The *Riemann integral* of a function over a smooth simply connected domain Ω_0, denoted as

$$\int_{\Omega_0} f(x)dx := \int_{\Omega_0} f(x)\Pi dx_i \tag{A2.1}$$

is defined in exactly the same manner as in the one dimensional case: cover Ω_0 with closed boxes with nonoverlapping interiors with sides of length ϵ and let $f_i^{\pm}$ denote the maximum resp. minimum value of f on the ith box, then

$$\int_{\Omega_0} f(x)dx := \lim_{\epsilon\downarrow 0} \epsilon^n \Sigma f_i^+ = \lim_{\epsilon\downarrow 0} \epsilon^n \Sigma f_i^- , \tag{A2.2}$$

provided the two expressions at the right give the same value.

Riemann's construction does not apply to unbounded functions. However, when f only grows beyond bounds for x going t o the boundary of Ω_0 we can extend the Riemann integral by calculating $\int_{\Omega_i} f(x)dx$ for a sequence of sets Ω_i growing to Ω_0 such that f is bounded on each Ω_i, and taking limits.

An alternative integral concept, due to Lebesgue, is essentially based on a partitioning of the range instead of the domain of f. It can be defined as, conditionally on the two expressions after the $:=$ sign giving the same value,

$$\int_{\Omega_0} f(x)dx := \lim_{\epsilon\downarrow 0} \sum_{i=-\infty}^{\infty} \epsilon i\lambda(\{x\in\Omega_0 \mid \epsilon(i-1)<f(x)\leq\epsilon i\}) \tag{A2.3}$$

$$= \lim_{\epsilon\downarrow 0} \sum_{i=-\infty}^{\infty} \epsilon(i-1)\lambda(\{x\in\Omega_0 \mid \epsilon(i-1)\leq f(x)<\epsilon i\})$$

where $\lambda(B)$ denotes the so-called *Lebesgue measure* ("volume") of the set B. The dx under the integral sign now also is interpreted as referring to this measure. The Lebesgue integral is more appropriate for our purposes as it allows a larger class of functions, *e.g.* the unbounded ones, to be integrated (but at the price of having to define the function λ for rather complicated sets). For piecewise continuous functions on bounded domains the two integrals coincide so that you can effectively think in terms of Riemann's construction if you find that more convenient.

The notation $L_1(\Omega)$, $\Omega\subset\mathbb{R}^n$, is used to indicate the space of functions $f:\Omega\to\mathbb{R}$ such that

$$\|f\|_1 := \int_{\Omega} |f(x)|dx < \infty , \tag{A2.4}$$

where the integral should be interpreted in the sense of Lebesgue, provided with $\|\cdot\|_1$ as a norm, with the understanding that any two functions f_1 and f_2 such that $\lambda(\{x\in\Omega \mid f_1(x)\neq f_2(x)\}) = 0$ will be considered identical objects. (The main advantage of Lebesgue's definition of the integral is that with this definition $L_1(\Omega)$ can be shown to be a Banach space.)

Under fairly general conditions which need not be detailed here (when doing specific calculations we shall always confine ourselves, possibly implicitly, to such classes of functions that these conditions are certainly fulfilled) we have the following rule for evaluating integrals

(i) $\displaystyle\int_{\Omega_0} f(x)dx = \int_{x_n\in B_n} \cdots \int_{x_1\in B_1(x_2,..,x_n)} f(x)dx_1 \cdots dx_n$ (*componentwise integration*)

where

$$B_i(x_{i+1},..,x_n) = \{z_i \mid z\in\Omega_0,\ z_{i+1} = x_{i+1},..,z_n = x_n\} .$$

Another rule which frequently comes in useful tells us how to deal with a change of variables. In the one-dimensional case the transformation $x = g(y)$ gives

$$\int_a^b f(x)dx = \int_{g^{-1}(a)}^{g^{-1}(b)} f(g(y))g'(y)dy \tag{A2.5}$$

where we have to assume that $g'(y) := Dg(y)\neq 0$ for $a<g(y)<b$. This formula applies independent of the sign of $g'(y)$ due to the information contained in the order of the integration boundaries $g^{-1}(b)$ and $g^{-1}(a)$. If we adhere to the notation of an integral over a domain Ω_0 we have to replace $g'(y)$ by its absolute value. This immediately generalizes to the n-dimensional case except that the role of $|g'(y)|$ is played by $|\det Dg(y)|$, *i.e.* the absolute value of the Jacobian of g. The rule then becomes

(ii) $\displaystyle\int_{\Omega_0} f(x)dx = \int_{g^{-1}(\Omega_0)} f(g(y)) \mid \det Dg(y)|dy$

for any invertible differentiable function g.

Apart from integrals over a domain Ω_0 of $\mathbb{R}^n$ we also need integrals over lower dimensional surfaces Σ in $\mathbb{R}^n$. An example of particular interest are subsets of the boundary $\partial\Omega$ of the i-state space. Usually these surfaces consist of a finite number of m-dimensional differentiable manifolds or "smooth surfaces", joined together by a finite number of edges, differentiable manifolds being defined as surfaces which locally can be represented as the image of some differentiable map $g:\mathbb{R}^m \to \mathbb{R}^n$ which on its range allows a differentiable inverse *. In that case we can use the connection with $\mathbb{R}^m$ to define a Lebesgue measure $d\sigma$ on Σ. Below we shall demonstrate the idea for Σ one dimensional as this is the only case for which we do concrete calculations. In that case one also speaks of *line integrals,* otherwise of *surface integrals.* The calculations for general surface integrals are analogous except for some technical details.

Consider a 1-dimensional curve Σ in $\mathbb{R}^n$ which can be represented as $g:\mathbb{R}\to\mathbb{R}^n$ with g differentiable and $Dg(t) \neq (0,..,0)^T$ for those values of t for which $g(t)\in\Sigma$. For Σ we define

$$\int_{\Sigma} f\, d\sigma = \int_{g^{-1}(\Sigma)} f(g(t))\, \|Dg(t)\| dt\,, \tag{A2.6}$$

where $\|\cdot\|$ is the Euclidean norm in $\mathbb{R}^n$. It is possible to show that this definition is independent of the choice of g. By varying g it is only possible to move more slowly or quickly along Σ, but the factor $\|Dg(t)\|$ makes that the points on Σ get a proportionally smaller or larger "weight". (The technical details of defining general surface integrals center around the calculation of the analogue of this factor $\|Dg\|$.)

EXERCISE A2.1: Prove the assertion above.
Hint: choose a function $\tilde{g}$ instead of g which can be written as $\tilde{g}(t) = g(h(t))$ and use the chain rule and integration rule (ii). Observe then that the assumption on Dg implies that any function $\tilde{g}$ satisfying the same assumption can indeed be written as $\tilde{g} = g(h)$.

* EXERCISE A2.2: What would you expect to be the analogue of $\|Dg(t)\|$ in the definition of a surface integral in $\mathbb{R}^3$ (Hint: first concentrate on linear maps so that Dg and g coincide and use your intuitive idea of surface area). Generalize to $(n-1)-$surfaces in $\mathbb{R}^n$.

Let Ω_0 be a domain in $\mathbb{R}^n$ with a smooth boundary $\partial\Omega_0$. Then we can define at every point of $\partial\Omega_0$ a *tangent (hyper) plane.* The outward *normal* vector to this tangent plane with (Euclidean) length 1 we shall always indicate as ν.

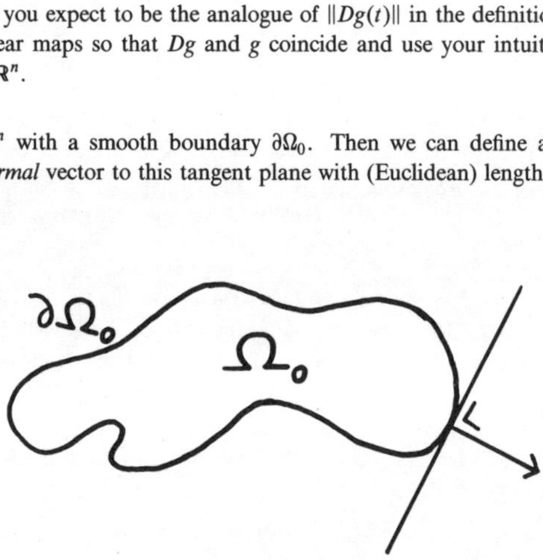

Fig. A.1

Using this notation the analogue of the $\mathbb{R}^1$ formula $\int_a^b f'(x)dx = f(b)-f(a)$ for differentiable functions $u:\mathbb{R}^n\to\mathbb{R}$ is

(iii) $\displaystyle\int_{\Omega_0} \nabla u\, dx = \int_{\partial\Omega_0} \nu u\, d\sigma.$

In this formula the requirement that $\partial\Omega_0$ is smooth can be relaxed to piecewise smoothness.

EXAMPLE A2.3: Let $\Omega_0 \subset \mathbb{R}^2$ be defined by $\Omega_0 = \{x \mid x_1^2 + x_2^2 < r\}$. We shall parametrize $\partial\Omega_0$ with the angle ϕ as $x_1 = r\cos(\phi)$, $x_2 = r\sin(\phi)$. In other words $\partial\Omega_0$ is the image of the function $g:[0,2\pi)\to\mathbb{R}^2:\phi\mapsto(r\cos(\phi),\, r\sin(\phi))^T$.

* With the notation $g:\mathbb{R}^m\to\mathbb{R}^n$ we do not necessarily mean that g is defined on the whole of $\mathbb{R}^m$ as we are interested only in local properties.

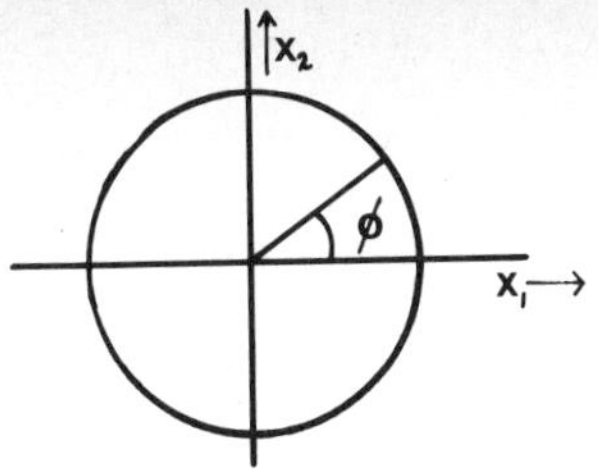

Fig. A.2

Then $Dg(\phi) = (-r\sin(\phi), r\cos(\phi))$ and $\|Dg(\phi)\| = r$. Now let $u(x) = x$, then $\nabla u(x) = (1,0)^T$ and

$$\int_{\Omega_0} \nabla u(x)dx = \begin{bmatrix} \pi r^2 \\ 0 \end{bmatrix}.$$

Since $\nu = (\cos(\phi), \sin(\phi))^T$ we have also

$$\int_{\partial\Omega_0} \nu u d\sigma = \int_0^{2\pi} \begin{bmatrix} \cos(\phi) \\ \sin(\phi) \end{bmatrix} r\cos(\phi) r d\phi = r^2 \int_0^{2\pi} \begin{bmatrix} \cos^2(\phi) \\ \sin(\phi)\cos(\phi) \end{bmatrix} d\phi = \begin{bmatrix} \pi r^2 \\ 0 \end{bmatrix}.$$

If we apply (iii) to each of the components of a function w: $\mathbb{R}^n \to \mathbb{R}^n$, and take the trace of the matrices on both sides of the equal sign we get

(iv) $\int_{\Omega_0} \nabla \cdot w(x)dx = \int_{\partial\Omega_0} \nu \cdot w d\sigma$ (*divergence* or *Gauss' theorem*).

(Note that this is a relation between the integrals of two scalar valued functions, one on the interior of Ω_0 and one on $\partial\Omega_0$.) The importance of Gauss' theorem is that it tells us how to deal with mass balances in Ω_0.

EXERCISE A2.4: Let $w: \mathbb{R}^2 \to \mathbb{R}^2$ be defined by $w(x) = (x_1, x_2)^T$ and let Ω_0 be as in the example above. Calculate both $\int_{\Omega_0} \nabla \cdot w(x)dx$ and $\int_{\partial\Omega_0} \nu \cdot w d\sigma$.

A3. Some useful relations from linear algebra: the differentiation of determinants

Let M denote the square matrix

$$M = [M_{ij}]_{j=1,..n}^{i=1,..n} = (m_{.1},..,m_{.n}) = \begin{bmatrix} m_{1.} \\ \cdot \\ \cdot \\ m_{n.} \end{bmatrix} \qquad \text{(A3.1)}$$

where a dot instead of an index means that index is still left free, so $m_{1.} = (m_{11},..,m_{1n})$. The i,j'th *minor* $|M_{ij}|$ of M is by definition the determinant of the $(n-1)\times(n-1)$ matrix obtained by deleting the i'th row and j'th column from M. The *cofactor* of the i,j'th component m_{ij} of M is defined as

$$\text{cofactor } m_{ij} := (-)^{i+j} |M_{ij}|, \qquad \text{(A3.2)}$$

So the cofactor of m_{ij} equals the determinant of the matrix obtained from M by replacing the j'th column of M by the j'th unit column vector (the i'th row of M by the i'th unit row vector), and

$$\det M = \sum_j m_{ij} \text{ cofactor } m_{ij} = \sum_i m_{ij} \text{ cofactor } m_{ij}. \qquad \text{(A3.3)}$$

Finally the *adjoint* of M is defined as

$$(\text{adj}M)_{ij} := \text{cofactor } m_{ji}. \qquad \text{(A3.4)}$$

If M is nonsingular Cramer's rule tells us that

$$M^{-1} = (\det M)^{-1} \operatorname{adj} M \,. \tag{A3.5}$$

The map det: $(\mathbb{R}^n)^n \to \mathbb{R}:(m_{\cdot 1},..,m_{\cdot n}) \to \det M$ is linear in its n vectorial arguments $m_{\cdot j}$. Therefore differentiating $\det M$ for $m_{\cdot j}$ gives

$$D_{m\cdot j}\det(m_{\cdot 1},..,m_{\cdot j},..,m_{\cdot n})\, h = \det(m_{\cdot 1},..,h,..,m_{\cdot n}) \tag{A3.6}$$

and from $\det M^T = \det M$

$$h^T D_{m_{i\cdot}} \det \begin{bmatrix} m_{1\cdot} \\ \cdot \\ \cdot \\ m_{i\cdot} \\ \cdot \\ \cdot \\ m_{n\cdot} \end{bmatrix} = \det \begin{bmatrix} m_{1\cdot} \\ \cdot \\ \cdot \\ h^T \\ \cdot \\ \cdot \\ m_{n\cdot} \end{bmatrix} . \tag{A3.7}$$

Finally, by setting h equal to the i'th unit column vector in (A3.6) or direct from (A3.3)

$$\frac{\partial \det M}{\partial m_{ij}} = \text{cofactor } m_{ij} \,. \tag{A3.8}$$

When our matrix M is a function of a single variable t, by the chain rule

$$\frac{d}{dt} \det M(t) = \sum_i \det \begin{bmatrix} m_{1\cdot} \\ \cdot \\ \cdot \\ \dfrac{dm_{i\cdot}}{dt} \\ \cdot \\ \cdot \\ m_{k\cdot} \end{bmatrix} . \tag{A3.9}$$

Now assume that

$$\frac{dM}{dt} = B(t)M \,. \tag{A3.10}$$

This relation can be rewritten as $\dfrac{dm_{ij}}{dt} = \sum_k b_{ik}\, m_{kj}$ or

$$\frac{dm_{i\cdot}}{dt} = \sum_k b_{ik}\, m_{k\cdot} \,,$$

and therefore, by the chain rule

$$\frac{d\det M}{dt} = \sum_i \sum_k b_{ik} \det \begin{bmatrix} m_{1\cdot} \\ \cdot \\ \cdot \\ m_{i-1,\cdot} \\ m_{k} \\ m_{i+1,\cdot} \\ \cdot \\ \cdot \\ m_{n\cdot} \end{bmatrix} = \sum_i b_{ii} \det M = \operatorname{trace} B \det M \,. \tag{A3.11}$$

B. Stochastic continuous *i*-state movements

So far we have excluded stochastic *i*-state movements from our considerations. The specification from first principles of *i*-models incorporating this type of movement requires a considerably greater insight in the detailed processes in and around the individual than we usually have, and we know of no model today in which a less than superficial connection with experimental data is made. Yet, as argued in remark 1.7, the incorporation of continuous stochastic *i*-state movements in our models may provide a description considerably closer to the actual biological mechanism than some of the *ad hoc* probabilistic postulates introduced earlier.

If we confirm ourselves to one-dimensional *i*-state spaces then, in the absence of jump contributions, the *p*-differential generators can be shown to be (see *e.g.* Cox & Miller, 1965; Goel & Richter-Dyn, 1979; Arnold, 1974; Karlin & Taylor, 1981 or Wong & Hajek, 1985)

$$\frac{\partial n}{\partial t} = \delta_{td} n + \delta_{ts} n = -\frac{\partial}{\partial x}(vn) + \frac{1}{2}\frac{\partial^2}{\partial x^2}(\sigma^2 n) \tag{B.1}$$

In this equation v corresponds to the infinitesimal mean and σ^2 to the infinitesimal variance of the stochastic x-movement. The *i*-state x itself follows a so-called diffusion process with drift v and infinitesimal variance σ^2.

One possible way to derive expressions for v and σ^2 is to start from a so-called stochastic differential equation for the *i*-state:

$$\frac{dx}{dt} = u(x,t) + \sigma(x,t)w(t)\,, \tag{B.2}$$

where w denotes an external stationary noise process with zero mean and

$$\lim_{T\to\infty} T^{-1}\, var(\int_0^T w(t)dt) = 1\,.$$

EXAMPLE B.1: *Stochastic cell growth.* Let x denote cell weight and let

$$\frac{dx}{dt} = za(x) - b(x)\,,$$

with z the nutrient concentration. Let z fluctuate around some mean value y and let

$$\alpha^2 = \lim_{T\to\infty} T^{-1}\, var(\int_0^T z(t)dt)\,,$$

then we can write

$$\frac{dx}{dt} = [ya(x) - b(x)] + [\alpha a(x)]w\,.$$

The next step is to assume that w shows negligible memory, or, in other words, that w approximates a so-called white noise. Then the *p*-equation becomes (Arnold, 1974; Karlin & Taylor, 1981; Wong & Hajek, 1985).

$$\frac{\partial n}{\partial t} = -\frac{\partial}{\partial x}(un) + \frac{1}{2}\frac{\partial}{\partial x}\left[\sigma^2\frac{\partial n}{\partial x}\right] = -\frac{\partial}{\partial x}((u - \frac{1}{2}\frac{\partial\sigma^2}{\partial x})n) + \frac{1}{2}\frac{\partial^2}{\partial x^2}(\sigma^2 n) \tag{B.3}$$

REMARKS B.2: (i) From (B.3) it appears that the noise also makes itself felt in the deterministic drift. However, this is partly a matter of convention. Clearly we cannot get away with the fact that the noise influences the infinitesimal mean. But we could have set

$$\delta_{td} n = -\frac{\partial}{\partial x}(un) \quad \text{and} \quad \delta_{ts} n = \frac{1}{2}\frac{\partial}{\partial x}(\sigma^2\frac{\partial n}{\partial x})$$

instead of the arrangement suggested by (B.1). The term drift has a less strict connotion than infinitesimal mean. Therefore it might be possible to make some case to call u from (B.3) the drift instead of v from (B.1). Only when σ^2 is independent of x, and in particular when $\sigma^2 = 0$ (so that $\delta_{ts} n = 0$ in either formalism), do we have coincidence of the two formalisms.
(ii) The assumption that x qualifies as *i*-state implies that the x-process is not allowed to show any memory other than that inherent in the present value of x itself. (This also was the reason that w should be white noise.) A direct consequence is that it is impossible for the x-process to be differentiable as long as $\sigma^2(x) > 0$. Therefore diffusion processes should at best be considered as approximations of considerably more complicated processes in which we leave out of

the consideration all additional short memory components.
(iii) A direct consequence of the non-differentiability of the x-process is that for w a white noise (B.2) cannot have its ordinary meaning. We have interpreted (B.2) here in the sense of Stratonovich (Arnold, 1974; Karlin & Taylor, 1981; Wong & Hajek, 1985). The reason is that we interpreted white noise as an approximation to a noise with a definite, even if short, memory. Only for the Stratonovich interpretation the map from w-process to x-process is continuous for w going to a white noise. The Ito interpretation which is more commonly adhered to by mathematicians on account of its other good properties, fails in this respect.
(iv) In general a simple differential equation like (B.2) will itself be an approximation, based on a time scale argument, of a much more complicated underlying process. In writing down (B.2) we therefore implicitly assumed that the fast time scales of that underlying process were much faster than the time scale of the fluctuations of w. This may be tricky when the noise fluctuations are fairly fast as well, as we assumed later in our argument. If the time scales of the noise and of the fast processes underlying the x process are of the same order of magnitude, we are forced to get the infinitesimal mean and variance of the limiting diffusion by a direct calculation, instead of by the two step strategy of first writing down (B.2) and then referring to the well-documented formula (B.3).

The k-dimensional analog of (B.1) reads

$$\frac{\partial n}{\partial t} = \delta_{td} n + \delta_{ts} n = \sum_i \frac{\partial}{\partial x_i}(v_i n) + \frac{1}{2} \sum_{ij} \frac{\partial^2}{\partial x_i \partial x_j}(c_{ij} n) \tag{B.4}$$

where $C = [c_{ij}]_{i,j=1,..k}$ denotes the, necessarily positive semi-definite, matrix of infinitesimal (co)variances. And the k-dimensional analog of (B.3) is

$$\frac{\partial n}{\partial t} = -\nabla \cdot \phi, \quad \phi = (u - \tfrac{1}{2} C \nabla) n \tag{B.5}$$

with

$$u_i = v_i - \frac{1}{2} \sum_j \frac{\partial}{\partial x_j} c_{ij}$$

ϕ again allows the interpretation of flux vector, just as in (3.1.8).

REMARK B.3: (B.4) and (B.5) show that our facile argument preceding (3.1.1) is considerably more tricky then it might seem. It certainly is possible to calculate the local mass change as the sum of the mass changes due to the *fluxes* in the direction of the various i-state variables. These fluxes themselves may be coupled, however, as is apparent from the occurrence of infinitesimal covariances.

In general (B.4) has to be complemented with side conditions. We shall only consider the nonsingular case and assume that $\det C$ is bounded away from zero and that v is bounded. We shall again denote as $\partial_- \Omega$ that part of the boundary transgression of which is known from biological considerations to lead to immediate local loss. (Due to our assumption of nonsingularity there is no possibility of individuals being pushed over some part of the boundary as a consequence of the processes in the interior of Ω as in the, singular, deterministic case.) The boundary condition expressing such loss can be shown to be

$$n(x) = 0 \quad \text{for } x \in \partial_- \Omega . \tag{B.6}$$

The local rate of mass loss can be calculated in the usual manner from the flux vector ϕ. In the same manner we shall again denote that part of the boundary where we know from biological considerations that there is influx of p-mass as $\partial_+ \Omega$. Balancing the arrival rate b_0 per unit of surface area with the local flux then gives

$$\nu(x) \cdot \phi(x) = b_\partial(x) \quad \text{for } x \in \partial_+ \Omega . \tag{B.7}$$

Finally we shall denote as $\partial_0 \Omega$ that part of the boundary on which we know that there is neither mass loss nor any influx, implying that

$$\nu(x) \cdot \phi(x) = 0 \quad \text{for } x \in \partial_0 \Omega . \tag{B.8}$$

REMARK B.4: (B.4) allows a lot more room for singularities than only the case of a deterministic mass flow. A complete classification for the one-dimensional case may be found *e.g.* in Karlin & Taylor (1981) or Goel & Richter-Dyn (1979).

Continuous stochastic i-state movements have the property that they immediately smooth away any delta-functions present in the initial data. This also applies to delta-function valued source and sink terms. We only give the formulae for the one-dimensional case.

A sink term of the form $-\beta\delta(x-a)n(x)$ gives rise to the transition condition

$$n(a+) = n(a-), \quad \phi(a-)-\phi(a+) = \beta n(a), \tag{B.9}$$

and a source term of the form $c\delta(x-a)$ to

$$n(a+) = n(a-), \quad \phi(a+)-\phi(a-) = c\,. \tag{B.10}$$

(For mathematicians: In this case it is fairly easy to give a distributional interpretation to a p-equation containing a delta-function valued sink term. Such in contrast to the case $\delta_{ts}n = 0$.)

EXAMPLE B.5: *Deterministic binary fission combined with stochastic individual growth.* The p-equation for this example is (remember x_1 was the value of x at which the cells divided and crossing x_0 resulted in death)

$$\begin{aligned}
&\frac{\partial n}{\partial t} = -\frac{\partial}{\partial x}(vn) + \frac{1}{2}\frac{\partial^2}{\partial x^2}(\sigma^2 n) \\
&n(x_1) = n(x_0) = 0 \\
&n(\tfrac{1}{2}x_1+) = n(\tfrac{1}{2}x_1-),\ \phi(\tfrac{1}{2}x_1+)-\phi(\tfrac{1}{2}x_1-) = 2\phi(x_1-)
\end{aligned} \tag{B.11}$$

with

$$\phi = vn - \frac{1}{2}\frac{\partial}{\partial x}(\sigma^2 n)$$

C. The p-equations for the examples from section 1

The invertebrate functional response

The p-equations for this example have already been given as (4.1.5). If there is a maximum gut capacity $\bar{s}<s_m$ then Ω has to be truncated at $\bar{s}$ and (4.1.5d) has to be replaced by

$$-f_0(\bar{s})n_0(t,\bar{s}) = \int_0^{\bar{\tau}} f_1(\bar{s})n_1(t,\bar{s},\tau)d\tau\,, \tag{4.1.5d'}$$

where

$$\bar{\tau} = \tau_e - a^{-1}\log\left[\frac{ac-u}{a\bar{s}-u}\right].$$

Size dependent reproduction in ectothermic animals

The p-equation for this example has already been given as (4.3.2). In laboratory experiments where we feed with food that does not reproduce we may set

$$\frac{dx}{dt} = \alpha - \beta x - \frac{\nu\xi x}{1+\xi x}\,p \tag{C.1}$$

with

$$p = \int_0^{a_{\max}} l^2(a)n(a)da\,. \tag{C.2}$$

A time scale argument based on the assumption that $\alpha,\beta,\nu>>1$ then gives

$$x \approx \frac{\alpha\xi-\beta-\nu\xi p + \sqrt{(\alpha\xi-\beta-\nu\xi p)^2+4\alpha\beta\xi}}{2\beta\xi}\,. \tag{C.3}$$

Reproduction by binary fission

The p--equation for this example is

$$\frac{\partial n}{\partial t} = -\frac{\partial n}{\partial a} - \frac{\partial Vn}{\partial x} - (\mu + b)n \tag{C.4}$$

$$n(0,x) = 4 \int_{a_0}^{\infty} b(a,2x)n(a,2x)da$$

If division occurs as soon as the size passes a stochastic threshold having a density g, a survivor function $\mathcal{G}$, $\mathcal{G}(x) = \int_x^{\infty} g(\xi)d\xi$, and a hazard rate γ, $\gamma = g / \mathcal{G}$, determined at birth, except that a cell needs to have passed the minimal age a_0 before it can divide, then the p-equation becomes

$$\frac{\partial n}{\partial t} = -\frac{\partial n}{\partial a} - \frac{\partial Vn}{\partial x} - (\mu + V\gamma)n$$

$$n(a_0+,x) = \mathcal{G}(x)n(a_0-,x) \tag{C.5}$$

$$n(0,x) = 4[(1-\mathcal{G}(x))n(a_0-,x) + \int_{a_0}^{\infty} V(2x)\gamma(2x)n(a,2x)da]$$

Colony size distribution in the diatom Asterionella

If we assume that bonds break exactly at age A the p-equations become

$$\frac{\partial n_k}{\partial t} = -\sum_{i=1}^{k} \frac{\partial n_k}{\partial a_i} - d(a_1)n_k + 2n_{k+1}(a_1,..,a_k,A), \quad k = 1,2, \cdots$$

$$\frac{\partial n_0}{\partial t} = -\frac{\partial n_0}{\partial a_0} - d(a_0)n_0$$

$$n_k(0,a_2,..,a_k) = d(a_2)n_{k-1}(a_2,..,a_k), \quad k = 2,3,.. \tag{C.6}$$

$$n_1(0) = \int_A^{\infty} d(a_0)n_0(a_0)da_0$$

$$n_0(A) = 2n_1(A)$$

for $0<a_1<..<a_k<A<a_0$, and n_k the bond-age density of colonies of size 2^k.

If bond breakage follows the rule from exercise (3.2.5), the p-equations are

$$\frac{\partial n_k}{\partial t} = -\sum_{i=1}^{k} \frac{\partial n_k}{\partial a_i} - d(a_1)n_k - b(a_k)n_k$$

$$+ \sum_{j=k+1}^{\infty} 2^{j-k} \int_{a_k}^{\infty} .. \int_{a_{j-1}}^{\infty} [b(a_{k+1}) - b(a_k)]n_j(a_1,..a_j)da_j..da_{k+1}, \quad k = 1,2, \cdots$$

$$\frac{\partial n_0}{\partial t} = -\frac{\partial n_0}{\partial a_0} - d(a_0)n_0 + \sum_{j=1}^{\infty} 2^j \int_{a_0}^{\infty} .. \int_{a_{j-1}}^{\infty} b(a_0)n_j(a_0,a_2,..,a_j)da_j..da_2 \tag{C.7}$$

$$n_k(0,a_2,..a_k) = d(a_2)\, n_{k-1}(a_2,..,a_k), \quad k = 2,3, \cdots$$

$$n_1(0) = \int_0^{\infty} d(a_0)n_0(a_0)da_0$$

$$n_0(0) = 0\,.$$

The case of independently breaking bonds has already been treated in example 5.4.2.

The PPP-problem

Figure C.1 depicts Ω. The upper right hand boundary of Ω is given by

$$\frac{\delta}{\alpha}[(x-1) + \frac{\beta}{\gamma}(y-1)] = r_0 \tag{C.8}$$

The p-equations are

$$\frac{\partial n_0}{\partial t} = -\zeta P n_0 + B \tag{C.9a}$$

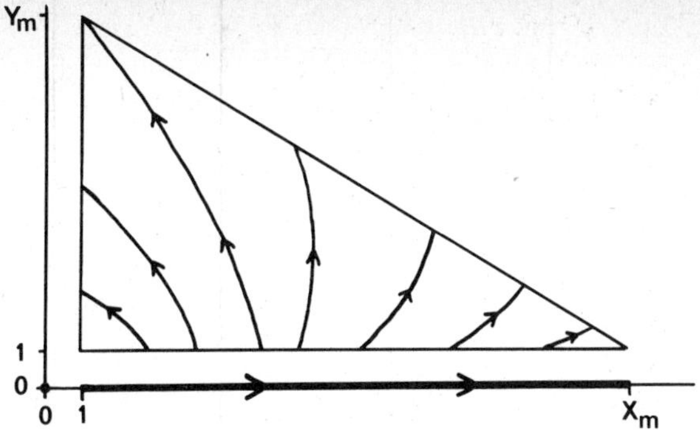

Fig. C.1. The i-state space of the PPP-problem

$$\frac{\partial n_1}{\partial t} = -\frac{\partial \alpha x n_1}{\partial x} - \eta Q n_1 \tag{C.9b}$$

$$\frac{\partial n_2}{\partial t} = -\frac{\partial(\alpha x - \beta y)n_2}{\partial x} - \frac{\partial \gamma y n_2}{\partial y} \tag{C.9c}$$

$$\alpha n_1(1) = \zeta P n_0 \tag{C.9d}$$

$$\gamma n_2(x,1) = \eta(x) Q n_1(x) \tag{C.9e}$$

The production rate of empty patches may be taken to be a function of the p-state as a whole. The simplest possibility is to set $B = f(n_0)$. P and Q have to be calculated from equations for the "air plancton". These have been discussed in example 6.2.9. The quantities R and S appearing there, corresponding to the production rates of "free" prey and predators respectively, can be calculated as

$$R = \alpha x_m^2\, n_1(x_m) + \int_1^{x_m} \frac{\alpha\gamma}{\beta} x^2\, n_2\left(x, \frac{\gamma}{\beta}\left(\frac{\alpha}{\delta} r_0 + 1 - x\right) + 1\right) dx \tag{C.10}$$

$$S = \int_1^{y_m} \alpha y n_2(1,y) dy + \int_1^{x_m} \frac{\alpha\gamma}{\beta} x\left(\frac{\gamma}{\beta}\left(\frac{\alpha}{\delta} r_0 + 1 - x\right) + 1\right) n_2\left(x, \frac{\gamma}{\delta}\left(\frac{\alpha}{\delta} r_0 + 1 - x\right) + 1\right) dx$$

If r_0 is not a fixed number but a random variable with hazard rate κ (*i.e.* $\kappa(r) = -d\log\mathcal{H}(r)/dr$ where $\mathcal{H}(r) = P(\{r_0 > r\})$) then (C.9b) and (C.9c) have to be replaced by

$$\frac{\partial n_1}{\partial t} = -\frac{\partial \alpha x n_1}{\partial x} - \eta Q n_1 - \delta x \kappa\left(\frac{\delta}{\alpha}(x-1)\right) n_1 \tag{C.9b'}$$

$$\frac{\partial n_2}{\partial t} = -\frac{\partial(\alpha x - \beta y)n_2}{\partial x} - \frac{\partial \gamma y n_2}{\partial y} - \delta x \kappa\left(\frac{\delta}{\alpha}((x-1) + \frac{\beta}{\gamma}(y-1)\right) n_2 \tag{C.9c'}$$

and (C.10) by

$$R = \int_1^{\infty} \delta x \kappa\left(\frac{\delta}{\alpha}(x-1)\right) n_1(x) dx + \int_1^{\infty}\int_1^{\infty} \delta x^2 \kappa\left(\frac{\delta}{\alpha}((x-1) + \frac{\beta}{\gamma}(y-1))\right) n_2(x,y) dx dy$$

$$S = \int_1^{y_m} \alpha y n_2(1,y) dy + \int_1^{\infty}\int_1^{\infty} \delta x y \kappa\left(\frac{\delta}{\alpha}((x-1) + \frac{\beta}{\gamma}(y-1))\right) n_2(x,y) dx dy \tag{C.10'}$$

Deterministic binary fission combined with stochastic individual growth

The p-equations for this example were already derived in example B.5, formula (B.11).

IV Age Dependence

J.A.J. Metz & O. Diekmann

1. Age as a substitute for comprehension

1.1. Why this special attention

On the whole the theory of structured population models is still *in statu nascendi.* We have a firm idea where the linear theory is heading, but a great deal of work remains to be done to get even a semblance of completeness, and our present understanding of nonlinear problems is scanty at best (but developing rapidly!). However, there is one specific area that is already well past puberty: that of purely age dependent problems. The deterministic linear theory, based on the so-called linear renewal (or Lotka's integral) equation, started with SHARPE'S and LOTKA'S work in 1911, still forms the backbone of human demography (KEYFITZ, 1968; COALE, 1972; POLLARD, 1973; a history of the early stages of the theory may be found in SAMUELSON, 1976), a corresponding stochastic theory of age dependent branching processes is nearing completion (JAGERS, 1975), and a great number of papers and one whole book (WEBB, 1985a) deal with nonlinear extensions.

It is certainly not our intention to add to this burgeoning specialist literature here especially since age is a variable with only minimal physiological connotations (see below and/or the discussion in III.2.2). However, as it turns out that often the easiest way to analyse a structured population model, at least in the linear case, is by transforming it into an equivalent renewal equation, we feel obliged to say at least a minimal amount about this technical device. In addition this gives us the opportunity to discuss in a concrete case how a special biological feature like "all individuals are born equal" can be exploited to arrive at clearcut results.

In this section we shall discuss the characterization of problems that allow an age type representation, either derived directly from observable *i*-behaviour or from some other more complicated model, as well as the related topic of using integral equations (as opposed to partial differential equations) as a modelling tool. The first section by necessity is fairly philosophical. The remaining sections will be more technical. In section 2 we shall introduce the basic linear theory centering around the so-called renewal theorem. In section 3 we shall consider some applications of that theorem in more complicated situations, and in section 4 we shall discuss some non-linear extensions. In these sections we shall be frankly parochial in that we concentrate mainly on examples from our own work. Section 5, finally, and also subsection 2.3.3 will be devoted to a totally different topic, the reduction of age dependent models to *p*-behaviourally equivalent differential equations in $\mathbb{R}^k$, $k<\infty$. Comparable techniques apply to other types of structured population models, but in the age dependent case we have by now a particularly good idea about their extent.

1.2. Which problems allow an age representation?

Two special biological properties set problems allowing an age representation apart from any other population problems. These are

(1) the population dynamical behaviour of any individual is in no way related directly to that of its parent(s) or the state of the environment immediately preceding its birth

(2) the average population dynamical properties of an individual, like mean rate of giving birth or probability of dying, do not depend on which circumstances it has experienced in its past. (NB There is no objection against a dependence of its birth and death rates on the current environment but the effect should be instantaneous.)

REMARK 1.2.1: We are referring here to what might be called generalized or long term population dynamically equivalent representations of the age type. This in contrast to strict *i*-state space representations which are supposed to

represent faithfully the actual *i*-behaviour in all its fine probabilistic detail. The difference is best explained by an example.
If a female human has just given birth we know that it will take her at least nine months before she can do so again. Therefore the information whether she is in the nine month *post partem* window or not allows us a better prediction of her future probabilities of giving birth. By the same argument the population dynamical future of a population artificially started with women who just gave birth should differ from that of a population with the same age distributions but in which none of the women are in the *post partem* window. However, under constant circumstances at least, the fraction of women in the *post partem* window in any cohort born into the population will just be a fixed function of the age of that cohort and so will be the *per capita* contribution of that cohort to the population birth rate. This *per capita* cohort birth rate should be calculated by taking the average over (the probability measure on the space of) all possible individual parturition histories.
An age representation will be strict if and only if in addition to conditions (1) and (2) we have that: (3) the individual birth process has so-called independent increments (which in the simplest case when multiple births are excluded means that it is a Poisson process) and the event of death is independent of the birth events, except for the possibility of a burst of births coinciding with the death event.

Properties (1) and (2) have counterparts referring to possible i-state representations of our individuals:

(1′) all individuals start identical in the sense that the probability distribution of their state at birth is identical, and

(2′) once born an individual's state process unfolds totally independent of its environment.

Here (2′) should be taken to imply that the differences between death rates in states that are attainable at the same age should remain constant at all time, for otherwise the environmental history may exert an influence on the relative likeliness of the various alternative *i*-state histories through selective deaths (see exercise 1.4.6 for an example). It can easily be seen that properties (1′) and (2′) imply conditions (1) and (2) respectively. Strictly speaking we cannot assert the reverse implication. However, as indicated in the next remark the counter examples are necessarily rather artificial.

REMARK 1.2.2: A sufficient condition for the reverse implication to hold true is that (a) the age representation is strict (see remark 1.2.1), and (b) we restrict the attention to minimal *i*-state representations (see III.2.2). Condition (a) in essence excludes the possibility that the environmental history experienced by an individual influences the fine detail of its actual birth process but not its average birth rate, and (b) in essence excludes the possibility that there exist components of the *i*-state vector which are for example related to the state of an individual's parents but which do not influence behaviour, as well as the presence of potential environmental influences on *i*-state behaviour in non-reachable parts of the *i*-state space.

If the physical environment is constant and there also is no direct interaction between individuals, condition (2) is satisfied by default. So for constant environmental conditions a great deal more problems allow an age representation than when conditions vary. (The *Daphnia* model from I.3 provides an example.)

REMARK 1.2.3: In the biological literature one frequently encounters a generalization of the age concept called physiological age. The most general and not very useful interpretation is that (i) an individual's average population dynamical properties can be represented fully by a one dimensional state process where the state variable can never decrease, and (ii) the state variable at birth can be set equal to zero (Note that when (i) is fulfilled (ii) is just equivalent to our earlier condition (1).) In this sense the *Daphnia* model from I.3 allows a physiological age representation.
In a more restricted interpretation the speed at which physiological age changes should be independent of physiological age itself. If we have some other *i*-state representation, say based on physiological insights, then for a physiological age representation to be possible it is sufficient that (1′) holds and that:

(2″) the only external influence allowed on the development of the *i*-state is through a multiplicative action on the speed of the *i*-state process as a whole.

(When the physiological age representation is strict and the other *i*-state representation is minimal (compare remark 1.2.2), these two conditions are also necessary.) If this multiplicative factor is called v, then we can introduce the function

$$p(t) := \int_0^t v(\tau)d\tau \tag{1.2.1}$$

called physiological time, and calculate the present physiological age a_p of an individual born at t_b as

$$a_p = p(t) - p(t_b) \tag{1.2.2}$$

(Compare the papers by Gurney, Nisbet & Blythe and by Goudriaan in part B.)

1.3. Integral equations as a natural modelling tool

For constant environmental conditions the observation that all individuals are born alike allows the derivation from first principles of an integral equation for the (total or population) birth rate b, alternatively called Lotka's equation by demographers and the renewal equation by mathematicians. Let $\phi(a)$ denote the mean number of offspring that an individual will beget per unit of time, a time units after it is born (dead individuals being assumed to have zero birth rate), and $g(t)$ the number of births per unit of time into the population which are not daughters of individuals born after $t=0$ (*i.e.* direct offspring from the founder population or possibly births from outside sources). Then the overall birth rate b can be written as the sum of g and the cumulation of the contributions from all individuals born after time zero:

$$b(t) = \int_0^t b(t-a)\phi(a)da + g(t). \tag{1.3.1}$$

The functions ϕ and g are known as the birth kernel and the forcing function respectively. Equation (1.3.1) has been the object of detailed mathematical study for a long time. Section 2 will give a rough summary of the main results. Subsection 1.4 will give some examples of the calculation of the birth kernel ϕ and the forcing function g.

Given the birth function b it is possible to calculate any linear functional (*i.e.* weighted cumulation) of the population state, like total number of individuals alive, or total biomass, as a function of time. For example, if $\mathcal{F}(a)$ denotes the probability that an individual survives at least to age a then the total population size N at time t equals

$$N(t) = \int_0^t b(t-a)\mathcal{F}(a)da + M(t), \tag{1.3.2}$$

where $M(t)$ denotes the number of survivors from the founder population.

If the environmental conditions vary dependent on the population development, then in the most general case there probably is no good alternative to writing down a full partial differential equation for the age distribution, along the lines laid out in chapters I and III. However, in many specific cases, for example when the total birth output is modified through mutual interaction of the neonates or by interaction with some specific segment of the population the size of which can be calculated as a linear functional of the birth history, it is still possible to derive a, now essentially non-linear, integral equation akin to equation (1.3.1), either directly or through the indirect route of first writing down a partial differential equation and then integrating it along the characteristics.

EXAMPLE 1.3.1: In III.6.2, example 6.2.4, we derived the following equation for the relation between the population birth rate b and recruitment h resulting from nursery competition

$$h = be^{-F(m)} \tag{1.3.3}$$

$$m = (b\,/\,F(m))(1-e^{-F(m)}) \tag{1.3.4}$$

where F is a function which tells how death rate in the nursery depends on crowding m, *i.e.* the number of competitors multiplied with their competitive strength. Therefore, if there are no other density dependent effects,

$$b(t) = \int_0^t \phi(a)h(b(t-a))da + g(t) \tag{1.3.5}$$

where $h(b)$ is defined by (1.3.3) with m the solution of (1.3.4). The dynamic properties of (1.3.3) - (1.3.5) will be examined in VI.3.

EXERCISE 1.3.2: Derive (1.3.5) by integrating the partial differential equation part of (III.6.2.9) along the characteristics.

The following example shows how equations like (1.3.1) and (1.3.5) can be derived direct from first principles in a slightly more rigorous manner. The main interest of this derivation is that the step of replacing detailed individual processes by just the average *pro capita* birth rate instead of being relegated to the verbal preparatory stages emerges as a consequence of the calculations.

Example 1.3.3: *Kermack's and McKendrick's general epidemic* In 1927 KERMACK and MCKENDRICK introduced a general epidemic model in which infectivity of an ill individual was assumed to depend in some general way on the time elapsed since it was infected. Further assumptions of the model are that all susceptibles are equally vulnerable to infection, and that the number of susceptibles can only alter due to them being infected, *i.e.* there are no new susceptibles born to the population, susceptibles do not die (ill individuals may) and there is no return to susceptibility at any time after an individual has been infected. Apart from that only the usual law of mass action assumption is made. Below we shall derive their basic equation from the most general assumptions along the lines laid out in METZ (1978).

As the course of illness v of a particular individual is assumed not to depend on either the way it got infected or on the circumstances the individual experiences after its infection we may proceed as if each individual is just labeled by the course its illness will take if ever it gets infected. If we wish we may think of any variability between individual courses of illness as being caused by genuine differences between the individuals themselves (provided those differences are in no way related to an individual's proneness to contracting an infection!) Let V denote the set of possible courses of individual illnesses. Then we can conceive our population of infected individuals, alive, recovered or dead, as a frequency distribution over this set. As V will in general be a very complicated set there is no direct way to represent our population of infecteds as a density over V. Therefore we shall represent it as a measure, to be denoted as m, *i.e.* if U denotes some sufficiently well behaved subset of V then $m(t)$ attributes to U just the number $m(t,U)$ (or rather the number per unit of area, but we shall omit this qualification from now on) of individuals infected up to time t whose courses of illness (are destined to) lie in U. Let $a(v,\tau)$ denote the infectivity contributed by an individual whose course of illness is v and who was infected τ time units ago (we assume that our parametrization is sufficiently detailed that an individual's infectivity depends deterministically on v and τ). Then the total infectivity y at time t equals

$$y(t) = \int_0^t \int_V a(v,\tau)\dot{m}(t-\tau,\{dv\})d\tau + g(t) \tag{1.3.6}$$

where $\dot{m} = \dfrac{dm}{dt}$, $m(t,\{dv\})$ denotes the number of infecteds up to time t whose course of illness lie in an infinitesimal set $\{dv\}$ around the point v, and g denotes the infectivity due to individuals already ill at $t = 0$ or derived from outside sources. If s is the density of susceptibles then the law of mass action gives

$$\dot{s} = -ys\,. \tag{1.3.7}$$

Finally our assumption that there is no relation between an individual organism's susceptibility to infection and the course of its illness if ever it gets infected gives us

$$\dot{m}(U) = -\dot{s}\mu(U) \tag{1.3.8}$$

where μ is the probability that an as yet uninfected organism's course of illness is destined to lie in U. Substituting (1.3.8) in (1.3.6) we get

$$y(t) = -\int_0^t \dot{s}(t-\tau) \int_V \mu(\{dv\})a(v,\tau)d\tau + g(t) \tag{1.3.9}$$

or

$$y(t) = -\int_0^t \dot{s}(t-\tau)\phi(\tau)d\tau + g(t) \tag{1.3.10a}$$

with

$$\phi(\tau) = \int_V a(v,\tau)\mu(\{dv\})\,. \tag{1.3.11}$$

Recalling that an average is nothing but the integral of a random variable over the corresponding probability measure, we see that ϕ is just the average infectivity of an individual who has been infected τ time units ago. And recalling (1.3.7) we see that our epidemic process can be modelled by (1.3.10a) together with

$$\dot{s}(t) = -y(t)s = -y(t)(\int_0^t \dot{s}(t-\tau)d\tau + s_0)\,. \tag{1.3.10b}$$

(1.3.10) is again a non-linear integral equation, this time relating the value of $\dot{s}$ at t to the behaviour of $\dot{s}$ during the interval $(0,t)$. In section 4 we shall deal in depth with the wealth of results that this equation permits one to derive. A related model in which infected individuals eventually return to the susceptibility class is treated in VI.2.

REMARK 1.3.4: Kermack's and McKendrick's paper is the classic of the mathematical theory of epidemics. Unfortunately it appears to be read considerably less frequently than it is cited. Apparently the idea got around that the paper only deals with an exceedingly special case which allows a simple ordinary differential equation representation in $\mathbb{R}^2$, of the kind discussed in 5.2. This is probably due to the following unfortunate combination of causes: a) the general equations proposed by Kermack & McKendrick were not in the mainstream of applied mathematics for some time to come, b) in his interesting and influential 1956 paper dealing with stochastic extensions of the simple model Kendall somehow gave the impression that this simple model essentially was what the Kermack & McKendrick paper was about, c) the original paper is not overly easy to obtain (this has recently been redressed by the reprint in OLIVEIRO PINTO & CONOLLY (1981)). As a result up to the present day papers get published purporting to analyse extensions of the Kermack & McKendrick model, whereas in fact they contain weaker results than those in the original paper. Therefore we would like to end with a commercial: Give Kermack's and McKendrick's paper a thorough try. It makes very rewarding reading!

The integral equations that result from our population models are always of the so-called Volterra type, *i.e.* the argument t of the unknown function b at the left hand side also delimits the integral occurring at the right. In our opinion there are two reasons for stressing the (re)formulation of models in terms of such integral equations. The first is entirely pragmatical: there are at present more results available in the mathematical literature for such integral equations than for the functional partial differential equations advocated in the previous chapters. The second reason is more of a philosophical nature. By writing down the integral equation direct from first principles, as was done in the last example, we circumvent the need to specify a detailed i-state space model, involving all sorts of possibly entirely unnecessary assumptions. Instead we concentrate on the minimal biological assumptions (1) and (2) from the previous subsection, thereby highlighting the range of applicability of our final results.

REMARK 1.3.5: The conditions allowing a population model to be replaced by an equivalent renewal equation are less restrictive than those allowing the construction of an age representation. In a constant environment the only condition is that individuals should pass through a stage in which they are all equal somewhere prior to giving birth. (For an age representation to apply this stage should occur immediately after birth.) For example, in the cell model from I.4 with $a \geqslant \frac{1}{2}$ (a was the size at which division could first occur) such a stage can be found in the reaching of size $\frac{1}{2}$, or any other fixed size between $\frac{1}{2}$ and a (compare II.10). The recruitment to the equalizing stage now replaces the birth rate function in equation (1.3.1).

1.4. The calculation of some birth kernels

In this subsection we shall illustrate the calculation of the birth kernel ϕ by means of some examples.

EXAMPLE 1.4.1: *Strictly age dependent reproduction and death* Assume that, provided an individual has not died yet, it gives birth at random, with birth events occurring in a Poisson process with rate $\lambda(a)$, where a denotes age, and that death occurs independent of an individual's history of giving birth, with a survivor function $\mathcal{F}(a)$. In that case the birth kernel can be written as

$$\phi(a) = \mathcal{F}(a)\,\lambda(a)\,. \tag{1.4.1}$$

The age distribution in this case also can be considered a p-state in the strict sense. The instantaneous death rate in a partial differential equation for the age distribution equals

$$\mu(a) = -\,d\log\mathcal{F}(a)\,/\,da\,. \tag{1.4.2}$$

EXERCISE 1.4.2: Write down both the integral and the partial differential equation versions of this model assuming constant environments, and transform the latter into the former by integration along the characteristics. In passing you will also have derived a formula for the forcing function g in equation (1.3.1). How would you approach the problem of constructing the reverse transformation?

EXAMPLE 1.4.3: *Finite i-state spaces.* Assume that the i-state space is finite, and represented as $\Omega = \{1,..,k\}$. Condition (2′) from 1.2 and the definition of the state concept together imply that the state process is a so-called continuous time Markov chain with killing. Such a process can be characterized by its constant transition rates $b_{ij}, i \neq j$, where b_{ij} is the probability per unit of time that the chain will jump to i whenever it happens to be in j, and the death rate μ_j which we shall assume to be constant as well. Let $b_{jj} := -(\sum_i b_{ij} + \mu_j)$, *i.e.* $-b_{jj}$ is the rate at which j is left, and

$B := [b_{ij}]_{i,j=1,..,k}$. B is called the differential generator of the chain. The definition of the state concept also implies that during any period the chain is in state j birth events should occur in a so-called process with time invariant independent increments. We shall assume the birth process to be independent of external influences implying that the mean number of births per unit of time produced by an individual in state j is a constant which we shall call a_j. We shall denote the column vector* $(a_1,..,a_k)^T$ as A. Finally condition (1′) from 1.2 compels us to assume that a neonate has a fixed probability c_j of being born in state j. We shall denote the column vector $(c_1,..,c_k)^T$ as C. Then

$$\phi(a) = A^T e^{Ba} C \tag{1.4.3}$$

and the probability that an individual survives to age a is

$$\mathcal{F}(a) = E^T e^{Ba} C \tag{1.4.4}$$

where $E=(1,..,1)^T$.

EXERCISE 1.4.4: Derive (1.4.3) by writing a differential equation for the population state vector N in the form

$$\frac{dN}{dt} = (B + CA^T)N = BN + (A^T N)C \tag{1.4.5}$$

(NB according to the rules of matrix multiplication $A^T N$ is a scalar, so that $A^T NC = CA^T N$, while CA^T is a matrix of rank one), solving this equation as if $b:=A^T N$ were known as a function of time, and substituting the solution in the defining formula for b. In passing you will have derived a formula for the forcing function g in equation (1.3.1).

EXERCISE 1.4.5: Calculate ϕ and $\mathcal{F}$ for the special model depicted in figure 1.4.1 and plot the results.

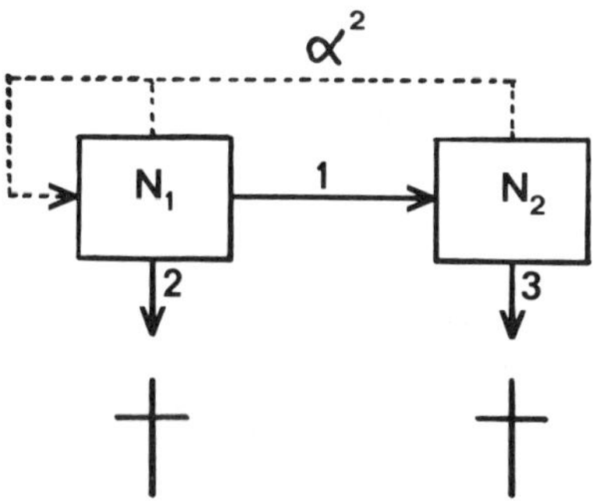

Fig. 1.4.1. Life cycle diagram corresponding to $B = \begin{pmatrix} -3 & 0 \\ 1 & -3 \end{pmatrix}$, $A^T=(0\ \ \alpha^2)$, $C=\binom{1}{0}$.

EXERCISE 1.4.6: Let the individual life cycle be given by the diagram of fig. 1.4.2 with α_1, α_2 and β constant. Assume that δ_1 and δ_2 each can take two values δ_i' and δ_i'' depending on the environment. First calculate the age dependent birth rate $\lambda(a) = \phi(a) / \mathcal{F}(a)$ for a constant environment. Show that $\lambda(a)$ is the same for both environments if (* and only if) $\delta_1' - \delta_1'' = \delta_2' - \delta_2''$. Also show that this condition on the death rates guarantees that the mean birth rate of a live individual is just a function of its age, no matter how the environment fluctuates. (Hint: Write down differential equations for the probabilities $p_j(t_0+a,\ a)$ that an individual born at t_0 is in state j at age a. Derive from this a differential equation for p_1+p_2. Finally show that $p_i / (p_1+p_2)$ satisfies a linear differential equation with constant coefficients.)

* ^{T} denotes transposition.

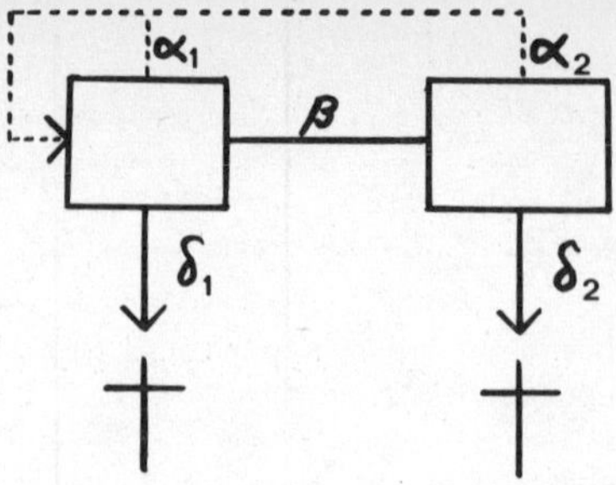

Fig. 1.4.2. Life cycle diagram corresponding to $B = \begin{bmatrix} -\beta+\delta_1 & 0 \\ \beta & \delta_2 \end{bmatrix}$, $A^T = (\alpha_1, \alpha_2)$ and $C = \binom{1}{0}$

REMARK 1.4.7: The main interest of example 1.4.3 is that the procedure outlined in exercise 1.4.4 can be reversed. Starting from a birth kernel like (1.4.3) it is always possible to derive an equivalent finite set of differential equations. This is especially helpful if we wish to do numerical studies on nonlinear problems. In section 5 we shall present this technique in more detail (and in section 2.3.3 we explore the connection with the representation theorems of linear systems theory).

EXAMPLE 1.4.8: *Deterministic binary fission combined with stochastic individual growth.* In III.1 and III.B we introduced a model in which an individual was assumed to grow according to a stochastic process of the diffusion type with size dependent infinitesimal mean and variance v and σ^2 respectively. Individuals were assumed to split into two equal parts on reaching size x_1, and to die on reaching size $x_0 < \frac{1}{2}x_1$. If f_1 denotes the (defective*) probability density of the time it takes to grow from size $\frac{1}{2}x_1$ to size x_1 and f_0 the (defective) probability density of the time it takes to reach size x_0 starting from $\frac{1}{2}x_1$, then

$$\phi(a) = 2f_1(a) \tag{1.4.6}$$

and

$$\mathcal{F}(a) = \int_a^\infty (f_0(\alpha) + f_1(\alpha))\, d\alpha\,. \tag{1.4.7}$$

(We assume here that there are no other causes of death. If the model also specifies a nonzero, possibly size dependent, death rate then we should add a third component in (1.4.7) denoting the (defective) probability density of the time till death from this cause.)

Calculating f_0 and f_1 is slightly complicated. In fact it is much easier to calculate their Laplace transforms

$$\bar{f}_i(s) := \int_0^\infty e^{-sa} f_i(a)da \tag{1..4.8}$$

As it turns out this is a bonus instead of a setback, as this Laplace transform is exactly the quantity we need in our calculations in section 2. We shall only give the procedure for $\bar{f}_1$.

Let $u_1(x,s)$ denote the Laplace transform of the (defective) probability density of the time an individual now sitting at x still needs to reach x_1, so that $\bar{f}_1(s) = u_1(\frac{1}{2}x_1, s)$. Then a well known result from the theory of diffusion processes (see e.g. COX & MILLER, 1965; GOEL & RICHTER-DYN, 1979; KARLIN & TAYLOR, 1981) tells us that u_1 solves the boundary value problem

$$\frac{1}{2}\sigma^2 \frac{d^2u_1}{dx^2} + v\,\frac{du_1}{dx} = su_1 \tag{1.4.9}$$

$$u_1(x_0,s) = 0, \quad u_1(x_1,s) = 1$$

In general we cannot solve (1.4.9) explicitly except for some special choices of v and σ^2. In 2.A we shall indicate how we can nevertheless use (1.4.9) to obtain useful bits of information about $\bar{f}_1$.

---

* A probability density f is called defective if $\int f(t)dt < 1$.

EXERCISE 1.4.9: Indicate how for a given initial size distribution we might calculate the Laplace transform of the forcing function g in equation 1.3.1.

2. Linear theory

We shall assume throughout this section that the physical environment is constant and that individuals interact neither directly nor indirectly. All the complexities of the (stochastic) i-behaviour then can be summarized by an age dependent mean birth rate $\lambda(a)$ and survival probability $\mathcal{F}(a)$, or even more concisely by the birth kernel $\phi(a)=\lambda(a)\mathcal{F}(a)$. We start with equation (1.3.1). In subsection 2.3 we shall also consider for the first time a p-equation for the age distribution.

2.1. An explicit expression for the population birth rate

To solve (1.3.1) we again use the biological interpretation. But first we introduce some notation to simplify our chores.

Given two locally integrable functions g_1 and g_2 on $\mathbb{R}^+$ we define their *convolution,* written as $g_1 * g_2$ to be the function*

$$(g_1 * g_2)(t): = \int_0^t g_1(t-\tau)g_2(\tau)d\tau\,. \tag{2.1.1}$$

The following relations hold (see also remark 2.1.4 for the precise interpretation of the equal sign)

$$g*(\lambda g_1 + \mu g_2) = \lambda g*g_1 + \mu g*g_2\,, \tag{2.1.2}$$

$$g_1*g_2 = g_2*g_1\,, \tag{2.1.3}$$

$$\delta * g = g\,, \tag{2.1.4}$$

where $\lambda,\mu\in\mathbb{R}$ and δ is the delta "function" introduced in III.5 (if you have skipped III.5 you may consider (2.1.4) as just the definition of δ). That is, functions under convolution behave pretty much like ordinary numbers under multiplication, with δ playing the role of 1. This is brought out even more sharply if we look at their Laplace transforms

$$\bar{g}(s) := \int_0^\infty e^{-st}\, g(t)dt\,, \tag{2.1.5}$$

for

$$\overline{g_1 * g_2}(s) = \bar{g}_1(s)\,\bar{g}_2(s)\,. \tag{2.1.6}$$

(Some general information about the Laplace transform for non-cognoscenti may be found in the appendix to this chapter.)

EXERCISE 2.1: Verify (2.1.2) to (2.1.4) and also (2.1.6). If necessary assume that g is continuous.

Using (2.1.1) we can write (1.3.1) as

$$b = \phi * b + g = b * \phi + g\,, \tag{2.1.7}$$

where, of course, we make the biologically natural assumption that both ϕ and g are nonnegative as well as locally integrable.

Now imagine an individual born at $t = 0$. By the same argument by which we derived (1.3.1) we can calculate the (mean) rates at which the various generations of its descendants are born as

$$\begin{array}{ll} 1^{st} \text{ generation births:} & \phi^{*1}: = \phi \\ 2^{nd} \text{ generation births:} & \phi^{*2}: = \phi * \phi^{*1} \\ \vdots & \vdots \\ n^{th} \text{ generation births:} & \phi^{*n}: = \phi * \phi^{*(n-1)} \end{array} \tag{2.1.8}$$

* The Fubini Theorem implies that $g_1 * g_2$ is a well-defined locally integrable function; see RUDIN (1974) 7.13.

Therefore the total (mean) birth rate of its clan as a function of time equals

$$\Phi := \sum_{n=1}^{\infty} \phi^{*n} \tag{2.1.9}$$

An elegant proof of the convergence of this sum may be found in JAGERS (1975) p. 105. Also see DOETSCH (1956) III.25. In the literature on Volterra integral equations *(e.g.* MILLER, 1971) Φ is known as the resolvent of (2.1.7), and the equation $\Phi = \phi\star\Phi + \phi$ is called the resolvent equation.

Remembering that g was the birth rate from individuals already present before $t=0$ and from the outside world, we infer that

$$b = g + \Phi\star g\,, \tag{2.1.10}$$

i.e. the total birth rate equals the starter birth rate g plus the cumulation of all the birth rates from clans started by it. (2.1.10) is an example of a so-called generation expansion, already referred to in I.5 (another example may be found in II.3).

EXERCISE 2.1.2: Verify that (2.1.10) satisfies (2.1.7)

EXERCISE 2.1.3: Simplify (2.1.10) by adapting the definition of Φ, using (2.1.4).

Since we derived (2.1.10) by the same reasoning by which we derived (2.1.7) anyway, there is not much sense in proving that (2.1.10) is the only solution: it is the only solution that matters. However, if we wish to use (2.1.7) as a basis for other calculations we have to make sure that the solution is unique. A somewhat formal proof goes as follows. Assume that b_1 and b_2 are two, possibly different, solutions. Then

$$b_1 - b_2 = (b_1 - b_2)\star\phi$$

and consequently, by induction,

$$b_1 - b_2 = (b_1 - b_2)\star\phi^{*n} \to 0$$

(for otherwise the sum in (2.1.9) would not converge), *i.e.* $b_1 = b_2$. This proof can be made rigorous in various ways (*i.e.*, under various assumptions on g).

(* Alternatively, let $\mathbf{1}$ denote the function which is identically one, then $\mathbf{1}$ $((b_1 - b_2) = \mathbf{1}\star\phi\star(b_1 - b_2)$ and consequently $(\mathbf{1} - \mathbf{1}\star\phi)\star(b_1 - b_2) = 0$. The Titchmarsh Theorem (Doetsch (1950) I.2.15 Satz 11 and 12) implies that necessarily either $b_1 - b_2 \equiv 0$ or $\mathbf{1} - \mathbf{1}\star\phi \equiv 0$. The latter is impossible since $\mathbf{1}\star\phi(t) \to 0$ for $t \downarrow 0$).

* REMARK 2.1.4: Some care is needed in interpreting the various equal signs. In general we only have equality almost everywhere. This is entirely sufficient for practical purposes. If g is continuous, then we can decide to restrict ourselves to continuous solutions only, in which case the equal signs in (2.1.7), (2.1.10) and that in $b_1 = b_2$ may be interpreted as referring to pointwise equality.
Formulae (2.1.9) and (2.1.10) provide an adequate means to calculate b for small t, simply by truncating (2.1.9) after a few terms. For large t the representation (2.1.9) has no practical value whatsoever. However, the large time behaviour is covered by a neat set of mathematical theorems reviewed in the next subsection. (But there is a price to pay: see subsection 2.3.)

2.2. Renewal theorems

We start with some heuristics. From previous experience we guess that b will eventually grow about exponentially. However, in contrast to the differential equation case dealt with in the previous chapters we cannot simply proceed by substituting a trial solution of the suspected form in (1.3.1) in order to find the possible growth rates supported by the dynamics, as in the integral equation case the representation of the dynamics comes in one package with the initial condition in the form of g. Apparently we have to get rid of the initial condition first. We can do this by considering an equation analogous to (1.3.1) but living on the whole time axis instead of just the positive half axis:

$$b(t) = \int_0^{\infty} b(t-a)\phi(a)da\,. \tag{2.2.1}$$

REMARK 2.2.1: We can derive (2.2.1) formally by substituting $b'(t) := b(t+t_0)/b(t_0)$ *(i.e.* we shift the time origin to

$-t_0$ and renormalize to $b'(0)=1$) in (1.3.1) and letting $t_0 \to \infty$. This indicates that information gleaned from (2.2.1) about the possible asymptotic behaviour of b can only be relevant in cases where $g(t)/b(t) \to 0$.

Substituting a trial solution of the form $b(t)=e^{rt}$ in (2.2.1) we find that r should satisfy

$$1 = \bar{\phi}(r) \tag{2.2.2a}$$

where

$$\bar{\phi}(s) := \int_0^\infty e^{-sa}\,\phi(a)\,da \tag{2.2.2b}$$

is the Laplace transform of the birth kernel ϕ.

Equation (2.2.2) is known as the *characteristic equation.* We can without loss of biological generality assume that ϕ increases at most exponentially, so that there exists a $\sigma \geqslant -\infty$ such that the integral in (2.2.2b) converges absolutely for $\mathrm{Re}(s) > \sigma$ and not for $\mathrm{Re}(s) < \sigma$. If we wish r to qualify as the asymptotic rate of exponential growth of our birth function b, then r should be real. So we concentrate on the behaviour of $\bar{\phi}$ for real s. Since $\phi(a) \geqslant 0$ the Laplace transform $\bar{\phi}$ is a monotonically decreasing smooth function of s with $\bar{\phi}(+\infty)=0$, and with $\lim_{s \downarrow \sigma} \bar{\phi}(s) = \theta \leqslant \infty$ (if $\sigma = -\infty$, which is the case when $\phi(a)$ goes to zero sufficiently fast, then necessarily $\theta = +\infty$). If $\theta < 1$ then (2.2.2) has no solution. Therefore we shall for the time being assume that $\theta \geqslant 1$. In that case (2.2.2) allows a unique solution. This solution r is called the *intrinsic rate of natural increase* associated with the birth kernel ϕ.

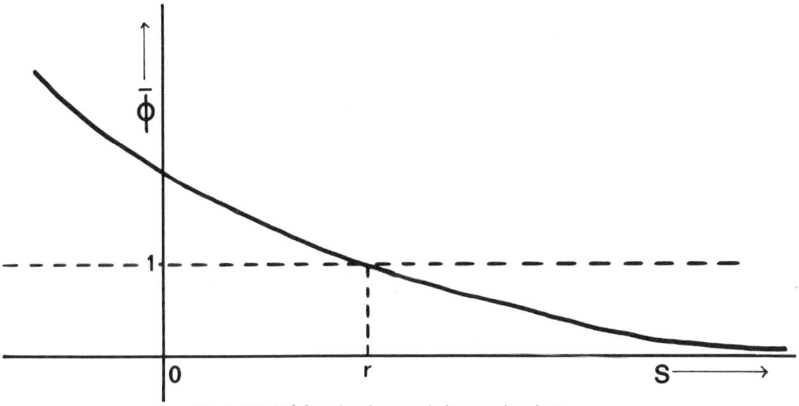

Fig. 2.2.1: Solving the characteristic equation for r

Some information about the whereabouts of r can be obtained from the observation that (see fig. 2.2.1)

$$\bar{\phi}(0) \gtrless 1 \Leftrightarrow r \gtrless 0 \tag{2.2.3a}$$

in combination with

$$\bar{\phi}(0) = R\,, \tag{2.2.3b}$$

where R is the so-called *net reproductive number, i.e.* the mean number of offspring a newborn individual is expected to beget during her lifetime. In general r has to be calculated numerically. Appendix 2A provides some convenient approximations.

EXERCISE 2.2.1: Check both (2.2.3a) and (2.2.3b).

EXERCISE 2.2.2: What does the characteristic equation from example 1.4.3 look like? What do you find for the characteristic equation of the linear differential equation (1.4.5)? Calculate r explicitly for the special case of exercise 1.4.5.

EXERCISE 2.2.3: Calculate R for example 1.4.3. Hint: Use $R = \bar{\phi}(0)$ and $\exp(Ba) = \frac{d}{da} B^{-1}\exp(Ba)$.

EXERCISE 2.2.4: Calculate R for example 1.4.8. Hint: Put $s=0$ in equation (1.4.9).

The statement that $b(t)$ will grow (or decrease) eventually exponentially at a relative rate r is the content of the so-called renewal theorem. This theorem comes in various disguises depending on what kind of assumptions one makes concerning ϕ and g. In a population context estimates for the integral of b are quite appropriate (indeed only the integral is a "number"), therefore we first present a theorem in this spirit and only thereafter give a variant which yields a much stronger conclusion, but at the expense of a much stronger assumption about g. But first of all we formulate the main technical tool as

LEMMA 2.2.5: *Assume that for some* $\delta>0$ *(i)* $\int_0^\infty e^{-(r-\delta)\tau}\,\phi(\tau)d\tau<\infty$ and *(ii)* $\bar{\phi}(s)\neq 1$ *for* $r-\delta\leq \mathrm{Re}\,s<r$, *then*

$$b(t) = \frac{e^{rt}\int_0^t e^{-r\tau}g(\tau)d\tau}{\int_0^\infty \tau e^{-r\tau}\phi(\tau)d\tau} + g(t) + \tilde{\Phi}*g(t) \tag{2.2.4}$$

where $\tilde{\Phi}$ *is such that* $\int_0^\infty e^{-(r-\delta)\tau}|\tilde{\Phi}(\tau)|d\tau < \infty$.

Note that the possible choices for δ in lemma 2.2.5 are restricted by two conditions. (i) refers essentially to the behaviour of ϕ at infinity; a sufficient condition for (i) is that $\phi(t)<ce^{(r-\delta-\epsilon)t}$ for all $t>T$, for some positive T and ϵ. (ii) refers to the position of any "other" roots of the characteristic equation in the complex plane; it can be shown that, for nonnegative ϕ, such roots always lie somewhat to the left of *a* the line $\mathrm{Re}\,s=r$, so there always exists a $\delta>0$ satisfying (ii). The next remark gives references for the proof of the lemma.

* REMARKS 2.2.6: (i) Lemma 2.2.5 is a generalization of a famous theorem of Paley and Wiener stating that the resolvent Φ is in L_1 whenever all roots of the characteristic equation lie in the left half plane (see *e.g.* MILLER, 1971). In lemma 2.2.5 the resolvent Φ is represented as $\Phi_0 + \tilde{\Phi}$, where $\Phi_0(t) = (\int_0^\infty \tau e^{-r\tau}\,\phi(\tau)d\tau)^{-1}e^{rt}$ corresponds to the pole of $\bar{\Phi}$ at r, and $\tilde{\Phi}$ lies in a weighted L_1 space. A proof that such a representation is possible may be found in JORDAN & WHEELER (1980a, b) or JORDAN, STAFFANS & WHEELER (1982).
(ii) In DIEKMANN & VAN GILS (1984) one can find indicated how lemma 2.2.5 fits in with the semi-group approaches sketched in subsection 2.3.2.
(iii) Of course one can generalize the representation (2.2.4) by calculating residues in additional poles of $\bar{\Phi}$.

The following two "renewal theorems" are direct consequences of lemma 2.2.5.

THEOREM 2.2.7: *If, in addition to the assumptions on* ϕ *from lemma 2.2.5,* $\int_0^\infty e^{-(r-\delta)\tau}g(\tau)d\tau<\infty$, *then*

$$b(t) = e^{rt}\frac{\int_0^\infty e^{-r\tau}g(\tau)d\tau}{\int_0^\infty \tau e^{-r\tau}\phi(\tau)d\tau} + f(t)\,, \tag{2.2.5}$$

where f *satisfies* $\int_0^\infty e^{-(r-\delta)\tau}\,|f(\tau)|\,d\tau<\infty$, *and consequently for any weighting function* h *with* $e^{-(r-\delta)a}h(a)$ *both bounded and integrable*

$$\int_0^t b(t-a)h(a)da = e^{rt}\frac{\int_0^\infty e^{-r\tau}g(\tau)d\tau\int_0^\infty e^{-ra}\,h(a)da}{\int_0^\infty \tau e^{-r\tau}\,\phi(\tau)d\tau} + O(e^{(r-\delta)t})$$

THEOREM 2.2.8: *If, in addition to the assumptions on* ϕ *from lemma 2.2.5* $e^{-(r-\delta)t}g(t)$ *is bounded, then*

$$b(t) = e^{rt}\frac{\int_0^\infty e^{-r\tau}g(\tau)d\tau}{\int_0^\infty \tau e^{-r\tau}\phi(\tau)d\tau} + O(e^{(r-\delta)t}) \tag{2.2.7}$$

REMARKS 2.2.9: (i) One can deal with systems of renewal equations in much the same manner as we did with one.
(ii) We have only used the nonnegativity of ϕ to guarantee the existence of a strictly dominant real root r. Appropriately modified forms of lemma 2.2.5 and theorems 2.2.7 and 2.2.8 apply when we dispense with the assumptions about the sign of ϕ and g.

In lemma 2.2.5 and theorems 2.2.7 and 2.2.8 we have concentrated on exponential estimates for the remainder term under the crucial hypothesis that for some positive δ the assumptions of lemma 2.2.5 hold. The following result which is essentially due to Feller and which with some right is referred to as *the* renewal theorem, makes no such hypothesis but has, as a consequence, a much weaker conclusion about the remainder term.

THEOREM 2.2.10: *Suppose that (i) g is continuous almost everywhere and (ii)* $\sum_{k=0}^{\infty} \sup_{0 \leq \tau < 1} e^{-rk} g(k+\tau) < \infty$ *then*

$$e^{-rt} b(t) \to \frac{\int_0^\infty e^{-r\tau} g(\tau) d\tau}{\int_0^\infty \tau e^{-r\tau} \phi(\tau) d\tau} \qquad \text{for } t \to \infty , \tag{2.2.8}$$

where the right hand side is interpreted as zero when the integral in the denominator diverges.

(A sufficient condition for (i) is that g is piecewise continuous and a sufficient condition for (ii) is that $e^{-rt} g(t) < f(t)$ with f decreasing and integrable.) A proof of this theorem may be found in JAGERS (1975).

REMARK 2.2.11: The renewal theorem essentially relates the asymptotic behaviour of the solution of the renewal equation (1.3.1) to the positive solutions of the "limiting equation" (2.2.1). It is this relationship which will be extended into the nonlinear realm in section 4.

We finish with two theorems which also apply when $\bar{\phi}(\sigma) < 1$. The first theorem provides some general insight in the behaviour of b when $R < 1$, the second one provides a general estimate for b. Proofs may be found respectively in JAGERS (1975) and HOPPENSTEADT (1975).

THEOREM 2.2.12: *If* $R < 1$ *and* $\lim_{t \to \infty} g(t) = \gamma$ *then*

$$\lim_{t \to \infty} b(t) = \gamma / (1 - R) \tag{2.2.9}$$

COROLLARY 2.2.13: *If* $R < 1$ *and* $g \to 0$ *then* $b \to 0$.

EXERCISE 2.2.14: Show that when $R < 1$

$$\int_0^\infty \Phi(\tau) d\tau = R / (1 - R) \tag{2.2.10}$$

and use this observation to interpret (2.2.9).

THEOREM 2.2.15: *For all* $\lambda \in \mathbb{R}$ *such that* $\bar{\phi}(\lambda) < 1$

$$e^{-\lambda t} b(t) \leq \frac{\sup_{0 \leq \tau < \infty} g(\tau) e^{-\lambda \tau}}{1 - \bar{\phi}(\lambda)} . \tag{2.2.11}$$

EXERCISE 2.2.16: Show that $\tilde{b}(t) := e^{-\lambda t} b(t)$ satisfies a renewal equation just like b, with $\tilde{\phi}(a) := e^{-\lambda a} \phi(a)$ and $\tilde{g}(t) := e^{-\lambda t} g(t)$.

* EXERCISE 2.2.17: Use the result of the previous exercise to prove theorem 2.2.15.

* EXERCISE 2.2.18: Derive the analogue of (2.2.11) for $\lambda \in \mathbb{C}$. (Hint: Take moduli where convenient.) Use this result to conclude that under the conditions of theorem 2.2.8 the Laplace transform of b exists for $\mathrm{Re}(s) > r$.

* EXERCISE 2.2.19: Use the complex inversion formula (A.13) together with a shift of contour as indicated in figure 2.2.2. to prove theorem 2.2.8.
Hint: Show that both $\overline{g}(s)$ and $\overline{\phi}(s) \to 0$ on the segments pp' and gg' for $\mathrm{Im}(p) = -\mathrm{Im}(q) \to \infty$ and that therefore $\overline{b}(s) \to 0$ on these segments. Therefore we can calculate b as the limit for $\mathrm{Im}(p) \to \infty$ of $\frac{1}{2\pi i} \int_q^p e^{st} \overline{b}(s) ds$ plus the sum of the residues of $\overline{b}(s) e^{st}$ within the contour. Show that the former term is $0(e^{(r-\delta)t})$.

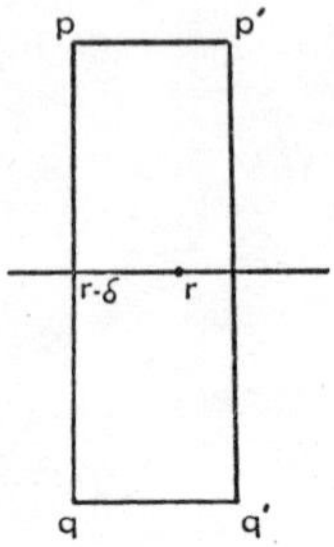

Fig. 2.2.2. Shift of contour used in exercise 2.2.19

2.3. *Semigroup approaches*

The renewal theorem provides an easy, cut and dried tool for most of our practical needs. The cost we pay for exclusively relying on it is that we lose contact with the semigroup and dynamical systems framework as expounded in chapter II and thereby with our easiest route towards nonlinear extensions, stability theory, bifurcation theory, *etc.*, as discussed in chapter VI. In this section we shall make up for this deficiency. In the spirit of chapter II we shall restrict ourselves throughout to the autonomous case, *i.e.* we shall assume that there are no births from outside sources. Moreover our approach will be almost entirely heuristic. Our aim is to sketch the various approaches, theories and results, and in particular their interrelationships. We shall not be very precise in the mathematical formulation, neither do we try to present the most general or sharpest results. More complete treatments may be found in WEBB (1984, 1985a) and in DIEKMANN (1980), VAN GILS (1984) and STAFFANS (1984).

2.3.1. *The age distribution*

The traditional biological way to define a dynamical system is in terms of the age density n, where* $n \in L_1(0,a_m)$, $a_m \leqslant \infty$, say. The semigroup transforming age distributions into each other will be called T_2. (The reason for the choice of index will become clear below.) Let the mean number of offspring produced per unit of time by an individual aged a be denoted as $\lambda(a)$, and the probability that individuals survive to age a as $\mathcal{F}(a)$. We shall assume that λ is (essentially) bounded. To calculate T_2 we observe that an animal aged a at $t=0$ has a probability $\mathcal{F}(a+t)/\mathcal{F}(a)$ to survive to time t and age $a+t$. Moreover, at time t an individual of age $a<t$ was born into the population at time $t-a$, and of the individuals born at that time only a fraction $\mathcal{F}(a)$ are still alive Therefore†

$$n(t,a) = \begin{cases} n(0,a-t)\mathcal{F}(a)/\mathcal{F}(a-t) & \text{for } a \geqslant t \\ b(t-a)\,\mathcal{F}(a) & \text{for } a<t \end{cases} \tag{2.3.1a}$$

where b is calculated from (2.1.10) with

---

* See chapter II section 1 or III.A.2 for a description of the function space L_1.

† We shall write $n(t,a)$ for $n(t)(a)$, the age density at t evaluated at a. Moreover, we shall often suppress either of the arguments depending on the context.

$$\phi(a) = \lambda(a)\mathcal{F}(a), \qquad g(t) = \int_0^{a_m} n(0,a)\,\lambda(a+t)\,\frac{\mathcal{F}(a+t)}{\mathcal{F}(a)}\,da\,. \tag{2.3.1b}$$

In the spirit of chapter II the asymptotic behaviour of T_2 should be deduced from the consideration of its differential generator A_2. For the full details of how this can be done you are referred to WEBB (1984, 1985a). We shall confine ourselves to deriving A_2 and calculating its dominant eigenvalue r.

To calculate A_2 we observe that $n(t+dt,\ a+dt) = n(t,a)\ \mathcal{F}(a+dt)/\mathcal{F}(a)$. Subtracting $n(t,\ a+dt)$ from both sides and writing $\mathcal{F}(a+dt)/\mathcal{F}(a) = 1-\mu(a)dt$ with $\mu = -d\log\mathcal{F}/da$, gives after division by dt

$$\frac{\partial n}{\partial t} = -\frac{\partial n}{\partial a} - \mu(a)n \tag{2.3.2}$$

and therefore

$$A_2 = -\frac{\partial}{\partial a} - \mu \tag{2.3.3a}$$

provided* $n \in AC$. Moreover $n(0)$ should equal b so the domain of A_2 is given by

$$\mathcal{D}(A_2) = \{n | n \in AC\ [0,a_m] \text{ and } n(0) = \int_0^{a_m} \lambda(a)n(a)da\} \tag{2.3.3b}$$

EXERCISE 2.3.1: Derive the explicit formulae for λ and μ from the model of fig. 1.4.1.

The first to derive the partial differential equation (2.3.2) was MCKENDRICK (1926). Later it was rediscovered by VON FOERSTER (1959) and many others. The explicit semigroup interpretation appeared in the literature only relatively recently in the work of WEBB (1979, 1981, 1983a,b, 1984, 1985a) and PRÜSS (1981, 1983a,b).

To calculate r we write $A_2\tilde{n} = r\tilde{n}$ to find

$$\tilde{n}(a) = \tilde{n}(0)e^{-ra}\ \mathcal{F}(a)\,. \tag{2.3.4}$$

The condition that $\tilde{n}$ should lie in $\mathcal{D}(A_2)$ then gives

$$1 = \int_0^{a_m} \lambda(a)e^{-ra}\ \mathcal{F}(a)da = \int_0^{\infty} e^{-ra}\phi(a)da \tag{2.3.5}$$

which is just our old acquaintance (2.2.2). Along the lines expounded in chapter II one can prove (WEBB 1984, 1985a) that under appropriate conditions

$$[T_2(t)n_0](a) = e^{rt}c(n_0)\mathcal{F}(a)e^{-ra} + 0(e^{(r-\epsilon)t}) \text{ for } t\to\infty\ , \tag{2.3.6}$$

where n_0 is the initial age density and $c(n_0)$ is a constant depending linearly on n_0. In appendix 2.B we shall discuss how c can be calculated direct from the interpretation.

EXERCISE 2.3.2: Derive (2.3.6) from the renewal theorem and calculate $c(n_0)$

* REMARK 2.3.3: When $a_m < \infty$ then it is easy to show that the linear (sub)space $F = \{n \in L_1(0,a_m) \mid \|n\|_B < \infty\}$ with $\|n\|_B = \int_0^{a_m} |\,n(a)\,|\ /\mathcal{F}(a)da$ is invariant under T_2. Moreover $L_1(0,a_m)$ gets mapped into F by any $T_2(t)$ with $t \geqslant a_m$. Therefore we may just as well choose F provided with the norm $\|\cdot\|_B$ instead of $L_1(0,a_m)$ as the space on which T_2 is supposed to act. As an alternative equivalent we may, on the analogy of the procedure followed in chapter II, make a change of variable $m(t,a) = n(t,a)/\mathcal{F}(a)$, with m assumed to be in $L_1(0,a_m)$. In this manner the death rate μ is eliminated from the differential generator.

---

* AC denotes the space of absolutely continuous functions, *i.e.* functions n such that $\int^a n(\alpha)d\alpha \in L_1$. See also chapter II definition 2.8 for an alternative characterization. The differential operator in (2.3.3a) should be interpreted as just the (left)inverse of the integration operator.

2.3.2. *Two semi-groups derived directly from the renewal equation itself*

There exist also less traditional ways to define semigroups for age dependent processes. Following DIEKMANN (1980) we shall discuss here two such semigroups based directly on two variants of the renewal equation.

The initial data enter (1.3.1) in the form of the function g. This therefore seems the obvious candidate as a basis for defining a semigroup. The question then is what becomes of g if we write down an equation for b from $\tau>0$ onwards and shift the time origin to τ. In the following we shall append an index τ to g to express the shift of time origin, g_0 being just our original g. From the dynamical viewpoint g_τ is the state of the system at time τ. Either by manipulating (1.3.1) or direct from the interpretation we find

$$g_\tau(t) = g_0(t+\tau) + \int_0^\tau b(s)\,\phi\,(t+\tau-s)ds \tag{2.3.7}$$

with b calculated from (2.1.10). Rewriting this in dynamical systems notation gives

$$T_1(\tau)g_0 = g_\tau \tag{2.3.8}$$

which may be considered the definition of the semigroup T_1. To complete this definition we still have to indicate which underlying function space for the "forcing" functions g we are dealing with. We shall postpone discussing this problem till after the introduction of yet another semigroup.

EXERCISE 2.3.4: Calculate the differential generator A_1 of T_1. Compare the result with that of exercises I.2.3.4 to 2.3.8. Calculate the characteristic equation corresponding to A_1 and make a "guess" at the asymptotic behaviour of $T_1(\tau)g_0$.

Usually the only sure way to know an animal's age is to observe when it was born. However, if our data include the history of the birth events then clearly there is no need to go through the intermediate stage of calculating an age distribution if all we wish to calculate anyhow are the future values of b. This can just as well be done from equation (2.2.1):

$$b(t) = \int_0^\infty \phi(a)b(t-a)da$$

The initial data for this equation are the values of b for $t\leq 0$. For later use we shall put $b(t)=:h_0(t)$ for $t\leq 0$, and rewrite (2.2.1) as

$$b(t) = \int_0^t \phi(a)b(t-a)da + \int_t^\infty \phi(a)h_0(t-a)da \tag{2.3.9}$$

Using the analogy with the previous case we can immediately rewrite this as the semigroup

$$(T_3(\tau)h_0)(t) = h_\tau(t) := b(t+\tau), \quad t\leq 0,\ \tau\geq 0\ , \tag{2.3.10}$$

where $b(t)=h_0(t)$ for $t\leq 0$ and for $t>0$ can be calculated from (2.1.10) with

$$g(t) = \int_t^\infty \phi(a)h_0(t-a)da\ . \tag{2.3.11}$$

EXERCISE 2.3.5: Calculate the differential generator A_3 of T_3. Compare the result with that of exercise I.2.3.9. Calculate also the characteristic equation corresponding to A_3, and make a "guess" at the asymptotic behaviour of $T_3(\tau)h_0$.

The (most obvious) relation between the semigroups T_1 to T_3 is indicated in the diagram in fig. 2.3.1 showing some of the relevant mappings between the, as yet unspecified, spaces on which they act. Here S_i are so-called intertwining maps which means that *e.g.* $S_1T_3(\tau)h_0=T_2(\tau)S_1h_0$ *etc.*

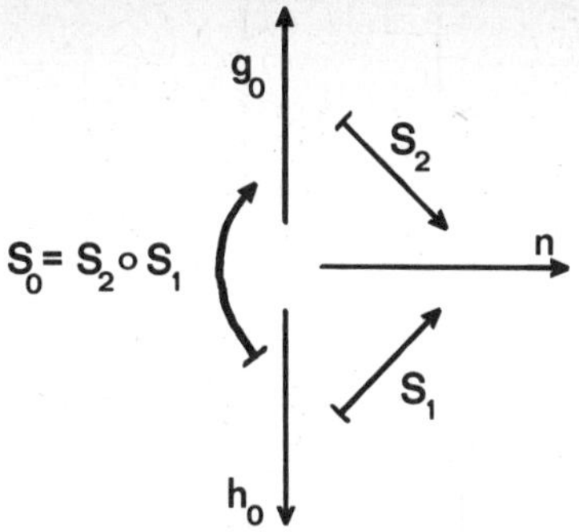

Fig. 2.3.1. Some maps connecting the semigroups T_1 to T_3.

The map S_0 is given by formula (2.3.11), showing that except for a choice of the spaces on which T_1 and T_3 act it is possible to reconstruct T_1 and T_3 from S_0 (provided H is sufficiently "rich" so that we may calculate the kernel ϕ from the action of S_0 on elements of H).

EXERCISE 2.3.6: Calculate S_1 and S_2 and discuss the degree to which the relation between T_1 to T_3 is unique.

As a last step in our definitions we still have to choose the spaces G and H on which T_1 and T_3 are supposed to act. In general this choice will depend on our practical needs. One possibility is to choose just $L_1(\mathbb{R}^-)$ for H and for G the closure of the image of H under the intertwining map S_0 (observe that $S_1 H$ is just equal to F from remark 2.3.3 and therefore also closure$(S_2 F) = G$). But this certainly is not the only possible choice.

The semigroup T_1 is studied in detail in DIEKMANN (1980) under the assumption that the kernel ϕ has compact support. One of the main results is that, with an appropriate choice of spaces, T_1 and T_3 are adjoints of each other. Many additional results are presented in VAN GILS (1984) (also see DIEKMANN (1982) and VERDUYN-LUNEL (1984, to appear)). STAFFANS (1984) gives a particularly nice overview covering intertwining as well as adjointness relations for both the cases of compact support, or finite delay, and infinite delay.

The semigroups T_1 to T_3 may be used as the basis for various nonlinear extensions. As none of them is based on a detailed conception of mechanism the choice between them is mainly a matter of convenience. A direct scientific argument for such a choice could be how the data are collected. Obviously T_3 and T_2 score best in this respect. From the mathematical point of view T_1 is especially pleasant to work with. However, the biological conditions allowing a problem to be formulated as a nonlinear extension of T_1 appear to be rather restrictive. Some examples may be found in section 4 and chapter VI (see also DIEKMANN & VAN GILS (1984)). T_2 certainly allows the largest scope for nonlinear extensions, the only (still rather strong!) restriction being that i-behaviour should fullfil conditions (1) and (2) from subsection 1.2. Mathematical references are PRÜSS (1983a,b), SCHAPPACHER (pers.comm.) and especially WEBB (1985a).

* 2.3.3. *Finite representability*

As a close to this subsection we shall say a few words about the possibility of representing our birth dynamics by means of a finite set of differential equations, as in the models covered by example 1.4.3 (see in particular exercise 1.4.4).

In the example we also have available a (semi)group T_4 acting on $\mathbb{R}^k$ obtained from solving the differential equation (1.4.5):

$$T_4(t) = \exp\,[(B + CA^T)t]\,, \tag{2.3.12}$$

B a square matrix, A and C single column matrices. Adding this setting to our intertwining diagram gives fig. 2.3.2,

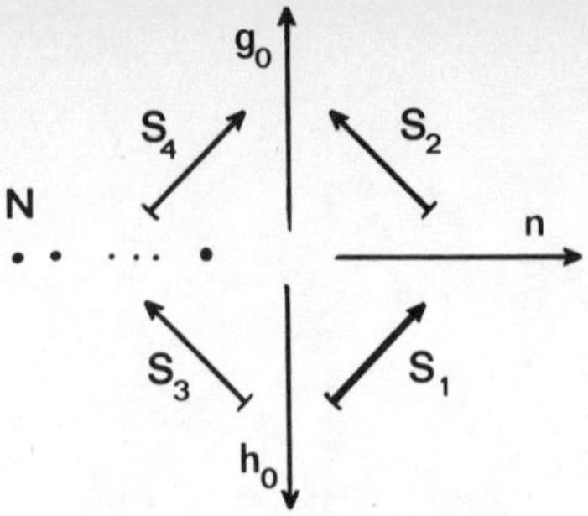

Fig. 2.3.2. The intertwining of the semigroups T_1 to T_4.

with

$$N = S_3\,h_0 = \int_0^t e^{Ba}Ch_0(-a)da\,. \tag{2.3.13}$$

and

$$g_0(t) = (S_4N)(t) = A^Te^{Bt}N\,, \tag{2.3.14}$$

while the map $S_0=S_4\circ S_3=S_2\circ S_1$ can, as always, be written as

$$S_0h_0 = \int_t^\infty \phi(a)h_0(t-a)da \tag{2.3.15a}$$

where now

$$\phi(a) = A^Te^{Ba}C\,. \tag{2.3.15b}$$

From the diagram it is clear that $S_0=S_2\circ S_1=S_4\circ S_3$ has a range of dimension $k'\leqslant k$. (2.3.15) moreover allows us to deduce that generically $k'=k$.

Given a linear map $S_0:H\to G$ which has k-dimensional range it is always possible to factorize this map into $S_0=S_4\circ S_3$ where $S_3:H\to\mathbb{R}^k$ and $S_4:\mathbb{R}^k\to G$. If this map is moreover of the form (2.3.15a) and H is sufficiently rich (*e.g.* $L_1(\mathbb{R}^-)$ will certainly do) then the results from systems theory* tell us that it is possible to calculate from S_0 a $k\times k$ matrix B and $k\times 1$ matrices A and C, such that (2.3.15b) holds. A, B and C are unique up to a choice of basis for the "intervening" space.

The previous results can be rephrased as: If (and only if) S_0 has a finite dimensional range it can be factorized into $S_0=S_4\circ S_3$ with $S_3:H\to\mathbb{R}^k$, $k=dim(S_0H)$. If such is the case S_3 induces a semigroup T_4 of the form (2.3.12) on $\mathbb{R}^k$ which intertwines with T_3 and T_1 in the manner of figure 2.3.2 with S_4 an isomorphism between T_4 and the restriction of T_1 to S_0H. The factorization and therefore also T_4 are unique up to a basis change. We shall refer to T_4 as a finite dimensional representation of our birth kernel ϕ.

From (2.3.15b) it can be seen immediately that a birth kernel allows a finite dimensional representation if and only if it can be written as a finite linear combination of polynomials times complex exponentials.

In general there is no guarantee that the components of the matrices A, B and C derived from a finite dimensionally representable kernel ϕ can be given signs compatible with a mechanistic model of the type described in example 1.4.3. Characterizing kernels which allow a mechanistically interpretable finite dimensional representation still seems to be an open problem.

In subsection 5.2 we shall consider the extension of the finite dimensional representability problem to the nonlinear case.

EXERCISE 2.3.7: Let $\phi(a)=pe^{-\alpha_1 a}$, $p,\alpha>0$. Calculate the map S_0 and show that it has finite dimensional range. Find

* A survey of these results with an eye towards animal behaviour may be found in METZ (1981); a more readily obtainable reference is KALMAN, FALB & ARBIB (1969).

a differential equation model generating the same birth process.

EXERCISE 2.3.8: Let $\phi(a)=p_1 e^{-\alpha_1 a}+p_2 e^{-\alpha_2 a}$, $p_1, p_2, \alpha_1, \alpha_2>0$. Same questions as exercise 2.3.7.

EXERCISE 2.3.9: Let $\phi(a)=pae^{-\alpha a}$, $p, \alpha>0$. Same questions as exercise 2.3.7. Hint: remember exercise 1.4.5.

EXERCISE 2.3.10: Let $\phi(a)=p_1 e^{-\alpha_1 a}+p_2 e^{-\alpha_2 a}$, $\alpha_1>\alpha_2$, $p_2>-p_1>0$. Same questions as exercise 2.3.7. Also try to find a mechanistically interpretable representation.

EXERCISE 2.3.11: What are all possible ϕ you can get from two differential equation representations of the birth law, without violating the constraint that ϕ be nonnegative? Do all these ϕ allow a mechanistically interpretable representation as well?

* EXERCISE 2.3.12: Let $\phi(a)=pe^{-\alpha a}+qe^{-\beta a}\cos(\omega a+\theta)$, $\beta>\alpha$, $p>q>0$. Show that S_0 has a three dimensional range. Find a (not necessarily interpretable) differential equation representation of the corresponding birth process.

** EXERCISE 2.3.13: Can you find an interpretable representation of the ϕ from the previous exercise as well?

* EXERCISE 2.3.14: Let

$$\phi(a) = \sum_{i=1}^{h} \sum_{j=0}^{k_i} p_{ij}\, a^j e^{-\alpha_i a} + \sum_{i=1}^{m} \sum_{j=1}^{n_i} q_{ij}\, a^j\, e^{-\beta_i a} \cos(\omega_i a+\theta_{ij})\,.$$

What is dim range S_0? Could you construct a (not necessarily interpretable) differential equation representation of the corresponding birth process?

** EXERCISE 2.3.15: Characterize all birth kernels ϕ allowing a mechanistically interpretable representation by finitely many differential equations.
Hint: The problem has some analogy with the problem of representing stochastic processes as functions of Markov chains (see e.g. ERICKSON, 1970).

2.A. Moments, cumulants and some approximations for r

In this appendix we shall derive some simple approximations for r. The coefficients appearing in these approximations allow nice probabilistic interpretations. We start with introducing some background information.

If f is a probability density of some nonnegative random variable T and $\bar{f}$ its Laplace transform, then

$$(-)^n \frac{d^n \bar{f}}{ds^n}(0) = \mathcal{E}T^n\,, \tag{2.A1}$$

where $\mathcal{E}$ denotes the expectation operator. Below we shall use the standard notation $\mu := \mathcal{E}T$ and $\sigma^2 := varT = \mathcal{E}T^2-(\mathcal{E}T)^2$.

EXERCISE 2.A1: Check (2.A1).

The cumulants κ_n of T are defined as

$$\kappa_n := (-)^n \frac{d^n \log\bar{f}}{ds^n}(0)\,. \tag{2.A2}$$

κ_1 and κ_2 just equal μ and σ^2 respectively. κ_3 / σ^3 is often used as a measure of the skewness of a distribution, and κ_4 / σ^4 as a measure of its kurtosis.

In renewal theory our main interest is $\bar{\phi}$, the Laplace transform of the birth kernel. We have already seen that $\bar{\phi}(0)$ can be interpreted as the mean number R of offspring of an individual. The normalized kernel

$$f := R^{-1}\phi \tag{2.A3}$$

can be interpreted as the probability density of a random variable, called "the age at child bearing".

REMARK 2.A2: For a biological interpretation of "the age at child bearing" consider the following thought experiment.

Start with a cohort of (potential) mothers. Wait till all their children are born. Sample at random one child. The age of its mother at the time of its birth is "the age at child bearing".

With our new notation in mind we now take a closer look at the characteristic equation (2.2.2):

$$1 = R\bar{f}(r)\,. \tag{2.A4}$$

Taking logarithms yields

$$0 = \log R + \log\bar{f}(r) = \log R - \mu r + \tfrac{1}{2}\sigma^2 r^2 + o(r^2)\,. \tag{2.A5}$$

We already know that $r=0$ for $\log R=0$. Therefore we may try to expend r in a power series in $\log R$: $r=a_1\log R+a_2(\log R)^2+\cdots$. Inserting this into (2.A5) and equating the coefficients of the successive powers of $\log R$ to zero gives

$$r = \frac{\log R}{\mu}\,(1+\tfrac{1}{2}(\sigma/\mu)^2\log R)+o(\log R))\,. \tag{2.A6}$$

(2.A6) highlights the effects of the mean number of offspring, mean age at childbearing and variance of the age at child bearing on the population growth rate.

Another approximation which is more accurate when the higher order cumulants are relatively small (or equivalently when the distribution of the age at child bearing is close to normal) comes from simply neglecting the $o(r^2)$ term in (2.A5), and solving the resulting quadratic equation for r:

$$r \simeq \frac{\mu-\sqrt{\mu^2-2\sigma^2\log R}}{\sigma^2}\,. \tag{2.A7}$$

The main usefulness of (2.A6) and (2.A7) is that they may do away with the need to measure the whole kernel ϕ. Generally simple statistics like R,μ and σ^2 are much easier to obtain experimentally!

REMARK 2.A3: If we are outside the range of applicability of the approximation formulae the only thing we can do is fit some low parameter family of curves to the observed individual reproduction data. A family with an exceptionally good record in this respect are the so-called shifted gamma densities

$$\phi(a) = \begin{cases} 0 & \text{for } a\leqslant\tau \\ R\dfrac{\alpha(\alpha(a-\tau))^{k-1}e^{-\alpha(a-\tau)}}{\Gamma(k)} & \text{for } a\geqslant\tau \end{cases} \tag{2.A8}$$

EXERCISE 2.A4: Let $\phi=Rf$ with f the gamma density

$$f(a) = \frac{\alpha(\alpha a)^{k-1}e^{-\alpha a}}{\Gamma(k)} \tag{2.A9}$$

Calculate lines of equal r/μ in the σ/μ versus $\log R$ plane. Compare the results with the approximate equal r/μ lines derived from either (2.A6) or (2.A7).

EXERCISE 2.A5: Let $\phi=Rf$ with f the homogeneous (α,β) density

$$f(a) = \begin{cases} (\beta-\alpha)^{-1} & \text{for } \alpha<a<\beta \\ 0 & \text{elsewhere} \end{cases}$$

Same questions as in the previous exercise.

EXERCISE 2.A6: Let ϕ be as in (2.A8). Calculate lines of equal r/τ in the $\log R$ versus $k/\alpha\tau$ plane for various values of $k^{-\frac{1}{2}}$. Compare the results with the approximate equal r/τ lines calculated from (2.A6) or (2.A7).

EXERCISE 2.A7: Devise a procedure to estimate the parameters of (2.A8) given a sample of ages at child bearing.

* EXERCISE 2.A9: Derive explicit expressions for R, μ and σ^2 for the model from example 1.4.8, using the directly available results for moments of the absorbtion time in diffusion processes (see *e.g.* GOEL & RICHTER DYN, 1974, or EWENS, 1969, 1979). (NB. The general result looks horrible. An interesting problem is whether these expressions can

be simplified for appropriate choices of v and σ^2. It would also be of interest to derive approximation formulae based on the assumption that σ^2 is relatively small.)

2.B. The dependence of the long run population size on the initial age distribution

In subsection 2.3.1 we argued that for $t\to\infty$

$$n(t,a) = [T_2(t)n_0](a) = e^{rt}c(n_0)\mathcal{F}(a)e^{-ra} + o(e^{rt}) , \tag{2.B1}$$

where n_0 is the initial age distribution. In this appendix we shall apply ourselves to calculating c direct from the interpretation. Before doing so we first introduce some notation. $n(t)$ will denote the age distribution at t. So $n(0)=n_0$. Moreover we shall normalize the eigenfunction $\tilde{n}$ corresponding to the dominant eigenvalue by setting $\tilde{n}(0)=1$, $i.e.$

$$\tilde{n}(a) = \mathcal{F}(a)e^{-ra} . \tag{2.B2}$$

(Please note that in general $\int n(a)da\neq 1$.) Further notation will be introduced as the need arises.

As T_2 is a linear semigroup c has to be a linear functional on the space of age distributions. Therefore we try writing

$$c(n) = \int_0^{a_m} \kappa v(a)n(a)da , \tag{2.B3}$$

with κ a constant introduced to absorb any future normalization.

* REMARK 2.B1: If, as we did in subsection 2.3.1, we choose our p-state space to be $L_1(0,a_m)$ any linear functional c can be written in the form (2.B3) with $v \in L_\infty(0,a_m)$. The explicit formula for v to be described below shows that when $\mathcal{F}$ is continuous, v is continuous as well. Therefore under this condition (2.B3) also applies to an enriched space containing in addition measures such as delta "functions".

Now consider a population started with one newborn individual. Let us call the age distribution of the clan she founds $n_{0,0}$. Eventually this clan will grow exponentially, so, with an appropriate definition of κ

$$n_{0,0}(t) \simeq \kappa\tilde{n}e^{rt} , \quad t\to\infty , \tag{2.B4}$$

where κ still has to be determined. Given (2.B4) we can calculate the asymptotic size of the clan founded by an individual aged a freshly introduced at time zero. Let $n_{\tau,a}$ denote the age distribution of the clan founded by a daughter born from her at age τ and time $\tau-a$. Then

$$n_{\tau,a}(t) \simeq \kappa \, \tilde{n} \, e^{r(t-\tau+a)} . \tag{2.B5}$$

The probability that an individual aged a ever reaches age τ is $\mathcal{F}(\tau)\,/\,\mathcal{F}(a)$, and the probability per time unit that she begets a daughter at that age is $\lambda(\tau)$. Therefore, if we add up all descendants of our individual aged a at time zero we get

$$n_{+,a}(t) = \int_a^{a_m} n_{\tau,a}(t)\lambda(\tau)\mathcal{F}(\tau)/\mathcal{F}(a)d\tau \simeq \kappa\tilde{n}v(a)e^{rt} , \tag{2.B6a}$$

with

$$v(a):=\frac{e^{ra}}{\mathcal{F}(a)} \int_a^{a_m} \lambda(\tau)\mathcal{F}(\tau)e^{-r\tau}d\tau . \tag{2.B6b}$$

Therefore

$$n(t) = \int_0^{a_m} n_{+,a}(t)n_0(a) \, da \simeq \kappa \int_0^{a_m} v(a)n_0(a)da \, \tilde{n}e^{rt} . \tag{2.B7}$$

The only thing that remains is finding an expression for κ. To that end we observe that necessarily

$$T_2(t)\tilde{n} = \tilde{n}e^{rt} = c(\tilde{n})\tilde{n}e^{rt} . \tag{2.B8}$$

Therefore we should have that $c(\tilde{n})=1$ or

$$\kappa = (\int_0^{a_m} v(a)\tilde{n}(a)da)^{-1} = (\int_0^{a_m} a\,\mathcal{F}(a)e^{-ra}\,\lambda(a)da)^{-1}\,. \tag{2.B9}$$

The function v introduced in (2.B6b) is known in the biological literature as the *reproductive value* of an individual aged a. The reason is that for large t the size of a clan founded by an individual aged a at time zero equals $v(a)$ times the size of the clan founded by a newborn individual.

EXERCISE 2.B2: Verify the last statement.

EXERCISE 2.B3: Confirm (2.B9). Hint: Use partial integration.

EXERCISE 2.B4: Derive the coefficient preceding e^{rt} in the renewal theorem 2.2.7, or in the right hand side of Feller's renewal theorem 2.2.8, by the same sort of procedure as used to derive $c(n_0)$.

EXERCISE 2.B5: Show that V defined by

$$V(t) := c(n(t)) \tag{2.B10}$$

satisfies

$$\frac{dV}{dt} = rV \tag{2.B11}$$

Hint: Differentiate under the integral sign in the definition of $c(n)$ and use the partial differential equation. Integrate partially and use the side condition.

* EXERCISE 2.B6: Show that the calculations of the previous exercise are equivalent to checking that

$$A_2^* c = rc \tag{2.B12}$$

where A_2^* is the adjoint of A_2.

3. Extensions of the linear theory

3.0. Introduction

In the introduction to this chapter we argued for the generality of the age dependent formalism in the linear case. The only thing that matters is a particular kind of "decoupling property": all individuals should be born equal, *i.e.* be independent of their forebears. This allows for a simple bookkeeping, in particular of the births, but also of derived quantities like total population size (compare formula (1.3.2)). It generally pays to keep an open mind for potential decoupling properties of this kind as their presence may be revealed or hidden depending on the way a population problem is formulated at the start. As all modelling involves an element of simplification it is even possible to introduce or do away with such properties before any equations are written down!

REMARK 3.0.1: We find it difficult to give explicit expression to the intuition underlying the general concept of "decoupling property" as it is used by us here. Roughly speaking we use the term to refer to various exploitable kinds of (stochastic) independencies inherent in i-behaviour.

In this section we present three examples of population problems which can be analyzed in great detail by relying on particular decoupling properties. In dealing with these examples we shall immediately use the relevant decoupling properties as our starting point. Especially in the last example, however, we only stumbled on the appropriate decoupling properties after a not inconsiderable number of false starts.

All three examples consider the distribution of some immediately observable discrete characteristic, like the number of scars on a cell derived from the division process or the size of small colonies of cells. As presented the examples may appear a trifle specific. However, they are representative of a large class of problems of their general kind. The specific details of each problem may depend on the very details of the biology of the organism under consideration, the method of approach, however, is (fairly) general.

3.1. Scar distributions in yeast

3.1.1. Budding yeasts

A full grown cell of the budding yeast *Saccharomyces cerevisiae* reproduces by forming small buds, which, after a short period of growth while remaining attached to the mother, detach themselves and start their lives as small independent cells. At the place where the bud was formed the mother cell retains a visible scar. Therefore the reproductive history of a cell can at least partially be traced by counting the number of such scars. (There even have been observed cells with over forty scars!) Given that we have available such an easy observable the questions arise (1) how the distribution of scar numbers develops over time, and (2) whether we might exploit any knowledge about the scar distribution in a population to make inferences about some aspect or other of the dynamics of the individual reproductive process. Models for scar distributions already have a not unconsiderable history in the literature. Some references are Adams *et al* (1981), Hamada *et al* (1982), Hamada & Nakamura (1982), Hjortso & Bailey (1983), Tuljapurkar (1983) and Gyllenberg (1985,a,b).

The most obvious decoupling moment in the life history of a budding yeast is the moment a young bud starts its life as an independent individual. Other decoupling moments of a different kind may occur when a yeast cell sheds a bud. We shall start concentrating on the first decoupling moment only as this allows us to write down just one renewal equation instead of a(n infinite) system of such equations. Let $\phi(a)$ denote the (mean) probability per unit of time that a yeast cell sheds a bud a time units after it started its independent life, then the rate of production of freshly detached buds b_0 satisfies

$$b_0 = \phi{\star}b_0 + g \,, \tag{3.1.1}$$

where g as usual denotes the rate of production of buds by cells that were already present at $t=0$. If we may assume, as we shall do from now on, that each time a bud is shed the only memory that remains of the reproductive history of a yeast cell is fully parametrized by the number of scars then ϕ can be calculated as

$$\phi = \sum_{n=0}^{\infty} \phi_n{\star}\cdots{\star}\phi_0 \,, \tag{3.1.2}$$

where ϕ_i denotes the, possibly defective (due to deaths), probability density of the time till the next bud is shed by a cell with i scars.

Exercise 3.1.1: Derive (3.1.2).

Our eventual aim is to calculate the scar distribution. Let b_i be the production rate of cells with i scars, $\mathcal{F}_i$ the survivor function of cells with i scars, *i.e.* $\mathcal{F}_i(\tau)$ is the probability that a cell with i scars is alive and moreover has not shed a new bud τ time units after it came into being as a cell with i scars, and let N_i denote the total number of scars, then for $i>0$

$$b_i = \phi_{i-1}{\star}\cdots{\star}\phi_0{\star}b_0 + g_i \,, \tag{3.1.3a}$$

and for all i,

$$N_i = \mathcal{F}_i \star b_i + M_i \,, \tag{3.1.3b}$$

where g_i is the rate at which the remnants of the initial population are still producing cells with i scars, and M_i is the number of cells initially present which still have i scars. (If cells are certain either to die or to produce daughters, M_i will eventually go to zero for any fixed i, and the same applies to any biologically acceptable g_i.) Finally let N denote the total population size and p_i the relative frequency of cells with i-scars, then

$$N = \sum N_i \,, \qquad p_i = N_i / N \,. \tag{3.1.4}$$

Exercise 3.1.2: Derive (3.1.3).

Exercise 3.1.3: Derive the following infinite system of renewal equations for the b_i

$$b_i = \phi_{i-1} \star b_{i-1} + g_i \,, \qquad b_0 = \sum_{j=0}^{\infty} (\phi_j \star b_j + g_j) \,, \tag{3.1.5}$$

where we conventionally put $g_0=0$. Use (3.1.5) to rederive (3.1.1) with

$$g = \sum_{n=1}^{\infty} (g_n + \sum_{j=0}^{n-1} \phi_n{\star}..{\star}\,\phi_{n-j}{\star}g_{n-j}) = \sum_{n=1}^{\infty} (g_n + \sum_{j=n}^{\infty} \phi_j{\star}..{\star}\,\phi_n{\star}g_n) \,. \tag{3.1.6}$$

Hint: Either exploit (2.1.2) and (2.1.3) or use Laplace transformation and ordinary algebra.

The characteristic equation corresponding to (3.1.1) is

$$1 = \bar{\phi}(r) = \sum_{n=0}^{\infty} \prod_{j=0}^{n} \bar{\phi}_j(r) . \tag{3.1.7}$$

The renewal theorem then tells us that our yeast population will eventually grow at a rate $r>0$, where r is the unique real solution of (3.1.7), if $\bar{\phi}(0)>1$. As the two problems which we set out to solve are only of interest for growing yeast populations we shall from now on make that assumption. Setting $b_0(t)=e^{rt}(c+o(1))$, for $t\rightarrow\infty$, in (3.13) gives

$$N_i(t) = e^{rt}(c\, \bar{\mathcal{F}}_i(r) \prod_{j=0}^{i-1} \bar{\phi}_j(r) + o(1)) , \tag{3.1.8}$$

and therefore, when $\tilde{p}_i$ denotes the asymptotic value of p_i for large t,*

$$\tilde{p}_i \propto \bar{\mathcal{F}}_i (r) \prod_{j=0}^{i-1} \bar{\phi}_j(r) , \tag{3.1.9}$$

where the product is interpreted as 1 when $i=0$. This solves our first problem.

If we are to make any progress on the inverse problem *(viz.* how can (3.1.9) be used to extract information about individual reproductive behaviour from the observed $\tilde{p}_i$) we have to make some further simplifying assumptions. Our first assumption, well entrenched in the microbiological literature, is that yeast cells do not die from old age or accumulation of scars, so that the only source of cell losses are random accidental deaths, or random washout from a culture vessel. Let μ denote the (age independent) death rate and ψ_i the probability density of the time till the next bud is produced by a cell already having i scars when there are no deaths, then

$$\phi_i(\tau) = e^{-\mu\tau}\, \psi_i(\tau) , \tag{3.1.10}$$

$$\mathcal{F}_i(\tau) = e^{-\mu\tau} \int_{\tau}^{\infty} \psi_i(\sigma)d\sigma , \tag{3.1.11}$$

and

$$\bar{\phi}_i(s) = \bar{\psi}_i(\mu+s) , \tag{3.1.12}$$

$$\bar{\mathcal{F}}_i(s) = \frac{1-\bar{\phi}_i(s)}{\mu+s} . \tag{3.1.13}$$

EXERCISE 3.1.4: Derive (3.1.10) and (3.1.11). Hint: Use the independence of the reproductive mechanism and cell loss.

EXERCISE 3.1.5: Derive (3.1.12) and (3.1.13).

Inserting (3.1.13) in (3.1.9) and setting the sum of the $\tilde{p}_i$ equal to one now gives

$$\tilde{p}_i = (1-\bar{\phi}_j(r)) \prod_{j=0}^{i-1} \bar{\phi}_j(r) . \tag{3.1.14}$$

where again the product is interpreted as 1 when $i=0$.

Using the explicit expression (3.1.14) in combination with the characteristic equation (3.1.7) it is easy to verify that the mean number of scars

$$\sum_{i=0}^{\infty} i\, \tilde{p}_i = 1 \tag{3.1.15}$$

in accordance with elementary biological reasoning: for every cell that is produced exactly one scar is produced as well. If a different mean scar number is found then we may conclude that cell death rate depends on scar number or cell age. (The other explanations that the sample is biased or that the stable value of the mean scar number has not yet been reached are both extremely unlikely, due to the basic nature of the biological mechanism of one new cell one scar. This acts equally, locally in non well stirred vessels, in growth limited populations, and by extension also in the population from which the inoculum was derived.)

---

* $\propto$ means proportional to.

EXERCISE 3.1.6: Check (3.1.15) algebraically.

To get some further idea of how (3.1.14) can be used assume temporarily that $\psi_i = \psi_0$ for all i. In that case the characteristic equation reduces to

$$1 = \sum_{n=0}^{\infty} [\bar{\phi}_0(r)]^{n+1} = \frac{\bar{\phi}_0(r)}{1-\bar{\phi}_0(r)} . \tag{3.1.16}$$

Therefore $\bar{\phi}_0(r) = \frac{1}{2}$ and consequently

$$\tilde{p}_i = (\tfrac{1}{2})^{i+1} . \tag{3.1.17}$$

If (3.1.17) is found to apply in our culture we may safely conclude that there is no relation between the number of scars and the time till the appearance of the next bud. In that case the scar distribution also cannot tell us anything further about the reproductive mechanism. This last conclusion may not come as a surprise. However, the main importance of (3.15) is as a null model, against which to test alternative hypotheses.

EXERCISE 3.1.7: Derive (3.1.17) by the following direct argument. Concentrate on the p_i. As deaths occur through random removal the dynamics of the p_i is not influenced by the deaths. Therefore any statement pertaining to the p_i should be independent of μ. Choose μ so that $r=0$. Then at equilibrium the N_i are constant so we can think in terms of the $\tilde{N}_i$ instead of the $\tilde{p}_i$. Therefore the transfer into the i-th scar class exactly balances the transfer and deaths from that scar class. When the ϕ_i are equal the relative transfer and death rates at equilibrium should be equal for all scar classes. This allows us to conclude that at equilibrium $\tilde{N}_i = \alpha \tilde{N}_{i-1}$ for some α, and therefore $\tilde{p}_i = \alpha \tilde{p}_{i-1}$. α can be determined from $\Sigma i \tilde{p}_i = 1$.

The next most simple model biologically is that $\psi_i = \psi_1$ for all $i \geqslant 1$ but that ψ_0 is different. In that case the characteristic equation reduces to

$$1 = \bar{\phi}_0(r) \sum_{n=0}^{\infty} [\bar{\phi}_1(r)]^n = \frac{\bar{\phi}_0(r)}{1-\bar{\phi}_1(r)} , \tag{3.1.18}$$

and we obtain

$$p_0 = 1-\bar{\phi}_0(r) = \bar{\phi}_1(r) \qquad p_i = [\bar{\phi}_0(r)]^2 \, [\bar{\phi}_1(r)]^{i-1} . \tag{3.1.19}$$

Therefore, if (and only if) newborn cells differ from full grown cells in their capacity to reproduce, but from the first reproduction act onwards cells stay essentially the same reproductively, the scar distribution is geometric with a modified zero'th term, and this zero'th term is related in a very specific way to the geometric parameter. Figure 3.1.1 shows the fit of this theoretical distribution to an observed one, drawn from HAMADA *et al* (1982). It can be seen that (3.1.17) certainly is out of the question, but that the fit of (3.1.19) is exceedingly good.

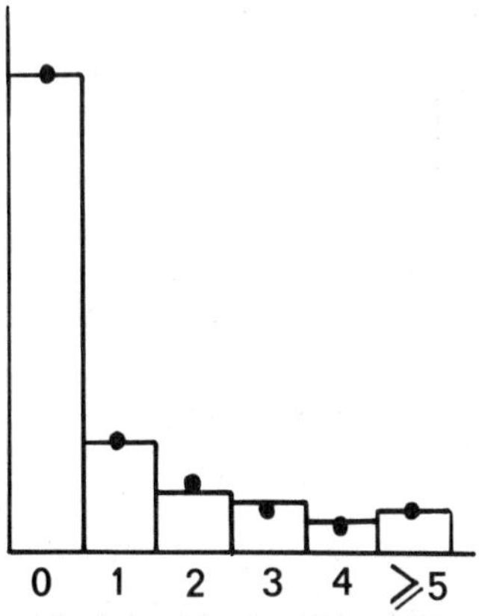

Fig. 3.1.1: Observed and theoretical ●, based on (3.1.19) with $\bar{\phi}_1(r)=0.6172$, scar number distribution in a sample of budding yeast *(Saccharomyces cerevisiae)* cells. Data from HAMADA *et al* (1982).

A final question is what kind of information about the difference between ψ_0 and ψ_1 can be obtained from

$\overline{\phi}_1(r)=1-\overline{\phi}_0(r)$ as estimated from the data. A biologically plausible hypothesis is that

$$\psi_0(\tau) = \begin{cases} 0 & \text{for } \tau<T \\ \psi_1(\tau-T) & \text{for } \tau\geqslant T \end{cases}, \tag{3.1.20}$$

where T is the time needed for the young bud to grow to an appropriate size. Then

$$\overline{\phi}_0(s) = e^{-(\mu+s)T}\overline{\phi}_1(s)\,, \tag{3.1.21}$$

and

$$T = (\mu+r)^{-1} \log\left[\frac{p_0}{1-p_0}\right]. \tag{3.1.22}$$

As r is just the population growth rate which can be relatively easily determined, and as generally random deaths can be neglected so that μ equals the dilution rate, we may conclude that the scar number distribution just enables us to calculate the time needed for a bud to grow to reproductive size, but no more than that.

Concluding remarks

We have shown how a judicious exploitation of various types of presumed "decoupling" moments in the life of a budding yeast cell allows us to derive relatively easily an explicit expression for the asymptotic distribution of the number of bud scars. The main assumption was that budding depends only on the time since the last bud was shed and the number of scars present, and not *e.g.* directly on cell age. When we moreover made the further (decoupling) assumptions that death rate μ is constant, independent of age or scar class, and that the interbudding times are identically distributed, independent of scar number, then the stationary scar distribution was found to be geometric with a modified zeroth term. The size of this zeroth term depends in a subtle manner on the distributions of both the interbudding times and the time till the production of the first bud. If the time till the first bud only differs from the interbudding times by the addition of a fixed delay T, the size of the zeroth term p_0 could be related to $(\mu+r)T$, where r is the population growth rate, allowing T to be inferred when p_0, r and μ are known (or, conversely, to infer r when T and μ are known *etc.*). These results indicate how a relatively easily obtainable population datum like the scar number distribution can be exploited to make inferences about the behaviour of the individuals constituting that population.

In all our calculations we concentrated on stationary distributions only, as this is the kind of datum that is most easily obtainable experimentally. Of course more information can be gleaned, at least in principle, from transient behaviour (see *e.g.* BERAN, STREIBLOVA & LIEBLOVA (1969), HJORTSO & BAILEY (1983) and TULJAPURKAR (1983)).

Finally we refer to GYLLENBERG (1985) for an analysis of more detailed models relating the scar distribution to cell sizes as the basic variable underlying the budding process in the manner of I.4 and II.

EXERCISE 3.1.8: Argue that under the assumptions made in this subsection the state of an individual can be parametrized by its scar number i and the time τ since it last shed a bud or was born. Write down the partial differential equations for the p-state (see chapter III for the methodology). Assume initially that the death rate μ may depend on i as well as on τ.

EXERCISE 3.1.9: Find an equation for the dominant eigenvalue and an expression for the corresponding stable distribution for the differential generator from exercise 3.1.8.

EXERCISE 3.1.10 Derive (3.1.9) from the result of exercise 3.1.9 with appropriate definitions of $\overline{\phi}$ and $\overline{\mathfrak{F}}$.

EXERCISE 3.1.11: Now assume that $\mu_i(\tau)=\mu$ and derive (3.1.14) to (3.1.19)

3.1.2. *Fission yeasts*

Fission yeasts, *Schizosaccharomyces pombe,* differ from budding yeasts in that they reproduce by cells dividing into two approximately equal parts. In the process two scars are formed, one on each sister cell. Moreover the scars that were already present on the mother are distributed over the daughter cells. The precise nature of the distribution rules is unknown. In fact the biological question is how the observed scar distribution can be used to make inferences about these rules and therefore by extension about the underlying process of cell growth and division (see *e.g.* CALLEJA *et al.* (1980) and HAMADA (1982)). Our main reason for including precisely this example, however, is technical: it shows how yet another type of biologically plausible decoupling property can be exploited to obtain more or less explicit mathematical results.

For a change we shall start this time from the partial differential equations for the p-state, assuming that the i-state is fully parametrized by the scar number i and cell age a:

$$\frac{\partial n_i(t,a)}{\partial t} = -\frac{\partial n_i(t,a)}{\partial a} - (d_i(a) + \mu_i(a))n_i(t,a)\,, \tag{3.1.23}$$

$$n_i(t,0) = 2 \sum_{j=max(1,i-1)}^{\infty} \pi_{ij} \int_0^{\infty} d_j(\alpha)n_j(t,\alpha)d\alpha\,,$$

where π_{ij} denotes the probability that a randomly chosen daughter of a cell with j scars has itself i scars, and $d_i(a)$ and $\mu_i(a)$ are the probabilities per unit of time that a cell with i scars aged a divides resp. dies.

The scar distribution can be found from

$$N_i(t) = \int_0^{\infty} n_i(t,a)da, \qquad N(t) = \sum_{i=1}^{\infty} N_i(t)\,, \qquad p_i(t) = N_i(t) \;/\; N(t)\,. \tag{3.1.24}$$

From the interpretation it follows immediately that

$$\pi_{ij} = \pi_{j+2-i,j}, \quad \text{for } i = 1, \cdots, j+1\,, \tag{3.1.25a}$$

$$\sum_{i=1}^{j+1} \pi_{ij} = 1\,. \tag{3.1.25b}$$

The values of the π_{ij} will be determined in some rough fashion by the processes of cell growth and division. In particular scar redistribution will be much more uneven when cell growth is unipolar than when it is bipolar *(vide* CALLEJA *et al,* 1980). Provided all μ_i are zero the π_{ij} together form the transition matrix of the Markov chain that results from following one cell line down through the generations. For later reference we shall call this chain M.

To determine the behaviour of the n_i for large time we turn, as usual, to the characteristic equation derived from (3.1.23):

$$\frac{d\tilde{n}_i}{da} = -\,(d_i(a) + \mu_i(a) + r)\,\tilde{n}_i(a) \tag{3.1.26a}$$

$$\tilde{n}_i(0) = 2 \sum_{j=\max(1,i-1)}^{\infty} \pi_{ij} \int_0^{\infty} d_j(\alpha)\tilde{n}_j(\alpha)d\alpha \tag{3.1.26b}$$

which on setting

$$\mathcal{F}_i(a) := e^{-\int_0^a (d_i(\alpha)+\mu_i(\alpha))d\alpha}\,, \qquad \phi_i(a) := d_i(a)\,\mathcal{F}_i(a) \tag{3.1.27}$$

(the similarity of the notation to that in the previous subsection is not incidental!) reduces to

$$\tilde{n}_i(0) = 2 \sum_{j=\max(1,i-!)}^{\infty} \pi_{ij}\, \bar{\phi}_j(r)\tilde{n}_j(0) \tag{3.1.28}$$

EXERCISE 3.1.12: Show that with the definitions (3.1.27) equation (3.1.28) is equivalent to (3.1.26).

It can be shown that under fairly general conditions on the π_{ij}, in particular that any state in the Markov chain M can be reached from every other state, there exists a unique real eigenvalue r with corresponding nonnegative eigenvector. Below we shall indicate how this eigenvector can be calculated explicitly if we make some special assumptions. Comparing the observed and predicted scar distributions then provides a test of our assumptions. If some more complicated assumptions are made one has to resort to numerical methods (see *e.g.* HAMADA, 1982). But even in that case the analytical reasoning makes it possible to concentrate on a relatively simple set of equations instead of having to solve the full partial differential equations (3.1.26).

We begin by making the biologically plausible assumption (compare the previous subsection) that $\mu_i(a)=\mu$, independent of i or a. In that case the biological mechanism "one additional cell → two more scars", leads us to infer that asymptotically the mean number of scars on a cell should be two:

$$\sum_{i=1}^{\infty} i\tilde{p}_i = 2 \tag{3.1.29}$$

This provides a test whether death rate is independent of scar number. (The fact that (3.1.29) is satisfied does not rule out that cell death depends on cell age as in this example cell age is not related to scar number!) An algebraic verification of (3.1.29) is provided by the exercises.

EXERCISE 3.1.13: Check (3.1.29). Hint: As a start show that

$$\tilde{N}_i = \int_0^\infty \tilde{n}_i(a)da = \frac{1}{r+\mu} \tilde{n}_i(0)\,(1-\bar{\phi}_i(r))\,,$$

then use (3.1.25) to show that

$$\sum_{i=1}^{j+1} i\,\pi_{ij} = \frac{j+2}{2}\,,$$

and use (3.1.28) to show that

$$\sum_{i=1}^\infty \tilde{n}_i(0) = 2\sum_{j=1}^\infty \bar{\phi}_j(r)\,\tilde{n}_j(0)$$

and

$$\sum_{i=1}^\infty i\,\tilde{n}_i(0) = \sum_{j=1}^\infty j\,\bar{\phi}_j\,(r)\,\tilde{n}_j(0) + \sum_{i=1}^\infty \tilde{n}_i(0)\,.$$

Use these identities to prove that

$$\tilde{N} = \sum_{i=0}^\infty \tilde{N}_i = \frac{1}{2(r+\mu)}\sum_{i=1}^\infty \tilde{n}_i(0)$$

and

$$\sum_{i=1}^\infty i\,\tilde{N}_i = \frac{1}{r+\mu}\sum_{i=1}^\infty \tilde{n}_i(0)\,.$$

Given the assumption of constant μ there are two types of scar redistribution rules for which we can get an expression for the p_i without making any further assumptions about the ϕ_j. The first one is

$$\pi_{1j} = \pi_{j+1,j} = \frac{1}{2}, \qquad \pi_{ij} = 0 \text{ for } i\neq 1,\ j+1\,, \tag{3.1.30}$$

or, in other words, one sister always ends up with one scar, the other gets all the old scars plus one new one. In essence this is the budding yeast model all over again, except that the number of scars is increased by one.

The second special case is that of equal sharing of the scar load:

$$j \text{ odd: } \pi_{ij} = \begin{cases} \frac{1}{2} & \text{for } i = (j+1)/2 \text{ and } i = (j+1)/2 + 1\,, \\ 0 & \text{for all other } i \end{cases} \tag{3.1.31}$$

$$j \text{ even: } \pi_{ij} = \begin{cases} 1 & \text{for } i = j/2 \\ 0 & \text{for all other } i \end{cases}$$

Clearly only a scar load of 2 reproduces itself in its daughters, and all cells with other scar loads give birth to daughters with scar loads nearer to 2. And indeed, from (3.1.28) we can deduce that

$$\tilde{n}_i(0) = \begin{cases} c & \text{(a free parameter) for } i = 2 \\ 0 & \text{for all other } i \end{cases} \tag{3.1.32}$$

and

$$1 = 2\,\bar{\phi}_2\,(r)\,, \tag{3.1.33}$$

i.e. the stable distribution consists of 2-scar cells only, and the characteristic equation equals that of a population consisting wholy of selfreproducing 2-scar cells.

EXERCISE 3.1.14: Derive (3.1.32) and (3.1.33) from (3.1.28).

Returning to general π_{ij} we make as our final simplification the biologically rather plausible decoupling assumption that time to cell division is independent of scar number so that the two processes of reproduction and scar redistribution can be considered separately. Summing over i in (3.1.28) then gives

$$1 = 2\,\bar{\phi}(r)\,, \tag{3.1.34}$$

the usual characteristic equation of the pure cell kinetics model. Substituting this in (3.1.28) again and using $\tilde{p}_i \propto \tilde{n}_i(0)$, results in

$$\tilde{p}_i = \sum_{j=\max(1,i-1)}^{\infty} \pi_{ij}\,\tilde{p}_j\,, \tag{3.1.35}$$

which is just the equation for the stationary distribution of the Markov chain M. Naturally in this case scar distribution does not contain any information about the division times. Using specific assumptions for the π_{ij} we can now calculate the $\tilde{p}_i$, either numerically or in some simple cases analytically. Translating hypotheses about the cell growth and division morphology into specific numerical values of the π_{ij} and subsequent calculation of the $\tilde{p}_i$ then makes possible a relatively easy empirical test of those assumptions. See CALLEJA *et al* (1980), and HAMADA (1982) for some further discussion.

EXERCISE 3.1.15: Let $\pi_{ij}=(j+1)^{-1}$ (all scar redistributions equally probable). Calculate the $\tilde{p}_i$ from (3.1.35) together with the conservation condition $\Sigma\tilde{p}_i=1$.
Hint: use (3.1.35) to derive the difference equation $\tilde{p}_i=\tilde{p}_{i-1}-(i-2)^{-1}\,\tilde{p}_{i-2})$ and use this to show that the $\tilde{p}_i=e^{-1}/(i-1)!$, *i.e.* scar number minus one is Poisson distributed with mean 1.

EXERCISE 3.1.16: Let $\pi_{ij}=q/2$, independent of j for $j \geqslant 2$ (check that necessarily $\pi_{11}=\pi_{21}=\frac{1}{2}$). In words: the probability that all the old scars end up on one of the two daughters equals q, independent of the number of old scars. Show that $\tilde{p}_1=\dfrac{q(1-\bar{\phi}_1(r))}{1+(q-1)\bar{\phi}_1(r)}$, and, therefore if $\phi_i=\phi$ independent of i, $\tilde{p}_1=q/(1+q)$.
Hint: exploit (3.1.28) and some of the identities from exercise 3.1.13.

EXERCISE 3.1.17: Write down an (infinite) system of renewal equations for $b_i(t)=n_i(o,t)$, either direct from the interpretation or by integrating (3.1.23) along the characteristics. Use the reasoning from the start of section 2.2 to arrive at (3.1.28) again.

3.2. *Colony size in the diatom* Asterionella

This example has already figured heavily in chapter III. However, except for some of the exercises, the development in this chapter has been kept completely independent, even to the extent of reintroducing the basic biology.

The planctonic diatom *Asterionella,* like many other such diatoms, occurs in small colonies. In the case of the, starshaped, colonies of *Asterionella* the main colony sizes are one, two, four, eight and sixteen. These colony sizes presumably derive from synchronous division of the cells in a colony, doubling colony size as newly formed sisters remain attached by inanimate bonds, and a breaking of the colonies into equal parts due to a weakening of those same bonds over time. In a colony of, say, size eight, one therefore has three types of bonds: four bonds linking sisters, two bonds that now link nieces, but originally linked their mothers, and one bond linking what have now become grand-nieces. (If you don't get the picture you might glance ahead to fig. 3.2.4). Below we shall refer to these bonds as respectively the child, parent and grandparent bonds. The age of any child bond equals colony cell age. If our colony of size eight does not double first, thereby transforming children into parents etc., it will in all probability fall apart into two colonies of size four each, due to a breaking of the grandparent bond.

In fast growing populations colony size generally is larger than in slow growing or declining populations. This is easily explained if under better circumstances interdivision times are shorter, but bond weakening always proceeds at the same pace. The nice thing about this is that it allows us to infer some population *growth* characteristic from a *single,* nonrepeated, sample from that population. The first mathematical problem then is to find out how exactly division and breakage rules are related to colony size distribution.

In chapter III we formulated three possible models for the division and breakage mechanisms. All these models had in common that 1) cell division rate only depended on cell age, and 2) that the probability per unit of time that a particular bond breaks depends only on the age of that bond. In the first model we assumed that a bond always breaks after a fixed time A. The second and third model allow for some variability in breakage time. In the second model bonds were assumed to break in complete dependence: if a particular bond breaks all its sisters break as well, and so do its forebears, thus keeping colony size strictly in the main sequence of 2^n, $n=0,1,2,...$ The third model specifically was designed to allow for strays. In it the opposite assumption was made that bonds are totally independent in their breaking behaviour. (Note that the first model is a special case of both the second and third models.) Studying more complicated models, allowing a partial dependence between breakage times in a colony, makes no sense at this stage as it will never be possible to distinguish between them on such meager data as colony size distributions alone. The situation might change though when more detailed culture data become available.

All the results in this subsection are our own and have not been published before. The initial impetus came from ideas and experiments of Marcel Donze (see KOOYMAN, 1976, and WESSEL, 1984). The work is certainly not finished: the doubly starred exercises represent routes we yet want to explore. However, we considered the story so appropriate to these lecture notes that we decided to incorporate it as just a snapshot of some of our current work.

Our assumptions about cell division imply that the total birth rate of cells satisfies the usual renewal equation. We shall assume that cells do not die. Then

$$b(t) = 2\int_0^t g(a)\, b(t-a)da + u(t)\,, \tag{3.2.1}$$

where g is the probability density of the time till division. (The reason for the somewhat aberrant notation is that in this way we are better equipped for the probabilistic arguments below.) And for large t

$$b(t) = ce^{rt} + o(e^{rt})\,, \tag{3.2.2}$$

with r the unique real root of

$$1 = 2\int_0^\infty g(a)e^{-ra}da\,. \tag{3.2.3}$$

EXERCISE 3.2.1: How is g related to the division rate d as introduced in chapter III formulae (5.4.2), (C.6) and (C.7)? ** How is u related to the quantities occurring in these equations?

We start considering the model in which bonds break at exactly age A. To calculate $N_k(t)$, the number of colonies with size 2^k, $k>0$, at time t, we observe that the oldest bond in the colony has age $a_k<A$, k the "generation-index", and that its parent was born at some time τ ago with $\tau>A$ (see fig. 3.2.1).

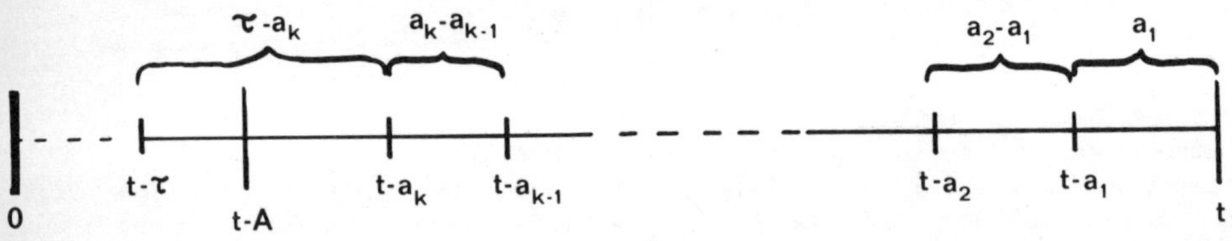

Fig. 3.2.1: Moments at which the bonds in a colony of size k and their last broken forebear were formed, and corresponding interdivision times.

The probability that exactly $k-1$ more divisions occurred between $t-a_k$ and t, so that colony size is 2^k, equals the probability that at least $k-1$ divisions occurred minus the probability that at least k divisions occurred:

$$P\,\{\text{exactly } k-1 \text{ divisions in } (t-a_k,\, t)\} = G_{k-1}(a_k) - G_k(a_k)\,,$$

with

$$G_0(a) = 1, \quad \text{and} \quad G_h(a) = \int_0^a g^{h*}(\alpha)d\alpha\,. \tag{3.2.4a}$$

Therefore, integrating over all possible combinations of birth time of the oldest bond and its parent, and observing that on breaking of the parent two colonies were formed, for $k>0$ and $t>A$,

$$N_k(t) = \int_0^A 2\int_A^t b(t-\tau)g(\tau-a_k)d\tau\,[G_{k-1}(a_k)-G_k(a_k)]da_k + R_k(t)\,, \tag{3.2.4b}$$

where R_k refers to all colonies of size 2^k in which the parent of the presently oldest bond was born before $t=0$. To calculate N_0, the number of single cells, we observe that these cells all have age $a_0>A$. Therefore, for $t>A$,

$$N_0(t) = 2 \int_A^t b(t-a_0)\, \mathcal{G}(a_0) da_0 + R_0(t) , \tag{3.2.4c}$$

where $\mathcal{G}(a_0)=1-G_1(a_0)$ is the probability that no further division has occurred after $t-a_0$.

EXERCISE 3.2.2: Argue that when g is not defective $R_k(t)\to 0$ for $t\to\infty$. Hint: observe that $R_k \leq \sum_m R_m$ and interpret this sum.

EXERCISE 3.2.3: Observe that for $t \geq A$ (3.2.4) can be written as

$$N_k = \psi_k \star b + R_k . \tag{3.2.5}$$

Give expressions for the kernels ψ_k.

For large t we find on setting $p_k=N_k / \sum_m N_m$ and $\tilde{p}_k = \lim_{t\to\infty} p_k$

$$\tilde{p}_0 = c \int_A^\infty e^{-r\tau}\, \mathcal{G}(\tau) d\tau , \tag{3.2.6a}$$

$$\tilde{p}_k = c \int_A^\infty e^{-r\tau} \int_0^A g(\tau-a)\, [G_{k-1}(a)-G_k(a)] da\, d\tau \quad \text{for } k>0 ,$$

with

$$c^{-1} = e^{-rA} \int_0^\infty e^{-r\tau}\, \mathcal{G}(\tau) d\tau . \tag{3.2.6b}$$

EXERCISE 3.2.4: Check (3.2.6). Hint: First observe that $p_k \propto N_k$, then take limits in (3.2.4) after some convenient renormalization. Finally check that for c given by (3.2.6b) $\Sigma\, \tilde{p}_k=1$.

For various special choices of g, like the gamma distributions from exercise 2.A4, it is possible to evaluate (3.2.6) explicitly. We shall leave this as an exercise, but for one exception. When $g(a)=\delta(a-D)$, we find relatively easily for $\alpha := A / D$, i.e. α^{-1} equals the population doubling time expressed in bond life times,

$$\left.\begin{aligned} \tilde{p}_h &= (\tfrac{1}{2})^{h-\alpha-1}-1 \\ \tilde{p}_{h-1} &= 2-(\tfrac{1}{2})^{h-\alpha-1} \end{aligned}\right\} \text{for } \alpha < h < \alpha+1 , \tag{3.2.7}$$
$$\tilde{p}_k = 0 \qquad \text{for } k \neq h,\ h-1 .$$

So given the $\tilde{p}_k$ we can immediately calculate α_1. However, some care is needed in interpreting (3.2.7). For g a delta function at D any births at t give rise to exactly twice that many births at $t+D$ without any "smearing" out. The renewal theorem as formulated in section 2 does not apply, and there is no convergence to a smooth exponential behaviour (also compare II.11). Therefore (3.2.7) cannot be interpreted as the long run values of the p_k for a process with this division time distribution. Instead it should be interpreted as the limit for the coefficient of variation going to zero of the long run limiting values of the p_k for some process in which there is still some variation in the interdivision times. In this sense (3.2.7) provides a useful approximation against which to check empirical data. The outcome from such a comparison not unexpectedly is that there clearly is a great deal of variation in either the interdivision or breakage times, as even in exponentially growing cultures the colony sizes certainly are distributed over a much larger number of classes than two only. So if we wish to fit real data we shall have to use one of the more elaborate families of distributions discussed in the exercises.

EXERCISE 3.2.5: Check (3.2.7). How is α related to the population growth rate r?
Hint: $g^{k*}(a)=\delta(a-kD)$. See further figure 3.2.2.

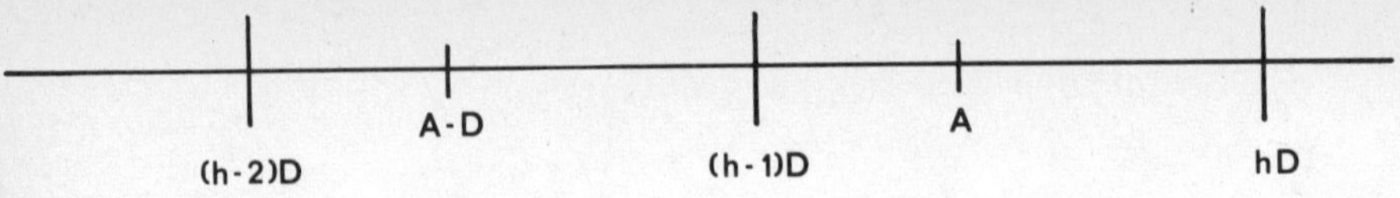

Fig. 3.2.2: The relative timing of breakage and division if the division time distribution is concentrated at D and the breakage time distribution at A.

** EXERCISE 3.2.6: Calculate explicit expressions for the $\tilde{p}_k$ assuming g to be a gamma density as in exercise 2.A4 with k an integer.
Hint: If $g_{k,\alpha}$ denotes the gamma density with parameters k and α as in exercise 2.A4, then $g_{k,\alpha}^{h*} = g_{hk,\alpha}$, and if $\mathcal{G}_{k,\alpha}$ denotes the corresponding survivor function, then for integer k

$$\mathcal{G}_{k,\alpha}(a) = \sum_{i=0}^{k-1} e^{-\alpha a} (\alpha a)^i / i\,! \tag{3.2.8}$$

** EXERCISE 3.2.7: Calculate explicit expressions for the $\tilde{p}_k$ on the assumption that g is a shifted gamma density, *i.e.* $g(a)=0$ for $a<\Delta$ and $g(a)=g_{k,\alpha}(a-\Delta)$ for $a>\Delta$ (notation as in the previous exercise), with k an integer.

When bonds do not break at an exactly fixed age we have some choice in making assumptions about the dependence structure of the breakage times of the different bonds in a colony. If we wish our colony sizes to remain in the main sequence only (2^k, $k=0,1,...$) we could *e.g.* assume that only the oldest bond can break. However, apart from the distinctly unbiological flavour of this assumption, it also makes us loose all nice decoupling properties, like the one we exploited above to such good avail. An assumption which leaves that decoupling property intact is the following: Assume that there exists a monotonically increasing function b of bond age, such that the hazard rate that all bonds of age a and older break equals $b(a)$. Under this assumption the survivor function of any specific individual bond equals

$$\mathcal{F}(a) = \exp[-\int_0^a b(\alpha)d\alpha]\,, \tag{3.2.9a}$$

independent of the ages of the other bonds it is associated with, and the hazard-rate of the i-th generation bonds in a colony breaking but not any younger ones, equals $b(a_i)-b(a_{i-1})$. Biologically such an assumption could be defended from the viewpoint that bonds weaken over time in a deterministic fashion and that bonds break due to random shocks in which all sufficiently weak bonds in a colony break together. This model we shall now develop in more detail.

To calculate the N_k we again look at the last broken forebear of the bonds present now. Let this forebear be born τ time units ago, and let the presently oldest bond have age a. This combination of events can come about in two ways (see fig. 3.2.3): Either the last forebear broke when the presently oldest bond was not born yet, which happens with probability

$$F(\tau-a):=1-\mathcal{F}(\tau-a) \tag{3.2.9b}$$

or it broke at some as yet unspecified time between α and $\alpha+d\alpha$ after the presently oldest bond was born, the presently oldest bond remaining intact at that time, which combination of events happens with probability $[b(\alpha+\tau-a)-b(\alpha)]\,\mathcal{F}(\alpha+\tau-a)d\alpha$. The probability that the presently oldest bond stayed intact till now in these two cases is respectively $\mathcal{F}(a)$ and $\mathcal{F}(a)/\mathcal{F}(\alpha)$. Combining the various possibilities gives

$$H(\tau,a):=P\{\text{ a bond born } \tau \text{ ago and having a child } a \text{ ago has broken itself but its child is still intact }\}$$

$$= F(\tau-a)\,\mathcal{F}(a) + \int_0^a [b(\alpha+\tau-a)-b(\alpha)]\,\mathcal{F}(\alpha+\tau-a)\,\frac{\mathcal{F}(a)}{\mathcal{F}(\alpha)}\,d\alpha\,. \tag{3.2.9c}$$

Using (3.2.9c) we may proceed just as in the previous case. The result is

$$\tilde{p}_k \propto \int_0^\infty e^{-r\tau} \int_0^\tau g(\tau-a)\,H(\tau,a)\,\{G_{k-1}(a)-G_k(a)\}\,da\,d\tau \tag{3.2.9d}$$

$$\tilde{p}_0 \propto \int_0^\infty e^{-r\tau} F(\tau)\,\mathcal{G}(\tau)\,d\tau\,.$$

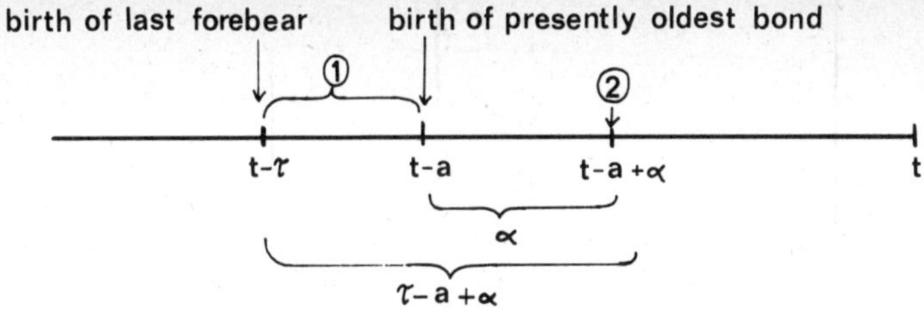

Fig. 3.2.3: Possible times at which the last forebear of the presently oldest bond could have been broken

EXERCISE 3.2.8: Check that when $\mathcal{F}(a)=1$ for $a<A$ and $\mathcal{F}(a)=0$ for $a>A$, $H(\tau,a)=1$ for $a<A<\tau$ and $H(\tau,a)=0$ elsewhere, so that for a degenerate breakage time distribution (3.2.9) indeed reduces to (3.2.6). Hint: Use delta-functions (and $0^2/0=0$).

** EXERCISE 3.2.9: Could you dream up a, non degenerate, family of distributions for which (3.2.9) can be evaluated explicitly?

In the final model to be discussed here we assume that all bonds break in complete independence. This necessarily upsets the decoupling conditions used in the previous two models. If we want to make any progress along the lines of this chapter we therefore have to introduce some other decoupling condition artificially, by restricting the supports of the probability densities g and f of respectively the time between divisions and bond lifetime. As we already know that colony sizes larger than 16 are never seen in practice, we shall assume that bonds can become at most great-grandmothers (4 generations in one colony), *i.e.* we assume that there exist a $D_{\min}$ and an $A_{\max}$, $A_{\max}<4D_{\min}$, such that

$$\left.\begin{aligned} g(a) &= 0\\ \mathcal{G}(a) &= 1\\ G(a) &= 0 \end{aligned}\right\} \text{ for } a<D_{\min}, \qquad \left.\begin{aligned} f(a) &= 0\\ \mathcal{F}(a) &= 0\\ F(a) &= 1 \end{aligned}\right\} \text{ for } a>A_{\max}, \tag{3.2.10}$$

where $\mathcal{G}$ and $\mathcal{F}$ and G and F are respectively the survivor functions and distribution functions corresponding to g and f, defined in the usual manner.

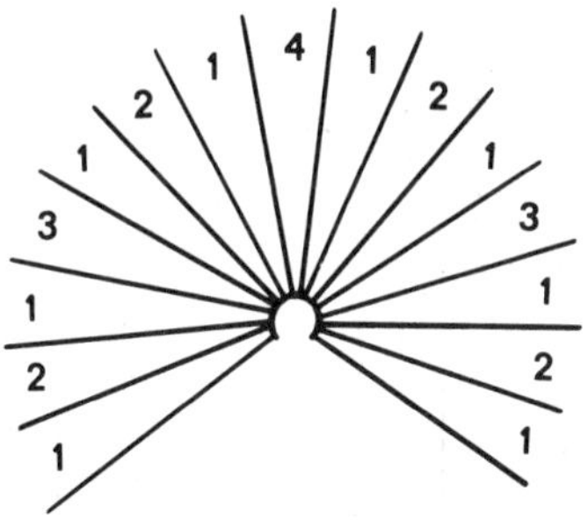

Fig. 3.2.4: (Hypothetical) colony of size 16 with the relative ages of the bonds (central) joining the radiating cells, 1 indicating the youngest bonds and 4 indicating the oldest one.

If bonds break independently colonies of all sizes may occur. Therefore we have to change our notation. M_i will denote the number of colonies of size i and q_i their relative frequency. (So $N_k=M_{2^k}$ and $p_k=q_{2^k}$). Due to the simplifying assumption (3.2.10) we can consider any colony as fragments of some hypothetical colony of size 16 (see figure 3.2.4). If the ages of the bonds in that colony are a_1 to a_4, where the numbering is as in fig. 3.2.4, then it is a simple combinatorial exercise to calculate the various ways in which a colony of size k can be obtained as a breakage product and their corresponding probabilities. Integrating over a_1 to a_4 then finally gives:

$$\tilde{q}_i \propto \int_0^\infty \int_0^{a_4} \int_0^{a_3} \int_0^{a_2} e^{-ra_4} g(a_4-a_3)g(a_3-a_2)g(a_2-a_1)\mathcal{G}(a_1)C_i^4(a_1,..,a_4)da_1..da_4 \tag{3.2.11a}$$

with

$$C_{16}^4 = \mathcal{F}(a_4)\, \mathscr{F}^2(a_3)\, \mathscr{F}^4(a_2)\, \mathscr{F}^8(a_1)$$

$$\cdot$$

$$\cdot \tag{3.2.11b}$$

$$\cdot$$

$$C_2^4 = \mathcal{F}(a_1)\, F(a_2)[2 + 4\, F(a_3) + 2\, F(a_4)] + F^2(a_1)[4\, \mathcal{F}(a_2) + 2\mathcal{F}(a_3) + \mathcal{F}(a_4)]$$

$$C_1^4 = F(a_1)[2 + 8F(a_2) + 4F(a_3) + 2F(a_4)]$$

EXERCISE 3.2.10: Calculate C_3^4 to C_{15}^4 by direct counting of the possibilities.

EXERCISE 3.2.11: Calculate the $\tilde{q}_i$ for g a delta function at D.

** EXERCISE 3.2.12: Let C_1^k to $C_{2^k}^k$ denote the coefficients analogous to C_1^4 to C_{16}^4 but now for the breakage products of a colony of size 2^k. Find a way to express the C_i^k, $i=1,..,2^k$ in the C_j^{k-1}, $j=1,..,2^{k-1}$. Use this recurrence relation to rederive the result of the previous exercise.
REMARK 3.2.13: Using the result of this exercise and computer methods it should be possible in principle to choose k very large, and in this way to relax the restricting conditions on which (3.2.11) was based.

** EXERCISE 3.2.14: In the three models discussed thus far we assumed that all individuals in a colony divide simultaneously. This is in accord with the observation that in a sample always either no cells in a colony are in the process of dividing or all of them are. However, models in which cells in a colony are less than fully dependent are thinkable as well. Again as an extreme one might assume that cells divide totally independently. (If the interdivision time distributions have but a small coefficient of variation and colony size is small, then cells in one colony would still tend to divide close to each other due to the trivial "synchronization" of their single common ancestor.) Try to divise models of this kind.

EXERCISE 3.2.15: How would the results obtained above be modified when a constant nonzero death rate is assumed. (Hint: Use the interpretation, there is almost no need to do any calculations!)

The three models described in this subsection represent the stage we have reached thus far with the *Asterionella* problem (or even more than that if you take account of the doubly starred exercises). The final stage of all this work should be the comparison of the calculated distributions for the various models with real, laboratory and field, data. After a model is found to fit sufficiently well the relationship of the estimated model parameters to the population growth rate should be worked out, *e.g.* by using an expansion procedure like that described in 2.A. We hope soon to find time to try our hand at all this.

You will no doubt have noticed that we did not refer even once to the differential equations for the *Asterionella* models derived in chapter III. The techniques used in this chapter simply are more powerful in this particular case. Yet it was precisely our investigation of those differential equations which gave us a sufficient grasp of the problem to come up with the solutions presented here. Moreover the fact that we could show (3.2.6) to match with the eigenfunction of (III.C.6) for the eigenvalue r gave us the confidence to proceed with the present approach.

* EXERCISE 3.2.16: Derive (3.2.6) from (III.C.6). Hint: First derive a formula for the stationary p-state distribution $\tilde{n}$ using the kind of reasoning which was used to derive (3.2.6). Check that the result satisfies $A\tilde{n}=r\tilde{n}$ where A is the differential generator from (III.C.6). Finally calculate the $\tilde{p}_i$ from the $\tilde{n}_i$.

** EXERCISE 3.2.17: Derive (3.2.9) from (III.C.7).

*** EXERCISE 3.2.18: Derive (3.2.11) from (III.5.4.2).

4. Some nonlinear extensions of the linear theory

In this section and the next one we shall consider various examples of nonlinear results which are directly based on the linear theory from section 2. The results in the present section discuss how in some cases p-behaviour far from the trivial equilibrium (the equilibrium at which $b = N = 0$) can be related to the behaviour of the linear problem derived by linearizing near that equilibrium. In the next section we deal with conditions which allow the reduction of the full p-equations of age dependent population dynamics to a differential equation on a finite dimensional space. A general survey of techniques for tackling nonlinear population problems is given in chapter VI. A good mathematical discussion of the general theory of nonlinear age dependent population dynamics can be found in WEBB (1985a). HOPPENSTEADT (1975) and especially NISBET & GURNEY (1982) provide some interesting biological perspectives.

In subsection 4.1 we discuss what might well be the oldest population dynamical model formulated as a nonlinear Volterra integral equation, the Kermack-McKendrick equation for the general epidemic. The main reason for treating this example, however, is that it allows us to introduce what might be considered the natural extension of the classic renewal theorem to the nonlinear realm. In subsection 4.2 we show for the *Daphnia* example from chapters I and III how the solution of the linear renewal equation may be used to construct an estimate for the solution of a nonlinear problem, and thereby to prove global stability of the trivial equilibrium.

4.1. Kermack's and McKendrick's (1927) general epidemic and the nonlinear renewal theorem

In (1927) KERMACK and MCKENDRICK introduced a general epidemic model, the assumptions of which have been set forth in example 1.3.3. For this model they proved the famous so-called threshold theorem, and also showed how to calculate the initial rate of increase of the epidemic from a linear renewal equation obtained by linearizing their general equation around the uninfected state. Oddly enough the general Kermack & McKendrick model was largely neglected by later applied mathematicians (see remark 1.3.4). Only in the 1970's KERMACK's and MCKENDRICK's work was again brought to the attention in the work of REDDINGIUS (1971) and WALTMAN (1974). As an answer to some of the questions raised by REDDINGIUS, METZ (1978) conjectured the "nonlinear renewal theorem" to be discussed below. The first part of this theorem was subsequently proved by DIEKMANN (1977; see also DIEKMANN & KAPER, 1978, and DIEKMANN & VAN GILS, 1984), the second part by GRIPENBERG (1983c) (but with a wrong reference as to the origin of the confecture!).

The answers to the exercises in this subsection can all be found in METZ (1978) or REDDINGIUS (1971).

Let s denote the number of susceptibles, $\dot{s}$ its time derivative, $\phi(\tau)$ the mean infectivity of an individual first infected τ time units ago (ϕ is called the infectivity kernel), and $g(t)$ the infectivity at time t due to outside sources and/or individuals that were already ill before $t=0$, then Kermack's and McKendrick's equation is (see example 1.3.3).

$$\dot{s}(t)/s(t) = \int_0^t \dot{s}(t-\tau)\phi(\tau)d\tau - g(t). \tag{4.1.1}$$

In words: The relative rate of change of the susceptible number equals minus the total infectivity, which in turn equals the cumulation of all the infectivities produced by individuals infected after $t=0$ (NB $\dot{s}$ is negative) plus the infectivity contributed by individuals already ill at $t=0$ and/or outside sources. Note that no new susceptibles are produced either through births or through previously infected individuals returning to the susceptible population.

EXERCISE 4.1.1: Assume that an infected organism stays ill for some random time with survivor function $\mathcal{F}$ after which it dies or recovers and becomes immune forever, and that an ill organism which contracted its illness τ time units ago has infectivity $a(\tau)$. Assume moreover that the susceptible population does not change due to other causes than individuals succumbing to disease. Write down joint differential equations for the "age" distribution of ill organisms (where "age" refers to the illness) and the susceptible number. Derive an equation of the form (4.1.1) by solving the partial differential equation for the age distribution by integration along characteristics on the supposition that $\dot{s}$ is known (Compare example 1.4.1 and in particular exercise 1.4.2).

EXERCISE 4.1.2: Assume that an individual illness follows a k-state, continuous time Markov chain with killing, where "killing" means either to recover and be immune forever or to die. Let the differential generator of this chain be B, let a newly infected organism enter the states of illness according to a probability vector C, and let the infectivity of the various states of illness be collected in a vector A. Write down joint differential equations for s and the distribution N of ill organisms. By solving the (linear) differential equation for N as if $\dot{s}$ were known and substituting the result in the equation for s derive an equation of the form (4.1.1). (Compare example 1.4.3 and in particular exercise

1.4.4.).

For the purpose of our further calculations it is easier to transform (4.1.1) into two more manegeable forms. Integrating once we find

$$\ln(s(t)/s_0) = \int_0^t s(t-\tau)\phi(\tau)d\tau - s_0 \int_0^t \phi(\tau)d\tau - h(t) \tag{4.1.2}$$

with

$$h(t) := \int_0^t g(\tau)d\tau \tag{4.1.3}$$

A rescaling

$$p := (s_0 - s)/s_0, \qquad \gamma := \int_0^\infty \phi(\tau)d\tau, \qquad \psi := \gamma^{-1}\phi \tag{4.1.4}$$

where p is the fraction of victims and γ is the total infective strength of the disease, then gives

$$-\ln(1-p(t)) = \gamma s_0 \int_0^t p(t-\tau)\psi(\tau)d\tau + h(t)\,. \tag{4.1.5}$$

This is the first of the two equations which we shall concentrate on below. The quantity

$$R = \gamma s_0 \tag{4.1.6}$$

will below, when we consider the linearized form of (4.1.5) around the uninfected state, be found to equal the mean number of secondary infections produced by one freshly infected individual introduced in a totally susceptible population of size s_0. R is known as the *net reproductive number.* Finally setting

$$x := -\ln(1-p), \qquad p = 1-e^{-x} \tag{4.1.7}$$

gives the *nonlinear renewal equation*

$$x(t) = \gamma s_0 \int_0^t f(x(t-\tau))\psi(\tau)d\tau + h(t) \tag{4.1.8}$$

with

$$f(x) := 1-e^{-x}\,. \tag{4.1.9}$$

With the definitions (4.1.3), (4.1.4), (4.1.7) and (4.1.9) equations (4.1.1), (4.1.2), (4.1.5) and (4.1.8) are all equivalent and we shall freely switch from one to the other depending on which is most convenient for a particular purpose.

EXERCISE 4.1.3: Check the calculations leading from (4.1.1) to (4.1.8). Also follow the route backwards to show the equivalence of all these equations.

As a first step in the analysis of our epidemic model we shall consider its large time behaviour. Since s can only decrease and cannot become negative, $s(t)$ has to go to a limit $s_\infty \geqslant 0$. It is easier, and more useful, to calculate this limit for the scaled equation (4.1.5). Letting $t \to \infty$ in this equation we obtain

$$-\ln(1-p_\infty) = \gamma s_0\, p_\infty + h_\infty\,. \tag{4.1.10}$$

Figure 4.1.1 shows p_∞ for various values of $\pi := f(h_\infty)$.

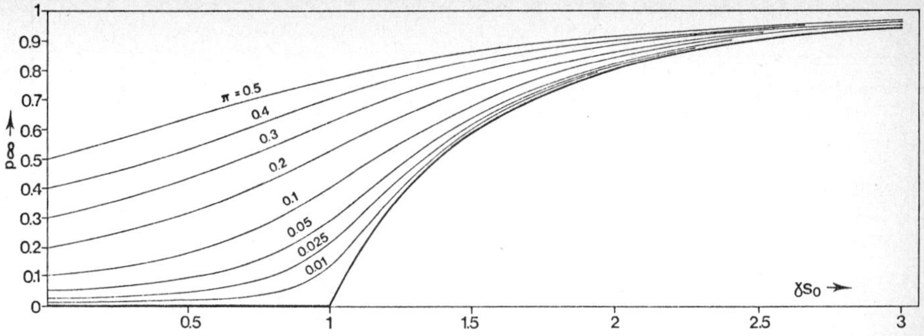

Fig. 4.1.1 The eventual fraction of (new) victims as a function of the number γs_0 of secondary infections produced by one freshly infected individual introduced into a totally uninfected population of size s_0 for various values of $\pi = 1 - e^{-h_\infty}$, where h_∞ is the total starting infectivity.

We see that for small total starting infectivity h_∞ the final epidemic size is negligible for $\gamma s_0 \leqslant 1$ but not for $\gamma s_0 > 1$. This observation is the content of Kermack's and McKendrick's celebrated *threshold theorem:*

$$\lim_{h_\infty \downarrow 0} p_\infty := \hat{p}_\infty \begin{cases} = 0 & \text{for } \gamma s_0 \leqslant 1 \\ > 0 & \text{if } \gamma s_0 > 1 \end{cases} . \tag{4.1.11}$$

The continuity of the solutions of (4.1.1) or its equivalent equations in the forcing functions g or h implies for all fixed $t > 0$ that $p(t) \downarrow 0$ for $h_\infty \downarrow 0$. Hence

$$0 = \lim_{t \uparrow \infty} \lim_{h_\infty \downarrow 0} p(t) \begin{Bmatrix} = \\ \\ < \end{Bmatrix} \lim_{h_\infty \downarrow 0} \lim_{t \uparrow \infty} p(t) = \hat{p}_\infty \begin{cases} \text{for } \gamma s_0 \leqslant 1 \\ \\ \text{for } \gamma s_0 > 1 . \end{cases} \tag{4.1.12}$$

As p necessarily increases monotonically with time, this may be interpreted as a stability of the uninfected state for $\gamma s_0 \leqslant 1$ versus an instability for $\gamma s_0 > 1$.

EXERCISE 4.1.4: Show by graphical means that $\hat{p}_\infty$ is a well-defined nonnegative solution of

$$-\ln(1 - \hat{p}_\infty) = \gamma s_0 \hat{p}_\infty . \tag{4.1.13}$$

Analyse how the solution set of this limit equation depends on γs_0. Which branches are the relevant ones?

EXERCISE 4.1.5: Show that for $\gamma s_0 >> 1$

$$p_\infty = 1 - e^{-(\gamma s_0 + h_\infty)} + o(e^{-(\gamma s_0 + h_\infty)}) \tag{4.1.14}$$

EXERCISE 4.1.6: Write for $\gamma s_0 - 1 << 1$ the relevant solution to (4.1.13) as $\hat{p}_\infty = a_1(\gamma s_0 - 1) + a_2(\gamma s_0 - 1)^2 + \cdots$. Calculate a_1 and a_2. (Hint: For $\gamma s_0 - 1 \gtrless 0$ you have to take different branches of the solution set, and therefore different values of the a_i.)

EXERCISE 4.1.7: Write for fixed $\gamma s_0 < 1$ the solution of (4.1.10) as $p_\infty = a_1 h_\infty + a_2 h_\infty^2 + \cdots$. Calculate a_1 and a_2. Could you extend this procedure to $\gamma s_0 > 1$?

EXERCISE 4.1.8: Analyse the simplest possible of all epidemic models given by

$$\frac{ds}{dt} = -\alpha s y \qquad \frac{dy}{dt} = \alpha s y - \beta y , \tag{4.1.15}$$

y the number of ill individuals. Start with an explicit calculation of the trajectories in the (s,y)-plane followed by an analysis of the equilibria and their stability properties.

EXERCISE 4.1.9: In the model from exercise 4.1.1 let

$$u(t) := \int_0^\infty n(t,\tau)w(\tau)d\tau\,, \qquad w(\tau) := \int_\tau^\infty a(\sigma)\,\mathcal{F}(\sigma)d\sigma\,/\,\mathcal{F}(\tau), \qquad \gamma = w(0)\,. \tag{4.1.16}$$

(What is the interpretation of w?) Show that

$$\frac{du}{dt} = \gamma sy - y \tag{4.1.17}$$

with y defined by

$$\frac{ds}{dt} = -ys\,. \tag{4.1.18}$$

Use this result to find a first integral of the p-equations in the form of a relation between s and u. Finally use this first integral to derive Kermack's and McKendrick's threshold theorem for this special case.

EXERCISE 4.1.10: In the model from exercise 4.1.2. let

$$u := WN, \qquad W := -A^T B^{-1}, \qquad \gamma = WC \tag{4.1.19}$$

Same questions as the previous exercise. Hint: The components of $-B^{-1}$ can be interpreted as the mean times that an individual presently in state of illness j will still spend in state of illness i.

So far we concentrated on the fraction of individuals ever to get infected. But we would like to know more. For example, does the maximum number ill at the same time also approach a positive limit for the initial infection going to zero, and if so, how can we characterize this limit? That is, we also want a characterization for small starting infections of the behaviour of the transients from an (almost) fully susceptible p-state to the final p-state after the epidemic has come to its natural end. Such a characterization is provided by what by some authors has been called the "nonlinear renewal theorem" (THIEME, 1985). Here we shall give a heuristic introduction in the context of the epidemic model. Before doing so we shall first have a look at the behaviour of the linearized equation near the noninfected state.

For small p (small x, see (4.1.7)) we have $p \simeq x \simeq f(x)$. Therefore p is approximated by the solution of the linear renewal equation

$$q(t) = \gamma s_0 \int_0^t q(t-\tau)\,\psi(\tau)d\tau + h(t)\,, \tag{4.1.20}$$

as long as both p and q are small. The characteristic equation corresponding to (4.1.20) is

$$1 = \gamma s_0\,\bar{\psi}(r) \tag{4.1.21}$$

Due to our normalization $\bar{\psi}(0)=1$, *i.e.* the net reproductive number R equals γs_0 as anounced previously. So the solution to (4.1.20) will stay bounded when $\gamma s_0 < 1$ and eventually grow exponentially when $\gamma s_0 > 1$. This is just the local linearized counterpart of the global stability result derived previously.

EXERCISE 4.1.11: For $\gamma s_0 < 1$ what tells (4.1.20) you about the behaviour of $q(t)$ for $t \to \infty$? Compare the result with the result from exercise 4.1.7. Hint: remember that $h(t) \to h_\infty$.

Now consider the full equation for $\gamma s_0 > 1$. If we start with a smaller and smaller starting infection it will take longer and longer before the epidemic reaches an appreciable level. This also means that we stay longer and longer in the region where the linear approximation applies. This in turn means that the detailed form of the initial conditions (the forcing function h) has less and less influence on the behaviour of p when finally the region is reached where the influence of the nonlinearity is felt: by the time p reaches that region the process has already stabilized to exponential growth. But we also have to look further and further into the future to see any interesting behaviour at all. (Fig. 4.1.2 shows an example of this phenomenon.) These observations form the basis for the following:

THEOREM 4.1.12: "Nonlinear renewal theorem, part I". *Assume that* $\gamma s_0 > 1$. *Choose some* $p_0 \in (0, \hat{p}_\infty)$ *and define* t_0 *by* $p(t_0) = p_0$ *(note that* t_0 *depends on* h*!). Then for all* t

$$\lim_{h_\infty \downarrow 0} p(t+t_0) = \hat{p}(t) \tag{4.1.22}$$

uniformly in t, *with* $\hat{p}(t)$ *defined as the unique positive solution of*

$$-\ln(1-\hat{p}(t)) = \gamma s_0 \int_0^\infty \hat{p}(t-\tau)\psi(\tau)d\tau \tag{4.1.23}$$

with

$$\hat{p}(0) = p_0 \,. \tag{4.1.24}$$

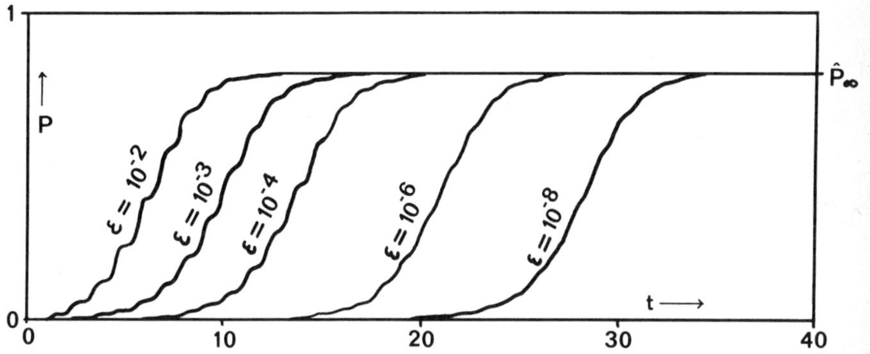

Fig.4.1.2 Fraction of victims as a function of time after an initial infection with a fraction ϵ freshly infected individuals for a scaled infectivity kernel $\psi(\tau)=0$ for $\tau\in[0,1)\cup[1.2,\infty)$ and $\psi(\tau)=5$ for $\tau\in[1,1.2)$, and a threshold parameter $\gamma s_0=2$.

The nonlinear renewal theorem part I is proved in DIEKMANN (1977) for the nonlinear renewal equation (4.1.18) on the general assumption that the function f is sublinear, *i.e.* $f(\alpha x)<\alpha f(x)$ for all $\alpha>0$, but the proof is flawed. DIEKMANN & KAPER (1978) use Tauberian methods to give a rigorous proof of the required estimates in a much more general setting. A completely different approach, using semi-group techniques and the idea of an invariant manifold, is adopted by DIEKMANN & VAN GILS (1984, section 7; for technical simplicity they assume $\psi(\tau)$ to be zero from some τ_0 onwards).

EXERCISE 4.1.13: Verify that f defined by (4.1.9) is sublinear.

* EXERCISE 4.1.14: Find the flaw in DIEKMANN (1977).

The fact that the limit in (4.1.22) is uniform allows us to infer that any quantity like the total number of individuals ill, that can be calculated as a function of time from a convolution of $-\dot{s}=s_0\dot{p}$ with a smooth bounded function h, also converges under the same limiting operation, and that their limits can be calculated from the convolution of h with $s_0\dot{\hat{p}}$. This in turn allows us to infer that also quantities like the maximum number of individuals ever to be ill at the same time converge, thereby answering the question posed earlier.

The "limit epidemics" $\hat{p}$ for different values of p_0 are all translates of each other. They increase monotonically and

$$\lim_{t\uparrow\infty} \hat{p}(t) = \hat{p}_\infty \,. \tag{4.1.25}$$

Moreover

$$\hat{p}(t) = e^{rt}\,(C(p_0) + o(1)) \quad \text{for } t\downarrow-\infty \,, \tag{4.1.26}$$

that is, the behaviour of our limit epidemics for $t\downarrow-\infty$ directly relates to the behaviour of the solution q to our linearized epidemic equation (4.1.20) for $t\uparrow\infty$. This allows us to calculate an estimate for t_0 by an appropriate matching of the two "limits". The result is provided by the following theorem first conjectured by METZ (1978) and subsequently proved by GRIPENBERG (1983c).

THEOREM 4.1.15: "Nonlinear renewal theorem, part II". *Let $\hat{p}_1(t)$ denote the unique solution to (4.1.23) for which*

$$\lim_{t\downarrow-\infty} e^{-rt}\,\hat{p}_1(t) = 1 \,, \tag{4.1.27}$$

then

$$\lim_{h_\infty\downarrow 0} (t_0 + r^{-1}\overline{h}(r)) = \hat{p}_1^{(-1)}(p_0) + r^{-1}\ln\overline{\Phi}(r) \tag{4.1.28a}$$

with

$$\bar{h}(s) = \int_0^\infty e^{-st}\, h(t)dt, \qquad \bar{\Phi}(s) = \int_0^\infty e^{-st}\, t\gamma s_0 \psi(t)dt \tag{4.1.28b}$$

and $\hat{p}_1^{(-1)}:(0,\hat{p}_\infty)\rightarrow\mathbb{R}$ *the inverse function of* $\hat{p}_1$.

The two parts of the nonlinear renewal theorem together totally cover the behaviour of the Kermack & McKendrick epidemic for small initial infections (and in practice most initial infections are small!): when h_∞ is small $p(t)$ apparently behaves as some suitable translate of the fixed curve $\hat{p}_1$ independent of the shape of h, where $\hat{p}_1$ is defined by (4.1.23) with (4.1.27) (an effective manner for actually calculating $\hat{p}_1$ is provided by exercise 4.1.17), an estimate of the amount by which $\hat{p}_1$ has to be translated being provided by (4.1.28). As proved, however, the theorem applies to a much wider class of models which can be (re)expressed in the form of the nonlinear renewal equation (4.1.8). See THIEME (1985) for further generalizations.

REMARKS 4.1.16: (i) So far the nonlinear renewal theorem has only been proved under rather restrictive technical assumptions on the function f in (4.1.8). It is to be expected that in the future other versions using other conditions and other convergence concepts will become available *(e.g.* THIEME, 1985).
(ii) METZ (1978) also gives a formulation of the theorem for the full stochastic model with formed the implicit basis of our deterministic equations. This is of some importance as our limiting procedure inevitably brings us into the regime where the number of ill individuals are so low that the law of large numbers on which the deterministic model is essentially based, is bound to fail.

EXERCISE 4.1.17: Calculating the limit epidemic is easiest starting from the scaled limiting form of equation (4.1.1):

$$\dot{\hat{p}}(t) = (1-\hat{p}_1(t))\,\gamma s_0 \int_0^\infty \dot{\hat{p}}_1(t-\tau)\psi(\tau)d\tau\,. \tag{4.1.29}$$

Assume that

$$\hat{p}_1(t) = \sum_{j=1}^\infty c_j\, e^{jrt} \tag{4.1.30}$$

By inserting this expression in (4.1.29) derive an equation for r as well as a recurrence relation for the c_j. In METZ (1978) it is shown that for bounded ψ the series expansion (4.1.30) converges on a left half line. Given the value of $\hat{p}_1$ on a left half line we can use any of the equations (4.1.23) or (4.1.29) to extend it numerically to the right.

EXERCISE 4.1.18: Use the result from the previous exercise to find the first few terms in a series expansion of $\ln\hat{p}_1^{(-1)}(p_0)$ for small p_0.

4.2. Population decline in ectotherms

In I.3 a model was presented describing the population growth in ectothermic animals. The main assumptions were that individual growth and reproduction depend on individual size l and food availability x and that the death rate μ was constant except for the possibility of death due to starvation of large animals when their food intake cannot keep pace with their maintenance metabolism. In III.4.3 and III.5.4 this model was extended to include general age dependent mortality as well. The resulting equations were (III.4.3.2):

$$\begin{aligned}
&\frac{\partial l(t,a)}{\partial t} = -\frac{\partial l(t,a)}{\partial a} + g(x,l(t,a)) && 0<a<a_m\leqslant\infty\\
&l(t,0) = l_b\\
&\frac{\partial n(t,a)}{\partial t} = -\frac{\partial n(t,a)}{\partial a} - \mu(a)n(t,a) && \text{for } l(t,a)\leqslant\bar{l}(x)\\
&n(t,a) = 0 && \text{for } l(t,a)>\bar{l}(x)\\
&n(t,0) = b(t)\\
&b(t) = \int_0^{a_m} \lambda(x,l(t,a))n(t,a)da\,.
\end{aligned} \tag{4.2.1a}$$

Equation (4.2.1a) updates the instantaneous age-length relation, (4.2.1b) updates the age distribution and (4.2.1c) is just the birth law. In this section we shall indicate how, given some natural monotonicity assumptions on g and β, the linear theory can give us some easy (but not very spectacular) results about the solutions to (4.2.1) when food availability is dynamically coupled to the population development. (In VI.2 and VI.4 similar results will be presented, but in a different context).

An inspection of the explicit formulae for g, λ and $\bar{l}$ from I.3 tells us that

g is nonnegative, Lipschitz continuous in x and l, increasing in x and decreasing in l, and $g(x,l) = 0$ for $l > \kappa \bar{l}(x)$, $\kappa < 1$,

$\bar{l}$ is nonnegative, and continuous and increasing in x.

λ is bounded, nonnegative, continuous and nondecreasing in x and piecewise continuous in l, and for x sufficiently small $\lambda(x,l) = 0$ for all l.

Moreover we shall assume that

μ is nonnegative, integrable on intervals bounded away from a_m and if $a_m = \infty$ then $\int_0^\infty \mu(a)da = \infty$, *i.e.* individuals cannot live on forever.

First we consider (4.2.1) for x a given, Lipschitz continuous, function of time. In that case (4.2.1a) can be solved simply by integration along the characteristics, giving us $l(t,a)$ for all a and t. Using the known $l(t,a)$ we can then solve (4.2.1b) by integration along the characteristics as

$$n(t,a) = \begin{cases} b(t-a)\exp[-\int_0^a \mu(\alpha)d\alpha]\, \chi(t,0,a) & \text{for } t \geq a \\ n(0,a-t)\exp[-\int_{a-t}^a \mu(\alpha)d\alpha]\, \chi(t,a-t,a) & \text{for } t < a \end{cases} \tag{4.2.2}$$

with

$$\chi(t,a_0,a) := \begin{cases} 1 & \text{if } l(t-a+\alpha,a) \leq \bar{l}(x(t-a+\alpha)) \text{ for all } \alpha \in [a_0,a] \\ 0 & \text{otherwise} \end{cases} \tag{4.2.3}$$

Finally $b(t)$ can be calculated from the usual generation expansion

$$b = \sum_{i=0}^{\infty} b_i \tag{4.2.4a}$$

with

$$b_0(t) = \begin{cases} \int_t^{a_m} \lambda(x(t),l(t,a))n(0,a-t) \exp[-\int_{a-t}^a \mu(\alpha)d\alpha]\, \chi(t,a-t,a)da & \text{for } t \leq a_m \\ 0 & \text{for } t \geq a_m \end{cases} \tag{4.2.4b}$$

$$b_i(t) = \int_0^{\min(t,a_m)} \lambda(x(t), l(t,a))b_{i-1}(t-a) \exp[-\int_0^a \mu(\alpha)d\alpha]\, \chi(t,0,a)da \,.$$

In the particular case that x is a fixed constant the death through starvation-mechanism operates at most once, on the initial condition, but after that no individual can grow sufficiently large for the mechanism to come into operation again. Moreover $l(t,a)$ does not depend on t for $a<t$. Therefore b satisfies just the usual linear renewal equation of sections 1 and 2. The various monotonicity properties imply that r is an increasing function of x, and that there exists an x_c such that $r(x_c)=0$ (*cf.* I.3). Moreover, if $x<x_c$ both $b(t) \to 0$ and $N(t) := \int n(t,a)da \to 0$ for $t \to \infty$, just as in the model with constant death rate discussed in I.3.

Interestingly the last result immediately extends to non constant x: if $x(t) \leq x_c - \epsilon$ for all t then $b(t) \to 0$ and therefore also $N(t) \to 0$ for $t \to \infty$. This can be seen as follows. First we observe that if we shut off the death through starvation mechanism, *i.e.* if we set $\chi = 1$ in (4.2.2) and (4.2.4), this can only make n and b larger. Next we observe that if for the process modified in this manner x' and x'' are two possible courses of the food availability such that $x'(t) \leq x''(t)$ for all t, then also for the corresponding birth rates and age densities $b'(t) \leq b''(t)$ and $n'(a,t) \leq n''(a,t)$

(assuming that the initial condition is the same). Setting $x''(t)=x_c-\epsilon$ then gives the desired result.

Finally we consider the case where the food availability is dynamically coupled to the population:

$$\frac{dx}{dt} = h(x)-u(x)\int_0^{a_m} v(l(t,a))n(t,a)da \tag{4.2.5}$$

with h Lipschitz continuous, positive on $(0,x_e)$ and negative on (x_e,∞), u and v both continuous nonnegative and increasing. The coupled system (4.2.1) and (4.2.5) has at most two equilibra, one in which $N=0$ and $x=x_e$, the other in which $x=x_c$ and N is calculated by setting $dx/dt=0$ in (4.2.5) (compare I.3.5). The latter equilibrium can only exist for $x_c<x_e$. The results derived thus far now allow us to prove that for $x_c>x_e$ the population will go extinct, independent of the initial conditions. To this end we observe that if x satisfies (4.2.5) then $\frac{dx}{dt}<h(x)$. Therefore there exists a t_0 such that $x(t)<x_e=x_c-\epsilon$ for all $t>t_0$. Combining with the result previously obtained for (4.2.1) we arrive at the desired result.

Remark 4.2.1: It is also possible to extend the argument to cover the case $x_c=x_e$. We shall only give a heuristic sketch. When $x_c=x_e$ there exists a t_0 such that $x(t)<x_c$ for all $t>t_0$. Therefore the mean number of offspring of individuals born after t_0 is bound to be smaller than one. Nonextinction would imply that with the advance of the generations this number converges to one. By (an extension of) the renewal theorem this implies that $b(t)\to\hat{b}\geqslant 0$. If $\hat{b}$ were positive this would imply that $x(t)<x_c-\epsilon$ from some t onwards for some $\epsilon>0$. But this in turn implies that the mean offspring number does not go to one contradicting the assumption of nonextinction.

5. Models allowing a reduction to a differential equation on $\mathbb{R}^k$.

In this final section we consider two classes of nonlinear age dependent models which allow a reduction of the full p-equation for the age distribution to a differential equation on a finite dimensional space. For the sake of the exposition we shall assume that we are dealing with an autonomous system on the space of age distributions. The extension to systems of p-equations coupled through their inputs and outputs as introduced in III is fairly straightforward. In the first subsection the transformation will be made through a direct application of the linear convergence result from 2.3.1. The second subsection extends the results from 2.3.3 to the nonlinear realm, (but the development is totally independent of that in 2.3.3!) Neither of the two classes of results are specific for age dependent models only. However the restriction to age dependent processes for the second class of problems allows us to give a fairly complete characterization of the nonlinear finite dimensional representability problem.

5.1. Models with a separable death rate

We shall say that a death rate $\mu(a,n)$ depending on age as well as on the p-state n as a whole, is separable when it can be written as

$$\mu(a,n) = \mu_1(a) + \mu_2(n) \tag{5.1.1}$$

If the only nonlinearity in the p-differential generator is through a separable dependence of the death rate on n then it is usually possible to capture the behaviour of the full population model asymptotically for large t by a differential equation in just one variable. This trick seems to be due independently to among others Gurtin, MacCamy, Coleman and Simmes (see Gurtin, 1980-1982), Prüss (1983a), Busenberg & Ianelli (1983, 1985) and Webb (198?a). We shall first illustrate the idea by a simple example, and thereafter put it in a general semigroup setting (thereby effectively lifting the restriction to age dependent models).

Example 5.1.1: Consider an age-structured population in which the only density dependence manifests itself through an increase of the age specific death rate $\mu_1(a)$ with an age independent factor $\mu_2(p)$, where p is the weighted average

$$p(t) = \int_0^\infty \gamma(a)n(t,a)da\,. \tag{5.1.2a}$$

The problem

$$\frac{\partial}{\partial t}n(t,a) = -\frac{\partial}{\partial a}n(t,a)-(\mu_1(a) + \mu_2(p))n(t,a) \qquad n(t,0) = \int_0^\infty \beta(a)n(t,a)da \tag{5.1.2b}$$

can be reduced to the linear problem

$$\frac{\partial}{\partial t} m(t,a) = -\frac{\partial}{\partial a} m(t,a) - \mu_1(a)m(t,a) \qquad m(t,0) = \int_0^\infty \beta(a)m(t,a)da \tag{5.1.3}$$

by the (implicit) transformation

$$m(t,a) := e^{\int_0^t \mu_2(p(\tau))d\tau} n(t,a)\,. \tag{5.1.4}$$

The linear theory from 2.3.1 now tells us that

$$m(t,a) = ce^{rt}\,(\tilde{m}(a) + o(1)) \qquad \text{for } t \to \infty\,, \tag{5.1.5}$$

where r is the dominant real eigenvalue corresponding to (5.1.3) and $\tilde{m}$ is the corresponding stable age distribution. Substitution in (5.1.4) and then into (5.1.2a) yields

$$p(t) = ce^{rt - \int_0^t \mu_2(p(\tau))d\tau} \left[\int_0^\infty \gamma(a)\tilde{m}(a)da + o(1)\right]. \tag{5.1.6}$$

Omitting the o (1) term we obtain by formal differentiation

$$\frac{dp}{dt} = (r - \mu_2(p))p\,. \tag{5.1.7}$$

Therefore we conclude that the asymptotic behaviour of n is revealed by (5.1.4), (5.1.5) and a study of (5.1.7).

A rigorous presentation of this useful trick in a more general setting may be found in WEBB (198?a).

A somewhat different approach to the same problem, again in a more general setting, has been taken by BUSENBERG & IANELLI (1983, 1985). They derive a (nonlinear) differential equation for the normalized age distribution

$$w(t,a) := n(t,a) \Big/ \int_0^\infty n(t,\alpha)d\alpha \tag{5.1.8}$$

and observe that one gets one and the same equation for all nonlinearities μ_2, including the special case $\mu_2 = 0$. Hence the normalized age distribution converges to the stable one as $t \to \infty$, leaving an ordinary differential equation problem for the total population size that is asymptotically autonomous.

In conclusion to this subsection we present yet another version of the trick, due to PRÜSS (1983), which closely mimicks the approach taken in II.14. Moreover, we shall, like Webb *(op. cit),* choose a formulation applying immediately to general structured population problems, not just age dependent ones.

Let A be the generator of a linear semigroup on a Banach space * X, and assume that A has a strictly dominant simple real eigenvalue r with corresponding eigenvector $\tilde{n}$, and that

$$X = \mathfrak{N}(A - rI) \oplus \mathfrak{R}(A - rI) \tag{5.1.9}$$

where the restriction of the semigroup to the second subspace satisfies an exponential estimate with exponent $r - \epsilon$ for some $\epsilon > 0$ Let $\mu_2 : X \to \mathbb{R}$ be smooth. Then the semilinear equation

$$\frac{dn}{dt} = An - \mu_2(n)n \tag{5.1.10}$$

has a solution for each initial condition. We assume that positivity arguments guarantee that the projection onto $\tilde{n}$ according to the decomposition (5.1.9) cannot vanish along a solution if we start with a "positive" initial condition. So for biologically relevant initial data we have the representation

$$n(t,a) = \rho(t)[\tilde{n}(a) + \bar{n}(t,a)] \tag{5.1.11}$$

where $\bar{n}(t,a) \in \mathfrak{R}(A - rI)$ and ρ is a real valued function. On substituting (5.1.11) into (5.1.10) and rearranging the terms we find

$$\left[\frac{d\rho}{dt} - r\rho + \mu_2(n)\rho\right]\tilde{n} = -\frac{d\rho}{dt}\bar{n} - \rho\frac{d\bar{n}}{dt} + \rho A\bar{n} - \mu_2(n)\rho\bar{n}\,. \tag{5.1.12}$$

Since $\mathfrak{N}(A - rI) \cap \mathfrak{R}(A - rI) = \{0\}$ both sides have to be zero. Therefore

* See II for the notion of a Banach space and the notation used in formula (5.1.9).

$$\frac{d\rho}{dt} = r\rho - \mu_2(n)\rho \tag{5.1.13}$$

and

$$\frac{d\bar{n}}{dt} = (A - rI)\bar{n} \tag{5.1.14}$$

(Note that (5.1.13) is used to obtain (5.1.14), and that in addition we have divided by ρ.) From (5.1.14) we conclude that

$$\bar{n}(t,\cdot) = 0(e^{-\epsilon t}) \qquad \text{for } t \to \infty . \tag{5.1.15}$$

Consequently the asymptotic behaviour of ρ for $t \to \infty$ is determined by the autonomous differential equation in $\mathbb{R}^1$.

$$\frac{d\rho}{dt} = (r - \mu_2(\rho\tilde{n}))\rho . \tag{5.1.16}$$

* EXERCISE 5.1.2: Formulate hypotheses about μ_2 which allow precise conclusions about the asymptotic behaviour of n. Pay special attention to the case of monotone (with respect to the cone X_+ which is left invariant by the semigroup) μ_2.

5.2. *Linear chain trickery*

In example 1.4.3 and exercise 1.4.4 and in exercise 4.1.2 we have met models which could be phrased as systems of ordinary differential equations, which could be transformed into an age representation. Interestingly enough, under certain conditions we can also effect a reverse transformation thereby providing ourselves with some relatively more easily studied examples of age dependent processes. This idea seems to be due to GURTIN & MACCAMY (1979) and has been used to good avail in diverse practical applications by NISBET & GURNEY and various coworkers (see the contribution by GURNEY, NISBET & BLYTHE in part B for references). Recently it has been discussed in a more abstract setting by HADELER (unpublished manuscript). In this subsection we shall deal with this idea in what we conjecture to be its most general form for age dependent models. Extensions to physiological as opposed to calendar age type models can be found in the contribution by Gurney, Nisbet & Blythe in part B, extensions to size structured models in MURPHY (1983a)).

We start with two examples essentially due to GURTIN & MACCAMY (1979). More examples on different levels of generality may be found in Gurtin's lecture notes (GURTIN, preprint 1982).

EXAMPLE 5.2.1: *A density dependent population model with two individual states.* Consider individuals which can be in two stages, reproductives and seniles, occurring in densities n_1 and n_2 respectively. Furthermore assume that the death rate of both groups equals $\mu(p)$, where p is the total population density, that the birth rates of the reproductive individuals equals $\beta(p)$ and that reproductive individuals become senile at a constant rate α, then

$$\frac{dn_1}{dt} = \beta(p)n_1 - \alpha n_1 - \mu(p)n_1 , \tag{5.2.1a}$$

$$\frac{dn_2}{dt} = \alpha n_1 - \mu(p)n_2 , \tag{5.2.1b}$$

$$p = n_1 + n_2 . \tag{5.2.1c}$$

Alternatively this equation may be rewritten in terms of the population density and the density of reproductives

$$\frac{dn_1}{dt} = \beta(p)n_1 - \alpha n_1 - \mu(p)n_1 , \tag{5.2.2a}$$

$$\frac{dp}{dt} = \beta(p)n_1 - \mu(p)p , \tag{5.2.2b}$$

where (5.2.2b) is obtained by adding (5.2.1a) and (5.2.1b).
The model embodied in equation (5.2.1) or (5.2.2) was specifically designed to allow an age representation: all individuals are born equal, the progression through the stages is independent of outside influences and the death mechanism

---

* We shall in this subsection revert to the notational convention of III in that we shall drop without warning arguments to which no direct reference is made.

is nonselective. To derive an equation for the age distribution we first observe that the death rate is independent of the stage an individual is in and therefore independent of age, so *

$$\frac{\partial n}{\partial t} = -\frac{\partial n}{\partial a} - \mu(p)n\,, \qquad n(t,0) = b(t)\,, \tag{5.2.3a}$$

where p is just the total density of individuals

$$p(t) = \int_0^\infty n(t,a)da\,. \tag{5.2.3b}$$

To calculate the birth rate b we note that the probability that an individual aged a is still in the reproductive state equals $e^{-\alpha a}$, so

$$b(t) = \beta(p)\int_0^\infty e^{-\alpha a}\, n(t,a)da\,. \tag{5.2.3c}$$

Only calculating *the* initial age distribution presents some difficulty if we have only $p(0)$ and $n_1(0)$ to go by. However, for practical purposes we may take just any age distribution compatible with there being $n_1(0)$ reproductives and $n_2(0)$ seniles at $t=0$ *i.e.* any $n(0,\cdot)$ satisfying

$$\int_0^\infty n(0,a)da = p(0)\,, \qquad \int_0^\infty e^{-\alpha a}\, n(0,a)da = n_1(0)\,. \tag{5.2.4}$$

Finally we shall effect the backward transformation from equation (5.2.3) into equation (5.2.2) again. The clue is provided by equation (5.2.4): what will do for the initial population state should do just as well for the population state at any later time. Therefore, for any age dependent process satisfying (5.2.3) we just define

$$n_1(t) := \int_0^\infty e^{-\alpha a}\, n(t,a)da\,, \qquad p(t) := \int_0^\infty n(t,a)da\,. \tag{5.2.5}$$

Then

$$\frac{dn_1}{dt} = \frac{d}{dt}\int_0^\infty n(t,a)e^{-\alpha a}\,da = \int_0^\infty \frac{\partial n(t,a)}{\partial t}\, e^{-\alpha a}\,da = -\int_0^\infty \left[\frac{\partial n(t,a)}{\partial a} + \mu(p)n(t,a)\right]e^{-\alpha a}\,da =$$

$$-\,n(t,a)e^{-\alpha a}\Big|_0^\infty - \alpha\int_0^\infty n(t,a)e^{-\alpha a}\,da - \mu(p)n_1 = \beta(p)n_1 - \alpha n_1 - \mu(p)n_1\,, \tag{5.2.6a}$$

where the last step is effected by using the side condition (5.2.3c) and analogously we find

$$\frac{dp}{dt} = \beta(p)n_1 - \mu(p)\,p\,. \tag{5.2.6b}$$

EXAMPLE 5.2.2: *A density dependent population model with three individual states.* This example is exactly equal to the previous one, except that we assume three stages: juvenile, reproductive adult, and senile, with equal transition rates between stages 1 and 2 and stages 2 and 3. The stage based equations are

$$\begin{aligned} \frac{dn_1}{dt} &= \beta(p)n_2 - \alpha n_1 - \mu(p)n_1\,, \\ \frac{dn_2}{dt} &= \alpha n_1 - \alpha n_2 - \mu(p)n_2\,, \\ \frac{dp}{dt} &= \beta(p)n_2 - \mu(p)p\,, \end{aligned} \tag{5.2.7}$$

with corresponding age based equations

$$\begin{aligned} &\frac{\partial n}{\partial t} = -\frac{\partial n}{\partial a} - \mu(p)n\,, \quad n(t,0) = b(t)\,, \\ &p(t) = \int_0^\infty n(t,a)da\,, \qquad b(t) = \beta(p)\int_0^\infty \alpha a e^{-\alpha a} n(t,a)da\,, \end{aligned} \tag{5.2.8}$$

and connecting rules

$$n_1(t) = \int_0^\infty e^{-\alpha a} n(t,a)da, \quad n_2(t) = \int_0^\infty \alpha a e^{-\alpha a} n(t,a)da, \quad p(t) = \int_0^\infty n(t,a)da\,. \tag{5.2.9}$$

EXERCISE 5.2.3: Check the transformation from (5.2.7) to (5.2.8) vice versa.

The stage models introduced in the examples are of no great interest in themselves, but the backward transformation rules are: suppose we were given some model which eventually led to equation (5.2.3) or equation (5.2.8), then it would always be possible to effect the transformation to equations (5.2.2) or (5.2.7) irrespective of the specific origin of the model. If we consider a particular age dependent mechanism like *e.g.* cannibalism, but we have some freedom of choice still in how we let the various quantities involved depend on the age distribution, then we could at least try whether a model of a form like (5.2.3) or (5.2.8) would be in our class, in order to find a relatively easily studied example. This then leaves us with two questions: (1) what does the ordinary differential equation counterpart tell us about the full partial differential equation, and (2) how far can the trick be extended?

Suppose we are given a partial differential equation of the form

$$\frac{\partial n(t,a)}{\partial t} = -\frac{\partial n(t,a)}{\partial a} - \mu(a,n)n(t,a), \quad n(t,a) = b(t) = \beta(n) \tag{5.2.10}$$

with $\mu(a,\cdot)$ and β functionals on the space of age distributions X. Equation (5.2.10) is supposed to define a dynamical system on X. Abstractly it can be rewritten as

$$\frac{dn}{dt} = F(n)\,. \tag{5.2.11}$$

The procedure in the examples then was (compare (5.2.5) and (5.2.6)) to seek a continuous linear map $P:X \to \mathbb{R}^k$ sending n into N such that

$$\frac{dN}{dt} = \frac{d}{dt}Pn = P\frac{dn}{dt} = PF(n) = G(Pn) = G(N) \tag{5.2.12}$$

for some function $G:\mathbb{R}^k \to \mathbb{R}^k$. If this procedure is really good it should also allow us to calculate $n(t,a)$ again at least for sufficiently large t. Integrating (5.2.10) along the characteristics shows that $n(t,a)$ for $a<t$ is fully determined by the solution of (5.2.12) if and only if both μ and β are so determined, *i.e.* it should be possible to write $\mu(a,n)=\mu_0(a,N)$ and $\beta(n)=\beta_0(N)$. This answers our first question, and also part of our second one.

The remaining part of the answer to the second question should be provided by analyzing what restrictions on μ_0, β_0 and P can be deduced from (5.2.12). As a first step we write the linear map P as

$$Pn = \int_0^\infty \Phi(a)n(a)da\,. \tag{5.2.13}$$

(Mathematically this implies that we have to make some assumptions about the space X.) Using (5.2.13) we find

$$PF(n) = \int_0^\infty \Phi(a)\left(-\frac{\partial n}{\partial a}(a) - \mu_0(a,Pn)n(a)\right)da = \Phi(0)\beta_o(Pn) + \int_0^\infty \left(\frac{d\Phi}{da}(a) - \Phi(a)\mu_0(a,Pn)\right)n(a)da\,, \tag{5.2.14}$$

(using that $\Phi(a)n(a)$ should go to zero for $a\to\infty$ from (5.2.13)). Inserting (5.2.14) in (5.2.12) gives

$$\int_0^\infty \left(\frac{d\Phi}{da}(a) - \mu_0(a,Pn)\Phi(a)\right)n(a)da = H(Pn) := G(Pn) - \Phi(0)\beta_0(Pn)\,. \tag{5.2.15}$$

Taking (Fréchet) derivatives for n at both sides and setting

$$\mu_1(a) := \mu_0(0,a), \qquad B := DH(0) \tag{5.2.16}$$

gives

$$\int_0^\infty \left(\frac{d\Phi}{da}(a) - \mu_1(a)\Phi(a)\right)n(a)da = B\int_0^\infty \Phi(a)n(a)da\,. \tag{5.2.17}$$

(5.2.17) can only be true for all n when

$$\frac{d\Phi}{da} = (B + \mu_1(a)I)\Phi(a)\,, \tag{5.2.18}$$

from which we conclude that

$$\Phi(a) = (\mathcal{F}(a))^{-1}e^{Ba}C \tag{5.2.19a}$$

with

$$C := \Phi(0), \qquad \mathcal{F}(a) := e^{-\int_0^a \mu_1(\alpha)d\alpha}\,. \tag{5.2.19b}$$

Finally inserting (5.2.18) in (5.2.15) gives.

$$\int_0^\infty (\mu_1(a)-\mu_0(a,Pn))\Phi(a)n(a)da = H(Pn)-BPn . \tag{5.2.20}$$

The right hand side of (5.2.20) is a function of Pn. Therefore the left hand side should also depend on n through Pn only. A sufficient (and we expect also necessary) condition for this is that

$$\mu_0(a,Pn) = \mu_1(a) + \mu_2(Pn) , \tag{5.2.21}$$

which is nothing but our old friend from 5.1, the separability condition.

Summarizing, we have found that if in (5.2.10)

$$\beta(n) = \beta_0(Pn), \qquad \mu(a,n) = \mu_1(a) + \mu_2(Pn) \tag{5.2.22}$$

where the linear map P is defined by

$$Pn = \int_0^\infty (\mathcal{F}(a))^{-1} e^{Ba} Cn(a)da \tag{5.2.23}$$

with $\mathcal{F}$ defined by (5.2.19b), then $N:=Pn$ satisfies

$$\frac{dN}{dt} = BN + C\beta_0(N)-\mu_2(N)N . \tag{5.2.24}$$

We also conjecture that this statement can be reversed: if there exists a continuous linear map $P:X\rightarrow\mathbb{R}^k$ such that (5.2.10) holds with $\beta(n)=\beta_0(Pn)$ and $\mu(a,n)=\mu_0(a,Pn)$ and $N:=Pn$ satisfies a differential equation $dN/dt=G(N)$, then P is necessarily of the form (5.2.23), μ of the form (5.2.22) and G of the form (5.2.24).

EXERCISE 5.2.4: Let B be the differential generator of a Markov chain with killing, *i.e.* $b_{ij}\geqslant 0$ for $i\neq j$, $b_{ii}\leqslant 0$ and $B^TE + U=0$ with all $u_i\geqslant 0$ and $E=(1,..,1)^T$, and let C be a probability vector, *i.e.* all $c_i\geqslant 0$ and $C^TE=1$. Now consider the population model

$$\frac{dN}{dt} = BN + C\beta_0(N)-\mu_2(N)N . \tag{5.2.25}$$

Derive the corresponding age representation and also the back transformation to (5.2.25) again.

EXERCISE 5.2.4: Let in (5.2.10)

$$\beta(n) = \frac{\theta\int_0^\infty \alpha_1 a e^{-\alpha_1 a} n(a)da}{1+\int_0^\infty n(a)da}$$

and

$$\mu(n,a) = \gamma_1+\gamma_2 \int_0^\infty e^{\alpha_2 a} n(a)da .$$

Find a k and a linear map $P:X\rightarrow\mathbb{R}^k$ such that $\beta(n)=\beta_0(Pn)$, $\mu(n,a) = \mu_2(Pn)$ and $N=Pn$ satisfies $dN/dt=G(N)$ for some function G.

We shall end this section with a few words about the practical implications of the result embodied in formula (5.2.22) to (5.2.24). Generally a model specification will begin with a specification of the functionals β and μ. If our reduction procedure is to apply, then it should be possible to write $\mu(a,n)$ as $\mu_1(a)+\mu_3(n)$. From μ_1 we can calculate $\mathcal{F}$ using its definition (5.2.19b). Next we observe that the components of $e^{Ba}C$ are necessarily all mixtures of polynomials times exponentials times sines and cosines. Conversely the realization theorems from systems theory (see e.g. KALMAN, FALB & ARBIB, 1969; BROCKETT, 1970; SILVERMAN, 1971; PADULO & ARBIB, 1974 or Wsillems, 1975) imply that any set of functions $\psi_i(a)$, $i=1,..h$ of that form can be expressed as $A_i e^{Ba}C$ for some $k\times k-$ matrix B and some vectors A_i and C. So we can find a representation of the form (5.2.24) if and only if $\beta(n)$ and $\mu_3(n)$ can both be calculated from some finite number of integrals $\int_0^\infty \psi_i(a)(\mathcal{F}(a))^{-1}n(a)da$ with ψ_i of the indicated form. For practical purposes it is moreover preferrable to choose k as small as possible. In most cases we dream up as examples, finding such a minimal representation of the $\psi_i(a)$ will not present a great problem. A guideline of how to proceed in general can be found in the references on systems realization mentioned earlier.

* EXERCISE 5.2.5: Discuss how the results from this subsection relate to those presented in 2.3.3.

** EXERCISE 5.2.6: Either prove that (5.2.21) is also a necessary condition for (5.2.20) to hold true for all n, or give a counter example.

A. *The Laplace transformation*

This appendix surveys the bare essentials about the Laplace transformation for non-cognoscenti. A detailed account may be found *e.g.* in WIDDER (1946), DOETSCH (1950, 1955, 1956) and VAN DER POL & BREMMER (1955). An extensive set of tables is Erdélyi *et al.* (1954).

Suppose that $f:\mathbb{R}^+ \to \mathbb{R}$ is some sufficiently well-behaved function, then its Laplace transform $\bar{f}$ is defined as

$$\bar{f}(s) := \int_0^\infty e^{-st} f(t)dt \tag{A.1}$$

for those complex s for which the integral converges. It can be shown that convergence of the integral for some $s_0=\sigma_0+i\tau_0$ entails the convergence for all $s=\sigma+i\tau$ with $\sigma>\sigma_0$, and moreover $\bar{f}$ is analytic in the half plane $\mathrm{Re}(s)>\sigma_0$. Generally we shall call by the same name the Laplace transform as defined by (A.1) and the analytic extension of $\bar{f}$ as a function of s. Finally, if f satisfies some exponential bound, $|f(t)| < ce^{\lambda t}$, then the integral certainly converges for $\mathrm{Re}(s)>\lambda$. These considerations are enough to justify many of our formal manipulations. The main usefulness of the Laplace transform, however, derives from the facts that 1) generally speaking the relation between f and $\bar{f}$ is one to one, and 2) we can deduce properties of f from $\bar{f}$ and vice versa.

Some examples of Laplace transform pairs are*

exponential density

$$f(t) = \alpha e^{-\alpha t} \quad \Leftrightarrow \bar{f}(s) = \frac{\alpha}{\alpha+s} \tag{A.2}$$

gamma density

$$f(t) = \frac{\alpha(\alpha t)^{k-1}e^{-\alpha t}}{\Gamma(k)} \quad \Leftrightarrow \bar{f}(s) = \left[\frac{\alpha}{\alpha+s}\right]^k \tag{A.3}$$

homogeneous density

$$f(t) = \begin{Bmatrix} (\beta-\alpha)^{-1} & \text{for } \alpha<t<\beta \\ 0 & \text{elsewhere} \end{Bmatrix} \Leftrightarrow \bar{f}(s) = \frac{e^{-\alpha s}-e^{-\beta s}}{(\beta-\alpha)s} \tag{A.4}$$

and in the limit, the delta function

$$f(t) = \delta(t-\alpha) \quad \Leftrightarrow \quad \bar{f}(s) = e^{-\alpha s} \tag{A.5}$$

Here we have chosen our f such that they integrate to one, *i.e.* $\bar{f}(0) = 1$. This is no restriction due to the linearity of the Laplace transform

$$f = \alpha f_1 + \beta f_2 \quad \Leftrightarrow \quad \bar{f} = \alpha \bar{f}_1 + \beta \bar{f}_2 \tag{A.6}$$

A class of functions which are not integrable but which allow a Laplace transform for $\mathrm{Re}(s)>0$ are

$$f(t) = t^\alpha,\ \alpha>-1 \quad \Leftrightarrow \quad \bar{f}(s) = \frac{\Gamma(\alpha+1)}{s^{\alpha+1}} \tag{A.7}$$

Finally two immensely useful properties of the Laplace transform, one in the context of integral equations, the other in the context of differential equations, are

$$f = f_1 * f_2 \quad \Leftrightarrow \quad \bar{f} = \bar{f}_1 \bar{f}_2 \tag{A.8}$$

$$f_2 = \frac{df_1}{dt} \quad \Leftrightarrow \bar{f}_2 = s\bar{f}_1 - \bar{f}_1(0)\,. \tag{A.9}$$

---

* The gamma function, Γ, is defined as $\Gamma(x) := \int_0^\infty t^{x-1}e^{-t}dt$. For integer x this reduces to (as can be shown by partial integration) $\Gamma(n) = (n-1)!$

EXERCISE A.1: Deduce the explicit solution (2.1.10) by Laplace transforming both sides of the renewal equation (2.1.7).

EXERCISE A.2: Prove (A.2) to (A.9). Hint: For (A.9) use partial integration.

Using (A.2) to (A.9) we can derive all other sorts of useful relations like

$$f_0(t) = \int_0^t f_1(\tau)d\tau \quad \Leftrightarrow \quad \bar{f}_0(s) = \bar{f}_1(s)/s \tag{A.10}$$

from (A.7) and (A.8) with $f_2(t) = t^0$,

$$f(t) = \int_t^\infty f_1(\tau)d\tau \quad \Leftrightarrow \quad \bar{f}(s) = (\int_0^\infty f_1(t)dt - \bar{f}_1(s))/s \tag{A.11}$$

from (A.10) and (A.6), and

$$f(t) = \begin{Bmatrix} f_1(t-\alpha) & \text{for } t>\alpha \\ 0 & \text{for } t<\alpha \end{Bmatrix} \Leftrightarrow \bar{f}(s) = e^{-\alpha s}\bar{f}_1(s) \tag{A.12}$$

from (A.5) and (A.8) with $f_2(t) = \delta(t-\alpha)$.

The problem of finding f from $\bar{f}$ is called the inversion problem. The crucial result is that for most classes of functions, in particular the continuous ones, the function f is uniquely determined by $\bar{f}$, with the understanding that two functions that differ only on a set of measure zero are counted as equivalent. Therefore it is possible to "calculate" some unknown f from its known Laplace transform $\bar{f}$ by looking $\bar{f}$ up, possibly after some rearrangement, in one of the extensive published tables of Laplace transform pairs. However, more often than not it is not possible for a given $\bar{f}$ to arrive at a pleasing formula for f. The main use of Laplace transform methods then is to deduce from $\bar{f}$ results about the behaviour of f for large t. (For example, the so-called Tauberian theorems relate the behaviour of f for large t to that of $\bar{f}$ near $s=0$.) Therefore Laplace transform methods are in a sense complementary to numerical methods which mainly concentrate on short time behaviour.

Various explicit formulae for f in terms of $\bar{f}$ are known, the most useful one being

$$f(t) = \frac{1}{2\pi i}\int_{\alpha-i\infty}^{\alpha+i\infty} e^{st}\bar{f}(s)ds \tag{A.13}$$

α being chosen such that all singularities of $\bar{f}$ lie to the left of the vertical line of integration $\mathrm{Re}(s)=\alpha$. The main utility of (A.13) comes from the fact that it allows us to use so-called contour integration to extract information about f from $\bar{f}$. We shall conclude this appendix with two examples in which we use contour integration to find (more or less) explicit expressions for f.

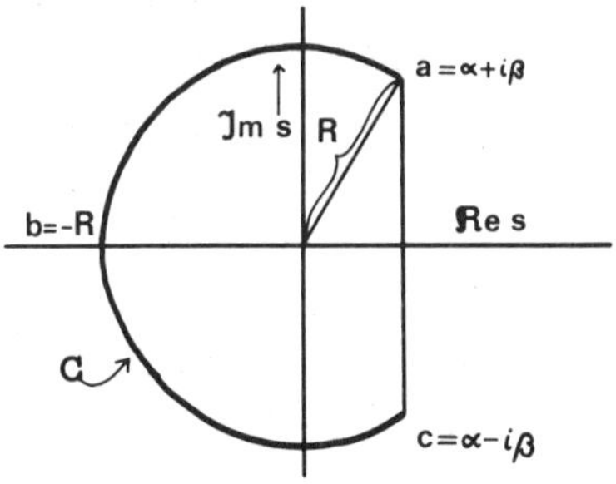

Fig. A.1. Integration contour used to evaluate (A.13).

EXAMPLE A.3: If $\bar{f}$ is meromorphic, *i.e.* its only singularities are poles, and if for the contour C depicted in fig. A.1

$$|\bar{f}(s)| < MR^{-\epsilon},\ M,\epsilon>0 \text{ for } s \text{ on the arc abc of } C \tag{A.14}$$

then for $R\to\infty$ the integral of $e^{st}\bar{f}(s)$ on the arc *abc* goes to zero, *i.e.* the integral around the contour C becomes equal to (A.13). But by the so-called residue theorem

$$\frac{1}{2\pi i}\oint_C e^{st}\bar{f}(s)da = \sum \text{ residues of } e^{st}\bar{f}(s) \text{ at poles inside } C \tag{A.15a}$$

where the residue of an n'th order pole at s_j equals

$$\lim_{s \to s_j} \frac{1}{(n-1)!} \frac{d^{n-1}}{ds^{n-1}} (s-s_j)^n e^{st}\bar{f}(s) = p(t)e^{s_j t} \tag{A.15b}$$

with p a polynomial of degree $n-1$.

The principal class of Laplace transforms satisfying (A.15) are the rational functions $\bar{f}(s)=p_1(s)/p_2(s)$ with the degree of the polynomial p_2 larger than that of p_1. In renewal theory the Laplace transform of the birth rate function

$$\bar{b} = \frac{\bar{g}}{1-\bar{\phi}} \tag{A.16}$$

is of this form if both ϕ and g are mixtures of polynomials times (possibly complex) exponentials like ϕ and g from example 1.4.3.

EXERCISE A.4: Calculate b for $\phi(a)=R\alpha^2 ae^{-\alpha a}$ and $g(t) = \gamma e^{\beta t}$

EXAMPLE A.4: When $\bar{f}$ is meromorphic but has infinetely many poles then (A.14) can never hold true. However, if we can find a sequence $R_i \to \infty$ such that (A.14) holds for those R_i, then we can still calculate f by taking the limit in (A.15). As an example of this procedure consider the renewal equation with $\phi(a)=R\beta^{-1}$ for $a \leqslant \beta$, $\phi(a)=0$ for $a>\beta$. Then

$$\bar{b}(s) = \frac{\beta s\bar{g}(s)}{\beta s - R + Re^{-\beta s}} . \tag{A.17}$$

If $g(t)=0$ for $t>\beta$ then $\bar{g}$ has no poles. Therefore the poles of $\bar{b}$ coincide with the solutions of

$$\frac{\beta s - R + \mathrm{Re}^{-\beta s}}{\beta s} = 0 \tag{A.18}$$

This equation has but one real solution r and all other solutions s_j have smaller real part. Moreover all the s_j are simple zeros of $(\beta s)^{-1}(\beta s - R + \mathrm{Re}^{-\beta s})$, so that all poles of $\bar{b}$ are of order 1. If we number the roots in order of decreasing real part, roots with positive imaginary part taking precedence over roots with negative imaginary part, then for $R \neq 1$ the results in BELLMAN & COOKE (1963) imply that for $t>\beta$

$$b(t) = \sum_{j=0}^{\infty} \frac{s_j \bar{g}(s_j)}{1-Re^{-\beta s_j}} e^{s_j t} \tag{A.19}$$

V. The Dynamical Behaviour of the Age-Size-Distribution of a Cell Population

H.J.A.M. Heijmans

Introduction

It is appropriate to think of the cell cycle as an ordered sequence of biochemical events, such as the synthesis of RNA and proteins and the replication of DNA, finally ending up in cell division. The rate at which these biochemical events, such as the increase of structural materials, proceed may heavily depend on volume (size) through such factors as diffusion times and surface to volume ratios. In section I.4 and chapter II we therefore considered the case that the position of a particular cell in its cycle was adequately described by its size. However, some of the biochemical reactions seem to proceed sequentially during a cell's life cycle and for such reactions cell age provides a better description. In this chapter we therefore consider the Bell-Anderson model for cell growth and division, the main assumption being that the *i*-state of a cell is given by a vector (a,x) where a and x represent age and size respectively. The *i*-state space Ω is a subset of $\mathbb{R}^+ \times \mathbb{R}^+$. In our study we restrict ourselves to the case that the growth rate of individual cells only depends on their size: see section 1. This study is certainly not the only one where age and size are considered to be the relevant parameters structuring microbial populations. Besides the paper by BELL & ANDERSON (1967), where one can find a derivation of the model discussed here, we mention SINKO & STREIFER (1967), BELL (1968), TRUCCO (1970), TRUCCO & BELL (1970), HANNSGEN, TYSON & WATSON (1985) and TYSON & HANNSGEN (1985b,) 198?, 1985a) for some related work in this area (see also section 9). Actually Sinko & Streifer independently derived the model discussed here, and they applied it to populations of the planarian worm *Dugesia tigrina.* In all the other papers variants of the model are investigated mathematically. In all of these papers, except the two last ones, one of the main restrictions is that the cell division probability only depends on age: TYSON and HANNSGEN (1985a) consider the so-called "tandem model" which is the transition probability model supplied with a critical size requirement. For a recent overview of some of this literature we refer to two articles written by JOHN TYSON (1985a,b).

In this chapter the division probability is allowed to depend on both age and size. After making some assumptions on the death, division, and growth rates we prove existence and uniqueness of solutions, after reducing the problem, by integration along the characteristics, to an abstract renewal equation for the birth function $B(t,x)$; see sections 2 and 3. Here x is "size at birth", and the word "abstract" means that the birth function is not just a scalar function of time, as it was in chapter IV, but takes its values in a Banach space; c.f. chapter II.

To determine the asymptotic behaviour of solutions, we first apply Laplace transformation to the renewal equation. This is done in section 4. Subsequently we write down the characteristic equation (which now, of course, is not of scalar type, but involves Banach space operators) and prove the existence of a dominant root, i.e. a root with largest real part (see section 6 for the details). This requires a certain amount of spectral theory. (In section 5 we give some results from the theory of positive operators, which are used in section 6.) It turns out that we need an extra condition on the growth rate to carry out this program, viz. assumption 6.4, which resembles the one made in chapter II. In the sections 7 and 8 we compute the residue associated with this dominant root, and apply the inverse Laplace transformation, which gives us the asymptotic behaviour of solutions. In section 9 it is argued that we cannot dispense with the extra condition on the growth rate. This disproves the supposition of BELL (1968) that, even in the case of exponential individual growth, a stable age-size distribution might exist if the division probability depends in an appropriate manner on age and size. BELL (1968) argued (and this indeed is correct) that this is impossible if the division probability depends on age only, and individual cells grow exponentially in size; a rigorous proof of this result is given in TRUCCO & BELL (1970), where the first and second moments of the distribution of birth sizes are computed.

The approach we have adopted in this chapter (i.e. reduction to a renewal equation) is not as different from the semigroup approach of chapter II as it seems at first sight; compare also section IV.2.3.

Finally we note that a reduction of the partial differential equation to a renewal equation for the birth function is generally possible. What one has to do is to replace the actual *i*-state vector $\underline{x}$ by the vector representing chronological age a plus state at birth $\underline{x}_b$: note that this new vector is one dimension higher if a was not already contained in $\underline{x}$. Then, by integration along the characteristics, one finds a renewal equation for $B(t,\underline{x}_b)$, representing the rate at which individuals with state $\underline{x}_b$ are born at time t.

1. The model

We assume that a cell is fully characterized by its age a and size x. Here size can mean volume, length, DNA-content or any other quantity which obeys a physical conservation law. Size increases with time and we assume that this process can be described by the ordinary differential equation

$$\frac{dx}{dt} = g(x). \tag{1.1}$$

This means in particular that the growth rate g does neither depend on age, which seems very reasonable from a biological point of view, nor on environmental factors (such as food density) which are influenced by the population itself, causing nonlinearities in the equation. Age also increases with time and obeys $\frac{da}{dt} = 1$. However our theory can be easily extended to the case where a denotes some physiological age, which does not necessarily increase linearly with time: $\frac{da}{dt} = f(a)$ where f is a bounded continuous positive function. We assume that if a cell divides, it produces two daughter cells, both having age zero and half the size of the mother. Let $n(t,a,x)$ be the cell density function, i.e. $\int_{x_1}^{x_2}\int_{a_1}^{a_2} n(t,a,x)dadx$ is the number of cells having age between a_1 and a_2, and size between x_1 and x_2. From the conservation principle it follows that the equation for the density function can be written as

$$\frac{\partial n}{\partial t} = -\nabla\cdot J - F - D, \tag{1.2}$$

where the flux $J = J(t,a,x)$ is given by $J = (n(t,a,x), g(x)n(t,a,x))^T$, and ∇ is the operator $(\frac{\partial}{\partial a}, \frac{\partial}{\partial x})^T$. The sinks F and D account for the individuals which "disappear" as a result of fission and death respectively. We refer to chapter III for a more general description how to derive balance equations such as (1.2).

Let fission and death be described by the per capita probabilities per unit of time $b(a,x)$ and $\mu(a,x)$ respectively,. then $F = F(t,a,x) = b(a,x)n(t,a,x)$ and $D = D(t,a,x) = \mu(a,x)n(t,a,x)$.

We shall now introduce a number of mathematical assumptions on the functions g, b and μ and discuss their biological meaning and/or mathematical motivation. With respect to the growth rate g we assume

(A_g) *g is a continuous function on* $[0,\infty)$ *and there exist constants* $g_{\min}$, $g_{\max}$ *such that* $0 < g_{\min} \leq g_{\max} < \infty$ and $g_{\min} \leq g(x) \leq g_{\max}$ *for all* $x \in [0,\infty)$.

It follows from this assumption that certain combinations of a and x are forbidden in the sense that cells with such a combination of age and size will never come into existence. More precisely there exists a (continuous) curve in the (a,x)-plane starting from $(a,x) = (0,0)$ and tending towards (∞,∞) below which no individual will ever dwell.

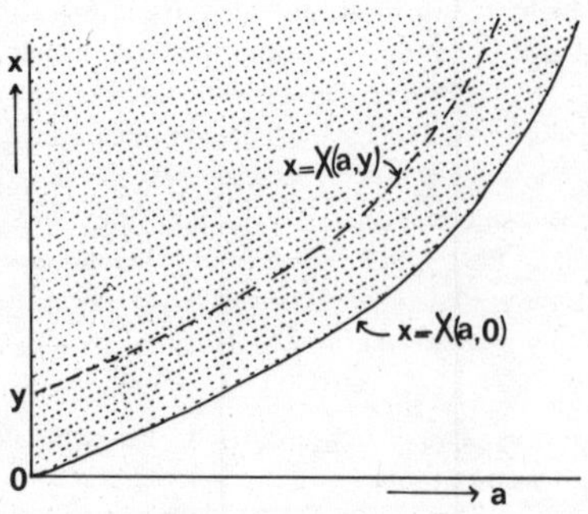

Figure 1.1. The set Ω. An individual with birth size y travels along the curve $\{X(a,y)|\ a \geq 0\}$ until it dies or divides.

We can compute this curve explicitly. Consider a cell whose size at birth is $x \geq 0$ (assuming that such cells indeed exist). Let $X(a,x)$ be its size at age a, if it has not died or divided before reaching that age. Then X is the solution of the initial value problem $\frac{dX}{da} = g(X)$, $X(0) = x$, which has a continuous (differentiable) solution tending to ∞ if a tends to ∞ because of assumption (A_g). The curve $\{(a,X(a,y))|a \geq 0\}$ is called the characteristic curve starting from $(0,y)$ (see figure 1.1). We refer to section 2 for more details.

Individuals can only exist in the shaded region $\Omega = \{(a,x)\in\mathbb{R}^+\times\mathbb{R}^+|x \geq X(a,0)\}$. The actual state space Ω_s (i.e. the subset of $\mathbb{R}^+\times\mathbb{R}^+$ in which indeed individuals do occur) is a subset of Ω, and in some cases Ω_s is smaller than Ω. (We refer to section 6 for an example.)

We impose the following conditions on b and μ:

(A_b) $\quad b \in L_\infty(\Omega)$ *(i.e. b is measurable and essentially bounded on* Ω)

$$b(a,x) = 0,\ a \leq a_0,\ (a,x) \in \Omega,$$

$$b(a,x) > 0,\ a > a_0,\ (a,x) \in \Omega,$$

$$\liminf_{a\to\infty} b(a,X(a,x)) \geq \underline{b} > 0 \text{ for every } x.$$

Here $a_0 > 0$ is some threshold below which cells cannot divide. The biological reason for this is that every cell has to go through a phase during which DNA is replicated, and the duration of this phase is more or less constant (see Bell & Anderson (1967), Eisen (1979)). Biologically, the last condition in (A_b) means that old individuals continue dividing at a positive rate.

(A_μ) $\quad \mu \in L_\infty^{loc}(\Omega)$, *i.e.* μ *is measurable and essentially bounded on compact subsets of* Ω,

$$\mu(a,x) \geq 0,\ (a,x) \in \Omega.$$

Let

$$d(a,x) = b(a,x) + \mu(a,x). \tag{1.3}$$

We assume

(A_d) $\quad$ *There exists a constant* d_∞ *with* $0 < d_\infty \leq \infty$ *such that* $\lim_{\sigma\to\infty} d(a+\sigma,X(\sigma,x)) = d_\infty$ *uniformly in a and x. Moreover, if* $d_\infty < \infty$, *there exists a constant* $M \geq 0$ *such that for all a and x:* $\int_0^\infty |d(a+\sigma,X(\sigma,x)) - d_\infty|d\sigma \leq M$.

Biologically assumption (A_d) means that the probability for a cell to reach age a without dying or dividing decreases more or less exponentially if a becomes large. In section 9 it is explained why this assumption is needed.

We can rewrite (1.2) as

$$\frac{\partial}{\partial t} n(t,a,x) + \frac{\partial}{\partial a} n(t,a,x) + \frac{\partial}{\partial x}(g(x)n(t,a,x)) = -(\mu(a,x) + b(a,x))n(t,a,x), \tag{1.4}$$

$$t \geq 0,\ (a,x) \in \Omega.$$

The fact that dividing mothers of age a and size $2x$ give birth to two daughters of age 0 and size x is accounted for by the boundary condition

$$n(t,0,x) = 4\int_{a_0}^\infty b(a,2x)n(t,a,2x)da. \tag{1.5}$$

See Bell & Anderson (1967) or chapter I for an explanation of the factor 4.

Remark 1.1. In (1.5) we only have to integrate over those ages a that satisfy $X(a,0) \leq 2x$.

We specify an initial condition

$$n(0,a,x) = n_0(a,x),\ (a,x) \in \Omega. \tag{1.6}$$

Biological considerations yield that n_0 should satisfy

$$n_0(a,x) \geq 0,\ (a,x) \in \Omega \text{ and } n_0 \in L_1(\Omega). \tag{1.7}$$

2. Reduction to an abstract renewal equation

Usually age-dependent population models are reduced to a renewal equation (which is a Volterra integral equation of convolution type) for the birth function (see chapter IV)). Here we will show that this can also be done for our age-size-structured model (1.4)-(1.6). In this case, however, we obtain an abstract renewal equation, in the sense that solutions take values in some function space.

Let $m(t,a,x)$ be defined by

$$m(t,a,x) = g(x)n(t,a,x), \tag{2.1}$$

then m satisfies the equation

$$\frac{\partial m}{\partial t} + \frac{\partial m}{\partial a} + g(x)\frac{\partial m}{\partial x} = -(\mu(a,x) + b(a,x))m(t,a,x), \tag{2.2a}$$

$$m(t,0,x) = \frac{4g(x)}{g(2x)}\int_{a_0}^{\infty} b(a,2x)m(t,a,2x)da, \tag{2.2b}$$

$$m(0,a,x) = m_0(a,x) \overset{def}{=} g(x)n_0(a,x). \tag{2.2c}$$

By the method of integration along characteristics (see COURANT & HILBERT (1962)) we can convert this system into an integral equation.

The characteristic curve through (t,a,x) is determined by $s \to (T(s,t),A(s,a),X(s,x))$, where s is an independent book-keeping variable and T,A,X are solutions of the ODE's $\frac{dT}{ds} = 1$, $T(0,t) = t$, $\frac{dA}{ds} = 1$, $A(0,a) = a$, $\frac{dX}{ds} = g(X)$, $X(0,x) = x$, thus $T(s,t) = s+t$, $A(s,a) = s+a$, and $X(s,x) = G^{-1}(s+G(x))$, where

$$G(x) = \int_0^x \frac{d\xi}{g(\xi)}, \quad x \geqslant 0, \tag{2.3}$$

can be interpreted as the time needed to grow from 0 to x and G^{-1} denotes its inverse. Observe that $G^{-1}(a) = X(a,0)$.

Now let t,a,x be fixed and let $\overline{m}(s) = m(T(s,t),A(s,a),X(s,x))$, then

$$\frac{d\overline{m}}{ds} = -d(A(s,a),X(s,x))\overline{m}(s), \tag{2.4}$$

where $d(a,x)$ is given by (1.3). Let

$$Q(s,a,x) \overset{def}{=} \exp\left[-\int_0^s d(A(\sigma,a),X(\sigma,x))d\sigma\right], \tag{2.5}$$

which can be interpreted as the probability that a cell with age a and size x reaches age $a+s$. From (2.4) we obtain that

$$\overline{m}(s) = \overline{m}(0)Q(s,a,x). \tag{2.6}$$

Let

$$t' = T(s,t),\ a' = A(s,a),\ x' = X(s,x). \tag{2.7}$$

(i) We choose $t = 0$. Then $a = a'-t'$, $x = X(-t',x')$. If we substitute this in (2.6) we obtain

$$m(t',a',x') = m(0,a'-t',X(-t',x'))\cdot Q(t',a'-t',X(-t',x')), \text{ if } a' > t'. \tag{2.8}$$

(ii) We choose $a = 0$. Then $t = t'-a'$, $x = X(-a',x')$, and we deduce from (2.6)

$$m(t',a',x') = m(t'-a',0,X(-a',x'))\cdot E(a',X(-a',x')), \text{if } a' < t', \tag{2.9}$$

where

$$E(a,x) \overset{def}{=} Q(a,0,x) = \exp\left[-\int_0^a d(\sigma,X(\sigma,x))d\sigma\right] \tag{2.10}$$

is the probability that a cell having size x at birth reaches age a.

If we drop the accents in (2.9) and (2.10), and use (2.1) and (2.2c) we find

$$n(t,a,x) = \frac{g(X(-t,x))}{g(x)} n_0(a-t, X(-t,x)) \cdot Q(t, a-t, X(-t,x)), \quad t < a, \tag{2.11}$$

$$n(t,a,x) = \frac{g(X(-a,x))}{g(x)} n(t-a, 0, X(-a,x)) \cdot E(a, X(-a,x)), \quad t > a. \tag{2.12}$$

Let the birth function B be defined by

$$B(t,x) = n(t,0,x). \tag{2.13}$$

If we substitute (2.11)-(2.12) into (1.5), then we obtain the following integral equation for B:

$$B(t,x) = \Phi(t,x) + \int_{a_0}^{t} k(a, 2x) B(t-a, X(-a, 2x)) da, \tag{2.14}$$

where

$$\Phi(t,x) = \frac{4g(X(-t, 2x))}{g(2x)} \int_{t}^{\infty} b(a, 2x) Q(t, a-t, X(-t, 2x)) \cdot n_0(a-t, X(-t, 2x)) da, \tag{2.15}$$

and

$$k(a,x) = \frac{4g(X(-a,x))}{g(x)} b(a,x) E(a, X(-a,x)). \tag{2.16}$$

$\Phi(t,x)$ is only defined for values of x satisfying $G(2x) \geqslant t$, and one should read $\Phi(t,x) = 0$ if $G(2x) \leqslant t$. Furthermore $k(a,x) = 0$ if $a \leqslant a_0$ or $a \geqslant G(x)$, and $k(a,x) \geqslant 0$ if $a_0 \leqslant a \leqslant G(x)$.

The integral equation (2.14) was also found by BELL (1968) but he only solved it for the special case that all cells divide at the same age (see also BEYER (1970)).

It follows from (2.11)-(2.12) that knowledge of the solution $B(t,x)$ of (2.14) yields the solution $n(t,a,x)$ of (1.4)-(1.6). Therefore we shall concentrate on (2.14) during the rest of this chapter. In section 9 we shall interprete some result in terms of the density $n(t,a,x)$.

We can rewrite (2.14) as the abstract renewal equation

$$B(t) = \Phi(t) + \int_{0}^{t} K(a) B(t-a) da, \tag{2.17}$$

where, for fixed $t \geqslant 0$, $\Phi(t) \in L_1[0,\infty)$ and $K(t)$ defines a bounded operator from $L_1[0,\infty)$ into itself:

$$(K(t)\psi)(x) = k(t, 2x)\psi(X(-t, 2x)), \quad \psi \in L_1[0,\infty), \tag{2.18}$$

where one should read $\psi(X(-t, 2x)) = 0$ if $G(2x) < t$.

REMARK 2.1. Throughout this chapter we call a Banach space-valued function integrable if it is Bochner-integrable. This means the following: let E be a Banach space with norm $\|\cdot\|_E$ and let $f: (a,b) \to E$, where $-\infty \leqslant a < b \leqslant \infty$. Then $f(t)$ is Bochner-integrable if and only if f is strongly measurable and $\|f(t)\|_E$ is Lebesgue integrable (see HILLE & PHILLIPS (1957)).

We call $B(t)$ a solution of (2.17) if and only if

i) $B(t) \in L_1[0,\infty)$, $t \geqslant 0$,

ii) $B(\cdot)$ is integrable on $[0,t_0]$ for all $t_0 \geqslant 0$,

iii) $B(t)$ obeys (2.17).

3. Existence and Uniqueness of solutions

It turns out that the proof of an existence and uniqueness result for the abstract renewal equation (2.17) is rather similar to the scalar case which has been extensively treated in the book of BELLMAN & COOKE (1963). First we shall prove a lemma.

LEMMA 3.1. (a) *Let d_∞ (of assumption (A_d)) be finite. Then there exist positive constants T_0, m_K, M_K and M_Φ such that for all $t \geqslant T_0$: $\|\Phi(t)\| \leqslant M_\Phi e^{-d_\infty t}$, and for all $\psi \in L_1[0,\infty)$: $m_K e^{-d_\infty t} \|\psi\| \leqslant \|K(t)\psi\| \leqslant M_K e^{-d_\infty t} \|\psi\|$.*
(b) Let $d_\infty = \infty$. For all $c > 0$ there exist constants $L_K(c), L_\Phi(c) > 0$ such that for all $t \geqslant 0$: $\|\Phi(t)\| \leqslant L_\Phi(c) e^{-ct}$, $\|K(t)\psi\| \leqslant L_K(c) e^{-ct} \|\psi\|$, for all $\psi \in L_1[0,\infty)$.

Proof. We shall only prove the second estimate in (a).

$$E(a,x) = \exp[-\int_0^a d(\sigma,X(\sigma,x))d\sigma] = \exp[-\int_0^a \{d(\sigma,X(\sigma,x))-d_\infty\}d\sigma]\cdot\exp[-\int_0^a d_\infty d\sigma].$$

Let M be the constant of assumption (A_d), then

$$e^{-M}e^{-d_\infty a} \leq E(a,x) \leq e^{M}e^{-d_\infty a}.$$

The second part of (a) now follows immediately from these estimates and the assumptions (A_g) and (A_b). In an analogous manner we can prove part (b). □

The following existence and uniqueness result can be proved.

Theorem 3.2. *Let $t_0 > 0$. There exists a unique bounded integrable solution $B(t)$ of (2.17) on $[0,t_0]$.*

The existence result can be established by the method of successive approximations. Uniqueness then follows from a Gronwall-type lemma. We refer to Bellman & Cooke (1963) where the scalar case has been worked out in great detail, and the reader will have no difficulty to see that all proofs can be carried over. Because t_0 can be chosen arbitrarily large, theorem 3.2 implies global existence of the solution $B(t)$.

Remark 3.3. Strictly speaking condition (A_b) and (A_μ) are sufficient to prove local existence and uniqueness.

In the next section we shall apply Laplace transformation to the integral equation (2.17). Therefore we need the following estimate.

Theorem 3.4. *There exists a $\beta \in \mathbb{R}$ such that $\|B(t)\| \leq M_B e^{\beta t}$, $t \geq 0$, where $M_B > 0$ is a constant.*

Proof. Let $\beta \in \mathbb{R}$ be such that $\|\Phi(t)\| \leq c_1 e^{\beta t}$ and $\int_0^\infty e^{-\beta t}\|K(t)\|dt = c_2 < 1$. From lemma 3.1 it is clear that such a β indeed exists. Then

$$\|B(t)\| \leq c_1 e^{\beta t} + \int_0^t \|K(a)\|\cdot\|B(t-a)\|da = c_1 e^{\beta t} + e^{\beta t}\int_0^t \{\|K(a)\|\cdot e^{-\beta a}\}\cdot\{\|B(t-a)\|\cdot e^{-\beta(t-a)}\}da.$$

Let $v(t) \stackrel{def}{=} \max_{0\leq a\leq t} \|B(a)e^{-\beta a}\|$, then $v(t) \leq c_1 + v(t)\int_0^t e^{-\beta a}\|K(a)\|da \leq c_1 + c_2 v(t)$, hence $v(t) \leq \frac{c_1}{1-c_2}$, from which we obtain that $\|B(t)\| \leq \frac{c_1}{1-c_2}e^{\beta t}$. □

4. Laplace Transformation

A technique which turned out to be extremely useful in the study of scalar renewal equations is Laplace transformation (*e.g.* Bellman & Cooke (1963), Hoppensteadt (1975) and chapter IV). This technique can also be employed in the study of abstract renewal equations such as (2.17). First we shall introduce some notations. Let $I \subseteq \mathbb{R}$ be an interval, and E a Banach space. We define by $L_p(I,E)$, $1 \leq p \leq \infty$, the Banach space consisting of all functions $f: I \to E$ satisfying $\|f\|_p \stackrel{def}{=} \{\int_I \|f(t)\|^p dt\}^{\frac{1}{p}} < \infty$, if $p < \infty$ and $\|f\|_\infty \stackrel{def}{=} \text{ess sup}\|f(t)\| < \infty$, if $p = \infty$. If $I = [0,\infty)$ we shall write $L_p(0,\infty;E)$ instead of $L_p([0,\infty);E)$.

Remark 4.1. We have to distinguish between the norm of $f(t)$, $t \geq 0$, as an element of E and the norm of f being an element of $L_p(I;E)$. In the first case we write $\|f(t)\|$, in the second case $\|f\|_p$.

Definition. Let f be a function from $[0,\infty)$ to some Banach space E, then its Laplace transform $\hat{f}$ is defined by $\hat{f}(\lambda) = \int_0^\infty e^{-\lambda t}f(t)dt$, whenever this integral is defined with respect to the norm topology.

The following result is standard (Hille & Phillips (1957)).

Lemma 4.2. *If $f \in L_1(0,\infty;E)$ then $\hat{f}(\lambda)$ is analytic in $\text{Re}\,\lambda > 0$ and continuous in $\text{Re}\,\lambda \geq 0$ (with respect to the norm-*

topology).

In what follows we need the Riemann-Lebesgue lemma (HILLE & PHILLIPS (1957), thm 6.4.2).

LEMMA 4.3 (Riemann-Lebesgue). *Let* $f \in L_1(0,\infty;E)$ *and* $\hat{f}$ *its Laplace transform. Then* $\lim_{|\eta|\to\infty} \hat{f}(\xi+i\eta) = 0$, *uniformly for* ξ *in bounded closed subintervals of* $(0,\infty)$.

Let the right-half-plane Λ be defined by

$$\Lambda \overset{def}{=} \{\lambda \in \mathbf{C} | \mathrm{Re}\,\lambda > -d_\infty\} \tag{4.1}$$

(where $\Lambda = \mathbf{C}$ if $d_\infty = \infty$). Then it follows from lemma 3.1 and lemma 4.2 that $\hat{K}(\lambda)$ and $\hat{\Phi}(\lambda)$ are defined and analytic in Λ. Moreover it follows from lemma 3.1 that $\hat{K}(\lambda)$ is not defined if $\mathrm{Re}\,\lambda < -d_\infty$.

REMARK 4.4. It is not a priori clear whether $\hat{K}(\lambda)$ is defined for λ on the vertical line $\mathrm{Re}\,\lambda = -d_\infty$. As to $\hat{\Phi}(\lambda)$ it depends on the initial age-size distribution $n_0(a,x)$ whether or not it is defined for values of λ satisfying $\mathrm{Re}\,\lambda \leqslant -d_\infty$. However this is not important for our purposes.

We define $\hat{B}(\lambda) = \int_0^\infty e^{-\lambda t} B(t)dt$ for those values of λ for which the integral converges. From theorem 3.4 we conclude that $\hat{B}(\lambda)$ exists if $\mathrm{Re}\,\lambda > \beta$. The convolution in (2.17) is converted by the Laplace transformation into a product of Laplace transforms. We wish to extend $\hat{B}(\lambda)$ to Λ minus some set Σ of singular points. More precisely

$$\hat{B}(\lambda) = \hat{\Phi}(\lambda) + \hat{K}(\lambda)\hat{B}(\lambda),\ \lambda \in \Lambda. \tag{4.2}$$

Let Σ be the set of all $\lambda \in \Lambda$ for which $I - \hat{K}(\lambda)$ is singular.

$$\Sigma = \{\lambda \in \Lambda | 1 \in \sigma(\hat{K}(\lambda))\}, \tag{4.3}$$

where $\sigma(\hat{K}(\lambda))$ denotes the spectrum of the operator $\hat{K}(\lambda)$. The condition $1 \in \sigma(\hat{K}(\lambda))$ is the usual precursor of a *characteristic equation* (HEIJMANS (to appear), HOPPENSTEADT (1975)).

For $\lambda \in \Lambda \setminus \Sigma$ we have

$$\hat{B}(\lambda) = (I - \hat{K}(\lambda))^{-1}\hat{\Phi}(\lambda). \tag{4.4}$$

In section 8 we shall prove that the element λ_d of Σ with largest real part determines the large time behaviour of the solution $B(t)$. Often λ_d turns out to be real, and the corresponding eigenvector of $\hat{K}(\lambda_d)$ to be positive: c.f. chapter II. The theory of positive operators is an important instrument to prove existence of λ_d, and has been succesfully exploited in a number of problems from population dynamics (DIEKMANN et al. (1984), HEIJMANS (1984a), HEIJMANS (1985b)). As an intermezzo we shall now present some results from positive operator theory with the emphasis on the existence and uniqueness of positive eigenvectors and eigenfunctionals.

5. Positive Operators

For the basic theory of order structures in a Banach space and positive operators, we refer to SCHAEFER (1974).

In the sequel E is some Banach space and E^* is it's dual, i.e. the space of all linear functionals (or linear forms) on E. We denote the duality pairing of $\psi \in E$, $F \in E^*$ with $\langle F,\psi \rangle$. A subset $E_+ \subseteq E$ is called a cone if the following conditions are satisfied

(i) E_+ is closed,

(ii) $\alpha\phi + \beta\psi \in E_+$ if $\phi,\psi \in E_+$ and $\alpha,\beta \geqslant 0$

(iii) $\psi \in E_+$ and $-\psi \in E_+$ implies that $\psi = 0$.

The reader can easily verify that by virtue of "$\phi \leqslant \psi$ iff $\psi - \phi \in E_+$" each cone $E_+ \subseteq E$ defines an order relation on E by which E becomes an ordered Banach space. We say that $\phi < \psi$ if $\phi \leqslant \psi$ and $\phi \neq \psi$. The cone E_+ is called total if the set $\{\psi - \phi | \psi,\phi \in E_+\}$ is dense in E. The dual set E_+^* is by definition the subset of E^* consisting of all positive functionals on E, i.e. $F \in E_+^*$ if and only if $F \in E^*$ and $\langle F,\psi \rangle \geqslant 0$ for all $\psi \in E_+$. If E_+ is total then E_+^* is a cone as well. A positive functional F is said to be strictly positive if $\langle F,\psi \rangle > 0$ for all $\psi \in E_+$, $\psi \neq 0$. A bounded linear operator T: $E \to E$ is called positive (with respect to the cone E_+) if $T\psi \in E_+$ for all $\psi \in E_+$. Notation $T \geqslant 0$. We denote the spectral radius of T by $r(T)$.

The first authors who systematically studied positive operators and their spectral properties were KREIN and RUTMAN (1948). In that paper they generalized the Frobenius theorem (which states that the spectral radius of a non-negative matrix is an eigenvalue of that matrix). They proved, among others, the following result.

THEOREM 5.1 (KREIN & RUTMAN (1948)). *Let $T: E \to E$ be compact and positive with respect to the total cone $E_+ \subseteq E$, and let $r = r(T) > 0$. Then there exists a $\psi \in E_+$, $\psi \neq 0$ such that $T\psi = r\psi$.*

They also introduced the notion of strong positivity. A positive operator $T: E \to E$ is called strongly positive if for all $\psi \in E_+$, $\psi \neq 0$ there is a natural number p such that $T^p\psi \in \overset{\circ}{E}_+$, where $\overset{\circ}{E}_+$ denotes the interior of the cone E_+ (assuming that E_+ has interior points). They proved that, if the assumptions of theorem 5.1 are fulfilled and, moreover, T is strongly positive, then

(a) T has (except for a constant) one and only one eigenvector $\psi \in E_+$. Moreover $\psi \in \overset{\circ}{E}_+$ and $T\psi = r\psi$.

(b) T^* has one and only one eigenvector $F \in E^*_+$, F is strictly positive and $T^*F = rF$.

(c) All other eigenvalues λ of T satisfy $|\lambda| < r(T)$.

Many years later their study was continued by a great number of authors, extending these ideas in several directions. Among others they weakend the condition that T has to be compact; in many cases it is sufficient that $\lambda = r(T)$ is a pole of the resolvent $R(\lambda,T) = (\lambda I - T)^{-1}$. Furthermore several different concepts generalizing the concept of strong positivity have been introduced. We mention three of these generalizations. SCHAEFER (1974) introduced in the early sixties the concept of irreducible positive operators. KRASNOSELSKII (1964) studied u_0-positive operators, and finally SAWASHIMA (1964) developed the theory of non-supporting operators. (Sawashima uses the terminology "non-support".) All three concepts have the advantage that the interior of the cone E_+ may be empty. The definitions of Schaefer and Sawashima are closely connected.

DEFINITION (SAWASHIMA (1964)). A bounded, positive operator $T: E \to E$ is called non-supporting with respect to E_+ if for all $\psi \in E_+$, $\psi \neq 0$, and $F \in F^*_+$, $F \neq 0$, there exists an integer p such that for all $n \geq p$ we have $\langle F, T^n\psi \rangle > 0$.

The following result, which was proved by SAWASHIMA (1964) is needed in the following section. The result can also be found in paper by MAREK (1970) which provides a comprehensive overview of some of the developments in positive operator theory between 1950 and 1970.

THEOREM 5.2. *Let the cone E_+ be total, let $T: E \to E$ be non-supporting with respect to E_+, and suppose that $r = r(T)$ is a pole of the resolvent, then*

(a) *$r > 0$ and r is an algebraïcally simple eigenvalue of T.*

(b) *The corresponding eigenvector ψ satisfies: $\psi \in E_+$ and $\langle H, \psi \rangle > 0$ for all $H \in E^*_+$, $H \neq 0$.*

(c) *The corresponding dual eigenvector is strictly positive.*

(d) *If, in addition, X is a Banach lattice and $\{\lambda \in \sigma(T): |\lambda| = r\}$ consists only of poles of the resolvent, then all remaining elements $\lambda \in \sigma(T)$ satisfy $|\lambda| < r$.*

6. Location of the singular points

From now on we let $X = L_1[0,\infty)$. In section 4 we defined the analytic operator family $\hat{K}(\lambda)$, $\lambda \in \Lambda$, being the Laplace transform of $K(t)$. Evidently $\hat{K}(\lambda)$ defines a bounded operator on X for all $\lambda \in \Lambda$.

$$(\hat{K}(\lambda)\psi)(x) = \int_{a_0}^{G(2x)} e^{-\lambda a} k(a, 2x)\psi(X(-a, 2x))da, \quad \psi \in X. \tag{6.1}$$

In the Appendix we shall prove the following result.

LEMMA 6.1. *For all $\lambda \in \Lambda$ the operator $\hat{K}(\lambda)$ is compact.*

We can now apply the following result, proved by STEINBERG (1968).

LEMMA 6.2. *Let E be a Banach space and Δ a subset of the complex plane which is open and connected. If $T(\lambda)$ is an* *analytic family of compact operators on E for $\lambda \in \Delta$, then either $(I-T(\lambda))$ is nowhere invertible in Δ or $(I-T(\lambda))^{-1}$ is meromorphic in Δ.*

(A function $\phi(\lambda)$ defined on a set $V \subseteq \mathbf{C}$ is called meromorphic if it is analytic on V except for an at most countable set of elements of V which are poles of finite order of ϕ.) It is clear that $\|\hat{K}(\lambda)\| \to 0$ if $\mathrm{Re}\,\lambda \to \infty$, implying that $I-\hat{K}(\lambda)$ is invertible if $\mathrm{Re}\,\lambda$ is large enough. Thus lemma 6.1 and lemma 6.2 yield:

THEOREM 6.3. *The function $\lambda \to (I-\hat{K}(\lambda))^{-1}$ is meromorphic in Λ.*

Therefore the set Σ defined by (4.3) is a discrete set whose elements are poles of $(I-\hat{K}(\lambda))^{-1}$ of finite order.

Now we shall employ positivity arguments to determine the so-called dominant singular point, i.e. the element of Σ with the largest real part. Before doing so we make an additional assumption on the growthrate g.

ASSUMPTION 6.4. *There exists a $\delta > 0$ such that $2g(x)-g(2x) \geqslant \delta$, all $x \in [0,\infty)$.*

In chapter II a similar assumption has been made to establish compactness of the semigroup. In section 9 we shall explain why assumption 6.4 is imposed. A consequence of this assumption is that a baby cell can not attain arbitrarily small sizes. We shall make this more explicit. If a cell is born with size x, then it can divide not earlier than a_0 time units later, and its daughers can not be smaller than

$$\gamma(x) \overset{def}{=} \tfrac{1}{2}X(a_0,x) = \tfrac{1}{2}G^{-1}(a_0+G(x)). \tag{6.2}$$

LEMMA 6.5. *γ has precisely one fixed point x_0. Let for arbitrary $x_1 \geqslant 0$ the sequence $\{x_n\}$ be defined recursively as $x_{n+1} = \gamma(x_n)$, $n \geqslant 1$ then: $x_1 < x_0$ implies $x_n < x_0$ and $x_1 > x_0$ implies $x_n > x_0$,* and $\lim_{n\to\infty} x_n = x_0$.

PROOF. The equation $\gamma(x)=x$ is equivalent to $G(2x)-G(x)=a_0$. The left hand side is zero when $x=0$ and its derivative $\dfrac{2g(x)-g(2x)}{g(2x)g(x)} > \dfrac{\delta}{g_{\min}} > 0$ by assumptions (A_g) and 6.4. This proves the first part of the lemma.
Next we consider the recurrence relation $x_{n+1}=\gamma(x_n)$. Since $\gamma(0)>0$, γ is continuous, and x_0 is the unique solution of $x_0=\gamma(x_0)$, we have $\gamma(x)>x$ if $0 \leqslant x < x_0$ and $\gamma(x)<x$ if $x>x_0$. In combination with $\gamma'(x)=\dfrac{g(a_0+G(x))}{2g(x)}>0$ this implies $x_n<x_{n+1}<x_{n+1}<x_0$ if $x_1<x_0$ and $x_n>x_{n+1}>x_0$ if $x_1>x_0$. Therefore $\lim_{n\to\infty} x_n$ exists. The continuity of γ implies that it is a fixed point. □

From this lemma and the observation that a baby cell attains the minimum birth size if all its ancestors have divided at age a_0, it follows that this minimum birth size is x_0 (which is positive if a_0 is positive), provided that there are infinitely many ancestors who all lived under the same growth regime.

REMARK 6.6. The state space Ω_s referred to in section 1 is given by $\Omega_s = \{(a,x) \in \mathbb{R}^+ \times \mathbb{R}^+ | x \geqslant X(a,x_0)\}$.

However, we do not want to restrict ourselves à priori to initial data defined on Ω_s only, but admit that $n_0(a,x)$ defined in (1.6) is positive on $\Omega \setminus \Omega_s$. We can prove the following result.

LEMMA 6.7. *If ψ is an eigenvector of $\hat{K}(\lambda)$, then $\psi(x) = 0$, $x < x_0$.*

PROOF. Let $\psi \in X$. It follows from (6.1) that $(\hat{K}(\lambda)^n\psi)(x) = 0$ if $x \leqslant x_n$, where $x_1 = \gamma(0)$ and $x_{n+1} = \gamma(x_n)$, $n \geqslant 1$. If ψ is an eigenvector of $\hat{K}(\lambda)$ then ψ is an eigenvector of $\hat{K}(\lambda)^n$ for every positive integer n. As a consequence $\psi(x) = 0$ if $x \leqslant x_n$, and now the result follows from lemma 6.5. □

We denote with Y the subspace of X containing all $\psi \in L_1[0,\infty)$ which are identically zero on $[0,x_0)$. Obviously $\hat{K}(\lambda)Y \subseteq Y$. We let $\hat{K}_0(\lambda)$ be the restriction of $\hat{K}(\lambda)$ to Y. It is clear immediately that lemma 6.1 and theorem 6.3 remain valid if $\hat{K}(\lambda)$ is replaced by $\hat{K}_0(\lambda)$. Moreover (4.3) can be replaced by $\Sigma = \{\lambda \in \Lambda | 1 \in \sigma(\hat{K}_0(\lambda))\}$. Let Y_+ be the subset of Y containing all elements which are non-negative a.c. (almost everywhere). The following result is straightforward.

THEOREM 6.8. *Y_+ defines a cone in Y which is total. Moreover $\hat{K}_0(\lambda)$ is positive with respect to Y_+ for all $\lambda \in \Lambda \cap \mathbb{R}$.*

We let Y_+^* be the dual of Y_+ and this defines a cone in Y^* because Y_+ is total. Clearly Y_+^* can be identified with $L_\infty^+[x_0,\infty)$, i.e. all measurable function on $[x_0,\infty)$ which are non-negative and essentially bounded.

The following lemma provides a useful characterization of the non-zero elements of Y_+^*.

LEMMA 6.9. *If $F \in Y_+^*$, $F \neq 0$, then there exists an $\epsilon > 0$ such that for all $f \in Y_+$ satisfying $f(x) > 0$ for almost every $x \in [x_0+\epsilon,\infty)$ the relation $<F,f> > 0$ holds.*

PROOF. $F \in Y_+^*$, $F \neq 0$ implies that there exists a measurable set $V \subset [x_0,\infty)$ with measure $\mu > 0$ such that $F(x) > 0$, $x \in V$. If we choose $\epsilon < \mu$, then the intersection $V \cap [x_0+\epsilon,\infty)$ has a measure which is greater than $\mu-\epsilon > 0$, and this yields the result. □

Now we can prove the following strong positivity result with respect to $\hat{K}_0(\lambda)$.

THEOREM 6.10. *For all $\lambda \in \Lambda \cap \mathbb{R}$ the operator $\hat{K}_0(\lambda)$ is non-supporting with respect to Y_+.*

PROOF. Let $\psi \in Y_+$, $\psi \neq 0$ and $\lambda \in \Lambda \cap \mathbb{R}$. If we substitute $z = X(-a, 2x)$ in (6.1) we obtain

$$(\hat{K}_0(\lambda)\psi)(x) = \int_{x_0}^{X(-a_0,2x)} e^{-\lambda(G(2x)-G(z))} \cdot k(G(2x)-G(z),2x) \frac{\psi(z)}{g(z)}\, dz.$$

Let $F \in Y_+^*$, $F \neq 0$ and let $\epsilon > 0$ be given by lemma 6.9. There exists an $x_1 > x_0$ such that $\int_{x_0}^{X(-a_0,2x_1)} \psi(z)dz > 0$. This yields that $(\hat{K}_0(\lambda)\psi)(x) > 0$ if $x \geq x_1$. Let $x_2 = \gamma(x_1)$, where γ is defined by (6.2). Then $(\hat{K}_0(\lambda)^2\psi)(x) > 0$, $x \geq x_2$. Recursively we find $(\hat{K}_0(\lambda)^n\psi)(x) > 0$, $x \geq x_n$, where $x_n = \gamma(x_{n-1})$, $n \geq 2$. We conclude from lemma 6.5 that there exists a $p \in \mathbb{N}$ such that $x_n < x_0 + \epsilon$ if $n \geq p$. Now we can apply lemma 6.9 which says that $<F,\hat{K}_0(\lambda)^n\psi > 0$ if $n \geq p$, and this proves the result. □

We can draw the following conclusions from theorem 5.2. Let $r_\lambda = r(\hat{K}_0(\lambda))$, $\lambda \in \Lambda$. If $\lambda \in \Lambda \cap \mathbb{R}$, then

(a) r_λ is an algebraïcally simple eigenvalue of $\hat{K}_0(\lambda)$.

(b) The corresponding eigenvector $\psi_\lambda \in Y_+$ satisfies $\psi_\lambda(x) > 0$, $x \in [x_0,\infty)$ a.e. We fix ψ_λ by the normalization $\|\psi_\lambda\| = 1$.

(c) The corresponding eigenfunctional $F_\lambda \in Y_+^*$ satisfies $F_\lambda(x) > 0$, $x \in [x_0,\infty)$ a.e., i.e. F_λ is strictly positive.

Hence, if $\lambda \in \Lambda$ is real and $r_\lambda = 1$, then $\lambda \in \Sigma$.

LEMMA 6.11. *There exists a unique $\lambda \in \Lambda \cap \mathbb{R}$ such that $r(\hat{K}_0(\lambda)) = 1$.*

PROOF. Let $\lambda,\mu \in \Lambda \cap \mathbb{R}$, $\lambda > \mu$ and $\psi \in Y_+$.

$$(\hat{K}_0(\mu)\psi)(x) = \int_{a_0}^{G(2x)} e^{-\mu a} k(a, 2x)\psi(X(-a, 2x))da$$

$$\geq e^{(\lambda-\mu)a_0} \int_{a_0}^{G(2x)} e^{-\lambda a} k(a, 2x)\psi(X(-a, 2x))da = e^{(\lambda-\mu)a_0}(\hat{K}_0(\lambda)\psi)(x).$$

If we substitute $\psi = \psi_\lambda$, then we obtain $\hat{K}_0(\mu)\psi_\lambda \geq e^{(\lambda-\mu)a_0} r_\lambda \psi_\lambda$. Taking duality pairings with F_μ on both sides yields

$$r_\mu \geq e^{(\lambda-\mu)a_0} \cdot r_\lambda \tag{6.3}$$

where we have used that $<F_\mu,\psi_\lambda>>0$. Thus $\lambda \to r(\hat{K}_0(\lambda))$ is strictly decreasing in $\Lambda \cap \mathbb{R}$. Moreover this function is continuous. It follows easily that $\lim_{\lambda\to\infty} r(\hat{K}_0(\lambda)) = 0$. If we can prove that $\lim_{\lambda\downarrow -d_\infty} r(\hat{K}_0(\lambda)) = \infty$ then the conclusion of the lemma follows. We have to distinguish between two cases.

(a) $d_\infty = \infty$. Then (6.3) implies that $\lim_{\lambda\to-\infty} r(\hat{K}_0(\lambda)) = \infty$.

(b) $d_\infty < \infty$. Since $\|\psi_\lambda\| = 1$,

$$r(\hat{K}_0(\lambda)) = \|\hat{K}_0(\lambda)\psi_\lambda\| = \int_{x_0}^{\infty} \{\int_0^{\infty} e^{-\lambda t}(K(t)\psi_\lambda)(x)dt\}dx = \int_0^{\infty} e^{-\lambda t}\{\int_{x_0}^{\infty} (K(t)\psi_\lambda)(x)dx\}dt$$

$$= \int_0^{\infty} e^{-\lambda t}\|K(t)\psi_\lambda\|dt \geq \int_{T_0}^{\infty} e^{-\lambda t}\|K(t)\psi_\lambda\|dt \geq \int_{T_0}^{\infty} m_K e^{-d_\infty t} e^{-\lambda t} dt = \frac{m_K}{\lambda+d_\infty} e^{-(\lambda+d_\infty)T_0},$$

where we have used lemma 3.1. The change of order of integration was permitted because of Fubini's theorem (DUNFORD & SCHWARTZ (1958)). It follows that $\lim_{\lambda \downarrow -d_\infty} r(\hat{K}_0(\lambda)) = \infty$. □

We denote the unique solution of $r(\hat{K}_0(\lambda)) = 1$ by λ_d, and we shall write ψ_d and F_d in stead of ψ_{λ_d} and F_{λ_d} respectively. We assume that ψ_d and F_d are normalized by

$$\|\psi_d\| = 1, \quad <F_d, \psi_d> = 1. \tag{6.4}$$

In order to prove that indeed λ_d is indeed the element of Σ with the largest real part, we need the following lemma (e.g. RUDIN (1966)).

LEMMA 6.12. *Let $f \in L_1[0,\infty)$ be a complex-valued function. Then $|\int_0^\infty f(x)dx| = \int_0^\infty |f(x)|dx$ if and only if there exists a constant $\alpha \in \mathbb{C}$, $|\alpha| = 1$ such that $|f(x)| = \alpha f(x)$ a.e. on $[0,\infty)$.*

THEOREM 6.13. *If $\lambda \in \Sigma$, $\lambda \neq \lambda_d$, then* $\mathrm{Re}\,\lambda < \lambda_d$.

PROOF. Suppose $\lambda \in \Sigma$ and $\hat{K}_0(\lambda)\psi = \psi$. Hence $|\hat{K}_0(\lambda)\psi| = |\psi|$, where $|\psi|(x) \stackrel{def}{=} |\psi(x)|$. This yields $\hat{K}_0(\lambda_R)|\psi| \geqslant |\psi|$, where $\lambda_R = \mathrm{Re}\,\lambda$. Taking duality pairings with F_{λ_R} on both sides yields $r_{\lambda_R} <F_{\lambda_R}, |\psi|> \geqslant <F_{\lambda_R}, |\psi|>$, from which we conclude that $r_{\lambda_R} \geqslant 1$. In the proof of lemma 6.11 we have shown that $\lambda \rightarrow r_\lambda$ is decreasing in $\lambda \in \Lambda \cap \mathbb{R}$, and this implies that $\lambda_R = \mathrm{Re}\,\lambda_d$. Now suppose that $\mathrm{Re}\,\lambda = \lambda_d$ and $\mathrm{Im}\,\lambda = \eta$. Thus $\hat{K}_0(\lambda_d)|\psi| \geqslant |\psi|$. Suppose that $\hat{K}_0(\lambda_d)|\psi| > |\psi|$. Taking duality pairings with F_d on both sides yields $<F_d,|\psi|> \; > \; <F_d,|\psi|>$ which is a contradiction. As a consequence $\hat{K}_0(\lambda_d)|\psi| = |\psi|$, from which we deduce that $|\psi| = c \cdot \psi_d$ for some constant c which we may assume to be one. Therefore $\psi(x) = \psi_d(x)e^{i\alpha(x)}$ for some real-valued function α. If we substitute this in $\hat{K}_0(\lambda_d)\psi_d = |\hat{K}_0(\lambda)\psi|$ we obtain

$$\int_{a_0}^\infty e^{-\lambda_d a} k(a,2x)\psi_d(X(-a,2x))da = |\int_{a_0}^\infty e^{-\lambda_d a - i\eta a} k(a,2x)\psi_d(X(-a,2x))e^{i\alpha(X(-a,2x))}da|.$$

From lemma 6.12 we conclude that $\alpha(X(-a,2x)) - \eta a = \beta$, for some constant β. If we substitute this in $\hat{K}_0(\lambda)\psi = \psi$ we obtain $e^{i\beta}\int_0^\infty e^{-\lambda_d a} k(a,2x))da = \psi_d(x)e^{i\alpha(x)}$, thus $\alpha(x) = \beta$ from which we conclude that $\eta = \mathrm{Im}\,\lambda = 0$. □

This result, combined with the Riemann-Lebesgue lemma (lemma 4.3) and theorem 6.3, implies among others that there exists a positive horizontal distance between λ_d and the other points in Σ.

COROLLARY 6.14. *There exists an $\epsilon > 0$ such that $\lambda_d - \epsilon > -d_\infty$ and* $\mathrm{Re}\,\lambda \leqslant \lambda_d - \epsilon$ *if $\lambda \in \Sigma$, $\lambda \neq \lambda_d$.*

Clearly $\hat{K}_0(\lambda)$ and $\hat{K}(\lambda)$ have the same eigenvectors (lemma 6.7). However $\hat{K}_0(\lambda)^*$ and $\hat{K}(\lambda)^*$ do not have the same eigenvectors. Let F_d' be the eigenvector of $\hat{K}(\lambda_d)^*$ corresponding to the eigenvalue one. Obviously, F_d' defines a positive functional on X. We can prove the following relation between F_d and F_d'. Let $<F_d', \psi_d> = 1$.

THEOREM 6.15. *For all $\psi \in Y$, the equality $<F_d,\psi> = <F_d',\psi>$ holds.*

PROOF. Let $\psi \in Y$, then $\psi = <F_d,\psi> \cdot \psi_d + \rho$, where $\rho \in \mathcal{R}(\hat{K}_0(\lambda_d) - I) \stackrel{def}{=} Z$, i.e. the range of $\hat{K}_0(\lambda_d) - I$. Since the spectral radius of the restriction of $\hat{K}_0(\lambda_d)$ to the subspace Z is strictly less than one (theorem 5.2d) it follows that $\|\hat{K}_0(\lambda_d)^n\rho\| < \theta^n\|\rho\|$ for all $\rho \in Z$, where θ is some constant strictly less than one. Since $\hat{K}(\lambda_d)\psi = \hat{K}_0(\lambda_d)\psi$ we have $<F_d',\psi> = <\hat{K}(\lambda_d)^{*n}F_d',\psi> = <F_d',\hat{K}_0(\lambda_d)^n(<F_d,\psi>\psi_d + \rho)> = <F_d,\psi> + <F_d',\hat{K}_0(\lambda_d)^n\rho>$. If we let $n \rightarrow \infty$ then the second term at the right-hand-side tends to zero yielding that $<F_d',\psi> = <F_d,\psi>$. □

7. Computation of the residue in λ_d.

Here we shall concentrate on the behaviour of $(I - \hat{K}(\lambda))^{-1}$ in a neighbourhood of $\lambda = \lambda_d$, which is a pole of finite order (cf. theorem 6.3). The techniques exploited in this section are very similar to those in a paper by SCHUMITZKY & WENSKA (1975). We define

$$R(\lambda) = (I - \hat{K}(\lambda))^{-1}, \quad \lambda \in \Lambda \setminus \Sigma. \tag{7.1}$$

Since $\hat{K}(\lambda)$ is analytic in a neighbourhood of λ_d we can write down its Taylor expansion.

$$\hat{K}(\lambda) = \sum_{n=0}^\infty (\lambda - \lambda_d)^n K_n, \tag{7.2}$$

where the series converges in the norm topology. Let $p \geqslant 1$ be the order of the pole of $R(\lambda)$ in $\lambda = \lambda_d$. In a neighbourhood of λ_d, $R(\lambda)$ can be represented by a Laurent series:

$$R(\lambda) = \sum_{n=-p}^{\infty} (\lambda-\lambda_d)^n R_n, \tag{7.3}$$

where by definition $R_{-p} \neq 0$. From

$$R(\lambda)(I-\hat{K}(\lambda)) = (I-\hat{K}(\lambda))R(\lambda) = I \tag{7.4}$$

if follows immediately that

$$R_{-p}(I-K_0) = (I-K_0)R_{-p} = 0. \tag{7.5}$$

From this relation and $K_0 = \hat{K}(\lambda_d)$ we obtain

$$\mathcal{R}(R_{-p}) = \{\psi_d\}, \tag{7.6}$$

where $\mathcal{R}(R_{-p})$ denotes the range of the operator R_{-p}, and $\{\psi_d\}$ stands for the span of the positive eigenvector ψ_d, i.e. $\{\psi_d\} = \{\gamma \cdot \psi_d | \gamma \in \mathbf{C}\}$. A relation similar to (7.4) is valid for the dual operators $K_0^* = \hat{K}(\lambda_d)^*$ and R_{-p}^*. Therefore

$$\mathcal{R}(R_{-p}^*) = \{F_d\}. \tag{7.7}$$

From (7.4) we also deduce that

$$-R_{-p}K_1 + R_{-p+1}(I-K_0) = 0, \quad \text{if } p > 1, \tag{7.8a}$$

$$-R_{-1}K_1 + R_0(I-K_0) = I, \quad \text{if } p = 1. \tag{7.8b}$$

Together with (7.5) this implies

$$R_{-p}K_1R_{-p} = 0, \quad \text{if } p > 1, \tag{7.9a}$$

$$R_{-1}K_1R_{-1} = -R_{-1}, \quad \text{if } p = 1. \tag{7.9b}$$

We can state our main result now.

THEOREM 7.1. *$R(\lambda)$ has a pole of order one in $\lambda = \lambda_d$ and the residue R_{-1} is given by*

$$R_{-1}\psi = \frac{<F_d',\psi>}{<F_d',-K_1\psi_d>} \cdot \psi_d, \quad \psi \in X. \tag{7.10}$$

Observe that $-K_1 = [-\frac{d}{d\lambda}\hat{K}(\lambda)]_{\lambda=\lambda_d}$ defines a positive non-supporting operator on Y and thus it follows from theorem 6.15 that $<F_d',-K_1\psi_d> = <F_d,-K_1\psi_d> > 0$.

PROOF OF THEOREM 7.1. Let ϕ_d and H_d be solutions of $R_{-p}\phi = \psi_d$ and $R_{-p}^*H = F_d$ respectively. On account of (7.6) and (7.7) such solutions indeed exist. If $p > 1$ then (7.9a) yields $0 = <H_d, R_{-p}K_1R_{-p}\phi_d> = <F_d,K_1\psi_d>$ which is a contradiction since F_d is strictly positive and $-K_1\psi_d$ is positive and nonzero. Therefore $p = 1$, and $\mathcal{R}(R_{-1}) = \{\psi_d\}$. Now let $R_{-1}\psi = f(\psi) \cdot \psi_d$ for some linear functional f. Then $<H_d,R_{-1}\psi> = <R_{-1}^*H_d,\psi> = <F_d,\psi> = <H_d,-R_{-1}KR_{-1}\psi> = <R_{-1}^*H_d, -K_1(f(\psi)\cdot\psi_d)> = f(\psi)\cdot<F_d,-K_1\psi_d>$, thus $f(\psi) = <F_d,\psi> / <F_d,-K_1\psi_d>$ which proves the result. □

It is not à priori clear whether or not $<F_d',\psi> > 0$ if $\psi \in X_+$, $\psi \neq 0$. This, however, is proved in the following lemma.

LEMMA 7.2. *If $\psi \in X_+$, $\psi \neq 0$ then $<F_d',\psi > 0$.*

PROOF. If the restriction of ψ to $[x_0,\infty)$ is not identically zero, then the result follows from theorem 6.15. Now suppose that ψ is positive on a subset of $[0,x_0]$ with positive measure. Thus

$$(\hat{K}(\lambda_d)\psi)(x) \geqslant \int_{G(2x)-G(x_0)}^{G(2x)} e^{-\lambda_d a}k(a,2x)\psi(X(-a,2x))da = \int_0^{x_0} e^{-\lambda_d(G(2x)-G(z))}\cdot k(G(2x)-G(z),2x)\frac{\psi(z)}{g(z)}dz > 0$$

for all $x \geqslant x_0$. Therefore $<F_d',\psi> = <\hat{K}(\lambda_d)^*F_d',\psi> = <F_d',\hat{K}(\lambda_d)\psi> > 0$. □

8. The inverse Laplace transform

Let E be a Banach space. The Hardy-Lebesgue class $H_p(\alpha;E)$ is the class of functions $g(\lambda)$ with values in E, which are analytic in $\mathrm{Re}\,\lambda > \alpha$ and satisfy the following conditions (cf. FRIEDMAN & SHINBROT (1967), HILLE & PHILLIPS (1957)).

$$\sup_{\zeta>\alpha}\left\{\int_{-\infty}^{\infty}\|g(\zeta+i\eta)\|^p d\eta\right\}^{\frac{1}{p}} < \infty, \tag{8.1a}$$

$$g(\alpha+i\eta) = \lim_{\zeta\downarrow\alpha} g(\zeta+i\eta) \text{ exists a.e. and is an element of } L_p(-\infty,\infty;E). \tag{8.1b}$$

The following inverse Laplace transform formula can be found in FRIEDMAN & SHINBROT (1967).

LEMMA 8.1. *Let $g(\lambda) \in H_1(\alpha;E)$, then the function*

$$f(t) = \frac{1}{2\pi i}\int_{\gamma-i\infty}^{\gamma+i\infty} e^{\lambda t} g(\lambda)d\lambda, \quad (\gamma \geqslant \alpha) \tag{8.2}$$

is defined and independent of γ, for all $t \in (-\infty,\infty)$. Moreover $f(t) = 0$, $t < 0$, $f(t)$ is continuous and $\hat{f}(\lambda) = g(\lambda)$.

We rewrite the abstract renewal equation (2.17) as

$$B = \Phi + K\star B, \tag{8.3}$$

where $K\star B$ denotes the convolution product, i.e. $(K\star B)(t) = \int_0^t K(a)B(t-a)da$. If we substitute

$$B = \Phi + v, \tag{8.4}$$

we obtain

$$v = \Psi + K\star v, \tag{8.5}$$

where

$$\Psi = K\star\Phi. \tag{8.6}$$

Taking Laplace transforms on both sides of (8.5) gives us

$$\hat{v}(\lambda) = (I-\hat{K}(\lambda))^{-1}\hat{\Psi}(\lambda). \tag{8.7}$$

We can prove the following result.

LEMMA 8.2. $\hat{v}(\lambda) \in H_1(\alpha;X)$, *if* $\alpha > \lambda_d$.

PROOF. Let $\lambda \in \mathbb{C}$ be such that $\mathrm{Re}\,\lambda \geqslant \alpha$. The function $\eta \to \hat{\Psi}(\zeta+i\eta)$ is an element of $L_1(-\infty,\infty;X)$ if $\zeta > -d_\infty$ (see section 6.3 of HILLE & PHILLIPS (1957)). Moreover we know from the Riemann-Lebesgue lemma (lemma 4.3) that $\|(I-\hat{K}(\zeta+i\eta))^{-1}\| \leqslant 2$ if $|\eta|$ is large enough, say $|\eta| \geqslant \eta_0$. From the continuity of the function $\eta \to (I-\hat{K}(\zeta+i\eta))^{-1}$ on $[-\eta_0,\eta_0]$ (if $\zeta \geqslant \alpha$) we conclude that there exists a constant $C > 0$ such that $\|(I-\hat{K}(\zeta+i\eta))^{-1}\| < C$ for all $\eta \in (-\infty,\infty)$. Thus $\|\hat{v}(\zeta+i\eta)\| \leqslant C\|\hat{\Psi}(\zeta+i\eta)\|$ where we have used (8.7). The positivity of $K(t)$ and $\Psi(t)$ yields that

$$\|\hat{\Psi}(\zeta+i\eta)\| \leqslant \|\hat{\Psi}(\alpha+i\eta)\|, \quad \zeta \geqslant \alpha,$$

and we conclude that condition (8.1a) is satisfied. The validity of condition (8.1b) follows from the analyticity of $(I-\hat{K}(\lambda))^{-1}$, $\hat{\Phi}(\lambda)$ and $\hat{K}(\lambda)$ on the region $\mathrm{Re}\,\lambda > \lambda_d$ and the fact that $\alpha > \lambda_d$. □

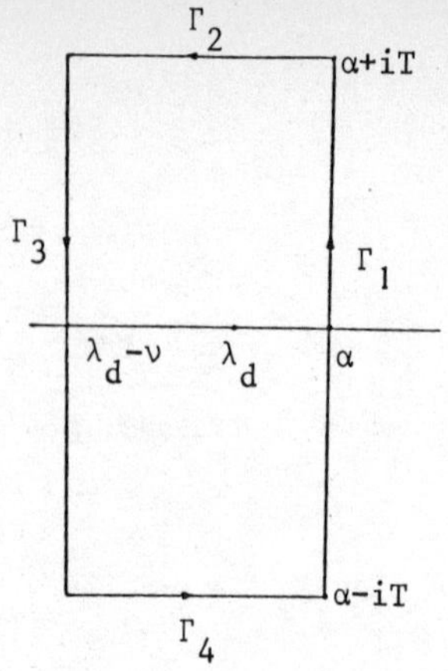

Figure 8.1. $\Gamma = \bigcup_{i=1}^{4} \Gamma_i$

Now let $\alpha > \lambda_d$, then lemma 8.1 yields that

$$v(t) = \frac{1}{2\pi i} \int_{\alpha - i\infty}^{\alpha + i\infty} e^{\lambda t} \hat{v}(\lambda) d\lambda \tag{8.8}$$

is well-defined. Some contributions to this integral can be evaluated by the method of residues. Therefore we shift the vertical integration curve Re $\lambda = \alpha$ to the left across the singularity $\lambda = \lambda_d$, such that it crosses no other elements of Σ (see fig. 8.1). Let $\epsilon > 0$ be given by corollary 6.14, and let $0 < \nu < \epsilon$. Let Γ be the rectangular contour in fig. 2. It follows immediately from the Riemann-Lebesgue lemma (lemma 4.3) that

$$\lim_{T \to \infty} \int_{\Gamma_i} e^{\lambda t} \hat{v}(\lambda) d\lambda = 0, \ i = 2,4.$$

Now it follows from Cauchy's theorem (which is also valid for vector-valued functions: see HILLE & PHILLIPS (1957)) that

$$v(t) = \frac{1}{2\pi i} \oint_{\Gamma} e^{\lambda t} \hat{v}(\lambda) d\lambda + \frac{1}{2\pi i} \lim_{T \to \infty} \int_{\lambda_d - \nu - iT}^{\lambda_d - \nu + iT} e^{\lambda t} \hat{v}(\lambda) d\lambda,$$

where we have used that the first integral does not depend on T. The residue theorem gives:

$$\frac{1}{2\pi i} \oint_{\Gamma} e^{\lambda t} \hat{v}(\lambda) d\lambda = \underset{\lambda = \lambda_d}{Res} \{ e^{\lambda t} \hat{v}(\lambda) \} = e^{\lambda_d t} R_{-1} \hat{\Psi}(\lambda_d) = e^{\lambda_d t} R_{-1} \hat{K}(\lambda_d) \hat{\Phi}(\lambda_d) = e^{\lambda_d t} \cdot \frac{<F_d', \hat{K}(\lambda_d) \hat{\Phi}(\lambda_d)>}{<F_d', -K_1 \psi_d>} \cdot \psi_d$$

$$= e^{\lambda_d t} \frac{<F_d', -K_1 \hat{\Phi}(\lambda_d)>}{<F_d', -K_1 \psi_d>} \cdot \psi_d,$$

where we have used theorem 7.1, (8.6) and (8.7). As in the proof of lemma 8.2 we have that the function $\eta \to \hat{v}(\lambda_d - \nu + i\eta)$ is an element of $L_1(-\infty, \infty; X)$. Now

$$\| \frac{1}{2\pi i} \int_{\lambda_d - \nu - i\infty}^{\lambda_d - \nu + i\infty} e^{\lambda t} \hat{v}(\lambda) d\lambda \| \leq M \cdot e^{(\lambda_d - \nu)t},$$

where

$$M \overset{def}{=} \frac{1}{2\pi} \int_{-\infty}^{\infty} \| \hat{v}(\lambda_d - \nu + i\eta) \| d\eta \text{ depends on } \nu \text{ and } \Phi.$$

We can state our main result now.

COROLLARY 8.3. *Let $\epsilon > 0$ be given by corollary 6.14, and let $0 < \nu < \epsilon$, then $\|e^{-\lambda_d t} B(t) - c \cdot \psi_d\| \leq L e^{-\nu t}$, $t \geq 0$, for* *some constant L, where* $c = \dfrac{<F'_d, \hat{\Phi}(\lambda_d)>}{<F'_d, -K_1\psi_d>}$ *is a constant depending linearly on* Φ.

PROOF. We have $B(t) = \Phi(t) + v(t)$, and $v(t) = e^{\lambda_d t}(c\cdot\psi_d + O(e^{-\nu t}))$. Now the result follows from lemma 3.1. □

REMARK 8.4. Observe from corollary 8.3 that if t has become infinite, no cells with size less than x_0 are born, although such cells may be present at time zero.

9. Interpretation, conclusions and final remarks

For the sake of converience we repeat (2.11) and (2.12)

$$n(t,a,x) = \frac{g(X(-t,x))}{g(x)} Q(t,a-t,X(-t,x))n_0(a-t,X(-t,x)),\ t \leqslant a,$$

$$n(t,a,x) = \frac{g(X(-a,x))}{g(x)} E(a,X(-a,x))B(t-a,X(-a,x)),\ t > a.$$

This does not define a classical solution of (1.4)-(1.6). However, it can be proved that n is differentiable along the characteristics of the partial differential operator $D = \frac{\partial}{\partial t} + \frac{\partial}{\partial a} + g(x)\frac{\partial}{\partial x}$, and in this sense indeed is a solution of (1.4)-(1.6).

Let

$$n_d(a,x) = e^{-\lambda_d a} \cdot \frac{g(X(-a,x))}{g(x)} E(a,X(-a,x))\psi_d(X(-a,x)). \tag{9.1}$$

Now we can restrate corollary 8.3 in terms of the solution n of (1.4)-(1.6).

COROLLARY 9.1. *Let* $\epsilon > 0$ *be given by corollary 6.14 and let* $0 < \nu < \epsilon$, *then the solution* $n(t,a,x)$ *of (1.4)-(1.6) satisfies* $\|e^{-\lambda_d t}n(t,\cdot,\cdot) - h(n_0)\cdot n_d\| \leqslant L'e^{-\nu t}\|n_0\|$, $t \geqslant 0$, *where* $\|\cdot\|$ *stands for the* $L_1(\Omega)$*-norm,* L' *is a positive constant, and* h *is a strictly positive linear functional on* $L_1(\Omega)$.

REMARK 9.2. h can be computed from $h(n_0) = \dfrac{<F'_d, \hat{\Phi}(\lambda_d)>}{<F'_d, -K_1\psi_d>}$.

Corollary 9.1 is a typical renewal result. The population grows (or decays) exponentially with exponent λ_d (which is sometimes called the Malthusian parameter). As time increases an asymptotically stable age-size distribution is reached. If $t = \infty$ the dependence on the initial condition is only reflected by the scalar $h(n_0)$.

If in our model the rates b and μ depend on age only then we can integrate (1.4)-(1.6) over all sizes x and we find the age-dependent problem

$$\frac{\partial N}{\partial t} + \frac{\partial N}{\partial a} = -(\mu(a)+b(a))N(t,a), \tag{9.2a}$$

$$N(t,0) = 2\int_0^\infty b(a)N(t,a)da, \tag{9.2b}$$

$$N(0,a) = N_0(a), \tag{9.2c}$$

where $N(t,a) \overset{def}{=} \int_0^\infty n(t,a,x)dx$. If the assumptions (A_b), (A_μ) and (A_d) of section 1 are satisfied then a stable age-distribution is reached as $t \to \infty$:

$$N(t,a) \sim e^{\lambda_d t}N_d(a),\ t \to \infty,$$

(this result can also be found in EISEN (1979)) and the growthrate $g(x)$ has no effect on this stable age-distribution. More details can be found in HANNSGEN et al. (1985).

Now we shall explain what can happen if assumption 6.4 is not fulfilled.

I. We expect that most of our result remain valid if $g(2x) < 2g(x)$, all x (but not necessarily $2g(x)-g(2x) > \delta$, for some $\delta > 0$). But probably one gets involved with great technical difficulties, which, however, do not provide additional insight.

II. If $g(2x) > 2g(x)$, for all x, then some sort of instability comes into the problem. Although γ defined by (6.2) again has a unique fixed point x_0, in this case it is unstable:

$$\frac{d\gamma}{dx}\Big|_{x=x_0} = \frac{g(2x_0)}{2g(x_0)} > 1.$$

For the sequence $\{x_n\}$ of lemma 6.4 this result in

$$x_n \to 0, \text{ if } x_1 < x_0,$$

$$x_n \to \infty, \text{ if } x_1 > x_0.$$

If we start with a population all of whose members have size $> \bar{x}(0)$, where $\bar{x}(0) > x_0$, then at time t all individuals have size $> \bar{x}(t)$, where $\bar{x}(t) \to \infty$. As a consequence there cannot exist a stable age-size distribution. A second problem arising in this case is caused by the fact that growth becomes very small if x tends to zero. As a consequence individuals can not grow away from zero.

III. Suppose that $g(2x) = 2g(x)$, all x. (Notice that this and also the former case is actually excluded by the boundedness condition on g. However, the same integral equation for the birth function $B(t)$ still holds.) Biologically this condition means that the time T needed to grow from x to $2x$ does not depend on x. We can prove that in this case the set of singular points Σ is periodic, i.e. there exists a $p > 0$ such that $\lambda \in \Sigma \Rightarrow \lambda + ikp \in \Sigma$, $k \in \mathbb{Z}$ (compare II.5).

LEMMA 9.3. *Let $g(2x) = 2g(x)$, for all x and let $T = G(2x) - G(x)$ (which does not depend on x), then Σ is periodic with period $p = \dfrac{2\pi}{T}$.*

PROOF. Suppose $\lambda \in \Sigma$ and let $\psi \in X$ be determined by $\hat{K}(\lambda)\psi = \psi$:

$$\psi(x) = \int_{a_0}^{\infty} e^{-\lambda a} k(a, 2x)\psi(X(-a, 2x))da.$$

Let $T = G(2x) - G(x)$ and $p = \dfrac{2\pi}{T}$. Let $\psi_k(x) = e^{-ikpG(x)} \cdot \psi(x)$, then

$$(\hat{K}(\lambda + ikp)\psi_k)(x) = \int_{a_0}^{\infty} e^{-\lambda a} e^{-ikpa} k(a, 2x)\psi(X(-a, 2x))e^{-ikp(G(2x)-a)}da$$

$$= e^{-ikpG(2x)} \int_{a_0}^{\infty} e^{-\lambda a} k(a, 2x)\psi(X(-a, 2x))da = e^{-ikp(T+G(x))} \cdot \psi(x) = \psi_k(x),$$

hence $\lambda + ikp \in \Sigma$. □

Now let $\psi_k(x) = e^{-ikpG(x)} \cdot \psi_d(x)$, where ψ_d is the positive eigenvector of $\hat{K}(\lambda_d)$ (assumed that a solution λ_d of $r(\hat{K}(\lambda)) = 1$ exists). Let

$$n_0^k(a,x) = e^{-\lambda_k a} \frac{g(X(-a,x))}{g(x)} E(a, X(-a,x)) \cdot \psi_k(X(-a,x)), \quad k \in \mathbb{Z},$$

where $\lambda_k = \lambda_d + ikp$ (see (9.1)). Choose $\gamma_k \in \mathbb{C}$, $k \in \mathbb{Z}$ such that $\Sigma_{k=1}^{\infty} |\gamma_k| < \frac{1}{2}$, $\gamma_{-k} = \bar{\gamma}_k$, and define the initial age-size-distribution $n_0(a,x)$ by

$$n_0(a,x) \overset{def}{=} n_0^0(a,x) + \sum_{\substack{k=-\infty \\ k\neq 0}}^{\infty} \gamma_k n_0^k(a,x) = (1 + 2Re \sum_{k=1}^{\infty} \gamma_k e^{-ikpG(x)}) \cdot n_0^0(a,x),$$

then $n_0(a,x) \geq 0$, $(a,x) \in \Omega$ and the solution $B(t,x)$ of the associated integral equation (2.14) is given by

$$B(t,x) = e^{\lambda_d t} \cdot \psi_d(x) \cdot \{1 + 2\,\mathrm{Re} \sum_{k=1}^{\infty} \gamma_k e^{ikp(t-G(x))}\} = e^{\lambda_d t} \cdot \psi_d(x) \cdot h(t,x)$$

where

$$h(t,x) \overset{def}{=} 1 + 2\,\mathrm{Re} \sum_{k=1}^{\infty} \gamma_k e^{ikp(t-G(x))}$$

satisfies

$$h(t+T,x) = h(t,x),$$

$$h(t,2x) = h(t,x).$$

This proves that there is no convergence to a stable age-size-distribution in this case (compare II.11).

This result disproves a remark of BELL (1968) which says that in case of exponential growth ($g(x)=c\cdot x$) there can exist a stable age-size-distribution if b depends in an appropriate manner on x and a. TRUCCO & BELL (1970) showed that in the case of dispersionless growth (i.e. $\frac{1}{x} X(a,x)$ depends on a only: this is satisfied if $g(x) = c\cdot x$) it is not possible that the first and second moments of the distribution of birth sizes both approach finite non-zero limits as $t \to \infty$, yielding that there does not exist a stable age-size distribution (see also TRUCCO (1970)). HANNSGEN, TYSON & WATSON (1985) proved that in case of exponential growth and under the assumption that the generation time (= age at which a cell divides) is a random variable with a given probability density function there cannot exist a stable, time-independent size distribution for the birth function.

IV. If $[0,\infty) = I_1 \cup I_2 \cup I_3$ such that $g(2x) < 2g(x)$, $x \in I_1$, $g(2x) = 2g(x)$, $x \in I_2$, $g(2x) > 2g(x)$, $x \in I_3$, then the question of convergence to a stable distribution is a very hard one, but also a very interesting and exciting one from the mathematical point of view.

The reason for making assumption (A_d) is a technical one. It guarantees the existence of a dominant element λ_d of Σ (see lemma 6.11).

Undoubtedly our theory is also valid if a less restrictive condition than (A_g) is imposed. However, our main purpose is not generality but to give an idea how abstract results from functional analysis can be used in the study of concrete structured population models. The results that we obtained here can also be found using semigroup methods, and readers who are trying to do so, will find out that the two approaches are more closely linked then it seems at first sight.

Appendix

Here we shall prove that for all $\lambda\in\Lambda$ the operator $\hat{K}(\lambda)$ is compact. We need the following result of KRASNOSELSKII et al. (1976, chapter 2, § 5. 6). They proved that a linear integral operator which has a compact majorant is compact itself. We shall make this more precise. Let $\Omega\subseteq\mathbb{R}$ be a measurable set and let the linear integral operator $T:L_1(\Omega)\to L_1(\Omega)$ be given by

$$(T\phi)(x) = \int_\Omega h(x,y)\phi(y)dy.$$

Suppose that

$$|h(x,y)| \leq h^+(x,y), \; x,y\in\Omega,$$

and let the operator T^+ be given by

$$(T^+\phi)(x) = \int_\Omega h^+(x,y)\phi(y)dy.$$

Then the following result holds (KRASNOSELSKII et al. (1976)):

LEMMA 1. *If T^+ is a bounded, compact operator from $L_1(\Omega)$ into itself then T is also compact.*

Now let $\lambda\in\Omega$, then

$$(\hat{K}(\lambda)\psi)(x) = \int_0^{X(-a_0,2x)} e^{-\lambda(G(2x)-G(z))}k(G(2x)-G(z),2x)\frac{\psi(z)}{g(z)}dz.$$

With (2.16), (A_g) and lemma 3.1 this yields

$$|e^{-\lambda(G(2x)-G(z))}.k(G(2x)-G(z),2x).\frac{1}{g(z)}|<e^{-(Re\lambda+d_\infty)(G(2x)-G(z))}.\frac{4}{g_{\min}}\|b\|_\infty\cdot e^M.$$

Let $p=\mathrm{Re}\lambda+d_\infty$, then $p>0$, since $\lambda\in\Lambda$. Let the operator $K^+(p)$ be defined as

$$(K^+(p)\psi)(x) = \int_0^{X(-a_0,2x)} e^{-p(G(2x)-G(z))}\psi(z)dz.$$

If we can prove that $K^+(p)$ is compact for all $p>0$ then it follows from Lemma 1 that $\hat{K}(\lambda)$ is compact for all $\lambda\in\Lambda$.

The following compactness criterium can be found in KUFNER et al. (1977).

LEMMA 2. *The bounded linear operator* $T:L_1(\Omega)\to L_1(\Omega)$ *is compact if for every* $\epsilon>0$ *there exists a* $\delta>0$ *such that* $\int_\Omega|(T\phi)(x+h)-(T\phi)(x)|dx<\epsilon\|\phi\|$ *for all* $\phi\in L_1(\Omega)$ *and* $|h|<\delta$.

We shall use this criterium to prove that $K^+(p)$ is compact for all $p>0$. For simplicity we assume that $g(x)=1$, for all x. The reader will have no difficulty to see that the proof can be carried through for more general g. Let $\psi\in L_1[0,\infty)$ and let $h>0$. Then

$$|(K^+(p)\psi)(x+h)-(K^+(p)\psi)(x)| = |e^{-2p(x+h)}\int_0^{2(x+h)-a_0} e^{pz}\psi(z)dz - e^{-2px}\int_0^{2x-a_0} e^{pz}\psi(z)dz|$$

$$\leq |e^{-2p(x+h)}-e^{-2px}|\cdot\int_0^{2x-a_0} e^{pz}|\psi(z)|dz + e^{-2p(x+h)}\int_{2x-a_0}^{2(x+h)-a_0} e^{pz}|\psi(z)|dz \overset{def}{=} f_1(x)+f_2(x),$$

where $f_1(x) = (1-e^{-2ph})(K^+(p)|\psi|)(x)$, $f_2(x)=e^{-2p(x+h)}\int_{2x-a_0}^{2(x+h)-a_0} e^{pz}|\psi(z)|dz$, and $|\psi|(x)\overset{def}{=}|\psi(x)|$. Thus

$$\|f_2\| = \int_0^\infty f_2(x)dx = \int_{\frac{1}{2}a_0}^\infty e^{-2p(x+h)}.\{\int_{2x-a_0}^{2(x+h)-a_0} e^{pz}|\psi(z)|dz\}dx$$

$$= \int_0^\infty e^{pz}|\psi(z)|\cdot\{\int_{\frac{1}{2}(z+a_0)-h}^{\frac{1}{2}(z+a_0)} e^{-2p(x+h)}dx\}dx = \frac{1-e^{-2ph}}{2p}e^{-pa_0}\|\psi\|.$$

From these two estimates and Lemma 2, the compactness of $K^+(p)$ and thus $\hat{K}(\lambda)$ follows immediately.

VI Nonlinear Dynamical Systems: Worked Examples, Perspectives and Open Problems.

O. Diekmann & H.J.A.M. Heijmans
(with contributions by F. van den Bosch)

1. Basic terminology and an outline of the program

1.1. Fundamental concepts of dynamical systems theory

A first and important step in the construction of a mathematical model of a (biological) system consists of the choice of a state space. The mathematical *state* should be a convenient representation of those physiological, chemical, physical and other relevant properties which in our conception of (or hypotheses about) reality uniquely determine the future, in the sense that for a given time course of experimental or environmental conditions (the input) we obtain a unique time course of the quantities we are interested in (the output). In section III.2 we made some remarks about the choice of a state space and about the construction of a state space from input-output data. Moreover, we presented a precise mathematical reformulation of the intuitive definition of "state" above in terms of a family of operators with a semigroup property.

When we restrict ourselves to time-independent inputs, things are a little simpler than in the general case. The input then is just a fixed parameter which we may suppress in the notation. In Chapter II we have treated some aspects of the theory of such autonomous (i.e., time-translation invariant) dynamical systems under the additional assumption of *linearity* and in the present chapter we shall discuss some aspects of autonomous *nonlinear* dynamical systems. Concerning outputs we confine ourselves to the remark that one may choose any (continuous) function of the state as output mapping.

We begin by introducing some terminology. Although certain states (for instance negative weight) might lack any biological interpretation, one can always define the dynamics on the whole linear state space if the system is linear, simply by using the linearity. This, of course, is no longer possible if the system is nonlinear. So let X be a Banach space and C a closed subset of X. A (semi) *dynamical system* on C is a mapping $u:\mathbb{R}^+ \times C \to C$ such that

(i) $u(\cdot,\phi):\mathbb{R}^+ \to C$ is continuous from above in $t=0$)

(ii) $u(t,\cdot):C \to C$ is continuous (for any $t \geqslant 0$)

(iii) $u(0,\phi) = \phi$

(iv) $u(t+s,\phi) = u(t,u(s,\phi))$, $\forall t,s \in \mathbb{R}^+$, $\phi \in C$.

REMARK 1.1.1. One can show that these properties imply that $u:\mathbb{R}^+ \times C \to C$ is continuous, *i.e.* u is continuous as a function of two variables.

A (one-parameter) *strongly continuous semigroup* on C is a family $\{T(t)\}_{t \geqslant 0}$ of (not necessarily linear) mappings from C into C such that

(i) $t \mapsto T(t)\phi$ is continuous from $\mathbb{R}^+$ to C

(ii) for each $t \geqslant 0$, $T(t):C \to C$ is continuous

(iii) $T(0) = I$ (where I denotes the identity operator on C)

(iv) $T(t+s) = T(t)T(s)$, $\forall t,s \in \mathbb{R}^+$.

The *infinitesimal generator* of $T(t)$ is the operator

$$A\phi = \lim_{t \downarrow 0} \frac{1}{t}(T(t)\phi - \phi)$$

for all $\phi \in \mathfrak{D}(A)$, the subset of C for which this limit exists.

The above notions of a dynamical system and a strongly continuous semigroup are equivalent in the sense that for a given closed subset C the identification

$$u(t,\phi) = T(t)\phi$$

gives a one-to-one correspondence between all dynamical systems on the one and all semigroups on the other hand. Therefore we can use these names interchangeably without causing confusion.

The outcome of the model building phase usually is a rather loosely defined "generator", and a first mathematical task is to *prove* that indeed one can associate a dynamical system with the model. Some results analogous to the Hille-Yosida Theorem II.2.6 exist for the nonlinear case (see PAZY, 1983b, for a nice survey), but as in the linear case one can usually follow a more direct and easy road. We shall demonstrate this road in section 2 by means of an example from mathematical epidemiology. The basic idea is to use integration along characteristics to convert the formal initial value problem into an integral equation and to use a so-called contraction mapping argument to prove existence and uniqueness of solutions of the integral equation on an appropriately small time interval. By repetition of the argument one can continue the solution as long as it remains bounded. Next one has to derive *a priori estimates,* guaranteeing that the solution cannot blow up in finite time, in order to obtain a solution defined for $0 \leq t < \infty$. Here the choice of C matters (of course C should also be adjusted to the biological interpretation; quite often, however, the definition of C involves only a precise mathematical formulation of obvious biological constraints; see section 2 for an example). Finally the circle may be closed by computing the infinitesimal generator of the semigroup. This amounts to a precise redefinition of the original "generator" and it provides us with a precise notion of "solution" (see Chapter II for a detailed example in the linear case).

After the existence and uniqueness of solutions to the initial value problem have been demonstrated, a more difficult task begins: one has to draw conclusions about biologically relevant aspects of the behaviour of the dynamical system. In the linear case we used spectral theory (Chapter II), Laplace transforms (Chapter IV) or both (Chapter V). Neither of the two has an analogue in the nonlinear case.

The behaviour of a nonlinear dynamical system can be essentially richer and more complicated than the behaviour of a linear dynamical system. In order to describe and classify various possibilities we need a lot more terminology, some of which we present below while referring to GUCKENHEIMER & HOLMES (1983), WALKER (1980) and SAPERSTONE (1981), for an introduction to the relevant literature.

Given the *initial state* ϕ at time zero, $T(t)\phi$ is the *state* at time t. The set

$$\Gamma^+(\phi) = \{T(t)\phi \mid t \geq 0\}$$

is called the *orbit* * starting at ϕ. Some rather simple orbits are those consisting of just one point or a closed curve of points. If

$$T(t)\hat{\phi} = \hat{\phi} \quad , \text{ for all } t \geq 0 \, ,$$

then $\hat{\phi}$ is called an *equilibrium* or a *steady state.* If

$$T(\tau)\phi_p = \phi_p \quad , \text{ for some } \tau > 0 \, ,$$

we call ϕ_p a *periodic point* with period τ and $\Gamma^+(\phi_p) = \{T(t)\phi_p | 0 \leq t < \tau\}$ is called a *periodic orbit.* The minimal $\tau > 0$ with the property above is called the minimal period, but usually the adjective "minimal" will be omitted.

We will be especially interested in those orbits or more complicated sets, which in some sense attract (or at least do not repel) orbits starting nearby. The distance dist (ϕ, V) between a point $\phi \in X$ and a non-empty subset V of X is defined as

$$dist(\phi, V) = \inf\{\|\phi - \psi\| \mid \psi \in V\} \, .$$

With respect to $\{T(t)\}$ we call a subset V of X

i) *stable* if $\forall \epsilon > 0 \exists \delta = \delta(\epsilon) > 0$ such that $\forall \psi \in C$ with dist$(\psi, V) < \delta$ necessarily dist $(T(t)\psi, V) < \epsilon$, $\forall t \geq 0$;

ii) *unstable* if V is *not* stable;

iii) *asymptotically stable* if V is stable and in addition $\exists \delta > 0$ such that $\forall \psi \in C$ with dist $(\psi, V) < \delta$ necessarily $\lim_{t \to \infty}$ dist $(T(t)\psi, V) = 0$.

---

* An orbit is sometimes also called a *trajectory,* but perhaps it is better to reserve the name trajectory for the subset $\{(t, T(t)\phi) | t \geq 0\}$ of $\mathbb{R} \times C$.

iv) *exponentially stable* if $\exists\delta>0$ such that $\exists\alpha = \alpha(\delta)>0$ and $\exists M = M(\delta)>0$ such that $\forall\psi\in C$ with dist $(\psi, V)<\delta$ necessarily dist $(T(t)\psi, V)\leq Me^{-\alpha t}$ dist (ψ, V), $\forall t\geq 0$.

If in the last two situations one may choose δ arbitrarily large one adds the adverb *"globally"*. If V is an orbit itself one speaks about *orbital stability* etc.. If an asymptotically stable set V consists of more than one point (for instance, V might be a periodic orbit), one says that orbits $\Gamma^+(\psi)$ converging to V have an *asymptotic phase* if $\phi = \phi(\psi)\in V$ exists such that $\lim_{t\to\infty}(T(t)\psi - T(t)\phi) = 0$. The *domain of attraction* of an asymptotically stable set V is the set $\{\psi\in C \mid \lim_{t\to\infty} \text{dist}\,(T(t)\psi, V) = 0\}$.

As a side-remark we mention that other stability concepts exist. The above definitions correspond to so-called *Lyapunov-stability.*

Not all stable sets are interesting. For instance, if $\hat{\phi}$ is a globally exponentially stable steady state, *any* set containing $\hat{\phi}$ is stable. We want to concentrate on those sets which are mapped into themselves by the semigroup and which, moreover, are related to the asymptotic (large time) behaviour of orbits. A subset M of C is called *positively invariant* if $T(t)\phi\in M$ for all $\phi\in M$ and all $t\geq 0$. We say that a point $\phi\in C$ has a *backward extension* if there exists a mapping $F:\mathbb{R}\times C\to C$ such that $F(0,\phi) = \phi$ and $F(t+s,\phi) = T(t)F(s,\phi)$ for all $s\in\mathbb{R}$ and $t\in\mathbb{R}_+$ (note that in the "forward" time direction F simply describes the action of $T(t)$ and that, at least in principle, several such backward extensions might exist since the semigroup operators might not be one-to-one; see *e.g.* section II.11). A subset M of C is called *invariant* if it is positively invariant and in addition each point of M has a backward extension which belongs to M.

EXERCISE 1.1.2. Show that (a) if $\hat{\phi}$ is an equilibrium then $\{\hat{\phi}\}$ is invariant, (b) if ϕ_p is a periodic point then $\Gamma^+(\phi_p)$ is invariant, (c) the union of two invariant sets is invariant, (d) a subset M of C is invariant if and only if $T(t)M = M$ for all $t\geq 0$.

For any $\phi\in C$ we define

$$\omega(\phi) = \{\psi\in C \mid \exists\{t_k\} \text{ such that } t_k\to\infty \text{ and } T(t_k)\phi\to\psi \text{ for } k\to\infty\}$$

and call it the ω- *limit set* of ϕ.

EXERCISE 1.1.3. Show that $\omega(\phi)$ is closed and positively invariant.

A much stronger result holds if the orbit $\Gamma^+(\phi)$ is precompact (i.e. the closure $\overline{\Gamma^+(\phi)}$ is compact). For the proof see, for instance, Walker (1980), page 167.

THEOREM 1.1.4. *If* $\Gamma^+(\phi)$ *is precompact, then* $\omega(\phi)$ *is nonempty, compact, connected and invariant. Moreover* $\lim_{t\to\infty}$ dist $(T(t)\phi, \omega(\phi)) = 0$.

REMARK 1.1.5. In the description of the orbit structure of dynamical systems it has proved useful to pay special attention to points which not necessarily belong to any ω-limit set, but nevertheless show some form of recurrent behaviour. This has led to the notions of nonwandering and chain recurrent points. See Guckenheimer & Holmes (1983).

The ideal second step in the analysis of a structured population model would be to trace all ω-limit sets and to determine their stability properties and their domain of attraction. As a rule it is impossible to obtain such a complete global overview of the orbit structure, but in special cases one of the following two methods might help to attain the end:

Lyapunov functions and the Invariance Principle.

Let $V:C\to\mathbb{R}$ be a continuous function. The derivative of V along the orbit starting in ϕ at $t = 0$ is given by

$$\lim_{t\downarrow 0}\frac{1}{t}\,(V(T(t)\phi)-V(\phi)) \,,$$

but this limit might not exist. Therefore we define

$$\dot{V}(\phi) = \liminf_{t\downarrow 0}\frac{1}{t}\,(V(T(t)\phi)-V(\phi)) \,, \tag{1.1}$$

where we allow that $\dot{V}(\phi) = -\infty$. The function V is called a *Lyapunov function* if $\dot{V}(\phi)\leq 0$ for all $\phi\in C$. So let V be a

Lyapunov function. We define

$$E = \{\phi \in C \mid \dot{V}(\phi) = 0\}$$

and

$$\mathcal{E} = \text{largest invariant subset of } E\,.$$

THEOREM 1.1.6. (Invariance Principle). *Let V be a Lyapunov function for $T(t)$ and assume that $\Gamma^+(\phi)$ is precompact. Then*

$$\lim_{t\to\infty} \operatorname{dist}(T(t)\phi, \mathcal{E}) = 0\,.$$

REMARKS 1.1.7. (i) We did not state the most general formulation of the Invariance Principle. See Walker (1980), section IV.4 or LaSalle (1976).
(ii) Theorem 1.1.6. is particularly useful if one can show that $\mathcal{E}$ is a discrete set of points. In section 4 we shall illustrate the use of the Invariance Principle in the simplest case in which $\mathcal{E}$ is just a singleton.
(iii) The difficulty with applying this strong and beautiful result is, of course, to find a Lyapunov function for the concrete problem at hand. Quite often this turns out to be an impossible task.

Monotonicity methods

Let X_+ be a closed convex cone in X (see Chapter V for relevant definitions, in particular of the $\leqslant$ symbol below). Assume that C is a subset of X_+. The semigroup $T(t)$ is called *monotone* if $\phi \leqslant \psi$ implies that $T(t)\phi \leqslant T(t)\psi$ for all $t \geqslant 0$. (So in the linear case monotonicity is nothing but positivity.) One can prove the following result.

THEOREM 1.1.8. *Suppose that*

(i) *$T(t)$ is a monotone semigroup*

(ii) *there exists precisely one equilibrium $\hat{\phi} \in C \setminus \{0\}$*

(iii) *for every $\phi \in C$, $\phi \neq 0$, there exist $t_0 > 0$ and $\underline{\phi}$, $\overline{\phi} \in C$ such that $0 < \underline{\phi} \leqslant T(t_0)\phi \leqslant \overline{\phi}$ and, moreover, $T(t)\underline{\phi}$ is increasing and $T(t)\overline{\phi}$ is decreasing with respect to t*

(iv) *orbits are precompact*

then $\lim_{t\to\infty} T(t)\phi = \hat{\phi}$ *for all* $\phi > 0$.

EXERCISE 1.1.9. Prove the above statement.

So under rather strong monotonicity assumptions the dynamics become rather simple. In section 4 we shall illustrate this general idea by means of an example from cell kinetics. Also see the contribution of Thieme to part B. We refer to Hirsch (1984a,b), Matano (1984) and Matano & Hirsch (in preparation) for a number of interesting results which hold without conditions like (ii) and (iii) above.

In general neither of these two methods will be applicable and one has to resort to a less ambitious program:

i) Trace all equilibria and determine their (local) stability.

ii) Find criteria for the existence and the (local) stability of periodic orbits.

iii) Find out whether there exist more complicated attracting invariant sets like tori or even so-called strange attractors with a Cantor set structure.

The last two items are fairly complicated already and computer experiments are usually an indispensable tool for such investigations.

It is frequently convenient to concentrate on qualitative changes of the orbit structure which may or may not occur when model parameters are varied. This method, which goes under the heading of *bifurcation theory,* is the mathematical counterpart of the experimental technique to study a system by slowly changing one of its controllable parameters while observing what happens.

In the next subsection we present some general principles from *local* stability and bifurcation theory in the context of ordinary differential equations, since for those the results are well-established and rigorously proved. In subsection

1.3 we speculate about some *global* aspects and about the practical meaning of stability. Then follow a number of examples from structured population dynamics which are intended to equip the reader with some feeling for the subject, and to illustrate various mathematical methods and techniques. All examples have in common that they can be reduced to an integral or an integro-differential equation, and as such are rather special. In a final section we mention several open problems for a more general class of first order functional partial differential equations. Most of these are concerned with generalizing the local stability and bifurcation results of subsection 1.2 to this class of equations. Finally we refer to Hale, Magalhaẽs & Oliva (1984) for an introduction to the global and generic theory of infinite dimensional dynamical systems, while noting that, as far as we know, no work in this spirit has been done for the kind of functional partial differential equations corresponding to structured population models.

So all together we try to present our wishful thinking about a comprehensive mathematical theory of *nonlinear* structured population models in the hope that our readers feel stimulated to contribute to its creation.

1.2. Linearized stability and bifurcation theory in the context of ordinary differential equations

In this subsection we shall sketch some of the basic results of nonlinear analysis without giving any proof. We refer to such sources as Hartman (1964), Hale (1969), Hirsch & Smale (1974), Guckenheimer & Holmes (1983), Iooss & Joseph (1980) and Chow & Hale (1982) for a detailed account of the much more comprehensive theory. Here our aim is just to help the mathematically uninitiated reader to build up some knowledge of the terminology and some intuition for the main results and problems.

In this subsection f denotes a function from $\mathbb{R}^N$ into $\mathbb{R}^N$ which is at least C^1-smooth and, moreover, is such that $f(0) = 0$. Hence $x = 0$ is an equilibrium for the ordinary differential equation

$$\frac{dx}{dt} = f(x) \tag{1.2.1}$$

or, in the terminology of the last subsection, for the dynamical system $x(t,x_0)$, where $x(t,x_0)$ denotes the solution of (1.2.1) with initial condition $x(0,x_0) = x_0$. In this context the function f is called a *vector field* on $\mathbb{R}^N$.

We shall concentrate on the behaviour of orbits in a neighbourhood of $x = 0$. Let $Df(0)$ denote the linearization of f at $x = 0$ (i.e., the Jacobi matrix of partial derivatives). An obvious question is: can one determine the stability of $x = 0$ from the stability of $x = 0$ with respect to the linear semigroup $e^{Df(0)t}$? Since the latter is completely determined by the eigenvalues of $Df(0)$ one can equivalently ask whether these have a decisive influence on the stability of $x = 0$. The positive answer is called the *Principle of Linearized Stability:*

THEOREM 1.2.1. (Poincaré-Lyapunov)

i) *If all eigenvalues of $Df(0)$ have negative real part, $x = 0$ is exponentially stable*

ii) *If at least one of the eigenvalues of $Df(0)$ has positive real part, $x = 0$ is unstable.*

If $Re\lambda \leq 0$ for all eigenvalues λ, with equality for at least one λ, the higher order terms come out of their subordinate position and control the situation.

The local equivalence between the semigroup and its linearization is described by the much stronger

THEOREM 1.2.2. (Hartman-Grobman) *If none of the eigenvalues of $Df(0)$ lies on the imaginary axis * then there exists a homeomorphism † h defined on a neighbourhood U of $x = 0$ which maps (restrictions to U of) orbits of the nonlinear semigroup onto orbits of the linear semigroup $e^{Df(0)t}$, preserving the direction in which they are traversed in the course of time.*

Let $S(t)$ denote the nonlinear semigroup corresponding to (1.2.1) and define the *local stable and unstable manifolds* by

$$W^s_{loc}(0) = \{x \in U \mid S(t)x \to 0 \text{ for } t \to \infty \text{ and } S(t)x \in U, \ \forall t \geq 0\} \tag{1.2.2}$$

$$W^u_{loc}(0) = \{x \in U \mid S(t)x \to 0 \text{ for } t \to -\infty \text{ and } S(t)x \in U, \ \forall t \leq 0\} \tag{1.2.3}$$

then we have

---

* An equilibrium with this property is called *hyperbolic.*

† A homeomorphism is simply a continuous map, having a continuous inverse.

Theorem 1.2.3. (the saddle point property) *If none of the eigenvalues of $Df(0)$ lies on the imaginary axis there exist local stable and unstable manifolds $W^s_{loc}(0)$ and $W^u_{loc}(0)$ of the same dimensions as the invariant generalized eigenspaces E^s and E^u of the linearized semigroup $e^{Df(0)t}$ corresponding to, respectively, the eigenvalues with negative real part and those with positive real part. The manifolds $W^s_{loc}(0)$ and $W^u_{loc}(0)$ are tangent to, respectively, E^s and E^u in $x = 0$ and they are as smooth as the function f.*

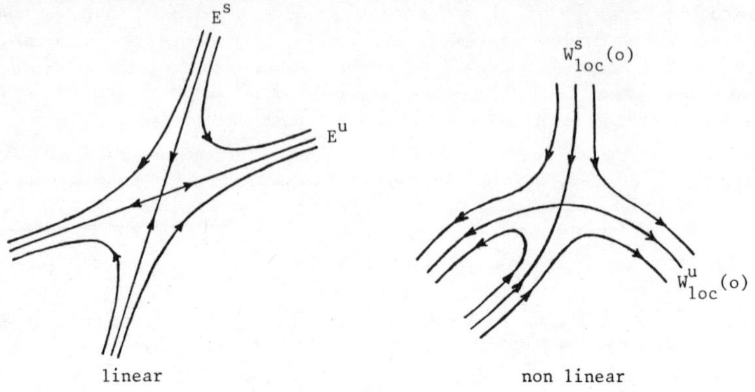

Homeomorphisms as in Theorem 1.2.2 generate a local equivalence relation among semigroups, which obviously can be extended to a global equivalence relation by omitting the special emphasis on a neighbourhood of an equilibrium point. A function f is called *structurally stable* if all functions in a C^1-neighbourhood of f belong to the same equivalence class. This notion might seem rather abstract at first sight, but actually it is suggested by the following practical considerations. If we model some real world phenomenon in terms of differential equations we always make idealizations and approximations while, moreover, leaving many minor effects completely out of consideration. So at best the real dynamics is close to the dynamics described by the differential equations. If the latter is "robust" (proof against small perturbations of the model assumptions or, in other words, of the differential equations themselves) we may trust its predictions, but otherwise we have to reexamine our modelling assumptions quite carefully.

In numerous applications f will naturally depend on *parameters* in such a way that the equivalence class changes when these parameters are varied (or, in other words, the parametrized family of vector fields cuts the "boundary" between two (or more) equivalence classes). As far as *local* equivalence is concerned, Theorem 1.2.2 tells us that this happens exactly when some eigenvalues of $Df(0)$ cross the imaginary axis for some specific value of the parameters. So if eigenvalues move from one halfplane into the other as parameters are varied, we expect to see qualitative changes in the (local) orbit structure. *Bifurcation theory* embraces the *classification* of possible qualitative changes and the development of constructive *algorithms* to determine the actual change that will occur in some concrete situation. In the simplest cases, viz.,

(i) a simple real eigenvalue crosses the imaginary axis with positive speed

(ii) a pair of complex conjugate simple eigenvalues crosses the imaginary axis with positive speed

everything is thoroughly understood.

Example 1.2.4. Let x denote the (one-dimensional) state variable and μ the (one-dimensional) parameter. We draw the zero set of $f = f(x, \mu)$ in the (x, μ)-plane for various functions f (see below).

Such pictures are called *bifurcation diagrams.* In (i)-(iv) the curve $x = 0$ is called the *trivial branch* and the other curves are called the *bifurcating branches.* The word *"bifurcation point"* refers to $\mu = 0$ in some texts and to $(x, \mu) = (0,0)$ in others.

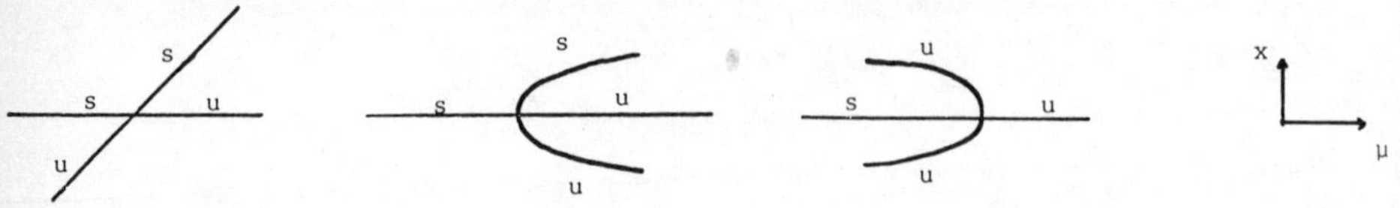

(i) $f = x(\mu - x)$ transcritical

(ii) $f = x(\mu - x^2)$ pitchfork

(iii) $f = x(\mu + x^2)$ pitchfork

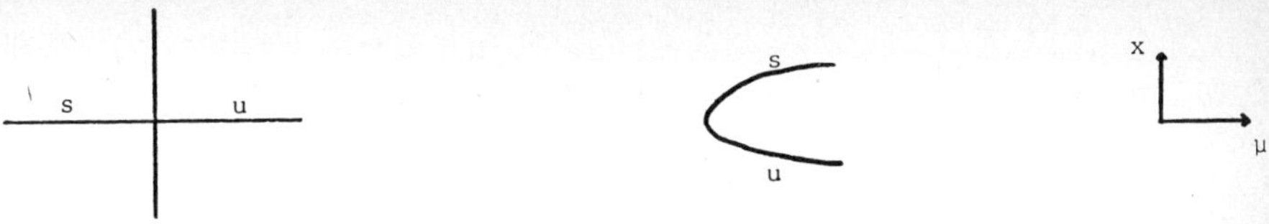

(iv) f = xμ
vertical

(v) f = $\mu - x^2$
saddle-node

Bifurcation theory tells us that the simple one-dimensional examples above are, locally near the bifurcation point, representative for the general case of a one-dimensional parameter μ and a higher (perhaps infinite) dimensional state variable x, under the stated condition (i) on the eigenvalue. Moreover, it provides us with a *constructive procedure* (based on a combination of the so-called Lyapunov-Schmidt reduction technique and the contraction mapping principle or, if you wish, the closely related implicit function theorem; an alternative approach uses reduction to normal forms; see Guckenheimer & Holmes, 1983) to calculate approximations to the bifurcating branches which cannot, of course, be calculated explicitly in general. Thus one can decide which of the pictures of Example 1.2.4 covers the situation at hand

EXAMPLE 1.2.5. In terms of polar coordinates $x = r\cos\theta$, $y = r\sin\theta$ the system

$$\dot{x} = \mu x - \omega y \pm x(x^2 + y^2)$$

$$\dot{y} = \omega x + \mu y \pm y(x^2 + y^2)$$

reduces to $\dot{r} = r(\mu \pm r^2)$, $\dot{\theta} = \omega$. So with the $-$ sign we have

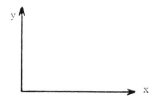

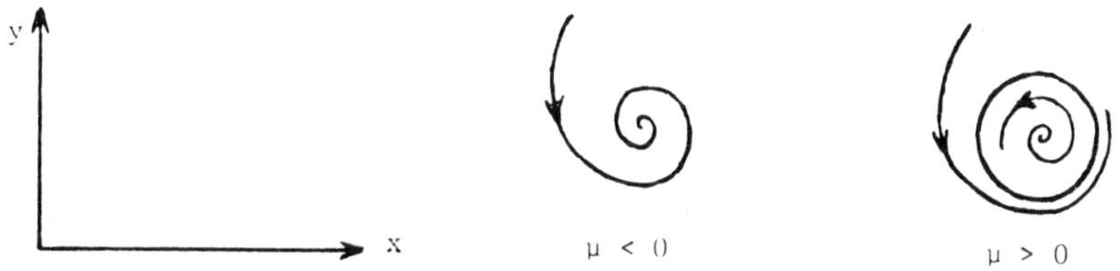

whereas with the $+$ sign we have

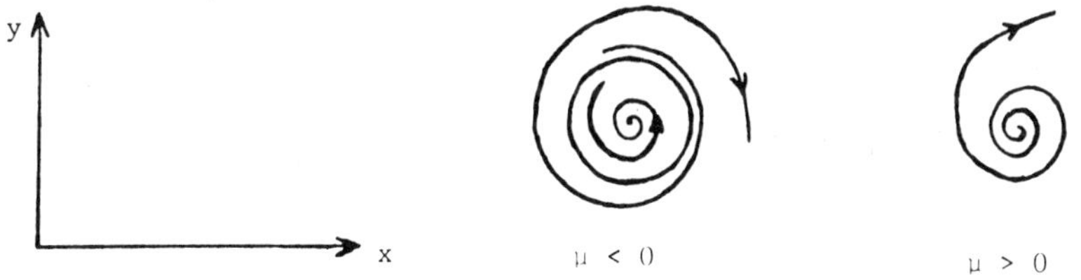

In other "pictures"

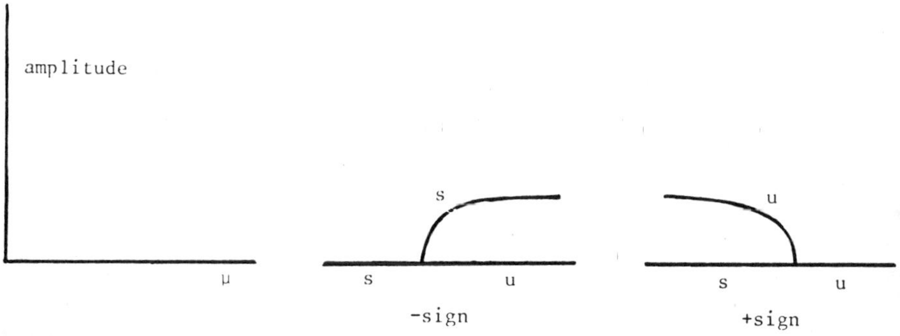

Again this simple example is representative for the general situation covered by condition (ii) above. The origination of a *periodic solution* from an equilibrium (which changes its stability type due to a pair of complex eigenvalues crossing the imaginary axis) is called a *Hopf bifurcation.*

EXERCISE 1.2.6. Check that the stability of the equilibria of $\dot{x} = f(x,\mu)$ with f as in $(i)-(v)$ in Example 1.2.4. is as indicated in the diagrams. Similarly determine the stability of the periodic solutions in Example 1.2.5.

If the trivial solution is stable for $\mu<0$ and unstable for $\mu>0$, then the bifurcating solutions which exist for $\mu>0$ are called *supercritical* and those which exist for $\mu<0$ are called *subcritical.* The outcome of Exercise 1.2.6 illustrates the general rule

supercritical bifurcating solutions are stable and subcritical bifurcating solutions are unstable

which is called the *principle of exchange of stability.*

The situation is essentially more complicated if eigenvalues have higher multiplicity, several eigenvalues cross simultaneously, eigenvalues hesitate or change their mind while lying on the imaginary axis (i.e. the derivative of the real part with respect to the parameter vanishes). The study of such situations is in full progress today and many results are known (CHOW & HALE, 1982, GUCKENHEIMER & HOLMES, 1983). Especially in the case of multiple eigenvalues or several (pair of) eigenvalues it is "natural" to consider families with more than one parameter in order to produce, as it is called in expressive language, a universal or partial unfolding of the singularity. Quite often it turns out that the local picture thus obtained gives, in fact, global information (see GOLUBITSKY and SCHAEFFER (1984) for a discussion of this phenomenon and the related concept of "organizing center").

EXERCISE 1.2.7. Draw the zero set of $x(x^2-\mu)(2x^2-\mu)$. What is the matter with the eigenvalue?

So far we concentrated on local results which can be proved in a constructive manner. But one can use topological non-constructive tools such as *degree theory,* to prove that the local branches cannot cease to exist and therefore should have a global continuation in some sense. More precisely they either have to set down in another bifurcation point or they should tend to infinity in the sense that either the x or the μ or both components are unbounded.

EXERCISE 1.2.8. Draw the zero sets of

(a) $x(x^2 + \mu^2 - 1)$,

(b) $x(x^2 + \mu^2 - 1)(x-\mu)$ *(secundary bifurcation),*

(c) $x(x + 1 - \frac{1}{\mu+1})$,

and interpret the results in the light of the remarks above.

Let $\xi\in\mathbb{R}^N$ be such that $\gamma(t) = x(t,\xi)$ is a periodic orbit with minimal period $p>0$. Linearization of the dynamical system $x(t,x_0)$ amounts to taking the derivative with respect to the initial condition x_0, and so we are led to consider the variational problem

$$\begin{cases} \dfrac{dH}{dt} = Df(\gamma(t))H \\ H(0) = I \end{cases}$$

where the $N\times N$ matrix $H(t)$ is $\frac{\partial x}{\partial x_0}(t,\gamma(0))$. Of particular interest are the eigenvalues of $H(p)$, which are called *Floquet multipliers* (or also *characteristic multipliers).* The fact that one of these eigenvalues has to be 1 reflects the translation invariance of the periodic orbit and can be proved by differentiation of the original nonlinear equation with respect to t. Indeed

$$\gamma''(t) = Df(\gamma(t))\gamma'(t)$$

implies that $\gamma'(t) = H(t)\gamma'(0)$ and consequently $f(\gamma(0)) = H(p)f(\gamma(0))$. The eigenvalue 1 of $H(p)$ is called the trivial multiplier.

THEOREM 1.2.9. *If all nontrivial Floquet multipliers lie strictly inside the unit circle in the complex plane, the periodic orbit γ is exponentially stable with asymptotic phase. If at least one of them lies strictly outside the unit circle, the periodic orbit γ is unstable.*

REMARKS 1.2.10. (i) One can show that H has a representation

$$H(t) = K(t)e^{Dt}$$

where K is p-periodic with $K(0) = I$, and D is constant. The eigenvalues of D are called *Floquet* (or *characteristic) exponents,* but one should realize that only the real parts of these are uniquely defined.
(ii) Let Π denote an $(N-1)$-dimensional subspace transverse to $f(\gamma(0))$, i.e. $\mathbb{R}^N = \Pi \oplus \text{span}\,\{f(\gamma(0))\}$. One can uniquely define a smooth function $\tau(x_0)$, for x_0 in a neighbourhood of $\gamma(0)$, such that $x(\tau(x_0),x_0)\in\Pi$ and $\tau(\gamma(0)) = p$.

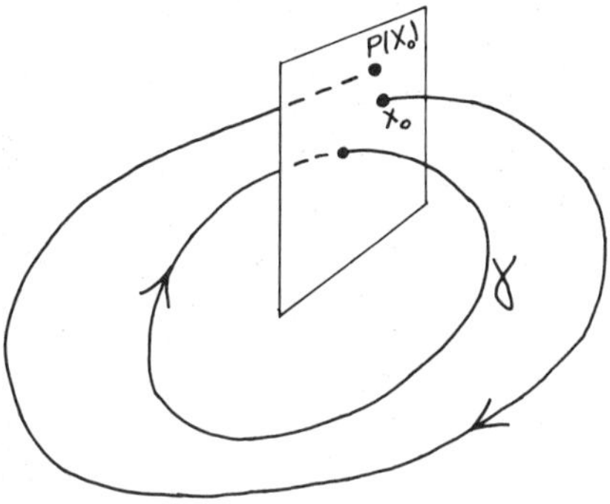

The mapping $P:\Pi\rightarrow\Pi$ defined by

$$P(x_0) = x(\tau(x_0), x_0)$$

is called the *Poincaré map* (or *return map).* Clearly $\gamma(0)$ is a fixed point of P. One can show that the linearization of P at $\gamma(0)$ has precisely the $(N-1)$ nontrivial Floquet multipliers as its eigenvalues.

Periodic solutions experience certain bifurcations when their stability character changes, but the situation is essentially more complicated than in the case of steady states. From a practical point of view there is the added difficulty that one usually cannot calculate the periodic solutions and their Floquet multipliers explicitly. Nevertheless it is useful to know that

i) if a multiplier leaves or enters the unit circle through -1 the old periodic solution gets company of a new one with a period which is about twice as large. Such *period doublings* may occur repeatedly and even accumulate.

ii) if a pair of complex multipliers leaves or enters the unit circle then, as a rule, an *invariant torus* comes into existence. Such a torus can be "filled" with quasi-periodic orbits.

Admittedly this is a very incomplete and vague description of a huge collection of fascinating mathematical results which have recently been established and we refer the interested reader to GUCKENHEIMER & HOLMES (1983) for a detailed account.

1.3. An impressionistic sketch of some global aspects.

Bifurcations may produce smooth transitions which are hardly recognizable as in

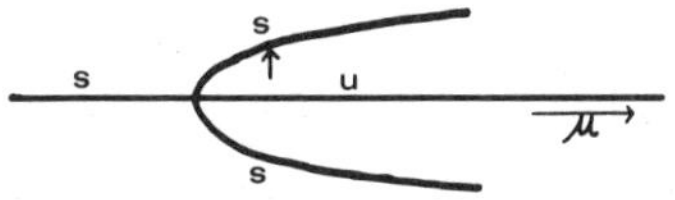

or they can lead to large *(catastrophic)* transition as in

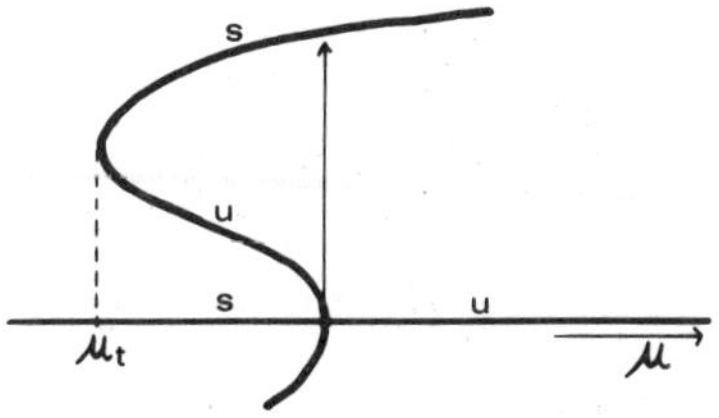

(Note that in *turning-points* such as μ_t the stability character of the steady states on one and the same branch may change; see Example 1.2.4 (v)). A situation which occurs rather frequently is an S - or Z - shaped branch:

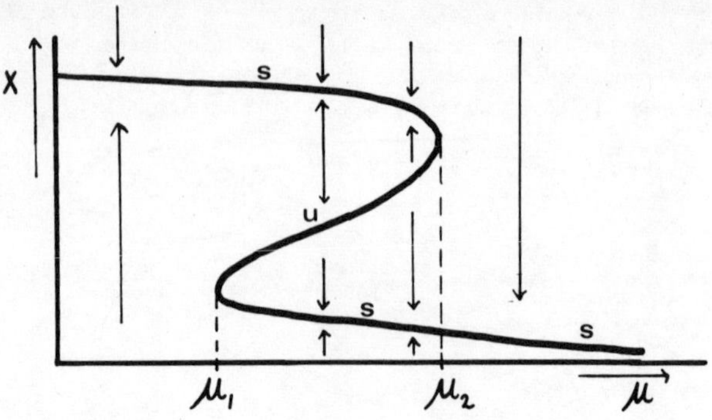

EXAMPLE 1.3.1. In the stimulating paper May (1977), R.M. May discusses the following simple pedagogical example. Consider the biomass of vegetation (say grass) as a dynamical variable x. In the absence of herbivores the dynamics of x is assumed to be given by

$$\frac{dx}{dt} = G(x)$$

with G a function that can be typified graphically as follows:

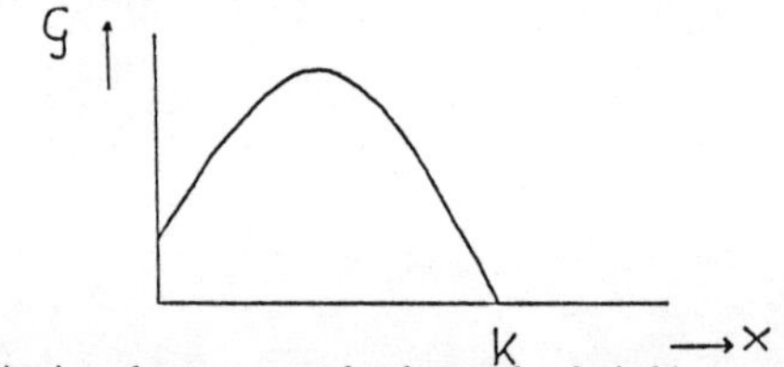

Here $G(0)$ is taken to be positive in order to account for the supply of wind-borne seeds from other areas; the decline is due to mutual shading and competition for nutrients and K is the carying capacity. Next, introduce a population of herbivores which is kept at a constant density μ. The grazing is described by the per capita consumption rate $c(x)$

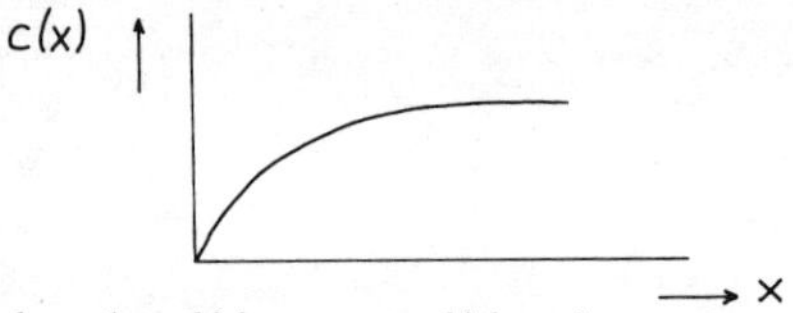

which is proportional to x at low x but which saturates at high x. So now

$$\frac{dx}{dt} = G(x) - \mu c(x)$$

and steady states have to satisfy the scalar equation

$$G(x) = \mu c(x).$$

The roots are easily found from graphical considerations:

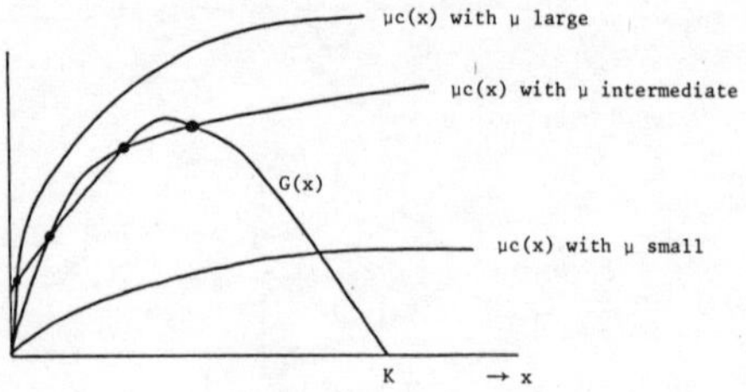

For large and small μ there exists only one steady state but for values in between we find three steady states (note that here we consider the case in which the c-curve saturates long before the G-curve reaches its top or, in other words, the nonlinearities have to be adjusted to each other in a specific manner in order to produce the Z).

For parameter values in between the turning points μ_1 and μ_2 the final state in which the system will settle down strongly depends on the *initial state* (say, at the beginning of spring). The domains of attraction of the stable upper and lower states are separated by the unstable state. In terms of a marble under the influence of gravity the landscape looks like:

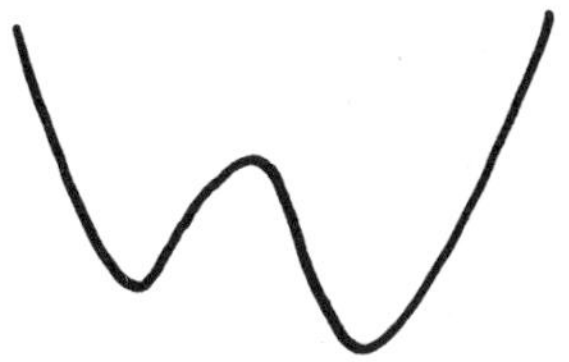

Alternatively one can say that sufficiently large perturbations of the state x may produce a transition from the lower to the upper state or vice versa. What exactly is "sufficient" depends on the exact value of μ: a typical feature of the Z-curve is the shrinking and swelling of the domains of attraction with variations in the parameter μ. Indeed, another way to produce such transitions is to make μ undergo variations which include a trip outside the (μ_1, μ_2) interval. Note that such μ-variations are accompanied by a *hysteresis effect:* the steady state in which we will find the system for a given $\mu \in (\mu_1, \mu_2)$ depends on the *road* along which μ came to its present value. If μ is itself a slowly varying dynamical variable (i.e., a time scale argument has been used to uncouple the μ-dynamics and the x-dynamics) this may even lead to oscillations in which slow, gradual changes are alternated by fast, catastrophic transitions.

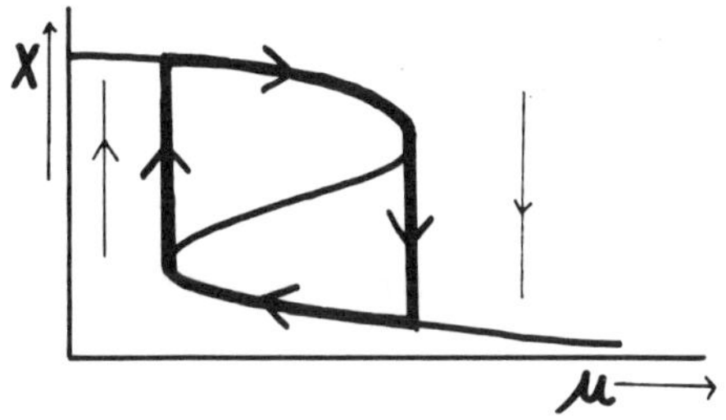

See the aforecited paper by MAY or see LUDWIG, JONES & HOLLING (1978) for an application of these ideas to the spruce- budworm (here x is the budworm density and μ^{-1} a measure for the leaf area).

The moral of the tale is that both chance and necessity play a leading part: the events are (assumed to be) predictable on the basis of deterministic relations between cause and effect, but nevertheless initial states and parameter values which, possibly, are subject to accidental perturbations (caused by forces or circumstances which we deliberately have chosen *not* to include in the model) have a decisive influence on what will actually happen.

Moreover the Z-curve illustrates the "grasp all, lose all" catastrophy: if in Example 1.3.1 one makes the cattle density μ larger and larger the domain of attraction of the economic desirable upper state becomes smaller and smaller until, for $\mu = \mu_2$, it is negligable and the least perturbation causes a catastrophic collaps of the grass-density. If, subsequently, the cattle density is lowered again the system will *not* return to the upper state but instead remain in the lower one! In other words: continuous changes in parameters may lead to discontinuous changes in steady states and the restoration of the old situation is *not* an immediate consequence of undoing the parameter change. The analogy with the consequences of over harvesting in a fishery suggests itself.

As a finale to this section we present some variations on a theme composed by C.S. Holling (HOLLING, 1973). The message is that putting a strong link between our mathematical stability concepts and our intuitive 'biological' ideas about ecosystem stability may be hazardous.

The mathematical stability of an equilibrium refers to the ability to recover from a temporary perturbation of the state variable(s). One can describe it in terms of the rate of return and the period of the oscillations that may accompany the return (or, in other words, the dominant eigenvalue). Therefore we are inclined to characterize a natural, real world, system that shows an almost constant appearance in the course of time as very stable. However, this constant appearance may be highly deceptive, by which we mean that it may be more the result of the almost complete absence of significant perturbations of both the state and the parameters.

As an example one can think of the great American lakes. Climatologic fluctuations are damped out by the water and the mobility of water animals guarantees a more or less homogeneous spatial distribution. For long times the composition of the inhabitant populations has been almost constant, a seemingly very stable situation. But human influence has led to radical changes. The supply of nutrients by waste water deposit and a strong harvesting pressure on certain fish populations has had a profound influence: some species became extinct and others were superseded by competitors which were better adapted to the new circumstances. There was no *resilience* and the disturbances created a new situation which was essentially different from the old.

On the other side of the spectrum one finds ecosystems which every now and then make large excursions through their state space (triggered by forest-fires, hurricanes or major pest outbreaks of short duration, like those of the spruce budworm) but which, maybe exactly while doing so, are very *resilient* in the sense that the overall pattern of relations between state-variables is essentially unaffected by many kinds of disturbances. Thus Holling argues that a strong time-inhomogeneity may enhance resilience and that the same may be true for spatial inhomogeneity (see the paper by SABELIS & LAANE in part B). Holling questions the appropriateness of the usual stability concepts for many types of ecosystems and promotes the resilience concept which heavily appeals to the intuition but seems hard to formalize in a strict mathematical sense (because it is neither clear what kind of disturbances one should consider (allow) nor what properties orbits should have). In the resilience concept at least two mathematical concepts, viz. dynamical stability and structural stability, are intertwined. The practical intuition on which it is based seems to be that in many (families of) systems a fast return rate to the equilibrium (or periodic solution) is linked to a large basin of attraction of that equilibrium (or periodic solution) and also to being far away (in parameter space) from a bifurcation point. Formalizing this notion presents one of the challenges of theoretical ecology!

2. An example of the construction of a dynamical system: an epidemic model with temporary immunity

2.1. The model.

In subsection IV.4.1 we considered the time course of an infectious disease within a closed population under the assumption that the disease conveys ever lasting immunity. We now drop this latter assumption and assume instead that an infected individual returns a fixed time T after its contagion to the class of susceptibles. After a scaling of time we may put $T = 1$. Hence $S(t)$, the number of susceptibles at time t, satisfies

$$S(t) = N - \int_0^1 i(t,a)da \tag{2.1}$$

where N denotes the total number of individuals and $\int_{a_1}^{a_2} i(t,\alpha)d\alpha, 0<a_1<a_2<1$, the number of individuals which were infected between $t-a_2$ and $t-a_1$. Let $A(a)$ denote the infectivity as a function of the time a elapsed since contagion. Of course we assume that $A(a) = 0$ for $a \geq 1$. The law of mass action assumption that the number of new cases equals the product of S and $\int_0^1 A(a)i(t,a)da$ leads to the problem

$$\begin{cases} \dfrac{\partial i}{\partial t} + \dfrac{\partial i}{\partial a} = 0 \\ i(t,0) = (N - \int_0^1 i(t,\alpha)d\alpha)\int_0^1 A(\alpha)i(t,\alpha)d\alpha \\ i(0,a) = \phi(a), \quad 0 \leq a \leq 1, \end{cases} \tag{2.2}$$

where the initial condition ϕ has to satisfy the *constraint*

$$\int_0^1 \phi(\alpha)d\alpha \leq N \tag{2.3}$$

(i.e., the number of infected individuals present at time $t=0$ is necessarily smaller than the total number present).

EXERCISE 2.1. Interpret the relation $\frac{dS}{dt}(t) = i(t,1) - i(t,0)$.

Let B_1 denote the indicator function of the interval [0, 1] i.e.

$$B_1(a) = \begin{cases} 1 & \text{for } 0 \leqslant a \leqslant 1 \\ 0 & \text{elsewhere} \end{cases} \tag{2.4}$$

The scaling

$$\begin{cases} i(t,a) = N\, n(t,a) \\ B_2(a) = \frac{N}{\gamma} A(a) \quad \text{with } \gamma = N \int_0^1 A(\alpha)d\alpha \end{cases} \tag{2.5}$$

yields

$$\begin{cases} \frac{\partial n}{\partial t} + \frac{\partial n}{\partial a} = 0 \\ n(t,0) = \gamma\,(1 - \int_0^1 B_1(\alpha) n(t,\alpha)d\alpha) \int_0^1 B_2(\alpha) n(t,\alpha)d\alpha \\ n(0,a) = \psi(a) \end{cases} \tag{2.6}$$

where the scaled initial condition ψ satisfies the constraint

$$\int_0^1 \psi(\alpha)d\alpha \leqslant 1\,. \tag{2.7}$$

EXERCISE 2.2. Give a biological interpretation of γ (see subsection IV.4.1).

Defining $y(t) = n(t,0)$ we find by integration along characteristics that y has to satisfy

$$y(t) = \gamma \left[1 - \int_0^t B_1(a) y(t-a)da - f_1(t)\right] \left[\int_0^t B_2(a) y(t-a)da + f_2(t)\right] \tag{2.8}$$

where

$$f_1(t) = \int_t^1 B_1(a)\psi(a-t)da = \begin{cases} \int_0^{1-t} \psi(a)da & , \text{ for } t \leqslant 1 \\ 0 & , \text{ elsewhere} \end{cases} \tag{2.9}$$

$$f_2(t) = \int_t^1 B_2(a)\,\psi(a-t)da = \begin{cases} \int_0^{1-t} B_2(a+t)\psi(a)da & , \text{ for } t \leqslant 1 \\ 0 & , \text{ elsewhere} \end{cases} \tag{2.10}$$

Readers who are not interested in the proof of the existence and the uniqueness of a solution may pass from here direct to Section 2.3.

The fact that naturally two forcing functions arise suggests to reformulate (2.8) as a two dimensional system. So define

$$x_i(t) = \int_0^t B_i(a) y(t-a)da + f_i(t)\,, \quad i = 1,2\,, \tag{2.11}$$

then

$$x(t) = \int_0^t K(a)\, g(x(t-a))da + f(t)\,, \tag{2.12}$$

where

$$K(a) = \begin{bmatrix} B_1(a) & 0 \\ 0 & B_2(a) \end{bmatrix} \tag{2.13}$$

and

$$g(x) = \begin{bmatrix} \gamma(1-x_1)\, x_2 \\ \gamma(1-x_1)\, x_2 \end{bmatrix} \tag{2.14}$$

Note that

$$y(t) = \gamma(1-x_1(t))\, x_2(t)\,. \tag{2.15}$$

2.2. Existence and uniqueness.

The main tool for the proof of the existence and uniqueness of solutions is the well known *contraction mapping principle.*

DEFINITION 2.3. Let X be a Banach space, U a closed subset and F a mapping from U to X. F is called a (strict) *contraction* on U if $\lambda \in [0, 1)$ exists such that

$$\|F(u)-F(v)\| \leq \lambda\, \|u-v\| \quad \text{for all } u,v\in U\,.$$

An element $u\in U$ is called a *fixed point* of F if $F(u) = u$.

THEOREM 2.4. (Banach-Cacciopoli). *If $F:U\to U$ is a contraction then F has a unique fixed point $\bar{u}\in U$; for any $u\in U$ the sequence $F^n(u)$ converges to $\bar{u}$ and, moreover,*

$$\|F^n(u)-\bar{u}\|\leq\lambda^n\, \frac{\|u-\bar{u}\|}{1-\lambda}\,.$$

DEFINITION 2.5. Let Λ be a set. A mapping $F:U\times\Lambda\to U$ is called a *uniform contraction* on U if $\lambda\in[0,1)$ exists such that

$$\|F(u,\rho)-F(v,\rho)\|\leq\lambda\|u-v\| \text{ for all } u,v\in U \text{and all } \rho\in\Lambda.$$

THEOREM 2.6. R (The uniform contraction principle). Let U and V be open subsets of Banach spaces X and Y and let $\bar{U}$ denote the closure of U. Suppose $F:\bar{U}\times V\to\bar{U}$ is a uniform contraction. Let $h = h(y)$ be the unique fixed point of $F(\cdot,y)$ in $\bar{U}$. If $F\in C^k(\bar{U}\times V,X), 0\leq k<\infty$ then $h\in C^k(V,X)$.

Assume that B_2 is a bounded measurable function. For any $\theta>0$ let $C_\theta = C[0,\theta] = C([0,\theta];\mathbb{R}^2)$ and define $F:C_\theta\to C_\theta$ by

$$F(x)(t) = \int_0^t K(a)g(x(t-a))da + f(t) \tag{2.16}$$

where f is a given element of C_θ.

EXERCISE 2.7. Verify that F indeed maps C_θ into itself.

We want to find a fixed point of F. Let

$$U = \{x|\ \|x-f\|_{C_\theta}\leq 1\}$$

Since $\|x\|_{C_\theta} \leq \|f\|_{C_\theta} + 1$ for all $x\in U$ we are led to introduce the Lipschitz constant L of g on the set $\{\xi|\ |\xi|\leq\|f\|_{C_\theta} + 1\}$, that is, $|g(\eta)-g(\xi)|_{\mathbb{R}_2} \leq L|\eta-\xi|_{\mathbb{R}_2}$, $\forall\eta,\xi$ in the ball of radius $\|f\|_{C_\theta} + 1$ in $\mathbb{R}_2$.

Putting

$$\|K\| = \max\{\sup B_1, \sup B_2\}$$

we clearly have for all $x,z \in U$

$$\|F(x)-F(z)\|_{C_0} \leqslant L\theta\|K\| \; \|x-z\|_{C_0} ,$$

and for θ sufficiently small $L\theta\|K\|<1$. Moreover,

$$\|F(x)-f\|_{C_0} \leqslant L\theta\|K\| \; \|x\|_{C_0} \leqslant L\theta\|K\| \; (1 + \|f\|_{C_0}) \leqslant 1$$

for θ sufficiently small. We conclude that for θ sufficiently small F is a contraction of U into U and consequently Theorem 2.4 implies that F has a unique fixed point in U.

EXERCISE 2.8. Use Theorem 2.6 to show that the fixed point depends continuously on the function f.

In the proof above the upper bound for θ depends on $\|f\|_{C_0}$. So if we repeat the argument to continue the solution to a larger time interval we have to investigate whether or not the new functions f will "blow-up" such that the intervals (the new upper bounds for θ) become smaller and smaller and do not cover the positive time axis as in the

EXAMPLE 2.9. The ordinary differential equation $\dot{y} = y^2$ has a solution $y(t) = -(t-c)^{-1}$ which explodes at the point $t = c$; the solution cannot be continued beyond the point $t = c$.

Now notice that the natural constraints $\int_0^1 \psi(a)da \leqslant 1$ and $\psi \geqslant 0$ imply that

$$0 \leqslant f_1(t) = \int_t^1 B_1(a)\psi(a-t)da \leqslant 1 \text{ and } 0 \leqslant f_2(t) = \int_t^1 B_2(a)\psi(a-t)da \leqslant \sup B_2 .$$

So we have *a priori bounds* for $\|f\|_{C_0}$ provided we demonstrate that the constructed solution, in terms of $n(t,\cdot)$, belongs to

$$C = \{\psi \in L_1(0,1) \mid \psi \geqslant 0 \text{ a.e. and } \int_0^1 \psi(a)da \leqslant 1\} . \tag{2.17}$$

Or, in other words, we have to show that C is (positively) *invariant* (as it should be on account of the biological interpretation). Since $\int_0^1 n(t,a)da = x_1(t)$ and $n(t,0) = y(t) = \gamma(1-x_1(t))x_2(t)$ we have to prove that $0 \leqslant x_1(t) \leqslant 1$ and $x_2(t) \geqslant 0$.

LEMMA 2.10. *If $f_1(0) \leqslant 1$ and $f_i(t) \geqslant 0$ for $i = 1,2$, then the solution of (2.12) satisfies $0 \leqslant x_1(t) \leqslant 1$ and $x_2(t) \geqslant 0$ as long as it is defined.*

PROOF. Assume first that $f_1(0)<1$. Let $\bar{t} = \sup\{t \mid x_1(t)<1\}$ and suppose that $\bar{t}<\infty$. By continuity $x_1(\bar{t}) = 1$. On $[0,\bar{t}]$, $x_2(t)$ is nonnegative and bounded from above (indeed, if we consider x_1 as known then x_2 can be obtained by monotone iteration: $x_2(t) = f_2(t) + \int_0^t B_2(a)\,(1-x_1(t-a))f_2(t-a)da + \cdots$, which shows nonnegativity; boundedness follows likewise from the inequality $x_2(t) \leqslant \gamma \int_0^t B_2(a)x_2(t-a)da + f_2(t)$). Differentiation of the equation for x_1 yields

$$\frac{dx_1}{dt} = \gamma(1-x_1)x_2 - h$$

where

$$h(t) = \begin{cases} \psi(1-t) & , t \leqslant 1 , \\ \gamma(1-x_1(t-1))x_2(t-1) & , t>1 . \end{cases}$$

Putting $z = 1-x_1$ we obtain

$$\frac{dz}{dt} = -\gamma z x_2 + h \geqslant -\gamma z x_2 \quad \text{for } t \in [0,\bar{t}]$$

$$\Rightarrow \ln\frac{z(t)}{z(0)} \geqslant -\gamma \int_0^t x_2(\tau)d\tau \Rightarrow z(t) \geqslant z(0)e^{-\gamma\int_0^t x_2(\tau)d\tau} > 0$$

for $t\in[0,\bar{t}]$ which is in contradiction with $z(\bar{t}) = 0$. So we conclude that $f_1(0)<1$ implies that $x_1(t)<1$. Invoking the continuous dependence on f we deduce that $f_1(0)\leqslant 1$ implies that $x_1(t)\leqslant 1$. □

SUMMARY: Take any $\psi\in C$. Choose θ such that $L\theta\|K\|(2 + \sup B_2)\leqslant 1$ where L is the Lipschitz constant of g in the ball of radius $2 + \sup B_2$. Then (2.12) has a unique solution in C_θ. Define for $0\leqslant t\leqslant\theta$

$$n(t,a) = \begin{cases} \psi(a-t) & ,a\geqslant t \\ \gamma(1-x_1(t-a))x_2(t-a) & , t>a. \end{cases} \tag{2.18}$$

Lemma 2.10 and equation (2.18) imply that $n(\theta,\cdot)\in C$. So with $n(\theta,\cdot)$ as a new initial condition we can construct a solution for $\theta\leqslant t\leqslant 2\theta$. Proceeding in this manner we obtain a unique solution defined for all $t\geqslant 0$. Thus we have associated a dynamical system with the problem (2.6) on C, the biologically interpretable subset of $L_1[0,1]$ (note that the continuity of the map $\psi\mapsto f$ together with the continuous dependence on f guarantees the continuous dependence on ψ).

REMARK. Without the restriction $\psi\in C$ solutions might indeed blow up in finite time.

EXERCISE 2.11. Verify that the infinitesimal generator is given by

$$A\psi = -\psi'$$

$$\mathfrak{D}(A) = \{\psi\in C \mid \psi \text{ is absolutely continuous and } \psi(0) = \gamma(1-\int_0^1 \psi(a)da)\int_0^1 B_2(a)\psi(a)da\}\,.$$

Herewith the first step of the program is completed. The next step is concerned with finding the steady states and determining their stability.

2.3. *The stability of the steady states.*

EXERCISE 2.12. Show that steady states are given by $\bar{n}(a) = \bar{y}$ where either $\bar{y} = 0$ or $\bar{y} = \dfrac{\gamma-1}{\gamma}$. The second one is biologically relevant (i.e. nonnegative) if and only if $\gamma>1$!

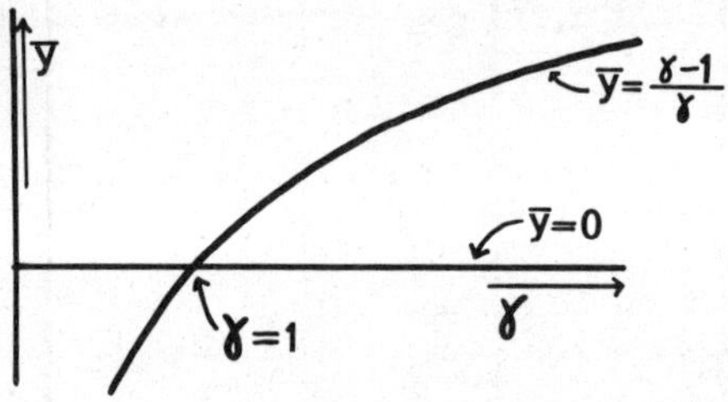

EXERCISE 2.13. Put $n(t,a) = \bar{y} + \rho(t,a)$ in (4.6) and neglect higher order terms in ρ to arrive at

$$\frac{\partial\rho}{\partial t} + \frac{\partial\rho}{\partial a} = 0$$

$$\rho(t,0) = \begin{cases} \gamma\int_0^1 B_2(a)\rho(t,a)da \quad \text{for } \bar{y} = 0 \\ \int_0^1 \{B_2(a) + (1-\gamma)B_1(a)\}\rho(t,a)da \text{ for } \bar{y} = \dfrac{\gamma-1}{\gamma}\,. \end{cases} \tag{2.19}$$

EXERCISE 2.14. Derive the characteristic equation for $\bar{y} = 0$ and show that all roots lie in the left half plane for $\gamma<1$ but that at least one root lies in the right half plane if $\gamma>1$ (consult subsection IV.4.1). Convince yourself that this result is trivial from a biological point of view by recalling the interpretation of γ and the threshold phenomenon discussed in subsection IV.4.1.

Since the principle of linearized stability is rigorously proved for systems of Volterra convolution integral equations like (2.12) (see, for instance, DIEKMANN & VAN GILS, 1984), Exercise 2.14 implies that $\bar{y} = 0$ is locally stable for $\gamma<1$ and unstable for $\gamma>1$. But in fact we can prove

THEOREM 2.15. *$\bar{y}$ is globally exponentially stable for $\gamma<1$.*

PROOF. Equation (2.8) and the fact that both y and f_1 are nonnegative imply that $y(t) \leq \gamma \int_0^t B_2(a) y(t-a) da + \gamma f_2(t)$.

Let z be the solution of the corresponding equation

$$z(t) = \gamma \int_0^t B_2(a) z(t-a) da + \gamma f_2(t).$$

We want to show that $y(t) \leq z(t)$ for all t. Let us first assume that $f_1(0)>0$. Then $y(0)<z(0)$. Let $\bar{t} = \sup\{t \mid z(t)>y(t)\}$ and assume that $\bar{t}<\infty$. Then necessarily $y(\bar{t}) = z(\bar{t})$. But on the other hand

$$z(\bar{t}) = \gamma \int_0^{\bar{t}} B_2(a) z(\bar{t}-a) da + \gamma f_2(\bar{t}) > \gamma \int_0^{\bar{t}} B_2(a) y(\bar{t}-a) da + \gamma f_2(\bar{t}) \geq y(\bar{t})$$

which is a contradiction. We conclude that $f_1(0)>0$ implies that $y(t)<z(t)$, $\forall t \in [0,\infty)$. The continuous dependence on f then implies that always $y(t) \leq z(t)$. In subsection IV.2.2 it was shown that z converges exponentially to zero for $t \to \infty$. Since $n(t,a) = y(t-a)$ for $t>1$ this proves the theorem. □

The principle of the exchange of stability tells us that $\bar{y} = \frac{\gamma-1}{\gamma}$ is stable for $\gamma>1$ but $\gamma-1$ small. In the present case this can also be easily verified by analysing the characteristic equation with the aid of the *implicit function theorem.*

THEOREM 2.16. *Let X, Y and Z be Banach spaces and $U \subset X$ and $V \subset Y$ open sets. Let F be a C^1-mapping from $U \times V$ into Z and let $(x_0,y_0) \in U \times Z$ be such that*

(i) $F(x_0,y_0) = 0$

(ii) *$D_x F(x_0,y_0)$ is an isomorphism (one-to-one and onto).*

Then there exist a neighbourhood $U_1 \times V_1$ of (x_0,y_0) and a function $f: V_1 \to U_1$ with $f(y_0) = x_0$ such that $F(x,y) = 0$ for $(x,y) \in U_1 \times V_1$ if and only if $x = f(y)$. If $F \in C^k$, $k \geq 1$, then $f \in C^k$.

EXERCISE 2.17. Show that the characteristic equation for $\bar{y} = \frac{\gamma-1}{\gamma}$ is

$$\bar{B}_2(\lambda) + (1-\gamma)\bar{B}_1(\lambda) = 1 \tag{2.20}$$

where $\bar{B}_i(\lambda) = \int_0^\infty e^{-\lambda\tau} B_i(\tau) d\tau$ is the Laplace transform of B_i.

EXERCISE 2.18. Verify that for $\gamma = 1$ (2.20) has the simple root $\lambda = 0$ while all other roots have strictly negative real part. Use the implicit function theorem to show that a unique root $\lambda(\gamma)$ exists with $\lambda(1) = 0$ and that $\lambda(\gamma)<0$ for $\gamma>1$.

Since roots can neither come out of the blue nor enter the right half plane at infinity (since $\bar{B}_i(\lambda) \to 0$ for $|\lambda| \to \infty$ with Re $\lambda \geq 0$) all roots have to lie in the left half plane when $0<\gamma-1<<1$. We conclude that $\bar{y} = \frac{\gamma-1}{\gamma}$ is *locally stable* for those values of γ. G. Gripenberg (1981) has derived sufficient conditions for the *global* stability of $\bar{y} = \frac{\gamma-1}{\gamma}$ (in the sense that any initial condition in C different from $\psi \equiv 0$ yields a solution which for $t \to \infty$ converges to $\bar{y} = \frac{\gamma-1}{\gamma}$).

QUESTION: Does $\bar{y} = \frac{\gamma-1}{\gamma}$ remain stable when γ is further increased?

As remarked above $\bar{y} = \frac{\gamma-1}{\gamma}$ can only loose its stability if roots cross the imaginary axis. The substitution $\lambda = 0$

into (2.20) leads to $\gamma = 1$ so in fact only couples of complex conjugated roots can possibly cross the imaginary axis and equivalently we may ask the

QUESTION: Do Hopf bifurcations occur if γ increases from 1 to, say, ∞?

In order to investigate whether for certain values of γ roots of (2.20) lie exactly *on* the imaginary axis one can take $\lambda = i\eta$ with $\eta \in \mathbb{R}$. From a real point of view we have two equations in two unknowns η and γ. Since the equations are linear in γ it is possible to eliminate γ and to derive one equation in the unknown η. By a miracle this nonlinear equation can be analysed completely in this special case (see DIEKMANN & MONTIJN (1982)). The outcome is summarized in

THEOREM 2.19. *When γ increases from 1 to ∞ exactly as many pairs of complex conjugated roots of (2.20) cross the imaginary axis as there are $n \in \mathbb{N}$ for which*

$$b_n = 2 \int_0^1 B_2(a)\sin(2\pi na)da > 0 .$$

These are simple and go from left to right with a positive speed at a "height" between $2n\pi$ and $(2n+1)\pi$.

The local Hopf bifurcation theorem for systems of Volterra convolution integral equations is proved in DIEKMANN & VAN GILS (1984).

Unfortunately Theorem 2.19 does not exclude the possibility that several pairs cross simultaneously and in resonance (i.e. one being an integer multiple of the other). Fiedler (preprint) has proved a *global* Hopf bifurcation theorem which applies to the present problem.

EXERCISE 2.20. Use the implicit function theorem to show that for $\gamma \to \infty$ the roots converge to the points $\pm 2n\pi i$ and that the sign of b_n determines whether they come from the right- or the left half plane.

REMARK. Only the first bifurcating periodic solution can possibly be stable (in other words: the later ones are necessarily unstable). Numerically the first usually corresponds to the root with the smallest η (i.e. the smallest frequency and the largest period; also see the next section) but there are exceptional cases in which a smaller period comes first.

EXERCISE 2.21. Show that $b_1 > 0$ if the support of B_2 is contained in $[0, \frac{1}{2}]$ and check the following interpretation: if the immunity period is long compared to the period of infectivity then enlargement of the population density leads to a destabilization of the stationary endemic state $\bar{y} = \frac{\gamma - 1}{\gamma}$; in this situation one can expect to see oscillations.

CONCLUDING REMARKS: In nonlinear problems one can usually exploit bifurcation theory to obtain some information about the dynamic behaviour. But especially questions concerning *global* aspects are very hard to answer by analytical means. The most promising approach seems to try to combine the outcome of *numerical experiments* with the rather abstract general (topological) *theory of dynamical systems and bifurcations.* Again we refer to Guckenheimer & Holmes (1983) and Golubitsky and Schaeffer (1984) for an outline of the main results and ideas. For the model of this section numerical experiments have not been performed and the results presented above are more or less all that is known to the present authors.

3. Hopf bifurcation in scalar nonlinear renewal equations and nursery competition.

3.1. Introduction to the theory.

In the preceeding subsection we encountered a system of two Volterra convolution equations (or, in other words, renewal equations) and we presented results concerning Hopf bifurcations. In this section we shall show in some more detail how such results can be derived by concentrating on the somewhat simpler problems which take the form of just one equation

$$x(t) = \int_0^\infty g(a)f(x(t-a))da \ , \quad x(t) \in \mathbb{R} \ , \tag{3.1}$$

where the (nonnegative) kernel g is normalized to have integral one, and where both g and the nonlinearity f may depend on parameters. Before becoming more specific we present some generalities.

Steady states of (3.1) are found from the equation

$$\bar{x} = f(\bar{x}). \tag{3.2}$$

The linearization around a steady state $\bar{x}$ is given by

$$y(t) = f'(\bar{x}) \int_0^\infty g(a)y(t-a)da \tag{3.3}$$

and the corresponding characteristic equation reads

$$1 = f'(\bar{x})\, \bar{g}(\lambda) \tag{3.4}$$

where, as before, $\bar{g}$ denotes the Laplace transform. In population problems g usually is nonnegative which allows us to draw the following conclusions (compare subsection IV.2.2; note that $\bar{g}(0) = 1$):

(i) if $f'(\bar{x})>1$ there is a dominant positive real root and $\bar{x}$ is unstable;

(ii) if $0<f'(\bar{x})<1$ there is a dominant negative real root and $\bar{x}$ is asymptotically stable; since the dominant root is real, typical trajectories approaching $\bar{x}$ will do so monotonically (i.e. without oscillating) and one calls $\bar{x}$ over-damped stable.

(iii) if $f'(\bar{x}) = 0$ there are no roots at all (if $f'(\bar{x})$ approaches zero, when some parameter is varied, roots will tend to ∞ in the left half plane of $\mathbb{C}$).

(iv) if $-1\leqslant f'(\bar{x})<0$ all roots are complex and have negative real part (indeed, $|\bar{g}(\lambda)|<1$ if $\mathrm{Re}\lambda\geqslant 0$, $\lambda\neq 0$, and therefore no root can lie in the right half plane); since now typical trajectories approaching $\bar{x}$ will oscillate one calls $\bar{x}$ under-damped stable.

(v) if $f'(\bar{x})<-1$ all roots are still complex but the possibility that they lie in the right half plane exists; so $\bar{x}$ is either under-damped stable or unstable.

Next consider the situation that $\bar{x}$ depends on some parameter(s). The stability properties of $\bar{x}$ can change in two ways when the parameters are varied. The first is connected with the bifurcation of steady states and occurs when $\lambda = 0$ is a root (which will be the case if and only if $f'(\bar{x}) = 1$). The second is connected with the bifurcation of periodic solutions and occurs if $\lambda = \pm i\omega$ is a root for some $\omega\in\mathbb{R}$. This will happen if and only if

$$\mathrm{Im}\, \bar{g}(i\omega) = 0 \tag{3.5}$$

and

$$f'(\bar{x}) = (\mathrm{Re}\, \bar{g}(i\omega))^{-1}. \tag{3.6}$$

In the following our approach will be to solve first (3.5) for ω (which will usually give countably many solutions ω_k) and subsequently analyse (3.6), with $\omega = \omega_k$, to find the parameter values for which roots lie exactly on the imaginary axis. Our strategy will be to concentrate on *two* parameters simultaneously, for the very simple reason that the curves in the plane which one gets as solutions of (3.6) suit the human physiological and psychological possibilities to take in information so very well.

As we have seen in subsection 1.2, the principle of the exchange of stability establishes a clear relation between the stability of the bifurcating periodic solutions and the direction of bifurcation. In DIEKMANN & VAN GILS (1984) an "explicit" formula is derived for the direction of bifurcation. The difference between "explicit" and explicit is that one still has to invert two matrices. In the present scalar case that is very easy and a straightforward application of Theorem 11.5 in DIEKMANN & VAN GILS (1984) yields:

THEOREM 3.1. *Consider a path in parameter space along which a simple root of (3.4) crosses the imaginary axis at ω from left to right with positive speed and assume that all other roots of (3.4) lie strictly in the left half plane. Let c_1 be defined by*

$$c_1 = -\frac{1}{2\,(f'(\bar{x}))^2\bar{g}'(i\omega)} \left\{ f'''(\bar{x}) + 2\,\frac{(f''(\bar{x}))^2}{1-f'(\bar{x})} + \frac{(f''(\bar{x}))^2\, \bar{g}(2i\omega)}{1-f'(\bar{x})\bar{g}(2i\omega)} \right\}. \tag{3.7}$$

Then $\mathrm{Re}\,c_1<0$ *implies that the bifurcating periodic solution exists supercritically and is stable, whereas* $\mathrm{Re}\,c_1>0$ *implies that the bifurcating periodic solution exists subcritically and is unstable.*

3.2. A first application.

Our first application is a caricature of the competition equation derived in subsection III.6.2 (exercise 6.2.7 with $F(\underline{m}) = -\log(1-m)$) in that we take explicit forms for both the kernel and the nonlinearity, basing ourselves on no other justification than the simplicity which results. So consider the equation

$$x(t) = \frac{\gamma}{2\epsilon} \int_{1-\epsilon}^{1+\epsilon} x(t-a) e^{-x(t-a)} da \, . \tag{3.8}$$

Here γ can be thought of as the expected number of offspring produced by a newborn individual during its entire life span in the absence of density dependent effects. The reproductivity is concentrated in a "window" of width 2ϵ centered at 1 (so time is scaled such that the midpoint of the reproductivity period, which can be thought of as a *generation time,* is reached exactly one time unit after birth). Finally, density dependence is incorporated in the exponential "correction" factor.

REMARK 3.2. Similar equations are used in human demography to describe the so-called Easterlin effect which says that women which are born during the bulge of a birth wave tend to get less children than those who are born in the lower valleys. See Swick (1981b).

Equation (3.8) has two steady states: $\bar{x} = 0$ and $\bar{x} = \ln\gamma$.

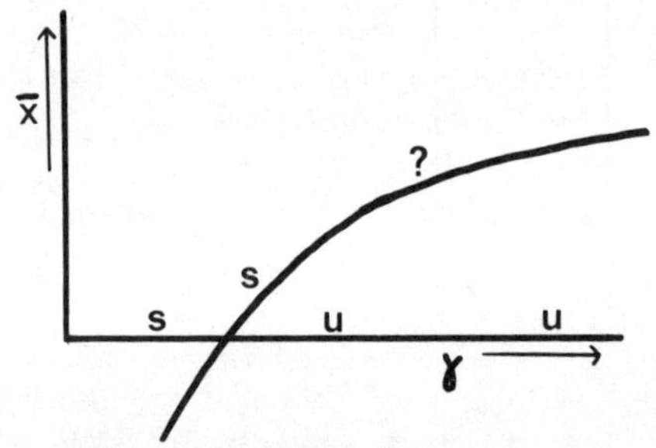

EXERCISE 3.3. Verify the stability assertions in the above diagram.

EXERCISE 3.4. Define $g(a) = \frac{1}{2\epsilon}$ for $1-\epsilon \leqslant a \leqslant 1+\epsilon$ and $g(a) = 0$ elsewhere. Show that the characteristic equation corresponding to the nontrivial steady state $\bar{x} = \ln\gamma$ is

$$1 = (1-\ln\gamma)\,\bar{g}(\lambda) \, . \tag{3.9}$$

EXERCISE 3.5. Show that $\bar{g}(i\omega) = \frac{\sin\epsilon\omega}{\epsilon\omega}[\cos\omega - i\sin\omega]$ and conclude that the roots of $\mathrm{Im}\bar{g}(i\omega) = 0$ are precisely the points $\omega = k\pi$, $k \in \mathbb{Z}$.

EXERCISE 3.6. Rewrite the equation $1 = (1-\ln\gamma)\,\mathrm{Re}\bar{g}(ik\pi)$ as

$$\ln\gamma = 1 + (-1)^{k+1} \frac{\epsilon k\pi}{\sin\epsilon k\pi} \tag{3.10}$$

Take $k = 1$ and draw $\ln\gamma$ as a function of ϵ (verify that $|\frac{\xi}{\sin\xi}| \geqslant 1$).

EXERCISE 3.7. Show that $|\frac{2\xi}{\sin 2\xi}| > |\frac{\xi}{\sin\xi}|$ for $\xi \neq 0$ and convince yourself that more generally $|\frac{k\xi}{\sin k\xi}| > |\frac{\xi}{\sin\xi}|$ for $\xi \neq 0$ and $k = 2,3,4, \cdots$. Now take $k = 2,3,4, \cdots$. and draw $\ln\gamma$ given by (3.10) as a function of ϵ for $0 \leqslant \epsilon < 1$. The result should look like figure 3.1.

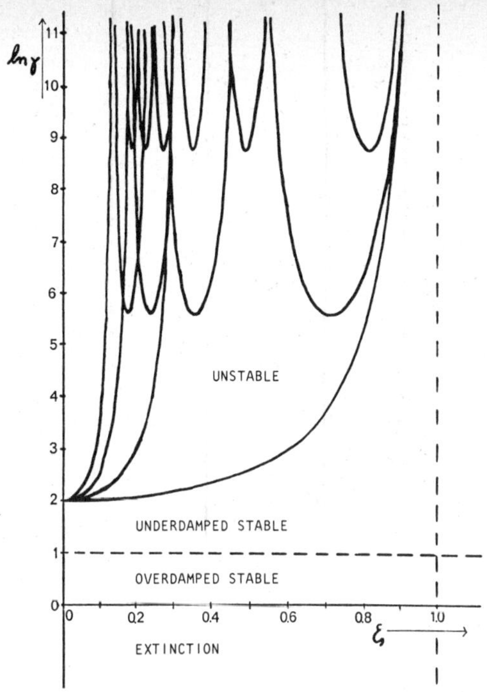

Figure 3.1. Parameter dependence of the qualitative behaviour of the solutions to (3.8).

CONCLUSION: There exists a curve in $(\ln\gamma,\epsilon)$-space given by $\ln\gamma = 1 + \frac{\epsilon\pi}{\sin\epsilon\pi}$ which is a *stability boundary* in the sense that it separates the region where $\overline{x}$ is stable from the region where $\overline{x}$ is unstable. One can pass this stability boundary either by shortening the reproductive period (decreasing ϵ) or by increasing the fertility (increasing γ)or by a combination of these two effects. At the stability boundary a Hopf bifurcation occurs and a periodic solution with a period of approximately two times the generation time arises.

We refer to Swick (1981b) for numerical experiments which indicate that the periodic solution may undergo a sequence of *period doublings* culminating in *chaotic behaviour* (in close analogy with the situation for the difference equation $x(t) = \gamma x(t-1)e^{-x(t-1)}$) when γ is further increased for some small value of ϵ.

In conclusion of our study of this simple example we shall try to determine the direction of the Hopf bifurcation.

EXERCISE 3.8. Show that

(i) $\overline{g}(2\pi i) = \frac{\sin\epsilon\pi}{\epsilon\pi}\cos\epsilon\pi$; note that $\overline{g}(2\pi i)\in\mathbb{R}$!

(ii) $\operatorname{Re}\overline{g}'(\pi i) = \frac{\sin\epsilon\pi}{\epsilon\pi} > 0$ for $0 \leqslant \epsilon < 1$.

(iii) $f'(\overline{x}) = 1-\ln\gamma$; $f''(\overline{x}) = \ln\gamma - 2$; $f'''(\overline{x}) = 3-\ln\gamma$ and conclude that

$$\operatorname{sign}\operatorname{Re}c_1 = \operatorname{sign}\left\{\frac{\epsilon\pi}{\sin\epsilon\pi} - 2 - 2\frac{\left(\frac{\epsilon\pi}{\sin\epsilon\pi}-1\right)^2}{\frac{\epsilon\pi}{\sin\epsilon\pi}+1} - \frac{\left(\frac{\epsilon\pi}{\sin\epsilon\pi}-1\right)^2\frac{\sin\epsilon\pi}{\epsilon\pi}\cos\epsilon\pi}{1+\cos\epsilon\pi}\right\}.$$

This expression is easily evaluated for $\epsilon = 0$ and $\epsilon = \frac{1}{2}$ and one finds $\operatorname{Re}c_1 < 0$, i.e. a supercritical stable bifurcation. But for $\epsilon\uparrow 1$, $\operatorname{Re}c_1 > 0$, so there must be at least one value of ϵ at which a change from supercritical to subcritical occurs. Numerically one finds there is exactly one at $\epsilon \approx 0{,}5536$. We have no biological interpretation for the switching from supercritical to subcritical.

3.3. Nursery competition.

Our second example is the model for competition in the nursery derived in section III.6.2. Consider

$$b(t) = \gamma \int_0^\infty g(a)b(t-a)e^{-(F(Q(t-a))-F(0))}da \tag{3.11}$$

where Q as a function of b is implicitly defined by

$$Q = b\frac{1-e^{-F(Q)}}{F(Q)} \tag{3.12}$$

and F is a strictly increasing function. In terms of the death rate ν and the fecundity function B the kernel g is

$$g(a) = \frac{1}{\gamma} B(a)e^{-\int_0^a \nu(\sigma)d\sigma} e^{-F(0)}, \tag{3.13}$$

where the parameter γ is chosen such that the integral of g is exactly one. Note that the seemingly superfluous insertion of the term $e^{-F(0)}$ allows us to give γ its usual interpretation of the expected number of offspring produced by a newborn individual during its entire lifespan in the absence of density dependent effects (the point is that the function F derived in Appendix III.6.A has $F(0)>0$ implying that even with all safe places available the rate of recruitment is different from the birth rate).

For $0<\ln\gamma<F(\infty)-F(0)$, (3.11) has a unique positive steady state

$$\bar{b} = \frac{\bar{Q}F(\bar{Q})}{1-e^{-F(\bar{Q})}} \tag{3.14}$$

where

$$\bar{Q} = F^{-1}(F(0) + \ln\gamma)\,. \tag{3.15}$$

After linearization one obtains the characteristic equation

$$1 = [1-\bar{b}F'(\bar{Q})Q'(\bar{b})]\,\bar{g}(\lambda)\,. \tag{3.16}$$

EXERCISE 3.9. Use (3.12) to deduce that

$$Q'(b) = \frac{1-e^{-F(Q)}}{F(Q) + F'(Q)[Q-be^{-F(Q)}]}$$

and use this expression to rewrite the coefficient of $\bar{g}(\lambda)$ in (3.16) as

$$\frac{F(\bar{Q}) + F'(\bar{Q})[\bar{Q}-\bar{b}]}{F(\bar{Q}) + F'(\bar{Q})[\bar{Q}-\bar{b}e^{-F(\bar{Q})}]} \tag{3.17}$$

We now consider two special functions F. The first is

$$F(Q) = \theta\frac{Q}{1+Q} \tag{3.18}$$

which is obtained from the cannibalism model discussed in the final remark of example III.6.2.3 by taking: (i) for Φ the Holling disk factor $\Phi(c) = \frac{1}{1+c}$; (ii) for c a constant times Q; (iii) a scaling of b and Q; (iv) $F(Q) = \theta Q\Phi(Q)$. The parameter θ is an indicator for the strength of the cannibalistic interaction (see DIEKMANN et al. (1986) for a more detailed interpretation).

EXERCISE 3.10. Show that now, for $0<\ln\gamma<\theta$,

$$\bar{Q} = \frac{\ln\gamma}{\theta-\ln\gamma} \tag{3.19}$$

$$\bar{b} = \frac{\gamma\ln^2\gamma}{(\gamma-1)(\theta-\ln\gamma)}$$

and derive for (3.17) the explicit expression

$$E(\theta,\gamma) = \frac{\theta(2\gamma-2-\gamma\ln\gamma)-\ln\gamma(\gamma-1-\gamma\ln\gamma)}{\theta(2\gamma-2-\ln\gamma)-\ln\gamma(\gamma-1-\ln\gamma)}. \tag{3.20}$$

EXERCISE 3.11. Solve the equation $E(\theta,\gamma) = -\Omega$ for θ in terms of γ and Ω to find

$$\theta = \ln\gamma\frac{\gamma-1-\gamma\ln\gamma + \Omega(\gamma-1-\ln\gamma)}{2\gamma-2-\gamma\ln\gamma + \Omega(2\gamma-2-\ln\gamma)} \tag{3.21}$$

We are interested in the function $\theta = \theta(\gamma)$ defined by (3.21) when

(i) $\Omega = 0$; this gives the boundary between the regions in parameter space where the steady state is overdamped stable and underdamped stable

(ii) $\Omega = -(\mathrm{Re}\bar{g}(i\omega))^{-1}>1$; with ω such that $\mathrm{Im}\bar{g}(i\omega) = 0$; this gives curves in parameter space where a root of the characteristic equation (3.16) lies on the imaginary axis.

EXERCISE 3.10. Show that

$$\frac{d\theta}{d\Omega} = \left[\frac{(\gamma-1)\ln\gamma}{2\gamma-2-\gamma\ln\gamma + \Omega(2\gamma-2-\ln\gamma)}\right]^2 >0$$

and conclude that the curve corresponding to the minimal positive value of Ω yields the boundary between the regions in parameter space where the steady state is stable and unstable.

EXERCISE 3.11. Show that the nominator of (3.21) has precisely one zero $\gamma^*(\Omega)$ in $(1,\infty)$ for $\Omega\geqslant 0$.

EXERCISE 3.12. Show that (3.21) satisfies the biological constraint $\theta>\ln\gamma$ if and only if $\gamma>\gamma^*(\Omega)$.

Figure 3.2A depicts the biologically relevant part of the graph of $\theta(\gamma)$ for $\Omega = 2.2618$. This value of Ω is obtained when $\nu(a) = \nu = 1$ and

$$B(a) = \begin{cases} 0 & \text{for } 0\leqslant a\leqslant 1 \\ \gamma e & \text{for } a>1. \end{cases}$$

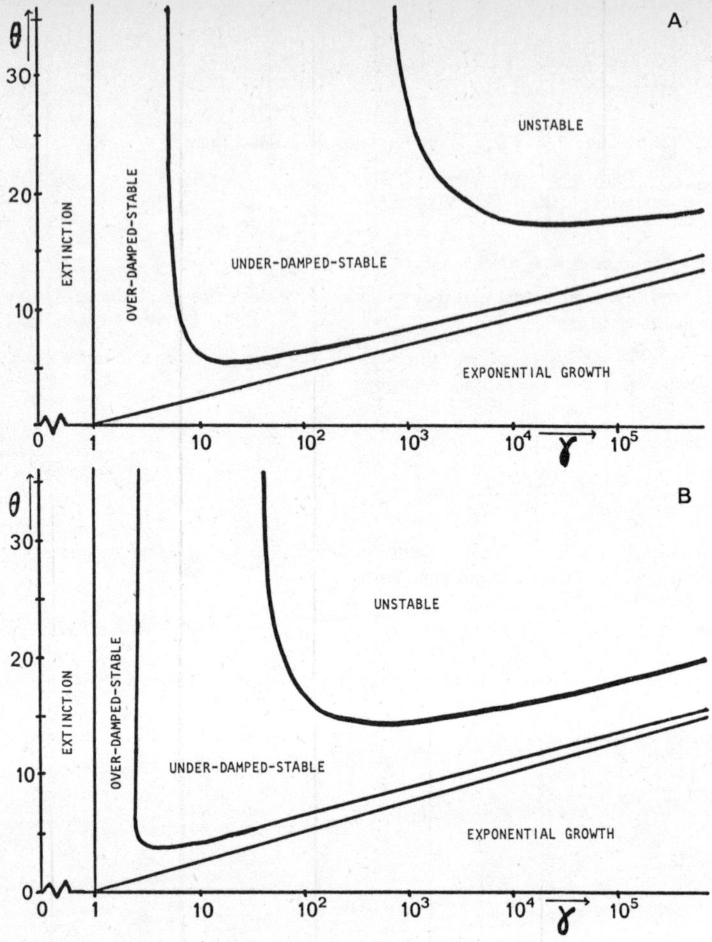

Figure 3.2. Stability diagram for the nursery competition model.
A: Cannibalism (formula (3.18)). B: Safe places (formula (3.22) with $\xi/\rho=2.75$.)

The second example is

$$F(Q) = \frac{\theta}{2Q}\{Q-\xi + \sqrt{(Q-\xi)^2 + \rho Q}\} \tag{3.22}$$

which was introduced in Appendix III.6.A to describe competition for safe places.

EXERCISE 3.13. Show that $F(0) = \lim_{Q\downarrow 0} F(Q) = \frac{\rho\theta}{4\xi}$ and $F(\infty) = \theta$.

EXERCISE 3.14. Show that now

$$\overline{Q} = \frac{\xi \ln\gamma}{\ln\gamma + \frac{\rho\theta}{4\xi} - \frac{1}{\theta}(\ln\gamma + \frac{\rho\theta}{4\xi})^2} \tag{3.23}$$

$$\overline{b} = \frac{\xi\gamma\ln\gamma}{(1-\frac{1}{\theta}(ln\gamma + \frac{\rho\theta}{4\xi}))(\gamma - \exp(-\frac{\rho\theta}{4\xi}).} \tag{3.24}$$

Subsequently one can derive for (3.17) the explicit expression

$$E(\theta,\gamma) = \frac{c_1\theta^2 + c_2\theta + c_3}{c_4\theta^2 + c_5\theta + c_6} \tag{3.25}$$

where

$$\begin{cases} c_1 = -\gamma \\ c_2 = 2\gamma(F(0) + \ln\gamma) + (1-4\frac{\xi}{\rho})(\gamma - e^{-F(0)} - \gamma(F(0) + \ln\gamma)) \\ c_3 = -4\gamma\frac{\xi}{\rho}(F(0) + \ln\gamma)^2 \\ c_4 = -e^{-F(0)} \\ c_5 = 2e^{-F(0)}(F(0) + \ln\gamma) + (1-4\frac{\xi}{\rho})(\gamma - e^{-F(0)} - e^{-F(0)}(F(0) + \ln\gamma)) \\ c_6 = -4e^{-F(0)}\frac{\xi}{\rho}(F(0) + \ln\gamma)^2 , \end{cases} \tag{3.26}$$

and the equation $E(\theta,\gamma) = -\Omega$ can be reduced to the equation

$$\theta^2 + d_1\theta + d_2 = 0 \tag{3.27}$$

where

$$\begin{cases} d_1 = -\{(1+\Omega)(1-4\frac{\xi}{\rho})\left[\frac{\gamma e^{F(0)} - 1}{\gamma e^{F(0)} + \Omega}\right] - (1 \quad 4\frac{\xi}{\rho})(F(0) + \ln\gamma) + 2(F(0) + \ln\gamma)\} \\ d_2 = 4\frac{\xi}{\rho}(F(0) + \ln\gamma)^2. \end{cases} \tag{3.28}$$

Although $F(0) = \frac{\rho\theta}{4\xi}$ depends on θ, we may interpret (3.27) as a quadratic equation in θ after a change of variable $\zeta = \gamma e^{F(0)}$. Thus we find, for given Ω,ξ and ρ,θ as a double valued function of ζ (the condition that the roots should be real and positive leads to a lower bound for the allowed values of ζ). Subsequently one can recover θ as a function of γ by inversion of the $\gamma \mapsto \zeta$ transformation (i.e. $\gamma = \zeta\exp(-\frac{\rho\theta}{4\xi})$). The outcome of the numerical calculations (with the same value of Ω as in Figure 3.2A) is presented in Figure 3.2B.

4. Lyapunov functions and monotone methods: the G-M model in cell kinetics

In this section we discuss a model for the growth of a cell population. One of the basic assumptions is that the cell cycle consists of two phases. The model is a slight adaption of a model first suggested by Kirk, Orr and Forest (1970), which describes the production of red blood cells by the bone marrow stem cell population. Later this model has also been investigated by Mackey (1978, 1981).

In this section we shall use two basic techniques to derive global stability results for trivial and nontrivial equilibria namely Lyapunov functions (and the invariance principle) and monotonicity methods. Although some of the computations can be simplified, we think that the formulation below most clearly illustrates the underlying idea.

4.1. The model

We consider a population of cells reproducing by division. We assume that within the cell cycle at least two phases can be distinguished: The *G*-phase or resting phase during which cells "just sit and wait", and the *M*-phase or mitosis phase. We assume that a cell which has entered the *M*-phase finally passes into mitosis and its daughters enter the *G*-phase again which they leave after an exponentially distributed time. For more biological details concerning the life cycle of cells we refer to Eisen (1979).

We assume that all individuals in the *G*-phase are identical and we denote their number at time t by $P(t)$. However cells in the *M*-phase can be distinguished from one another according to some one-dimensional quantity x, which we shall call maturity, but which can be anything such as age, or the concentration of some chemical substance (like

DNA) within the cell. We let $n(t,x)$ be the maturity distribution, i.e. $\int_{x_1}^{x_2} n(t,x)dx$ is the number of M-cells with maturity between x_1 and x_2. A cell entering the M-phase has maturity $x = 0$. As in section I.4 and chapter II we conceive of fission as a stochastic process which can be described by a function $b(x) \geqslant 0$. We assume that the maximal maturity is $x = 1$, which is achieved mathematically by assuming that

$$\int_0^1 b(x)dx = \infty .$$

As in section I.4 and chapter II we assume that the maturity of an individual in the M-phase increases deterministically according to the ODE

$$\frac{dx}{dt} = V(x) ,$$

where $V(x)$ is called the growth rate. For simplicity we assume a constant death rate $\mu > 0$ for all individuals in the G-phase as well as in the M-phase. Finally we let γ be the transition probability, i.e. the chance per unit of time that cells in the G-phase enter the M-phase.

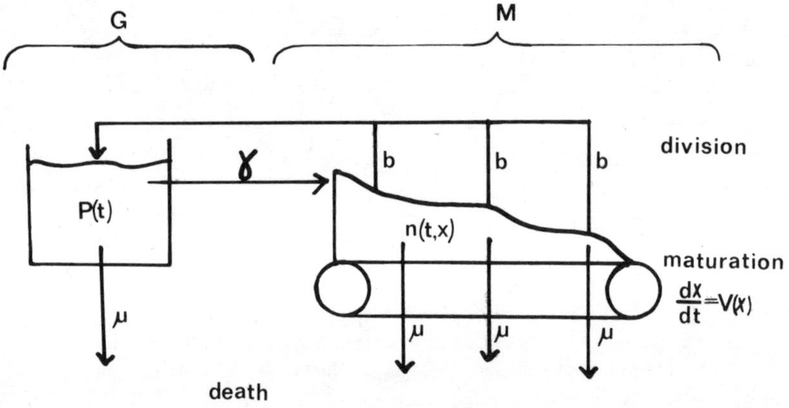

Figure 4.1.1. Schematic representation of the G-M model.

When μ, γ, V and b are all constant the dynamics of $P(t), n(t,x)$ is governed by the following linear system:

$$\frac{dP}{dt} = -\mu P - \gamma P + 2 \int_0^1 b(x)n(t,x)dx \tag{4.1}$$

$$\frac{\partial n}{\partial t} + \frac{\partial}{\partial x}(V(x)n(t,x)) = -\mu n(t,x) - b(x)n(t,x) \tag{4.2}$$

$$V(0)n(t,0) = \gamma P. \tag{4.3}$$

EXERCISE 4.1. Convince yourself that (4.1) - (4.3) are correct and interprete the separate terms.

We supply (4.1) - (4.3) with the initial conditions

$$P(0) = P_0 \geqslant 0 \tag{4.4}$$

$$n(0,x) = n_0(x) \geqslant 0, \ 0 \leqslant x \leqslant 1 . \tag{4.5}$$

As in chapter II, we make the following

ASSUMPTIONS 4.2:

H_V: V is a strictly positive continuous function

H_b: b is non-negative and continuous on [0, 1) and $\lim_{x \uparrow 1} \int_a^x b(\xi)d\xi = \infty$.

Let $G(x), E(x)$ be defined as

$$G(x) = \int_0^x \frac{d\xi}{V(\xi)}, \tag{4.6}$$

$$E(x) = \exp(-\int_0^x \frac{\mu + b(\xi)}{V(\xi)} d\xi). \tag{4.7}$$

EXERCISE 4.3. Show that the distribution $n(t,x)$ can be expressed in terms of $P(t)$ in the following way

$$n(t,x) = \gamma P(t - G(x)) \frac{E(x)}{V(x)}.$$

EXERCISE 4.4. Use this result to show that the characteristic equation associated with (4.1) - (4.3) looks as follows:

$$\lambda + \mu + \gamma = 2\gamma \int_0^1 \frac{b(x)E(x)}{V(x)} e^{-\lambda G(x)} dx.$$

Show that this equation has exactly one real solution λ_d. Show that λ_d is strictly dominant, i.e. for all other roots of the equation the inequality $\text{Re}\lambda < \lambda_d$ holds.

Of course, in practice the quantities μ, γ, V and b will depend on environmental factors such as food density, concentration of toxic chemicals etc. In this section we consider the very simple case where μ, V and b do not depend on such factors, and γ depends on the number of individuals in the G-phase $P(t)$.

ASSUMPTION 4.5. $\gamma = \gamma(P)$, γ is a Lipschitz-continuous function on $[0,\infty)$ which is decreasing, and moreover $\lim_{P\to\infty} \gamma(P) = 0$.

This means among other things that the chance that a cell in the G-phase enters the M-phase decreases if the size of the total G-population increases and that it approaches zero for $P \to \infty$.

It was first suggested by Kirk, Orr and Forest (1970), see also Mackey (1981), that the transition probability γ might depend on the concentration of some mitotic inhibitory substance. At the end of this section we shall indicate how this problem is related to assumption 4.5. See also Mackey (1978).

Assumption 4.5 makes (4.1) - (4.3) nonlinear, but keeps the system autonomous.

4.2. *Existence and uniqueness*

First we have to make some remarks on the problem of existence and uniqueness of solutions. As in chapter II we impose the following condition on the initial function n_0 (compare this to the result in Exercise 4.3):

ASSUMPTION 4.6. $n_0(\cdot)/E(\cdot)$ is a continuous function on $[0,1]$.

Now let X be the space of all pairs (ρ,ν) such that $\rho \in \mathbb{R}$ and $\frac{\nu(\cdot)}{E(\cdot)} \in C[0,1]$, and let X be supplied with the norm

$$\|(\rho,\nu)\| = |\rho| + \sup_{0 \leq x \leq 1} \frac{|\nu(x)|}{E(x)}.$$

EXERCISE 4.7. Verify that with this norm X defines a Banach space.

We define the cone X_+ in the following way:

$$\phi = (\rho,\nu) \in X_+ \text{ if and only if } \rho \geq 0 \text{ and } \frac{\nu(x)}{E(x)} \geq 0,\ x \in [0,1].$$

Now let $\phi_1 = (\rho_1,\nu_1)$, $\phi_2 = (\rho_2,\nu_2) \in X_+$ then $\phi_1 \leq \phi_2$ if $\phi_2 - \phi_1 \in X_+$, $\phi_1 < \phi_2$ if $\phi_2 - \phi_1 \in X_+$ and $\phi_2 \neq \phi_1$ and finally $\phi_1 << \phi_2$ if $\rho_1 < \rho_2$ and $\frac{\nu_1(x)}{E(x)} < \frac{\nu_2(x)}{E(x)}$ for all $x \in [0,1]$.

We define $X(t,x)$ as the maturity of an individual at time t given that its maturity at time zero was x. In other words $X(t,x)$ is the solution of

$$\frac{dX}{dt} = V(X(t,x)),\ X(0,x) = x\ ,$$

and $X(t,x)=G^{-1}(t+G(x))$, with $G(x)=\int_0^x \frac{d\xi}{V(\xi)}$ (compare chapter II), provided that $0 \leq t < G(1)-G(x)$.

Let the initial pair $(P_0,n_0) \in X_+$ then $(P(t),n(t))$ is called a solution of (4.1) - (4.5) (with $\gamma = \gamma(P(t))$ substituted) if and only if

(i) $(P(t),n(t)) \in X_+,\ t \geq 0,$

(ii) $P(t)$ is differentiable for $t>0$ and

$$\frac{dP}{dt}(t) = -\mu P(t) - \gamma(P(t))P(t) + 2\int_0^1 b(x)n(t,x)dx\ ,$$

(iii) $\lim_{h\to 0}\frac{1}{h}\{V(X(h,x))n(t+h,\ X(h,x)) - V(x)n(t,x)\}$ exists for all $t>0$, $0<x<1$ and

$$\frac{1}{V(x)}\lim_{h\to 0}\frac{1}{h}\{V(X(h,x))n(t+h,X(t,x)) - V(x)n(t,x)\} = -\mu n(t,x) - b(x)n(t,x),\quad t>0,\ 0<x<1\ ,$$

(iv) $V(0)n(t,0) = \gamma(P(t))P(t),\ t>0\ ,$

(v) $P(0) = P_0\ ,\ n(0,x) = n_0(x),\ 0 \leq x \leq 1\ .$

Condition (iii) means that n has to be differentiable along the characteristics.

As in Exercise 4.3 we can deduce

$$n(t,x) = \gamma(P(t-G(x)))P(t-G(x))\frac{E(x)}{V(x)}\ , \tag{4.8}$$

and if we substitute this in (4.1) we obtain

$$\frac{dP}{dt}(t) = -\mu P(t) - f(P(t)) + 2\int_0^1 k(x)f(P(t-G(x)))dx\ ,$$

where

$$k(x) = \frac{b(x)}{V(x)}E(x),\ 0 \leq x < 1\ , \tag{4.9}$$

$$f(P) = \gamma(P)P\ ,\ P \geq 0\ . \tag{4.10}$$

From (4.8) we conclude that

$$n(t,x) = \frac{E(x)}{V(x)}\ f(P(t-G(x)), t>G(x)\ .$$

A similar calculation shows that

$$n(t,x) = \frac{E(x)}{V(x)}\frac{V(X(-t,x))}{E(X(-t,x))}\ n_0(X(-t,x)),\ t<G(x)\ .$$

Therefore at time $t = G(x)$, where $0<x<1$, $n(t,x)$ is discontinuous in x unless $f(P_0) = V(0)n_0(0)$. We define the subset C of X as

$$C = \{\phi = (\rho,\nu) \in X_+ | f(\rho) = V(0)\nu(0)\}\ .$$

Since the definition of a solution of (4.1) - (4.5) involves that $n(t,x)$ has to be continuous for $t>0$ we must start with initial pairs (P_0,n_0) belonging to C. We can then prove the following existence and uniqueness result.

THEOREM 4.8. *Let $\phi_0 = (P_0,n_0) \in C$, then there exists a unique solution $\phi(t) = (P(t),\ n(t)) \in C$ of the system (4.1) - (4.5).*

One way to obtain this result is to apply standard local existence and uniqueness results for retarded functional differential equations to the integro-differential equation obtained above (see Hale (1977)). As in section 2 of this chapter, global existence follows if one can give an a priori estimate on the solution: see lemma 4.10 below.

Now we can define a nonlinear semigroup (or dynamical system) $T(t)$ on C in the following standard way:

$$\phi(t) = T(t)\phi_0,\ t \geq 0,\ \phi_0 \in C\ .$$

It is incorporated in the definition of solutions that $T(t)$ is nonnegativity-preserving. However, it is an easy task to show that $\phi_0 \in C$, $\phi_0 \neq 0$, implies that $\phi(t) = T(t)\phi_0 >> 0$ for t large enough. This fact will be exploited in the proof of theorem 4.27.

4.3. Boundedness of solutions

We conclude from (4.8) that it suffices to show that $P(t)$ remains bounded for all $t>0$.

First we integrate (4.2) from 0 to 1. Let

$$N(t) = \int_0^1 n(t,x)dx ,$$

then

$$\frac{dN}{dt}(t) = -\mu N(t) + \gamma(P(t))P(t) - \int_0^1 b(x)n(t,x)dx .$$

EXERCISE 4.9. Check this.

Combining this with (4.1) yields

$$\frac{dM}{dt} = -\mu M + \gamma(P)P , \tag{4.11}$$

where $M(t) = 2N(t) + P(t)$, $t \geqslant 0$.

LEMMA 4.10. *$M(t)$ is bounded, $t \geqslant 0$.*

PROOF. Suppose that $M(t)$ is not bounded. Then we can choose a strictly increasing sequence $t_n > 0$, $n \in \mathbb{N}$ such that $M(t_n) \to \infty$ if $n \to \infty$ and $\dot{M}(t_n) \geqslant 0$.

(i) Suppose that $P(t_n)$ is a bounded sequence, $P(t_n) \leqslant P_{\max}$, $n \in \mathbb{N}$, then $\dot{M}(t_n) \leqslant \gamma(0)P_{\max} - \mu M(t_n)$, and therefore $\dot{M}(t_n) < 0$ if n is large enough which is a contradiction.

(ii) Suppose that $P(t_n) \to \infty$ as $n \to \infty$. Then $\gamma(P(t_n)) < \frac{1}{2}\mu$ as n is large enough. Since $\dot{M}(t_n) \leqslant (\gamma(P(t_n)) - \mu)P(t_n)$ this contradicts $\dot{M}(t_n) \geqslant 0$. □

Since $0 \leqslant P(t) \leqslant M(t)$, lemma 4.10 implies boundedness of $P(t)$ and hence of the solutions of (4.1) - (4.5).

THEOREM 4.11. *Every solution $(P(t), n(t))$ of (4.1) - (4.5) is bounded and precompact.*

EXERCISE 4.12. Show that precompactness follows from (4.8).

4.4. Extinction of the population

It is intuitively clear that, if the population even under the most favourable growth conditions (i.e. $\gamma(P(t)) = \gamma(0)$, for all $t \geqslant 0$) does become extinct, then there is no hope for survival under the actual circumstances. Below we shall make this intuitive idea more precise.

EXERCISE 4.13. Use Exercise 4.4 to show that for every fixed $\gamma \geqslant 0$ there exists a $\mu(\gamma) \geqslant 0$ such that for $\mu \geqslant 0$ the dominant eigenvalue λ_d is given by $\lambda_d = \mu(\gamma) - \mu$. Show that $\mu(\gamma)$ is increasing with γ.
Hint: The characteristic equation can be rewritten as

$$\lambda + \mu + \gamma = 2\gamma \int_0^1 k_0(x) e^{-(\lambda+\mu)G(x)} dx \text{ ,where } \int_0^1 k_0(x)dx = 1 .$$

Let for $\gamma \geqslant 0$ and $\lambda \in \mathbb{C}$

$$\pi_\gamma(\lambda) = \frac{2\gamma}{\lambda + \mu + \gamma} \int_0^1 k(x) e^{-\lambda G(x)} dx .$$

EXERCISE 4.14. Compare this expression to the characteristic equation obtained in Exercise 4.4. Interpret $\pi_\gamma(0)$ as the net reproduction rate (i.e. the average number of offspring of a newborn individual) if the transition probability is γ.

Now we can state our "extinction result".

THEOREM 4.15. *If $\pi_{\gamma(0)}(0) \leq 1$, then the population becomes extinct, i.e. the solution $(P(t), n(t))$ of (4.1) - (4.5) satisfies*

$$(P(t), n(t)) \to 0 \text{ as } t \to \infty .$$

EXERCISE 4.16. Show that the condition in this theorem is satisfied iff $\mu \geq \mu(\gamma(0))$ (see Exercise 4.13).

In order to prove theorem 4.15 we shall construct a suitable Lyapunov function (see section 1 of this chapter). Let

$$r(x) = \frac{2}{E(x)} \int_x^1 k(y) dy . \tag{4.12}$$

EXERCISE 4.17. Interpret $r(x)$ as the expected number of offspring of a cell in the M-phase with maturity x. Show that $r(x) = 2$ if $\mu = 0$.

We define the continuous function $\mathcal{V}$ on X by:

$$\mathcal{V}(\rho,\nu) = \rho + \int_0^1 r(x)\nu(x)dx, \ (\rho,\nu) \in X . \tag{4.13}$$

We can give the following intuitive interpretation of the function $\mathcal{V}$. Obviously an individual in the M-phase has a greater chance to divide eventually than an individual in the G-phase. Since we can interpret $r(x)$ as the expected number of offspring of an individual in the M-phase this function assigns a value to every individual, representing its (expected) future contribution to the population.

Let $(P_0,n_0) \in C$ and $(P(t), n(t)) = T(t)(P_0,n_0)$, $t \geq 0$, then

$$\frac{d}{dt}\mathcal{V}(P(t),n(t)) = ((r(0)-1)\gamma(P(t))-\mu)P(t) .$$

EXERCISE 4.18. Use $r'(x) = \dfrac{\mu + b(x)}{V(x)} r(x) - \dfrac{2b(x)}{V(x)}$ to prove this.

Hence $\dot{\mathcal{V}}(\rho,\nu) = ((r(0)-1)\gamma(\rho)-\mu)\rho$, $(\rho,\nu) \in C$.

EXERCISE 4.19. Prove that $\dot{\mathcal{V}}(\rho,\nu) \leq 0$, for all $(\rho,\nu) \in C$ iff $\pi_{\gamma(0)}(0) \leq 1$.

As in section 1 of this chapter we define

$$E = \{(\rho,\nu) \in C \mid \dot{\mathcal{V}}(\rho,\nu) = 0\} ,$$

and we let $\mathcal{E}$ be the largest invariant subset of E. Suppose that $\pi_{\gamma(0)}(0) \leq 1$. Then

$$E = \{(\rho,\nu) \in C \mid \rho = 0\} .$$

Let $(\rho,\nu) \in \mathcal{E}$. Then $\rho = 0$. Since $\mathcal{E}$ is invariant it follows from (4.8) that $\nu = 0$, hence

$$\mathcal{E} = \{(0,0)\} .$$

Since moreover, for every $(P_0,n_0) \in C$ the orbit $\Gamma^+(P_0,n_0)$ is precompact (theorem 4.11) we obtain from the invariance principle (theorem 1.1.6) that $(P(t),n(t)) \to (0,0)$ as $t \to 0$, and this proves theorem 4.15.

4.5. *Existence of a nontrivial equilibrium and monotonicity on an invariant subset*

It follows from theorem 4.15 and exercise 4.16 that the trivial equilibrium is (globally) stable if the deathrate μ is large enough.

EXERCISE 4.20. Show that there exists a unique nontrivial equilibrium $(\hat{P},\hat{n})$ of (4.1) - (4.3) if $\pi_{\gamma(0)}(0)>1$, where $\hat{n}(x) = f(\hat{P})\frac{E(x)}{V(x)}$, and $\hat{P}$ is determined by

$$\gamma(\hat{P}) = \frac{\mu}{r(0)-1} \quad (\text{i.e } \pi_{\gamma(\hat{P})}(0) = 1)\,.$$

(Recall the corresponding result for the Daphnia-model in section I.3.)

We assume for the rest of this section that $\pi_{\gamma(0)}(0)>1$. This is equivalent to

$$\frac{r(0)\gamma(0)}{\mu+\gamma(0)} > 1 \tag{4.14}$$

From assumption 4.5 we conclude that $f(P) = P\gamma(P)$ is increasing for small values of P. From a biological point of view the following assumption means no restriction of generality.

ASSUMPTION 4.21. There exists a P_m, $0<P_m\leqslant\infty$ such that f is increasing on $[0,P_m)$ and nonincreasing on (P_m,∞).

We recall that γ is decreasing. We also make the following

ASSUMPTION 4.22. $\hat{P}<P_m$.

For future use we note that assumption 4.22 can be reformulated as

$$(r(0)-1)\gamma(P_m)<\mu\,. \tag{4.15}$$

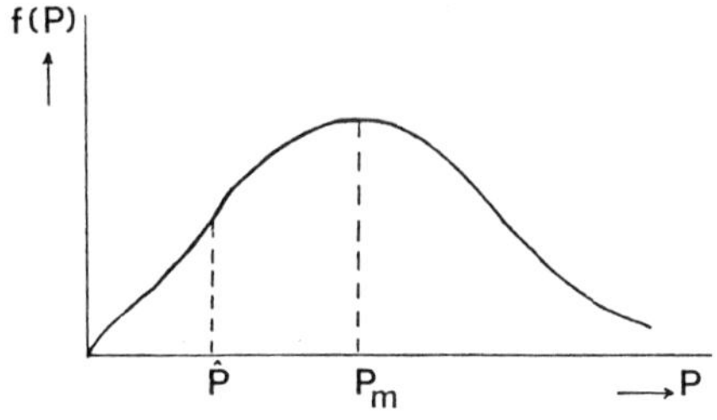

For $P\geqslant 0$ we define $\Phi_P\in C$ by

$$\Phi_P = (P, f(P)\frac{E(\cdot)}{V(\cdot)})\,. \tag{4.16}$$

Let $\hat{P}<\tilde{P}<P_m$ and let the bounded subset $\tilde{C}\subseteq C$ be given by

$$\tilde{C} = \{\phi\in C \mid \phi<<\Phi_{\tilde{P}}\}\,. \tag{4.17}$$

LEMMA 4.23. $\tilde{C}$ is positively invariant under the action of $T(t)$.

PROOF. Let $\phi_0\in\tilde{C}$ and suppose that $T(t)\phi_0\notin\tilde{C}$ for some $t>0$. Let t_0 be the smallest t for which this is so, and let $T(t)\phi_0 = (P(t), n(t))$. There are three possibilities:

i) There exists an $x, 0<x\leqslant 1$, such that $\frac{V(x)}{E(x)}n(t_0,x) = f(\tilde{P})$. Let $h>0$ be such that $t-h>0$ and $X(-h,x)>0$, then $\frac{V(X(-h,x))}{E(X(-h,x))}n(t_0-h, X(-h,x)) = \frac{V(x)}{E(x)}n(t_0,x) = f(\tilde{P})$ as follows directly by integration of (4.2) along characteristics. But this yields $T(t_0-h)\phi\notin\tilde{C}$ which is contradictory with the definition of t_0.

ii) Let $V(0)n(t_0,0) = f(\tilde{P})$. Then $f(P(t_0)) = f(\tilde{P})$ and this implies that $P(t_0) = \tilde{P}$.

ii) The third possibility is $P(t_0) = \tilde{P}$.

Therefore we may assume $P(t_0) = \tilde{P}$. Since $P(t_0-h)<\tilde{P}$ for $0<h\leqslant t_0$ we find that $\dot{P}(t_0)\geqslant 0$. On the other hand

$$\dot{P}(t_0) = -\mu\tilde{P}-f(\tilde{P}) + 2\int_0^1 b(x)n(t_0,x)dx < -\mu\tilde{P}-f(\tilde{P}) + 2\int_0^1 k(x)f(\tilde{P})dx = (r(0)-1)f(\tilde{P})-\mu\tilde{P}<0$$

because $\tilde{P}>\hat{P}$. Thus we have obtained a contradiction and the result is proved. □

In the sequel we shall need the following technical lemma.

LEMMA 4.24.

a) *Let* $\phi\in C$, $\phi>>0$, *then there exists a sequence* $\{\phi_k\}_{k\in\mathbb{N}}$ in C *such that* $0<<\phi_k<<\phi$, $k\in\mathbb{N}$ *and* $\lim_{k\to\infty}\phi_k = \phi$.

b) *Let* $\phi\in\tilde{C}$, *then there exists a sequence* $\{\phi_k\}_{k\in\mathbb{N}}$ *in* $\tilde{C}$ *such that* $\phi_k>>\phi$, $k\in\mathbb{N}$ *and* $\lim_{k\to\infty}\phi_k = \phi$.

PROOF. We shall only give a proof of b).
Let $\phi = (\rho,\nu)\in\tilde{C}$. Then $\rho<\tilde{P}$ and $\frac{V(x)}{E(x)}\nu(x)<f(\tilde{P})$, $0\leq x\leq 1$. Let $\{a_k\}_{k\in\mathbb{N}}$ be a sequence in $\mathbb{R}$ such that $a_k\to 0$, $k\to\infty$ and for all $k\in\mathbb{N}$: $a_k>0$ and $\frac{V(x)}{E(x)}\nu(x) + a_k<f(\tilde{P})$, $0\leq x\leq 1$. Let $\nu_k(x) = \nu(x) + a_k\frac{E(x)}{V(x)}$, and let $\rho_k\in(\rho,\tilde{P})$ be determined by $f(\rho_k) = V(0)\nu_k(0) = V(0)\nu(0) + a_k = f(\rho) + a_k$. Then $\phi_k = (\rho_k,\nu_k)$, $k\in\mathbb{N}$, satisfies the conditions of the lemma. □

We shall now prove two monotonicity results.

THEOREM 4.25. *$T(t)$ is monotone on $\tilde{C}$, i.e. $\phi,\psi\in\tilde{C}$ and $\phi\leq\psi$ imply that $T(t)\phi\leq T(t)\psi$, $t\geq 0$.*

PROOF. Let $\phi,\psi\in\tilde{C}$, $\phi\leq\psi$ and let $\{\psi_k\}_{k\in\mathbb{N}}$ be a sequence in $\tilde{C}$ such that $\psi_k\to\psi$, $k\to\infty$ and $\psi_k>>\psi$, $k\in\mathbb{N}$ (cf. lemma 4.24b). We show that $T(t)\phi<<T(t)\psi_k$ for all $t>0$ and $k\in\mathbb{N}$. Suppose there is a $k\in\mathbb{N}$ for which this is not true, and let $t_0>0$ be the smallest t for which the strict inequality is not satisfied. Let $T(t)\phi = (P(t;\phi),n(t;\phi))$ and $T(t)\psi_k = (P(t;\psi_k), n(t;\psi_k))$. As in the proof of lemma 4.23 we can show that $P(t_0;\phi) = P(t_0;\psi_k)$. Since $P(t_0-h;\phi)<P(t_0-h;\psi_k)$, $0<h\leq t_0$ we conclude that

$$\dot{P}(t_0;\phi)\geq\dot{P}(t_0;\psi_k)\,.$$

This, however, implies that

$$2\int_0^1 b(x)n(t_0,x;\phi)dx \geq 2\int_0^1 b(x)n(t_0,x;\psi_k)dx\ ,$$

which is a contradiction. Now let $t>0$ be fixed. Then $T(t)\phi<<T(t)\psi_k$, $k\in\mathbb{N}$. Letting $k\to\infty$ and using the continuity of $T(t)$ we find

$$T(t)\phi\leq T(t)\psi$$

and the result follows. □

THEOREM 4.26.

a) *If $0<P<\hat{P}$ then $T(t)\Phi_P$ is increasing in t.*

b) *If $\hat{P}<P<\tilde{P}$ then $T(t)\Phi_P$ is decreasing in t.*

PROOF. We shall only prove a). The proof of b) proceeds along the same lines. Let $0<P<\hat{P}$, and let Q be such that $P<Q<\hat{P}$. Suppose we can show that

$$\Phi_P<<T(t)\Phi_Q,\ t>0\,.$$

Then, letting Q approach P, we obtain

$$\Phi_P\leq T(t)\Phi_P,\ t>0\ ,$$

and now the monotonicity of $T(t)$ gives

$$T(s)\Phi_P\leq T(s)T(t)\Phi_P = T(s + t)\Phi_P,\ s\geq 0,\ t\geq 0$$

which would imply the result. Therefore we shall prove that indeed $\Phi_P<<T(t)\Phi_Q$ for all $t>0$. Suppose not. Again let t_0 be the smallest t such that the strict inequality is not satisfied. As in the proof of lemma 4.23 we can show that

$$P = P(t_0;\Phi_Q)\,.$$

Here $T(t)\Phi_Q = (P(t;\Phi_Q),\ n(t;\Phi_Q))$. Since $P<P(t;\Phi_Q)$, $0\leq t<t_0$, we obtain

$$\dot P(t_0;\Phi_Q)\leq 0\,.$$

On the other hand

$$\dot P(t_0;\Phi_Q) = -\mu P - f(P) + 2\int_0^1 b(x)n(t_0,x,\Phi_Q)dx > -\mu P - f(P) + 2\int_0^1 b(x)f(P)\frac{E(x)}{V(x)}dx$$

$$= -\mu P - f(P) + r(0)f(P) > 0\,,$$

since $P<\hat P$. This is a contradiction, and the result is proved. □

4.6. Global stability of the nontrivial equilibrium

In this section we make again the assumptions 4.21 and 4.22. Let $\tilde P$ satisfy $\hat P<\tilde P<P_m$ and let the invariant bounded subset $\tilde C$ be given by (4.17). Suppose that the initial condition $\phi_0\neq 0$ is contained in $\tilde C$. From the remark following theorem 4.8 we obtain that there exists a $t_1>0$ such that

$$T(t_1)\phi_0 >> 0\,.$$

A straightforward calculation shows that there exist $\underline{P},\bar P$ such that

$$0<\underline{P}\leq\hat P\leq\bar P<\tilde P \text{ and } \Phi_{\underline{P}}\leq T(t_1)\phi_0\leq\Phi_{\bar P}\,.$$

Since $\{T(t)\Phi_{\underline{P}}\}_{t\geq 0}$ and $\{T(t)\Phi_{\bar P}\}_{t\geq 0}$ define a precompact increasing and decreasing net respectively, we may conclude that both nets converge to a limit which defines a fixed point of $T(t)$. But the only fixed point is $\hat\phi = (\hat P,\hat n)$ (see exercise 4.20) and therefore

$$\lim_{t\to\infty} T(t)\Phi_{\underline{P}} = \lim_{t\to\infty} T(t)\Phi_{\bar P} = \hat\phi\,.$$

We conclude from

$$T(t-t_1)\Phi_{\underline{P}}\leq T(t)\phi_0\leq T(t-t_1)\Phi_{\bar P}$$

that

$$\lim_{t\to\infty} T(t)\phi_0 = \hat\phi\,,$$

and we have proved the following result

THEOREM 4.27. *Let $\phi_0\in\tilde C\setminus\{0\}$, then $\lim_{t\to\infty} T(t)\phi_0 = \hat\phi$.*

We can prove our main result now.

THEOREM 4.28. *Let $\phi_0\in C$, $\phi_0\neq 0$, then $\lim_{t\to\infty} T(t)\phi_0 = \hat\phi$.*

PROOF. i) Suppose $P_m = \infty$ Let $\phi_0\in C$, $\phi_0\neq 0$. If $f(P)\to\infty$ as $P\to\infty$ then the proof follows from the fact that $\phi_0\in\tilde C$ if $\tilde P$ is large enough. If $f(P)\to f_\infty<\infty$ as $P\to\infty$, then we conclude from (4.8) that for $t>G(1)$ we have $T(t)\phi_0\in\tilde C$ if $\tilde P$ is large enough.
ii) Let $P_m<\infty$. Let $\phi_0\in C$, $\phi_0\neq 0$, and $(P(t),n(t)) = T(t)\phi_0$. Suppose $P(t)\geq P_m$ for all $t\geq t_0$ where $t_0>0$. Now let $(\rho,\nu)\in\Omega(\phi_0)$ (i.e. the omega limit set of ϕ_0, cf. section 1) then $\rho\geq P_m$, and $\dot{\mathcal{V}}(\rho,\nu) = (r(0)-1)f(\rho)-\mu\rho<0$ which is impossible.

We may conclude that there exists a $t_1\geq G(1)$ such that $P(t_1)<P_m$. Let $\tilde P$, $P(t_1)<\tilde P<P_m$ be such that

$$-\mu\hat P - f(\tilde P) + r(0)f(P_m)<0$$

(note that such a $\tilde P$ exists since $-\mu P_m + (r(0)-1)f(P_m)<0$). We show that $P(t)<\tilde P$ for all $t\geq t_1$. Suppose not. Let t_2 be the smallest value of t greater than t_1 such that $P(t_2) = \tilde P$. Then $\dot P(t_2)\geq 0$. On the other hand

$$\dot P(t_2) = -\mu\tilde P - f(\tilde P) + 2\int_0^1 b(x)n(t_2,x)dx \leq -\mu\tilde P - f(\tilde P) + 2\int_0^1 k(x)f(P_m)dx = -\mu\tilde P - f(\tilde P) + r(0)f(P_m)<0\,,$$

which is a contradiction. Therefore

$$P(t)<\tilde{P},\ t\geq t_1\ ,$$

and from (4.8) we conclude that

$$T(t)\phi_0 = (P(t),n(t))\in\tilde{C},\ t\geq t_1 + G(1)\ ,$$

where $\tilde{C}$ is given by (4.17). This proves the result. □

4.7. *Final remarks*

KIRK, ORR and FORREST (1970) present a model describing the control of the bone marrow stem cell population, which supplies the circulating blood population. They assumed that the production process was controlled by some stem cell specific mitotic inhibitor (one of the family of chalones), and that the mitotic phase was of constant duration. In terms of our G-M-model this last assumption is equivalent to the supposition that all cells divide at reaching maturity x = 1. Using our notation their model is described by

$$\frac{dP}{dt} = -\mu P - \tilde{\gamma}(c)P + 2V(1)n(t,1)\ , \tag{4.18}$$

$$\frac{\partial n}{\partial t} + \frac{\partial}{\partial x}(V(x)n(t,x)) = -\delta n(t,x)\ , \tag{4.19}$$

$$V(0)n(t,0) = \tilde{\gamma}(c)P\ , \tag{4.20}$$

$$\frac{dc}{dt} = \tilde{\rho}P - \tilde{\sigma}c\ . \tag{4.21}$$

Here c denotes the concentration of the mitotic inhibitor produced by G-cells at a rate $\tilde{\rho}$, and desintegrating at a rate $\tilde{\sigma}$. μP is due to loss from the G-phase via differentiation into the various channels. KIRK, ORR and FORREST (1970) assumed $\delta = 0$ (MACKEY (1978) calls this the 'normal' situation) and they 'solved' the system by using analogue computer techniques. This model was also studied by MACKEY (1978, 1981) and he calls it the *pluripotential stem cell model.* Mackey extensively examines the case that the dynamics of c is much faster than the dynamics of P and n, which can be modelled by assuming that $\tilde{\rho} = \rho/\epsilon$, $\tilde{\sigma} = \sigma/\epsilon$,, where ϵ is a small parameter. Keeping ρ and σ constant and letting ϵ tend to zero (4.21) can be replaced by

$$\rho P = \sigma c$$

(the so-called quasi-steady state situation). Now we may substitute in (4.18) and (4.20)

$$\tilde{\gamma}(c) = \tilde{\gamma}(\frac{\rho}{\sigma}P) = \gamma(P)\ ,$$

which happens to be the case that we considered. Additionally Mackey assumes that the transition probability γ depends on the total G-population in a decreasing manner, to be precise

$$\gamma(P) = \frac{\gamma_0\theta^q}{\theta^q + P^q}\ ,$$

for some integer q and positive constants γ_0 and θ.

EXERCISE 4.29. Assume the quasi-steady state situation (i.e. $\tilde{\gamma}(c) = \gamma(P)$) and let τ be the duration of the M-phase (i.e. $\tau = G(1)$). Let $P(t),n(t,x)$ be a solution of the system thus obtained and $N(t) = \int_0^1 n(t,x)$. Show that $P(t)$, $N(t)$ obey the delay-differential equations.

$$\frac{dP}{dt} = -\mu P - \gamma(P)P + 2\gamma(P_\tau)P_\tau e^{-\delta\tau}\ ,\ t>\tau\ , \tag{4.22}$$

$$\frac{dN}{dt} = -\delta N + \gamma(P)P - \gamma(P_\tau)P_\tau e^{-\delta\tau}\ ,\ t>\tau\ , \tag{4.23}$$

where $P_\tau(t) = P(t-\tau)$ and $N_\tau(t) = N(t-\tau)$.

It is easily seen that under some conditions on the parameters a unique non-trivial equilibrium $\hat{P}$, $\hat{N}$ of (4.22) - (4.23) exists. Numerical experiments suggest local stability of this equilibrium in some subset of the parameter space;

these experiments also indicate the occurrence of Hopf bifurcations (and ultimately the existence of solutions of a more 'complex' nature). As a matter of fact this local stability and the onset of Hopf bifurcation can be determined from the characteristic equation which can be obtained by linearizing around the equilibrium $\hat{P}$.

EXERCISE 4.30. Convince yourself that this characteristic equation takes the form

$$\lambda + A + Be^{-\lambda\tau} = 0 ,$$

where A,B are constants depending on the parameters.

The situation becomes completely different (as well in our $G-M$-model as in the variant described above) if the growthrate V depends on time, caused by environmental factors. Let us, as a simple example, consider (4.18) - (4.20) where $\tilde{\gamma}(c)$ is assumed to be a constant γ and where the growthrate V depends only on P (and not on x).

$$\frac{dP}{dt} = -\mu P - \gamma P + 2V(P)n(t,1) \tag{4.24}$$

$$\frac{\partial n}{\partial t} + V(P)\frac{\partial n}{\partial x} = -\delta n(t,x) \tag{4.25}$$

$$V(P)n(t,0) = \gamma P . \tag{4.26}$$

This system cannot be reduced to a two-dimensional system of delay-differential equations with fixed delay, because the duration of the M-phase depends on the moment of entry. Let us define $\tau(t)$ by

$$1 = \int_{t-\tau(t)}^{t} V(P(s))ds .$$

EXERCISE 4.31. Interpret $\tau(t)$ as the time which an individual reaching maturity 1 at time t has spent in the M-phase.

We can reduce (4.24) - (4.26) to the system

$$\frac{dP}{dt}(t) = -(\mu+\gamma)P(t) + 2\gamma e^{-\delta\tau(t)}P(t-\tau(t))$$

$$\frac{dN}{dt}(t) = -\delta N(t) + \gamma P(t) - \gamma e^{-\delta\tau(t)}P(t-\tau(t)) ,$$

$$\frac{d\tau}{dt}(t) = 1 - \frac{V(P(t))}{V(P(t-\tau(t)))} ,$$

where $N(t) = \int_0^1 n(t,x)dx$, i.e. a three-dimensional system of variable-delay differential equations, and such a reduction has been extensively investigated by NISBET and GURNEY (1983). See also the contribution by Gurney, Nisbet and Blythe to part B of these lecture notes.

EXERCISE 4.32. Show that the above reduction to the three-dimensional system of variable delay-differential equations is indeed correct.

5. Reduction to an ODE-system: a chemostat model for a cell population reproducing by unequal fission

In this section we consider a model for the growth of a cell population living in a chemostat (cf. section I.4) and reproducing by unequal fission. The main feature of the model is that it can be reduced to an ODE-system in the sense that the dynamics of substrate and biomass (represented by the total population) are governed by the classical two dimensional chemostat system (cf. HERBERT et al. (1956), WALTMAN (1983), and section I.4). The knowledge of the behaviour of solutions of this system makes it possible to characterize the omega limit sets of solutions of the full problem.

5.1. The model

As in section I.4 and chapter II it is assumed that a cell is fully characterized by its size x. However in the present model fission not necessarily results in equal parts, but the ratio p between the birth size of a daughter and the division size of her mother is a random variable described by a smooth probability density function $d(p)$, which does not depend on the division size of the mother: since size is conserved at division, d is symmetric around $p = \frac{1}{2}$. Moreover

$$\int_0^1 d(p)dp = 1 .$$

For more biological details we refer to Koch and Schaechter (1962).

Now we shall first write down the equations:

$$\frac{\partial n}{\partial t}(t,x) + \frac{\partial}{\partial x}(\beta(S(t))xn(t,x)) = -Dn(t,x)-b(x)n(t,x) + 2\int_0^1 \frac{d(p)}{p} b(\frac{x}{p})n(t,\frac{x}{p})dp , \tag{5.1}$$

$$\frac{dS}{dt}(t) = D(S^i - S(t)) - \alpha\beta(S(t))\int_{x_{\min}}^1 xn(t,x)dx . \tag{5.2}$$

See section III.3.3 for a derivation of the corresponding linear equation (see also Remark I.4.3.4, Exercise I.4.3.6 and Exercise II.11.3). Here $n(t,x)$ denotes the size distribution (per unit of volume), $S(t)$ is the concentration of the limiting nutrient in the chemostat (see section I.4.5 for a description of the chemostat device). D is the dilution rate, $b(x)$ is the division rate, S^i is the concentration of the limiting substrate in the inflowing fluid, α is a conversion factor describing the relation between size-units and substrate units ($1/\alpha$ is sometimes called the yield constant) and $x_{\min}$ is the minimum size. The growth rate V of the individuals depends on S, but is at every fixed time proportional to the individual's size

$$V = \beta(S(t))x .$$

Among others this means that we assume the 'Structural Nutrient Hypothesis': see DIEKMANN, LAUWERIER, ALDENBERG & METZ (1983) and § I.4.5. We make the following assumptions on β, d and b.

ASSUMPTIONS 5.1.

H_β: $\beta(S) = \dfrac{k_1 S}{1+k_2 S}$, where $k_1, k_2 > 0$.

H_d: $d(p)>0$, $p \in (\frac{1}{2}-\Delta, \frac{1}{2}+\Delta)$ and $d(p) = 0$ outside this interval. $d(p)$ is symmetric around $p = \frac{1}{2}$, $\int_0^1 d(p)dp = 1$ and d is continuously differentiable on $[\frac{1}{2}-\Delta, \frac{1}{2}+\Delta]$.

H_b: b is continuous on $[0,1)$, $b(x)>0$, $x \in (a,1)$ and $b(x) = 0, x \leq a$. Moreover $\lim_{x \uparrow 1} \int_a^x b(\xi)d\xi = \infty$, and the function $x \to \frac{b(x)}{\bar{\beta}x}\exp[-\int_a^x \frac{b(\xi)}{\bar{\beta}\xi}d\xi]$ is bounded, where $\bar{\beta} = \lim_{S\to\infty}\beta(S) = \frac{k_1}{k_2}$.

The function β in H_β is called the Monod-Michaelis-Menten function (cf. Remark I.4.5.2). H_β is only made for simplicity: essential is that $\beta(0) = 0$, β is increasing and $\lim_{S\to\infty}\beta(S)$ exists and is finite.

The last condition in H_b says that the function $\phi_b(x)$ given by (I.4.1.5) for $V(x) = \bar{\beta}x$ is bounded. Note that it follows from these assumptions that

$$x_{\min} = a\,(\tfrac{1}{2}-\Delta) ,$$

and we have to supply (5.1) - (5.2) with the boundary condition

$$n(t,x_{\min}) = 0 . \tag{5.3}$$

Additionally we impose the initial conditions

$$n(0,x) = n_0(x) \geq 0 \tag{5.4}$$

$$S(0) = S_0 \geq 0 . \tag{5.5}$$

In this section we shall try to avoid as many technical details as possible: an exhaustive investigation of the problem can be found in Heijmans (1984b).

5.2. The linear equation

It appears sensible to start with an investigation of the linear problem that is obtained if we assume that β does not depend on $S(t)$, but is a constant. The anålysis of this linear problem proceeds along the lines of chapter II, and this we consider a justification for omitting the technical details. Let

$$E(x) = \exp(-\int_a^x \frac{b(\xi)}{\beta\xi}\, d\xi) .$$

If we assume

$H_{0,\text{linear}}$: $\quad n_0(\cdot)\,/\,E(\cdot)$ belongs to $L^1[x_{\min},1]$,

(a very similar condition is imposed in chapter II, the main difference being that we considered continuous functions at that place) then we can prove that there exists a unique solution $n(t,x)$ of

$$\frac{\partial n}{\partial t}(t,x) + \frac{\partial}{\partial x}(\beta x n(t,x)) = -Dn(t,x) - b(x)n(t,x) + 2\int_{\frac{1}{2}-\Delta}^{\frac{1}{2}+\Delta} \frac{d(p)}{p}\, b(\frac{x}{p})n(t,\frac{x}{p})dp \tag{5.6}$$

$$n(t,x_{\min}) = 0 \tag{5.7}$$

$$n(0,x) = n_0(x) . \tag{5.8}$$

We can associate a strongly continuous linear semigroup T_β with the problem in the following way

$$T_\beta(t)n_0 = n(t),$$

where $n(t) = n(t,.)$ is the solution of (5.6) - (5.8) and the subindex β indicates the dependence on β. For a function $\phi \in L^1[x_{\min},1]$ we define

$$W[\phi] = \int_{x_{\min}}^{1} x\phi(x)dx . \tag{5.9}$$

This quantity can be interpreted as the biomass of a population whose size distribution is described by ϕ, and it plays a major role in the subsequent analysis.

EXERCISE 5.2. Let n_0 obey $H_{0,\text{linear}}$ and let $n(t) = T_\beta(t)n_0$, then $W(t) \overset{def}{=} W[n(t)]$ obeys the ordinary differential equation:

$$\frac{dW}{dt} = (\beta - D)W(t),\ W(0) = W[n_0].$$

Prove this.

It can be shown rigorously that the infinitesimal generator associated with (5.6) - (5.7) has a strictly dominant eigenvalue λ_d,

$$\lambda_d = \beta - D,$$

which is a pole of the resolvent and algebraically simple. The corresponding spectral projection is given by

$$P\phi = W[\phi]\cdot n_d , \tag{5.10}$$

where n_d is the positive eigenvector of A belonging to eigenvalue λ_d normalized by $W[n_d] = 1$. n_d is called the stable size distribution. Sometimes we shall write $n_d(\beta)$ to indicate the dependence on β.

EXERCISE 5.3. See whether you can find a relation between this result and the outcome of Exercise 5.2!

Now the following result holds:

THEOREM 5.4. *There exist constants* $\epsilon, M > 0$ *such that*

$$\|T_\beta(t)n_0 - W[n_0]e^{(\beta-D)t}n_d\| \leqslant Me^{(\beta-D-\epsilon)t}\|n_0\|,$$

for all $t \geqslant 0$ *and initial functions satisfying* $H_{0,\text{linear}}$.

Note that there exists a stable size distribution even though $V(2x) = 2V(x)$ for all x. This, of course, is due to the fact that d is not a delta function but smooth.

5.3. *An O.D.E. system related to the nonlinear problem.*

If we multiply (5.1) by x and integrate (as in Exercise 5.2) we obtain

$$\frac{dW}{dt}(t) = (\beta(S(t)) - D)W(t) \tag{5.11}$$

where now $W(t) = W(t;S_0, n_0) = W[n(t;S_0,n_0)]$, if a solution $S(t;S_0,n_0)$, $n(t;S_0,n_0)$ of (5.1) - (5.5) exists. Furthermore we can rewrite (5.2) as

$$\frac{dS}{dt}(t) = D(S^i - S(t)) - \alpha\beta(S(t))W(t)\,, \tag{5.12}$$

and we have obtained a two-dimensional O.D.E. system, called the Monod equations which has been extensively investigated in the literature (e.g. HERBERT et al. (1956), HSU et al. (1977)). For the following results we refer to this last reference. The system (5.11) - (5.12) supplied with $W(0) = W_0 \geq 0, S(0) = S_0 \geq 0$ has a unique bounded solution. Obviously (5.11) - (5.12) always has the trivial equilibrium $W = 0, S = S^i$. Let

$$D_{crit} = \frac{k_1 S^i}{1 + k_2 S^i}$$

where k_1, k_2 are defined by H_β.

EXERCISE 5.5. Show that there exists a unique nontrivial equilibrium $\hat{S} = \frac{D}{k_1 - k_2 D}$, $\hat{W} = \frac{1}{\alpha}(S^i - \frac{D}{k_1 - k_2 D})$ if $D < D_{crit}$.

Moreover it can be proved that this equilibrium is globally asymptotically stable if $D < D_{crit}$.

EXERCISE 5.6. Let $D < D_{crit}$. Let for $S, W > 0$, $\mathcal{V}$ be defined as:

$$\mathcal{V}(S,W) = (S - \hat{S} - \hat{S}\log\frac{S}{\hat{S}}) + c(W - \hat{W} - \hat{W}\log\frac{W}{\hat{W}})\,,$$

where

$$c = \alpha k / (k_1 - k_2 D).$$

Show that $\dot{\mathcal{V}}(S,W) = -\frac{(S - \hat{S})^2}{S(1 + k_2 S)}(-(k_1 - k_2 D)S^i - k_2 DS) \leq 0$ and that global stability of $\hat{S}$, $\hat{W}$ follows from the invariance principle (theorem 1.6).

5.4. *The nonlinear problem.*

Now let us return to our original nonlinear problem (5.1) - (5.5). The observation that the nonlinear function $\beta(S(\cdot))$ can be defined a priori as a function of time t makes the existence and uniqueness proof a relatively easy one. More precisely: it can be shown that for every initial pair (S_0, n_0) satisfying $S_0 \geq 0$, $n_0(x) \geq 0$ and some condition $H_{0,\text{nonlinear}}$, which we shall not write down explicitly, but which is the nonlinear analogue of $H_{0,\text{linear}}$, there exists a unique solution which we denote with $(S(t;S_0,n_0), n(t;S_0,n_0))$, or briefly $(S(t), n(t))$.

Before we proceed we shall give some definitions. Let X be the space of pairs (σ,ν) with norm $\|(\sigma,\nu)\|_X = |\sigma| + \|\nu\|$. Let C be the subset of X consisting of all pairs (σ,ν) such that $\sigma \geq 0$, $\nu \geq 0$ and (σ,ν) satisfies $H_{0,\text{nonlinear}}$.

EXERCISE 5.7. If $D \geq D_{crit}$ then the only equilibrium is the trivial equilibrium $(S^i, 0)$. If $D < D_{crit}$ then there exists a unique nontrivial equilibrium $(\hat{S}, \hat{n})$ where $\hat{n} = \hat{W} n_d(D)$ and $\hat{S}$, $\hat{W}$ are determined by Exercise 5.5.

Now we can state the main result of this section.

THEOREM 5.8. *Let $(S_0,n_0)\in C$, $(S_0,n_0)\neq(0,0)$ and let $(S(t),n(t))$ be the associated solution of* (5.1) - (5.5). *If $D\geqslant D_{crit}$ then* $\lim_{t\to\infty} S(t) = S^i$, $\lim_{t\to\infty} n(t) = 0$. *If $D<D_{crit}$ then* $\lim_{t\to\infty} S(t) = \hat{S}$, $\lim_{t\to\infty} n(t) = \hat{n}$.

The remainder of this section is concerned with a sketch of the proof of this global stability result. The missing details can be found in Heijmans (1984).

Let $D<D_{crit}$. We can associate a dynamical system with the problem in the standard way: $T(t)(S_0,n_0) = (S(t;S_0,n_0), n(t;S_0,n_0))$, for $(S_0,n_0)\in C$. Now $T(t)$ leaves C invariant. Let $(S_0,n_0)\in C\setminus\{(0,0)\}$ and let $\Gamma^+(S_0,n_0)$ be the orbit through (S_0,n_0). Then $\Gamma^+(S_0,n_0)$ is bounded and precompact. (The proof of this result is rather technical and we shall omit it.) Therefore the omega-limit set $\Omega(S_0,n_0)$ is non-empty, compact and invariant and, moreover, $(S(t), n(t))\to\Omega(S_0,n_0)$ as $t\to\infty$. It can be shown that $\Omega(S_0,n_0)\subseteq C$. Now let $(\Sigma,\nu)\in\Omega(S_0,n_0)$, then it must be that $\Sigma = \hat{S}$ and $W[\nu] = \hat{W}$. Note that the function $\mathcal{V}$ of Exercise 5.6 also defines a Lyapunov function for the original system (5.1) - (5.3). The invariance of $\Omega(S_0,n_0)$ yields that for all $t\geqslant 0$ there exists an element $(\Sigma^{-t},\nu^{-t})\in\Omega(S_0,n_0)$ such that $T(t)(\Sigma^{-t}, \nu^{-t}) = (\Sigma,\nu)$. Since $S(s;\Sigma^{-t},\nu^{-t}) = \hat{S}$ for all $s\geqslant 0$ we have that $n(s;\Sigma^{-t},\nu^{-t}) = T_D(s)\nu^{-t}$. In words: for initial pairs belonging to the omega-limit set the nonlinear problem reduces to the linear problem with $\beta(S(t))$ replaced by $\beta(\hat{S}) = D$. From theorem 5.4 we conclude that

$$\|T_D(s)\nu^{-t}-\hat{n}\| \leqslant Me^{-\epsilon s}\,\|\nu^{-t}\| \leqslant M'e^{-\epsilon s},\ s\geqslant 0\,,$$

where M' can be chosen independent of ν^{-t}, since $\Omega(S_0,n_0)$ is precompact. Substituting $s = t$ and using that $T_D(t)\nu^{-t} = \nu$ yields

$$\|\nu-\hat{n}\| \leqslant M'e^{-\epsilon t}\,,$$

and from the fact that this estimate is valid for all $t\geqslant 0$ we conclude that $\nu = \hat{n}$. Thus we have shown that

$$\Omega(S_0,n_0) = (\hat{S},\hat{n})\,,$$

and this proves the result.

EXERCISE 5.9. Let $D\geqslant D_{crit}$. Show that $\lim_{t\to\infty} (S(t),n(t)) = (S^i,0)$.

Hint: $\|n(t)\| = \int_{x_{min}}^{1} n(t,x)dx \leqslant \frac{1}{x_{min}} \int_{x_{min}}^{1} xn(t,x)dx$.

Observe that contrary to the so-called "Stochastic Threshold" model discussed in section I.4.5, a change of the dilution rate D (which is a control parameter) does cause a deformation of the shape of the stable size distribution $\hat{n}$.

6. Interaction through the environment: some open problems.

The Daphnia model from section I.3 and variants of the cell proliferation models from the preceeding section (see the remarks in subsection 4.7 and remark 6.5 below) have in common that the individuals interact with each other only indirectly: they all consume from a common resource pool. A similar situation arises if the individuals produce a toxic chemical substance which has a restraining effect on their growth or an inhibiting effect on mitosis.

So consider the system of differential equations

$$\frac{dn}{dt} = A(S)n \tag{6.1}$$

$$\frac{dS}{dt} = F(S,L[n])\,, \tag{6.2}$$

where n takes values in an infinite-dimensional space X and S in $\mathbb{R}$, and where for each positive S the unbounded operator $A(S)$ generates a strongly continuous semigroup of bounded *linear* operators on X, which leaves a cone X_+ invariant. L denotes a continuous linear functional on X and F a smooth function of two real variables with $F(0,y)\geqslant 0$ for all $y\geqslant 0$. The idea is that the dynamical system generated by (6.1) - (6.2) should leave the cone $X_+\times\mathbb{R}_+$ invariant.

REMARK 6.1. Useful generalizations are immediate: i) S may take values in $\mathbb{R}^m$; ii) $\frac{dS}{dt} = F(S,L(S)[n])$, where for each $S\in\mathbb{R}^m$, $L(S)$ is a continuous linear function from X into $\mathbb{R}^k$ and where the mapping $S\mapsto L(S)$ from $\mathbb{R}^m$ into

$\mathfrak{L}(X;\mathbb{R}^k)$ as well as $F:\mathbb{R}^m \times \mathbb{R}^k \to \mathbb{R}^m$ are smooth. But we don't want to complicate the formulation with technical details. Instead we present the simplest example which displays the essential features.

PROBLEM 6.2. Prove that the initial value problem for (6.1) - (6.2) is well-posed. That is, prove the existence and uniqueness of a solution for given initial data

$$n(0) = n_0 \in X_+,\ S(0) = S_0 \geqslant 0$$

and prove that the solution depends continuously on (n_0, S_0), uniformly for t in compact sets. Here it seems crucial to formulate suitable hypotheses about the dependence of A on S. The *unbounded* generator A should depend smoothly on S, but in what sense? Actually, we may only need that the corresponding semigroups of *bounded* operators depend smoothly on S. So the Trotter-Kato ideas (PAZY, 1983, III.4) suggest to make assumptions about the dependence of the resolvent of A on S. Note that when S is a *given* function of time we may consider (6.1) as a non-autonomous *linear* evolution equation, and therefore the existing theory for these (PAZY, 1983, V) may prove useful. The crux of the matter is to find hypotheses which can be *easily* checked in concrete applications.

Let us assume that for fixed S the semigroup generated by $A(S)$ has a strictly dominant real eigenvalue $\lambda_d(S)$. So if $(\hat{n}, \hat{S})$ is a nontrivial steady state then necessarily

$$\lambda_d(\hat{S}) = 0\,, \tag{6.3}$$

and

$$\hat{n} = k n_d(\hat{S})\,, \tag{6.4}$$

where $n_d(\hat{S})$ is the positive eigenvector corresponding to $\lambda_d(\hat{S})$, normalized in some convenient way. In many applications $\lambda_d(S)$ is a strictly monotone continuous function of S and in that case (6.3) has a unique solution $\hat{S}$ if $\lambda_d(0) < 0 < \lambda_d(\infty)$. Subsequently k has to be determined from the scalar equation

$$F(\hat{S}, kL[n_d(\hat{S})]) = 0\,. \tag{6.5}$$

One finds a unique solution if F is monotone in its second argument, which it usually is. Note that biological relevance requires that $k \geqslant 0$ (cf. I.3.5).

In order to linearize formally around an equilibrium $(\hat{n}, \hat{S})$ we put

$$n = \hat{n} + u \quad , \quad S = \hat{S} + \xi\,,$$

and obtain, upon neglecting higher order terms,

$$\frac{du}{dt} = A(\hat{S})u + \xi A'(\hat{S})\hat{n} \tag{6.6}$$

$$\frac{d\xi}{dt} = \frac{\partial F}{\partial S}\xi + \frac{\partial F}{\partial y}L[u]\,. \tag{6.7}$$

where both partial derivatives of F are evaluated at $(\hat{S}, L[\hat{n}])$. (Here we deliberately avoid to specify the sense in which the derivative $A'(\hat{S})$ is taken; so everything is purely formal and part of the problem will be to make this precise). For $\lambda \notin \sigma(A(\hat{S}))$ we can solve the first equation of the eigenvalue problem

$$(\lambda I - A(\hat{S}))u = \xi A'(\hat{S})\hat{n} \tag{6.8}$$

$$(\lambda - \frac{\partial F}{\partial S})\,\xi = \frac{\partial F}{\partial y}\,L[u]\,, \tag{6.9}$$

to obtain

$$u = \xi(\lambda I - A(\hat{S}))^{-1} A'(\hat{S})\hat{n}\,. \tag{6.10}$$

Upon substitution of this expression into (6.9) we find the *characteristic equation*

$$\lambda - \frac{\partial F}{\partial S} = \frac{\partial F}{\partial y} L((\lambda I - A(\hat{S}))^{-1} A'(\hat{S})\hat{n})\,. \tag{6.11}$$

Note that $\lambda \in \sigma(A(\hat{S}))$ cannot belong to the point spectrum corresponding to (6.8) - (6.9) if $A'(\hat{S})\hat{n} \notin R(\lambda I - A(\hat{S}))$.

PROBLEM 6.3. Prove the Principle of Linearized Stability. That is, show that the stability of an equilibrium $(\hat{n}, \hat{S})$ is determined by the position of the roots of (6.11) in the complex plane (giving due attention to the exceptional points $\lambda \in \sigma(A(\hat{S}))$).

Note that in general (6.1) - (6.2) is not semi-linear (i.e., it is not necessarily a bounded perturbation of a fixed unbounded generator) and hence the standard theory does not apply (in particular there is no immediate meaningful variation of constants formula). It is to be expected that the techniques which have to be developed for solving Problem 6.3 are useful as well (if not sufficient) for proving the Hopf bifurcation theorem for (6.1) - (6.2).

Of course it is a difficult problem in itself to determine whether all roots of (6.11) lie in the left half plane, or whether at least one root lies in the right half plane. Hence it is very important to develop analytical and numerical methods to decide about this question. Known results and approaches from the theory of delay equations (BELLMAN & COOKE 1963, COOKE 1985) could be taken as a starting point.

As argued in the preceding sections one can sometimes use monotonicity arguments to derive strong conclusions. Thus it is natural and important to pay attention to

PROBLEM 6.4. Find conditions on A,F and L which guarantee that the nonlinear semigroup corresponding to (6.1) - (6.2) is monotone (perhaps after restriction to a closed invariant subset).

Finally it would be advantageous to have a (partial) description of the special class of equations for which one can construct a Lyapunov function.

REMARK 6.5. Sometimes time scale arguments are used to replace (6.2) by the quasi-steady state equation

$$F(S,L[n]) = 0 .$$

If one can solve from this for S as a function of $L[n]$, substitution of the result into (6.1) leads to an S-independent problem. The cell proliferation model of section 4 is of this type. For this class of equations one can pose the same problems as above.

In the above we have concentrated on interaction through the environment which excludes (among other things) predator-prey and parasitoid-host interactions. Inclusion of such interactions leads to more complicated nonlinearities. For the corresponding class of equations (which is more difficult to describe abstractly in full generality) one has many open problems analogous to those described above. For age-dependent problems a reasonably complete mathematical theory is coming into existence. WEBB (1985a), PRÜSS (1983b) and SCHAPPACHER (in preparation) managed to prove the Principle of Linearized Stability, despite the fact that the problem is not semi-linear, but their proofs are still rather complicated and it will be very nice if a simple proof could be given. CUSHING (1983a) and PRÜSS (1983a) treat the Hopf bifurcation theorem (also see DIEKMANN & VAN GILS (1984) for those problems which can be reduced to Volterra convolution integral equations).

Notwithstanding that a lot of interesting and difficult mathematical problems arising in the context of structured populations have been solved, we can safely end this chapter, and in fact Part A of these lecture notes, by concluding that the field of *nonlinear* structured population models is largely a collection of open mathematical problems.

Note added in proof:
Very recently, Ph. CLÉMENT, O. DIEKMANN, M. GYLLENBERG, H.J.A.M. HEIJMANS and H.R. THIEME developed a perturbation theory for dual semigroups which yields a natural and very convenient framework for the local stability and bifurcation theory.

Bibliography

* : early papers which we found particularly stimulating.
m : mathematical survey paper or book.
b : biological survey paper or book.
mb : survey paper or book on mathematical biology.
s : paper dealing with one or more structured population models.
sm : s-paper with the stress on the mathematics.
sb : s-paper with the stress on the biology.

s ADAMS, E.D., ROTHMAN, E.D. and BERAN, K. (1981). *The age structure of populations of Saccharomyces cerevisiae.* Math. Biosc. **53:** 249-263.

s ALDENBERG, T. (1979). *The calculation of production, reproduction and total growth in the autonomous Van Sickle equation.* Hydrobiol. Bull. **13:** 3-12.

s ALDIS, G., TOGNETTI, K. and WINLEY, G. (1981). *The relationship between the age-position dependent integral formulation and the continuity equation for one dimensional growth.* Math. Biosc. **57:** 191-209.

s ANDERSON, E.C., BELL, G.I., PETERSEN, D.F. and TOBEY, R.A. (1969). *Cell growth and division IV. Determination of volume growth rate and division probability.* Biophys. J. **9:** 246-263.

s ANDERSON, E.C. and PETERSEN, D.F. (1967). *Cell growth and division II. Experimental studies of cell volume distribution in mammalian suspension cultures.* Biophys. J. **7:** 353-364.

s ARGENTESI, F., de BERNARDI R. and DI COLA, G. (1978a). *Some mathematical methods for the study of population dynamics.* Report III. 77. Istituto per le Applicazioni del Calcolo 'Mauro Picone', Roma.

s ARGENTESI, F., de BERNARDI, R. and DI COLA, G. (1978b). *Single species population dynamics.* Report III. 78 Istituto per le Applicazioni del Calcolo 'Mauro Picone', Roma.

s ARINO, O. and KIMMEL, M. (198?). *Asymptotic analysis of a cell-cycle model based on unequal division.*

m ARIS, R. (1978). *Mathematical Modelling Techniques.* Research Notes in Math. 24, Pitman, London.

m ARNOLD, L. (1974). *Stochastic Differential Equations.* Wiley, New York.

* AUSLANDER, D.M., OSTER, G.F. and HUFFAKER, C.B. (1974). *Dynamics of interacting populations.* J. Franklin Inst. **297:** 345-376.

m BALAKRISHNAN, A.V. (1976). *Applied Functional Analysis.* Springer, Berlin.

m BARDOS, C. (1970). *Problèmes aux limites pour les équations aux dérivées partielles du premier ordre.* Ann. Scient. Ec. Norm. Sup. 4e Serie t. **3:** 185-233.

s BEDDINGTON, J.R. and FREE, C.A. (1976). *Age structure effects in predator-prey interactions.* Theor. Pop. Biol. **9:** 15-24.

* BELL, G.I. (1968). *Cell growth and division. III. Conditions for balanced exponential growth in a mathematical model.* Biophys. J. **8:** 431-444.

* BELL, G.I. and ANDERSON, E.C. (1967). *Cell growth and division. I. A mathematical model with applications to cell volume distributions in mammalian suspension cultures.* Biophys. J. **7:** 329-351.

m BELLENI-MORANTE, A. (1979). *Applied Semigroups and Evolution Equations.* Clarendon Press, Oxford.

m BELLMAN, R. and COOKE, K.L. (1963). *Differential-Difference Equations.* Academic Press, New York.

BERAN, K., STREIBLOVA, E. and LIEBLOVA, J. (1969). *On the concept of the population of the yeast Saccharomyces cerevisiae.* In: Proceedings of the Second Symposium on Yeasts (A. Kochova-Kratochivilova, ed.). Publ. House Slovak. Acad. Sci., Bratislava, 353-363.

s BERNARDI, M.L., CAPELO, A.C. and PERITI, P. (1984). *A mathematical model for the evolution of cell populations under the action of mutagenic agents.* Math. Biosc. **71:** 19-39.

BERTALANFFY, L. von (1934). *Untersuchungen über die Gesetzlichkeit des Wachstums, I Teil Allgemeine Grundlagen der Theorie: mathematische und physiologische Gesetzlichkeiten der Wachstums bei Wassertieren.* Arch. Entwicklungsmech. Org. **131:** 613-652.

s BERTUZZI, A., GANDOLFI, A. and GIOVENCO, M.A. (1981). *Mathematical models of the cell cycle with a view to tumor studies.* Math. Biosc. **53:** 159-188.

sm BEYER, W.A. (1970). *Solution to a mathematical model of cell growth, division and death.* Math. Biosc. **6:** 431-

436.

sm DI BLASIO, G. (1979). *Nonlinear age-dependent population growth with history-dependent birth rate.* Math. Biosc. **46:** 279-291.

sm DI BLASIO, G., IANNELLI, M. and SINESTRARI, E. (1981). *An abstract partial differential equation with a boundary condition of renewal type.* Boll. U.M.I. Anal. Funz. e Appl. (V) XVIII-C, 259-274.

s DI BLASIO, G., IANNELLI, M. and SINESTRARI, E. (1982). *Approach to equilibrium in age structured populations with an increasing recruitment process.* J. Math. Biol. **13:** 371-382.

s BLYTHE, S.P., NISBET, R.M. and GURNEY, W.S.C. (1982). *Instability and complex dynamic behaviour in population models with long time delays.* Theor. Pop. Biol. **22:** 147-176.

s BLYTHE, S.P., NISBET, R.M. and GURNEY, W.S.C. (1983). *Formulating population models with differential aging.* In: Population Biology (H.I. Freedman & C. Strobeck, eds.). Springer Lect. Notes in Biomath. **52:** 133-140.

s BLYTHE, S.P., NISBET, R.M. and GURNEY, W.S.C. (1984). *The dynamics of population models with distributed maturation periods.* Theor. Pop. Biol. **25:** 289-311.

BLYTHE, S.P., NISBET, R.M., GURNEY, W.S.C. and MACDONALD, N. (1985). *Stability switches in distributed delay models.* J. Math. Anal. Appl. **109:** 388-396.

s BOTSFORD, L.W. (1981a). *The effects of increased individual growth rates on depressed population size.* Amer. Nat. **117:** 38-63.

s BOTSFORD, L.W. (1981b). *Optimal fishery policy for size-specific, density-dependent population models.* J. Math. Biol. **12:** 265-293.

s BOTSFORD, L.W. and WICKHAM, D.E. (1978). *Behavior of age-specific density-dependent models and the Northern California Dungeness Crab (Cancer magister) Fishery,* J. Fish. Res. Board Can. **35:** 833-843.

s BRAUER, F. (1983). *Nonlinear age-dependent population growth under harvesting.* Int. J. Comput. Math. Appl. **9:** 345-352.

s BRITTON, N.F. and WHEALS, A.E. (preprint). *Mathematical models for a G_0 phase in* Saccharomyces cerevisiae

m BROCKETT, R.W. (1970). *Finite Dimensional Linear Systems.* Wiley, New York.

sm BROKATE, M. (1985). *Pontryagin's principle for control problems in age-dependent population dynamics.* J. Math. Biol. **23:** 75-101.

sm BRUNOVSKÝ, P. (1983). *Notes on chaos in the cell population partial differential equation.* Nonlinear Analysis. Theory, Methods & Applications **7:** 167-176.

sm BRUNOVSKÝ, P. and KOMORNÍK, J. (1984). *Ergodicity and exactness of the shift on $C[0,\infty)$ and the semiflow of a first order partial differential equation.* J. Math. Anal. Appl. **104:** 235-245.

sb BURDETT, I.D.J. and KIRKWOOD, T.B.L. (1983). *How does a bacterium grow during its cell cycle?* J. Theor. Biol. **103:** 11-20.

s BUSENBERG, S.N. and COOKE, K.L. (1980). *The effect of integral conditions in certain equations modelling epidemics and population growth.* J. Math. Biol. **10:** 13-32.

s BUSENBERG, S.N. and COOKE, K.L. (eds). (1981). *Differential Equations and Applications in Ecology, Epidemics and Population Problems.* Academic Press. New York.

sm BUSENBERG, S.N. and IANNELLI, M. (1983). *A class of nonlinear diffusion problems in age-dependent population dynamics.* Nonlinear Analysis, Theory, Methods & Applications **7:** 501-529.

s BUSENBERG, S.N. and IANNELLI, M. (1985). *Separable models in age-dependent populations dynamics.* J. Math. Biol. **22:** 145-173.

m BUTZER, P.L. and BERENS, H. (1967). *Semi-Groups of Operators and Approximation.* Springer, Berlin.

CALLEJA, G.B., ZUKER, M. and JOHNSON, B.F. (1980). *Analysis of fission scars as permanent records of cell division in Schizosaccharomyces pombe* J. Theor. Biol. **84:** 523-544.

s CAPASSO, V., GROSSO, E. and PAVERI-FONTANA, S.L. (eds.) (1985). *Mathematics in Biology and Medicine.* Springer Lect. Notes in Biomath. **57.**

sm CASTILLO CHÁVEZ, C. (preprint). *Nonlinear character-dependent models with constant time delay in population dynamics.*

sm CASTILLO CHÁVEZ, C. (preprint). *Linear character-dependent models with constant time delay in population dynamics.*

s CHARLESWORTH, B. (1980). *Evolution in Age-Structured Populations.* Cambridge Univ. Press, Cambridge.

s CHIANG, A.S. and THOMPSON, R.W. (1981). *The age distribution from continuous biochemical reactors with cell reproduction by mitosis.* J. Theor. Biol. **89:** 321-333.

sm CHIPOT, M. (1983). *On the equations of age-dependent population dynamics.* Arch. Rat. Mech. Anal. **82:** 13-26.

s CHIPOT, M. and EDELSTEIN, L. (1983). *A mathematical theory of size distributions in tissue culture.* J. Math. Biol. **16:** 115-130.

CHOW, S-N., DIEKMANN, O. and MALLET-PARET, J. (1985). *Stability, multiplicity and global continuation of symmetric periodic solutions of a nonlinear Volterra integral equation.* Japan J. Appl. Math. **2:** 433-469.

m CHOW, S-N. and HALE, J.K. (1982). *Methods of Bifurcation Theory.* Springer, New York.

s COALE, A.J. (1972). *The Growth and Structure of Human Populations, a Mathematical Investigation.* Princeton University Press.

s COFFMAN, C.V. and B.D. COLEMAN (1978, 1979). *On the growth of populations with narrow spread in reproductive age.* II. Conditions of convexity, J. Math. Biol. **6:** 285-303. III. Periodic variations in the environment, J. Math. Biol. **7:** 281-301.

* COLE, L.C. (1954). *The population consequences of life history phenomena.* Quart. Rev. Biol. **29:** 103-137.

s COLEMAN, B.D. (1978). *On the growth of populations with narrow spread in reproductive age.* I. General theory and examples. J. Math. Biol. **6:** 1-19.

s COLEMAN, C.S. and FRAUENTHAL, J.C. (1983). *Satiable egg eating predators.* Math. Biosc. **63:** 99-119.

COOKE, K.L. (1985). *Stability of delay differential equations with applications in biology and medicine.* In: Mathematics in Biology and Medicine (V. Capasso, E. Grosso and S.L. Paveri-Fontana, eds.) Springer Lect. Notes in Biomath. **57:** 439-446.

COOKE, K.L. and GROSSMAN, Z. (1982). *Discrete delay, distributed delay and stability switches.* J. Math. Anal. Appl. **86:** 592-627.

s COOKE, K.L. and YORKE, J.A. (1973). *Some equations modelling growth processes and gonorrhea epidemics.* Math. Biosc. **16:** 75-101.

Cornish-Bowden, A. (1979). *Fundamentals of Enzyme Kinetics.* Butterworth, London.

m COURANT, R. and HILBERT, D. (1962). *Methods of Mathematical Physics.* Interscience, New York.

m COX, D.R. (1962). *Renewal Theory.* Methuen, London.

m COX, D.R. and MILLER, H.D. (1965). *The Theory of Stochastic Processes.* Methuen, London.

s Curry G.L. et al. (1981). *Approximating a closed-form solution for cotton fruiting dynamics.* Math. Biosc. **54:** 91-113.

m CURTAIN, R.F. and PRITCHARD, A.J. (1977). *Functional Analysis in Modern Applied Mathematics.* Academic Press, London.

s CUSHING, J.M. (1977). *Integro-differential Equations and Delay Models in Population Dynamics.* Springer Lect. Notes in Biomath. **20.**

s CUSHING, J.M. (1980). *Model stability and instability in age structured populations.* J. Theor. Biol. **86:** 709-730.

s CUSHING, J.M. (1981). *Stability and maturation periods in age structured populations.* In: Differential Equations and Applications in Ecology, Epidemics, and Population Problems. (S.N. Busenberg and K.L. Cooke eds.). Academic Press, New York.

s CUSHING, J.M. (1983). *Bifurcation of time periodic solutions of the McKendrick equations with applications to population dynamics.* Comp. & Maths. with Appls. **9:** 459-478.

s CUSHING, J.M. (1984). *Existence and stability of equilibria in age-structured population dynamics.* J. Math. Biol. **20:** 259-276.

s CUSHING, J.M. (1985a). *Global branches of equilibrium solutions of the McKendrick equations for age-structured population growth.* Comp. & Math. with Appl. **11:** 459-478.

s Cushing, J.M. (1985b). *Equilibria in structured populations. J. Math. Biol.* **23:** 15-39.

s CUSHING, J.M. and SALEEM, M. (1982). *A predator prey model with age-structure.* J. Math. Biol. **14:** 231-250. (Erratum **16** (1983) 305) mb D'ANCONA, U. (1954). *The Struggle for Existence.* Brill, Leiden.

m DAVIES, E.B. (1980). *One-Parameter Semigroups.* Academic Press, London.

s DEANGELIS, D.L. and COUTANT, C.C. (1982). *Genesis of bimodal size distributions in species cohorts.* Trans. Amer. Fish. Soc. **111:** 384-388.

s DEANGELIS, D.L., COX, D.K. and COUTANT, C.C. (1979). *Cannibalism and size dispersal in young-of-the-year large mouth bass: experiment and model.* Ecological Modelling **8:** 133-148.

s DEANGELIS, D.L., HACKNEY, P.A. and WEBB, J.C. (1980). *A partial differential equation model of changing sizes and numbers in a cohort of juvenile fish.* Env. Biol. Fis. **5:** 261-266.

s DEANGELIS, D.L. and MATTICE, J.S. (1979). *Implications of a partial differential equation cohort model.* Math. Biosc. **47:** 271-285.

sm DEIMLING, K. (1981) *Equilibria of an age-dependent population model.* In: Nonlinear Differential Equations: Invariance, Stability and Bifurcation (P. de Mottoni & L. Salvadori, eds). Academic Press, New York.

s DEMETRIUS, L. (1983). *Statistical mechanics and population biology.* J. Stat. Phys. **30:** 709-753.

sm DIEKMANN, O. (1977). *Limiting behaviour in an epidemic model.* Nonlinear Analysis, Theory, Methods and Applications **1:** 459-470.

DIEKMANN, O. (1978). *Thresholds and travelling waves for the geographical spread of infection.* J. Math. Biol. **6:** 109-130.

DIEKMANN, O. (1979). *Integral equations and population dynamics.* In: Numerical treatment of integral equations (H.J.J. te Riele, ed.) CWI Syllabus **41:** 117-149.

DIEKMANN, O. (1980). *Volterra integral equations and semigroups of operators.* MC Report TW 197.

DIEKMANN, O. (1982). *A duality principle for delay equations.* In: Equadiff 5 (M. Gregas, ed.). Teubner Texte zur Math. **47:** 84-86.

s DIEKMANN, O. (1985). *The dynamics of structured populations: some examples.* In: Mathematics in Biology and Medicin (V. Capasso, E. Grosso and S.L. Paveri-Fontana eds.). Lect. Notes in Biomath. **57:** 7-18.

DIEKMANN, O. and GILS, S.A. VAN (1984). *Invariant manifolds for Volterra integral equations of convolution type.* J. Diff. Equ. **54:** 139-180.

s DIEKMANN, O., HEIJMANS, H.J.A.M. and THIEME, H.R. (1984). *On the stability of the cell size distribution.* J. Math. Biol. **19:** 227-248. Part II: *Time periodic developmental rates.* To appear in Comp. & Maths. with Appls.

DIEKMANN, O. and KAPER, H.G. (1978). *On the bounded solutions of a nonlinear convolution equation.* Nonlinear Analysis, Theory, Methods & Applications **2:** 721-737.

s DIEKMANN, O., LAUWERIER, H.A., ALDENBERG, T. and METZ, J.A.J. (1983). *Growth, fission and the stable size distribution.* J. Math. Biol. **18:** 135-148.

s DIEKMANN, O., METZ, J.A.J., KOOIJMAN, S.A.L.M. and HEIJMANS, H.J.A.M. (1984). *Continuum population dynamics with an application to Daphnia magna.* Nieuw Archief voor Wiskunde (4) **2:** 82-109.

DIEKMANN, O. and MONTIJN, R. (1982). *Prelude to Hopf bifurcation in an epidemic model: analysis of a characteristic equation associated with a nonlinear Volterra integral equation.* J. Math. Biol. **14:** 117-127.

s DIEKMANN, O., NISBET, R.M., GURNEY, W.S.C. and BOSCH, F. VAN DEN (to appear). *Simple mathematical models for cannibalism: a critique and a new approach.* Math. Biosc.

s DIETZ, K. and D. SCHENZLE (1985). *Proportionate mixing models for age-dependent infection transmission.* J. Math. Biol. **22:** 117-120.

m DOETSCH, G. (1937). *Theorie und Anwendung der Laplace Transformation.* Springer, Berlin.

m DOETSCH, G. (1950,1955,1956). *Handbuch der Laplace-Transformation* I. Theorie der Laplace-Transformation, II, III. Anwendung der Laplace-Transformation, 1,2. Abt. Birkhäuser, Basel.

DOUCET, P.G. and STRAALEN VAN, N.M. (1980). *Analysis of hunger from feeding rate observations.* Anim. Behav. **28:** 913-921.

m DUNFORD, N. and SCHWARTZ, J.T. (1958). *Linear Operators.* Part I: General Theory. Interscience, New York.

s EDELSTEIN, L. (to appear) *Models for plant-herbivore systems.* J. Math. Biol.

s EDELSTEIN, L. and HADAR, Y. (1983). *A model for pellet size distributions in submerged mycelial cultures.* J. Theor. Biol. **105:** 427-452.

mb EISEN, M. (1979). *Mathematical Models in Cell Biology and Cancer Chemotherapy.* Springer Lecture Notes in Biomathematics 30, Berlin.

s ELDERKIN, R.H. (1982). *Seed dispersal in a patchy environment with global age dependence.* J. Math. Biol. **13:** 283-303.

m ERDÉLYI, A., MAGNUS, W., OBERHETTINGER, F. and TRICOMI, F.G. (1954). *Tables of integral transforms.* II Vols. McGraw-Hill, New York.

ERICKSON, R.V. (1970). *Functions of Markov chains.* Ann. Math. Statist. **41:** 843-850.

m ERWE, F. and PESCHL E. (1972). *Partielle Differentialgleichungen erster Ordnung.* Bibliographisches Institut, Mannheim.

mb EWENS, W.J. (1969). *Population Genetics.* Methuen, London.

mb EWENS, W.J. (1979). *Mathematical Population Genetics.* Springer Verlag, Berlin.

m FATTORINI, H.O. (1983). *The Cauchy Problem.* Addison-Wesley, London.

s FELDMANN, U. (1979). *Wachstumskinetik.* Springer Medizinische Informatik und Statistik **11.**

FELLER, W. (1941). *On the integral equation of renewal theory.* Ann. Math. Stat. **12:** 243-267.

m FELLER, W. (1966). *An Introduction to Probability Theory and Its Applications.* Vol. II, Wiley, New York.

FIEDLER, B. (preprint) *Global Hopf bifurcation for Volterra integral equations.*

* FISHER, R.A. (1958). *The Genetical Theory of Natural Selection.* Dover, New York.

m FLEMING, W.H. (1977). *Functions of Several Variables.* 2nd edition. Springer, New York.

b FOX, L.R. (1975). *Cannibalism in natural populations.* Ann. Rev. Ecol. Systems **6:** 87-106.

sb FRANSZ, H.G. (1974). *The functional response to prey density in an acarine system.* PUDOC, Wageningen.

s FRAUENTHAL, J.C. (1983). *Some simple models of cannibalism.* Math. Biosc. **63:** 87-98.

s FREDRICKSON, A.G. (1971). *A mathematical theory of age structure in sexual populations: random mating and monogamous marriage models.* Math. Biosc. **10:** 117-143.

* FREDRICKSON, A.G., RAMKRISHNA, D. and TSUCHIYA, H.M. (1967). *Statistics and dynamics of procaryotic cell populations.* Math. Biosc. **1:** 327-374.

s FRIED, J. (1973). *A mathematical model of proliferating cell populations: further development and consideration of the resting state.* Math. Biosc. **18:** 397-408.

m FRIEDMAN, A. and SHINBROT, M. (1967). *Volterra integral equations in Banach space.* Trans. Am. Math. Soc. **126:** 131-179.

FUJII, K. (1978). *Computer simulation studies of the cyclities of Tribolium population dynamics.* Res. Popul. Ecol. **19:** 155-169.

s FUNAKOSHI, H. and YAMADA, A. (1980). *Transition phenomena in bacterial growth between logarithmic and stationary phases.* J. Math. Biol. **9:** 369-387.

s GAGE, T.B., WILLIAMS, F.M. and HORTON, J.B. (1984). *Division synchrony and the dynamics of microbial populations: a size-specific model.* Theor. Pop. Biol. **26:** 296-314.

sm GAJEWSKI, H. and ZACHARIAS, K. (1982). *On an initial-value problem for a transport equation in polymer chemistry.* Math. Nachr. **109:** 135-156.

GAUSE, G.F. (1934, 1971). *The Struggle for Existence.* Williams & Wilkins, New York; Dover, New York.

m GELFAND, I.M. and SHILOV, G.E. (1964). *Generalized Functions.* Acad. Press, New York.

GILS, S.A. VAN (1984). *Some studies in dynamical system theory:* I *Volterra integral equations of convolution type,* II *Hopf bifurcation and symmetry.* Thesis, Technical University of Delft.

mb GOEL, N.S. and RICHTER-DYN, N. (1979). *Stochastic Models in Biology.* Acad. Press, New York.

m GOLUBITSKY, M. and SCHAEFFER, D. (1984). *Singularities and Groups in Bifurcation Theory.* Vol. I. Springer, New York.

s GOPALSAMY, K. (1978). *Dynamics of maturing populations and their asymptotic behaviour.* J. Math. Biol. **5:** 383-398.

s GOPALSAMY, K. (1982). *Age specific coexistence in two species competition.* Math. Biosc. **61:** 101-122.

GREINER, G. (1981). *Zur Perron-Frobenius-Theorie stark stetiger Halbgruppen.* Math. Z. **177:** 401-423.

GREINER, G. (1984). *A typical Perron-Frobenius theorem with applications to an age-dependent population equation.* In: Infinite-dimensional Systems (F. Kappel and W. Schappacher, eds.). Springer Lect. Notes in Math. **1076:** 86-100.

GREINER, G. (to appear). *Perturbing the boundary conditions of a generator.* Houston J. Math.

GREINER, G., VOIGT, J. and WOLFF, M. (1981). *On the spectral bound of the generator of semigroups of positive operators.* J. Operator Th. **5:** 245-256.

s GRIFFEL, D.H. (1976). *Age dependent population growth.* IMA J. Appl. Math. **17:** 141-152.

sm GRIPENBERG, G. (1980). *Periodic solutions of an epidemic model.* J. Math. Biol. **10:** 271-280.

sm GRIPENBERG, G. (1981). *On some epidemic models.* Quart. Appl. Math. **39:** 317-327.

s GRIPENBERG, G. (1983a). *Stability analysis of a distributed parameter model for the growth of micro-organisms.* Comp. & Maths. with Appls. **9:** 431-442.

s GRIPENBERG, G. (1983b). *A stationary distribution for the growth of a population subject to random catastrophes.* J. Math. Biol. **17:** 371-379.

sm GRIPENBERG, G. (1983c). *An estimate for the solution of a Volterra equation describing an epidemic.* Nonlinear Analysis, Theory, Methods & Applications **7:** 161-165.

sm GRIPENBERG, G. (1983d). *On a nonlinear integral equation modelling an epidemic in an age-structured population.* J. reine u. angew. Math. **341:** 54-56.

sm GRIPENBERG, G. (1983e). *Stability of periodic solutions of some integral equations.* J. reine u. angew. Math. **331:** 16-31.

m GUCKENHEIMER, J. and HOLMES, Ph. (1983). *Nonlinear Oscillations, Dynamical Systems and Bifurcation of Vector Fields.* Springer, New York.

s GURNEY, W.S.C. and NISBET, R.M. (1980). *Age- and density-dependent population dynamics in static and variable environments.* Theor. Pop. Biol. **17:** 321-344.

s GURNEY, W.S.C. and NISBET, R.M. (1983). *The systematic formulation of delay-differential models of age or size structured populations.* In: Population Biology (H.I. Freedman & C. Strobeck, eds.). Springer Lect. Notes in Biomath. **52:** 163-172.

s GURNEY, W.S.C. and NISBET, R.M. (1985). *Fluctuation periodicity, generation separation and the expression of larval competition.* Theor. Pop. Biol. **28:** 150-180.

s GURNEY, W.S.C., NISBET, R.M. and LAWTON, J.H. (1983). *The systematic formulation of tractable single species population models incorporating age structure.* J. Anim. Ecol. **52:** 479-495.

s GURTIN, M.E. (1980, second draft 1982). *The Mathematical Theory of Age-Structured Populations.* Manuscript.

sm GURTIN, M.E. (1983). *Some questions and open problems in continuum mechanics and population dynamics.* J. Diff. Equ. **48:** 293-312.

s GURTIN, M.E. and LEVINE, D.S. (1979). *On predator-prey interactions with predation dependent on age of prey.* Math. Biosc. **47:** 207-219.

s GURTIN, M.E. and LEVINE, D.S. (1982). *On populations that cannibalize their young.* SIAM J. Appl. Math. **42:** 94-108.

* GURTIN, M.E. and MACCAMY, R.C. (1974). *Nonlinear age-dependent population dynamics.* Arch. Rat. Mech. Anal. **54:** 281-300.

s GURTIN, M.E. and MACCAMY, R.C. (1979a). *Some simple models for nonlinear age-dependent population dynamics.* Math. Biosc. **43:** 199-211, and 213-237.

s GURTIN, M.E. and MACCAMY, R.C. (1979b). *Population dynamics with age dependence.* In: Nonlinear Analysis and Mechanics. Herriot-Watt Symposium III. Pitman, Boston.

s GURTIN, M.E. and MACCAMY, R.C. (1981). *Diffusion models for age-structured populations.* Math. Biosc. **54:** 49-59.

s GURTIN, M.E. and MURPHY, L.F. (1981a). *On the optimal harvesting of age-structured populations: some simple models.* Math. Biosc. **55:** 115-136.

s GURTIN, M.E. and MURPHY, L.F. (1981b). *On the optimal harvesting of persistent age-structured populations.* J. Math. Biol. **13:** 131-148.

s GYLLENBERG, M. (1982). *Nonlinear age-dependent population dynamics in continuously propagated bacterial cultures.* Math. Biosc. **62:** 45-74.

s GYLLENBERG, M. (1983). *Stability of a nonlinear age-dependent population model containing a control variable.* SIAM J. Appl. Math. **43:** 1418-1438.

s GYLLENBERG, M. (1985a). *The age structure of populations of cells reproducing by asymmetric division.* In: Mathematics in Biology and Medicine. (V. Capasso, E. Grosso and S.L. Paveri-Fontana, eds.) Springer Lect. Notes in Biomath. **57:** 320-327

s GYLLENBERG, M. (1985b). *The size and scar distributions of the yeast Saccharomyces cerevisiae,* Centre for Mathematics and Computer Science Report AM-R8509, Amsterdam.

s GYLLENBERG, M. and HEIJMANS, H.J.A.M. (to appear). *An abstract delay-differential equation modelling size-dependent cell growth and division.* SIAM J. Math. Anal.

s HADELER, K.P. (198?). *Reduction of integral equations and Lotka age structure models to ordinary differential equations.*

s HADELER, K.P. and DIETZ, K. (1984). *Population dynamics of killing parasites which reproduce in the host.* J. Math. Biol. **21:** 45-66.

s HAIMOVICI, A. (1979a). *On the growth of a population dependent on ages and involving resources and pollution.* Math. Biosc. **43:** 213-237.

s HAIMOVICI A. (1979b). *On the age dependent growth of two interacting populations.* Boll. Un. Mat. Ital. **15:** 405-429.

m HALE, J.K. (1969). *Ordinary Differential Equations.* Wiley, New York.

m HALE, J.K. (1977). *Theory of Functional Differential Equations.* Springer, Berlin.

m HALE, J.K., MAGALHAES, L.T. and OLIVA, W.M. (1984). *An Introduction to Infinite Dimensional Dynamical Systems-Geometric Theory.* Springer, New York.

HAMADA, T. (1982). *Stationary scar-class structure of populations of Schizosaccharomyces pombe: letter to the editor.* J. Theor. Biol. **99:** 835-838.

HAMADA, T., KANNO, S. and KANO, E. (1982). *Stationary stage structure of yeast populations with stage-dependent generation time.* J. Theor. Biol. **97:** 393-414.

HAMADA, T. and NAKAMURA, Y. (1982). *On the oscillatory transient stage structure of yeast populations.* J. Theor. Biol. **99:** 797-805.

s HANNSGEN, K.B. and TYSON, J.J. (1985). *Stability of the steady-state size distribution in a model of cell growth and division.* J. Math. Biol. **22:** 293-301.

s HANNSGEN, K.B., TYSON, J.J. and WATSON, L.T. (1985). *Steady-state size distributions in probabilistic models of the cell division cycle.* SIAM J. Appl. Math. **45:** 523-540.

s HARA, T. (1984). *A stochastic model and the moment dynamics of the growth and size distribution in plant populations.* J. Theor. Biol. **109:** 173-190.

m HARTMAN, P. (1964). *Ordinary Differential Equations.* Wiley, New York.

s HASTINGS, A. (1977). *Spatial heterogeneity and the stability of predator-prey systems.* Theor. Pop. Biol. **12:** 37-48.

s HASTINGS, A. (1978). *Spatial heterogeneity and the stability of predator-prey systems: predator-mediated coexistence.* Theor. Pop. Biol. **14:** 380-395.

s HASTINGS, A. (1983). *Age-dependent predation is not a simple process.* I. Continuous time models. Theor. Pop. Biol. **23:** 347-362.

s HASTINGS, A. (1984a). *Age-dependent predation is not a simple process. II. Wolves, Ungulates, and a discrete time model for predation on juveniles with a stabilizing tail.* Theor. Pop. Biol. **26:** 271-282.

s HASTINGS, A. (1984b). *Simple models for age-dependent predation.* In: Mathematical Ecology (S.A. Levin & T.G. Hallam, eds.). Springer Lect. Notes in Biomath. **54:** 114-119.

s HASTINGS, A. (1984c). *Delays in recruitment at different trophic levels: Effects on stability.* J. Math. Biol. **21:** 35-44.

s HASTINGS, A. and WOLLKIND, D. (1982). *Age structure in predator-prey systems. A general model and a specific example.* Theor. Pop. Biol. **21:** 44-56.

sm HEIJMANS, H.J.A.M. (1984a). *Structured populations, linear semigroups and positivity.* CWI Report AM-R8417. To appear in Math. Zeitschr.

s HEIJMANS, H.J.A. M. (1984b). *On the stable size distribution of populations reproducing by fission into two unequal parts.* Math. Biosc. **72:** 19-50.

sm HEIJMANS, H.J.A.M. (1984c). *Holling's 'hungry mantid' model for the invertebrate functional response considered as a Markov process.* Part III: Stable satiation distribution. J. Math. Biol. **21:** 115-143.

s HEIJMANS, H.J.A.M. (1985a). *Dynamics of Structured Populations.* Ph.D. Thesis, Amsterdam.

sm HEIJMANS, H.J.A.M. (1985b). *An eigenvalue problem related to cell growth.* J. Math. Anal. Appl. **111:** 253-280.

sm HEIJMANS, H.J.A.M. (1986). *Markov semigroups and structured population dynamics.* to appear in proceedings of the symposium "Aspects of positivity in Functional Analysis", Tübingen, 1985.

sm HEIJMANS, H.J.A.M. and METZ, J.A.J. (in prep.). *Small parameters in structured population models and the Trotter-Kato theorem.*

HEINEKEN, F.G., TSUCHIYA, H.M. and ARIS, R. (1967). *On the mathematical status of the pseudo-steady state hypothesis of biochemical kinetics.* Math. Biosc. **1:** 95-114.

HERBERT, D. ELSWORTH, R. and TELLING, R.C. (1956). *The continuous culture of bacteria, a theoretical and experimental study.* J. Gen. Microbiol. **14:** 601-622.

m HILLE, E. and PHILLIPS, R.S. (1957). *Functional Analysis and Semigroups.* Amer. Math. Soc. Coll. Publ., Providence.

m HIRSCH, M.W. (1984a). *The dynamical systems approach to differential equations.* Bull. AMS **11:** 1-64.

HIRSCH, M.W. (preprint). *Stability and convergence in strongly monotone flows.*

HIRSCH, M.W. (1984b). *Differential equations and convergence almost everywhere in strongly monotone flows.* Contemporary Math. **17,** Amer.Math. Soc., Providence: 267-285.

m HIRSCH, M.W. and SMALE, S. (1974). *Differential Equations, Dynamical Systems, and Linear Algebra.* Academic Press, New York.

HJORTSO, M.A. and BAILEY, J.E. (1983). *Transient responses of budding yeast populations.* Math. Biosc. **63:** 121-148.

b HOLLING, C.S. (1959). *Some characteristics of simple types of predation and parasitism.* Canad. Entomol. **91:** 385-398.

* HOLLING, C.S. (1966). *The functional response of invertebrate predators to prey density.* Mem. Ent. Soc. Canada **48.**

HOLLING, C.S. (1973). *Resilience and stability of ecological systems.* Ann. Rev. Ecol. Syst. **4:** 1-23.

s HOPPENSTEADT, F. (1974). *An age dependent epidemic model.* J. Franklin Inst. **297:** 325-333.

* HOPPENSTEADT, F. (1975). *Mathematical Theories of Populations: Demographics, Genetics and Epidemics.* SIAM.

sm HOPPENSTEADT, F. (1976). *A nonlinear renewal equation with periodic and chaotic solutions.* SIAM-AMS Proceedings **10:** 51-60.

s HORWOOD, J.W. and SHEPHERD, J.G. (1981). *The sensitivity of age-structured populations to environmental variability.* Math. Biosc. **57:** 59-82.

s HSU, P-H. and FREDRICKSON, A.G. (1975). *Population-changing processes and the dynamics of sexual populations.* Math. Biosc. **26:** 55-78.

HSU, S.B., HUBBELL, S. and WALTMAN, P. (1977). *A mathematical theory for single-nutrient competition in continuous cultures of micro-organisms.* SIAM J. Appl. Math. **32:** 366-383.

HUFFAKER, C.B. (1958). *Experimental studies on predation: dispersion factors and predator-prey oscillations.* Hilgardia **27:** 343-383.

b HUTCHINSON, E.G. (1978). *An Introduction to Population Ecology.* Yale Univ. Press, London.

m HUTSON, V. and PYM, J.S. (1980). *Applications of Functional Analysis and Operator Theory.* Academic Press, London.

sm IANNELLI, M. (1985). *Mathematical problems in the description of age structured populations.* In: Mathematics in Biology and Medicine (V. Capasso, E. Grosso and S.L. Paveri-Fontana, eds.) Springer Lect. Notes in Biomath. **57:** 19-32.

s IMPAGLIO, J. (1985). *Deterministic Aspects of Mathematical Demography.* Springer, Berlin.

m IOOSS, G. (1979). *Bifurcation of Maps and Applications.* North-Holland.

m IOOSS, G. and JOSEPH, D.D. (1980). *Elementary Stability and Bifurcation Theory.* Springer, New York.

b IVLEV, V.S. (1955, 1961). *Experimental Ecology of the Feeding of Fishes.* Pishchepromizdat, Moscow (in Russian). Yale Univ. Press, New Haven.

* JAGERS, P. (1975). *Branching Processes with Biological Applications.* Wiley, London.

JAGERS, P. (1982). *How probable is it to be first born? and other branching-process applications to kinship problems.* Math. Biosc. **59:** 1-15.

JAGERS, P. (1983). *On the Malthusianness of general branching processes in abstract type spaces.* In: Probability and Mathematical Statistics, essays in honour of Carl-Gustav Esseen (A. Gut and L. Holst, eds.) Dept. of Math., Uppsala University.

s JAGERS, P. (1983). *Stochastic models for cell kinetics.* Bull. Math. Biol. **45:** 507-519.

JAGERS, P. and NERMAN, O. (1984). *The growth and composition of branching populations.* Adv. Appl. Prob. **16:** 221-259.

JAGERS, P. and NERMAN, O. (1985). *Branching processes in periodically varying environment.* Ann. Prob. **13:** 254-268.

s JÖNSSON, T. (unpublished). *Remarks on the Koch-Schaechter cell cycle model and the use of branching processes.*

JORDAN, G.S. and WHEELER R.L. (1980a), *Structure of resolvents of Volterra integral and integro-differential systems.* SIAM J. Math. Anal. **11:** 119-132.

JORDAN, G.S, and WHEELER R.L. (1980b), *Weighted L^1-remainder theorems for resolvents of Volterra equations.* SIAM J. Math. Anal. **11:** 885-900.

JORDAN, G.S., STAFFANS, O.J. and WHEELER, R.L. (1982). *Local analyticity in weighted L^1-spaces and applications to stability problems for Volterra equations.* Trans. Amer. Math. Soc. **274:** 749-782.

m KALMAN, R.E., FALB, P.L. and ARBIB, M.A. (1969). *Topics in Mathematical Systems Theory.* McGraw-Hill, New York.

m KAMPEN, N.G. VAN (1981). *Stochastic Processes in Physics and Chemistry.* North Holland, Amsterdam.

s KAPUR, J.N. (1982). *Age-structured population models with density dependence.* Bull. Can. Math. Soc. **74:** 207-215.

m KARLIN, S. and TAYLOR, H.M. (1981). *A Second Course in Stochastic Processes.* Acad. Press, New York.

m KATO, T. (1976). *Perturbation Theory for Linear Operators.* 2nd ed. Springer, Berlin.

KENDALL, D.G. (1956). *Deterministic and stochastic epidemics in closed populations.* In: Proceedings of the Third Berkeley Symposium on Mathematical Statistics and Probability (J. Neyman, ed.) IV: 149-165.

* KERMACK, W.O. and MCKENDRICK, A.G. (1927). *A contribution to the mathematical theory of epidemics.* Proc. Roy. Soc. A **115:** 700-721.

* KERMACK, W.O. and MCKENDRICK, A.G. (1932). *Contributions to the mathematical theory of epidemics. II. The problem of endemicity.* Proc. Roy. Soc. A **138:** 55-83.

* KERMACK, W.O. and MCKENDRICK, A.G. (1933). *Contributions to the mathematical theory of epidemics. III. Further studies on the problem of endemicity.* Proc. Roy. Soc. A **141:** 94-122.

m KERSCHER, W. and NAGEL, R. (1984). *Asymptotic behavior of one-parameter semigroups of positive operators.* Acta Appl. Math. **2:** 297-310.

s KEYFITZ, N. (1968). *Introduction to the Mathematics of Population.* Addison-Wesley, Reading.

s KEYFITZ, N. (1977). *Applied Mathematical Demography.* Wiley, New York.

s KEYFITZ, N. and BEEKMAN, J.A. (1984). *Demography through Problems.* Springer.

s KIMMEL, M., DARZYNKIEWICZ, Z., ARINO, O. and TRAGANOS, F. (1984). *Analysis of a model of cell cycle based on unequal division of mitotic constituents to daughter cells during cytokinesis.* J. Theor. Biol. **110:** 637-664.

KIRK, J., ORR, J.S. and FORREST, J. (1970). *The role of chalone in the control of the bone marrow stem cell population.* Math. Biosc. **6:** 129-143.

* KOCH, A.L. and SCHAECHTER, M. (1962). *A model for statistics of the cell division process.* J. Gen. Microbiol. **29:** 435-454.

s KOOIJMAN, S.A.L.M. (1983a). *De dynamica van populaties onder chemische stress.* TNO Report R83/24.

s KOOIJMAN, S.A.L.M. (1983b). *Toxicity at the population level.* TNO Report P83/27.

s KOOIJMAN, S.A.L.M. (to appear). *What the hen can tell about her eggs: egg development on basis of energy budgets.* J. Math. Biol.

s KOOIJMAN, S.A.L.M. and METZ, J.A.J. (1984). *On the dynamics of chemically stressed populations: the deduction of population consequences from effects on individuals.* Ecotox. Env. Saf. **8:** 254-274.

s KOOYMAN, Chr. (and DONZE, M.) (1976). *Populatiedynamica van Asterionella in de planktonproeven bij Harculo.* KEMA Report IV 6886-76.

m KRASNOSEL'SKII, M.A. (1964). *Positive Solutions of Operator Equations.* Noordhoff, Groningen.

m KRASNOSEL'SKII, M.A., ZABREIKO P.P., PUSTYLNIK, E.I. and SBOLEVSKII, P.E. (1976). *Integral Operators in Spaces of Summable Functions.* Noordhoff, Leiden.

m KREIN, M.G. and RUTMAN, M.A. (1948). *Linear operators leaving invariant a cone in a Banach space.* Usephi Mat. Nauk **3:** No. 1 (23), 3-95 [Russian]; English transl.: Am. Math. Soc. Translations (1), **10:** 199-325. (1950).

s KUCZEK, T. (1984). *Stochastic modelling for the bacterial life cycle.* Math. Biosc. **69:** 159-169.

m KUFNER, A., JOHN, O. and FUCIK, S. (1977). *Function Spaces,* Noordhoff, Leiden.

sm KUTTLER, K.L. HILGERS, J.W. and COURTNEY, T.H. (1985). *The solution of an evolution equation describing certain types of mechanical and chemical interaction.* Appl. Anal. **19:** 75-88.

m LADAS, G.E. and LAKSHMIKANTHAM, V. (1972). *Differential Equations in Abstract Spaces.* Academic Press, New York.

s LANGHAAR, H.L. (1972). *General population theory in the age-time continuum.* J. Franklin Inst. **293:** 199-214.

m LASALLE, J.P. (1976). *The Stability of Dynamical Systems.* SIAM, Philadelphia.

sm LASOTA, A. (1981). *Stable and chaotic solutions of a first order partial differential equation.* Nonlinear Analysis, Theory, Methods & Applications **5:** 1181-1193.

s LASOTA, A. and MACKEY, M.C. (1984). *Globally asymptotic properties of proliferating cell populations.* J. Math. Biol. **19:** 43-62.

s LASOTA, A., MACKEY, M.C. and WAZEWSKA-CZYZEWSKA, M. (1981). *Minimizing therapeutically induced anemia.* J. Math. Biol. **13:** 149-158.

s LEBOWITZ, J.L. and RUBINOW, S.I. (1974). *A theory for the age and generation time distribution of a microbial population.* J. Math. Biol. **1:** 17-36.

sb LEE, K.Y., BARR, R.O., GAGE, S.H. and KZHARKAR, A.N. (1976). *Formulation of a mathematical model for insect pest ecosystems - the cereal leaf beetle problem.* J. Theor. Biol. **59:** 33-76.

s LEVIN, S.A. and GOODYEAR, C.P. (1980). *Analysis of an age-structured fishery model.* J. Math. Biol. **9:** 245-274.

LEVIN, S.A., COHEN, D. and HASTINGS, A. (1984). *Dispersal strategies in patchy environments.* Theor. Pop. Biol. **26:** 165-191.

* LEVIN, S.A. and PAINE, R.T. (1974). *Disturbance, patch formation and community structure.* Proc. Nat. Acad. Sc. (USA) **71:** 2744-2747.

s LEVIN, S.A. and PAINE, R.T. (1975). *The role of disturbance in models of community structure.* In: Ecosystem analysis and prediction (S.A. Levin, ed.) SIAM: 56-67.

s LEVIN, S.A. and SEGEL, L.A. (1982). *Models of the influence of predation on aspect diversity in prey populations.* J. Math. Biol. **14:** 253-284.

s LEVINE, D.S. (1981). *On the stability of a predator-prey system with egg-eating predators.* Math. Biosc. **56:** 27-46.

s LEVINE, D.S. (1983a). *Bifurcating periodic solutions for a class of age-structured predator-prey systems.* Bull. Math. Biol. **45:** 901-915.

s LEVINE, D.S. (1983b). *Some age-structured effects in predator-prey models.* In: Population Biology (H.I. Freedman & C. Strobeck, eds.) Springer Lect. Notes in Biomath. **52:** 304-316.

m LIN, C.C. and SEGEL, L.A. (1974). *Mathematics Applied to Deterministic Problems in the Natural Sciences.* Macmillan, New York.

s LOPEZ, A. (1961). *Problems in Stable Population Theory,* Office of Population Research, Princeton.

sm LOSKOT, K. (1985). *Turbulent solutions of a first order partial differential equation.* J. Diff. Equ. **58:** 1-14.

* LOTKA, A.J. (1907). *Relation between birth rates and death rates.* Science N.S. **26:** 21-22.

* LOTKA, A.J. (1922). *The stability of the normal age distribution.* Proc. Nat. Acad. Sci. **8:** 339-345.

* LOTKA, A.J. (1925). *Elements of Physical Biology.* Williams & Wilkins, Baltimore.

s LOTKA, A.J. (1939). *On an integral equation in population analysis.* Ann. Math. Stat. **10:** 1-25.

* LOTKA, A.J. (1956). *Elements of Mathematical Biology.* Dover, New York. (reprint of Lotka 1925).

LUDWIG, D., JONES, D.D. and HOLLING, C.S. (1978). *Qualitative analysis of insect outbreak systems: the spruce budworm and forest.* J. Anim. Ecol. **47:** 315-332.

MACDONALD, N. (1978). *Time Lags in Biological Models.* Springer Lect. Notes in Biomath. **27.**

* MCKENDRICK, A.G. (1926). *Application of mathematics to medical problems.* Proc. Edinb. Math. Soc. **44:** 98-130.

MCMACON, J.W. and RIGLER F.H. (1963). *Mechanisms regulating the feeding rate of Daphnia magna* Straus. Canad. J. Zool. **41:** 321-332.

MACKEY, M.C. (1978). *A unified hypothesis for the origin of aplastic anemia and periodic haematopoiesis.* Blood **51:** 941-956.

MACKEY, M.C. (1981). *Some models in hemopoiesis: predictions and problems.* In: Biomathematics and Cell Kinetics (M. Rotenberg, ed.). Elsevier-North Holland: 23-38.

MACKEY, M.C. and DÖRMER, P. (1982). *Continuous maturation of proliferating erythroid precursors.* Cell Tissue Kinet. **15:** 381-392.

sm MCLEOD, J.B. (1964). *On the scalar transport equation.* Proc. London Math. Soc. **14:** 445-458.

MALTHUS, T.R. (1798, 1970). *An essay on the principle of population (and: A summary view of the principle of population)* Penguin, Harmondsworth, Middlesex.

sm MARCATI, P. (1981). *Asymptotic behavior in age-dependent population dynamics with hereditary renewal law.* SIAM J. Math. Anal. **12:** 904-916.

sm MARCATI, P. (1982). *On the global stability of the logistic age-dependent population growth.* J. Math. Biol. **15:** 215-226.

sm MARCUS, M. and MIZEL, V.J. (1980). *Semilinear hereditary hyperbolic systems with nonlocal boundary conditions.* J. Math. Anal. Appl. **76:** 440-475. **77:** 1-19.

m MAREK, I. (1970). *Frobenius theory of positive operations, comparison theorems and applications.* SIAM J. Appl. Math. **19:** 607-620.

m MARTIN, R.H. (1976). *Nonlinear Operators and Differential Equations in Banach Spaces.* Wiley. New York.

MATANO, H. (1984). *Existence of nontrivial unstable sets for equilibriums of strongly order-preserving systems.* J. Fac. Sc. Univ. Tokyo **30:** 645-673.

MATANO, H. and HIRSCH, M.W. (in prep.). *Existence theorem for stable equilibria in strongly order-preserving systems.*

mb MAY, R.M. (1973). *Stability and Complexity in Model Ecosystems.* Princeton University Press.

MAY, R.M. (1977). *Thresholds and breakpoints in ecosystems with a multiplicity of stable states.* Nature **269:** 471-477.

mb MAYNARD-SMITH, J. (1974). *Models in Ecology.* Cambridge Univ. Press.

s MEE, C.V.M. VAN DER and ZWEIFEL, P.F. (198?). *A Fokker-Planck equation for growing cell populations.*

METZ, J.A.J. (1977). *State space models for animal behaviour.* Ann. Syst. Res. **6:** 65-109.

s METZ, J.A.J. (1978). *The epidemic in a closed population with all susceptibles equally vulnerable; some results for large susceptible populations and small initial infections.* Acta Biotheor. **27:** 75-123.

mb METZ, J.A.J. (1981). *Mathematical representations of the dynamics of animal behaviour (an expository survey).* Thesis, University of Leiden.

s METZ, J.A.J. and BATENBURG, F.H.D. (1984). *Holling's 'hungry mantid' model for the invertebrate functional response considered as a Markov process.* Part 0: *A survey of the main ideas and results.* In: Mathematical Ecology (S.A. Levin & T.G. Hallam, eds.) Lect. Notes in Biomath. **54:** 29-41.

s METZ, J.A.J. and BATENBURG, F.H.D. (1985a, b). *Holling's 'hungry mantid' model for the invertebrate functional response considered as a Markov process.* Part 1: The full model and some of its limits. Part 2: Negligible handling time. J. Math. Biol. **22:** 209-238, 239-257.

m MIKUSINKSKY, J. (1959). *Operational Calculus.* Pergamon.

m MILLER, R.K. (1971). *Nonlinear Volterra Integral Equations,* W.A. Benjamin, Menlo Park Ca.

MIYATA, H., MIYATA, M. and ITO, M. (1978). *Cell-cycle in fission yeast, Schizosaccharomyces pombe. I. Relationship between cell size and cycle time.* Cell Struct. Funct. **3:** 39-46.

m MODE, Ch. J. (1971). *Multitype Branching Processes.* Elsevier, New York.

m MODE, Ch.J. (1985). *Stochastic Processes in Demography and their Computer Implementation.* Springer, Berlin.

s MURDOCH, W.W., NISBET, R.M., BLYTHE, S.P., GURNEY, W.S.C. and REEVE, J.D. (preprint). *An invulnerable age class and stability in delay-differential parasitoid-host models.*

s MURPHY, L.F. (1983a). *A nonlinear growth mechanism in size structured population dynamics.* J. Theor. Biol. **104:** 493-506.

s MURPHY, L.F. (1983b). *Density dependent cellular growth in an age-structured colony.* Comp & Maths. with Appls. **9:** 383-392.

mb MURRAY, J.D. (1977). *Lectures on Non-linear Differential Equation Models in Biology.* Clarendon Press, Oxford.

NAGEL, R. (1984). *What can positivity do for stability?* In: Functional Analysis: Surveys and Recent Results III (K.D. Bierstedt & B. Fuchssteiner, eds.) North-Holland: 145-153.

NERMAN, O. (preprint) *The growth and composition of supercritical branching populations on general type spaces.*

NIIZEKI, S. (1984). *On the Cauchy problem for Volterra-Lotka's competition equations with migration effect and its travelling wave like solutions.* Funk. Ekv. **27:** 1-24.

mb NISBET, R.M. and GURNEY, W.S.C. (1982). *Modelling Fluctuating Populations.* Wiley, New York.

s NISBET, R.M. and GURNEY, W.S.C. (1983). *The systematic formulation of population models for insects with dynamically varying instar duration.* Theor. Pop. Biol. **23:** 114-135.

s NISBET, R.M. GURNEY, W.S.C., BLYTHE, S.P. and METZ J.A.J. (1985). *Stage structure models of populations with distinct growth and development processes.* IMA J. Math. Appl. Biol. Med. **2:** 57-68.

s NISHIMURA, Y. and BAILEY, J.E. (1980). *On the dynamics of Cooper-Helmstetter-Donachie procaryote populations.* Math. Biosc. **51:** 305-382.

s NÖBAUER, W. and TIMISCHL, W. (1983). *Mathematical models of the sterile-insect technique.* In: Recent Trends in Mathematics (Kurke et al., eds.) Teubner **50:** 224-233.

NÜSSBAUM, R.D. (1970). *The radius of the essential spectrum.* Duke Math. J. **38:** 473-478.

NUSSBAUM, R.D. (1984). *A folk theorem in the spectral theory of C_0 semigroups.* Pacific J. Math. **113:** 433-449.

mb OKUBO, A. (1980). *Diffusion and Ecological Problems.* Mathematical Models. Springer, Berlin.

s OLDFIELD, D.G. (1966). *A continuity equation for cell populations.* Bull. Math. Biophys. **28:** 545-554.

mb OLIVEIRO PINTO, F. and CONOLLY, B.W. (1981). *Applicable Mathematics of Non-physical Phenomena.* Ellis Horwood, Chichester.

s OSTER, G. (1976). *Internal variables in populations dynamics.* In: Some Mathematical Questions in Biology VII. AMS, Providence.

s OSTER, G. (1977). *Lectures in Population Dynamics.* In: Modern Modeling of Continuum Phenomena (R.C.DiPrima, ed.) AMS, Providence: 149-190.

s OSTER, G., IPAKTCHI, A. and ROCKLIN, S. (1976). *Phenotypic structure and bifurcation behavior of population models.* Theor. Pop. Biol. **10:** 365-382.

s OSTER, G. and TAKAHASHI, Y. (1974). *Models for age-specific interactions in a periodic environment.* Ecol. Monogr. **44:** 483-501.

m PADULO, L. and ARBIB, M.A. (1974). *Systems Theory, a Unified Approach to Continuous and Discrete Systems.* Saunders, Philadelphia.

s PAINE, R.T. and LEVIN, S.A. (1981). *Intertidal landscapes: disturbance and the dynamics of pattern.* Ecol. Monogr. **51 (2):** 145-178.

* PAINTER, P.G. and MARR, A.G. (1968). *Mathematics of microbial populations.* Ann. Rev. Microbiol. **22:** 519-548.

m PAZY, A. (1983a). *Semigroups of Linear Operators and Applications to Partial Differential Equations.* Springer, New York.

m PAZY, A. (1983b). *Semigroups of operators in Banach spaces.* In: Equadiff 82 (H.W. Knobloch & K. Schmitt, eds.) Springer Lect. Notes in Math. **1017:** 508-523.

m POL, B. VAN DER and BREMMER, M. (1955). *Operational Calculus,* 2nd. ed., Cambridge Univ. Press.

b POLIS, G.A. (1981). *The evolution and dynamics of intraspecific predation.* Ann. Rev. Ecol. Syst. **12:** 225-251.

s POLLARD, J.H. (1973). *Mathematical Models for the Growth of Human Populations.* Cambridge Univ. Press., Cambridge.

POWELL, E.O. (1964). *A note on Koch & Schaechter's hypothesis about growth and fission of bacteria.* J. gen. Microbiol. **37:** 231-249.

sm POZIO, M.A. (1980). *Behavior of solutions of some abstract functional differential equations and applications to predator-prey dynamics.* Nonl. Anal. TMA **4:** 917-938.

sm PRÜSS, J. (1981). *Equilibrium solutions of age-specific population dynamics of several species.* J. Math. Biol. **11:** 65-84.

sm PRÜSS, J. (1983a). *On the qualitative behaviour of populations with age-specific interactions.* Comp. & Maths. with Appls. **9:** 327-339.

sm PRÜSS, J. (1983b). *Stability analysis for equilibria in age-specific population dynamics.* Nonlinear Analysis, Theory, Methods & Applications **7:** 1291-1313.

s RAMKRISHNA, D. (1979). *Statistical models of cell populations.* In: T.K. Chose, A. Fiechter, eds., Advances in biochemical engineering - II, Springer, Berlin.

s RAMKRISHNA, D., FREDRICKSON, A.G. and TSUCHIYA, H.M. (1968). *On relationships between various distribution functions in balanced unicellular growth.* Bull. Math. Biophys. **30:** 319-323.

s RANTA, J. (1982). *On the mathematical modelling of microbial age dynamic and some control aspects of microbial growth processes.* Acta Polytechnica Scandinavica Math **35.**

RASHEVSKY, N. (1959). *Some remarks on the mathematical theory of the feeding of fishes.* Bull. Math. Biol. **21:** 161-182.

REDDINGIUS, J. (1971). *Notes on the mathematical theory of epidemics.* Acta Biotheor. **20:** 125-157.

s RODOLPHE, F., SHISHINY, H.E. and ONILLON, J.C. (1977). *Modelisation de deux populations d'aleurodes ravageurs des cultures.* C.R. Cong. AFCET Modélisation et maîtrise des systèmes **1:** 527-535.

s RORRES, C. (1976). *Stability of an age-specific population with density-dependent fertility.* Theor. Pop. Biol. **10:** 26-46.

s RORRES, C. (1979). *A nonlinear model of population growth in which fertility is dependent on birth rate.* SIAM J. Appl. Math. **37:** 423-432.

s RORRES, C. (1979). *Local stability of a population with density-dependent fertility.* Theor. Pop. Biol. **16:** 283-300.

s ROTENBERG, M. (1972). *Theory of population transport.* J. Theor. Biol. **37:** 291-305.

s ROTENBERG, M. (1975). *Equilibrium and stability in populations whose interactions are age-specific.* J. Theor. Biol. **54:** 207-224.

s ROTENBERG, M. (ed.) (1981) *Biomathematics and Cell Kinetics,* Elsevier/North-Holland, Amsterdam.

s ROTENBERG, M. (1982). *Theory of distributed quiescent state in the cell cycle.* J. Theor. Biol. **96:** 459-509.

s ROTENBERG, M. (1983). *Transport theory for growing cell populations.* J. Theor. Biol. **103:** 181-199.

mb ROUGHGARDEN, J. (1979). *Theory of Population Genetics and Evolutionary Ecology:* An Introduction. MacMillan, New York.

ROUGHGARDEN, J. (1983). *Competition and theory in community ecology.* Amer. Nat. **122:** 583-601.

s RUBINOW, S.I. (1968). *A maturity-time representation for cell populations.* Biophys. J. **8:** 1055-1073.

mb RUBINOW, S.I. (1975). *Introduction to Mathematical Biology.* Wiley, New York.

mb RUBINOW, S.I. (1975). *Mathematical Problems in the Biological Sciences.* Regional Conference Series in Applied Mathematics 10. SIAM, Philadelphia.

s RUBINOV, S.I. (1978). *Age-structured equations in the theory of cell populations.* In: Studies in Mathematical Biology II (S.A. Levin ed.). Math. Assoc. Amer., Washington.

s RUBINOW, S.I. and LEBOWITZ, J.L. (1974). *A mathematical model of neutrophil production and control in normal man.* J. Math. Biol. **1:** 187-225.

s RUBINOW, S.I. and OPENHEIM BERGER, R. (1979). *Time-dependent solution to age-structured equations for sexual populations.* Theor. Pop. Biol. **16:** 35-47.

m RUDIN, W. (1973). *Functional Analysis.* McGraw-Hill, New York.

m RUDIN, W. (1974). *Real and Complex Analysis,* 2nd ed. McGraw-Hill, New York.

sb SABELIS, M.W. (1981). *Biological control of two-spotted spider mites using phytoseiid predators.* Part I. Modelling the predator-prey interaction at the individual level. Agricultural Research Reports 910, Pudoc, Wageningen, the Netherlands.
Part 2. Modelling the predator-prey interaction at the population level. Agricultural Research Reports, ??, Pudoc, Wageningen, the Netherlands.

s SAIDEL, G.M. (1968). *Bacterial cell populations in a continuously changing environment.* J. Theor. Biol. **19:** 287-296.

s SALEEM, M. (1983). *Predator-prey relationships: egg-eating predators.* Math. Biosc. **65:** 187-197.

s SALEEM, M. (1984a). *Egg-eating age-structured predators in interaction with age-structured prey.* Math. Biosc. **70:** 91-104.

s SALEEM, M. (1984b). *Predator-prey relationships: indiscriminate predation.* J. Math. Biol. **21:** 25-34.

SAMUELSON, P. (1976). *Resolving a historical confusion in population analysis.* Human Biol. 559-580.

s SANCHEZ, D. (1978). *Linear age-dependent population growth with harvesting.* Bull. Math. Biol. **40:** 377-385.

m SAPERSTONE, S.H. (1981). *Semidynamical Systems in Infinte Dimensional Spaces.* Springer, New York.

SAWASHIMA, I. (1964). *On spectral properties of some positive operators.* Nat. Sci. Dept. Ochanomizu Univ. **15:** 53-64.

m SCHAEFER, H.H. (1974). *Banach Lattices and Positive Operators.* Springer, Berlin.

m SCHAPPACHER, W. (1983). *Asymptotic behavior of linear C_0-semigroups.* Quaderni 83/1. Universita' degli Studi di Bari.

SCHUMITZKY, A. and WENSKA, T. (1975). *An operator residue theorem with applications to branching processes and renewal type integral equations.* SIAM J. Math. Anal. **6:** 229-235.

m SCHWARTZ, L. (1950, 1951). *Théorie des Distributions.* Parts I, II. Hermann, Paris.

m SEGEL, L.A. (1977a). *An introduction to continuum theory.* In: Modern Modelling of Continuum Phenomena (R.C. DiPrima, ed.) AMS, Providence: 1-60.

m SEGEL, L.A. (1977b). *Mathematics Applied to Continuum Mechanics.* Macmillan, New York.

mb SEGEL, L.A. (ed. 1980). *Mathematical Models in Molecular and Cellular Biology.* Cambridge University Press.

* SHARPE, F.R. and LOTKA, A.J. (1911). *A problem in age-distributions.* Phil. Mag. **21:** 435-438.

m SILVERMAN, L. (1971). *Realization of linear dynamical systems.* IEEE Trans. Automatic Control **16:** 554-567.

s SILVERT, W. and PLATT, T. (1978). *Energy flux in the pelagic ecosystem: a time-dependent equation.* Limnol. Oceanogr. **23:** 813-816.

m SIMMONS, G.F. (1963). *Introduction to Topology and Modern Analysis.* McGraw-Hill, New York.

sm SINESTRARI, E. (1979). *Asymptotic behaviour of solutions of a nonlinear model of population dynamics.* Rend. Acc. Naz. Lincei LXVII: 186-190.

sm SINESTRARI, E. (1980). *Nonlinear age-dependent population growth.* J. Math. Biol. **9:** 331-345.

sm SINESTRARI, E. and WEBB, G.F. (198?). *Nonlinear hyperbolic systems with nonlocal boundary conditions.*

* SINKO, J.W. and STREIFER, W. (1967). *A new model for age-size structure of a population.* Ecology **48:** 910-918.

* SINKO, J.W. and STREIFER, W. (1971). *A model for populations reproducing by fission.* Ecology **52:** 330-335.

s SMITH, D. and KEYFITZ, N. (1977). *Mathematical Demography (Selected Papers).* Biomathematics Vol. 6, Springer, Berlin.

s SMITH, J.L. and WOLLKIND, D.J. (1983). *Age structure in predator-prey systems: intraspecific carnivore interaction, passive diffusion and the paradox of enrichment.* J. Math. Biol. **17:** 275-288.

SOLOMON, M.E. (1949). *The natural control of animal populations.* J. Anim. Ecol. **18:** 1-35.

s SONG JIAN and CHEN RENZHAO. (1983). *Dynamic characteristics of nonstationary population systems and computational formulas of several important demographic indices.* Scientia Sinica (A) XXVI (12) 1314-1325.

s SOWUNMI, C.O.A. (1976). *Female dominant age-dependent deterministic populations dynamics.* J. Math. Biol.

3: 9-17.

STAFFANS, O.J. (1984). *Semigroups generated by a convolution equation.* In: Infinite-dimensional Systems (F. Kappel & W. Schappacher, eds.) Springer Lect. Notes in Math. **1076:** 209-226.

STEINBERG, S. (1968). *Meromorphic families of compact operators.* Arch. Rat. Mech. Anal. **31:** 372-380.

STRAALEN, N.M. VAN (1983a). *Vergelijkende demografie van Springstaarten.* Thesis, Free University, Amsterdam.

STRAALEN, N.M. VAN (1983b). *Physiological time and time-invariance.* J. Theor. Biol. **104:** 349-359.

s STRAALEN, N.M. VAN (1985). *Production and biomass turnover in stationary stage-structured populations.* J. Theor. Biol. **113:** 331-352.

s STRAALEN, N.M. VAN (preprint). *Turnover of accumulating substances in populations with age-structure.*

s STREBEL, D.E. (1985). *Environmental fluctuations and extinction - single species.* Theor. Pop. Biol. **27:** 1-26.

* STREIFER, W. (1974). *Realistic models in population ecology.* In: A. Mac Fadyen (ed.) Advances in Ecological Research **8:** 199-266.

s STREIFER, W. and ISTOCK, C.A. (1973). *A critical variable formulation of population dynamics.* Ecology **54:** 392-398.

s SUBRAMANIAN, G. and RAMKRISHNA, D. (1971). *On the solution of statistical models of cell populations.* Math. Biosc. **10:** 1-23.

s SUDBURY, A. (1981). *The expected population size in a cell-size dependent branching process.* J. Appl. Prob. **18:** 65-75.

s SWICK, K.E. (1981a). *A nonlinear model for human population dynamics.* SIAM J. Appl. Math. **40:** 266-278.

s SWICK, K.E. (1981b). *Stability and bifurcation in age-dependent population dynamics.* Theor. Pop. Biol. **20:** 80-100.

s SWICK, K.E. (1985). *Some reducible models of age dependent dynamics.* SIAM J. Appl. Math. **45:** 256-267.

m TAKÁCS, L. (1960). *Stochastic Processes: Problems and Solutions.* Methuen, London.

m TANABE, H. (1979). *Equations of Evolution.* Pitman, London.

m TAYLOR, A.E. and LAY, D.C. (1979). *Introduction to Functional Analysis.* Wiley. New York.

sm THIEME, H.R. (1984a). *Renewal theorems for linear discrete Volterra equations.* J. reine & angew. Math. **353:** 55-84.

sm THIEME, H.R. (1984b). *Renewal theorems for linear periodic Volterra integral equations.* J. Int. Equ. **7:** 253-277.

s THIEME, H.R. (1985). *Renewal theorems for some mathematical models in epidemiology.* J. Int. Equ. **8:** 185-216.

s THOMPSON, R.W., DIBIASIO, D. and MENDES, C. (1982). *Predator-prey interactions: egg-eating predators.* Math. Biosc. **60:** 109-120.

m TITCHMARSH, E.C. (1979). *The Theory of Functions,* 2nd edition. Oxford University Press.

s TOGNETTI, K. (1975). *The two stage integral population model.* Math. Biosc. **24:** 61-70.

s TOGNETTI, K. and WINLEY, G. (1980). *The growth of a column of age and position dependent cells.* Math. Biosc. **50:** 59-74.

sm TOSKOT, K. (1985). *Turbulent solutions of a first order partial differential equations.* J. Diff. Equ. **58:** 1-14.

m TREVES, F. (1967). *Topological Vector Spaces, Distributions and Kernels.* Acad. Press, New York.

s TRUCCO, E. (1965). *Mathematical models for cellular systems. The von Foerster equation.* Bull. Math. Biophys. **27:** 285-305, 449-471.

s TRUCCO, E. (1967). *Collection functions for non-equivivant cell populations.* J. Theor. Biol. **15:** 180-189.

s TRUCCO, E. (1970). *On the average cellular volume in synchronized cell populations.* Bull. Math. Bioph. **32:** 459-473.

s TRUCCO, E. and BROCKWELL, P.J. (1968). *Percentage labeled mitoses curves in exponentially growing cell populations.* J. Theor. Biol. **20:** 321-337.

s TRUCCO, E. and BELL, G.I. (1970). *A note on the dispersionless growth law for single cells.* Bull. Math. Biophys. **32:** 475-483.

s TSCHUMY, W.O. (1982). *Competition between juveniles and adults in age-structured populations.* Theor. Pop. Biol. **21:** 255-268.

s TULJAPURKAR, S.D. (1982). *Population dynamics in variable environments IV: Weak ergodicity in the Lotka equation.* J. Math. Biol. **14:** 221-230.

s TULJAPURKAR, S.D. (1983). *Transient dynamics of yeast populations.* Math. Biosc. **64:** 157-168.

s TYSON, J.J. (1985a). *The coordination of cell growth and division - intentional or incidental?* Bio Essays **2:** 72-77.

s TYSON, J.J. (1985b). *The coordination of cell growth and division: a comparison of models.* In: Temporal Order (L. Rensing and N.I. Jaeger, eds.) Springer, Berlin: 291-295.

s TYSON, J.J. and DIEKMANN, O. (to appear). *Sloppy size control of the cell division cycle.* J. Theor. Biol.

s TYSON, J.J. and HANNSGEN, K.B. (1985a). *The distributions of cell size and generation time in a model of the cell cycle incorporating size control and random transitions.* J. Theor. Biol. **113:** 29-62.

s TYSON, J.J. and HANNSGEN, K.B. (1985b). *Global asymptotic stability of the size distribution in probabilistic models of the cell cycle.* J. Math. Biol. **22:** 61-68.

s TYSON, J.J. and HANNSGEN, K.B.(preprint). *Cell growth and division: a deterministic/probabilistic model of the cell cycle.*

s VALERON, A.J. and MACDONALD, P.J.M. eds., (1978). *Biomathematics and Cell Kinetics.* Elsevier/North-Holland, Amsterdam.

s VANSICKLE, J. (1977). *Analysis of a distributed-parameter population model based on physiological age.* J. Theor. Biol. **64:** 571 586

s VENTURINO, E. (1984). *Age-structured predator-prey models.* Math. Mod **5:** 117-128.

VERDUYN-LUNEL, S.M. (1984). *Linear autonomous retarded functional differential equations: A sharp version of Henry's theorem.* Report AM-R8405, CWI, Amsterdam.

Verduyn-Lunel, S.M. (to appear). *A sharp version of Henry's theorem on small solutions.* J. Diff. Equ.

VERHULST, P.F. (1838). *Notice sur la loi que la population suit dans son accroisement.* Correspondence Mathématique et Physique Publiée par A. Quetelet **10:** 113-121.

sb VILLARREAL, E., AKCASU, Z. and CANALE, R.P. (1976). *A theory of interacting microbial populations: multi-group approach.* J. Theor. Biol. **58:** 285-317.

m VICHNEVETSKY, R. and BOWLES, J.B. (1982). *Fourier analysis of numerical approximations of hyperbolic equations.* SIAM.

VOIGT, J. (1980). *A perturbation theorem for the essential spectral radius of strongly continuous semigroups.* Mn. Math. **90:** 153-161.

VOLTERRA, V. (1926). *Variazioni e Fluttuazioni del Numero d'Individui in Specie Animali Convisenti.* Mem. accad. Lincei (6) **2:** 31-113.

VOLTERRA, V. (1927). *Variazioni e Fluttuazioni del Numero d'Individui in Specie Animali Convisenti.* R. Comitato Talassografico Italiano, Memoria **131:** 1-142.

* VONFOERSTER, H. (1959). *Some remarks on changing populations.* In: The Kinetics of Cellular Proliferation (F. Stohlman, ed.) Grune and Stratton, New York.

s VOORN, W.J. (1983). *Statistics of cell size in the steady-state with applications to* Escherichia Coli. Thesis, University of Amsterdam.

m WALKER, J.A. (1980). *Dynamical Systems and Evolution Equations.* Plenum Press, New York.

s WALTMAN, P. (1974). *Deterministic threshold models in the theory of epidemics.* Springer Lect. Notes in Biomath. **1.**

mb WALTMAN, P. (1983). *Competition Models in Population Biology.* Regional Conference Series in Appl. Math **45.** SIAM.

s WANG, Y., GUTIERREZ, A.P., OSTER, G. and DAXL, R. (1977). *A population model for plant growth and development: coupling cotton-herbivore interaction.* Can. Ent. **109:** 1359-1374.

s WANG, F.S.J. (1980). *Stability of an age-dependent population.* SIAM J. Math. Anal. **11:** 683-689.

WEBB, G.F. (1979). *Compactness of bounded trajectories of dynamical systems in infinite dimensional spaces.* Proc. Roy. Soc. Edinburgh **84a:** 19-33.

sm WEBB, G.F. (1981). *Nonlinear semigroups and age-dependent population models.* Ann. Mat. Pura & Appl.

CXXIX: 43-55.

sm WEBB, G.F. (1982). *Nonlinear age-dependent population dynamics with continuous age distributions.* In: Evolution Equations and their Applications. Research Notes in Mathematics **68:** Pitman, Boston.

sm WEBB, G.F. (1983a). *Nonlinear age dependent population dynamics in* L^1. J. Int. Equ. **5:** 309-328.

sm WEBB, G.F. (1983b). *The semigroup associated with nonlinear age-dependent population dynamics.* Comp. & Maths. with Appls. **3:** 487-497.

sm WEBB, G.F. (1984). *A semigroup proof of the Sharpe-Lotka theorem.* In: Infinite-dimensional Systems (F. Kappel & W. Schappacher, eds.) Springer Lect. Notes in Math. **1076:** 254-268.

sm WEBB, G.F. (1985a). *Theory of Nonlinear Age-Dependent Population Dynamics.* Marcel Dekker, New York.

s Webb, G.F. (1985b). *Dynamics of populations structured by internal variables.* Math. Z. **189:** 319-336.

s WEBB, G.F. (preprint a). *Logistic models of structured population growth.*

s WEBB, G.F. (preprint b). *An operator-theoretic formulation of asynchronous exponential growth.*

s WEBB, G.F. (to appear). *A model of proliferating cell populations with inherited cycle length.* J. Math. Biol.

s WEBB, G.F. and GRABOSCH, A. (preprint). *Asynchronous exponential growth in transition probability models of the cell cycle.*

WESSEL, W.W. (and DONZE, M.) (1984). *Aanzet tot kwantitatieve behandeling van de populatie-dynamica van Asterionella formosa in micro-ecosystemen.* KEMA Report III 5395-84.

s WICHMANN, H.-ERICH. (1984). *Regulationsmodelle und ihre Anwendung auf die Blutbildung.* Medizinische Informatik und Statistik (Springer). **48.**

m WIDDER, D.V. (1946). *The Laplace Transform.* Princeton University Press.

m WILLEMS, J.C. (1975). *Minimal realizations in state space form from input/output data.* In: Systeemleer (H.F.J.M. Buffart & J.M.L. Ouds, eds.) Stenfert Kroese, Leiden.

sb WILLIAMS, F.M. (1971). *Dynamics of microbial populations.* In: Systems analysis and simulation in ecology (B.C. Patten, ed.) Vol. I, Academic Press: 197-267.

s WINLEY, G. and TOGNETTI, K. (1981). *A growth function for a column of cells.* Math. Biosc. **56:** 209-216.

s WITTEN, M. (1981, 1982). *Modeling cellular systems and aging processes:* I. Mathematics of cell system models - a review. Mechanisms of Ageing and Development **17:** 53-94. II. Some thoughts on describing an asynchronously dividing cellular system. In: Nonlinear Phenomena in Math. Sc. (V. Lakshmikantham, ed.) Academic Press 1023-1035.

s WITTEN, M. (guest editor). (1983). *Hyperbolic partial differential equations: populations, reactors, tides and waves: theory and applications.* Comp. & Maths. with Appls. **9** no. 3 (special issue).

s WITTEN, M. (guest editor). (1985). *Hyperbolic PDE II.* Comp. & Maths. with Appls. **11** no. 1-3 (special issue).

s WOLLKIND, D.J., HASTINGS, A. and LOGAN, J. (1980). *Functional response, numerical response and stability in arthropod predator-prey ecosystems involving age structure.* Res. Popul. Ecol. **22:** 323-338.

s WOLLKIND, D., HASTINGS, A. and LOGAN, J. (1982). *Age structure in predator-prey systems. II. Functional response and stability and the paradox of enrichment.* Theor. Pop. Biol. **21:** 57-68.

m WONG, E. and HAJEK, B. (1985). *Stochastic Processes in Engineering Systems.* Springer Verlag, New York.

WONG, J. Tze Fei (1975). *Kinetics of Enzyme Mechanisms.* Academic Press, London.

WULFF, F.V. (1980). *Animal community structure and energy budget calculations of a Daphnia magna (Straus) population in relation to the rock pool environment.* Ecol. Modell. **11:** 179-225.

s YAMADA, A. and FUNAKOSKI, H. (1982). *On a mixed problem for the McKendrick-Von Foerster equation.* Quart. Appl. Math. XL: 165-192.

s YELLIN, J. and SAMUELSON, P.A. (1977). *Comparison of linear and nonlinear models for human population dynamics.* Theor. Pop. Biol. **11:** 105-126.

m YOSIDA, K. (1980). *Functional Analysis.* 6th edition. Springer, Berlin.

Index of examples

Invertebrate functional response
- only handling time based (disk equation): 5-12.
- purely satiation based: 6, 12-20, 45 (E 5.1), 86 (E 2.1.3), 116 (E 5.3.2 - E 5.3.4), 119 (E 6.2.2).
- satiation and handling time: 78 (E 1.1), 82, 101 (E 3.3.3), 105 (E 4.1.2), 106 (E 4.1.3), 110 (E 4.2.4), 119 (E 6.2.2), 133.

Finite state animal: 6-8, 140-142 (E 1.4.3 - R 1.4.7), 151 - 153, 169 (E 4.1.2), 178 - 182, 184.

Strict age dependence: 113 (R 1.2.1), 140 (E 1.4.1), 145 (E 2.2.2 - E 2.2.4), 171 (E 4.1.8), 172 (E 4.1.10).

Egg eating predators: 119-121 (E 6.2.3), 224.

Nursery competition: 121 (E 6.2.4 - E 6.2.8), 99-100, 138 (E 1.3.1), 222-225.

Kermack's & McKendrick's general epidemic: 139-140 (E 1.3.3), 169-174.

Epidemic with temporary immunity: 214-220.

Size dependent reproduction in ectotherms (Daphnia)
- age independent: 21-31, 85 (E 2.1.1), 87 (E 2.2.1), 102 (E 3.4.1) 103 (E 3.4.3), 109 (E 4.2.2), 116 (E 5.3.5), 118 (E 6.2.1), 123-124 (E 6.3.1 - 6.3.3), 133.
- with age dependent death rate: 79 (E 1.2), 83, 109 (E 4.2.2), 111 (E 4.3.2), 116 (E 5.4.1), 133, 174-176.

Size dependent cell kinetics
- pure size dependent and equal division: 31-43, 46-77, 103 (E 3.5.1), 106 (E 4.1.5), 114 (E 5.1.3), 118 (E 6.1.1), 140 (R 1.3.5).
- pure size dependent, unequal division: 37, 38 (E 4.3.6), 73 (E 11.3, E 11.4), 100 (E 3.3.1), 114 (E 5.1.3), 118 (E 6.1.1), 237-241.
- with additional age dependence: 80 (E 1.4), 82, 112 (E 4.3.5), 115 (E 5.2.2), 133, 185-202.
- stochastic growth: 82 (E 1.13), 82, 87 (E 2.2.1), 87 (R 2.2.2), 131 (EB.1), 133 (EB.5), 135, 142-143 (E 1.4.8 - E 1.4.9), 154 (E 2.A9).

G-M model in cell kinetics: 227-237.

Scar distribution in yeasts
- budding yeasts: 157-160.
- fission yeasts: 160-163.

Colony size in the diatom *Asterionella*: 80 (E 1.6), 83, 101 (E 3.3.2, E 3.3.4), 117 (E 5.4.2), 134, 163-169.

Prey-Predator-Patch problem: 81 (E 1.10), 82, 106 (E 4.1.4), 110 (E 4.2.5 - E 4.2.8), 122 (E 6.2.9), 134-135.

Part B

From Physiological Ecology to Population Dynamics a Collection of Papers

I. Individuals and laboratory populations

The basic strategy of the structured approach to population dynamics, as advocated in these notes, starts from mechanistic considerations on the individual level incorporating various amounts of detail about the underlying physiology. The resulting *i*-model * is used as the substrate for subsequent *p*- modelling. In the end we may say that we have explained population phenomena from considerations on the *i*-level.

The only place where this approach may be found in pure undiluted form is in the study of laboratory populations. It is only under the carefully controlled conditions of the laboratory that we can come to grips with the relevant physiological processes, and it is only in the laboratory that we can sufficiently control the environmental influences experienced by a population.

This section on individuals and laboratory populations contains four papers, all of which bear on topics related to models introduced as examples in part A. But, as befits papers centering on a biological theme, the treatment here is less stylized and the development is pursued in continual contact with real data.

In the paper by Kooijman the problem is addressed how the simple *i*-size based models from part A should be extended to incorporate the often appreciable effects of stored energy or nutrient reserves. The paper stays on the *i*-level in that no explicit *p*-model is formulated, even if various population consequences are derived directly from the *i*-model. Another aspect of the paper that deserves mentioning, is the interesting but possibly confusing alternation between forward and inverse arguments: A partial mechanistic specification is completed by forcing the mechanism to generate a well established growth law and a scaling law under constant feeding conditions. Combining the thus inferred mechanism with various fairly innocuous further mechanistic assumptions then leads to a host of predictions which are compared with experimental data, known phenomenological relationships etc.

The suite of three papers by Sabelis and coworkers deal with the interaction between a predatory mite and its spider mite prey. The first paper considers the predator's functional response, paying great attention to the detailed interaction between theory and experiment. The second paper goes on to consider the dynamics of predator and prey populations within one patch of rose leaves. The final paper considers the dynamics of a population of such patches. All three papers go direct from biological considerations to the computer implementation of discretized versions of the functional partial differential equations describing population behaviour. The first model is treated analytically in chapters I and III of part A. An analytical formulation of the third model may be found in chapter III of part A.

* The prefixes *i*- and *p*- refer respectively to individual and population.

POPULATION DYNAMICS ON BASIS OF BUDGETS

S.A.L.M. Kooijman
Division of technology for society TNO
P.O.Box 217, 2600 AE DELFT, THE NETHERLANDS

SUMMARY

This paper describes a simple and general model for the feeding, storage, growth and reproduction of an ectotherm as functions of a possibly fluctuating food density at constant temperature. The model assumes a hyperbolic functional response, a fixed ratio between ingestion and assimilation rate, and a storage that is proportional to assimilation rate and weight when food supply has been constant for some time; it further assumes, a fixed ratio between energy spent on reproduction and growth plus maintenance at constant food density, a von Bertalanffy growth at constant food density, and a juvenile stage that ends as soon as the animal attains sufficient weight. In the present formulation, the storage dynamics is central. It describes the gradual increase in respiration when food density is suddenly increased, as well as the gradual decrease in respiration rate and time remaining until death when the animal is starved. The model is shown to fit quite well the available data on feeding, respiration, growth, reproduction of female Daphnia magna and their survival time when deprived of food. It explains the occurrence of males in this parthenogenetically reproducing species. The model can also be used to describe microbial dynamics, if it is assumed that division occurs as soon as a certain cell size has been attained. The relationship between the present model and existing descriptions on the substrate-limited growth of bacteria and the nutrient and light-limited growth of algae has been evaluated. They turn out to be special cases of the present model, which explains some observed deviations from existing theories.

INTRODUCTION

The purpose of this paper is to describe the growth and reproduction behaviour of individuals by a model that can serve as a basis for studies in population dynamics. The literature on this subject, although vast, largely falls into two categories. One concerns models that incorporate a great deal of biological detail, and so do not allow of studies in population dynamics other than through computer simulation. The other concerns simple models allowing of some mathematical analysis, but conflicting with known biological facts. The suitability of computer simulation as a research tool being limited, the first type of model can hardly be expected to be capable of tracking down general phenomena in population dynamics, and phenomena predicted by the second type of model may not be relevant.

More progress in the understanding of population dynamics is to be expected from relatively simple models involving just enough biology to fit the data in the literature on physiological ecology. In this paper an attempt has been made to formulate such a model on basis of a simple energy budget of an ectotherm. It extends the ideas given in Kooijman & Metz, 1984, to include storage considerations. Evidence from experiments with the water flea, Daphnia magna, will be adduced in support of the model, the formulation of which has nevertheless been kept as general as possible. However, no attempt has been made to cover all literature on the subject. The literature on marine poikilotherms has been reviewed by Conover, 1978, who cites some 1100 references, and more generally by Bradfield & Llewellyn, 1982. In the literature, several attempt have been made to model the growth process in Daphnia, but no description covers storage in addition to being explicit c.f. Wulff, 1980 and Paloheimo et al., 1982.

A further feature of the present model is that with minor adaptations it can also be used for unicellulars. The understanding of the substrate-limited growth of bacteria and the nutrient and light-limited growth of algae has recently made rapid progress. It will be shown how the proposed formulation relates to some of this work, and how it explains some experimental results deviating from existing theories on the dynamics of microbes. The successful concept of cell quota in the description of nutrient contents of algal cells (see Droop, 1983, for a review) is a special case of the wider concept of storage introduced here. The discussion will be restricted to the situation of constant temperature. A list of frequently used symbols is given in the appendix.

ENERGY BUDGETS OF ECTOTHERMS

This discussion is based on those two state variables of the organism which seem to be the most relevant viz. storage and 'weight'. The latter term is intended to be the measure of the size of the organism, e.g. volume, a cubic length measure, or wet weight; for small aquatic animals, wet weight is hard to measure directly, and most of the literature on these animals uses dry weight. The biomass of microbes is often indicated by their carbon content. These two weight measures are usually related to wet weight by power laws. Porter et al., 1983, state that the dry weight of daphnids is proportional to length to the power 2.39. Strathmann, 1966, state that the carbon content of diatoms is proportional to their volume to the power 0.758, and for other algae to the power 0.866. Energy storage materials usually consist of proteins, lipids and carbohydrates. When food is abundant, they contribute considerably to carbon content and to dry weight. Since the amount of storage materials depends on the availability of food, the above-mentioned conversions to size do not make sense if they are not related to the feeding status of the organism. This problem is less relevant for the wet weight of aquatic animals, because storage materials usually replace water. For animals that do not change shape very much during their life, like daphnids, a cubic length measure seems to be the most appropriate measure of size, because it can be measured rapidly without harm to the animals. We shall assume, therefore, that energy and nutrient storage are related to the chemical composition of an individual, and weight to its size.

Figure 1 shows a diagram of the energy flow, in it simplest form, through an ectothermal individual. The pathways indicated by the numbered arrows will be discussed briefly in the sections numbered identically.

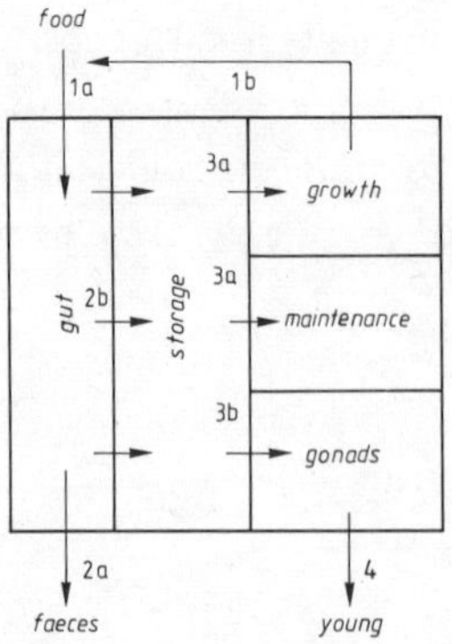

Fig. 1. The energy or nutrient flow through an individual

1. The ingestion rate, $\dot{I}$, tends to increase with the food density, X, and weight, W, of the animal. When starved animals are fed, they often ingest at a higher rate (see e.g. Watts & Young, 1980, for Daphnia feeding on algae), so ingestion rate may also depend on gut content. However, this fast process (,starved daphnids are able to fill their guts within 7.5 minutes, see Geller, 1975,) will not be incorporated in our model.

1a. For a fixed weight, the ingestion rate as a function of food density is known as the functional response, which frequently takes the form of a hyperbolic function. See Kooijman & Metz, 1984 for Daphnia. In symbols, we have $\dot{I} = X/(1/\dot{F}_m + X/\dot{I}_m)$, where X is the food density, $\dot{I}_m$ the maximum ingestion rate, and $\dot{F}_m$ the filtering or searching rate in the absence of food.

When offered different food items, individuals usually select for type and size. We shall briefly focus on the latter, because it provides an argument for the maximum ingestion being determined by the digestion rate, i.e. expressed as carbon content, the maximum ingestion rate is independent of the size of food items, provided that their chemical composition is similar. Frost, 1972, found this to be the case for copepods fed on species of diatoms of different cell sizes. On the other hand, Geller, 1975, found that, when the maximum ingestion rate is expressed in terms of ingested food volume instead of its carbon content, it is independent of the cell size of six widely different species of food algae. This may be a coincidence, merely due to one species of alga being less digestible than another. If X(V) denotes the density of food particles of volume V, and P(V) the ingestion probability of an item of volume V in the filtered water in the area searched, the ingestion rate $\dot{I}(V)$ of particles of volume V becomes $\dot{I}(V) = PX/[1/\dot{F}_m + \int_V PX/\dot{I}_m]$, where the maximum ingestion rate $\dot{I}_m$ is proportional to a weighted integral over the size distribution, of ingested particles, $\int_V PX/\int_V PXC$, where C(V) is the carbon content of a particle of volume V. An efficient filter feeder (e.g. Daphnia) feeding on a suitable algal or bacterial species will ingest all particles, so P(V)=1. For copepods, on the other hand, which capture their food particles more actively and tend to select the larger algae (Strickler, 1982), the results of Frost, 1972, can be interpreted to mean that the catching probability function increases with particle volume. If such is the case, it will probably also be a function of the size of the animal itself.

When the food contains several types of particles, the catch function may also depend on the (relative) abundances of the types, as has been found by DeMott, 1982, for Bosmina feeding on mixtures of algae and bacteria.

1b. Obviously, the ingestion rate increases with the size of organism. For daphnids, which do not change very much in shape during growth, it is about proportional to $W^{2/3}$ (Kooijman & Metz, 1984). This relation is plausible when we realize that the maximum filtering rate, i.e. $\dot{F}_m = (d\dot{I})/(dX)$ for X = 0, probably depends on the surface area of the filtering apparatus, and the maximum ingestion rate, $\dot{I}_m$, on the surface area of the gut. For microbes, $\dot{F}_m$ is related to the probability of a nutrient particle coming into contact with a free site on its surface, and $\dot{I}_m$ to the maximum number of binding sites; both magnitudes are related to surface area.

2a. The gut has the function of a buffer, frequently fortified by the presence of a stomach. The nutritional gains from the ingested food can be assumed to be about proportional to a moving average of the ingestion rate. For small-particle feeders like daphnids, the residence time of a food particle has been found to be between 28 and 54 minutes at 15°C, depending on the algal type (Geller, 1975). Because of these short residence times, we will neglect the buffer function of the gut, which is justified if food density does not change too rapidly. The energetic gain from the food, the so-called assimilation energy $\dot{A}$ will, for this type of animal, assumed to be proportional to ingestion rate, and we shall disregard the possibility of digestion being less efficient at high ingestion rates as has been postulated by Paloheimo et al., 1982.

2b. So $\dot{A} = [\dot{A}_m]\, W^{2/3} f$, where $f = X/(K+X)$, the shape parameter of the functional response curve, $K = \dot{I}_m/\dot{F}_m$, being independent of the weight, W, of the animal, and $[\dot{A}_m]$ being proportional to $\dot{I}_m/W^{2/3}$, where the proportionality factor involves the energy gain per unit weight of particles. In microbiological studies K has become known as the saturation constant.

3. The observation that at a constant food density X, the growth curve for Daphnia magna closely resembles a von Bertalanffy curve (Kooijman & Metz, 1984) suggests that the decline in growth with age is due to the increasing metabolic needs of the animal. This is confirmed by the results of the feeding experiments with Daphnia reported in Fig. 2. In these experiments the length of the animals was monitored during growth at two levels of chlorella densities which were alternated after one, two or three weeks at 20°C. The curves without shift represent least squares adaptations of the von Bertalanffy growth curves. The curves in the other figures represent the expected growth for animals changing

their growth regime momentarily to the same parameters as the ones without shift. We can conclude that daphnids retain their ability to grow, and that larger daphnids adapt more gradually to a new growth regime. This will be explained by a weight related storage.
Full-grown animals still reproduce abundantly, suggesting that at constant food density the storage utilization rate $\dot{C}$ falls into a part $\kappa\dot{C}$ spent on growth and routine metabolism, and a part $(1-\kappa)\dot{C}$ spent on reproduction. The balance equation for the storage S is $\dot{S} = \dot{A} - \dot{C}$.

3a. Growth is given by $\eta\dot{W} = \kappa\dot{C} - \dot{M}$, where η is the energy requirement per unit increase of weight, and $\dot{M} = \xi W$ the routine metabolic rate, which is taken to be proportional to weight. Since $\dot{C}$ is an unknown function of S and W, we substitute $\dot{C} = \dot{A} - \dot{S}$ and assume that, at constant food density, the storage, which depends on W, is in a pseudo-equilibrium S*, i.e. it can be written as a function of f and W, so $\dot{W} = (\kappa\dot{A} - \dot{M}) / (\eta + \kappa(dS^*)/(dW))$.
If growth is of the von Bertalanffy type, i.e. $\dot{W}$ is a weighed difference between $W^{2/3}$ and W, S* has to be proportional to W. We assume that it is also proportional to f, giving $S^* = [S_m]\, fW$, where the parameter $[S_m]$ can be interpreted as the weight-specific maximum storage. Substitution of this equilibrium storage in the growth equation gives the growth rate as a function of f and W. Following the diagram of Fig. 1, however, we do not want it as a function of f, but as one of the state variable S. The observations in Fig. 2 also show that a sudden change in f does not produce a sudden change in $\dot{W}$. We therefore substitute $f = [S]/[S_m]$, where [S] is the weight-specific storage, i.e. $[S] = S/W$, which gives

$$\dot{W} = \frac{[S]}{[S] + \eta/\kappa} \frac{[\dot{A}_m]}{[S_m]} W^{2/3} - \frac{\xi/\kappa}{[S] + \eta/\kappa} W$$

So the maximum weight W_m of an adult is given by $W_m^{1/3} = \kappa[\dot{A}_m]/\xi$, and the maximum growth rate by $\dot{W}_m = (4/27)\, W_m\, (\xi/\kappa)/([S_m] + \eta/\kappa)$, which is reached in animals of weight W_m 8/27 for f = 1.
The storage utilization rate now becomes

$$\dot{C} = \frac{[S]}{[S] + \eta/\kappa} \left\{ \frac{\eta}{\kappa} \frac{[\dot{A}_m]}{[S_m]} W^{2/3} + \frac{\xi}{\kappa} W \right\}$$

If we substitute this in the storage balance equation, we have

$$\dot{S} = [\dot{A}_m] W^{2/3} \left\{ f - \frac{1}{[S_m]} \frac{[S]\eta/\kappa}{[S] + \eta/\kappa} \right\} - W \frac{[S]\; \xi/\kappa}{[S] + \eta/\kappa}$$

Fig. 2. Length development at 20°C of Daphnia magna in two constant Chlorella densities (a), in a shift up at 1(b), 2(c) or 3(d) weeks, and in a shift down at 1(e), 2(f) or 3(g) weeks in these two densities. For further explanation see text.

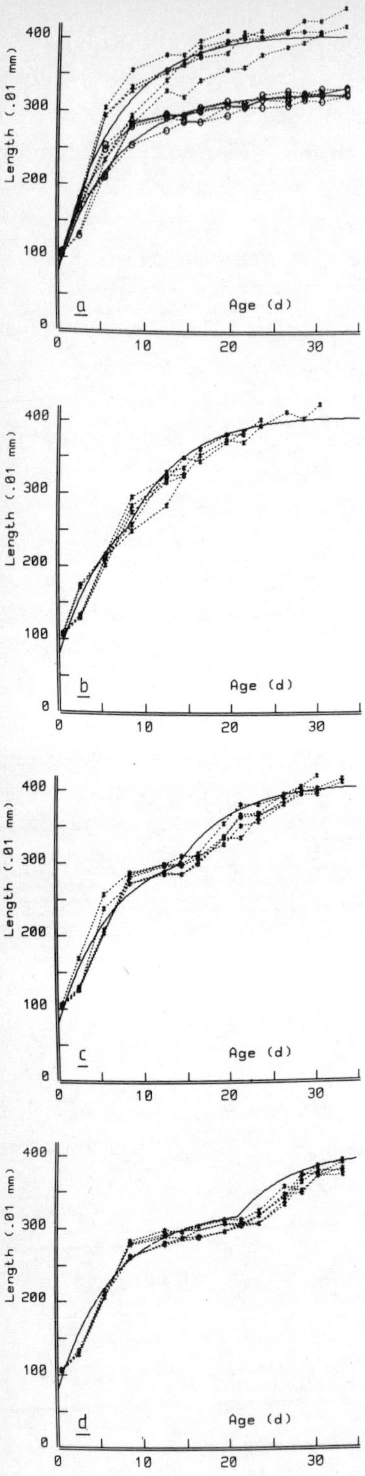

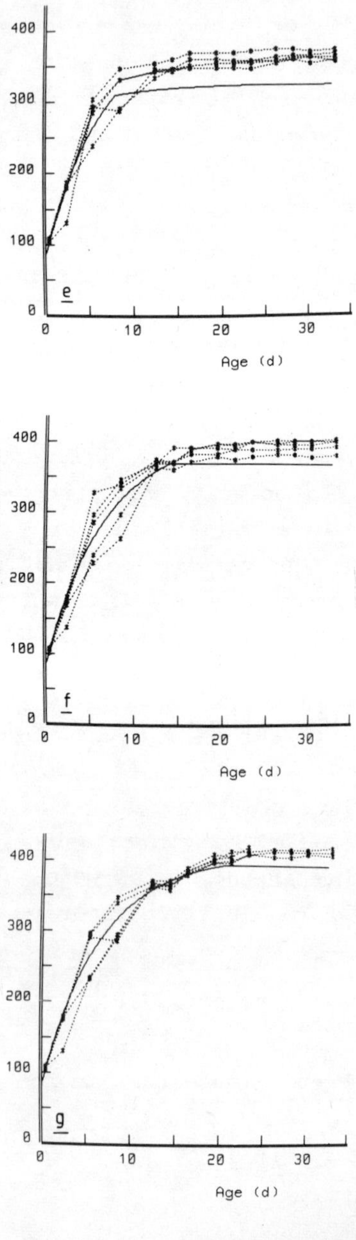

When given in the dynamics of the weight-specific storage, $[\dot{S}] = \dot{S}/W - S\dot{W}/W^2$, this simplifies to $[\dot{S}] = [\dot{A}_m]W^{-1/3} \{f - [S]/[S_m]\}$.
If storage decreases to less than $S = W^{4/3} ([S_m]/[\dot{A}_m]) \dot{\zeta}/\kappa$ the animal can no longer fulfil its metabolic needs in this regime, because $\kappa \dot{C} < \dot{M}$. The results presented in Fig. 2 suggest that κ will increase to $\kappa \dot{C} = \dot{M}$, from which follows that $\kappa = (\dot{\zeta}/[S]) ([S_m]/[\dot{A}_m])W^{1/3}$, and so $\dot{C} = [S] ([\dot{A}_m]/[S_m])W^{2/3}$.
Under poor feeding conditions, therefore, the animal decreases its utilization rate and ceases growing. Without any food uptake, the storage is emptied by a first-order process down to $S = W^{4/3} \dot{\zeta} [S_m]/[\dot{A}_m]$, corresponding to $\dot{C} = \dot{M}$. It will then die of starvation. In the non-growth region of the storage, where W is a constant, its balance equation therefore is $\dot{S} = [\dot{A}_m]W^{2/3} \{f-[S]/[S_m]\}$.
When given in the dynamics of the weight-specific storage, this again gives $[\dot{S}] = [\dot{A}_m]W^{-1/3} \{f - [S]/[S_m]\}$, which means that the dynamics of the weight-specific storage is the same whether the animals are growing or not. The only difference is that maintenance is at the expense of growth when the animals are growing, and of reproduction when they are not.

3b. As stated above, the energy spent on reproduction equals $(1-\kappa)\dot{C}$. In daphnids, the energy seems to be converted into young in female animals if the weight of the animals exceeds a treshold value W_J, (Kooijman & Metz, 1984). At this weight, there is no obvious change in growth. (The von Bertalanffy curves fit well in the entire weight range). This suggests that during the pre-reproductive period, the gonads receive an inflow of energy for their ripening.

4. Daphnids normally reproduce parthenogenetically; female diploid adults beget female diploid offspring without intervention by males (Taub, 1982). The occurrence of males will be discussed in the section on starvation. The reproduction process in daphnids is coupled with the moulting cycle. The latter depends on temperature, but not on the feeding status. At 20°C Daphnia magna moults every 2 or 3 days. Just after moulting, eggs are deposited in the brood pouch and develop without food supply (Green, 1956) into young, which are released just before moulting. During this period the adult restores its energy reserves, so the energy chanelled into reproduction can be regarded as being constant (Tessier & Goulden, 1982). If, at constant food density, growth is of the von Bertalanffy type from birth onwards the weight-specific storage of the young should equal that of the adult.

Tessier et al., 1983, actually observed that the storage in the form of triglycerides in young depends on the adult's feeding success; young born of well-fed adults survived for twice as long when starved as did the offspring of starved adults. So the energy investment per young is $W_b\ (\omega + [S])$, where the parameter ω can be interpreted as the weight specific energy requirement for the formation of offspring tissue, W_b being weight of a young at birth (The length of Daphnia magna at birth is 0.8 mm).

The reproduction rate for growing animals now becomes

$$\dot{R} = \frac{1-\kappa}{W_b} \frac{1}{[S] + \omega} \frac{[S]}{[S] + \eta/\kappa} \left\{ \frac{[\dot{A}_m]}{[S_m]} \frac{\eta}{\kappa} W^{2/3} + \frac{\zeta}{\kappa} W \right\}$$

For non-growing ones, we have

$$\dot{R} = \frac{1}{W_b} \frac{1}{[S] + \omega} \left\{ [S] \frac{[\dot{A}_m]}{[S_m]} W^{2/3} - \zeta W \right\}$$

The maximum reproduction rate, which is reached for animals of weight W_m for $f = 1$, then becomes $\dot{R}_m = (1-\kappa)(\zeta/\kappa)(W_m/W_b)/([S_m]+\omega)$.

We have now completed the quantitative description of the arrows in Fig. 1, in terms of the state variables weight, W, and storage S. We can summarize this description in the state space representation, given in Fig. 3. At constant food density, an individual has weight W_b at birth and moves along a line through the origin, as indicated, but does not leave the growth region. This situation will be discussed in the next section. More generally, we can state that, if a population experiences a period of constant food density, all individuals will gather on a line through the origin, regardless of their food history. If the population has experienced higher food densities in the past, there may be a group weighing more than $(f\kappa[\dot{A}_m]/\zeta)^3$, unable to grow, but still able to reproduce. For low food densities, $f < \kappa$, there may be another group weighing more than $(f[\dot{A}_m]/\zeta)^3$ that will eventually die of starvation. This situation will also be briefly discussed. On basis of the (ultimate) reproduction and survival behaviour at fixed food densities, we can classify them into the categories mentioned in Fig. 3. Lines of equal growth and reproduction in the state space are given in Fig. 4.

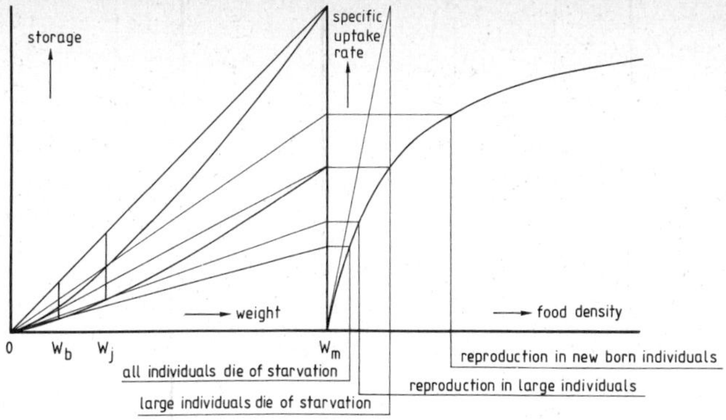

Fig. 3 State space representation of an ectotherm. For an increasing storage at given weight, there is a region in which an animal cannot exist due to starvation (upperbound $S=\kappa[S_m]W^{4/3}/W_m^{1/3}$), a no-growth region (upperbound $S=[S_m]W^{4/3}/W_m^{1/3}$), a growth region (upperbound $S=[S_m]W$) and a region an animal can not reach. For further explanation see text. In this figure, the value for κ has been chosen 1/2.

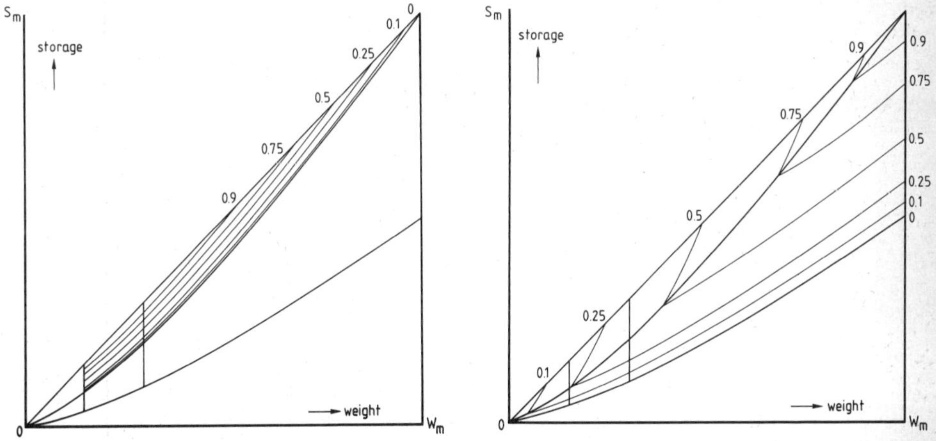

Fig. 4. Lines of equal growth (left) and reproduction (right). In this figures, the following choices have been made: $\kappa=1/2$, $[S_m]\ \kappa/\eta=10$ and $[S_m]/\omega=5$. The numbers indicated along the lines are fractions of maximum growth and reproduction.

SOME PHENOMENA UNDER CONDITIONS OF CONSTANT FOOD DENSITY

Utilization rate can be measured indirectly as respiration rate. It is not at all clear, however how oxygen consumption and carbon dioxide production rates exactly relate to the energy flows under consideration, even for animals with empty guts. Storage materials are usually classified as proteins, glycogen and triglycerides. Growth and reproduction result in formation of tissues also consisting of these components, possibly in other proportions. Although the preservation of energy and oxygen, nitrogen and carbon may be negligible in comparison with the amounts of storage materials used during the measurement of the respiration rate, the energy involved in this preservation process is not. The energy channelled into reproduction can be quite substantial, as has been found for crustaceans (Kmeleva, 1972 gives 0.5 time the utilization rate) and especially for daphnids (Richman, 1958, gives up to 0.8 times the utilization rate).

The energy gain from the utilization of storage materials depends on their composition. For aquatic animals Brafield & Llewellyn, 1982, give the conversion heat loss in joules =

(11.16x mg O_2 cons.) + (2.62x mg CO_2 prod.) - (9.41x mg NH_3 prod.).

If the composition of storage material remains the same, the conversion of oxygen consumption into energy involves a constant factor.

This means that at constant food density, the respiration rate can be written as $aW^{2/3} + bW$, where the quotient of the regression coefficients b/a equals $(\dot{\xi}/\eta)\ [S_m]/[\dot{A}_m]$, see Fig. 5. The fit is quite satisfactory.

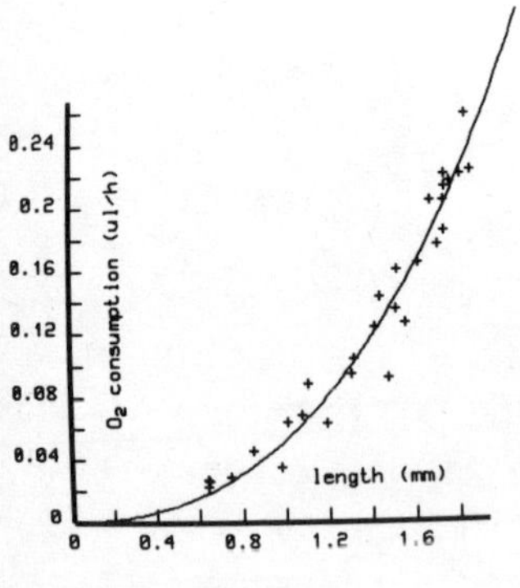

Fig. 5. Respiration rate of Daphnia pulex with few eggs at 20°C as a function of length. Data are from Richman, 1958, table 5. The fitted curve is $0.0336\ L^2 + 0.01845\ L^3$, and is indistinguishable from the curve $0.0516\ L^{2.437}$.

The curve usually fitted to respiration size data is the allometric one, aW^b, see e.g. Richman, 1958, who found b=0.88 and Kersting & V.d. Leeuw-Leeghwater, 1976, who found b=0.82 for Daphnia pulex. As is obvious from Fig. 5, the data can never tell the difference between the two curves. See Kooijman, 1984, for a more extended discussion.

At constant food density, the equilibrium storage equals $[S_m]fW$, for which the growth equation can be solved, resulting in

$$W^{1/3}(a) = W_\infty^{1/3}\,(1-be^{-\dot{\gamma}a}),\ \text{where } W_\infty^{1/3} = \kappa\ f\ [\dot{A}_m]/\ \dot{\zeta},\ b = 1-W_b^{1/3}/W_\infty^{1/3},$$

$$\dot{\gamma} = (\dot{\zeta}/\kappa)/\{3(\eta/\kappa + [S_m]f)\}\quad \text{and a the age.}$$

The rate $\dot{\gamma}$ therefore decreases with increasing food density, owing to the presence of a storage. Reanalysis of the data on daphnid growth at different chlorella densities of Kooijman & Metz, 1984, reveals a significantly increased goodness of fit, compared with the situation without storage, $[S_m]=0$; see Fig. 6. Note that the inverse rate $\dot{\gamma}$ depends linearly on the asymptotic length: $1/\dot{\gamma} = 3\eta/\dot{\zeta} + 3([S_m]/[\dot{A}_m])W_\infty^{1/3}$. The quotient of the slope parameter and the intercept, $(\dot{\zeta}/\eta)\ [S_m]/[\dot{A}_m]$, should equal the quotient of the regression coefficients corresponding to W and $W^{2/3}$ respectively for the respiration rate as a function of weight, as we have seen before.

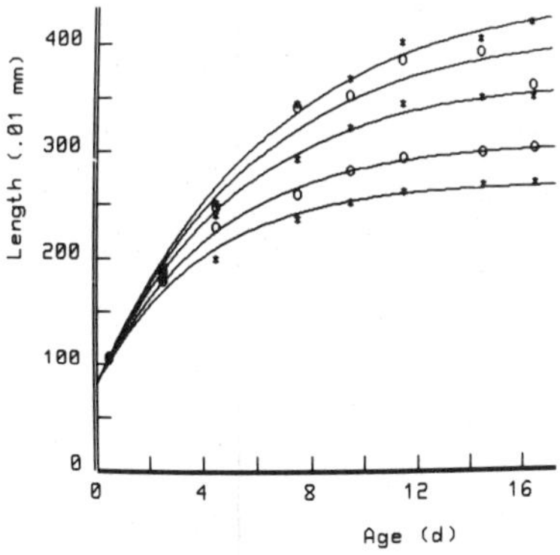

Fig. 6. Lenght L of Dapnia magna as a function of age a for various Chlorella densities X at 20°C. The fitted curves have the form $L = fL_m - (fL_m-L_b)\exp\{-a\dot{\gamma}\}$, where f = .407, .465, .554, .625 and .682 respectively, L_b = .80 mm, L_m = 6.60 mm, $\dot{\gamma}= 1/\{3\eta/\dot{\zeta} + 3f[S_m]\kappa/\dot{\zeta}\}$ with $\eta/\dot{\zeta}$= 0.672 d and $[S_m]\kappa/\dot{\zeta}$= 3.06 d. All parameters have been estimated by nonlinear simultaneous regression, except L_b and L_m, which have been determined from other data.

The duration of the pre-reproductive period, i.e. the age at which weight W_J is reached, is found form $W(J) = W_J$ to be

$$J = 3\ (\eta/\kappa + [S_m]f)(\dot{\zeta}/\kappa)^{-1}\ \ln\{(W_\infty^{1/3} - W_b^{1/3})/(W_\infty^{1/3} - W_J^{1/3})\}$$

for $f \geqq (\dot{\zeta}/\kappa)W_J^{1/3}/[\dot{A}_m]$. At lower food densities, the weight W_J will not be reached.

Like the respiration rate, the reproduction rate can be written as $aW^{2/3} + bW$ at constant food density, where b/a also has the interpretation $(\dot{\zeta}/\eta)\,[S_m]/[\dot{A}_m]$. Khmeleva, 1972, found for crustaceans that the relation between weight and reproduction is similar to that between weight and respiration. For Daphnia magna it is illustrated in Fig. 7.

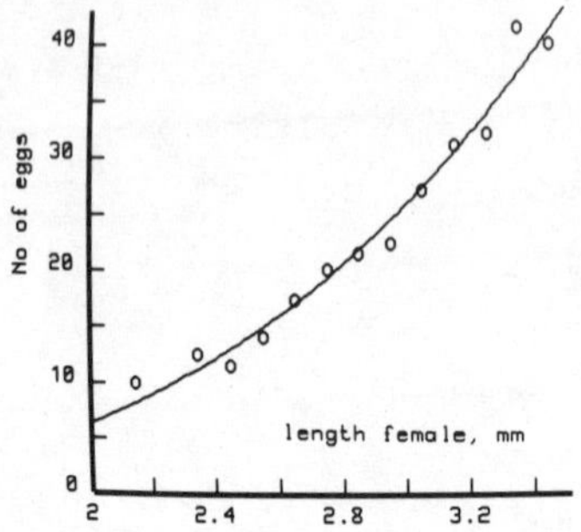

Fig. 7. Number of eggs in brood pouches of Daphnia magna, sampled from a wild population. The data are from Green, 1954. The fitted curve is $-1.01\ L^2 + 1.295\ L^3$.

Although the fit is satisfactory, particularly because nothing is known about the feeding history of the wild population sampled, this regression cannot contribute to the estimation of the energy budget parameters; even a small change in the shape of the curve has a large effect on the regression parameters. In Fig. 7, these parameters even fall outside the relevant range.

At constant food density, but for unknown food history of the population, an optimum relationship of reproduction rate to individual weight will, in principle be established. For $\kappa < 2/3$, the maximum reproduction for $f < \kappa 3/2$ is $(W_\infty/W_b)\ (\dot{\zeta}/\kappa)(\omega+[S_m]f)^{-1}\ (4/27)\kappa^{-2}$ for $W^{1/3} = (2/3)\ [\dot{A}_m]f/\dot{\zeta}$, i.e. it lies on the curve $S = (3/2)\ ([S_m]/[\dot{A}_m])W^{4/3}\ \dot{\zeta}$ in the state space representation. For $f > \kappa 3/2$, the maximum reproduction occurs for W_m. For $\kappa > 2/3$, the maximum reproduction is $(W_\infty/W_b)(\dot{\zeta}/\kappa)(\omega + [S_m]f)^{-1}\ (1-\kappa)$ for $W^{1/3} = \kappa f[\dot{A}_m]/\dot{\zeta}$, i.e. it lies on the curve separating the growth and the no-growth regions in the state space.

Up to now, we have compared the reproduction rates of animals on a line through the origin in the state space. Comparing the reproduction rate of full-grown animals, which are on the curve separating the growth and the no-growth regions, we find that the reproduction rate is linearly proportional to weight, with a proportionality factor of $\dot{R}_m/W_m$.

At constant food density, given the explicit expression for weight as a function of age, we can write the reproduction rate explicitly as a function of age. Although the formulae now become lengthy, it is easy to write out the primary production efficiency on basis of wet weight, defined by $(W_b\dot{R} + \dot{W})/\dot{I}$, and to show that it decreases with increasing food density, because storage increases with food density. This finding may be the solution to the problem mentioned by Conover, 1978, in his review on this topic, namely that he fails to see why production efficiencies should decrease with increasing food density.

SOME PHENOMENA ACCOMPANYING STARVATION

The storage development at constant food density in an animal of weight W and storage S_o at time zero, in the no-growth region of its state space, is easily found from the balance equation to be

$$S(t) = S_o \exp\left\{-\frac{[\dot{A}_m]t}{[S_m]W^{1/3}}\right\} + \frac{\dot{A}_1[S_m]W^{2/3}}{[\dot{A}_m]}\left(1-\exp\left\{-\frac{[\dot{A}_m]t}{[S_m]W^{1/3}}\right\}\right)$$

Where $\dot{A}_1$ is the influx of assimiliation energy. So in case of starvation, i.e. $\dot{A}_1=0$, the storage decays exponentially: $S(t) = S_o \exp.\{-t/\alpha\}$, where the time constant α can be interpreted as $W^{1/3}[S_m]/[\dot{A}_m]$. Figure 8 shows that the dry weight and lipid content of starved Daphnia magna at 20°C actually follows such a decay.

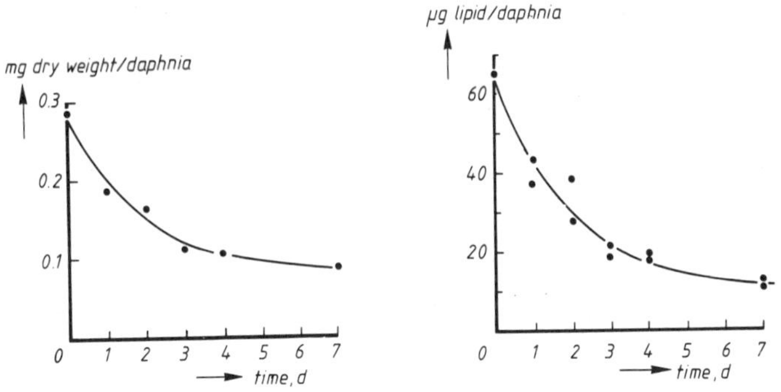

Fig. 8. Dry weight and mean lipid content of starved Daphnia magna at 20°C, measured in groups of 25 animals with a mean length of 3.4 mm. The fitted curves starting from 0.27 mg and 64 μg respectively are exponential decay functions with time constants of 1.99 d and 1.88 d, respectively, with asymptotes of 0.19 mg and 54 μg resp.

From these curves we estimate $[S_m]/[\dot{A}_m]$ to be 0.586 d/mm and 0.552 d/mm respectively. (This type of decay functions has also been found by Richman, 1958, and Lemke & Lampert, 1975, for Daphnia pulex, with twice as large values for $[S_m]/[\dot{A}_m]$). The ratio of the storage at the entry into the no-growth region of the state space and the storage at death by starvation equals κ. The time to death by starvation for the animals was 7 d, so we estimate κ to be about exp. $\{-7/1.99\} = 0.024$ and exp.$\{-7/1.88\} = 0.03$, both of which seem to be incredibly small. However, errors in the estimation of the asymptote of the dry weight and lipid content strongly influence this estimation of κ. Estimations based on oxygen consumption or carbon dioxide production rate do not suffer from such errors, because the asymptote is zero.

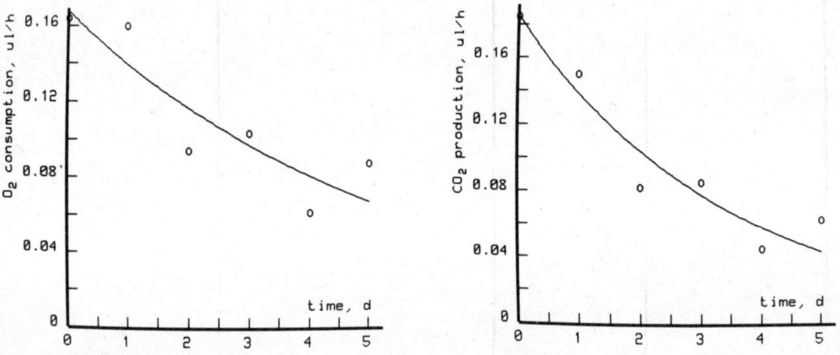

Fig. 9. Oxygen consumption and carbon dioxide production rate in starved Daphnia pulex of 1.62 mm at 20°C. Data from Richman, 1958. The fitted curves starting from 0.167 µl/h and 0.185 µl/h, respectively, are exponential decay functions with time constants of 5.48 d and 3.43 d.

From the curves in Fig. 9, we estimate $[S_m]/[\dot{A}_m]$ to be 3.38 d/mm and 2.11 d/mm. For a starvation time of 7 d, we have κ = exp.$\{-7/5.48\}=0.28$ and κ = exp.$\{-7/3.43\}=0.13$, which seem to be more credible.

If the influx of assimilation energy happens to be less than the routine metabolic rate, the animal will eventually die by starvation. If it is assumed that the storage S_0 has been in equilibrium with a higher influx $\dot{A}_0$ experienced earlier, the time until death by starvation is found from $S(t_\dagger) = W^{4/3}\,\dot{\zeta}[S_m]/[\dot{A}_m]$ to be

$$t_\dagger = W^{1/3}\,([S_m]/[\dot{A}_m])\,\ln\{(\dot{A}_0 - \dot{A}_1)/(\dot{M} - \dot{A}_1)\},$$

where as we have stated before, the influxes $\dot{A}$ are assumed to be proportional to $W^{2/3}$, and the routine metabolic rate $\dot{M}$ to W. In case

of starvation, i.e. $\dot{A}_1=0$, the starvation period reduces for $[S_\dagger] = S(t_\dagger)/W$ to $t_\dagger = ([S_\dagger]/\dot{\xi})\ \ln\{[S_o]/[S_\dagger]\}$.
The number of young born in this period is
$[[S_\dagger]\ \ln\{[S_\dagger]/[S_o]\} + (\omega + [S_\dagger])\ \ln\{(\omega + [S_o])/(\omega + [S_\dagger])\}]\ W/(\omega W_b)$.
Depending on the values of the parameters and especially of κ, the survival time can decrease as well as increase with weight of the animals. This result holds for animals differing in weight, but starting at the equilibrium storage $[S_m]f_oW$, which represents a line in the state space representation. Comparing animals fullgrown at constant food density, we have $t_\dagger = -W^{1/3}\ ([S_m]/[\dot{A}_m])\ \ln\kappa$. Such animals are on the curve, separating the growth and no-growth region in the state space representation. So the latter starvation time is the time needed by the animals to cross the no-growth region of the state space. This relation for starvation times can be of use in estimating the parameter κ, if the value for $[S_m]/[\dot{A}_m]$ has been obtained from the analysis of respiration rates or growth curves, as outlined in the previous section.
The starvation times and total number of young born in this period have been determined for individuals of Daphnia magna at 20°C kept without food in two kinds of water derived from groundwater. This had been supplemented with salts to arrive at the major ion concentrations in mmol: Na: 1.19; K: 0.2; Ca: 1.36; Mg: 0.73; Cl: 2.72; SO_4: 0.73; HCO_3: 1.39. Before use, the media were filtered over charcoal and bacterial filters. The two media differed in that one had been kept in stock for several weeks and had been aerated with compressed air from a central supply, whereas the other had not. The results, shown in Fig. 10, indicate that in the water used directly, the animals survived for about 5.5 days irrespective of their length, while in the aerated water, the small ones ($\leqq$ 2 mm) survived for up to three weeks. The difference may have been caused by the presence of organic matter, possibly arising from microbial degradation of oil residues in the compressed air. This conjecture is supported by the finding that the aerated medium contained 2 mg/l of total organic carbon, whereas the other medium contained only 1.8 mg/l (Most, if not all of which is not biodegradable).
The theory predicts that animals weighing more than $([\dot{A}_m]\ f_s/\dot{\xi})^3$ will fail to survive at input $f_s = X_s/(K+X_s)$. The food density allowing an animal just to survive is known as the threshold food density, and is given by $X_s = KW^{1/3}/([\dot{A}_m]/\dot{\xi}-W^{1/3}) = KW^{1/3}/(W_m{}^{1/3}/\kappa - W^{1/3})$. For Daphnia magna fed on the alga Chlorella pyrenoidosa, K equals 1.4×10^5 cells/ml and the maximum length is 6.4 mm, see Kooijman & Metz, 1984.

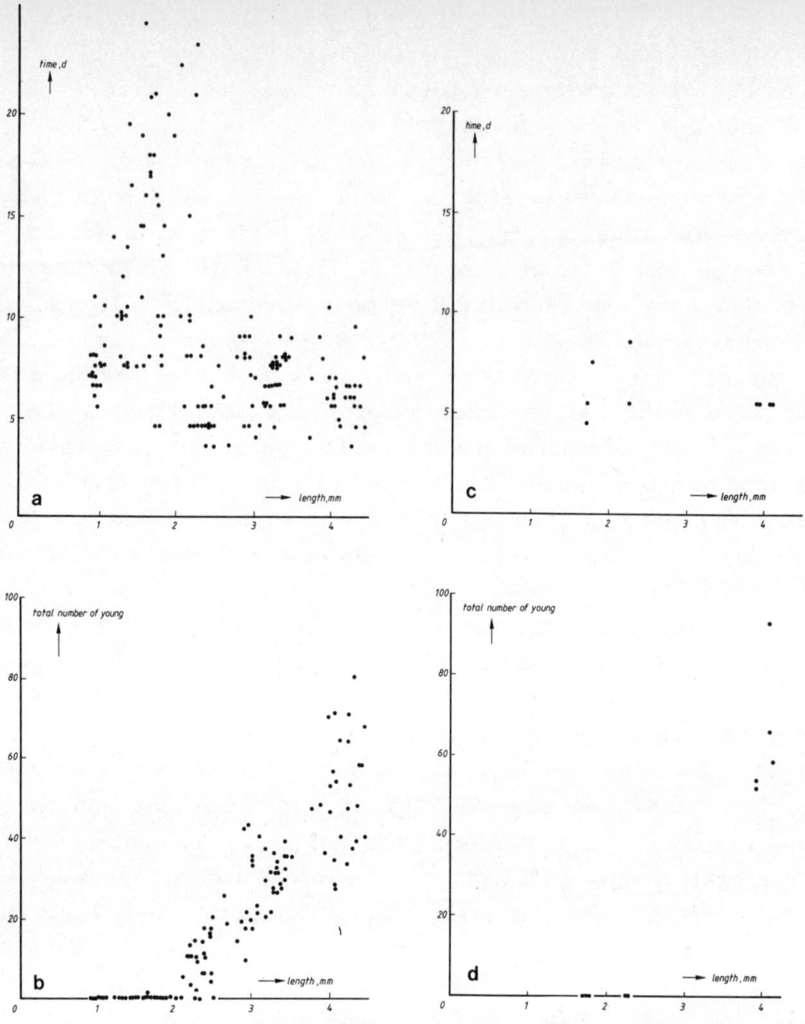

Fig. 10. The survival time of Daphnia magna at starvation after a period of abundant food (Chlorella) and the total number of young of each daphnid born in this period, as a function of length in stocked (a, b) and freshly prepared (c, d) media at 20°C.

If we take κ to be 1/4, we have for an animal of 2 mm length: $X_s \cong 10^4$ cells/ml, which corresponds to 0.1 mg C/l, well within the observed difference in carbon content. In fact, such low threshold food densities pose a practical problem in starvation experiments.
Substituting $W=W_J$ in the formula for the threshold food density, we obtain the threshold food density at which a population can rejuvenate

itself. It can be regarded as a lower bound for the threshold food density for the population. It was introduced by Lampert, 1977, who defined it to be such that the biomass of a population remains constant, and so just compensates mortality.
The ingestion rate at the threshold food density is known as the maintenance ration, and is given by $\dot{I}_s = \dot{\zeta} W\ [\dot{I}_m]/[\dot{A}_m]$; see Kooijman & Metz, 1984. For daphnids about 3 mm long, it equals some 6 chlorella cells a second at 20°C, see Kooijman, 1983. The utilization rate at this ingestion rate eventually becomes equal to the routine metabolic rate, so $\dot{I} = \dot{M} = \dot{\zeta} W$.
The quotient $\dot{M}/\dot{I}_s$ equals $[\dot{I}_m]/[\dot{A}_m]$, i.e. the conversion factor for food into assimilation energy. Kersting, 1983, inverted this relation by converting the maintenance ration to an energy measure and used the respiration/ingestion ratio to determine the threshold food density. This procedure is valid if it is known how the conversion to an energy measure should be performed for the food and animal species under consideration.
Some species show remarkable adaptations in their life histories to the size dependent effects of a drop in food density. As has been mentioned before, daphnids normally reproduce parthenogenetically. However, large female daphnids start producing males when the food density declines rapidly. This situation usually involves a large variance of weights in the population, because large animals (older ones, grown up at high food density) as well as small ones (new-born animals due to the high reproduction rate) are abundant. They produce no males when food density declines slowly, usually corresponding to a smaller variance of weights, see Kooijman, 1983. After copulation, those females produce winter eggs, which represent a resting stage, see Fig. 11. This behaviour can be understood when we realize that at non-increasing food densities, the small individuals will outcompete the large ones (because of their lower threshold food density) if the variance in weights is large, whereas this mechanism is much less operative if the variance is small.

Fig. 11. Life cycle of Daphnia magna. The female normally produces eggs in a brood pouch (a). The eggs developing into young females, which are released just before moulting (b). Under certain circumstances, eggs develop into males (c), which will copulate with females with empty brood pouches (d). These females then produce two winter eggs (e) in an ephippium, which represents a resting stage (f). For further details, see text.

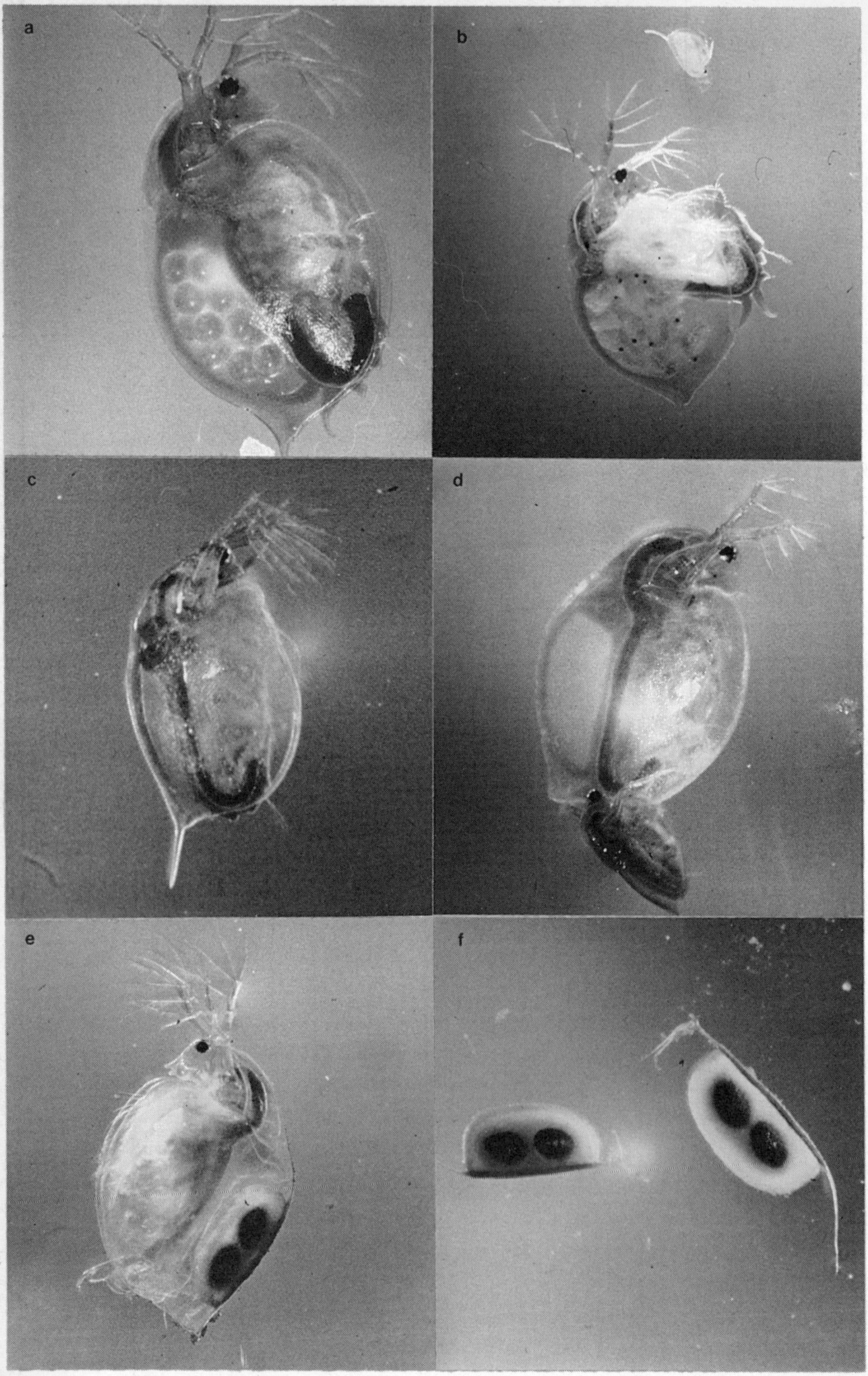
a
b
c
d
e
f

UNICELLULAR DYNAMICS

The food items ingested by multicellulars usually satisfy all their nutritional requirements although many exceptions to this general rule have been noted. For instance, Checkley, 1980, reports that copepods ingest many more diatoms if these are starved of nitrogen, in order to fulfill their requirement for this element. Scott, 1981, reports that the marine rotifer Brachionus plicatilis requires dissolved vitamin B_{12}. In unicellulars the uncoupling of the various nutritional and energy sources is much more widely spread. The interaction between these inputs will not be discussed here; we will assume that the input of the non-limiting nutrients remains constant. A further essential difference between the dynamics of multicellulars and unicellulars is that in the latter, proliferation is coupled to growth of individuals more directly: the asexual reproduction of unicellulars can usually be well described by a division following attainment of a certain division weight, W_D. For a division into p parts, we have $W_D = pW_b$. In this section we shall consider the implications of the present model for unicellular dynamics, and we shall discuss its relations to relevant results in the extensive literature on this subject. Most of this literature focuses on the description of equilibrium states of chemostats, where the dilution rate, and eventually the population growth rate is constant, in terms of measurable quantities like yield, uptake rate, and cell quota. We shall evaluate these relations on basis of the present model, first for non-conservative substrates, i.e. substrates whose degradation provides the energy necessary for maintenance and proliferation, and secondly for the special case of conservative substrates, here called nutrients, i.e. substrates which provide the chemical elements to be incorporated in the biomass.
The dynamics turns out to be dependent on the scaling of uptake rate of substrates with cell size. We shall consider two scaling relations:

- the uptake is proportional to $W^{2/3}$ because this has been observed to be relevant for ectotherms and
- uptake is proportional to W, because this is the assumption usually made in the literature on unicellular dynamics.

This gives four categories to be studied.

1. Unicellulars feeding by phagocytosis like ciliates probably resemble multicellulars most closely in their dynamics. At constant food density the division interval, D, is easily found from the equation $W(D) = W_D$.

It corresponds to the expression for the length of the juvenile stage in multicellulars, given before for $\kappa = 1$, because no reproduction is involved. In situations where death is not important, the population growth rate at constant food densities is found from $\dot{r} = (\ln p)/D$ for division into p. parts. It results in

$$\dot{r} = \frac{[\dot{A}_m]f - \dot{\zeta}}{[S_m]f + \eta} (\ln p)/\{3\,(W_\infty^{1/3} - 1) \ln \frac{W_\infty^{1/3} - W_b^{1/3}}{W_\infty^{1/3} - (pW_b)^{1/3}}\}$$

where $W_\infty^{1/3} = f[\dot{A}_m]/\dot{\zeta}$. For $f < (\dot{\zeta}/[\dot{A}_m])(pW_b)^{1/3}$ the cells will not reach their threshold weight for division. We will see in the next section that in case of an uptake rate proportional to weight instead of surface area, the population growth rate reduces to the first factor.

2. The literature on bacterial dynamics actually assumes that the uptake rate of substrates is proportional to weight, so in the present notation $\dot{I} = [\dot{I}_m]fW$. For bacilli it may be argued that the surface area scales approximately with weight, because the rod diameter is more or less constant. The balance equation for the storage then becomes

$$\dot{S} = W\,\{[\dot{A}_m]f - \frac{[S]}{[S] + \eta}\,(\frac{[\dot{A}_m]\,\eta}{[S_m]} + \dot{\zeta})\} \text{ for } [S] \geq \frac{[S_m]\,\dot{\zeta}}{[\dot{A}_m]}$$

Growth is given by $\dot{W} = W\,\frac{[S][\dot{A}_m]/[S_m] - \dot{\zeta}}{[S] + \eta}$. Expressed in the dynamics of the weight-specific storage, $[\dot{S}] = \dot{S}/W - S\dot{W}/W^2$, we have $[\dot{S}] = [\dot{A}_m]\,(f - [S]/[S_m])$. For $[S] \leqq [S_m]\,\dot{\zeta}/[\dot{A}_m]$ the cell dies of starvation.

The population growth rate at constant substrate density is found from $\dot{r} = (\ln 2)/D$, and the division interval D from $W(D) = 2\,W_b$, as before. The result is $\dot{r} = ([\dot{A}_m]f - \dot{\zeta})/([S_m]f + \eta)$.
Substitution of $f=X/(K+X)$, reveals the hyperbolic relation between $\dot{r}$ and the substrate density X, which for $\dot{\zeta}=0$ reduces to the well-known Monod equation $\dot{r}=\dot{r}_m X/(X+K')$ for $\dot{r}_m=[\dot{A}_m]/([S_m] + \eta)$ and $K'=K\eta/([S_m] + \eta)$. It still plays a central role in the literature on microbial dynamics, in spite of the existence of maintenance in bacteria growing on non-conservative substrates having been known for a long time. Pirt, 1965 noticed that the slope of the line in a plot of inverse yield vs inverse population growth rate equals the so-called 'maintenance coefficient' in equilibrium situations.
The yield, Y, defined as the quotient of biomass formed and substrate consumed, is

$$Y = \dot{r}/([\dot{I}_m]f) = \dot{r}([\dot{A}_m] - [S_m]\dot{r})/\{\,[\dot{I}_m]\,(\dot{\zeta} + \eta\dot{r})\}.$$

The present model simplifies to the formula in Pirt, 1965, for zero storage, $[S_m] = 0$. In our notation, we have

$\frac{1}{Y} = \eta \frac{[\dot{I}_m]}{[\dot{A}_m]} + \frac{[\dot{I}_m]\ \dot{\zeta}}{[\dot{A}_m]\ \dot{r}}$. The parameter $\frac{[\dot{A}_m]}{\eta\ [\dot{I}_m]}$ has been called the "true yield" and corresponds to the yield for $\dot{\zeta} = 0$. Pirt calculated the "maintenance coefficient", corresponding to $\dot{\zeta}\ [\dot{I}_m]/[\dot{A}_m]$, for two bacterial species growing on two substrates, aerobically and anaerobically, and obtained a range of 0.083 to 0.55 h^{-1} on basis of dry weight (apart from a selomonad that behaved erratically). This wide range is due to differences in the ability of the bacteria to convert the substrate into energy. The corresponding range for the product of the "maintenance coefficient" and the "true yield", $\dot{\zeta}/\eta$, confusingly called the "maintenance rate constant" (Gons & Mur, 1975), is 0.0393 - 0.0418 h^{-1}. So, whereas $[\dot{A}_m]$ is highly dependent on the environment, this indicates that the internal parameters $\dot{\zeta}$ and η are essentially constant.

Later, Stouthamer, 1979, 1980 analysed the dependence of the "specific rate of consumption" $\dot{U}$ on the population growth rate $\dot{r}$. The "specific rate of consumption" also known as the "specific uptake rate", is defined as the total substrate (or nutrient) uptake by the biomass, per unit biomass. In the equilibrium situation of a constant substrate density, it is given by $\dot{U}=\dot{r}/Y$. Stouthamer derived a linear relationship between $\dot{U}$ and $\dot{r}$, but he observed deviations which he explained by the growth rate influencing the fermentation pattern. Our model provides an alternative explanation, which can be regarded as a further interpretation of the explanation by Stouthamer, resulting in $\dot{U} = [\dot{I}_m](\dot{\zeta} + \eta\dot{r})/([\dot{A}_m] - [S_m]\dot{r})$. This hyperbolic relation between $\dot{U}$ and $\dot{r}$ is meaningful only on the interval $\{0,([\dot{A}_m] - \dot{\zeta})/([S_m] + \eta)\}$; its asymptote on $\dot{r} = [\dot{A}_m]/[S_m]$ lies outside this interval.

In their study on light-limited growth in algae, Gons & Mur, 1975 and van Liere, 1979, noticed that in an equilibrium situation the intercept of the linear relation between specific light-energy uptake rate and growth rate can be interpreted to be the maintenance rate constant. This relation follows from our model for zero storage, $[S_m] = 0$ and a volume-related input: $\dot{I} = [\dot{I}_m]f\ W$. This relation may be realistic if chlorophyll content is volume related, and the cell is optically thin.

In that case, the volume increase of a cell is exponential, as has been found by Donze & Nienhuis, 1973. See Fig. 12. Gons & Mur, 1975 also observed a deviation from linearity for high uptake rates, which they explained by assuming that light energy utilization is less

efficient at higher growth rates. Their data closely obey the hyperbolic relation $\dot{I}/W =$
$[\dot{I}_m]\ (\dot{\zeta} + \eta\dot{r})/([\dot{A}_m] - [S_m]\dot{r})$ obtained from our model for $[S_m] \neq o$.

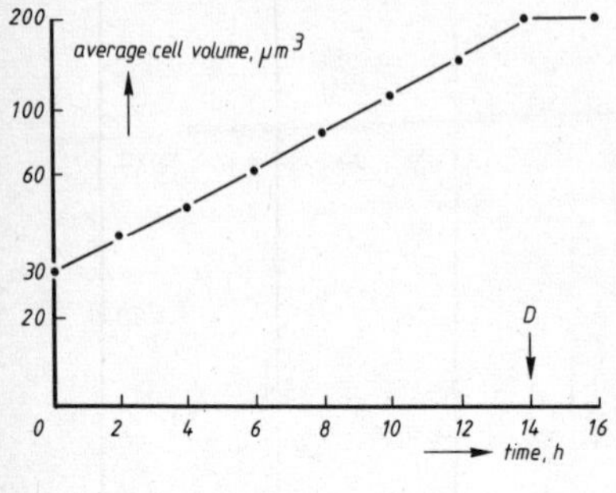

Fig. 12. Exponential volume increase during the light period in Scenedesmus obliquus growing in a synchroneous culture at 20°C. From Donze & Nienhuis, 1973.

See Fig. 13. So this hyperbolic relation affords an alternative explanation for the deviation.

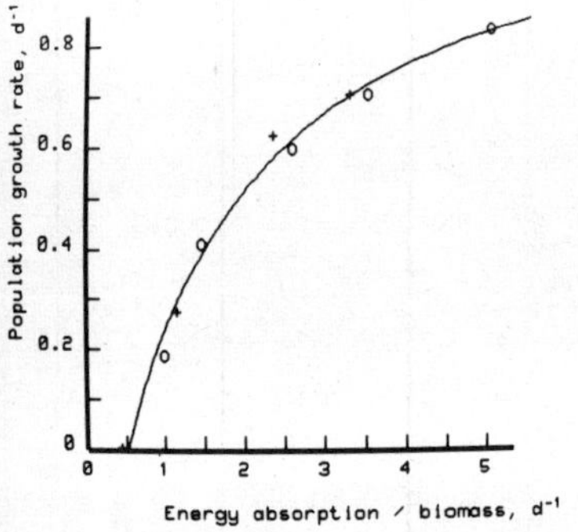

Fig. 13. Relation between biomass-specific energy absorption and population growth rate in Scenedesmus protuberans at 20°C. Data from Gons & Mur, 1975. The dots refer to an illumination of 30 kJ/d and the crosses to 13 kJ/d. The curve is $X = (1+2.284\ Y)/(1.921-1.636Y)$.

3. The literature on nutrient-limited growth of algae is very extensive. As far as I know, it assumes that input is proportional to W, i.e. $\dot{I} = [\dot{I}_m]fW$. There being no routine metabolism for nutrients, we may put $\dot{\zeta} = o$. It has, however, often been noted that nutrients are excreted. If we assume that the rate of excretion is a fixed proportion, $(1-\kappa)$, of the utilization rate, the resulting balance equations for the storage becomes

$$\dot{S} = [\dot{I}_m]fW - \frac{[S]\ \eta/\kappa}{[S]+\eta/\kappa}\ \frac{[\dot{I}_m]}{[S_m]}\ W.$$

Growth is given by $\dot{W} = W\,[S][\dot{I}_m]\{[S_m]([S]+\eta/\kappa)\}^{-1}$. Expressed in the dynamics of the weight-specific storage, we have $[\dot{S}] = [\dot{I}_m](f-[S]/[S_m])$. These equations correspond to the formulae presented by Nyholm, 1976, 1977. At constant nutrient concentration, the storage is at its equilibrium value $S^* = [S_m]fW$. If, in this situation, death is not important, the population growth rate can be obtained as before, and results in $\dot{r} = [\dot{I}_m]f/([S_m]f + \eta/\kappa)$.
The first investigator to recognize the importance of storage in nutrient-limited growth of algae was Droop, 1968.
He introduced the concept of cell quota Q, defined as the nutrient concentration in the cell. In the equilibrium situation of a chemostat, this concentration equals the quotient of the substrate consumed and the biomass formed, and so we have $Q=1/Y = [\dot{I}_m]\ f/\dot{r}$. Substitution of $\dot{r}$ gives $Q = [S_m]\ f+\eta/\kappa$. Droop considered the situation of zero loss, i.e. $\kappa=1$, from which it follows that $Q=[S]+\eta$. The parameter η, which we took to be the amount of nutrient required for formation of a unit of biomass, has been called the subsistence quota, or the minimum quota needed for growth to proceed. This is obvious, because $Q=\eta$ implies $[S]=0$. Droop observed a linear relationship between the cell quota and the specific uptake rate $\dot{U}$. In the equilibrium situation of a constant nutrient density, we have $\dot{U} = \dot{r}Q$. Substitution of $\dot{r}$ and $f = (Q-\eta/\kappa)/[S_m]$ gives $\dot{U} = (Q-\eta/\kappa)\ [\dot{I}_m]/[S_m]$, which is linear in Q.

4. The parameter η can be interpreted as the quotient of the amount of nutrients built into the structure of the algae and cell size. Comparing different species, Shuter, 1978, found that the nitrogen and phosphorus contents of algae that do not grow because they are starved of these nutrients, scale with volume to the power 0.709 ± 0.066. This finding suggests that the amount of these structural nutrients is essentially related to surface area. The weight-specific amount of nutrient needed for growth therefore decreases with cell weight. If we assume the uptake of nutrients to be also surface related, i.e. $\dot{I} = [\dot{I}_m]fW^{2/3}$ we have the balance equation $\dot{S} = [\dot{I}_m]fW^{2/3} - \dot{C}$. The utilization rate $\dot{C}$ now relates to growth $\dot{W}$ as $\kappa\dot{C} = 2/3\ \eta W^{-1/3}\ \dot{W}$. Nutrient absorbtion being much more rapid than growth, the storage is at its equilibrium value, S^*, at constant input. Cell growth is usually assumed to be exponential. This implies that S^* scales with $W^{2/3}$, and suggests that $S^* = [S_m]fW^{2/3}$. Writing $\{S\}$ for the surface area-specific storage, i.e. $\{S\}=S/W^{2/3}$, the balance equation is

$$\dot{S} = [\dot{I}_m]fW^{2/3} - (\eta/\kappa)W^{2/3}([\dot{I}_m]/[S_m])\ \{S\}/(\{S\}+ \eta/k) \text{ and}$$

$$\dot{W} = W(3/2)([\dot{I}_m]/[S_m])\ \{S\}/(\{S\} + \eta/\kappa)$$

When expressed in the dynamics of the surface-specific storage, we have $\{\dot{S}\} = \dot{S}/W^{2/3} - 2/3\ S\dot{W}/W^{5/3}$, so $\{\dot{S}\} = [\dot{I}_m]/(f-\{S\}/[S_m])$.
If death is not important, the population growth rate at constant nutrient concentration is $\dot{r} = 1.5\ [\dot{I}_m]f\{[S_m]f + \eta/\kappa\}^{-1}$.
The yield is given by $Y=\dot{r}\ \{[\dot{I}_m]fA\}^{-1}$ where A is the quotient of the average value of $W^{2/3}$ and of W in the population of cells. If there is any scatter in the threshold weight for division, the distribution of cell sizes at constant nutrient density usually converges rapidly to the stable cell size distribution. This has density $2W_b/W^2$, see Voorn, 1983, so $A = (2/3)\ W_b^{2/3}\ (1-2^{-1/3})\ \{W_b 2\ln 2\}^{-1} \cong 0.1\ W_b^{-1/3}$
The specific uptake rate $\dot{U}$ is $\dot{U} = \dot{r}/Y = ([\dot{I}_m]/[S_m])(Q3/2 - A\eta/\kappa)$ which means that Droop's linearity between $\dot{U}$ and Q still holds.

DISCUSSION

The course of the weight, storage and reproduction rate in an ectotherm containing the parameters $P = \{W_b,\ W_J,\ K,\ [\dot{I}_m],\ [\dot{A}_m],\ [S_m],\ \kappa,\ \dot{\zeta},\ \eta,\ \omega\}$ has been described with a set of two coupled differential equations. The parameters involving energy, viz. $[\dot{A}_m]$, $[S_m]$, $\dot{\zeta}$, η and ω, only occur as ratios in the model, so only nine parameters have to be estimated, and no actual conversion to energy is necessary. Its various aspects have been checked for Daphnia magna feeding on Chlorella. At constant food density the equations for weight, storage and reproduction rate can be written explicitly as a function of age. Even in this case, the storage has an effect on growth and reproduction behaviour, resulting in, e.g., a decrease in production efficiency for increasing food densities. This result contradicts a statement made in Kooijman & Metz, 1984. For zero storage, $[S_m]=0$, the present equations for growth and reproduction reduce to those in Kooijman & Metz, 1983. Apart from the considerations on starvation, no attempt has yet been made to analyse the implications of the model in dynamic environments. For this purpose it is necessary to develop techniques for handling two-dimensional, or, if age is to be included, three-dimensional state spaces for individuals with survival boundaries. As we have shown above, our model can be used to describe nutrient- and light-limited algal growth, and substrate-limited bacterial growth, as well as being able to describe growth in multicellulars. For unicellulars it affords explanations of some observed deviations from current theories. For nutrient-limited algal growth, the model becomes much simpler involving only the parameters $\{W_b$, K, $[\dot{I}_m]$, $[S_m]$, κ, $\eta\}$, where W_b disappears when input is weight-related.

A remarkable feature of the reasoning here presented is that, at constant food density, the respiration rate can be written as a weighted sum of a surface area and a weight measure, this sum approximately scaling with $W^{0.8}$ whereas the routine metabolic rate scales with weight for individuals of the same species. The dependence of respiration rate on weight has caught the attention of many research workers in ecophysiology, who usually compare (widely) different species. Their work prompts a study of the implications the model has for the comparison of respiration rates and related variables between species of different sizes. The key to such comparisons is the maximum size a species can reach, which can be written, as has been shown, as a simple function of the parameters. These parameters must therefore vary in a systematic way between species of different size. It has proved to be possible to explain in this way the relations reported in the literature between, on the hand, size of a species and, in the other, ingestion rate, animal abundance, maximum growth rate, respiration rate, reproduction rate, duration of pre-reproductive period, starvation time as well as maximum population growth rate. For further discussions, see Kooijman, 1984.

Our model formulations shows some promise for studies on phytoplankton-zooplankton interactions. Some progress may be expected from a study of population dynamics on a basis of budgets.

ACKNOWLEDGEMENTS

I would like to thank Mrs A. de Ruiter for her assistance in the experimental work.

LITERATURE

Brafield, A.E. & Llewellyn, M.J., 1982. Animal energetics. Balckie, Glascow.

Checkly, D.M.Jr. 1980. The egg production of a marine planktonic copepod in relation to its food supply: Laboratory studies. Limnol. Oceanogr. 25, 430-446.

Conover, R.J. 1978. Transformation of organic matter In: Kinne, O. (ed) Marine Ecology, Vol 4: Dynamics, 221-499. J. Wiley & Sons, Chichester.

DeMott, W.R., 1982. Feeding selectivities and relative ingestion rates of Daphnia and Bosmina. Limnol. Oceanogr. 27, 518-527.

Donze, M. & Nienhuis, H. 1973. Studies on photosynthesis in synchronous cultures of Scenedesmus. In M. Donze (ed.). Thesis, University of Leiden, 85 pp.

Droop, M.R. 1968. Vitamin B_{12} and marine ecology. IV. The kinetics of uptake, growth and inhibition in Monochrysis lutheri. J. Mar. Biol. Ass. U.K. 48, 689-733.

Droop, M.R., 1983. 25 Years of Algal Growth Kinetics. Botanica Marina 26, 99-112

Frost, B.W., 1972. Effects of size and concentration of food particles on the feeding behaviour of the marine copepod Calanus pacificus. Limnol Oceanogr. 17, 805-815.

Geller, W.K., 1975. Die Nahrungsaufnahme von Daphnia pulex in Abhängigkeit von der Futterkonzentration, der Temperatur, der Körpergrösse und dem Hungerzustand der Tiere. Arch Hydrobiol/ Suppl 48, 47-107.

Gons, H.J. & Mur. L.R., 1975. An energy balance for algal populations in light-limiting conditions. Verh. Internat. Verein. Limnol. 19, 2729-2733.

Green, J., 1954. Size and reproduction in Daphnia magna (Crustacea: Cladocera). Proc. Zool. Soc. Lond. 124, 535-546.

Green, J., 1956. Growth, size and reproduction in Daphnia (Crustacea: Cladocera). Proc. Zool. Soc. Lond. 126, 173-204.

Kersting, K., 1983. Direct determination of the "threshold food concentration" for Daphnia magna. Arch. Hydrobiol. 96, 510-514.

Kersting, K. & v.d. Leeuw-Leegwater, C., 1976. Effects of food concentration on the respiration rate of Daphnia magna. Hydrobiologia 49: 137-143.

Khmeleva, N.N., 1972. Intensity of generative growth in crustaceans. Dokl.Acad.Sci. U.S.S.R. (Biol Sci) 207, 633-636.

Kooijman, S.A.L.M., 1983. Toxicity at population level. Proceedings of the ESA/SETAC Symposium on Multispecies Toxicity Testing, (to be published)

Kooijman, S.A.L.M., 1984. Energy budgets can explain body size relations. Submitted for publication.

Kooijman, S.A.L.M. & Metz, J.A.J., 1984. On the dynamics of chemically stressed populations: the deduction of population consequences from effects on individuals. Ecotoxicology and Environmental Safety 8, to appear.

Lampert, W. 1977. Studies on the carbon balance of Daphnia pulex de Geer as related to environmental conditions III. Production and production efficiency. Arch. Hydrobiol. Suppl 48, 336-360.
IV. Determination of the "threshold" concentration as a factor controlling the abundance of zooplankton species. Arch. Hydrobiol. Suppl. 48, 361-368.

Lemke, H.W. von & Lampert, W., 1975. Veränderungen in Gewicht und der Chemischen Zusammensetzung von Daphnia pulex im Hunger. Arch. Hydrobiol. Suppl. 48 (1) 108-137.

Liere, E. van, 1979. On Oscillatoria agardhii Gomont, experimental ecology and physiology of a nuisance bloomforming cyanobacterium. Univ. of A'dam Ph.D. thesis.

Nyholm, N, 1977. Kinetics of Phosphate Limited Algal Growth. Biotechnol. Bioeng. 19, 467-492.

Nyholm, N. 1976. A Mathemathical Model for Micribial Growth under Limitation by Conservative Substances. Biotechnol. Bioeng. 18, 1043-1056.

Paloheimo, J.E., Crabtree, S.J. & Taylor, W.D., 1982. Growth model for Daphnia. Can.J.Fish.Aquat.Sci. 39, 598-606.

Pirt, S.J., 1965. The maintenance energy of bacteria in growing cultures. Proc. Roy. Soc. London (B) 163, 224-231.

Porter, K.G., Gerritsen, J. & Orcutt, J.D.Jr., 1982. The effect of food concentration on swimming patterns, feeding behaviour, ingestion, assimilation and respiration by Daphnia. Limnol. Oceanogr. 27, 935-949.

Richman, S., 1958. The transformation of energy by Daphnia pulex. Ecological Monographs 28, 273-291.

Scott, J.M. 1981. The vitamin B_{12} requirement of the marine rotifer Brachionus plicatilis. J. mar. biol. Ass. U.K.61, 983-994.

Shuter, B.J. 1978. Size dependence of phosphorus and nitrogen subsistence quotas in unicellular microorganisms. Limnol Oceanogr. 23, 1248-1255.

Stouthamer, A.H., 1979. The search for correlation between theoretical and experimental growth yield. Microbial Biochemistry 21, 1-47.

Stouthamer, A.H. 1980. Energy regulation of microbial growth. Vierteljahrschrift der Naturforschenden Gesellschaft in Zürich 125, 43-60.

Strathmann, R.R. 1966. Estimating the organic carbon content of phytoplankton from cell volume or plasma volume. Limnol. Oceanogr. 12, 411-418.

Strickler, J.R., 1982. Calanoid copepods, feeding currents and the role of gravity. Science 218, 158-160.

Taub, F.B., 1982. A critical review of Daphnia ecology and culture. Unpublished report. Office of pesticides and toxic substances U.S. EPA 401 M Street, S.W. Washington, D.C. 20460.

Tessier, A.J., Henry, L.C., Goulden, C.E. and Durand, M.W., 1983. Starvation in Daphnia: Energy reserves and reproductive allocation.

Tessier, A.J. and Goulden, C.E., 1982. Estimation of food limitation in Cladoceran populations. Limnol. Oceanogr. 27, 707-717.

Voorn, W.J., 1983. Statistics of cell size in the steady-state with applications to Escherichia coli. Ph-D thesis, Amsterdam.

Watts, E. & Young, S., 1980. Components of Daphnia feeding behaviour. J. Plankton Res. 2, 203-212.

Wulff, F.V., 1980. Animal community structure and energy budget calculations of a Daphnia magna (Straus) population in relation to the rock pool environment. Ecological Modelling 11, 179-225.

LIST OF FREQUENTLY USED SYMBOLS

Symbol	dimension	interpretation
X, X_s	food (nutrient).length^{-3}	food or nutrient density, threshold-
f		functional response as a portion to its maximum
$\dot{I}$, $\dot{I}_m$, $\dot{I}_s$	food (nutrient).time^{-1}	ingestion rate, maximum-, threshold-
$[\dot{I}]$	food (nutrient).length^{-2}.time^{-1}	size specific-
$\dot{A}$, $\dot{A}_m$,	energy.time^{-1}	assimilation energy rate, maximum-
$[\dot{A}_m]$	energy.length^{-2}.time^{-1}	size specific-
S, S_m	energy (nutrient)	storage, maximum-
$[S]$, $[S_m]$	energy (nutrient).length^{-3}	weight specific-, maximum weight specific-
$\dot{S}$	energy (nutrient).time^{-1}.	storage change rate
W, W_b, W_J	length^3	weight,-at birth,-at the end of the juvenile stage
W_D, W_∞, W_m	length^3	-at division, adult-, maximum-
$\dot{W}$	length^3.time^{-1}	growth rate
$\dot{R}$, $\dot{R}_m$	time^{-1}	reproductive rate, maximum-
$\dot{M}$	energy.time^{-1}	routine metabolic rate
$\dot{C}$	energy (nutrient).time^{-1}	utilization rate
$\dot{F}_m$	length^3.time^{-1}	maximum filtration or searching rate
K	food (nutrient).length^{-3}	saturation constant
κ		proportion of utilized energy spent on growth and routine metabolism
η	energy.length^{-3}	energy requirement for a unit increase in weight
ζ	energy.length^{-3}.time^{-1}	maintenance energy consumption rate per unit of weight
ω	energy.length^{-3}	energy requirement for a unit increase in offspring tissue

<u>APPENDIX</u> The idea behind the κ-rule.

Blood has a low take-up capacity for energy (or nutrient), but a relatively high transportation rate; Many times an hour it is pumped through the body. Passing along the gut, it takes up any energy delivered there, which is rapidly circulated through the whole body. At separated sites along the vessels, two types of cells are waiting to pick up energy from the blood, the (many) somatic and (few) ovary cells. These cells are not able to react to each other's activities other than through the concentration of energy in the blood. The carriers that remove energy from the blood across the cell membrane, into the two types of cells, have the same activity dependence on the concentration of energy in the blood, but they may have different efficiencies. One might assume that all carriers are identical, but that the numbers of carriers in the membranes of the two types of cells differ. This is the basis of the κ-rule. The efficiency (a number) of the carriers in the ovary cells is controlled by hormones, depending on age, size and environment. For Daphnia, this efficiency seems to be constant as long as feeding states permits. Inside each somatic cell, the energy partition occurs within the same cell, maintenance and growth are natural competing processes. It is therefore reasonable to assume that maintenance is at the expense of growth, not at reproduction as long as energy permits. The hormonal system has to intervene for maintenance to be at the expense of reproduction. It will do so in poor conditions. The main part of the maintenance energy is involved in the course grain regulation of the enzyme system of the cell; i.e. a continuous process of breaking down and building up. This process is closely related to those occurring during cell growth and division. (Observing the oxygen consumption pattern, the energy spent in movement in Daphnia is only of minor importance.) At still other sites along the blood vessels, carriers regulate the energy content of the blood. They can not observe which type of cell removed the energy from the blood nor the energy influx along the gut.

They only see the energy content of the blood, the amount of stored energy and the size of the animal. Therefore, $[\dot{S}]$ cannot depent on κ. The size dependence of the energy regulation is plausible, because the stored energy is chemically represented by more or less massive solid lipids, which are deposited on certain surfaces inside the animal. Since these surfaces increase with $W^{2/3}$ and the volume with W in growing animals, the blood has a decreasing ability to reach the stored energy; the lipid layers grow thicker and blood can only reach the outer surface. It is therefore quite natural that $[\dot{S}]$ appears to be proportional to $W^{-1/3}$.

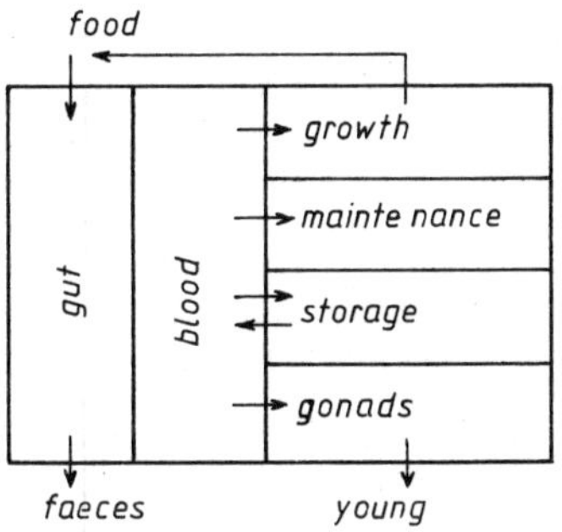

The Functional Response of Predatory Mites to the Density of Two-spotted Spider Mites

M.W. Sabelis

Department of Population Biology*
University of Leiden, P.O. Box 9516
Kaiserstraat 63, 2300 RA Leiden
The Netherlands

ABSTRACT

Stochastic queueing theory was applied to construct a model that calculates the rate of prey killing by a predatory mite from the underlying food conversion processes and the feeding-state dependent searching behaviour, as well as from prey behaviour and prey density. The model was provided with data derived from experiments on food conversion and searching behaviour of the predatory mite, *Phytoseiulus persimilis,* in prey colonies with either eggs or (preovipositional) females of the two-spotted spider mite, *Tetranychus urticae.* Predictions of the killing rate did not differ from measurements on predation at different prey densities in monocultures of each of both prey stages and in mixtures of the two prey stages. The agreement between model prediction and measurements in the prey stage mixture indicates that searching behaviour of the predatory mite is not modified as a result of the prey stages being presented together. Moreover, it was shown that partial consumption of the prey's food content is of crucial importance to an understanding of the number of prey eaten at high prey densities.

Sensitivity analysis of the model showed that 10% changes in the rate of food resorption from the gut affected the killing rate to an extent equivalent to ca. 50% changes in the (feeding state dependent) rate of successful attack. To put it into an evolutionary perspective selection to speed up the physiological processes underlying food conversion into egg biomass will be much more intense than selection for improved ability to capture prey. Because the rate of food resorption is linearly related to the food content of the gut (g) with coefficient r_{rs}, selection may lead to physiological improvements (in r_{rs} and $g_{\max}$) and to behaviour directed to keep g at its maximum. The latter evolutionary path may have resulted in the predatory mites having a strong tendency to feed to capacity by feeding on whatever prey stage they can seize given their feeding-state dependent effort in attack (time spent handling is small compared to the time needed to clear food from the gut). Partial prey consumption by well fed predators may also be viewed as a way to maximize g, but it is not clear why feeding is not continued until the prey is fully utilized. Possible reasons are discussed.

Introduction

The functional response has been defined by Solomon (1949) in terms of the relationship between the number of prey consumed per unit of time per predator and the prey density. Solomon stressed that it is basically an individual response of predators to the density of their prey, thereby implicitly recognizing the importance of the predatory behaviour for the functional response. In the context of population dynamics the emphasis should be on the response of all predators together. This entails the need to construct models that calculate the average rate of predation from the underlying individual behaviour of the predators and prompted a large number of theoretical studies (Holling, 1959a,b, 1961, 1965, 1966; Royama, 1971; Rogers, 19672; Nakamura, 1964; Fransz, 1974; Oaten and Murdoch, 1975;

* Part of this work was performed at the Department of Animal Ecology, Agricultural University, Wageningen.

Hassell et al., 1976; Van Lenteren and Bakker, 1976; Taylor, 1976; Hassell et al., 1977; Curry and DeMichele, 1977; Real, 1977; Hassell, 1978; Sjöberg, 1980; Sabelis, 1981; Mills, 1982; Arditi, 1983; Metz and Van Batenburg, 1984a,b,c). The model discussed below has been developed to calculate predation rates of the predatory mite, *Phytoseiulus persimilis* Athias-Henriot, in relation to the density of the two-spotted spider mite, *Tetranychus urticae* Koch (Sabelis, 1981).

Holling (1959ab) first developed a model, known as the 'disc' equation, which focussed on the time budget of the predator. The time available for foraging is considered to be the limiting variable. To seize and consume a prey item takes some time, thereby reducing the time available to search for other prey. When prey is abundant, the maximum number of prey caught would be equal to the predator's life time divided by the time spent handling a prey item. For predatory mites this maximum is much higher than realized; the time taken to handle a prey egg is about 5 minutes so that a maximum of 12 eggs per hour would be possible, whereas the maximum predation rate found in reality is slightly more than 1 egg per hour. The most important factor causing this difference is predator satiation; the motivation to seize prey decreases with the feeding state of the predator and when completely satiated with food the predator does not respond to prey stimuli. Hence, to model prey consumption by a predatory mite the predator's foraging behaviour should be related to its motivational state and the dynamics of this state should be taken into account. Such models have been developed by Holling (1966), Nakamura (1974), Fransz (1974), Taylor (1976), Curry and DeMichele (1977), Sjöberg (1980), Sabelis (1981), Mills (1982) and Metz and Van Batenburg (1984 a, b, c). The term 'prey density' requires further explanation. Spider mites are phytophagous arthropods that feed on the leaf parenchyma of their host plants. They give rise to patchy infestations, each patch consisting of a group of colonized leaves of one or more plants that are very close to each other. Within the patch the distribution of spider mites is much less heterogeneous, but yet it varies from leaf to leaf and depends on plant architecture. Hence, if there are more or less homogeneous distributions of spider mites, these are to be found on the leaves. Clearly, prey density cannot be defined solely in terms of the number of spider mites per unit area. Its definition requires a notion of the spatial scale and, hence, of the prey distribution. Here, however we will focus on the part of a leaf that is colonized by the spider mites. This part is readily recognized by the labyrinth of silken threads stretched between the cuticle surface and the prominent structures (e.g. ribs) on the underside of the leaves. Within this self-made web the adult females feed on the leaf and deposit their eggs, which develop into adults via a series of mobile feeding stages interrupted by moulting periods. Prey density should therefore be specified in terms of the prey stages that are present in the webbed area of the leaf. Of course, the developmental stage or the age of the predator should be specified as well.

To simplify the problem it is first assumed that one young mature predator female forages in a large webbed leaf area where all resident prey are in the same stage of development. Two prey stages will be considered below, i.e. either the egg or the female in the preoviposition phase, representing opposite extremes of mobility and size of the prey. After modelling the functional response to the density in monocultures of each of both prey stages, the model is extended to include prey consumption in mixed cultures of the prey, while assuming that all components of foraging behaviour are set by the feeding state of the predator and are not directly affected by the stage composition of the prey supply in the webbed leaf area. As a next step, the monoculture models are validated by comparing predicted and experimentally established values of the predation rate. If these validations are successful, then predictions of the mixed culture models are validated by experiment. This in fact constitutes a test of the assumption that the predator's behaviour is determined by its feeding state and not by the two prey stages being presented together. Discrepancies between prediction and measurement are an indication that foraging behaviour and/or prey stage preference have changed as a consequence of the simultaneous presence of two prey stages. This method was first put forward by de Wit (1960) in the context of competition between plant species. The first researchers to apply the method to the analysis of prey (stage) preference were Rabbinge (1976; see also De Wit and Goudriaan, 1978) - who used De Wit's yield equation - and Cock (1978) - who used Rogers' variant of Holling's disc equation (see also Chesson, 1983). But neither of these equations take the feeding state of the predator into account. The models discussed below do take this state into account and are an extension of models previously developed by Fransz (1974), Curry and DeMichele (1977), Taylor (1976) and Sjöberg (1980). Analytic formulations of these models are presented by Metz and van Batenburg (1984 a, b, c).

Clearly, the main goal of modelling in this study is to predict the predation rate from the underlying behaviour and food conversion physiology of the predatory mite, so that predictions can be compared with actual measurements and the explanatory value of the model can be assessed. In this way, model validation may help to detect gaps in our knowledge of the predator-prey system. Furthering our understanding of how the system works, is not the only goal, however. Using the validated model it is possible to assess the relative importance of each of the predator's traits to the predicted value of the predation rate. This so-called sensitivity analysis paves the way to speculate on which of the traits will be subject to the most severe selection pressure. Assuming that the ovipositional rate strongly depends on the rate of predation, selection is likely to act most intense on the predator's property that is crucial in determining the rate of predation. Hence, it is possible to formulate hypotheses on how natural selection will have moulded the behavioural and physiological properties of the predator. By reconsidering what is known of the predator's biology it may then be possible to trace the properties that may have been favoured by selection. The question of interest in this chapter is whether an increase in searching ability is more important in determining the predation rate, than a proportionally equivalent increase in the ability to convert food into egg biomass. The answer to this question will shed some new light on the evolutionary ecology of predatory mites.

The final goal of the modelling effort is to end up with a simple predation model for use as a building block (a subroutine) in models of the population dynamics of predator and prey. Rather than to select a simple predation model on a-priori grounds, it may be worthwile first to study the predation process in detail and then to decide on which model captures the essence of the predation process and is still simple enough for use in models of population dynamics. However, the specification of such a model will not be discussed here, but in the next chapter

Dynamics of the predator's feeding state

It may be reasonable to assume that the dynamics of the predator's motivational state parallel the dynamics of the food content of its gut. Changes in motivation can then be simulated by integrating the rate of ingestion and the rate of gut emptying (digestion, resorption and egestion) over sufficiently short time intervals. This simple approach allows for the calculation of an indicator of the motivational state during observation of predatory behaviour, so that this state variable and the behavioural components of prey searching can be related to each other. These relations can then be used in predation models to calculate the rate of predation.

The postulation of one intervening variable, indicating the hunger drive, may be too simple a hypothesis to account for all hunger related behaviour (Dethier, 1976). For example, it may be questioned, whether the motivational state adapts itself instantaneously to a change in the food content of the gut. Or to put it in the form of a testable hypothesis: does a certain level of gut fullness induce the same searching behaviour whether it is achieved via ingestion or gut emptying starting from large or small amounts of food in the gut respectively? The answer is yes, but - as may be expected - exceptions have been found (Sabelis, 1981). For example, 6 days of starvation at 20 ° C can lead to irreversible loss of strength or to prolonged periods of very aggressive attacks even after the gut is completely filled. In the latter case it appeared that the predators recovered from their state of metabolic imbalance after some hours. Notwithstanding these noteworthy exceptions the simple hypothesis formulated above holds under a wide range of conditions and it was not rejected at starvation periods shorter than 2 days (Sabelis, 1981).

Another important assumption underlying the above approach is that the quality of the food is constant. The existence of specific hungers for proteins, sugars or water has been shown for various organisms (Dethier, 1976). If the food quality differs among the developmental stages of spider mites, diet composition may be relevant in defining the motivational state of the predator. It is therefore important to test the model in such a way that deviating responses of the predator have at least some chance of being detected. One possible test method is to predict the predation rate in mixed cultures of prey stages from data on physiological and behavioural components obtained in monocultures of each prey stage. Discrepancies between model prediction and actual measurements of the predation rate may indicate - among others - that food quality differs among the prey stages. This approach is followed here. Another method is to compare e.g. the rate of oviposition of predatory mites in monocultures of each prey stage and in mixed cultures. The problem, however, is that such comparisons are only justified if the level of gut fullness is the same in each of these oviposition experiments. The (rather small) differences in reproduction reported in the literature are therefore certainly no proof of the developmental stages of the prey having differential food quality (Sabelis, 1981).

Under the above set of conditions the food content of the predator's gut can be quantified by measuring food ingestion and the removal of food from the gut by resorption and egestion, further referred to as gut emptying. To do this, a Cahn elektrobalance was found to be indispensable for weighing the tiny mites.

Ingestion

By interrupting feeding periods and measuring either the predator's weight increase (in the case of large prey) or the fraction of ^{32}P ingested from its ^{32}P-labelled prey (in the case of small prey stages) it was possible to show that the fraction of ingested food is related to feeding time t according to the negative exponential, $1- \exp(-r_i t)$, where r_i is the relative rate of ingestion. Estimates of r_i are presented in Table 1. Because the product of these ingestion constants and total feeding time per prey stage always exceeds 2, it is clear that most of the food (note that 1 - exp (-2) = 0.865) is ingested in the first few minutes of the feeding period. Because - as we will see below - the time constant of gut emptying exceeds the time constant of ingestion ($=1/r_i$) by a factor 100 to 3000, it is justified to conceive of the

Table 1

The relative rate of ingestion (r_i) and the maximum amount of food (f_{max}) ingested from prey eggs or prey females by single females of *Phytoseiulus persimilis* that had been starved for 2 days at 20°C. The maximum ingestible food content per prey stage is indicated by the difference (D) between fresh and dry weight and the summed amount of food ingested by three predators feeding consecutively on the same prey individual (see *).

Prey stage	r_i (min^{-1})	f_{max} (μg)	D (μg)
egg	1.8	1.0	0.8
female	0.05	7.8	17.9
female*		17.1*	17.9

ingestion process as a single gulp of the prey's food content.

To determine the maximum amount of ingestible food per prey stage starved predators were weighed before and after feeding on a particular prey stage. The results are presented in Table 1, together with the fresh and dry weight of the prey stages. Apparently, all prey stages were sucked dry except for the adult female prey. The latter was only utilized completely until after three consecutive predators had fed on it. Clearly, the adult female contains much more food than a single predator can stow in its gut. The same conclusion was drawn from feeding experiments with the other prey stages, using predators that were not severely starved, but had been deprived of food for short periods. Therefore, partial ingestion of the prey's food content should be taken into account. If $g_{\max}$ is the maximum food content of the gut and g the actual food content, then the amount of ingested food, f, is obtained from the following formula:

$$f = \min(f_{\max}, g_{\max} - g) .$$

Thus, the predator consumes either all the food in its prey ($f_{\max}$) or stops feeding, when satiated.

Gut emptying

The food ingested will be cleared from the gut by resorption and egestion. It is possible to estimate the food deficit of the gut by weighing a starved predator before and after a short period (ca. 1 hour) in which it is allowed to feed until satiation. Hence, the food deficit of the predator's gut can be measured in relation to the period for which the predator was deprived of food after being in a satiated state. The results presented in Figure 1 demonstrate that the maximum food content of the gut ($g_{\max}$) is equal to 8.1 μg and that the gut is emptied in an exponential fashion:

$$g_t = g_{\max} \exp(-r_g t)$$

where g_t is the food content of the gut after starvation for a period t starting from total satiation ($g_0 = g_{\max}$), and where r_g is the relative rate of gut emptying, which was ca. 0.81 day^{-1} at 15 ° C, ca. 1.84 day^{-1} at 20 ° C and ca. 3.11 day^{-1}at 25 ° C. That r_g depends on temperature, is to be expected in poikilotherms, such as mites. This method, however, may be biased because (1) it is difficult to define precisely when the predator is satiated, (2) a resorption 'leak' is inevitable during the period needed to satiate the predator and (3) the manipulations necessitated by the weighing procedure may affect the results. Therefore, an alternative method was developed to estimate r_g. It involves using a model of the predator's food expenditure (Figure 2), which is based on the assumption that predatory mites are 'weight watchers'; as soon as the amount of food available for maintenance processes exceeds the needs (modelled as the body weight exceeding an optimum), all food resorbed into the haemolymph is allocated to the terminal ocyte, but whenever an energy deficit arises because food for maintenance is used up by respiration and transpiration, the food is allocated to the soma to meet the needs. The relative rate of resorption r_{rs} can be estimated with the aid of this model if the food content of the gut and the weight loss via respiration, transpiration and oviposition is known. When prey eggs are abundant, the gut can be considered to be almost completely filled with food, and the concurrent ovipositional rate can be measured, which when multiplied by the weight of a predator egg gives the egg biomass produced per unit time. The weight loss via respiration and transpiration can be estimated from the weight loss of predators during a starvation period. Hence, when prey eggs are abundant and the gut is filled to its capacity, the relative rate of resorption r_{rs} is the only unknown in the following 'balance' equation

$$r_{rs} g_{\max} = r_{rt} b + R_{ov} E$$

$g_{\max}$ is the maximum food content of the gut (μg);r_{rt} is the relative rate of respiration and transpiration (day^{-1}); b is the somatic body weight (total weight minus gut content, weight of integument and terminal oocyte; all in μg);R_{ov} is the rate of oviposition (eggs/day); E is the weight of an egg at deposition (μg). An approximate formula to calculate r_{rs} is given by:

$$r_{rs} = (0.04(T-11)7.1 + 0.3(T-11)4.7) / 8 ,$$

where T, the temperature, should exceed 11 ° C. To find r_g the value r_{rs} should be raised by the relative rate of egestion r_{eg}. Approximate estimates of the defaecation rate and the diameter of the faecal droplets were obtained by observing well fed female predators: $r_{eg} = 0.01(T-11)$. Clearly, most of the ingested food is resorbed and then utilized for egg production. In conclusion, the alternative method based on the balance equation resulted in similar but somewhat higher values for r_g, i.e. 0.9 day^{-1} at 15 ° C, 2.07 day^{-1} at 20 ° C, and 3.22 day^{-1} at 25 ° C. The results obtained by both the direct and the indirect method were evaluated by supplying them to the model presented in Figure 2 for calculations of the rate of oviposition under various regimes of food ingestion. Comparison with actual data on oviposition under these food regimes showed that the estimates obtained via the indirect method are preferable

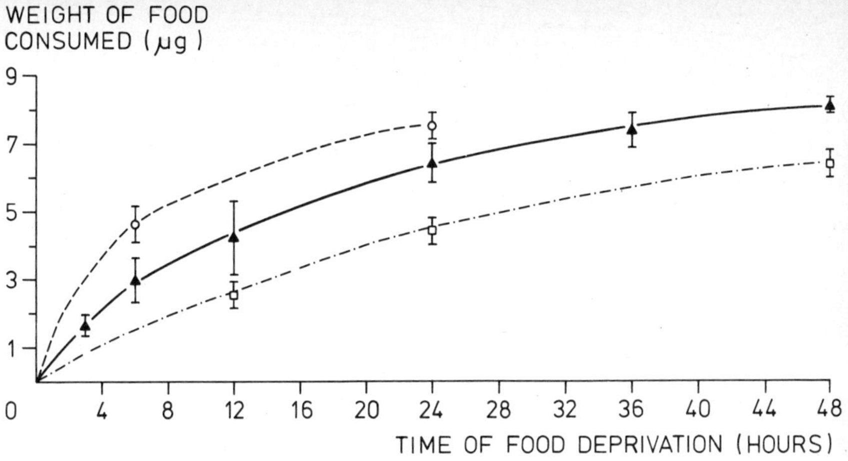

Figure 1 The weight of the food consumed until the female predators were satiated, and its relation to the duration of the period for which the predator was deprived of food prior to *ad lib* feeding. Results are presented for three temperature treatments during the food deprivation period: 15° C - □ , 20° C - ▲ , 25° C - 0 . These symbols are positioned to indicate mean values; the standard deviations above and below the means are indicated by bars; each experiment was replicated 20 times.

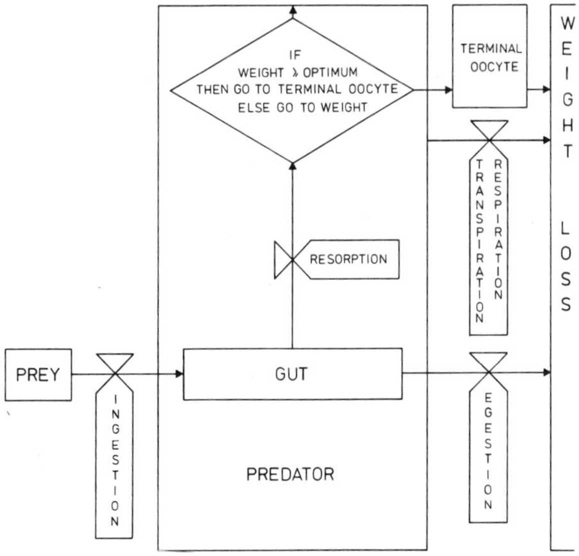

Figure 2 Forrester-diagram of food mass expenditure by a female predatory mite. The state variables are represented by rectangles and the rate variables by valve symbols. See text for explanation.

(Sabelis, 1981).

The rate of successful attack and its relation to the feeding state

The relation between the rate of successful attack and the feeding state of the predator can be derived from a time series of recorded behavioural events and from a simultaneous record of the dynamic changes in the predator's gut fullness. Because direct measurements of the latter are not feasible, an indirect method is the only possible solution. To this end the dynamic changes in gut fullness were calculated from the time series of ingestion events, given the initial feeding state. These calculations were based on the assumption that all predators were identical, that their food conversion is a deterministic process and that the food content of the prey is constant. In order to fulfil these requirements the following precautions were taken:

(1) temperature and humidity were kept constant at 20 ° C and 70% RH (r_g constant);

(2) prey items were standardized with respect to stage, size and feeding history (f_{max} constant);

(3) they were supplied only in monocultures and only at high prey density;

(4) the predators were young females 3 to 6 days old after their final moult (r_g not age dependent);

(5) they were in a satiated state at the start of the experiment.

Initial satiation was guaranteed by rearing predators at high prey supply and by sampling only those predators that has just finished feeding on a prey item. To enable behaviour to be studied at any feeding state desired, the predator was deprived of food for a known period before being allowed access to the prey. Then, the predator was released on a rose leaf on which a colony of spider mites was established. Hence, by allowing the prey to settle, predatory behaviour could be studied in spider mite colonies formed in a natural way, i.e. all possible cues relevant to prey searching (web, faeces. exuviae, chemical clues) and the prey's defense (web) were present. How the behavioural observations were analysed so as to obtain an estimate of the attack rate, will be discussed below.

Calculation of the attack rate

The attack rate σ is equal to the reciprocal of the time between successive prey catches. This time interval can be subdivided in the time spent searching and the time spent handling per captured prey. In formula:

$$\sigma = (t_{\text{searching}} + t_{\text{handling}})^{-1} .$$

Predatory mites do not spend much time in unsuccessful attacks. The attack response occurs after predator-prey contact and lasts only a few seconds. However, the time spent handling a captured prey takes several minutes and may therefore be relevant for calculating the time spent per successful capture. The handling period includes all time involved in attacking, feeding, palpating, cleaning and return feeding on the same prey individual. The time spent searching or its reciprocal, the searching rate, is determined by several factors, which will be discussed consecutively below.

First, consider the general case in which both predator and prey are walking non-stop on a plane surface. Then, the rate of encounter is determined by the prey density, the walking velocity and the distance of prey detection. If the predator does not detect its prey unless they make physical contact, then the predator searches along a path having a width equal to the sum of the diameters of predator and prey (w). Hence, the rate of encounter e equals:

$$e = d\, w\, v$$

d is the prey density and v is the resultant walking velocity, which is obtained from its component velocities of predator and prey (v_p and v_s) by simple vector analysis:

$$\vec{v} = \vec{v}_p + \vec{v}_s, \text{ so that } v^2 = v_p^2 + v_s^2 - 2v_p\, v_s \cos(\theta)$$

θ is the angle between the momentary walking directions of predator and prey. If the walking directions are mutually independent, all values of θ occur with equal probability. Hence, by taking expectations of the resultant walking velocity, it can be shown (Skellam, 1958) that:

$$E(v^2) = E(v_p^2) + E(v_s^2), \text{ since } \int_0^{2\pi} \cos(\theta)\, d\theta = 0 .$$

The resultant walking velocity v can be calculated using the following approximate formula:

$$v = (v_p^2 + v_s^2)^{\frac{1}{2}} .$$

In the extreme case of predator and prey having the same walking speed (which is tantalizing if they are walking in the same direction), Skellam (1958) showed that this formula underestimates $E(v^2)$ by only 10%.

The presence of a web poses several specific problems that must be solved before realistic calculations of e are possible. When the predator walks tortuously as is usual in the web of their prey, then local depletion of prey decreases the rate of encounter with prey. This local depletion will hardly affect the overall prey density provided that the prey colony is sufficiently large. To estimate the effect of recrossing areas already searched, a numerical simulation model was made which simulates the walking behaviour of predator and prey on the basis of the assumption that the direction of each step deviates from the direction of the previous step by an angle chosen at random from an experimentally defined frequency distribution, autocorrelations being accounted for. The mites were represented as circles and during simulation contacts between the circles were registered. To estimate the maximum possible decrease in e the searching paths were simulated for the extreme case that the predator consumed each prey encountered and the prey was immobile.

Another problem related to the presence of web arises from the three-dimensional web structure which allows the predator and prey to pass over and under each other without mutual contact. As the spider mites tend to concentrate on the leaf surface on which they feed, they are not randomly distributed in the web layer, whereas their perpendicular projections on the leaf surface tend to be so. Therefore, the reduction of e due to this lack of coincidence is most readily quantified by measuring the fraction of real contacts out of all coinciding projections on the leaf surface. Thus, the rate of encounter e should be multiplied by a fraction c, further referred to as the coincidence in the webbed space.

Not all encounters result in predation. Either the predator is not motivated to attack despite contacting a prey, or the prey is able to escape from predator attack. Hence, to calculate the rate of successful attack the encounter rate e should be multiplied by the success ratio s, which is measured as the fraction of successful attacks out of the total number of prey contacted by the predator (with its pedipalps or first legs).

Of course, the mites do not walk non-stop; they spend considerable time 'resting' (= no displacement). Thus, the rate of successful attack also depends on the proportion of the time that the mites spend in actual displacement. This fraction is termed the walking activity a. Assuming that the activity is non-periodic, that active and non-active periods alternate frequently (time constant $\gg \sigma^{-1}$) and that the active periods of predator and prey are independent of the momentary activity of neighbouring mites, the rate of encounter can be divided into sub-rates that pertain to the three relevant combinations of predator and prey activity; encounters between a walking predator and a walking prey occur with probability a_p times a_s, those between a walking predator and a 'resting' prey with probability a_p times $(1-a_s)$, and those between a walking prey and a 'resting' predator with probability a_s times $(1-a_p)$. The values of some of the behavioural components discussed earlier, depend on the particular combination of the state of predator and prey activity. For example, in a state of complete rest the predator draws up its first legs close to its mouthparts, so that its effective size becomes smaller. Another example is that a resting predator does not usually attack a prey touching the predator's rear. Consequently, the success ratio should be halved in this particular case.

In conclusion, the time spent searching is approximated by:

$$t_{\text{searching}}^{-1} = d[w_{ww} (v_p^2 + v_s^2)^{\frac{1}{2}} a_p a_s c_{ww} k_{ww} s_{ww} + w_{wr} v_p a_p(1-a_s) c_{wr} k_{wr} s_{wr} + w_{rw} v_s a_s(1-a_p) c_{rw} k_{rw} s_{rw}] .$$

The reduction factor resulting from the effect of recrossing areas already searched is represented by k. This factor k, the width of the searching path w (sum of diameters of predator and prey), the coincidence factor c and the success ratio s are provided with the subscripts ww, wr and rw to indicate their possible dependence on the combination of activity states of predator and prey respectively (w = walking; r = resting). The walking speed and the walking activity are not provided with these subscripts assuming that active and non-active periods of predator and prey alternate frequently and are independent of the momentary activity of other predator or prey mites. The latter assumptions are approximately correct when predator and prey forage in the prey's web and the prey's activity is not increased due to a deteriorating leaf condition.

The above approximate equation will be used to calculate the time spent searching per captured prey individual. To correctly estimate the average time interval between prey catches the time spent per successful attack should be calculated as the sum of the searching time and the handling time, the reciprocal of which gives the rate of successful attack σ.

Behavioural components

Components of the attack rate σ were estimated by analysing continuous observations of predatory behaviour in a spider-mite colony, and their relation with the feeding state was established using calculations of the dynamics of gut filling based on the actual time series of ingestion events. The observations were done by use of binocular microscope and video equipment. Details of experimental methods are given elsewhere by Sabelis, (1981). The results of the behavioural component analysis are summarized in Tables 2 and 3, Figures 3 and 4. A brief explanation is given below.

Table 2

The lateral reach and distal length of females of *Phytoseiulus persimilis* and of two stages of *Tetranychus urticae.*

		distal length (cm)	lateral reach (cm)	half the sum of width and length (cm)
female predator	walking	0.098	0.084	0.091
	resting	0.071	0.042	0.056
female prey (in preoviposition phase)	-	0.075	0.044	0.060
prey egg	-	0.014	-	0.014

Table 3

Parameters of the distribution of angular deviations obtained from analysing the walking paths of females of *Phytoseiulus persimilis* and *Tetranychus urticae* in the webbed leaf area.

	Mean (radians)	Standard Deviation (radians)	λ	Critical level(1)
satiated predator	0.000	0.301	-0.047	0.80
starved predator	-0.001	0.319	-0.035	0.85
female prey	0.000	0.259	-0.190	0.45

(1) $P(\chi^2 > X^2)$ where n=c-1-3 , c=no. of frequency class and X^2 = test statistic

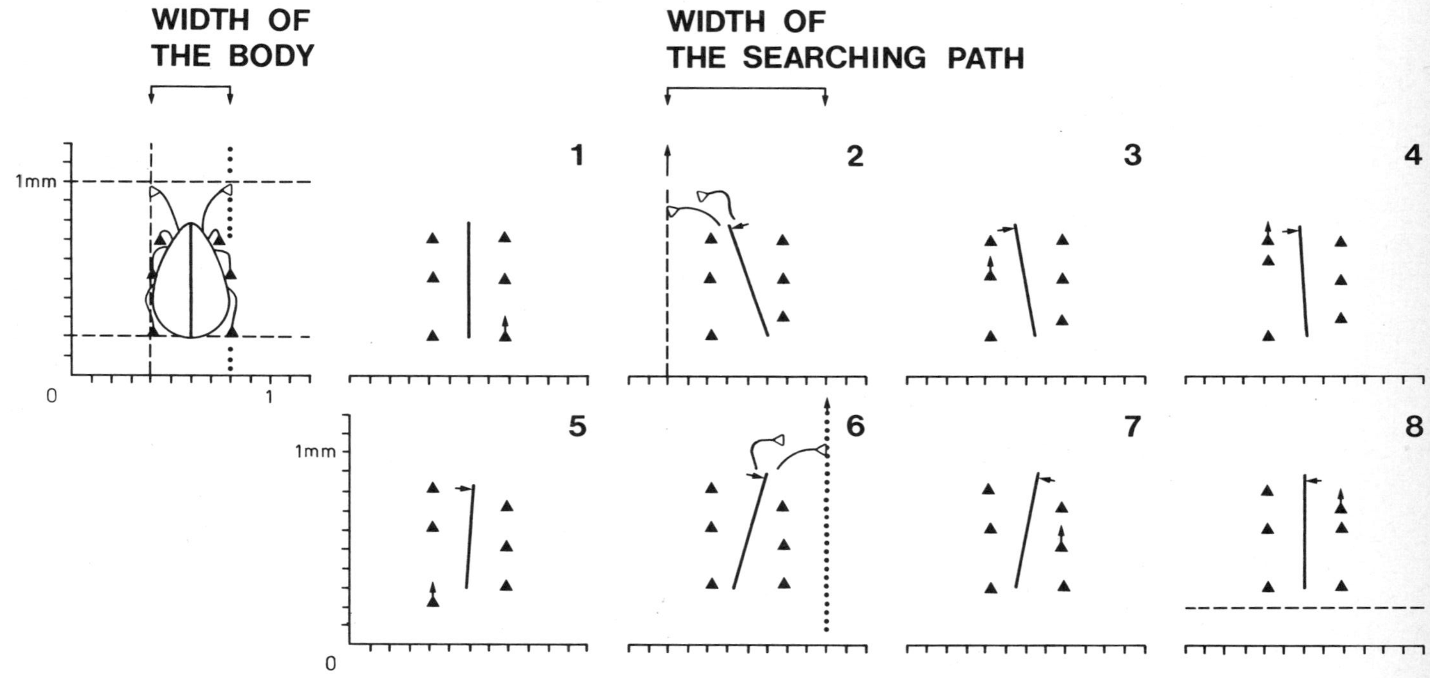

Figure 3 The width of the searching path, as determined by body-axis and front-leg movements of the predatory mite; ▲, tarsi of legs, 2, 3 and 4; Δ, tarsus of leg 1; arrows indicate leg-movement; the straight line represents the position of the body-axis. Eight successive phases of forward locomotion are depicted.

Width of the searching path. The first legs of predatory mites are equipped with various mechanoreceptors and chemosensors. These legs have no function in locomotion, but are solely involved in prey detection. They are moved to and fro in a way resembling the use of antennae in hymenopterous parasitoids. Together with concurrent left-right movements of the body-axis the lateral reach of the predatory mite is extended in the way illustrated in Figure 3. The increased width of the searching path appears to be similar to the length of the mite (Table 2). In a state of rest the width of the mite is much smaller than the length, even though the latter is decreased because the resting predator keeps its first legs drawn up close to its mouthparts (Table 2). The adult female of the two-spotted spider mite does not increase the width of its path, because it makes no to and fro movements. Its length and width are approximately the same as those of the predator in a state of rest (Table 2). The egg of the spider mite is 4 to 5 times smaller than the female, and is round (Table 2). In calculations of the attack rate the predator and prey mites are conceived as circular units with a diameter equal to half the sum of the real length and width.

Walking speed. Most published data on walking speed have demonstrated that the walking speed of predatory mites increases with the state of food deprivation. These data were obtained, however, on leaves or artificial substrates without web. In the web, walking speed is dramatically reduced because of the tangled structure of the web which impedes forward locomotion. The predator inspects the web intensively with its first legs and once a passage is found, it moves forward. Therefore, moments of standstill and displacement alternate frequently. Because the web impedes locomotion in this way and because the short instants of standstill were included in the record of the time spent walking (due to lags in reaction time of observer), walking speed in the web is much lower than on a leaf without web. For the same reasons it is likely that in the web the effects of hunger on walking speed are levelled off (Figure 4). Similarly, the effect of temperature on locomotion, usually found when predators are walking on unwebbed substrates, was not found from the data recorded in the web (Sabelis, 1981). The walking speed of the spider mites is affected largely in the same way as that of the predatory mites. The adult females of the spider mite usually move over very short distances to change their feeding site, walking at a speed of ca. 0.011 cm/s.

Walking pattern and recrossing. The effect of recrossing local areas previously freed of prey was quantified by simulating the predator's walking path in a large area with eggs as prey. Walking paths obtained from video-recordings were analysed by measuring angular deviations over fixed step lengths (0.04 cm). Time series analysis showed that sequential angular deviations were only slightly autocorrelated. For the simulation angular deviations were sampled at random from the frequency distribution of measured angular deviations (x). Ignoring autocorrelations, this distribution can be described in its cumulative form by a three-parametric model:

$$\underline{x} \simeq E(\underline{x}) + (\mathrm{var}(\underline{x}))^{\frac{1}{2}} \, (p^{\lambda} - (1-p)^{\lambda}) / \lambda$$

x is a stochastic variate with expectation $E(x)$, variance var (x) and a symmetric distribution with kurtosis related to λ, the values of which are presented in Table 3. To sample a random value of p from a uniform distribution ($0 \leqslant p \leqslant 1$), random values of the angular deviation x can be obtained (in radians). These deviations were consecutively added to the actual walking direction. From the cosine and the sine of the new direction the changed position of the mite can be calculated, taking the step length as the hypotenuse. Whenever the distance between the mite's position and the egg positions was found to be less than half the sum of the diameters of predator plus prey egg, the prey egg was discarded. The tortuous path of the predator causes local depression of prey density, but in the total area prey density was hardly affected, as the area taken was large (and the simulation period sufficiently short). From simulations of 100 replicate searching paths, recrossing resulted in a rate of prey encounter that was 19% lower than estimated for the case of predators avoiding areas already visited. Additional simulations with mobile prey showed that the effect of recrossing becomes very small as a consequence of prey moving into sites previously cleared of prey by the predator. In this way provisional estimates of the effect of recrossing were obtained by simulation.

Coincidence in the webbed space. Webbing not only reduces walking speed, but also the chance of actual contact, because predator and prey may pass over and under each other without making tarsal contact. The fraction of real contacts out of all coinciding perpendicular projections on the leaf surface were measured for well fed predators and starved predators. Both estimates were close to 0.5, irrespective of the prey stage supplied (Figure 4). Apparently, starved predators do not modify their search so as to increase the chance of contact with prey.

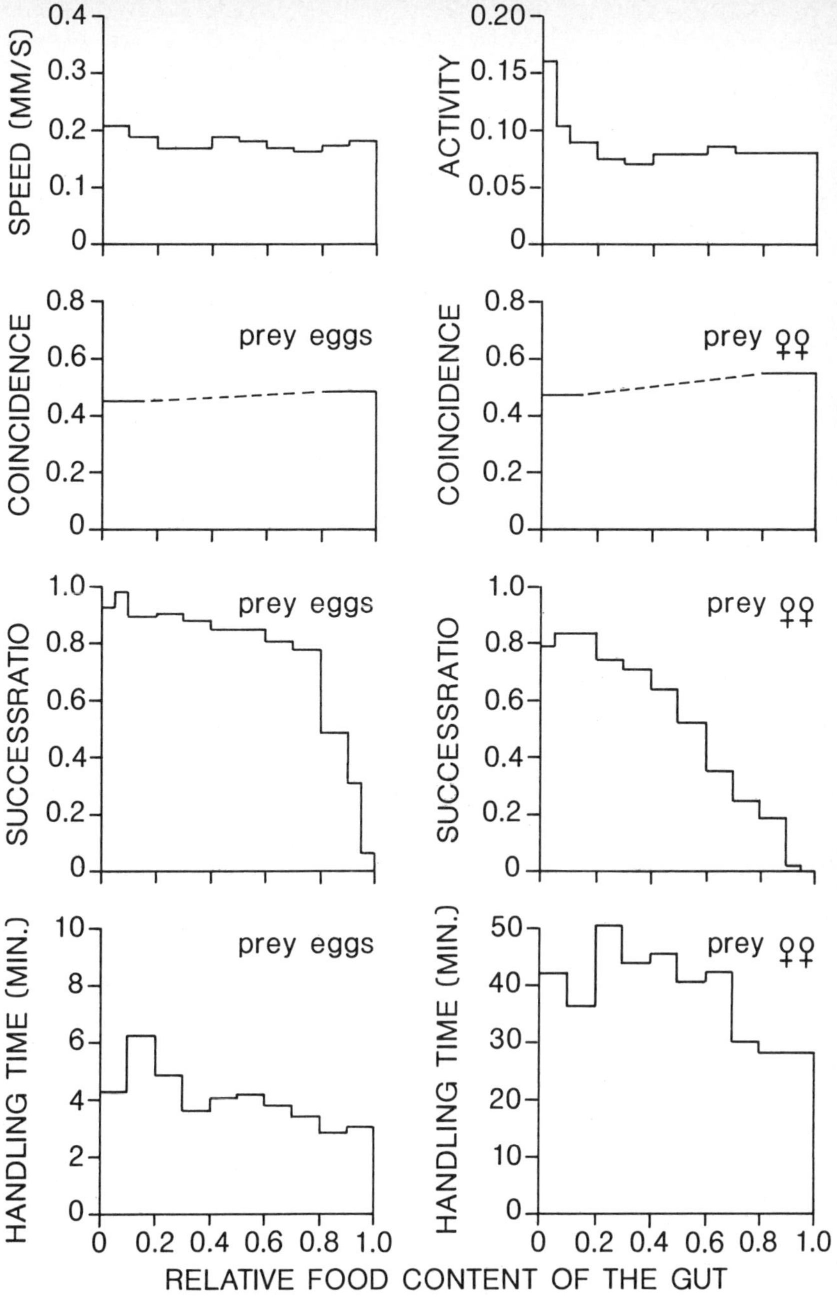

Figure 4 Components of the searching behaviour of *P. persimilis* for either eggs or adult females of *T. urticae,* and the relation of these components with the food content of the predator's gut. Behaviour was observed at 20° C and 70% RH, using standardized predator females and rose leaves with a prey colony consisting of spider mites that were in the egg stage or in the preoviposition phase.

Walking activity. By conceiving of the walking state as a Bernoulli variable (1 = walking, 0 = resting) and by random sampling of moments at which the activity state was evaluated, the fraction of the time spent in forward locomotion was estimated. This procedure was carried out for several levels of gut fullness of the predator. The results shown in Figure 4 reveal that walking activity increases from 0.08 to 0.16 when the food content of the gut falls below 20% of its maximum. Compared with the predator, the female spider mite spends less time in walking (about 4% of the observation time).

Success ratio. During the behavioural observations contacts between predator and prey were scored when the first tarsi of the predator touched the integument of its prey. If these contacts were followed by an attack and by feeding, the encounter was considered to have been successful. A series of independent experiments showed that the prey rarely survive after being fed upon by the predator. The term 'success' therefore refers both to successful food acquisition and to prey death. The records of predator-prey contacts were classified according to the (calculated) level of gut fullness at the moment of first contact with a prey item. For each class of gut fullness the fraction of successful attacks out of all predator-prey contacts were calculated. These so-called success ratios and their relation to the feeding state of the predator are shown in Figure 4. Clearly, starved predators are more motivated than satiated predators to attack their prey. Moreover, at the same level of gut fullness adult female spider mites are attacked with much less chance of success than are spider-mite eggs.

Handling time

The time spent handling a captured prey generally increases with the state of food deprivation of the predator (Sandness and McMurtry, 1972: Fransz, 1974; Sabelis, 1981). This also applies to the data pertaining to the time spent handling by *P. persimilis* (Figure 4). The handling time also depends on the prey stage involved. Apparently, the predator spends less time to consume the food of a spider mite egg than to consume food from a spider-mite female.

Disturbance

A note on disturbance effects is needed, because these have played an important role in the explanations given for the various shapes of the functional responses in acarine predator-prey systems. Mori and Chant (1966) attributed the domed shape of the functional response of *P. persimilis* to the disturbing effect of female prey 'bumping into' predators that had just started attacking, handling or feeding on a prey. Sandness and McMurtry (1972) found a second plateau in the functional response of another predatory mite and attributed this to a stimulating effect on predators that were disturbed while feeding, to attack the prey that caused the disturbance. Predator-prey and prey-prey encounters may also stimulate walking activity and walking speed (Sabelis, 1981). All these effects, however, were recorded under rather artificial conditions. For example, the above mentioned observations were done on substrates of paper or Perspex, where walking speed and walking activity are much higher than in the web of the spider mites. Therefore, the occurrence of the above mentioned effects were scored for predatory mites foraging in the webbed area of a spider-mite colony (Sabelis, 1981). However all these effects were observed rarely on this substrate, so that their role will not be considered in the models to be discussed below.

Models of predation processes

In principle, the model to be discussed below calculates the mean rate of prey killing in a large population of predators. The predators are assumed to have identical properties and each predator is assumed to forage in a separate prey colony. The predation process consists of a sequence of prey searching and prey handling activities. The time required for both activities is subject to stochastic variation as a consequence of random search, random prey distribution etc., but is also determined by the motivational state of the predator, the dynamics of which are assumed to be related to the food content of the predator's gut. At satiation the predator will not attack prey mites (or will attack without harming the prey), but as time proceeds, the consumed food will be digested, resorbed and egested, so that the food content of the gut decreases and the motivation to attack increases. Because of the discrete nature of feeding events and the stochastic variation in searching periods each predator goes through a unique time series of feeding states and feeding events. By considering a large number of replicate predation experiments the expected rate of prey killing can be calculated. This clearly shows the need to develop stochastic models of the predation process.

To bring mathematical treatment within the realm of queueing theory (or stochastic theory of birth-death processes) (Gross and Harris, 1974) three simplifying assumptions must be made. Firstly, the food content of the prey and that of the predator's gut are assumed to consist of *discrete* food units. But, note that this assumption has to be made anyway in the experimental analysis of predatory behaviour. The second assumption is that ingestion of the prey content is conceived of as a single gulp. This is a reasonable assumption, because the time constant of ingestion ($1/r_i$) is less than 1% of the time constant of gut emptying ($1/r_g$). The third assumption is that the time spent handling a caught prey has a negligible influence on the time available to find and seize prey. Several facts are in support of this assumption:

(1) the total time spent handling is less than 10% of the experimental period (under steady state conditions)

(2) the handling time per prey is short, the food content of the prey is large relative to the maximum food content of the gut and most of the food is consumed during a short initial part of the handling period, so that the predator will often reach satiation early in a handling period and gain little by attacking a prey after or during the handling period.

Thus, under the above assumptions the predation process is reduced to a sequence of searching periods segregated from each other by discrete ingestion events. Handling time, however, will not be completely ignored below; its length will be included in the calculation of the time spent per successful attack when the predator were to remain in the same motivational state ($1/\sigma_g$). The analogy with solved problems in queueing theory has been recognized first by Taylor (1976) and Curry and DeMichele (1977); in their view, prey mites in the predator's gut could be considered as a line of customers waiting to be served by a service facility, i.e. waiting to be digested by the predator. By analogy with the gut capacity the line of customers has a maximum length and consists of a finite number of customers. Usually stochastic queueing models assume that inter-arrival times and service times obey the exponential distribution or, equivalently, that the arrival rate and the service rate follow a Poisson distribution. Essential to the predation process is that the intensity parameter of the Poisson distribution depends on the length of the waiting line; the longer the line, the lower the arrival rate and the higher the service rate. With respect to the customer-prey analogy one problem still remains: as a consequence of partial ingestion and differences in the food content of the prey stages the prey is not equivalent to the customer. This can be solved by dividing the food content of the prey and that of the predator's gut into a large number of discrete food units. In this way the prey is analogous to a group of customers that may simultaneously join the waiting line. The model, thus, described will be further referred to as the finite state Markov model.

One may wonder why the satiation-axis is discretized already in the process of model building. According to Metz and Van Batenburg (1984abc) it is preferable to write down the full continuous model in the form of partial differential equations (see Part A). However, as these equations can only be solved by numerical means in which the satiation axis is divided up in discrete compartments anyway, it is decided to skip the intermediate'posh'mathematics and go direct from the biology to a discrete approximating Markov model. Nevertheless, the results obtained by Metz and Van Batenburg (1984abc) by studying the asymptotic behaviour of the full continuous version and some relevant approximations will prove very important in interpreting the results obtained by using the finite state Markov model. The stochastic Markov model will be presented first, followed by a short discussion of some deterministic versions of the predation process. Subsequently, the results of the stochastic and deterministic models will be compared to illustrate some relevant properties of the predator-prey interaction under consideration.

Predation approximated by a finite state Markov process

Assume that the food content of the prey (f_{max}) consists of h discrete food units of size u, that the food content of the predator's gut consists of n food units and varies between 0 (empty gut) and N (at satiation), and that gut emptying is mimicked by random removal of single food units from the gut. The resulting finite Markov chain approximation has as its state space the integers from 0 to N where $N=[g_{max}/u]$ and $[x] \overset{def}{=}$ integer part of x. Such a process is fully characterized by its transition rates. First consider the downward transitions from n to $n-1$. These are assumed to occur at such a rate that the mean time needed to make a transition in the absence of any prey capture exactly equals the time the deterministically decreasing satiation needs to cross the same distance. This was supposed to give the best approximation to the real continuous process for sufficiently small u. Let ρ_n denote the downward transition rate and the holding time of any state of a continuous time Markov chain be exponentially distributed with the sum of the outward transition rates as a parameter. The inverse of this parameter then equals the mean of the exponential distribution. Therefore, ρ_n should be set equal to $r_g^{-1}\ln(un/(un-u))$, or equivalently for transition from n to $n-1$:

$$\rho_n = r_g(ln(\frac{n}{n-1}))^{-1} \text{ for } n = [g/u] = 2,3,.....,N$$

For $n=0$ or $1, \rho_n$ was conventionally put:

$$\rho_1 = r_g(ln(\frac{1}{1-z}))^{-1}, \text{ where } z = 0.99$$

$$\rho_0 = 0$$

If the prey are all in the same developmental stage and contain $h=[f_{max}/u]$ food units, the upward transitions are from n to $n+h$, if $n+h$ is less than N, otherwise the transition will end at N. These upward jumps occur at a rate σ_n, which is the rate of successful attack σ calculated for a predator at state n. Given the upward and downward transition rates the differential equations for the probabilities p_n that the predator is in state n are:

$$\frac{dp_0}{dt} = -\sigma_0 p_0 + \rho_1 p_1$$

$$\frac{dp_n}{dt} = -(\rho_n + \sigma_n)p_n + \rho_{n+1}p_{n+1} + \sigma_{n-h}p_{n-h} \text{ for } n = 1,2,....N-1$$

$$\frac{dp_N}{dt} = -\rho_N p_N + \sum_{i=N-h}^{N-1} \sigma_i p_i$$

with the convention that $p_{n-h}=0$ for $n-h<0$. If there are two types of prey (e.g. prey eggs and prey females) with maximum food contents f_1 and f_2 (or h_1 and h_2 food units per prey type) and with catching rates $\sigma_{n,1}$ and $\sigma_{n,2}$, the set of differential equations for p_n is:

$$\frac{dp_0}{dt} = -(\sigma_{0,1} + \sigma_{0,2})p_0 + \rho_1 p_1$$

$$\frac{dp_n}{dt} = -(\sigma_{n,1} + \sigma_{n,2} + \rho_n)p_n + \rho_{n+1}p_{n+1} + \sigma_{n-h_1,1}\,p_{n-h_1} + \sigma_{n-h_2,2}p_{n-h_2} \text{ for } n=1,2,..,N-1$$

$$\frac{dp_N}{dt} = -\rho_N p_N + \sum_{i=N-h_1}^{N-1} \sigma_{i,1}p_i + \sum_{i=N-h_2}^{N-1} \sigma_{i,2}p_i \,.$$

When prey density is constant, the probability distribution $p_n(t)$ will enter into a steady state p_n. Under this condition p_n can be found by selecting an arbitrary starting value for p_0, solving p_1 from $p_0'(t)=0$, p_2 from $p_1'(t)=0$, etc., and finally by rescaling the p_n values such that all p_n sum to unity. The mean killing rate for each prey type (K_1 and K_2) is obtained from:

$$K_1 = \sum_{n=0}^{N-1} \sigma_{n,1}\,p_n \text{ and } K_2 = \sum_{n=0}^{N-1} \sigma_{n,2}p_n \,.$$

The mean rate of food ingestion I is calculated from:

$$I = \sum_{n=0}^{N-h_1} \sigma_{n,1}\,p_n\,h_1 + \sum_{n=N-h_1+1}^{N-1} \sigma_{n,1}p_n(N-n) + \sum_{n=0}^{N-h_2} \sigma_{n,2}\,p_n\,h_2 + \sum_{n=N-h_2+1}^{N-1} \sigma_{n,2}\,p_n(N-n)\,.$$

This completes the description of the Markov model. For the calculations of the mean killing rate and the mean ingestion rate, the unit food size u was set equal to $0.25\mu g$, so that N equals 32 and the transition rates, ρ_n, $\sigma_{n,1}$ and $\sigma_{n,2}$, can be computed using the data presented in the previous sections.

Deterministic approaches to the predation process

Stochastic variation in searching periods and discrete ingestion events are two salient elements of the stochastic model discussed above. To elucidate their role in determining the overall predation rate it is worthwhile to consider some deterministic alternatives. These alternative models are deterministic in that the gut fullness of each predator can only be in one and the same state at one moment in time. Such a simplification can be achieved in two ways: (1) by conceiving of predation as a process of continuous consumption of infinitely small portions of prey food (the model of the prey food sucker) or (2) by conceiving of predation as a sequence of searching periods of fixed length, separated from each other by ingestion events (the model of ingestion cadences). These models will be discussed briefly and only for the case of a monoculture of prey stages.

In the model of the prey food sucker ingestion is no longer a discrete event, but is conceived of as a continuous process; the predators keep on consuming infinitely small portions of food at a rate equal to the product of σ_g and f_{max}. In the steady state the amount of food ingested balances the amount of food cleared from the gut:

$$\sigma_g f_{max} = r_g g \,.$$

Under these conditions the prey killing rate K equals the value of σ_g satisfying the above equality. This model has been discussed in more detail by Fransz (1974), Sabelis (1981) and Metz and Van Batenburg (1984ab) and it was referred to as the Rashevsky limit in Part A.

In the model of the ingestion cadences stochastic variation in searching periods is ignored as an aspect of the foraging process, so that each predator undergoes the same series of searching periods and ingestion events. Under steady state conditions ingestion occurs after fixed time intervals of searching and the amount of food ingested must then be equal to the amount of food cleared from the gut during the searching period t^* (the asteriks refers to the steady state situation). If g^* is defined in the steady state, as the fullness of the gut at the start of a searching period just after the ingestion event, then it follows that:

$$f = g^*(1 - \exp(-r_g t^*)), \text{ where } f = f_{max} \text{ if } g^* < g_{max} \text{ and } f \leqslant f_{max} \text{ for } g^* = g_{max}$$

Hence, the length of the searching period t^* between successive ingestion events can be obtained given the values of g^*, f_{max} and r_g:

$$t^* = \ln(g^* / (g^* - f)) / r_g .$$

The reciprocal of t^* is equal to the sum of products of the rate of attack σ_g and the fraction of the time t^* spent in gut content class $g(=t_g / t^*)$. Because t^* cancels, it follows that: $\sum_{g=g^*-f}^{g^*} \sigma_g t_g = 1$, where $t_g = \ln(g / (g-1)) / r_g$ (the residence time in class g for $g > 1$). Because σ_g is a function of prey density, a value of prey density (and thus a series of σ_g's) should be found that satisfies the above equality. Hence, for any given value of g^* it is possible to trace back the prey density at which this steady state cadence of ingestion is achieved. This inverse method resembles to some extent the simulation method used by Holling (1966) to calculate the prey killing rate under steady conditions.

Comparison of calculations with stochastic and deterministic models

The deterministic models are simplified versions of the Markov model. To elucidate the effect of each simplifying step the predictions of these models must be compared. In Figure 5, calculations with the stochastic and deterministic versions are presented for the case of young female predators of *P. persimilis* foraging in a webbed leaf area with a constant density of *T. urticae* eggs as prey.

Clearly, the deterministic sucker model predicts a much lower maximum for the overall predation rate than the models in which predation is conceived of as a series of discrete ingestion events, but in the range of low prey egg densities the predictions of all three models show negligible differences. Two possible causes should be taken into consideration:

1. the convexity of the relation between σ_g and g (or σ_n and n); Metz and Van Batenburg (1984bc) showed that the deterministic sucker model will underestimate the overall predation rate if the relation between σ_g and g is convex, whereas it will overestimate the overall predation rate, if the relation between σ_g and g is concave. At first sight the data on searching behaviour in Figure 4 indicate a concave $\sigma_g - g$ relationship, so that an overestimation would have been expected. But, when these data are scrutinized, it will be clear that in the range of g near to full satiation the relationship between σ_g and g is of the convex type. This may explain why at low prey egg densities the deterministic sucker model (slightly) overestimates the overall predation rate, whereas at high prey egg densities the overall predation rate is underestimated.
2. No account of partial ingestion. Because the deterministic sucker model assumes that the predator continuously consumes infinitely small portions of food (at a rate equal to $\sigma_g f_{max}$), this model is by definition incapable of accounting for partial ingestion. Hence, at high prey egg densities the predators from the latter model have to digest much more food per captured prey than the predators from the event-oriented models.

Because the predictions of the event-oriented models differ only slightly, it may be concluded that the absence of partial ingestion in the deterministic sucker model is the most important cause.

The deterministic and stochastic models that are based on the event-oriented concept give similar predictions of the overall predation rate, but the deterministic model of ingestion cadences predicts a slightly lower predation rate. Three possible causes should be taken into account:

1. The convexity of the relationship between σ_g and g (at high g). Due to stochastic variation in the time intervals between successive prey catches the state of the predator is represented by the probability distribution p_n, which after multiplication with σ_n and subsequent summation gives the expected rate of predation. Because of the curvilinear relationship between σ_g and g the expected rate of predation is likely to differ from that predicted by the model of ingestion cadences. Maybe, the concavity if the $\sigma_g - g$ relationship at high values of g has the same effect as discussed above.
2. Stochastic variation in time intervals between successive prey catches will result in different amounts of food ingested at each prey catch event. In the cadence model the searching periods are of fixed length (under steady state conditions) and, thus, the amount of food ingested per catch event is equal. Hence, the total amount of food ingested in the stochastic case may very well differ from that in the deterministic case.

3. The stochastic model assumed random removal of food units from the gut. However, when the food unit *u* is taken small and the maximum number of gut content classes is large, the gut emptying process is approximately deterministic. It was found that enlargement of the number of gut classes had hardly any effect on the predictions of the Markov model.

Though I favour the second cause, it is not possible to show which of these causes is the most important.

For the calculations discussed below the finite state Markov model is used, because the deterministic sucker model is obviously not adequate and the deterministic cadence model requires very unconvenient numerical calculation procedures.

Experimental validation of the finite state Markov model

To validate the performance of the Markov model in predicting the functional response, comparison with measurements of the actual rate of predation was needed. Therefore, a series of predation experiments was carried out in monocultures of prey eggs or of prey females and in mixed cultures of prey eggs and prey females. To prevent egg deposition prey females were used that were in their preoviposition phase. The following features of the experimental procedure are relevant here:

i) *the experimental arena* consisted of a leaf disc with a diameter of 3.7 cm, floating upside down on water-soaked cotton wool in a petri dish; the leaf disc was punched out of the centre of a rose leaf (cv Sonia); the water barrier prevented the mites from escaping from the leaf disc;

ii) *prey egg density at the start of the predation experiments* was established by releasing ovipositing spider-mite females in adequate numbers one day before, by goading the prey females at the appropriate time with a needle, to drive them to the edge of the area, and removing them once they had got there and finally, by correcting prey density (i.e. by puncturing eggs with a thin needle) until prey density was brought to the desired level; care was taken not to harm the structure of web;

iii) *initial prey female density* was established by releasing the desired number of prey females on to the leaf disc ca. half a day before the start of the experiment; these females had moulted only a few hours before they were released and hence were definitely in the pre-oviposition phase for some 2 days more;

iv) *mixtures of prey eggs and prey females* were established in essentially the same way, but the pre-ovipositional females were not released onto the disc until a few hours before the predation experiment started; the mixtures of the prey stages were established so as to obtain approximately the same amount of ingestible food for the predator ($16-20\mu g$ per cm^2 webbed leaf area);

v) *the condition of the predator* was standardized by using young female predators whose last moult had occurred 3-8 days previously and by allowing them an adaptation period of 8 to 12 hours prior to the actual predation experiment: in this adaptation period the predator was placed on a leaf disc with prey at exactly the same density and stage composition as on the disc used for the predation experiment. Just before the predation experiment the predator was placed on a new disc by transfer on a small excised portion of the leaf disc on which it was allowed to enter steady state conditions;

vi) *the experimental period* was set at 10 hours; if the predator dispersed from the disc into the water before the 10 hour period had elapsed, the replicate was discarded because the exact moment of departure was not known; this loss of replicates never exceeded 20% of the total number of replicates used;

vii) *constancy of prey density* could not be ensured by replacement of the prey killed, but the decrease of prey density was kept within acceptable limits as a consequence of the short experimental period and the sufficiently large size of the experimental arena and the webbed leaf area; even if the predator had consumed prey eggs at a maximum rate of 1 per hour and prey females at a maximum rate of 1 per 3 hours, the decrease in overall prey density in the webbed area (ca. 10 cm^2) would not have exceeded 1 unit of prey egg density or 0.33 unit of prey female density during the experimental period of 10 hours;

viii) *the number of prey killed* was measured by counting the remaining live prey; control experiments showed that abiotic mortality (shrivelled eggs or lifeless prey females) was negligible over a period of 10 hours; the remnants of the killed prey females were used to check the mortality assessed; no such check was feasible for egg mortality because of the presence of the eggs that had been punctured with a needle; calculations of the mortality of prey females were corrected whenever some prey females were recorded as having drowned in the water surrounding the leaf disc;

ix) *the number of replicates* of each experiment was invariably 20;

x) *environmental conditions* were identical to those present during the behavioural observations, i.e. 20°C and 70%

RH.

This completes the description of the experimental procedure.

The measurements in monocultures of each prey stage are presented in Figure 5 (prey eggs) and Figure 6 (prey females), as the mean and the standard deviation of the number of prey consumed per 10 hours. The standard error is within the range of 0.24 and 0.52 prey, so that half the length of the two-sided confidence interval is within the range of 0.5 and 1.1 prey (α=0.05;t_{19}^{α}=2.09). Assuming steady state conditions the finite state Markov model was used to predict the functional response from the underlying behavioural and physiological parameters and functions. It is clear that these predictions were very close to the measured values of the prey consumption. To ascertain the robustness of the model its sensitivity to errors in the input parameters and functions need to be assessed. Here, it is sufficient to consider errors in σ_n and ρ_n only. The sensitivity to changes in σ_n can be evaluated from the predictions of the queueing model in Figure 5 and 6, because σ_n is approximately a linear function of prey density (the time spent handling a prey is short!). Clearly, even if + 50% or - 50% changes in prey density are considered, predictions still fall within the confidence interval of the measurements. The sensitivity to changes in ρ_n can be summarized most conveniently in terms of the relation between prey consumption at high prey egg density (K_{egg} in prey eggs per day; d_{egg} = 40 prey eggs per cm^2)and the relative rate of gut emptying r_g (day^{-1}):

$$K_{egg} = 13.58(r_g)^{0.69} .$$

Thus, errors in r_g larger than 10% would have resulted in rejection of the model. Clearly, the performance of the model in predicting the functional response largely hinges upon a correct estimation of r_g, rather than of σ_n.

The measurements in the mixtures of prey eggs and prey females are presented in Figure 7. Again assuming steady state conditions the predictions of the Markov model lie well within the 95% confidence intervals of the average prey consumption measured. This leads to the conclusion that there is no reason to assume that prey stage preference changes as a result of the predator experiencing the two prey stages simultaneously. Another important conclusion revealed by the simulation is that the mean food content of the gut is not the same in each of the prey stage mixtures (see Figure 7). Even if they were the same, the mean gut fullness in the mixed culture would differ from that in the monoculture of each prey stage at the same prey stage density. For example, the mean fraction of the gut filled with prey food at a density of 1.5 female per cm^2 is equal to ca. 0.74 and this fraction equals 0.80 at a density of 5 eggs per cm^2, whereas in the mixture of 5 prey eggs plus 1.5 female per cm^2 the mean fraction of the gut filled with food is equal to 0.85. Because the success ratio of the predator in relation to prey eggs and prey females is very much dependent on the degree of gut fullness and, thus, so are the relative prey preferences as predicted from the behavioural observation in the monocultures, it appears to be essential to standardize the food content of the gut, or - as elaborated here - to include the feeding state of the predator in the analysis of prey stage preferences. Cock (1978) and Fernando & Hassell (1980) also analysed the prey stage preference of *P. persimilis* in relation to different developmental stages of *T. urticae.* They used the 'random predator' equation with two parameters, i.e. the catch rate and the handling time (Rogers, 1971; Royama, 1972) to describe the functional responses to prey density in monocultures of the prey stages. These equations were then used to predict the diet composition of the predator in mixtures of prey stages. As the 'random predator' equation of Rogers and Royama does not take the feeding state of the predator into account, the measured diet composition in the culture of mixed prey stages is likely to be different from the predicted diet, because at least part of discrepancies result from ignoring the effects of the predator's feeding state on the rate of successful attack. This may explain why Cock (1978) and Fernando & Hassell (1980) found marked deviations from the prediction, indicating that there is some change in predator searching behaviour arising from both prey types being presented together. The analysis of prey stage preference, as presented here, shows that these changes in searching behaviour may be the result of changed feeding states rather than of changes in prey stage preference *per se.*

The agreement between measured and predicted diet is certainly no proof of the absence of behavioural changes due to the predator experiencing both prey stages together. In theory there may be behavioural changes compensating each other's effect on the diet composition. From an evolutionary point of view the selective advantage of such a combination of behavioural changes are obscure unless it would allow the predator to spend more of its food in reproduction.

If the observed diets had differed from those predicted, it would still be too premature to infer that prey stage preferences had changed as a result of the prey types being presented together. Several other explanations are to be considered. For example, the behaviour of the prey types may have changed as a result of being together, or the predator's walking speed and/or activity may be stimulated leading to increased rates of encounter with both prey types, or the rate of food conversion may have changed under conditions of a mixed diet (without affecting prey stage preference). But, if these alternative explanations can be ruled out after detailed observations on the behaviour of predator and prey in the mixed culture or can be considered to be insufficient to explain the differences between observed and predicted diets, prey preference (in terms of the success ratio s or the width of the searching path w) has changed as a result of the prey types being presented together. This would be the kind of phenomenon that deserves special

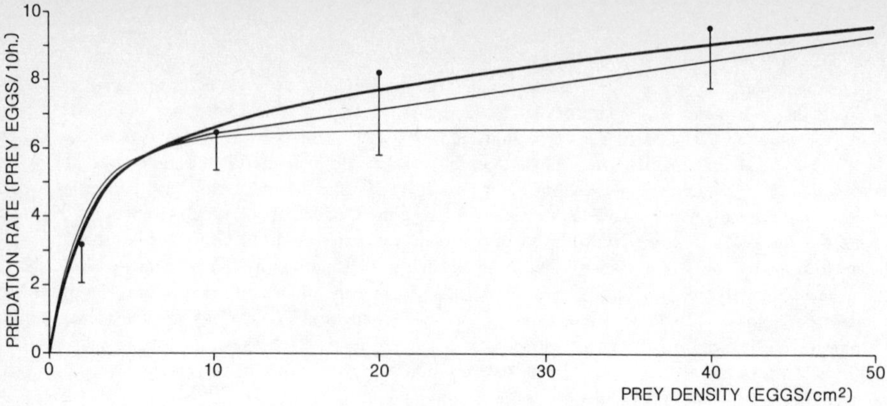

Figure 5 The functional response of *Phytoseiulus persimilis* to the density of eggs of *Tetranychus urticae.* Lines of different thickness indicate the steady state calculations: thin line = the deterministic 'sucker' model, the line of intermediate thickness = the deterministic model of ingestion cadences, the thick line = the finite state Markov model. The mean of the measured functional response is indicated by a black dot; the bar above or below the dot indicates the standard deviation.

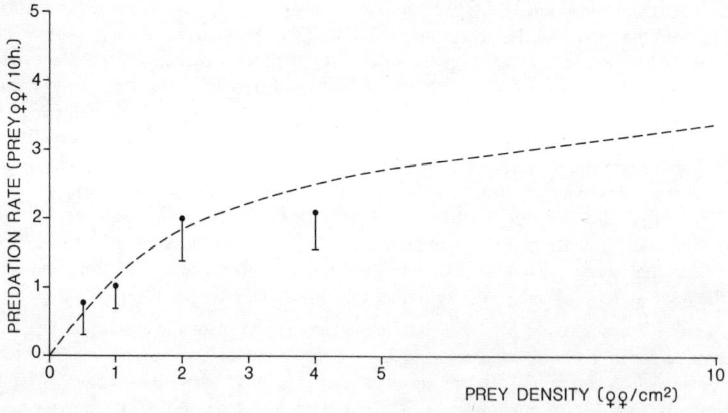

Figure 6 The functional response of *Phytoseiulus persimilis* to the density of preovipositional females of *Tetranychus urticae.* The dashed line represents steady state calculations using the finite state Markov model. The mean of the measured predation is indicated by a dot; the bar indicates the standard deviation.

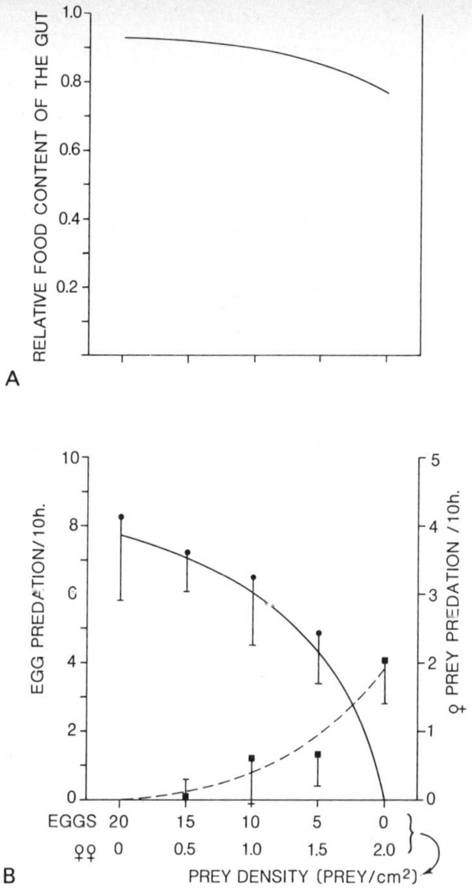

Figure 7 A. Mean values of the relative food content of the predator's gut at different combinations of prey egg and prey female densities (see x-axis in Figure 7B). These values are obtained by steady state calculations using the finite state Markov model, i.e. mean = $\sum_{n=0}^{N} p_n \, n / N$.

B. The functional response of *Phytoseiulus persimilis* to the density of eggs and of preovipositional females of *Tetranychus urticae,* as a result of the prey stages being presented together in different quantities. The line represents steady state calculations of prey egg consumption using the finite state Markov model. The black dots give the mean of the measured prey consumption and the bars give the standard deviation of the measured prey egg consumption. The dashed line represents steady state calculations of prey female consumption using the finite state Markov model. The black squares and associated bars indicate the mean and standard deviation of the measured consumption of prey females.

attention from the biologist; it raises the question *why* it is more profitable to the predator to change its prey selection behaviour.

On prey preference and reproductive success

The profitability of each prey type or mixture of prey types should be expressed in terms of the reproductive success that accrues to the predator from including them in its diet. A prey type may be more profitable than others, because its effective food mass is larger, because its composition of nutrients or other chemical compounds is superior, because less energy (or time) is involved in digesting nutritive compounds, in detoxifying poisoned compounds, or in attacking and catching the prey. (Remember that the success ratio is determined by the attack strategy of the predator as well as by the defensive strategy of the prey!). Natural selection will favour predator genotypes coding for prey selection behaviour that maximizes reproductive success. This prediction can be tested by comparing reproductive performance in monocultures of the prey types and in mixtures of the prey types. If the theoretically best strategy agrees with the actual prey selection behaviour, the predator behaves optimally in terms of its reproductive success (which fits in the selectionist's view, but is not a proof, of course).

Charnov (1976) elaborated a simple model of optimal prey selection. He assumed that prey types only differed in the amount of food (or energy) available to the predator and in the time required to find, catch and consume the food of the prey types. Moreover, he assumed that not the feeding time nor the searching time, but only the prey selection behaviour has been subject to natural selection. Under this set of assumptions he calculated the ratio of energy gained to the feeding and searching time required, either for each of the prey types alone or for combinations of the prey types. The largest ratio predicts the best prey selection strategy.Charnov's model of optimal prey selection may apply to the predator-prey system under consideration, when the rate of food intake by the predator is equal to the rate at which the food is leaving its gut. However, as a consequence of the random prey distribution and the random predator search within the leaf area webbed by the spider mites each predator goes through a unique set of state transitions. Even when the prey density is kept constant and the predator has been foraging for a considerable time, fluctuations in the feeding state of the predator may still be considerable. Because both the amount of food ingested from the prey stages and the handling time depend on the feeding state of the predator, it is difficult to see how Charnov's prey selection model applies to the prey selection behaviour of a predatory mite. Moreover, the total time spent handling prey items by *P. persimilis* was found to be less than 10% of the experimental period and numerous prey items were contacted by the predator without subsequent attack. Hence, it is unlikely, that selection to minimize the time spent handling per unit of food consumed has an important bearing on the choice between prey stages of *T. urticae.* It makes more sense to analyse the prey selection behaviour of predatory mites in terms of the rate of resorption, because the resorbed food is spent in egg production or in maintenance processes, increasing the chances of survival to reproduce in times to come. The rate of resorption (R) depends on the relative rate of resorption (r_{rs}) and on the food content of the gut (g) according to the following equation:

$$R = r_{rs} g \ .$$

Assuming that r_{rs} and $g_{\max}$ have been driven to their physiological maximum by natural selection (or that the time scale of evolutionary change in r_{rs} is much larger than the time scale of dynamic changes in g), and that the prey types do not differ in food quality, prey selection behaviour will maximize reproductive success, if it maximizes g:$R_{\max} = r_{rs} g_{\max}$. This may explain why predatory mites have sacrificed efficient use of the available food in their prey and practise partial ingestion of the food as an effective way of maximizing g. It may also explain why the predatory mite does not modify its searching behaviour in response to the two prey stages being offered together. Clearly, because the time constant of gut emptying ($1/r_g$) exceeds the time constant of ingestion ($1/r_i$) by a factor 100 to 3000, the time needed to find and catch a prey is always shorter when both prey stages are searched for, than when one of the prey stages is selected and the other is being ignored. Therefore, if the searching behaviour is determined only by the feeding state of the predator, the rate of resorption R is maximized by attacking both prey stages with probabilities equal to those in the monocultures of prey (i.e. σ_g's are equal). In this respect prey stage selection by *P. persimilis* can be considered as optimal. However, why the predator does not maximize g by staying near to a captured prey and continue to feed until the prey is emptied, is still an open question. When a well fed predator attacks and feeds on a spider-mite female, she usually leaves a considerable portion of food behind and does not return later on to feed on the same prey. In an attempt to find possible explanations some considerations are given below:

- the food in the partially consumed prey may deteriorate and therefore become unprofitable in due time; however, a decrease in food quality is unlikely because hungry predators have been observed to feed on killed prey, but a decrease in food ingestibility does occur since after killing the prey looses much water by evaporation, which influences the viscosity of the prey's food content.

- other activities, such as selection of oviposition sites or the deposition of marking pheromones (to inform other female predators that the local area contains less prey and to give them the choice to invade and compete or to continue searching for a more suitable prey area).

These hypotheses are amenable to experimental investigation and may shed some new light on the optimality of the foraging behaviour. The most salient result of the preference analysis is that prey selection behaviour seems not to be affected by the prey stage composition offered to the predator. It was shown that behavioural observations on predators foraging in monocultures of the prey stages could be used to predict the diet of the predators foraging in mixed cultures of these prey stages. If there would have been important differences in food quality between the prey stages or in the energy involved in attacking, seizing and feeding on the prey stages, an optimal predator would have changed its prey selection behaviour when both prey stages are offered together. Thus, the absence of such a behavioural change suggests that the prey stages do not differ in food quality and are easily seized (assuming optimality). This suggestion has some intuitive appeal, because (1) a considerable part of the spider-mite female, i.e. the reproduction organ, contains the same food as present in the eggs suggesting the absence of large differences in food quality, and (2) attacks by predatory mites are of short duration and do not involve chasing or struggling, which suggests that little energy is spent in attacks on prey.

Concluding remarks

The main concern of a population biologist is (1) to elucidate which factors are of predominant importance in determining the reproductive success of living organisms, (2) to trace how natural selection has molded the traits of these organisms (and which are the constraints to evolutionary change), and (3) to determine the consequences of the individual's traits for population persistence and stability (in time and space). To return to the predatory mites questions pertaining to (3) will be discussed elsewhere (Sabelis; this volume), but some of the questions pertaining to (1) and (2) can be answered, as summarized below:

- What determines reproductive success of a predatory mite foraging in a colony of spider mites? Clearly, foraging behaviour and food conversion are of crucial importance; *Phytoseiulus persimilis* relies entirely on *Tetranychus* mites as prey, females spend most of their food (up to 70%) in egg production, and they are capable to deposit an egg mass per day equal to their own body weight. To give a full answer, however requires more information on patch time allocation and the control of offspring quality (e.g. sex allocation), which is beyond the scope of this paper.
- Which of the determinants, foraging behaviour or food conversion (i.e. σ_g or r_g), contributes most to reproductive success? Or to place the question in an evolutionary context: which of these determinants is subject to the most severe selection pressure? By modelling the predation process and quantifying the input parameters and functions it was shown that the overall predation rate is much more affected by standard changes in r_g than by proportionally equivalent changes in σ_g. Hence, it is to be expected that selection to increase r_g will have been (and will be) the most intense.
- How important is the time spent in handling the prey caught in determining the overall rate of predation? As also emphasized by Metz and Van Batenburg (1984bc) handling time is not a very important factor in determining the overall predation rate. It was found that the time constant of ingestion ($1/r_i$) is negligibly small compared to the time constant of gut emptying ($1/r_g$). Moreover, under steady state conditions the total time spent handling comprises less than 10% of the inter-catch time intervals, even at high prey densities. Hence, the time taken to clear food from the gut (and motivate the predator to attack a subsequent prey) is of overriding importance.
- Why do predatory mites partially ingest the food content of their prey? It was shown that the gut is emptied exponentially. The time constant of gut emptying is almost equal to the time constant of food resorption because the defaecation rate is low. Hence, the rate of resorption (and the rate of oviposition) is maximized by keeping the food content of the gut at its maximum. Apparently, the predators achieve this by consuming small portions of each prey caught, thereby sacrificing efficient use of prey resources. However, why partial ingestion of the prey is the rule and return feedings on the same prey are rarely observed, is a question to be solved in the future.
- Do predatory mites change their searching behaviour in response to the prey stages being presented together? Based on the system analysis presented in this paper there is no reason to conclude that prey selection behaviour is affected by the prey stage composition offered to the predator. This suggests that there is little difference in the food quality of the prey stages and in the energy (or time) involved in attacking and handling the prey stages. If this is true, it follows that the rate of resorption (and oviposition) is maximized by

attacking and feeding on both prey stages instead of selecting one particular stage.

As will be discussed elsewhere in this Volume, several of the answers to the above questions are used in constructing population models of the interaction between predatory mites and spider mites.

It would be interesting to apply the method of preference analysis advocated in this paper to the case of prey species selection by predatory mites. Of course, direct observations to detect changes in prey selection behaviour in response to the prey type composition are indispensable as these may provide definite proof of changed searching behaviour. However, there is always a danger in relying completely on the visual analysis, because one may overlook aspects hitherto unknown. Hence, to check the behavioural interpretations in an independent way and to detect whether unknown factors play a role, it is necessary to construct and validate predation models. The degree of detail to be included in the models depends on the questions to be asked. For example, Chesson (1978) states that selective predation occurs when the relative frequencies of prey types in a predator's diet differ from the relative frequencies in the environment. She derived a simple preference index that does not depend on the number of each prey type present, and advocates its use to detect change in consumer behaviour in relation to prey density and prey stage composition (Chesson, 1978, 1983) (but note that changes in *prey* behaviour may be detected as well). However, with respect to predators such behavioural changes are commonplace. More interesting questions are whether predators change their searching behaviour in response to the prey type composition in the environment given they are in a certain feeding state. Because changes in prey preference may not occur until after a series of prey captures, it is very difficult to standardize the feeding state so as to make predation experiments in mono and mixed prey cultures directly comparable. Hence, predation models such as the finite state Markov model may provide a useful tool in detecting interesting changes in prey preference of predators.

ACKNOWLEDGEMENTS: I thank Hans Metz for several thought provoking comments on the first version of the manuscript, Odo Diekmann for many helpful suggestions, and Piet Ruardij for his contribution in the implementation of path analysis on a computer. The figures were drawn by Martin Brittijn and Gerard Wever.

REFERENCES

R. ARDITI, 1983. *A unified model of the functional response of predators and parasitoids.* Journal of Animal Ecology 52: 293-303.

E.L. CHARNOV, 1976. *Optimal foraging: attack strategy of a mantid.* American Naturalist 110: 141-151.

J. CHESSON, 1978. *Measuring preference in selective predation.* Ecology 59: 211-215.

J. CHESSON, 1983. *The estimation and analysis of preference and its relationship to foraging models.* Ecology 64: 1297-1304.

M.J.W. COCK, 1978. *The assessment of pereference.* Journal of Animal Ecology 47: 805-816.

G.L CURRY, and D.W. DEMICHELE, 1977. *Stochastic analysis for the description and synthesis of predator-prey systems.* Canadian Entomologist 109: 1167-1174.

V.G. DETHIER, 1976. *The Hungry Fly.* Harvard University Press.

C.T. DE WIT, 1960. *On Competition.* Verslagen van Landbouwkundige Onderzoekingen (Agricultural Research Reports) 66.8, Pudoc. Wageningen, pp. 82.

C.T. DE WIT, and J. GOUDRIAAN, 1978. *Simulation of Ecological Processes.* Simulation Monographs, Pudoc, Wageningen, pp. 175.

M.J.H.P. FERNANDO, and M.P. HASSELL, 1980. *Predator-prey responses in an acarine system.* Researches on Population Ecology 22: 301-322.

H.G. FRANSZ, 1974. *The functional Response to Prey Density in an acarine System* Simulation Monographs, Pudoc, Wageningen, pp. 143.

D. GROSS, and C.M. HARRIS, 1974. *Fundamentals of Queueing Theory.* Wiley, New York, pp. 55.

M.P. HASSELL, 1978. *The Dynamics of Arthropod Predator-Prey Systems.* Monographs in Population Biology 13, Princeton University Press, pp. 237.

M.P. HASSELL, J.H. LAWTON and J.R. BEDDINGTON, 1976. *The components of arthropod predation. I. The prey death rate.* Journal of Animal Ecology 45: 135-164.

M.P. HASSELL, J.H. LAWTON, and J.R. BEDDINGTON, 1977. *Sigmoid functional responses by invertebrate predators and parasitoids.* Journal of Animal Ecology 46: 249-262.

C.S. Holling, 1959a. *The components of predation as revealed by a study of small mammal predation of the European pine saw fly.* Canadian Entomologist 91: 293-320.

C.S. Holling, 1959b. *Some characteristics of simple types of predation and parasitism.* Canadian Entomologist 91: 385-398.

C.S. Holling, 1961. *Principles of insect predation.* Annual Review of Entomology 6: 163-182.

C.S. Holling, 1965. *The functional response of predators to prey density and its role in mimicry and population regulation.* Memoirs of the entomological Society of Canada 45: 3-60.

C.S. Holling, 1966. *The functional response of invertebrate predators to prey density.* Memoirs of the entomological Society of Canada 48: 1-86.

J.A.J. Metz and F.H.D. Van Batenburg, (1984a). *Holling's "hungry mantid" model for the invertebrate functional response considered as a Markov process.* Part 0: A survey of the main results. In: Mathematical Ecology (Eds S.A. Levin and T.G. Hallam), Proceedings Trieste, 1982. Springer Lecture Notes in Biomathematics 54: 29-41.

J.A.J. Metz and F.H.D. Van Batenburg (1984b). *Holling's "hungry mantid" model for the invertebrate functional response considered as a Markov process.* Part I: The full model and some of its limits. Report AM-R8419, Centre for Mathematics and Computer Science.

J.A.J. Metz, and F.H.D. Van Batenburg, (1984c). *Holling's "hungry mantid" model for the invertebrate functional response considered as a Markov process.* Part II: Negligible handling time. Report AM-R8420, Centre of Mathematics and Computer Science.

N.J. Mills, 1982. *Satiation and the functional response: a test of a new model.* Ecological Entomology 7: 305-315.

H. Mori and D.A. Chant, 1966. *The influence of prey density, relative humidity and starvation on the predaceous behaviour of Phytoseiulus persimilis Athias-Henriot (Acarina: Phytoseiidae).* Canadian Journal of Zoology 44: 483-491.

K. Nakamura, 1974. *A model of the functional response of a predator to prey density involving the hunger effect.* Oecologia 16: 265-278.

A. Oaten and W.W. Murdoch, 1975. *Switching, functional response and stability in predator-prey systems.* American Naturalist 109: 299-318.

R. Rabbinge, 1976. *Biological control of the fruit-tree red spider mite.* Simulation Monographs, Pudoc, Wageningen, pp. 228.

L.A. Real, 1977. *The kinetics of the functional response.* American Naturalist 111: 289-300.

D.J. Rogers, 1972. *Random search and insect population models.* Journal of Animal Ecology 41: 369-383.

T. Royama, 1971. *A comparative study of models for predation and parasitism.* Researches in Population Ecology, Supplement 1, 1-91.

M.W. Sabelis, 1981. *Biological control of two-spotted spider mites using phytoseiid predators.* Agricultural Research Reports 910, Pudoc, Wageningen, pp. 242.

J.N. Sandness and J.A. McMurtry, 1972. *Prey consumption behaviour of Amblyseius largoensis in relation to hunger.* Canadian Entomologist 104: 461-470.

S. Sjöberg, (1980). *Zooplankton feeding and queueing theory.* Ecological Modelling 10: 215-225.

J.G. Skellam, 958. *The mathematical foundations underlying line transects in animal ecology,* Biometrics 14: 385-400.

M.E. Solomon, 1949. *The natural control of animal populations.* Journal of Animal Ecology 18: 1-35.

R.J. Taylor, 1976. *Value of clumping to prey and the evolutionary response of ambush predators.* American Naturalist 110: 13-29.

J.C. Van Lenteren and K. Bakker, 1976. *Functional responses in invertebrates.* Netherlands Journal of Zoology 26: 567-572.

Local Dynamics of the Interaction between Predatory Mites and Two-spotted Spider Mites

M.W. Sabelis & J. van der Meer

Department of Population Biology,*
University of Leiden,
Kaiserstraat 63,
P.O. Box 9516,
2300 RA Leiden, The Netherlands

ABSTRACT

The interactions between the predatory mite, *Phytoseiulus persimilis,* and the two-spotted spider mite, *Tetranychus urticae,* take place in patchy infestations caused by spider mites on their host plants. A model was developed to simulate the predator-prey dynamics within a spider-mite patch. It was provided with experimental data on the properties of individual predators and prey, i.e. data on development, reproduction, survival, predation and dispersal. The model serves to extrapolate from the individual level to the population level. To investigate whether the existing knowledge of the acarine system provides an explanation for the observed local dynamics, population experiments were carried out in a wind tunnel to provide the conditions for aerial dispersal, and the results were compared with simulations. Validation of the model was successful when it was assumed that the predators do not disperse until after prey extermination. System analysis revealed that the dispersal rate of the predators is not exclusively determined by their feeding state, but that their take-off behaviour can be suppressed in response to cues left by their prey. Evolutionary explanations for the strong tendency to stay until the local prey population is eliminated, are discussed.

Introduction

The two-spotted spider mite, *Tetranychus urticae* Koch (Acarina: Tetranychidae), is a phytophagous arthropod with a length of ca. 1 mm. It causes damage to various crops of agricultural importance (Helle and Sabelis,1985). Reduction of the photosynthetically active leaf area is brought about by the spider mites feeding on the cells of the leaf parenchyma (at a rate of ca. 20 cells per minute). Per individual mite the damage is little, but as a consequence of their high intrinsic rate of increase, r_m, (see Sabelis, 1985a) populations of spider mites can virtually destroy the leaf area of a standing crop. The high r_m is caused by high rates of survival, development and oviposition and by a female biased sex ratio among the offspring. The eggs are deposited in a labyrinth of silken threads stretched between the leaf surface and the prominent parts (e.g. ribs, leaf edges) on the underside of the leaves. Within this self-made web the mites feed and juveniles develop into adults via a series of mobile feeding stages interrupted by moulting periods. The young females mate and then disperse to found new colonies on the same or other leaves in the vicinity. Due to their tendency to form colonies on leaves that are close to each other, spider-mite infestations have a patchy character. A group of colonized leaves on one plant or several neighbouring plants is further referred to as a spider-mite patch.

The predatory mite, *Phytoseiulus persimilis* Athias-Henriot (Acarina: Phytoseiidae), has proven to be effective in controlling spider-mite outbreaks in several agricultural crops (McMurtry, 1982). This predator is of the same size as

---

* Part of this work was performed at the Department of Animal Ecology, Agricultural University, Wageningen, The Netherlands

the spider mite and passes through the same series of stages during development. Adult females are the most active prey searchers. Moving along the stems and petioles of the plant they may move onto the leaf surface to intrude the webbed part of the leaf area in search for spider mites. Well fed female predators invest c. 70% of the ingested food in producing their own eggs and they are capable of producing an egg mass per day equal to their own body weight (Sabelis, 1981; Sabelis, this Volume). The capacity of the predator to control spider mites hinges largely on its strong numerical response. Under normal conditions the predatory mites have an intrinsic rate of population increase that is close to that of the spider mites. Hence, it is not surprising that in a small and *closed* environment the spider mites are usually completely eliminated, after which the predators die from starvation. Under realistic conditions spatial units, such as a colonized leaf or a spider-mite patch, do not form a closed environment. The question, therefore, is what type of local predator-prey dynamics is revealed when the predator and its prey are free to leave these spatial units on their own accord and by their own means.

To find a suitable set-up for such an experiment one should first consider how mites disperse. They can leave their food source by active locomotion, but this will not bring them far due to their small size and the heterogeneous structure of the vegetation and the soil surface. Long-distance dispersal is possible via air currents or by phoresy, i.e. by hitchhiking on other organisms with a better dispersal capacity. Not much is known of phoretic transport of the mites under study, but aerial dispersal is well documented in field studies and the conditions under which take-off occurs are known from observations on individual mites in the lab (see Kennedy & Smitley (1985) and Sabelis & Dicke (1985) for review). The most salient result of the behavioural studies is that spider mites and predatory mites can decide when to take-off in response to particular environmental conditions (Sabelis & Afman, 1984; Smitley & Kennedy, 1985). They are certainly not passively blown of a leaf, but actively determine when to be carried aloft.

To elucidate the effect of dispersal on the predator-prey interaction in a patch experiments were carried out in which local predator-prey dynamics and dispersal were measured simultaneously. These population experiments were done in a wind tunnel to provide the conditions for aerial dispersal to occur. The results of these experiments were analysed by means of a simulation model. The model was provided with data on the properties of individual predators and prey. It served to extrapolate from the individual level to the population level. In this way it was possible to investigate whether our biological knowledge of the acarine system provides and explanation for the observed dynamics in a spider-mite patch.

A summary of the main features of this simulation model precedes the description and analysis of population experiments in a wind tunnel.

Simulation of local predator-prey dynamics

The framework of the simulation model consists of a series of state variables representing the number of individual mites in each consecutive stage or age class. Development and ageing processes are simulated by use of the so-called BOXCARTRAIN-method developed by Goudriaan (1973) and discussed elsewhere in this Volume. The residence times of the mites in each stage or age class are updated each time step in 'hour glass' integrals. After elapse of the stage or age-specific period the mites are transferred to the next stage or age class. Dispersion in developmental time is mimicked by means of the fractional repeated shift method (Goudriaan, 1973). The adult phase is divided into periods of equal length (age classes) and the ageing process is simulated by complete shifts at the appropriate moments. Each time step the number of eggs produced during the preceding time interval by mothers of all age classes are added to obtain the state variable representing the number of new eggs. In this way population growth of the predatory mites and the spider mites can be simulated starting from a specified age distribution.

Development and reproduction depend on the amount of food acquired. In the case of the spider mites this aspect is not taken into account because plant food was amply available throughout the population experiments in the wind tunnel. However, in the case of the predatory mites prey availability should be accounted for because prey numbers appeared to be subject to large fluctuations during the population experiments. The finite state Markov model for prey search behaviour discussed in the preceding paper (Sabelis; this Volume) is used to calculate the amount of food ingested and the 'weight watcher' model is used to calculate the conversion of ingested food into body tissue and egg biomass. However, before these (sub)models can be used, it should be pointed out first what is meant by prey density. This requires a clear notion of the predator-prey interaction space. By definition, this space is confined to a spider-mite patch, i.e. a collection of neighbouring leaves colonized by spider mites. Because behavioural observations showed that the predators use little time in moving between colonies in one patch (usually much less than one hour per day), it is allowed to consider the separate spider-mite colonies as parts of one coherent large super colony. Hence, the leaf area webbed by the spider mites constitutes the battle field for predator-prey interaction and prey density may be defined as the number of spider mites per unit webbed leaf area. Within this webbed area the spider mites are not randomly distributed and the predator tends to stay longer in the areas with many prey and few conspecific predators (Sabelis, 1981; Bernstein, 1984). However, these aspects are ignored in the simulation model because the major part of the prey density variation between sites within the webbed leaf area would give rise to predation rates close to the maximum anyway and because local depressions of prey density are often associated with an overall decrease in prey density within the webbed leaf area, especially when the predators have become abundant. Although spatial variation in prey density is largely ignored, there is one aspect of this variation taken into account. This relates to the webbed area that has become devoid of prey, because the spider mites have matured and moved away. Because the predators spend little time in searching in these 'empty' areas, the interaction space is calculated as the webbed leaf area minus the 'empty' or exploited leaf area.

As will be discussed later, there are several predator components that are influenced by predator crowding. Therefore, predator density is calculated, as the actual number of predators per unit of webbed leaf area (corrected for the empty area).

In contrast to leaf-to-leaf dispersal within the spider-mite patch dispersal out of the patch is explicitly taken into account. Two versions of the model will be compared: one in which the predators are assumed to stay within the webbed area until the prey is exterminated, and one in which dispersal rates are calculated from functions derived from experiments on take-off behaviour of individual predators in a small wind tunnel (Sabelis and Afman, 1984). Except for the initial release of predators into the spider-mite patch no further predator invasions (or returns) are taken into account as these were prevented by the set-up of the population experiments.

As mites are poikilothermic arthropods, the ambient temperature determines the kinetics of biochemical processes underlying development, reproduction and food conversion. Therefore, temperature is made a driving variable in the population model. It is assumed that all temperature related rate variables react instantaneously to any change in temperature. This hypothesis is supported by the available experimental data on the effect of alternating temperatures on predator and prey components (Rabbinge, 1976; Sabelis, 1981). Effects of humidity are not considered in the model, which is only justified if relative humidity is (kept) within the range of 50 to 90%.

As will be clear, the model is of the state variable type (Forrester, 1961; De Wit and Goudriaan, 1978). State and rate variables are distinguished and mathematical expressions are given to calculate the value of each rate variable from the state of the system. The state variables are updated by rectilinear integration of the rate variable over short time intervals. Unfortunately, integration methods that continuously adjust the time interval of integration on the basis of some accuracy criterion could not be used. Rectilinear integration was required because some rate variables in the BOXCAR subroutine involve a division by the time interval of integration (De Wit and Goudriaan, 1978, p. 20). Therefore the size of the time interval has to be fixed in advance. As a rule of thumb it has to be smaller than 20% of the time constant of the integration process (Ferrari, 1978, p.26), where the time constant associated with a state variable N equals $|d\log N / dt|^{-1}$. In the case of more state variables, the time constant of the system is governed by the integration with the smallest time constant. However, the time constants of different processes treated in a simulation model often differ considerably. Hence, would the above rule of thumb be applied, the computing effort will be very inefficient with respect to the slow processes. One way to alleviate this difficulty is to assume that the fast processes are in a steady state. This assumption is made with respect to the predation rate. Thus, each time step prey density is calculated first and then the predation rate under the assumption that the steady state is instantaneously reached. This approach is justified when the rate of food conversion by the predator is much faster than the rate of prey density change. This condition seems to be fulfilled as simulations without the steady state assumption (with time intervals of some seconds instead of one hour) did not reveal important deviations from simulations that were based on the assumption of steady state predation.

Numerous models of acarine predator-prey interactions have been published (Logan, 1982). These are either simulation models (Rabbinge, 1976; Dover et al., 1979; Fujita et al., 1979; Rabbinge and Hoy, 1980; Johnson and Wellington, 1984; Shaw, 1984) or composite models (composed of analytic and simulation models) (Wollkind and Logan, 1978; Hastings and Wollkind, 1982; Wollkind et al., 1980, 1982). The model discussed here is mainly of the simulation type and it differs conceptually from published models in the following respects:

(1) the explicit choice for a local spatial scale wherein predator-prey dynamics are considered,

(2) the corresponding operational definition of prey density,

(3) the direct coupling of functional and numerical responses by modelling the conversion of ingested food into predator egg biomass,

(4) explicit account of evasive movements of predator and prey, i.e. their take-off to aerial voyages,

(5) incorporation of hitherto unknown functions and parameters (e.g. the influence of physiological age on the ovipositional rate of the predator, sex ratio control by the predator in response to prey and predator density, the effect of prey density on aerial dispersal of the predator).

Functional relationships and parameters

The structure of the model comprises several functions and parameters, that were obtained by observation of individual mites in the laboratory. However, it was not possible to obtain data covering all possible factorial combinations. In these cases methods were developed to extrapolate to the range of conditions where little or no measurements were available. A brief summary of the data and extrapolation methods is given below. A more detailed discussion is to be found in Sabelis (1981; in press).

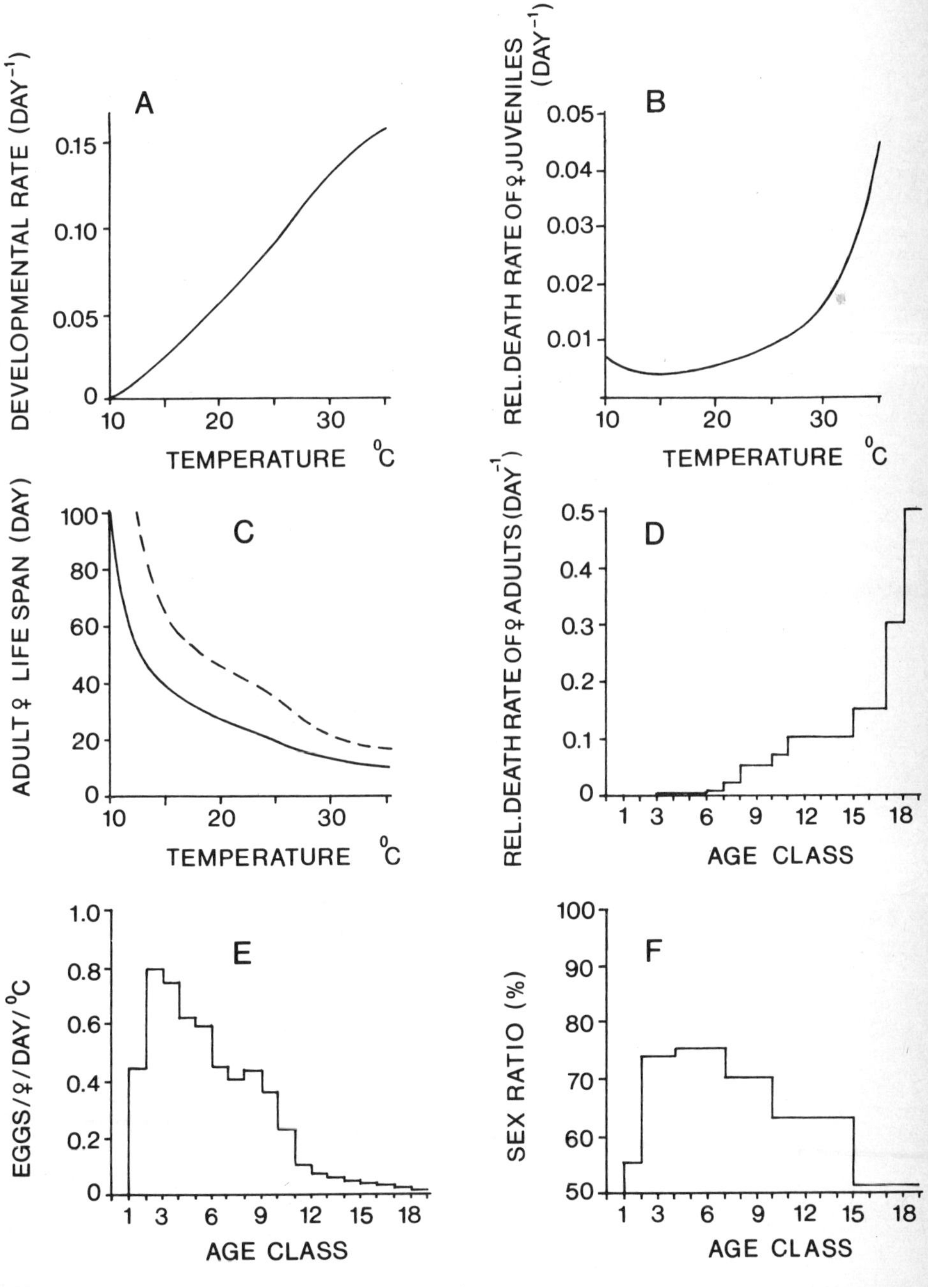

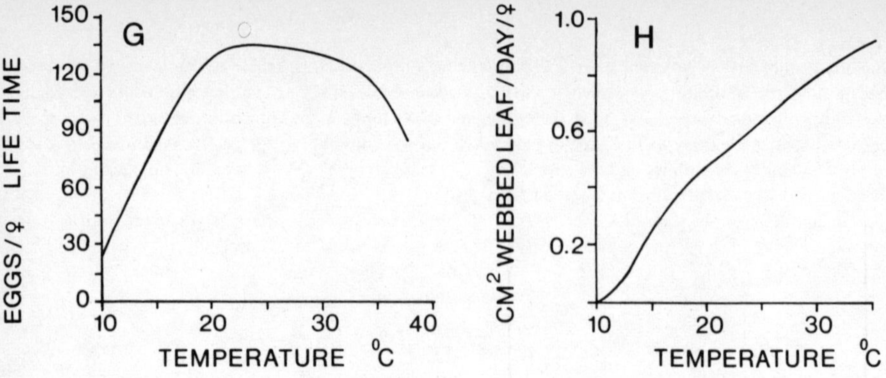

FIGURE 1

Life history components (A-G) and web production (H) of the two-spotted spider mite, *Tetranychus urticae*. Single ♀ mites were studied and provided with ample supply of plant food and sperm

- A The rate of 'egg-to-egg' development (i.e. the inverse of the time between birth and the first egg produced) and its relation to temperature.
- B The relative death rate of juvenile females (in the egg-to-egg period) and its relation to temperature.
- C The 50% points (drawn line) and the 98% points (dashed line) of the cumulative distribution of female life spans (from last moulting to death) and their relation to temperature.
- D The relative death rate of adult females in relation to their age classified in periods of equal length (3.5 days at 15°C; 2.5 days at 20°C; 2 days at 25°C; 1.25 days at 30°C; 1 day at 35°C)
- E The slope of the regression line of the ovipositional rate against temperature and its relation to age (since last moulting). The age-axis is divided in age classes (see D).
- F The proportion daughters (♀♀ / (♂♂ + ♀♀)) among the offspring and its relation to the age age class of the mother (see D).
- G The total number of eggs produced in a female's life time and its relation to temperature.
- H The bean leaf area covered by web per female per day and its relation to temperature.

Web production and life history of the two-spotted spider mite

All functioanl relationships were measured in experiments where single female spider mites were studied and provided with ample plant food and sufficient males (to ensure sperm supply). These experiments should have been done preferably on Lima bean plants, as this was the host plant used in the population experiments to be analysed. However, results of such experiments were not available. Instead, results of experiments on leaves of another bean variety (cv Noord-Hollandse Bruine) and on rose leaves (cv Sonia) were used, thus assuming absence of any host plant effect. Though this seems a reasonable assumption for the life history data (Sabelis, 1981), it is not correct when web production is considered. On roses the expansion of the web over the leaf surface proceeds at a lower rate, than usually observed on bean (Sabelis, unpublished data). Therefore additional experiments were carried out to measure the web expansion rate on Lima bean plants.

Development. The inverse of the time between birth and first oviposition is referred to as the rate of egg-to-egg development. It is linearly related to temperature in the range of 12°C to 35°C (Figure 1A). In the simulation model a distinction is made between each developmental stage in the egg-to-egg period. The egg stage comprises c. 39% of the egg-to-egg period, the larval stage 9.5%, the first moulting stage 8%, the first nymphal stage 7%, the second moulting stage 8%, the second nymphal stage 8.5%, the last moulting stage 11% and the preoviposition phase 9%. The standard deviation of stage-specific periods is in the range of 5 to 18% of the mean. The egg-to-adult period of the males is 3-7% shorter than that of the females.

Juvenile survival. The fraction of juvenile spider mites that reach adulthood, is used to calculate the relative death rate r_d from: $r_d = -\ln(N_t/N_0)/t$, where $N=$ the number of mites and t is the developmental period. The relation between r_d and temperature is given in Figure 1B. No distinction between male and female mortality is made.

Life span. In Figure 1C the 50% point and the 98% point of the cumulative distribution of female life spans from last moulting to death) are presented in relation to temperature. The maximum life span, taken to be equal to the 98% point, was divided into 18 periods of equal length. For each of these age classes the relative death rate r_d was calculated. It turned out that these r_d's varied little with temperature. Therefore, it suffices to present the relation between (smoothed estimates of) r_d and the age class number, as shown in Figure 1D. The life span of the males was estimated to be c. 70% of that of the ♀♀.

Oviposition. The oviposition rate depends on the age of the mother and on temperature. These relations can be simplified by fitting the following regression equation for each age class separately: $0_x = a_x(T-11)$, where 0_x is the oviposition rate in age class x, a_x is the slope of the regression line, T is the temperature and the threshold temperature for oviposition is 11°C. By estimating the slope a_x for each age class the relation in Figure 1E was obtained. Similar simplifications were reported to be successful by Rabbinge (1976) and Shaw (1984). In Figure 1G it is shown additionally that the total number of eggs produced during a female's life time is rather constant between 18°C and 35°C, but decreases above and below this temperature range.

Sex ratio. The proportion of daughters among the offspring appeared to depend on the age of the mother. It was only measured at 25°C and it was assumed to be similarly related to the age class number at all other temperatures (Figure 1F). Sex ratio control in response to the condition of the host plant has been reported by Wrensch and Young (1983), but this was not included in the model (in agreement with the assumption that the host plant remains in a good condition throughout the simulated period).

Web production. Experiments designed to quantify web production showed that the amount of web per unit of colonized leaf area rapidly approached a constant level (Sabelis, 1981). It only exceeded this level when the food plant was overexploited as a food source (and the mites spent much time in walking and webbing). Hence, it is reasonable to assume that the webbed leaf area is a homogeneous substrate. The expansion rate of the leaf area webbed per female (cm^2/♀/ day) is linearly related to temperature, just as the oviposition rate and the developmental rate (Figure 1H).

Predation and life history components of the predatory mites

The life history traits were measured using the experimental design discussed in the preceding paper (Sabelis; This Volume), in which prey density was kept almost constant so that the predation process approached a steady state. The predation experiments were all done with fertilized adult females of *P. persimilis.* The predator-prey arenas consisted of rose leaves (cv Sonia) on which females of *T. urticae* had constructed a web and deposited eggs (Lima bean leaves instead of rose leaves were used in the experiments on the effect of predator and prey density on sex ratio among the offspring). Life history components and predation were measured in three types of experiments differing with respect to prey density, predator density and temperature level:

(1) single predator per arena, high prey density (40 eggs/ cm^2 webbed leaf area) and different temperature levels

(10, 15, 20, 25, 30, 35°C) (Figure 2 and 4)

(2) single predator per arena, different prey density levels and one constant temperature (25°C - Figure 5)

(3) different numbers of predators per arena, high prey density (40 eggs/ cm^2) and one constant temperature (25°C)-(Figure 6)

Thus, one factor was changed while keeping the others constant. The assumptions underlying extrapolations to multifactorial changes will be discussed below. Some of these assumptions were substantiated by experimental tests.

Predation. By means of the finite state Markov model the rate of predation can be calculated from (1) the relative rate of gut emptying (r_g) and (2) the relation between the rate of successful attack (σ) and the fullness of the gut (g). Both input data were obtained from measurements at high prey densities, in monocultures of either eggs or preovipositional females of the two-spotted spider mite, at one constant temperature level and with one young female predator per arena. For the model of predator-prey dynamics the predation rate needs to be calculated for predators at any prey density, prey stage mixture, temperature, age and predator density. This extended range of conditions can be covered most easily if the following simplifying assumptions are made:

(1) the components of searching behaviour are determined only by gut fullness and the prey stage (in accordance with observations in prey stage monocultures),

(2) the rate of successful attack (σ) is determined only by these behavioural components and (independently of these components) by prey density,

(3) the relative rate of food conversion (as well as r_g) depends on temperature, (st)age and, as will be discussed later, also on physiological age and predator density.

The arguments for these assumptions and the elaboration of their consequences are given below;

- The predation model can be extended with additional data on predatory behaviour with respect to other prey stages than eggs and females, i.e. larvae, nymphs, moulting stages and adult males. These data are available from Sabelis (1981). The values of σ calculated from these data are intermediate between the σ values for prey eggs and prey females.
- Validation of the predation model in mono- and mixed cultures of prey eggs and prey females suggested that the behavioural components and their relation to gut fullness did not change as a consequence of the prey stages being presented together and in different densities. (See Sabelis, this Volume). These findings are in support of assumptions 1 and 2, but validation with respect to other (combinations of) prey stages are lacking.
- The kinetics of biochemical processes underlying food conversion is determined by ambient temperature. Because the developmental rate and the oviposition rate of most phytoseiids are linearly related to temperature, the rate of food conversion is expected to be influenced accordingly. The threshold temperature for development and reproduction is c. 11°C for *P. persimilis.* Hence, the relation between r_g and temperature (T) is of the form: $r_g = C(T-11)$, where C is the gut emptying constant in day $^{-1}$ ° C $^{-1}$. In the preceding paper (Sabelis; this Volume) measurements are presented of r_g at three temperatures so that C can be determined. Based on this relation (assumption 3) and the assumption that the rate of successful attack is not affected by temperature, the predation rate can be calculated at different constant temperatures. To validate the hypothesis that the effect of temperature on the conversion rate is of decisive importance in determining the effect of temperature on the predation rate, calculations and measurements of the predation rate were compared. In Figure 2 the results are shown for young female predators foraging at high prey egg density and 4 different temperature levels. Clearly, calculated and measured values differ little, so that there is no good reason to reject the hypothesis.
- Predation rate and oviposition rate are not constant during adult life. For example, females, which are constantly well fed and females whose oviposition was interrupted by a long starvation period turn out to have the same life time egg production. Apparently, total egg production is limited in some rough sense. Energetic and mass balance considerations imply that a slackening of the daily egg production should coincide with a smaller intake rate. Indeed, this relationship was confirmed by several independent experiments (Sabelis, 1981). In the context of the model this was implemented by letting the relative rate of food resorption (r_{rs}) depend on the total number of eggs produced in the past. The appropriate relation can be established by means of the 'weight watcher' model (see preceding paper; Sabelis, this Volume). This model was used to estimate the value of r_{rs} from measurements of the oviposition rates over the full life span of a female predator. The latter procedure was followed for the case of high prey egg density so that r_{rs} could be estimated under the assumption that the gut is filled to capacity ($g = g_{\max}$). The resulting estimates of r_{rs} and their relation to the number of eggs produced is presented in Figure 4E. To validate the above hypothesis on the role of past egg production (as a measure of physiological age) experiments on life time reproduction at low prey egg

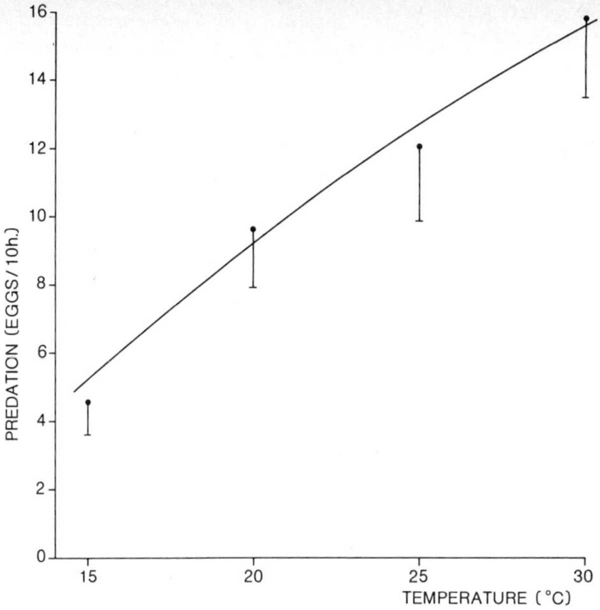

FIGURE 2

The number of eggs of *Tetranychus urticae* killed by young females of *Phytoseiulus persimilis* under steady state conditions during a 10 hour period ands its relationship with the environmental temperature. Measurements of the number of prey eggs killed at four constant temperatures (15°C, 20°C, 25°C and 30°C) are indicated by black dots (means) and bars (standard deviations). The drawn line represents the steady state calculations of the predation on eggs using the finite state Markov model. The model is provided with data on the effect of temperature on the relative rate of gut emptying, but *not* on the components of searching behaviour. Experiments and calculations are carried out at the prey density of 40 eggs per cm^2 webbed leaf area.

densities were carried out and these were simulated assuming that (physiological) age has no influence on the rate of successful attack (σ). The simulation results were encouraging in that the time delays in decreasing r_{rs} values are correctly predicted (Figure 3). However, after c. 50 days most predators stopped producing eggs irrespective of their oviposition history. Obviously, ageing processes take a heavy toll after the critical age of 50 days.

- Although the ovipositing females are the most voracious of all, the juveniles (esp. the nymphs; larvae do not successfully attack prey) and the males cannot be ignored as predators. Their predation rate is approximately equal to that of a non-reproductive female with the exception that the nymphs usually do not attack adult female spider mites (Sabelis, 1981).

- It has been shown that the oviposition rate (and therefore also the predation rate) decreases with predator density (Eveleigh & Chant, 1982). This is not caused by a waste of time available for searching prey due to struggles between predators, nor by changes in behavioural components, such as walking speed, walking activity and success ratio (Sabelis; unpublished data). Hence, as there is no indication of the predator density affecting the relative rate of defaecation, it follows that the decreased rate of oviposition should be caused by a decrease of the relative rate of food resorption r_{rs} in response to predator density. Using the 'weight watcher' model, values of r_{rs} were estimated from life time reproduction experiments at various predator densities. The results are presented in Figure 6C.

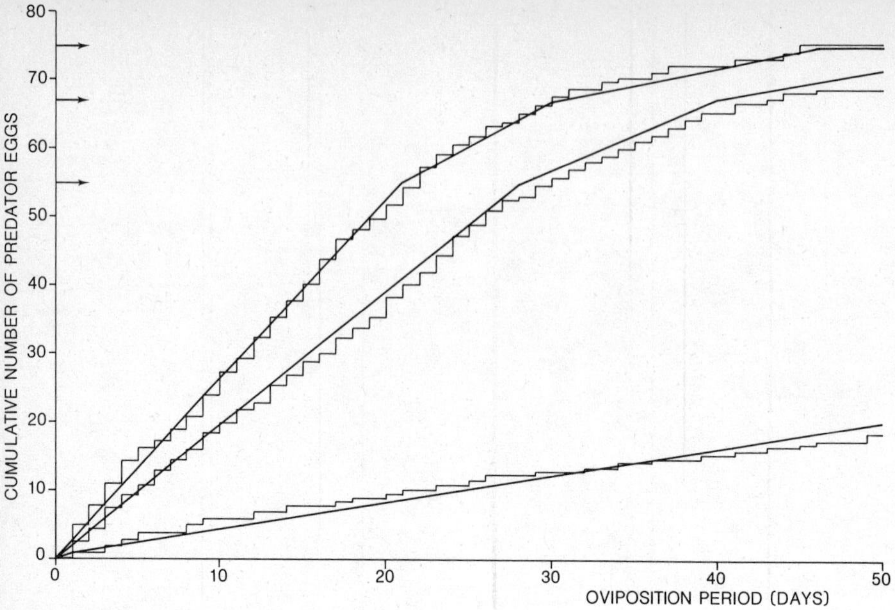

FIGURE 3

The cumulative number of eggs deposited by females of *Phytoseiulus persimilis* in the course of the oviposition period. Experimental results are presented in histograms for three situations differing with respect to prey density: 40 eggs/cm^2 webbed leaf area (upper histogram), 4 eggs/cm^2 (intermediate histogram) and 1 egg/cm^2 (lower histogram).

Experiments were done at 20°C and 70% RH. The straight lines represent steady state calculations using the finite state Markov model and the 'weight watcher' model . (See text for explanation). Above the egg production levels indicated by arrows along the *y*-axis one parameter of the model, i.e. the relative rate of gut emptying r_g was adapted so as to make a good fit between calculated and measured egg production *at high prey egg density*. Using the relation between r_g thus obtained, and egg production was calculated for the case of *low* prey egg densities.

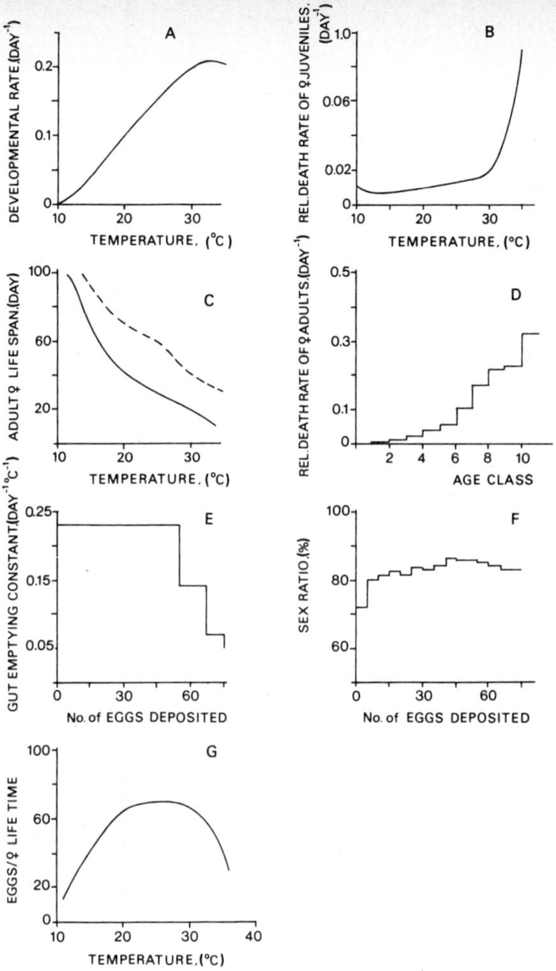

FIGURE 4

Life history components of the predatory mite, *Phytoseiulus persimilis.* Single ♀ predators were studied under conditions of ample prey supply and continuous presence of males.

A The rate of egg-to-egg development (i.e. the inverse of the time elapsed between birth and the first egg produced) and its relation to temperature.

B The relative death rate of juvenile and preovipositional females (i.e in the egg-to-egg period) and its relation to temperature.

C The 50% points (drawn line) and the 98% points (dashed line) of the cumulative distribution of female life spans (from last moulting to death) and their relation to temperature.

D The relative death rate of adult females in relation to their age, classified in periods of equal length (9 days at 15°C; 7 days at 20°C; 6 days at 25°C; 4.5 days at 30°C; 3.3 days at 35°C).

E The relation between the gut emptying constant C of predators in the oviposition phase and the cumulative number of eggs deposited (as a measure of physiological age). For explanation, see text.

F The proportion daughters (♀♀/(♂♂+♀♀)) among the offspring and its relation to the cumulative number of eggs deposited by the mother (see E).

G The total number of eggs produced by female predators in their life time and its relation to temperature.

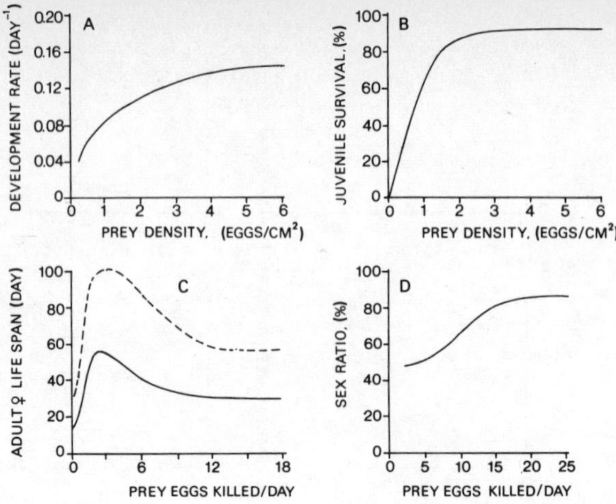

FIGURE 5

The influence of the level of prey supply on some life history components of the predatory mite, *Phytoseiulus persimilis* Athias-Henriot. Experimental conditions: 25°C, 70% RH.

A The rate of egg-to-egg development and its relation to prey density (eggs per cm^2 webbed leaf area).

B Survival of the juveniles (esp. the nymphal stages) and its relation to prey density (eggs per m^2 webbed leaf area)

C The 50% points (drawn line) and the 98% points (dashed line) of the cumulative distribution of female life spans (from last moulting to death) and their relation to the rate of prey egg consumption (eggs per day).

D The proportion daughters among the offspring and its relation to the prey consumption rate (eggs per day) of the mother.

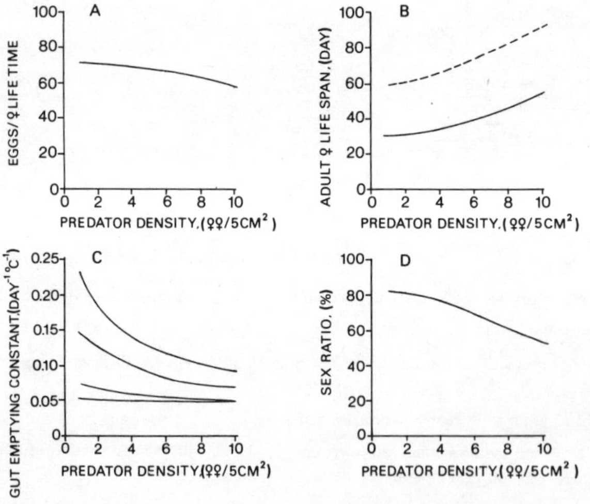

FIGURE 6

The influence of predator density (number of predators per 5 cm^2 webbed leaf area) on the rate of food conversion and some life history components of the predatory mite, *Phytoseiulus persimilis* Athias-Henriot. Experimental conditions: ample prey, 25°C, 70% RH.

A The total number of eggs produced per female predator during her life time and its relation to predator density.

B The 50% points (drawn line) and 98% points (dashed line) of the cumulative distribution of female life spans and its relation to predator density.

C The gut emptying constant C and its relation to predator density. For explanation, see text.

D The proportion daughters among the offspring produced at different levels of the predator density.

Development. The rate of egg-to-egg development increases linearly with temperature, but above 32 °C it levels off (Figure 4A). In the simulation model each stage in the egg-to-egg period is distinguished in order to account for differential rates of development, survival and predation. The egg stage comprises 34% of the egg-to-egg period, the larval stage 12%, the first nymphal stage 15.5%, the second nymphal stage 16.5% and the preovipositional stage 22%, whereas moulting takes a negligible amount of time. The standard deviations are within 7-12% of the mean developmental time. The egg-to-adult period of the males is slightly shorter than that of the females. In Figure 5A the effect of prey density on the developmental rate of the predator is shown. The critical prey density level below which the developmental rate decreases, is c. 3 prey eggs per cm ^{2}(Sabelis, unpublished data, but see also Eveleigh and Chant, 1982). Whether the density of immature predators affects the developmental rate independently of prey density, has not been investigated yet.

Survival. The relative death rate r_d is usually very small except when temperature rises above 32°C or falls below 5°C (Figure 4B), and when prey density decreases below 2 prey eggs per cm ^{2} (Figure 5B). According to Eveleigh and Chant (1982) juvenile survival is not significantly affected by the density of conspecific immatures.

Life span. Age and temperature dependence of survival are presented in Figure 4C for the case of well fed predators. The maximum life span was taken to be equal to the 98% point of the cumulative frequency distribution. This period was divided into 10 subperiods of equal length. For each of the 10 age classes the relative death rate r_d was calculated. It turned out that the calculated values of r_d varied little with temperature in the range of 12°C to 32°C, so that within this range it suffices to relate r_d to the age class only (Figure 4D).

Prey and predator density are factors influencing the life span. When prey density decreases, the life span initially increases, but below the consumption rate of 2 prey eggs/day a steep decline takes place (Figure 5C; Sabelis, 1981 and unpublished data). In absence of prey and presence of free water for drinking the mean life span is approximately 14 days, which is half the life span of well fed predators. The effect of the density of conspecific females on female life span is shown in Figure 6B. Under conditions of ample prey supply the mean life span increases with predator density. The extension of the life span at moderately low prey density and high predator density coincides with a lower rate of food turnover into egg biomass. This suggests that, like the oviposition rate, the life span depends on the amount of food converted in the past. In the context of the model this was implemented by means of a physiological time scale, i.e. age was measured in terms of the fraction of the maximum life span completed, rather than in terms of the actual time alive. Thus, the adults were classified in 10 age classes and shifted to the next class as soon as the class boundaries were passed. The residence time per age class (t_x) was dependent on prey density, predator density and temperature. Life span data measured under various conditions revealed a rather constant ratio of the mean to the standard deviation, implying that survival is constant per age class. Therefore, it suffices to divide the class-specific relative death rates, $r_d(x)$, (see Figure 4D) by the proportional change in the life span. At very low prey density such a correction cannot be justified, because the decreased life span results from starvation and not from a shorter residence residence time per age class. For example, interruption of the oviposition period by a long starvation period increases the life span, relative to that of a continuously well fed female. Therefore, in the context of the model the effect of very low prey density was implemented by adapting the value of the relative death rate, while maintaining a long residence time per age class (i.e. equal to the maximum measured at a consumption rate of 3 prey eggs per day (see Figure 5C). Thus, if $Q(y)$ is defined as the ratio of the expected life span at prey density y to that at high prey density, then $Q(y)$ never drops below unity and $Q(y)=Q(y_3)$ when y is below the prey density at which the prey egg consumption rate equals 3 eggs per day (y_3).

Because the effect of predator density was measured only at high prey density, extrapolations to low density regimes are to be based on the assumption of proportionality. Thus, if $\bar{L}(y,z)$ is defined as the expected mean life span at prey density y and predator density z, then the proportionality factor $R(z)$ is obtained from:

$$R(z)=\bar{L}(\text{high},z)/\bar{L}(\text{high},1)$$

Mean and maximum life span can now be calculated from:

$$\bar{L}(y,z) = \bar{L}(y,1)R(z) \text{ and } L_{\max}(y,z)=L_{\max}(y,1)R(z)$$

Because the available experimental data revealed a rather constant ratio of the mean life span to its standard deviation, the relative death rate for each age class x was adapted by multiplication with the inverse of $R(z)$.

The above extrapolation seems not to make sense when prey is absent or very scarce. Starvation is likely to affect the predators independent of predator density. It is therefore provisionally assumed that $R(z)=1$ when the prey consumption rate drops below the level required for maintenance processes (c. 2 prey eggs per day at 25°C). Another gap to be filled by extrapolation arises from the fact that the role of prey and predator density has been studied at one particular temperature only (25°C). Again, proportionality was assumed in the context of the model. The effect of predator density was taken into account by multiplication with $R(z)$, whereas the effect of prey density on life span necessitated an extra multiplication with $Q(y)$:

$$Q(y) = \bar{L}(y,1)/\bar{L}(\text{high},1)$$

Like $R(z)$, the factor $Q(y)$ influences $\bar{L}(x,y), L_{max}(x,y)$ and the relative death rate $r_d(x)$. The life span of well fed males is equal to ca. 75% of that of well fed females. Data on the effect of prey density and predator density are not available.

Oviposition. Because the predation rate is largely determined by the relative rate of food resorption and most of the ingested and resorbed food is allocated to egg production, it follows that the predation rate and the oviposition rate are similarly affected by temperature, prey density and predator density. Hence, as discussed above, the 'weight watcher' model can be used to calculate the rate of oviposition from the rate of predation. The key factor in the food conversion process is the relative rate of gut emptying $r_g(\text{day}^{-1})$. The effect of temperature on r_g is linear with $r_g = C(T-11)$. Prey density is assumed not to affect C (the gut emptying constant in day $^{-1}$°C^{-1}), but physiological age (or more precisely the number of eggs produced; see Figure 4E) and predator density (Figure 6C) have pronounced effects on C.

Sex ratio. The sex ratio among the offspring is little influenced by the age of the mother (Figure 4F), except that females usually start the oviposition period by producing a son. In contrast to what has been suggested by Charnov (1982) and Bull (1983), phytoseiid females are capable to control the sex ratio in response to environmental conditions such as prey and predator density (Sabelis, 1985b). At high prey density and low predator density sib-mating is likely to be more frequent, whereas low prey density and h igh predator density tend to promote random mating. As predicted by evolutionary theory of sex allocation (Charnov, 1982), phytoseiid females produce female biased sex ratios under the former conditions and invest equally in sons and daughters under the latter conditions (Sabelis, 1985b). Preliminary results of experiments with *P. persimilis* are shown in Figure 5D for variable predator density and high prey density. The combined effect of low prey density and high predator density has not been measured yet. In the context of the model it was assumed that the effect of predator density (z) on the proportion females ($S(y,z)$) can be represented by a relative effect on the female bias:

$$S(y,z) = 0.5 + (S(y,1)-0.5)S(\text{high},z)/S(\text{high},1)$$

Dispersal of spider mites and predatory mites

Mites disperse by phoretic means, by walk or via wind currents. Reports on phoretic transport of the mites under study are absent, but locomotory dispersal (Hussey & Parr, 1963; Sabelis, 1981; Bernstein, 1984) and aerial dispersal (Fleschner et al., 1956; Sabelis & Afman, 1984; Smitley & Kennedy, 1985) have been studied more thoroughly. In the population experiment to be described below phoretic and locomotory dispersal are made irrelevant due to the experimental set-up, which was specifically arranged to study aerial dispersal in the course of the predator-prey interaction in a spider-mite patch. Therefore, it suffices here to present data on aerial dispersal in response to food availability and crowding. These observations were made in a mini wind tunnel, specifically designed to observe take-off behaviour of individual mites on a small leaflet (Sabelis & Afman, 1984).

Two-spotted spider mites showed little tendency to take off when the host plant leaflet was in good condition. Because the latter situation prevailed throughout all population experiments, it suffices to mention that the relative rate of spider-mite dispersal ($-\ln(N_t/N_0)/t$; N_t = number of mites on leaflet at t; t = observation period) is 0.005 to 0.03 day $^{-1}$. Among the few dispersers adult females were the most frequent. No experimental data were available to elucidate the effect of predator density on take-off behaviour of their prey (but see Bernstein, 1984, for the effect on locomotory dispersal of spider mites).

The dispersal tendency of predatory mites increased with the time spent without food. Well fed females of *P. persimilis* were starved for different periods and then placed on a leaflet in the mini-wind tunnel for observation during c. 10 min. At 25°C and 35% RH 18% of the predators dispersed after 2 hours starvation, 54% after 9 hours and 83% adter 23 hours. To estimate the relative rate of aerial dispersal r_a in relation to the state of food deprivation well fed predators were placed on a leaflet in the mini-wind tunnel and then, using time-lapse video techniques, they were continuously observed until they dispersed aerially. From the cumulative distribution of take-off events the relative rate of dispersal r_a was calculated for different starvation classes and, using the data on the rate of gut emptying, the food content of the predator's gut was related to r_a under the assumption that the feeding state, and not the feeding history, is the factor determining the dispersal rate. The resulting relation is given in Figure 7.

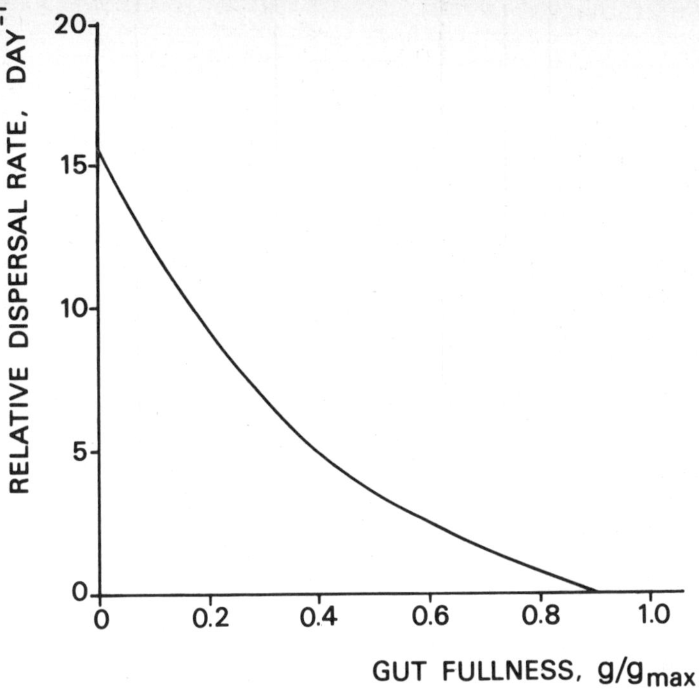

FIGURE 7

The relation between the relative rate of aerial dispersal and the relative food content of the gut, the latter being calculated from the time spent without prey (since the predator was in a satiated state) and the relative rate of gut emptying (at 25°C).

Caution should be exercised, however, in using this function in the population model, because the dispersal rates were measured *in absence of prey!* Clearly, a hungry predator is not expected to disperse when there is plenty of prey around. This was corroborated by experiments in the mini-wind tunnel. Starved females of *P. persimilis* dispersed much less frequently when the leaflet was contaminated by residues left by spider mites while feeding during the day preceding the experiment. In that case, only 7% of the predators dispersed during 10 min. observations, whereas 83% dispersed when spider-mite cues were absent. The suppression of take-off behaviour was mediated, at least partly, by chemical residues, because the suppression was also elicited when the leaflet was treated with a water extract from spider-mite infested leaves that were brushed before extraction in order to remove the mites and their web. Due to oxidation and/or evaporation of these chemicals the suppressive effect on take-off behaviour was not of a permanent nature; a few days after removal of the mites and their web dispersal was almost as frequent as in experiments with clean leaflets. Therefore, it can be concluded that the starved predators are not simply blown from the leaflet due to loss of strength to withstand the wind, but that they can decide whether to stay or not. How these decisions relate to prey and predator density, has not yet been investigated, but the system analysis and the population experiment to be discussed below, will shed some new light on these aspects.

The effect of aerial dispersal on local predator-prey dynamics

To investigate whether the biological information summarized in Figure 1-7 provides an explanation for the dynamics of predator and prey in a spider-mite patch the system was analysed by simulation and experiments at the population level. A group of small Lima bean plants was provided with spider mites and predatory mites and then placed in a large wind tunnel to enable the mites to disperse by wind, as soon as they are motivated to do so. Under field conditions wind streams will bring the mites over distances long enough to make the chance of return negligibly small. The dispersers were therefore caught and then removed from the wind tunnel. In this way, it was possible to study local predator-prey dynamics under conditions where mite invasions on the plants were controlled by the experimenter and the evasions were recorded in the course of the experiment. The population experiments and their analysis are discussed below.

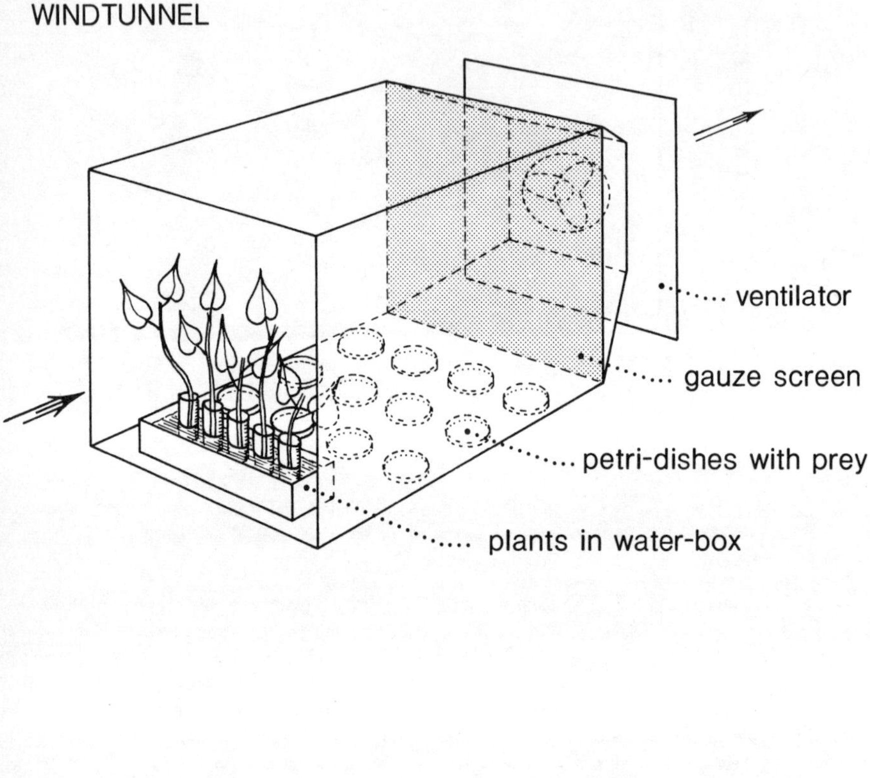

FIGURE 8

A The set-up of the wind tunnel used for the experiments on predator-prey dynamics. For explanation, see text (Note, however, that the transversal strips on the walls of the tunnel are *not* indicated). The arrows indicate the prevailing wind direction.

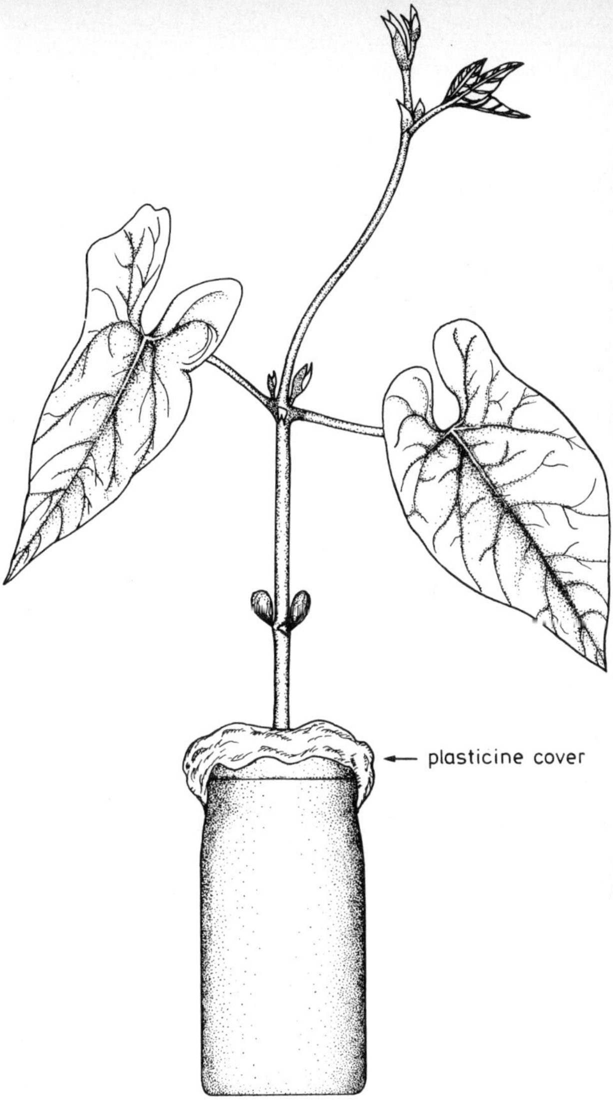

B The plant unit used in the experimnents on predator-prey dynamics in a wind tunnel. For explanation, see text.

Population experiments in a wind tunnel: materials and methods

The following features of the experimental procedure are relevant:

(i) *Wind;* the experiments were carried out in a wind tunnel (1 m length, 0.55 m width and 0.7 m height) of the type presented in Figure 8A. A wind stream of c. 2 m/s was brought about by a ventilator in the centre of a cone-shaped connection to the leeward side of the tunnel. Near to the opposite end of the tunnel (the upwind side) a box with young bean plants was placed. The mites dispersing from the plants were intercepted in several ways. As the plant-box was filled with water, they could drop onto the water surface where they were easily detected while floating. They could be intercepted by a gauze with a mesh size fine enough to prevent the female mites from moving out of the wind tunnel. The intercepted predatory mites become hungry and under this condition they are known to move upwind and to be very efficient in locating upwind spider-mite colonies. Therefore, 20 petri dishes with spider-mite infested leaves on moistened cotton wool were placed at the bottom side of the tunnel in-between the plant-box and the gauze. Once the predators arrived on these leaves, they stayed there due to the ample supply of prey. To avoid confusion with the two-spotted spider mites used in the population experiment a red-coloured tetranychid mite *(Tetranychus pacificus* McGregor) was used as 'bait' for the dispersed predatory mites. As the predatory mites have a strong tendency to move along leaf ribs and edges, the walls of the tunnel were provided with rib-like structures positioned transversal to the wind direction to increase the probability of finding the prey-bait dishes at the bottom side of the tunnel. Baits for entrapment of dispersing prey mites were also installed in the wind tunnel. These consisted of petri-dishes with bean leaf discs without pacific spider mites.
To study the effect of wind on dispersal two replicate experiments at very low wind speed (c. 0.15 m/s) were done in addition to those at the wind speed of 2 m/s (See Figure 9A).

(ii) *Host plant;* Young Lima bean plants were grown in small bottles until the first two leaves were full grown (Figure 8B). The leaf area per plant was kept constant (ca. 400 cm^2) by removing the apical parts of the plant throughout the population experiment. A total of ten plant units were placed in the water-filled box. Care was taken that the leaves touched each other so as to enable the mites to move from plant to plant without any hindrance.

(iii) *Climate;* All experiments were carried out in a climate room at 23°C and 55% RH.

(iv) *Initial number of mites;* Young fertilized females were selected for release. They were in the oviposition phase, but had deposited only a few eggs before. At the start of the experiment females of *T. urticae* were released and after a period of prey population build-up females of *P. persimilis* were transferred to the spider-mite patch. Three experiments were done with ten replicates each:
- release of 8 female spider mites at day 0 and 4 female predators at day 9 (Fig. 9A)
- release of 8 female spider mites at day 0 and 4 female predators at day 12 (Fig. 9B)
- release of 16 female spider mites at day 0 and 4 female predators at day 9 (Fig. 9C)

During the first 3 days after release missing females were replaced.

(v) *Population census;* counts of mites on the bean plants and the dispersed mites were made at intervals of 3 days. The mites on the bean plants were counted while inspecting each bean plant separately under the binocular microscope. To limit the amount of work only the adult females were recorded. The dispersed mites were counted while inspecting the water-filled plant-box and the tunnel walls by naked eye, whereas the leaf discs with or without prey were inspected under the binocular microscope.

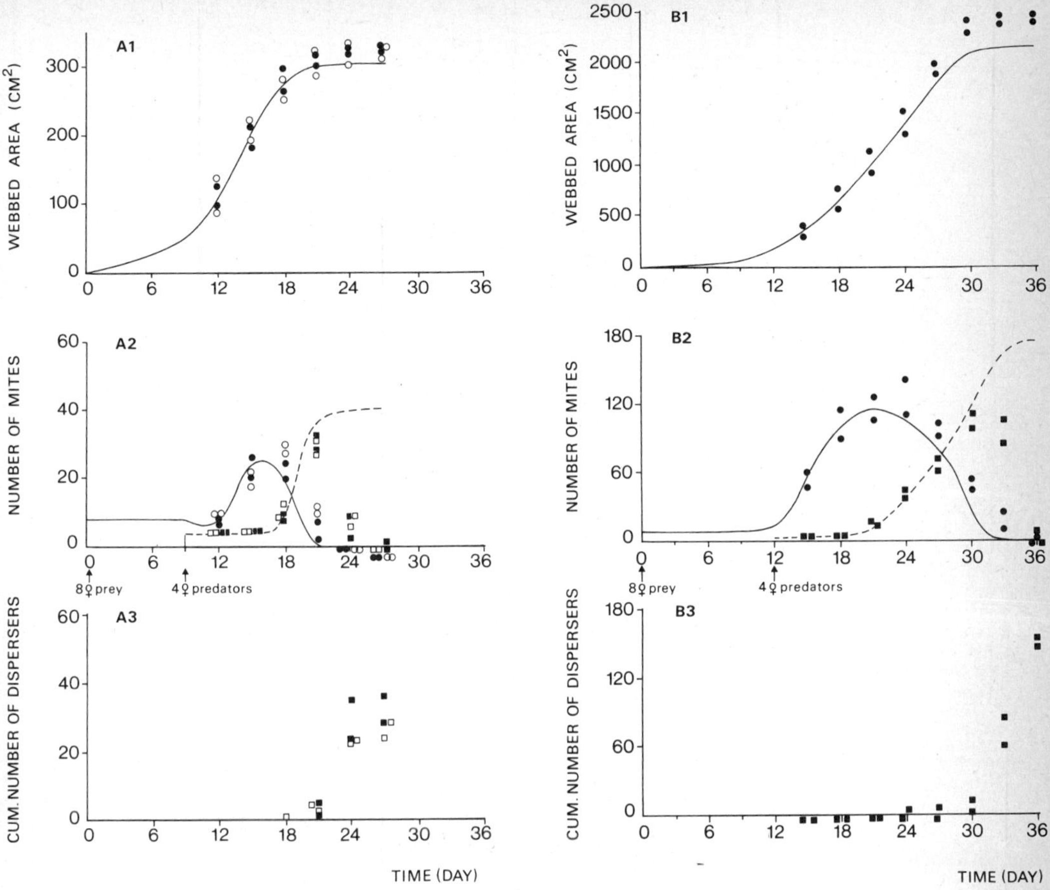
A1
WEBBED AREA (CM^2)
A2
NUMBER OF MITES
8♀ prey
4♀ predators
A3
CUM. NUMBER OF DISPERSERS
TIME (DAY)
B1
WEBBED AREA (CM^2)
B2
NUMBER OF MITES
8♀ prey
4♀ predators
B3
CUM. NUMBER OF DISPERSERS
TIME (DAY)

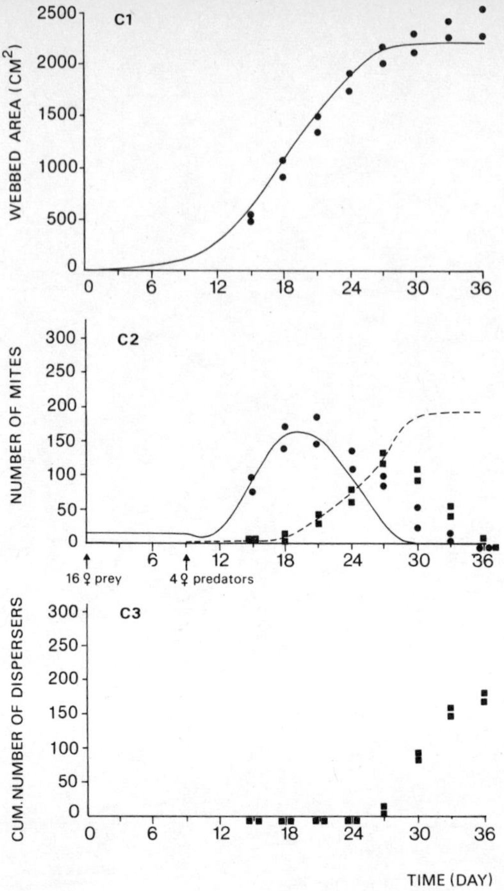

FIGURE 9

Predator-prey dynamics on bean plants placed in a windtunnel (23°C, 70% RH). The experimental results are indicated by dots (prey) and squares (predator). Black dots and squares relate to experimental conditions of 2 m/s wind speed, whereas open dots and squares relate to 'still air' conditions (ca. 0.2 m/s wind speed). Each measurement represented by these signs is based on one single replicate. Simulations of the predator-prey dynamics are indicated by drawn lines (prey) and dashed lines (predator).

For each experiment the results are summarized by presenting the time series of three variables: (1) the leaf area covered by web of the two-spotted spider mites, (2) the number of adult female mites, and (3) the cumulative number of adult female mites that had dispersed from the bean plants (i.e. caught on the petri dishes provided with prey food or on the water surface underneath the bean plants).

A Predator-prey dynamics after releasing 8 young female spider mites on day 0 and 4 young female predatory mites on day 9.

B Predator-prey dynamics after releasing 8 young female spider mites on day 0 and 4 young female predatory mites on day 12.

C Predator-prey dynamics after releasing 16 young female spider mites on day 0 and 4 young female predators on day 9.

Note, that the number of dispersed spider mites are *not* indicated, being very low (total number of dispersers ranged from 1 to 8 per replicate experiment).

Results of population experiments

The results of the population experiments are shown in Figure 9 and the following conclusions can be drawn:

(1) the predator-prey interaction ended in all but one case in spider-mite elimination,

(2) the period between predator release and prey extermination increased with the initial pre-predator ratio and the time interval between prey and predator introduction,

(3) dispersal of the spider mites occurred at a very low rate, whereas dispersal of the predatory mites peaked around the moment of prey extermination,

(4) in still air all dispersers were found on the water surface of the plant box, whereas in the presence of an air current of 2 m/s most dispersers were trapped in the baits positioned at the downwind size of the plants (bean leaves for the spider mites and colonies of the Pacific spider mites for the predatory mites),

(5) predator-prey dynamics in 'still air' were very similar to those in the air current,

(6) the majority of the dispersers were adult females, immatures and males were not found to be dispersing until after prey extermination.

Analysis of measured predator-prey dynamics by simulation

To assess the explanatory value of our biological knowledge of the predator-prey interaction, the population experiments were simulated after providing the model with appropriate additional data on initial numbers of mites, date of release and temperature. With respect to the dispersal process three versions of the model were created and tested. In the first version the dispersal rate was assumed to be determined by the actual food content of the gut (See Figure 7). In the second version it was assumed to be determined by the steady state expectation value of the gut fullness. In the third version the predators were assumed not to disperse at all. The latter is equivalent to assuming that the predators do not disperse until after the moment of prey extermination. Hence, the three versions suffice to cover the conceivable cases in terms of their extremes. Both the first and the second version of the simulation model produced a dynamical pattern that was in sheer contradiction to the observed pattern; the spider-mite population continued to increase indefinitely and was little influenced by predation due to the predator's high rate of dispersal. The third version, however, produced a dynamical pattern that was remarkably similar to the observed pattern (See Figure 9A, B and C), except for the following aspects:

1) In the experiments the predatory mites obviously did not delay the moment of dispersal as strictly as assumed in the third version of the model

2) The model consistently predicted the moment of prey extermination at a slightly earlier time than observed.

It is reasonable to assume that these deviations were interrelated; prey elimination was slightly postponed due to the predator exodus starting some days before complete prey extermination had actually occurred. Notwithstanding the early take-off of some predatory mites it is clear that the majority of aerial voyages occur during a few days before and after the moment of prey extermination. If starvation would have been the only factor influencing dispersal, 95% of the predators would have left the plants within a day and the departure period would be much shorter than observed. This strongly suggests that the predators were arrested in the webbed leaf area, even when there was little or no prey. The arrestment may well be caused by the predator's response to prey residues. A second point suggested by the system analysis is that predator density had no effect on aerial dispersal, as stimulation would be the only biologically conceivable influence of predator crowding. Note, however, that cannibalism among the predatory mites was not taken into account in the model. It may therefore be that the actual feeding state of the resident predators was better than simulated on the basis of spider-mite supply. This may have caused the actual predators to be less motivated to disperse.

Although the simulations with the third version of the model were very successful, their explanatory value cannot be clear without a sensitivity analysis of the model. This, however, will not be treated in this paper, as it will be discussed at length by Sabelis (in press). It suffices here to mention that (1) among the relatively most important components are the developmental rate of predator and prey (See also Caswell & Hastings, 1982; Sabelis, 1985c,d) and the relative rate of gut emptying and that (2) the effects of prey density were relevant to the outcome of the simulations, but that (3) the effects of high predator density could just as well be ignored (predator density did not increase above 0.1 per cm ^{2}, both in the experiments and the simulations presented in Figure 9).

Discussion

The most salient result of the system analysis presented in this paper is that local prey extermination is predominantly caused by the tendency of the predators to stay in the prey patch until the prey population's very end. Whether this phenomenon is realistic under field conditions, is still an open question. It may be that if the population experiments were to be repeated in a larger and more complex environment, the prey population would not be exterminated, but persisted through repeated local temporary escapes from the predators. However, all population experiments carried out in cucumber and rose crops under greenhouse conditions (i.e. where wind speed is low (< 0.5 m/s) and hence locomotory dispersal prevails) invariably resulted in local prey extermination (Sabelis et al., 1983; Sabelis, in press). Moreover, practical experience of workers in biological control of spider mites shows that spider-mite populations are decimated by predatory mites to the extent that reintroduction of the predators is required as soon as the spider-mite population resurges (Hussey and Scopes, 1985). Certainly, the action of predatory mites is more close to that of a pesticide than to that of a regulatory agent that keeps prey density permanently at a low level. On the other hand, it may well be that the uniformity of crops, such as cucumber and rose, does prevent locomotory dispersal of the prey from being a successful way of escaping from predators. Undoubtedly, there will be situations in nature and there can be situations created by experimenters (see e.g. Huffaker, 1958), in which spatial heterogeneity is such that locomotory dispersal does provide a chance of (temporary) escape.

From an evolutionary point of view a relevant question is why predatory mites tend to stay in a spider-mite patch until the prey population is eliminated. A set of alternative dispersal strategies would be conceivable, all of which have in common that the dispersal rate may exceed zero even when the local prey population is not exterminated. An interesting special case among these alternatives is that of 'the prudent disperser', a variant of the well known foraging strategy of 'the prudent predator' (Slobodkin, 1974; Gilpin, 1975; Slatkin & Maynard Smith, 1979). The dispersers are called prudent when they expose themselves to the risks associated with passive aerial dispersal, thereby alleviating the predation pressure on the local prey population so as to prevent the prey's extermination and providing predators, that stay behind, i.e. their sibs, with better reproductive prospects. When a prey patch is colonized exclusively by predators employing one single dispersal strategy, predators of the 'prudent' type would initially reproduce less because part of them disperse, aerially and incur high risks of death. But — later on — the overall number of (dispersed) predators (and its ratio to the within-patch time period of predator-prey interaction) exceeds that produced in a prey patch with predators that disperse only after prey extermination. Hence, provided that the prey does not overexploit its food source at an early moment, predators of the prudent type seem to have a selective advantage to predators of the exterminator type in terms of the total number of dispersers produced per time unit. However, when predators of each of the two types enter the same prey patch, then the prey left by the prudent dispersers is not only to the advantage of their relatives, but also to that of the exterminator type. Which of the two predator types will be the winners, is an open question, unless a third more flexible strategy exists whereby individuals employ the 'prudent' strategy when alone and the exterminator strategy when other predators in the prey patch do the same. Such an ability to discriminate between the opponent's strategies seems however unlikely for predatory mites, so that we are left with the question which of the above mentioned dispersal strategies is evolutionarily stable. This will be the subject of a forthcoming paper.

Another relevant question in the context of (co-)evolution is why the (aerial) dispersal rate of two-spotted spider mites is so low even when predatory mites tend to drive local spider-mite populations to extinction. If predation pressure imposes a strong selective force, one would expect selection for escape behaviour of the prey in response to predator density. Bernstein (1984) found increased locomotory dispersal of spider mites when predator density was high. The effectiveness of such an escape response remains to be assessed, however. More information on escape strategies of the prey are needed.

Acknowledgements:
We thank Hans Metz for constructive comments and several suggestions for improvement of the text. Rijndert de Fluiter constructed the wind tunnels. The figures were drawn by Jan Herzberg.

References

C. Bernstein, (1984). *Prey and predator emigration responses in the acarine system Tetranychus urticae - Phytoseiulus persimilis.* Oecologia **61,** 134-142.

J.J. Bull, (1983). *The Evolution of Sex Chromosomes and Sex Determining Mechanisms.* Benjamin and Cummings, Menlo Park.

H. Caswell & A. Hastings (1980). *Fecundity, developmental time, and population growth rate: an analytical solution.* Theoretical Population Biology **17,** 71-79.

E.L. Charnov (1982). Monographs in Population Biology **18,** Princeton University Press, 355 pp.

C.T. De Wit & J. Goudriaan (1978). *Simulation of Ecological Processes.* Pudoc, Wageningen, 175 pp.

M.J. Dover, B.A. Coft, S.M. Welch and R.I. Tummala, (1979). *Biological control of Panonychus ulmi (Acarina: Tetranychidae) by Amblyseius fallacis (Acarina: Phytoseiidae)* Environmental Entomology **8,** 282-292.

E.S. Eveleigh & D.A, Chant (1981). *Experimental studies on acarine predator-prey interactions: the numerical response of immature and adult predators (Acarina: Phytoseiidae).* Canadian Journal of Zoology 59 **7,** 1407-1418.

E.S. Eveleigh & D.A. Chant (1982). *Experimental studies on acarine predator-prey interactions: the effects of predator density on immature survival, adult fecundity and emigration rates, and the numerical response to prey density (Acarina: Phytoseiidae).* Canadian Journal of Zoology 60 **4,** 630-638.

Th.J. Ferrari, (1978). *Elements of System Dynamics Simulation.* Pudoc, Wageningen, 86 pp.

C.A. Fleschner, M.E. Bedgley, D.W. Richter & J.C. Hall (1956). *Air drift of spider mites.* Journal of Economic Entomology **49:** 624-627

J.W. Forrester, (1961). *Industrial Dynamics.* Massachusetts Institute of Technology Press, Cambridge, Massachusetts, U.S.A.

K. Fujita, T. Inoue & A. Takafuji, (1979). *Systems analysis of an acarine predator-prey system.* I. Researches on Population Ecology **21,** 105-119.

M.E. Gilpin (1975). *Group Selection in Predator-Prey Communities.* Princeton University Press, Princeton, New Yersey, 108 pp.

J. Goudriaan, (1973). *Dispersion in simulation models of population growth and salt movement in the soil.* Netherlands Journal of agricultural Science **21,** 269-281.

A. Hastings & D. Wollkind, (1982). *Age structure in predator prey systems.* Theoretical Population Biology **21,** 44-56.

W. Helle & M.W. Sabelis (1985). *Spider Mites: Their Biology, Natural Enemies and Control.* Elsevier Science Publishers, Amsterdam.

C.B. Huffaker, (1958). *Experimental studies on predation: dispersion factors and predator-prey oscillations.* Hilgardia 27 **14,** 343-383.

N.W. Hussey & W.J. Parr, (1963). *Dispersal of the glasshouse red spider mite Tetranychus urticae Koch (Acarina: Tetranychidae).* Entomologia experimentalis et applicata **6,** 207-214.

N.W. Hussey & N.E.A. Scopes (1985). *Control of Tetranychidae in greenhouse vegetables (Britain).* In: Spider Mites (Eds W. Helle & M. Sabelis), World Crop Pest Series, Vol. 1B, Elsevier, Amsterdam.

D.L. Johnson & W.G. Wellington, (1984). *Simulation of the interactions of predatory Typhlodromus mites with the European red mite, Panonychus ulmi (Koch.* Researches on Population Ecology **26,** 30-50.

G.G.Kennedy & D.R. Smitley (1985). *Dispersal.* In: Spider Mites: Biology, Natural Enemies and Control (Eds. W. Helle and M. Sabelis), World Crop Pest Series, Vol. 1A, Elsevier, Amsterdam.

J.A. Logan (1982). *Recent advances and new directions in phytoseiid population models.* In: Recent Advances in Knowledge of the Phytoseiidae (M.A. Hoy, Editor), University of California, Publication no. 3284.

J.A. McMurtry (1982). *The use of phytoseiids for biological control: progress and future prospects.* In: Recent Advances in Knowledge of the Phytoseiidae (M.A. Hoy, Editor), University of California, Publication no. 3284.

R. Rabbinge (1976). *Biological control of fruit-tree red spider mite.* Pudoc, Wageningen, 228 pp.

R. Rabbinge & M.A. Hoy (1980). *A population model for the two-spotted spider mite Tetranychus urticae and its predator Metaseiulus occidentalis.* Entomologia experimentalis et applicata **28,** 64-81.

M.W. Sabelis (1981). *Biological control of two-spotted spider mites using phytoseiid predators. Part I: Modelling the predator-prey interaction at the individual level.* Agricultural Research Reports 910, Pudoc, Wageningen, 242 pp.

M.W. Sabelis (1985a). *Reproductive strategies of spider mites.* In: Spider Mites: Biology Natural Enemies and Control. (Eds. W. Helle and M. Sabelis), World Crop Pest Series, Vol. 1A, Elsevier, Amsterdam.

M.W. Sabelis (1985b). *Sex allocation.* In: Spider Mites: Biology, Natural Enemies and Control (Eds. W. Helle and M. Sabelis), World Crop Pest Series Vol. 1B, Elsevier, Amsterdam.

M.W. Sabelis (1985c). *Capacity for increase.* In: Spider Mites: Biology, Natural Enemies and Control (Eds. W. Helle and M. Sabelis), World Crop Pest Series Vol. 1B, Elsevier, Amsterdam.

M.W. Sabelis (1985d). *Development.* In: Spider Mires; Biology, Natural Enemies and Control (Eds. W. Helle and M. Sabelis), World Crop Pest Series Vol. 1B, Elsevier, Amsterdam.

M.W. Sabelis (in press). *Biological control of two-spotted spider mites using phytoseiid predators. Part 2: Modelling the predator-prey interaction at the population level.* Agricultural Research Reports, Pudoc, Wageningen.

M.W. SABELIS & B.P. AFMAN (1984). *Factors initiating or suppressing aerial dispersal of the predatory mite, Phytoseiulus persimilis.* Abstracts of the Entomological Congress, Hamburg, 1984.

M.W. SABELIS, F. VAN ALEBEEK, A. BAL, J. VAN BILSEN, T. VAN HEIJNINGEN, P. KAIZER, G. KRAMER, H. SNELLEN, R. VEENENBOS & J. VOGELEZANG (1983). *Experimental validation of a simulation model of the interaction between Phytoseiulus persimilis and Tetranychus urticae on cucumber.* SROP/WPRS Bulletin: 207-229.

M.W. SABELIS & M. DICKE (1985). *Long-range dispersal and searching behaviour.* In: Spider Mites: Biology, Natural Enemies and Control (Eds. W. Helle and M. Sabelis), World Crop Pest Series, Vol. 1B, Elsevier, Amsterdam.

P.B. SHAW (1984). *Simulation model of a predator-prey system comprised of Phytoseiulus persimilis Athias-Henriot (Acari: Phytoseiidae) and Tetranychus urticae (Acari: Tetranychidae).* I. Structure and validation of the model. Researches on Population Ecology **26,** 235-259.

M. SLATKIN & J. MAYNARD SMITH (1979). *Models of coevolution.* Quarterly Review of Biology 54 **3,** 233-263.

L.B. SLOBODKIN (1974). *Prudent predation does not require group selection.* American Naturalist **108,** 665-678.

D.R. SMITLEY G.G. KENNEDY (1985). *Photo-oriented aerial dispersal behaviour of Tetranychus urticae (Acari: Tetranychidae) enhances escape from the leaf surface.* Annals of the Entomological Society of America 78 **5,** 609-614.

D.J. WOLLKIND & J.A. LOGAN (1978). *Temperature-dependent predator-prey mite ecosystem on apple tree foliage.* Journal of Mathematical Biology **6,** 265-283.

D.J. WOLLKIND, A. HASTINGS & J.A. LOGAN (1980). *Functional response, numerical response and stability in arthropod predator-prey ecosystems involving age structure.* Researches on Population Ecology **22,** 275-299.

D.J. WOLLKIND, A. HASTINGS & J.A. LOGAN (1982). *Age structure in preator-prey systems. II. Functional response and stability and the paradox of enrichment.* Theoretical Population Biology **21,** 57-68.

D.J. WOLLKIND & J.L. SMITH (1983). *A mathematical model of the paradox of enrichment in arthropods: comparison between theory and experiment.* Northwest Science 57 **4,** 241-248.

D.L. WRENSCH & S.S.Y. YOUNG (1983). *Relationship between primary and tertiary sex ratio in the two-spotted spider mite (Acarina: Tetranychidae).* Annals of the Entomological Society of America **76:** 786-789.

Regional Dynamics of Spider-Mite Populations that Become Extinct Locally because of Food Source Depletion and Predation by Phytoseiid Mites (Acarina: Tetranychidae, Phytoseiidae)

M.W. Sabelis & W.E.M. Laane

Department of Population Biology*
Division of Ecology
University of Leiden
Kaiserstraat 63
P.O. Box 9516
2300 RA Leiden
The Netherlands

ABSTRACT

Two-spotted spider mites are tiny, phytophagous arthropods. They disperse by passive transport in air currents (or by hitchhiking on other organisms). After landing they may find and colonize a suitable host plant. Local spider-mite populations rapidly increase until they have exterminated their host plants or until they become exterminated by predatory mites. After invasion of a predator into a spider-mite patch the predator population increases at the expense of the spider mites. The predators do not disperse until after the local spider-mite population has crashed. When their food source is depleted they disperse by being carried away in air currents. They may accidentally land or walk into a spider mite patch or into the odour plume that is emitted by the spider mites and betrays their upwind presence to the predatory mite. Hence, the risk to a spider-mite patch of being invaded by a predator depends on the area covered by the prey patch and the associated odour plume and on the number of predators dispersing and searching for prey.

Although both prey and predator populations are locally transient, their regional populations do persist. Simulations showed that regional populations enter a stable limit cycle under a wide range of conditions. Population cycling is caused by the delayed numerical response resulting from the fact that predators do not disperse from a spider-mite patch until after prey extermination. Sensitivity analyses of the simulation model suggest that the stability of the limit cycle hinges on a complex set of factors contributing to the asynchronization of local predator-prey cycles, whereas the stability does not rely on (predator/prey) mortality factors that are positively related to prey density at a local and/or regional scale. Some traits of the simulated dynamics are indeed found in published experiments on acarine population dynamics, but especially the stability of the limit cycle dynamics needs to be further substantiated by experimental work.

Introduction

One of the most salient properties of the interaction between the predatory mite *Phytoseiulus persimilis* Athias-Henriot (Acarina: Phytoseiidae), the two-spotted spider mite *Tetranychus urticae* Koch (Acarina: Tetranychidae) and the host plants is that local populations of prey and predator sooner or later become extinct. After a female spider mite has found a suitable host plant she starts feeding, converts plant food to eggs, and gives rise to a population that rapidly expands over the plant (or group of plants in close vicinity) in a coherent way as if it were a drop of oil that has fallen on water. Such a local spider-mite infestation is further referred to as a spider-mite patch (or prey patch). If predators are absent the host plant will eventually be overexploited and exhausted as a food source to the spider mites. If no other host plants are available in the vicinity, the local spider-mite population will crash because of starvation and dispersal. As shown in the previous chapter (Sabelis, this volume), the predator-prey subsystem exhibits similar dynamics. After a female predatory mite invades a spider-mite patch, she starts feeding, reproduces and gives rise to a

* Part of this work was performed at the Department of Animal Ecology, Agricultural University, Wageningen.

population that rapidly increases and sooner or later causes the prey population to decrease. Eventually, the prey population is suppressed to such low levels that most - if not all - predators would die if they would have stayed put instead of dispersing and searching for new food sources elsewhere.

To find new food sources the tiny and wingless mites should cover larger distances than possible by locomotion only. This is done by passive transport in air currents and possibly also by passive transport on larger organisms with better dispersal capacities than mites, a phenomenon called phoresy. Aerial dispersal by spider mites has been demonstrated to occur in the field (Hoy, 1982; Brandenburg & Kennedy, 1982). In windtunnel experiments, however, it was found that only a small proportion of the spider-mite population disperses aerially even though the host plant was overexploited (Laane, unpublished data). It is possible that the sticky web produced by spider mites serves as a means to attach to larger animals that happen to pass by (Sabelis, unpublished data), but this hypothesis needs further substantiation. How spider mites find suitable host plants after passive transport has not been investigated yet. With respect to the predatory mites, experiments in a large wind tunnel (Sabelis, this Volume) showed that the majority disperses aerially when the prey is more or less exterminated. In other wind-tunnel experiments it has been shown that take-off is suppressed by the predatory mites in response to chemical cues produced by the spider mites while feeding on their host plant (Sabelis & Afman, 1984). As soon as the spider mites are eliminated, the production of these cues is stopped, the predatory mites become hungry and prone to be taken up by air currents. Aerial dispersal has been demonstrated to occur in the field for several species of predatory mites (Johnson & Croft, 1981; Hoy, 1982), whereas reports on phoretic dispersal of predatory mites are very scarce. After passive transport over long distances the predators are able to locate upwind spider-mite patches that are close to the landing spot. Odour emitted by the spider mites may be an important clue in finding the prey patches. It indicates the upwind position of the patch, it may provide gradients that help to keep the predator moving along the odour plume and it may function as a signal to the predator to continue moving upwind instead of starting another aerial voyage (Sabelis & van de Baan, 1983; Sabelis, Afman & Slim, 1984; Sabelis, Vermaat & Groeneveld, 1984; Sabelis, Schippers, Gunther & van der Weel, in prep.). In summary, the predator-prey system is characterized by a sequence of local population outbreaks, local extinction, passive long-distance dispersal and short-distance search to find suitable food sources in the vicinity of the landing spot.

Because the tritrophic system described above is locally transient, it may be questioned which factors promote its apparent persistence on a large spatial scale. Both population modelling and population experiments have provided several hypotheses that may help explain the persistence and the stability of the system. We will first briefly comment upon these hypotheses and also add some hypotheses that emerged from considering the particular predator-prey system described above. The relative importance of all these hypotheses will be evaluated by simulations with a model provided with realistic biological data as far as available.

Theories on persistence of tritrophic systems

Theories on persistence and regional stability can be divided in two categories. The first set of theories emphasizes the importance of (positive) density-dependent processes that regulate the number of predators and prey by negative feedback responses (e.g. Hassell, 1978). The second set of theories emphasizes the importance of factors that promote asynchrony of local predator-prey cycles and diminish the chances of regional extinction (e.g. Den Boer, 1968; Reddingius & Den Boer, 1970, Reddingius, 1971; Maynard Smith, 1974; Levin, 1974, 1976; Crowley, 1977, 1978, 1981).

Regulation by intraspecific competition among the prey

When spider mites have overexploited their host plants, the population will decline because of lack of food. This may in turn give the host plants a chance to recover either by regenerative growth or by germination of seeds from the seed bank. If the host plants regenerate soon enough to provide food for the surviving spider mites, the population of spider mites may start to increase until the host plants are overexploited again. Huffaker (1958) created a persistent mite-plant system by providing new host-plant material at regular intervals. Takafuji et al. (1983) found large oscillations of spider mites and uninfested leaves on rose plants over a period of half a year without adding new host plants. The rose apparently recovered soon enough to get sustained oscillations of the number of spider mites. This type of regulation of prey numbers may also effect regulation of predator numbers, as discussed by Hassell (1978). Even an unstable predator-prey model, such as the Nicholson-Bailey model, can be stabilized by including intraspecific competition among the prey.

Regulation of the prey population by predators

Predators may regulate the prey population below the level where intraspecific competition for food affects the prey's rate of increase. This requires an adequate functional and numerical response of the predators to increasing prey densities so as to prevent the prey from overexploiting the host plants. Moreover, the response of the predators at low prey densities should allow the prey population to resurge sooner or later. This may be achieved in various ways;

(1) A few prey may be left in the patch at the end of the predator-prey interaction period, because the predator spends less time in searching at low prey density to save energy for survival, because the predators become starved and less able to attack their prey, or because the predators leave the patch at low prey density. Though behavioural observations (Sabelis, 1981; Eveleigh & Chant, 1982) have given some evidence for each of these causes, there is no good reason to assume that these are effective at the patch level, because the spider mites seem to be eliminated according to most published reports (Chant, 1961; Bravenboer & Dosse, 1962; Laing & Huffaker, 1969; Takafuji, 1977; Takafujii et al., 1981; Sabelis, in press; Sabelis, this volume). Presumably, the number of predators that stay in the patch until prey elimination is too large to give the prey any chance of escape. Moreover, juvenile predators which have low food requirements compared to adult females and less tendency to disperse away from a patch, may be responsible for clearing away the last remnants of the prey population.

(2) The predators may spend more time in searching for prey patches when average densities are low rather than high (Murdoch & Oaten, 1975; Oaten, 1977; Hassell & May, 1974). These increased travel times at low prey density may also cause higher risks of dying because of starvation (Fujita et al. 1979). The result of both effects of low prey density may be that the undetected prey patches get more chance to expand and to become an important source of dispersing spider mites.

(3) The predators may tend to aggregate in high density patches, thus increasing the chance of low density patches to escape from predation (Hassell & May, 1974). The interception of dispersing predators will certainly increase with patch size and the size of the prey-odour plume emitted. Hence, it is possible that dispersing predators invade and aggregate in large prey patches, thereby decreasing the chances of the small prey patches to be invaded by predators.

(4) A fixed number of prey or a fixed proportion of the prey population may reside in refuges (Hassell, 1978). For example, diapausing spider-mite females hide in crevices of the bark of trees or in litter. They are less preferred prey to the predatory mites and they may be effectively protected from predation, being in a refuge. At first sight host plants growing at sites with low humidity represent refuges to the spider mites, because the spider mites thrive best at c. 30% RH and because all eggs of the predatory mites will shrivel and die at relative humidities below c. 50%. However, these conditions lead to unlimited growth of the spider mite numbers until the host plant is overexploited and henceforth the refuge is eliminated. Such a refuge is therefore only of temporary value.

The feedback mechanisms described above are in no way a guarantee for the stabilization of predator-prey interactions. The conditions under which these are stabilizing, are discussed by Hassell (1978).

Asynchronization of local predator-prey cycles

Dispersal has been thought of as having a synchronizing effect on local predator-prey cycles. When the cycles are in phase, local extinctions become synchronized too, so that the predator-prey system will not persist regionally (Maynard Smith, 1974; Levin, 1974, 1976). Phase differences between local cycles are therefore essential to its persistence, but the question is by what factors these are promoted. A discussion of these factors follows below:

(1) If dispersing prey have better capacities to reach new areas than their predators, the prey patches may have a start of their invaders. Huffaker (1958) emphasized the importance of the temporary escapes accomplished in this way. He found self-perpetuating predator-prey cycles only when he created a complex micro-environment in which the spider mites could reach new sites more easily than their predators. The spider mites are capable of spinning silken ropes and used these to cross barriers that cannot be surmounted by predatory mites because these are not capable of producing such 'life lines'. This difference in capacity of displacement may be relevant in short-distance dispersal only, but on a small spatial scale its effects are not apparent because local prey extermination seems to be the rule rather than the exception. Hence, it may have been relevant in Huffaker's experiments and perhaps also in certain situations in the field, but we fail to see the generality of this principle. This is because both spider mites and predatory mites disperse passively by air currents and spider mites do not use the silken ropes for aerial dispersal, such as wolf spiders that use it for 'ballooning' (Coates, 1974; Richter, 1971). As a consequence of their larger weight ballooning is very useful to the spiders to be taken up by wind currents. But spider mites and predatory mites are much smaller and certainly light enough to do it without ballooning! The important point to note is that dispersal being passive a predatory mite has the same chances of reaching a certain point in space than a spider-mite. Hence, though Huffaker's

hypothesis may have been relevant in his laboratory experiments, it does not provide a plausible explanation for the apparent persistence of the acarine predator-prey populations in the field.

(2) Fugitive responses of the prey upon encounter with predators may also increase the chances of prey patches to keep ahead of their waylayers. Bernstein (1984) reported increased short-distance (!) dispersal of spider mites in response to predator density, but whether this response is an effective escape at a larger spatial scale, remains to be investigated.

(3) If colonization of host plants by spider mites is a matter of chance, simultaneous foundation of prey patches is an unlikely event. The same applies to the invasions of predators into prey patches. Hence, even in the case all predator-prey cycles are of equal length, simultaneous extinction is unlikely too.

(4) That random termination of local cycles also promotes phase differences between local cycles, follows from the preceding point (Crowley, 1981). However, the literature on acarine predator-prey interactions does not provide concrete examples to illustrate its relevance.

(5) The risk of a prey patch to be invaded by predators is related to the time elapsed since colonization, i.e. the life time of that patch. Most probably, there will always be a certain time-lag between a new host plant being colonized by a female spider-mite and the resultant spider-mite patch being discovered by a predatory mite. This is because spider mites need some time to make a web and deposit other clues that are important for a predatory mite to locate its prey. Sabelis and Van de Baan (1983) and Sabelis et al. (1984b) showed that *P. persimilis* is able to locate one-day old spider-mite patches from some distance after perceiving volatile chemicals emitted by spider mites. However, younger spider-mite colonizations did not elicit such a response. Apparently, prey odour production requires some time before concentrations are sufficient to help the predator find its prey. The web is also an important clue to locating a spider-mite patch. Upon tarsal contact with the silk the predator decreases its walking speed and intensifies chemosensory inspection of the webbed leaf area in its search for spider mites. Moreover, the web and associated clues have a positive influence on the predation success. Once a predator invades a web, the probability of actual prey consumption after tarsal contact with a prey is much higher than on an unwebbed substrate (Sabelis, 1981). Thus, the predator will not be very successful until after the spider mites have had the opportunity of making a web and of producing other clues. However small the time-lag between colonization and discovery, the spider mites do get some time to reproduce without predator influence. Another reason why prey-patch risk is not constant with the age of the patch, follows from the fact that the leaf area exploited by the spider mites and covered by their web increases with the age of the patch. The chance of a predator landing in the patch after aerial dispersal increases linearly with patch size. Moreover, the larger the patch, the higher the chance that it will be encountered by predators searching by locomotory means after aerial dispersal and landing. Increased risk of discovery also results from the increasing production of volatile chemicals that can be perceived by the predatory mites and that are used in combination with wind direction to find any upwind spider-mite patch in the neighbourhood of their position on the ground (Sabelis & Van de Baan, 1983; Sabelis, Vermaat & Groeneveld, 1984; Sabelis, Schippers, Gunther & Van de Weel, in prep.). Thus, as the spider-mite population grows, both the leaf area covered by web and the odour plume area in which the prey odour concentration is sufficiently high to elicit a searching response in the predators, expand, thereby increasing the risk of discovery by the dispersing predators.

(6) The length of the predator-prey interaction period depends on the size of the prey patch at the moment of first predator invasion (Sabelis, this volume). The first predator invading a prey patch initiates a rapid population build-up so that successive invaders contribute less and less, the later their date of arrival is. Hence, the larger the prey patch, the more prey it contains and the longer the predator-prey interaction will last (unless prevented by some cause, such as overexploitation of the host plant by the spider mites). Even when the risk of prey patch detection would be constant with the age of the prey patch, variation in patch age implies variation in the length of the interaction periods, which promotes phase differences between local predator-prey cycles. Moreover, genetic variation in properties of the colonizing prey and the invading predators also contributes to phase differences.

(7) Spatial variation in abiotic factors (climate, soil) will certainly be present to some extent. It may influence colonization success and modify the length of interaction periods. It may therefore help to keep local cycles out of phase.

(8) Variation in local host-plant quality and quantity is likely to be an important factor causing phase differences. Host plant range, host plant distribution and especially, host plant selection by spider mites are poorly understood. These aspects need more scrutiny before they can be included in models in a meaningful way.

An important assumption underlying the theory of asynchronization of local cycles relates to the size of the environment. The number of sites available for colonization and the number of prey patches available for predator invasion should be large to minimize the chance of regional extinction. Several theoretical studies (Maynard Smith, 1974; Levin, 1974; Hilborn, 1975; Zeigler, 1977; Crowley, 1977, 1978 and 1981) have indicated that multi-patch systems tend to be more stable and persist longer.

The relative importance of the hypothesized mechanisms

To determine the relative importance of the above hypotheses the regional interaction between predatory mites and spider mites has been modelled and the sensitivity of the model to parameter changes and structural modifications has been tested. The model calculates regional dynamics from descriptive functions of local prey or predator-prey cycles, as proposed by Maynard Smith (1974). These functions relate the number of mites, the patch size (leaf area damaged and covered by the prey's web), the area covered by the plume of prey odour and the number of mites taking off for aerial dispersal with the time elapsed since the start of the cycle. The chance of food source colonization depends on the available food source area and the number of dispersing mites, whereas the number of dispersing mites depends on their ability to survive, their searching success and dispersal from the existing colonized food sources. Consequently, the model contains several feedback relationships, so that modelling is the only way out to elucidate the potential role of the proposed mechanisms in stabilizing the number of predators, prey and host plants and, hence, in ensuring the persistence of these three populations. The complexity of the predator-prey interaction makes formal mathematical analysis very difficult (see e.g. partial differential equation no. 14 discussed by Levin (1976)), so that a simulation approach is needed to provide some new insight. Such an approach would probably be much too comprehensive, if it were not confined to the parameter space set by our current knowledge of the interaction between the two-spotted spider mite and its predator *P. persimilis.* Hence, the emphasis in this study is to provide the model with realistic data as far as available. The descriptive functions of within-patch dynamics constitute the most realistic part of the model. These functions are derived from a validated simulation model designed to simulate the predator-prey interaction in a single patch (Sabelis, 1981; Sabelis et al., 1983). Also, the number of mites that leave the patch and disperse aerially, can be approximately estimated using this model (Sabelis, this volume). The 'fate and fortune' of the dispersing mites are, however, based on no more than best guesses because of incomplete information on the ability to find distant food sources. Adequate information on host-plant dynamics is also lacking. Host plant quality and quantity is therefore considered to be constant for all patches and not influenced by external conditions (weather and soil). The influence of the weather, especially the temperature, on the predator-prey interaction is considered to be constant too. This is reasonable because the rate of population increase and the rate of food consumption are approximately linearly related to temperature in the range from 13° C to 32° C (Sabelis, 1981).

Very recently, a related model of regional predator-prey dynamics has been described by Takafuji et al. (1983). These authors used a simulation approach developed by Maynard Smith (1974), just like we did. In Takafuji's and Maynard Smith's models patches have a fixed size and are either empty, or colonized by prey and/or invaded by predators. The risk of a prey patch to be invaded by a predator is therefore considered to be independent of its life time. The most salient difference with our approach is that our model takes into account the age-dependent risks of prey patches to be invaded by predators. This age-dependency ensues from the fact that prey patch size, as well as the size of the odour plume emitted, increase with the life time of the prey patch, and from the fact that founders of a prey patch have a start of their predators, thus constituting an initial refuge. These extensions of the model are essential if one intends to assess the relative importance of hypotheses concerning regulation with those concerning asynchronization.

Modelling host-plant area and age-structured prey and predator patches

The model does not take into account the detailed spatial structure of the environment, but instead considers two subunits only: the leaf area suitable for colonization by spider mites and the leaf area that is not suitable. The fraction of the leaf area covered by host plants was roughly estimated to be 0.03 in a natural environment. The maximum area covered by host and non-host plants was arbitrarily taken to be 3 x 10^{10} m^2 and to cover 10^{10} m^2 soil surface. The non-host plant area was assumed to be constant, but the host-plant area was assumed to obey the logistic growth law with parameters g_h and H_{max} with an additional leaf consumption term taking damage by spider mites into account. The relative rate of plant growth g_h was set equal to 0.15 m^2 leaf area per m^2 soil surface per day and H_{max} follows from the data presented above. The actual host-plant leaf area (H) and the total leaf area of the environment (host plus non-host) served to calculate the fraction of the plant area available for colonization by spider mites (f_h). The host-plant area H was assumed to recover readily by regenerative growth or by germination of seeds from the seed bank. Therefore, in the simulation model H was prevented from dropping to zero. In this way host-plant dynamics was made to be a persisting component of the tritrophic system which is regulated by intraspecific competition in the absence of phytophagous mites.

As a logical consequence of the neglect of environmental spatial structure but in contrast to the models used by Maynard Smith (1974), Hilborn (1975) and Zeigler (1977) the spatial distribution of prey and predator was not considered explicitly. Instead, the prey were assumed to form distinct patches, the so-called prey patches which upon detection by predatory mites are referred to as predator patches, and only the frequencies of these patch types were taken into consideration (F_s and F_p,respectively). Within-patch dynamics of prey and predator were described by a series of state transitions according to the ideas originally put forward by Maynard Smith (1974) and followed by many others (Hilborn, 1975; Zeigler, 1977; Hastings, 1977, 1979; Gurney & Nisbet, 1978; Takafuji et al., 1983). The prey patches went through a pre-set series of state transitions depending on the time (t in days) elapsed since colonization of the host plant by one spider-mite female. It was assumed that later arrivals of spider mites in the already established prey patches hardly contributed to the exponential population growth initiated by the founder female. Each state in the series was characterized by the number of spider-mite females ($n_s(t)$) and the finite rate of spider-mite dispersal (q_s). In our model each state was additionally characterized by the area covered by the spider-mite's web (the leaf area $w_s(t)$ or its projection on the soil surface $a_s(t)$) and the soil surface covered by its odour plume ($a_0(t)$) to enable the risk of predator invasions to be calculated. It was assumed that in the absence of predators, population growth continued unchecked until $t = 100$ days, an arbitrarily chosen age at which the population crashes due to deaths as a result of food scarcity and for a minor part because of increased dispersal. The mean crash period lasts 5 time units, during which predators are rendered ineffective by the excessive webbing produced by actively wandering spider mites. If all available hostplants are colonized by the spider mites, the same crashing procedure was started. In this way within-patch dynamics of the spider mites can be described as a function of the life time of the patch. The frequency of prey patches of age $t(F_s(t))$ is the dynamic quantity calculated by the model, whereas all properties of these prey patches can be derived from their relation with the age of the patch. To reduce the bookkeeping of the frequencies the life time of a prey patch is divided in classes of one day. During the simulation the frequencies are shifted to the next class at the end of each day.

After the first predator invades a prey patch, this patch is transformed to a predator patch. For each age of the prey patch at invasion (age t_p) there is a pre-set series of state transitions for the predator patch. Thus the dynamics of the predator patch are determined at its creation. This is a reasonable approach, because later invasions will not contribute much to the population explosion initiated after arrival of the first female predator. Bookkeeping of the frequencies of predator patches (F_p) was rigorously simplified to economize computing time. At their creation predator patches are characterized for their life time by a number of state variables, the values of which are set by the age of the prey patch at predator invasion (t_p). These state constants are the length of the interaction period (l_p), the mean number of spider-mite females during the interaction period ($\overline{n}_{s,p}(t_p)$) and the finite rate of dispersal ($q_{s,p}(t_p)$) (to calculate the number of dispersing spider mites), the mean area covered by the spider-mite's web ($\overline{w}_{s,p}(t_p)$) and the mean area covered by the prey odour plume ($\overline{a}_{o,p}(t_p)$) (to calculate the risk of dispersing predators to be intercepted in predator patches), and finally, the number of female predators that will disperse after prey elimination ($\overline{n}_p(t_p)$). The advantage of this simplified representation of the dynamics within predator patches is that all predator patches that still have a time $l(0 \leq l \leq l_p)$ to live can be grouped into one big superpatch with a number of state variables labelled by l. As soon as l is zero, the female predators join the population of dispersing predators, whereas the predator patch vanishes implying that no prey, no predators and no webbed leaf area remains.

All state-age relations pertaining to prey patches ($n_s(t)$), $w_s(t)$ or $a_s(t)$) except for $a_o(t)$ and $q_s(t)$ were derived from a model that simulates the population growth of spider mites in a patch. It was assumed that the population grows unlimited until $t = 100$ days and decreases quickly thereafter. The state-interaction-period relations ($\overline{n}_{s,p}(t_p)$, $\overline{w}_{s,p}(t_p)$ or $\overline{a}_{s,p}(t_p)$ except for $\overline{a}_{o,p}(t_p)$ and $q_{s,p}(t_p)$) were derived from a model that simulates the population growth of spider mites and predatory mites, as well as their interaction within a patch. This model was developed and validated by Sabelis (1981; this Volume and in press) and Sabelis et al. (1983). It was assumed that prey population growth stops if the webbed area attains the same size as a predator-free patch at $t = 100$ days, whereafter it crashes. By fixing this upper limit to the predator's food, both the number of predators dispersing after prey extermination (n_p) and the length of the interaction period have a bell-shaped relation with t instead of a progressive course. The area covered by the prey-odour plume ($a_o(t)$ and $a_{o,p}(t_p)$) was calculated from the Gaussian plume model (Bossert & Wilson, 1963; Fares et al., 1980), from the minimum number of prey needed to elicit a searching response of the predator (Sabelis & Van de Baan, 1983) and from the number of spider mites present in the patch ($n_s(t)$ or $\overline{n}_{s,p}(t_p)$). The calculations and underlying assumptions are presented in Appendix 1. The finite rates of spider-mite dispersal were obtained from wind-tunnel experiments with potted bean plants in various degrees of exploitation by the spider mites (Laane, unpublished data). A gauze screen was placed at the downwind end of the tunnel to intercept the dispersing mites. The fraction of the mite population that dispersed aerially during a period of 6 hours, was used to calculate the relative rate of dispersal ($d_s = -\ln(q_s) / 0.25 \text{ day}^{-1}$) and this was related to the state of host-plant exploitation (= fraction of bean leaf area covered by webbing). The latter relation was used in our model, taking $w_s(t) / w_s(100)$ as a relative measure of the state of exploitation of the host plant. It was assumed that spider-mite dispersal from predator patches could be found from the same relation, i.e. it was assumed to be independent of the presence of predators. All relations have been smoothed or simplified: they are shown in Figures 1 and 2.

Prey patches were classified according to age t, which is the time elapsed since a single spider-mite female colonized a group of suitable host plants. The dynamics of the prey in the patch was described over 100 age classes of one day each. During simulation the frequencies of prey patches $F_s(t)$ are shifted simultaneously to the next class at the end of each day. Because the frequencies are labelled by age t the state variables ($n_s(t), q_s(t), w_s(t), a_s(t)$ and $a_o(t)$) are readily obtained from the relations shown in Figures 1 and 2.

The predator patches were not classified by age, but the values of their state variables ($n_p(t_p)$, $\bar{n}_{s,p}(t_p)$, $q_{s,p}(t_p)$, $\bar{a}_o(t_p)$) were set to a value that depends on the age of the prey patch at first predator invasion (t_p) according to the relations shown in Figures 1 and 2. The values thus obtained were added to overall state variables ($F_p(l)$, $N_{s,p}(l)$, $D_{s,p}(l)$, $A_{s,p}(l)$, $A_{o,p}(l)$ and $N_p(l)$) of all predator patches together (the so-called superpatch) which are labelled with an index l corresponding to the length of the predator-prey interaction period l_p. The latter period was given by its relation with t_p shown in Figure 2. The subsequent procedure was to diminish the value of the label l by one unit each day until l is zero, as if it were a rocket countdown. In this way the total number of dispersing prey ($D_{s,p}$) the total number of dispersing predators ($N_p(l=0)$), the total area covered by web ($A_{s,p}$) and by prey odour ($A_{o,p}$) could be monitored, starting from the age of the prey patches at first predator invasion t_p.

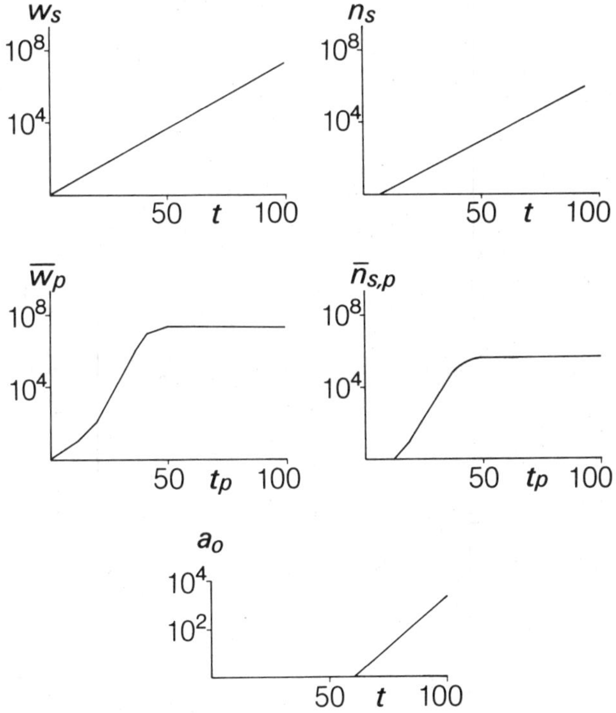

Figure 1 Within-patch state variables of the spider mites in prey and predator patches and their relation to the age of the prey patch (t in days) or the age of the prey patch at first predator invasion (t_p in days). These variables are:

(a) The leaf area (cm^2) covered by a spider-mite patch: $w_s(t) = \exp(0.171t)$,

(b) the number of spider-mite females: $n_s(t) = \max(1, \exp(-1.2 + 0.16t))$,

(c) the mean leaf area (cm^2) covered by the predator patch ($\bar{w}_p$) during the interaction period and its relation with t_p,

(d) the mean number of spider-mite females ($\bar{n}_{s,p}$) during the interaction period and its relation with t_p,

(e) the area (cm^2) covered by the odour plume emitted by the spider-mite patch: $a_0(t) = \exp(0.19t - 11.8)$. Note that $\bar{a}_{o,p}(t_p)$ was obtained from this equation by first finding a value for t in Figure 1b that would correspond to $n_s = \bar{n}_{s,p}$, as if the patch under consideration in Figure 1e were a prey patch without predators.

Figure 1a, 1b, 1c and 1d were obtained from within-patch simulations with a model described in Sabelis (1981 and in press) using data on the spider mite *Tetranychus urticae* and the predatory mite *Phytoseiulus persimilis.* The calculations leading to Figure 1e are presented in Appendix 1.

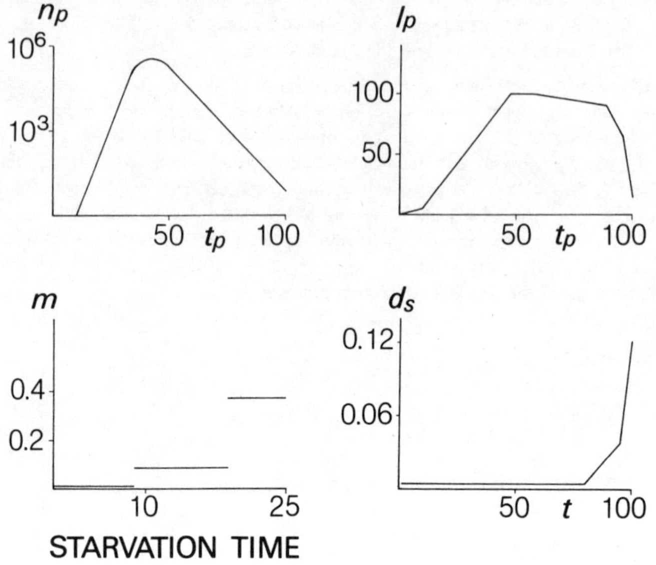

Figure 2 Dispersal characteristics of predatory mites and spider mites:

(a) the number of female predators that will disperse after elimination of their prey (n_p) and its relation with the age of the patch at first invasion (t_p),

(b) the length of the predator-prey interaction period (days) after which the predators will disperse because of prey scarcity (l_p) and its relation with t_p,

(c) the relative death rate of dispersing predators (*m* in day^{-1}) and its relation with the time of food deprivation (days) according to data from Sabelis (1981),

(d) the relative rate of spider-mite dispersal (d_s in day^{-1}) and its relation with the age of the spider-mite patch (*t* in days) (Laane, unpublished data). Assuming that predators do not influence prey dispersal, the relative rate of spider-mite dispersal from predator patches ($d_{s,p}$ can be obtained from this relation after finding a value for *t* in Figure 1b that would correspond to $n_s = \bar{n}_{s,p}$, as if the patch under consideration in Figure 2d were a prey patch without predators. Figure 2a and 2b were obtained from within-patch simulations with a model described by Sabelis (1981; this Volume and in press) using data on the predatory mite *P. persimilis.*

Modelling dispersal, colonization and invasion

Prey and predator must disperse aerially if they are to have some chances of reaching new resources. In our model dispersal distances were not considered. Instead, aerial dispersal was conceived as a process that distributes the mites randomly over the environment, and the environment was assumed to be closed (i.e. no emigration nor immigration of the mites). All locations in the environment were assumed to be equally accessible to prey and predator. Dispersal is either passive by the wind or active by walk. The time needed for aerial voyages was assumed to be infinity small and resources can only be detected after landing. Locomotory search starts after landing, it is a random walk and its success depends on the time allotted, the rate of movement and the area covered by resources and their associated cues.

Spider mites seem not to be good in locomotory search, but this still awaits more investigations. For the time being we assumed locomotory search for host plants to be absent. Moreover, as they seem to die soon in absence of host plants, we arbitrarily assumed that only one aerial voyage is made, resulting either in successful colonization or death. The number of new prey patches ($F_s(1)$) is obtained from the product of the total number of dispersing spider mites ($D_s + D_{s,p}$) and the fraction of the total leaf area available for colonization. Thus, the number of prey patches increases by females colonizing available host plants, thereby assuming that colonizations do not overlap and prey patches will never fuse in times to come. The number of prey patches decreases as a consequence of resource overexploitation (at $t = 100$) and first predator invasions.

Calculating the risk that a prey patch will be detected by dispersing predators is the most crucial part of the model. The following assumptions were made:

(1) predators may disperse aerially repeatedly, but only once per day;

(2) if landed in a prey patch or associated odour plume, predator invasion is considered to be certain and to take place immediately;

(3) if not, locomotory search is started until a prey patch or its odour plume is found;

(4) dispersing predators have no shortage of water, which allows them to survive for about 3 weeks. The latter assumption was modelled by putting the number of predators dispersing from the predator patches after extermination of their prey ($N_p(l=0)$) into the first of a series of state variables representing the number of dispersing predators in subsequent age classes. There were 25 age classes, each one day long, and the appropriate relative death rates (m) are shown in Figure 2. Finally, it was assumed that

(5) predator-prey dynamics in a patch is not affected by the state of food deprivation of the first predator arriving in the patch. There is some truth in this assumption because the adult females of predatory mites retain their fecundity even after long periods of starvation (Sabelis, 1981).

Following the above tentative reasoning the probability R that at least one predator will invade a prey patch of age t, is obtained from the following equation:

$$R = 1 - \exp(-r.D_p.\Delta\tau)$$

$\Delta\tau$ = time step of integration (day)
D_p = total number of dispersing predators
r = relative rate of prey patch discovery (day^{-1})

The probability that predators dispersing aerially land in a prey patch of age t or its odour plume, is equal to f_s, which is the fraction of the total leaf area covered by these prey patches and their plumes. If the dispersing predators make aerial voyages once per day, the relative rate of successful landings in prey patches is also given by f_s. The probability that predators dispersing aerially do not land in a prey patch nor in the associated plume area, is equal to $1-f_s$. This group of predators continues to search by walk and encounter prey patches and their plumes at rate e (day^{-1}). Hence, the relative rate of prey patch discovery can be calculated by the following equation:

$$r = (1-f_s).e + f_s$$

If the predators would not make one aerial voyage at the start of each day (as assumed in the above equation), but several voyages in the course of a day (ν/day), the equation can be modified as follows:

$$r = e + f_s.\nu$$

where for simplicity e is not corrected by a factor which takes care of the happy few which hit a prey patch right away. The rate of encounter with prey patches and their odour plumes depends on the rate of movement of the predator (walking speed (m/day) times fraction of time spent walking $= s$), on the size of the prey patches, their plume areas and the predator (i.e. the sum of their diameters assuming these are all circular objects $= \varnothing_s + \varnothing_o + \varnothing_p$) and on the density of prey patches in the total environment ($= F_s(t)/A$, where A = the area of the environment in m^2). In formula:

$$e = s.(\varnothing_s + \varnothing_o + \varnothing_p).F_s(t)/A$$

$\varnothing_s = 2.(a_s(t)/\pi)^{\frac{1}{2}}$ = diameter of prey patch (m)

$\varnothing_0 = 2.(a_o(t)/\pi)^{\frac{1}{2}}$ = diameter of the odour plume (m)

$\varnothing_p$ = diameter of the predator (m)

In this way the risk of prey patches of age t to be invaded by at least one predator was calculated. To obtain the number of prey patches invaded by predators, the risk R was multiplied by the frequency of prey patches of age t ($F_s(t)$). In Figure 3 a typical relation between the prey patch risk and the age of the prey patch is presented for the particular case that $D_p = 10^8$ and $F_s(t) = 1$. It is shown that the risk increases exponentially with age and size of the prey patches until it becomes high and gradually approaches unity.

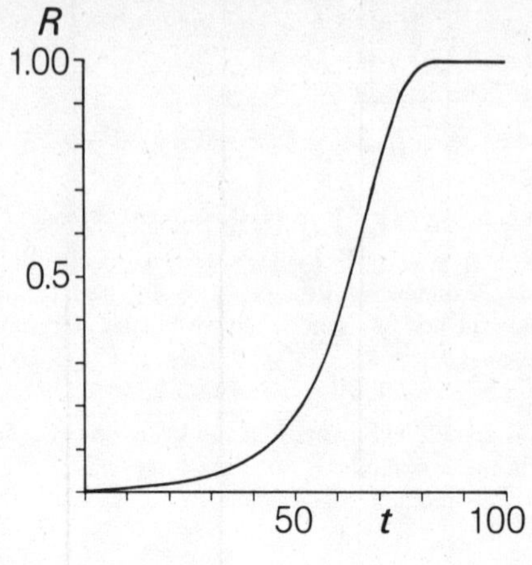

Figure 3 The risk of prey patch discovery by predators in relation to the age of the prey patch (t in days) using the formula for R (see text) with $D_p = 10^8$ and $F_s(t) = 1$ for $1 \leqslant t \leqslant 100$.

For biological reasons it is unlikely for spider-mite patches to be discovered immediately after the patch has been founded. The colonizing spider-mite females require some time to produce sufficient web and kairomones to elicit a searching response from the predators (Sabelis, 1981; Sabelis & Van den Baan, 1983). Therefore, it was assumed that spider-mite patches run no risk of predator invasion before the age of one day. This constitutes a temporary refuge, albeit of very short duration.

The number of dispersing predators will decrease because predators detect prey patches and also because they detect prey patches that have been detected before, i.e. the predator patches. To account for this decrease the fraction of predators that have not detected prey or predator patches was calculated first and then multiplied by D_p to obtain the number of predators, that continue to disperse and search for prey. The process of predators being catched in prey and predator patches is called interception. The probability of interception I was calculated following the same way of reasoning as discussed above with respect to R, thereby assuming that interception is independent of the number of predators in the predator patches and that it is determined by the mean size of the patch and the associated odour plume (as derived from Figure 1 using the mean number of prey present during the interaction period ($\overline{n}_{s,p}$) in the case of predator patches).

To understand the sensitivity analysis presented in the next section it is important to realize that each process is modelled in a separate Fortran D0-loop, in which calculations are made for each age class of the prey patch, each class of the predator-prey interaction period, or for each class of the starvation period to which dispersing predators are condemned. These processes are successively:

1. ageing of prey and predator patches
2. crashing of the prey patches at overexploitation of the host plant
3. dispersal of spider mites from prey patches and predator patches
4. host plant dynamics and calculation of leaf surface available for colonization
5. colonization of the available leaf area by spider mites
6. dynamics of dispersing predators (mortality, interception, take-off)
7. risk to prey patches of being invaded by predators
8. formation of new predator patches
9. decrease of prey patches due to invasion by predators

The important point to note is that several variables can be changed independently although in the real world they are inextricably linked. For example, all prey patches can be made identical in size when calculating risks of predator invasion, whereas when calculating interception of predators in prey and predator patches,variable patch size is taken into account. Another point to note is that the properties of predator patches are determined by the age of the prey patch at invasion (t_p) and their values are kept apart in variables labelled by the countdown index l until l becomes equal to zero. Therefore, it is possible e.g. to make all predator patches identical in size without affecting the mean number of spider mites in the predator patches or the number of predators dispersing after elapse of the predator-prey interaction period. In this way the role of each process can be studied much more analytically than would be possible in real world experiments on true predator-prey interactions.

The above description shows that the interaction between host plant, spider mites and predatory mites was modelled taking into account logistic growth of the host plants, age structure of the prey patches, the life time of the predator patches, colonization of host plants by prey, detection of prey patches by predators and interception of dispersing predators in prey and predator patches. The model output consisted of the total number of spider-mite females in prey and predator patches (N_s and $N_{s,p}$), the frequency of prey patches (F_s), the frequency of predator patches (F_p), the total number of dispersing prey (D_s) and dispersing predators (D_p) and the fraction of leaf area in the environment available for colonization by prey (f_h). Moreover, the mean age of the prey patches and the mean future life span of the predator patches were calculated, to summarize within-patch dynamics in the output files. Although within-patch dynamics were simulated fairly realistically, the dynamics of the host plants and the dispersal and searching processes were built on numerous assumptions and best guesses. Future experiments should aim to bridge the most important gaps in our biological knowledge.

Simulation and initialization

After formalization of the conceptual model and specification of the causal links the model was implemented on a computer in a Fortran-based simulation language (CSMP). Simulation consists of updating a massive bookkeeping scheme of state variables over small time increments ($\Delta\tau$). A time step of 0.1 day proved to be a reasonable compromis between output precision and investment in computer time.

To start the program initial values were set to the patch-type frequencies and the number of dispersing mites. We only used one way of fixing initial values: a low number of prey patches of given age and a large number of dispersing predators enter the simulations at $\tau = 0$. This is done to cause a severe perturbation of the interaction. If the initial number of dispersing predators is taken too high, the prey is suppressed to such low levels that the predator population crashes. This happened when 10^{10} or more predators (i.e. more than one predator per m^2) enter the simulation as dispersing mites. Figure 4A shows such a simulation in which prey survives after the predators have crashed. The number of prey patches increases until most of the host plants are colonized and thereafter the prey population enters a limit cycle with a period and amplitude set by the rate of host-plant regeneration and the rate of prey population growth. In Figure 4B the initial number of dispersing predators is taken to be lower, i.e. one predator per 100 m^2($D_p = 10^8$). In this case the host-plant area available for colonization remains close to the maximum attainable, because the predators exert effective control over the phytophagous mites. Though the prey population increases after initial perturbation, it is soon followed by an increase in predator patches until most of the prey patches have been invaded by predators. According to Figure 4C the number of dispersing predators keeps increasing, whereas the number of dispersing prey has already reached a constant level. Eventually, the number of dispersing predators reaches a level of one predator per 10 m^2, causing the invasion risk to the prey patches to become very high. In this period the predators discover the prey patches very soon after the elapse of the built-in temporary refuge period that follows after a host plant has been colonized by a spider-mite female (see e.g. day 370 in Figure 4D, 4F and 4G). The result is a decrease in dispersing prey soon followed by a decrease in dispersing predators. The invasion risks to the prey patches become low again and the prey population resurges after the density of dispersing predators has decreased to below one per 500 m^2. The mean age of the prey patches increases (see e.g. day 430 to day 485 in Figure 4D, 4H and 4I), thus offering the predators new chances to discover prey patches, to multiply and, ultimately to produce dispersing predators. Consequently, the life span (Figures 4E and 4I) and the number of predator patches (Figure 4B) increases, soon followed by an increase in the number of dispersing predators (Figure 4C). The subsequent course of events is identical to that described above and shown in Figures 4B-E. It is important to note that the fluctuations of prey and predators converge to a fixed cyclic pattern: **a limit cycle.**

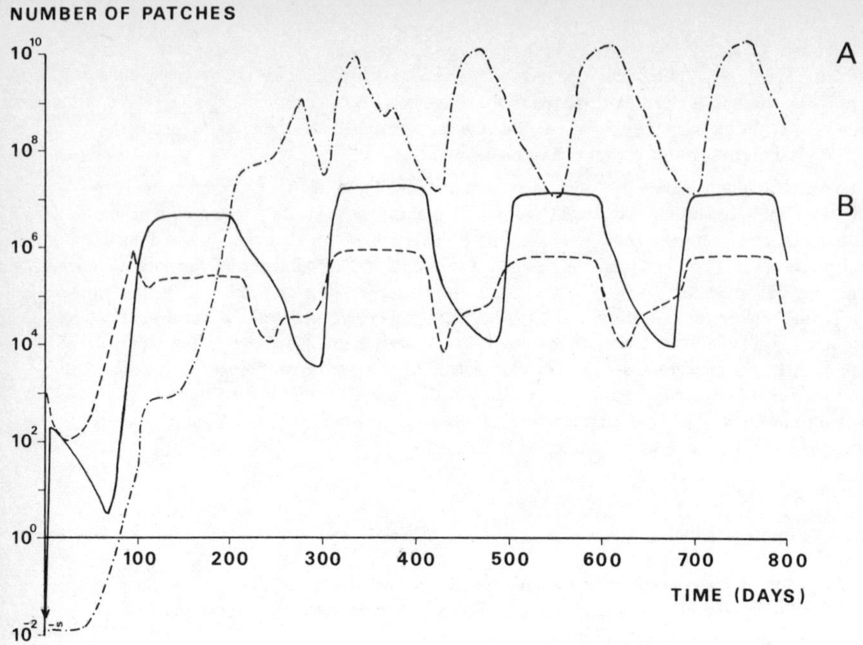

Figure 4A Regional dynamics of the number of prey patches, when the initial number of (dispersing) predators was large ($D_p = 10^{10}$) causing prey scarcity, regional extinction of predators and subsequently unlimited growth of the prey population until the availability of uncolonized host-plant area became limiting (—·—·—·—).

Figure 4B Regional dynamics of the number of prey and predator patches, when the initial number of (dispersing) predators was lower ($D_p = 10^8$) than in Figure 4A, so that the predators did not become extinct in the initial phase of predator-prey interaction. (number of prey patches: - - - - - ; number of predator patches: ———)

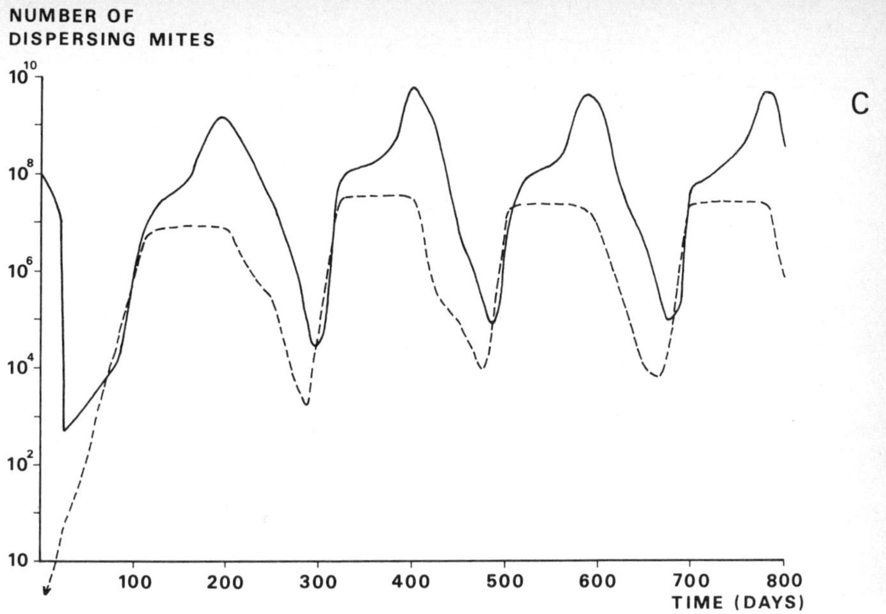

Figure 4C Regional dynamics of the number of dispersing prey and dispersing predators, associated with the simulation presented in Figure 4B.

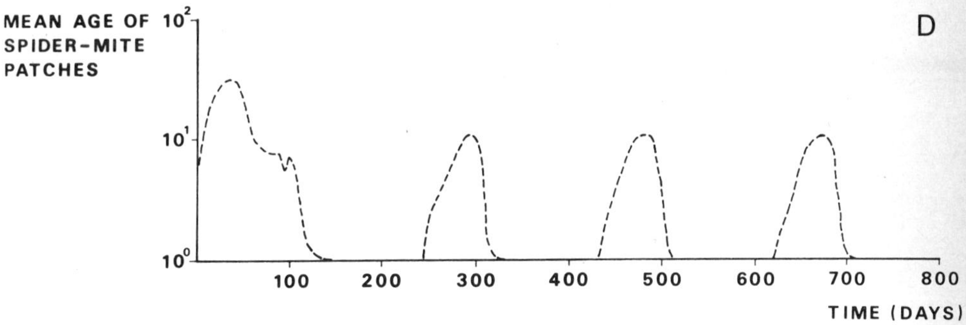

Figure 4D Mean age of the prey patches in the course of the simulation presented in Figure 4B.

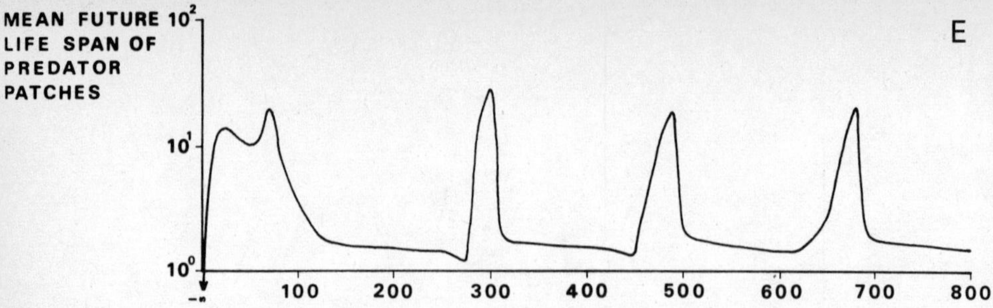

Figure 4E Mean future life span of the predator patches in the course of the simulation presented in Figure 4B.

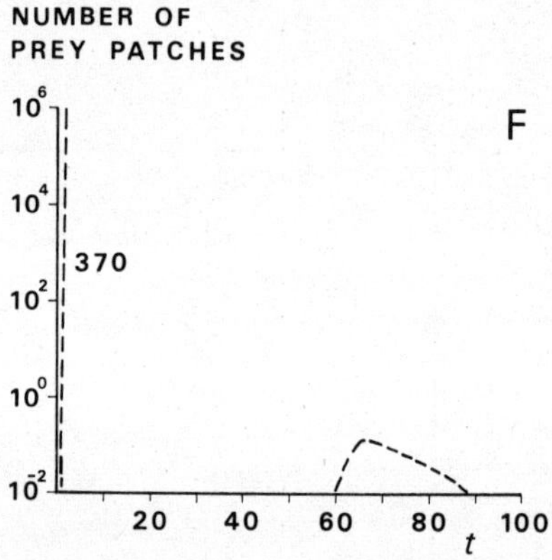

Figure 4F Frequency distribution of prey patches in relation to prey patch age (t) on day 370 of the simulation presented in Figure 4B; this day is in a period when the number of dispersing predators are high causing the prey patches to be invaded by predators soon after new prey patches are founded.

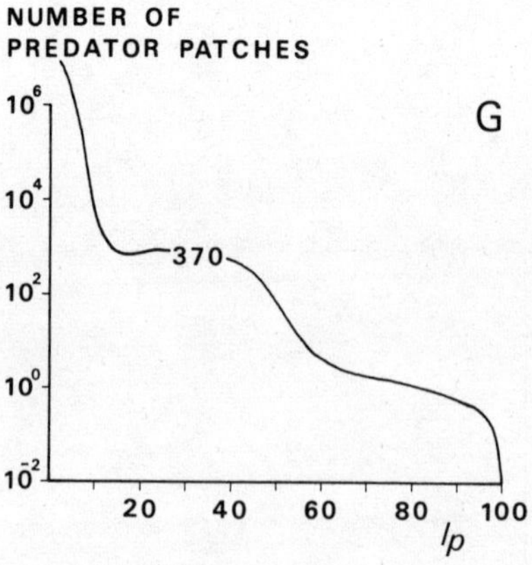

Figure 4G Frequency distribution of predator patches in relation to their future life span (l_p) on day 370 (see also Figure 4F).

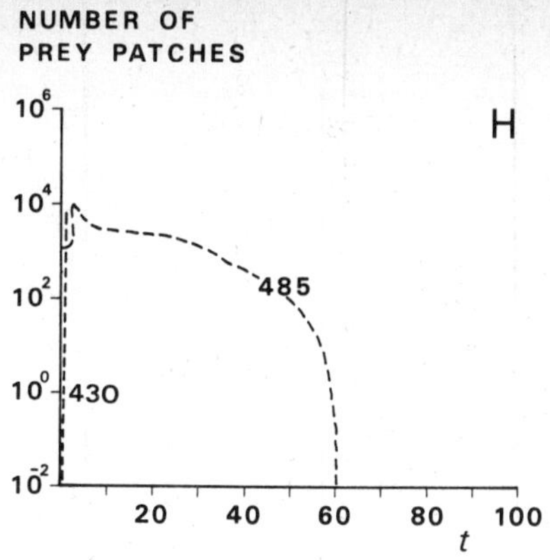

Figure 4H Frequency distribution of prey patches in relation to prey patch age (t) on days 430 and 485 of the simulation presented in Figure 4B; these days mark a period in which the number of dispersing predators reach the lowest level, causing the prey patches to expand and to be invaded by predators at a later time (compare with Figure 4F)

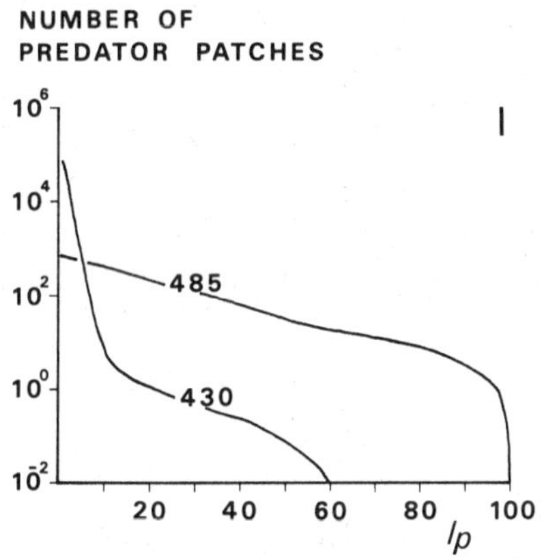

Figure 4I Frequency distribution of predator patches in relation to their future life span (l_p) on days 430 and 485 (see also Figure 4H)

Specifications of the simulations		Log (No. prey patches)		Log (No. predator patches)	
In words	In formulas	Lowest value	Highest value	Lowest value	Highest value
Standard conditions at $\tau = 0$ (Figure 4B)	$D_p = 10^8$ $F_s(t) = 100$ for $1 \leq t \leq 10$, else $F_s(t) = 0$	3.92	5.76	3.91	7.06
all prey patches of age 1	$F_s(1) = 1000$ $F_s(t) = 0$, if $t > 1$	3.90	5.77	3.91	7.06
equal number of prey patches in each age class	$F_s(t) = 10$, for $1 \leq t \leq 100$	3.78	5.93	4.06	7.24
Increased number of prey patches	$F_s(t) = 10^4$ $1 \leq t \leq 10$, else $F_s(t) = 0$	3.92	5.76	3.91	7.06
equal predator-prey ratio, but decreased initial number	$D_p = 10^7$ $F_s(t) = 10$ for $1 \leq t \leq 10$, else $F_s(t) = 0$	3.91	5.76	3.91	7.06
ditto, but increased initial number	$D_p = 10^9$ $F_s(t) = 1000$ for $1 \leq t \leq 10$ else $F_s(t) = 0$	3.88	5.76	3.91	7.06
ditto, but increased even more (Figure 4A)	$D_p = 10^{10}$ $F_s(t) = 10^4$, for $1 \leq t \leq 10$ else $F_s(t) = 0$	7.17	10.14	extinction!	

Table 1

Dynamics of the number of patches, as affected by initial number and initial age distribution of the prey patches ($F_s(t)$), as well as by initial number of dispersing predators (D_p).

When the initial number of dispersing predators is not too high, the simulated fluctuations invariably converge to a limit cycle at the same level, with the same period and with the same amplitude. These simulations are listed in Table 1 and their results are summarized by presenting the log-transformed values of the number of prey and predator patches at the peak of the third cycle, as well as the trough of the subsequent cycle. These values give sufficient information on the position and amplitude of the limit cycle, as the fluctuations converge very rapidly to it. The values that deviate from the majority of the simulation results, indicate either that the initial perturbation has been severe, causing convergence to take longer or that the initial number of dispersing predators was taken too high, resulting in predator extinction because of prey scarcity. Evidently, the predator-prey-hostplant system, as we have modelled it, tends to enter into a stable limit cycle over a wide range of initial values. The mechanisms underlying this dynamic behaviour were analysed by investigating the system's response to structural changes in the model and to modifications of the parameters.

Sensitivity analysis

The analysis is based on a series of simulations over periods of 800 days. The results are summarized by log-transformed third peak and subsequent trough values of the number of prey and predator patches, as in Table 1. But these values have only been presented, if the simulated fluctuations converged to a limit cycle. Whenever deviating types of dynamic behaviour were found, verbal descriptions are given or the fluctuations are presented in a figure. Though - as shown in Figure 4B-I - other output variables could have been presented as well, only patch frequency dynamics are given as conclusive evidence for the response of the system to structural changes in the model and to parameter modifications. This reduces the amount of information without affecting the conclusions. The simulations presented in Figure 4B serve as a basis for comparison and are henceforth referred to as the standard simulation.

Prey dispersal and host-plant colonization

Increased dispersal of the spider mites contributes to their escape from predator attack. We tested the effect of different dispersal rates that were chosen within or close to the range of actually measured values (Laane, unpublished data; Figure 2D). The simulated patch frequencies of both prey and predator patches entered a limit cycle at a slightly higher or lower level, depending on whether dispersal was increased or decreased (Table 2).

Specifications of the simulations		Log (No.prey patches)		Log (No. predator patches)	
In words	In formulas	Lowest value	Highest value	Lowest value	Highest value
Standard conditions	$\max(f_h) = 0.03$ d_s and $d_{s,p}$ as in Figure 2D	3.92	5.76	3.91	7.06
maximum dispersal	$d_s = d_{s,p} = 0.122$	3.98	5.84	4.10	7.14
minimum dispersal	$d_s = d_{s,p} = 0.0048$	3.92	5.76	3.91	7.06
below-minimum dispersal from young patches and above-maximum dispersal from from old patches	$d_s = d_{s,p} = 0$, at $t = 1$ and gradually approaches $d_s = d_{s,p} = 0.244$ at $t = 100$	4.01	5.93	4.11	7.23
maximum dispersal from predator patches only	$d_{s,p} = 0.122$	3.95	5.85	4.09	7.15
maximum area with suitable hostplants increased	$\max(f_h) = 0.5$	4.41	6.35	4.53	7.66

Table 2

Dynamics of the number of patches, as affected by dispersal of spider mites.

This is probably because the escape resulting from dispersal is only temporarily effective. The same applies to increased dispersal of spider mites from predator patches. Bernstein (1984) presented experimental evidence for increased emigration of spider-mite females as a response to predator density. Such a response may lead to a temporary escape from their enemies which has clear advantages to the individual prey. However, our simulations do not point to important effects at the population level.

Colonization chances of dispersing spider mites depend largely on the host-plant area available. In the standard simulation this area remained close to the maximum attainable, which is equal to 3% of the total leaf area present in the environment. However realistic this small percentage of hostplant area may be under natural conditions, in agricultural crops the chances for colonization may be much higher, which may also increase the chance of pest outbreaks. However, this fear was not substantiated by a simulation in which the maximum hostplant area was taken to be 50% of the total leaf area in the environment. The peak and trough values of the patch frequencies increased only fourfold, whereas the uncolonized hostplant area remained close to the maximum attainable, as with the standard simulation (Table 2). Moreover, it appears that the dynamic behaviour of the predator-prey-hostplant system does not hinge upon the colonization chances of the prey since the limit cycle character was not affected.

Prey refuges

Absolute temporary refuges were built into the standard model in two phases of prey-patch life. The first phase is the one-day period following colonization of the host plant. The second is the period of host-plant exhaustion after t = 100 days, when the prey prevents the predators from being effective by producing a labyrinth of web. Prolonging the first phase from 1 to 20 days had a pronounced effect on the level and amplitude of the fluctuations, which converged to a limit cycle (Table 3).

Specifications of the simulations		Log (No. prey patches)		Log (No. predator patches)	
In words	In formulas	Lowest value	Highest value	Lowest value	Highest value
Standard conditions	$r(t) = 0$ $r(t)$, see text	3.92	5.76	3.91	7.06
Young prey patches without risk of being discovered by predators	$r(t) = 0$, for $1 \leqslant t \leqslant 20$	4.27	7.76	4.43	7.94
Old prey patches without risk of predation	$r(t) = 0$, for $80 \leqslant t \leqslant 100$	3.89	5.77	3.87	7.07
initial prey coloniz-zations run a risk of (instantaneous) discovery (Fig. 5)	$r(1) \neq 0$, but calculated in the same way as $r(t)$	-0.62	5.32	4.39	6.36
predators unable to exterminate prey patch completely (Fig. 6)	$F_s(3)$is increased each day with $F_p(1)$	no limit cycle, very slow but steady increase			

Table 3

Dynamics of the number of patches, when young and old prey patches are designated as refuges, and when prey survive locally.

It caused the prey to escape from predator attack at low prey density; the escapees helped the prey population to resurge when the number of dispersing predators became low. Prolonging the second phase did not affect the dynamics, which is undoubtedly because prey patches rarely get the chance to grow old (Table 3).

Because of the apparent importance of the temporary refuges of young prey patches, we also investigated how the system responds to the absence of these refuges. Recall from the program description that colonization of the host plant by a spider mite and invasion by a predator are calculated in this order at each time step. Hence, removing the temporary refuge from the model allows the predators to attack colonizing spider mites instantaneously. Simulations without the initial refuge resulted in fluctuations converging to a limit cycle (Figure 5A). As expected from the above results the level of the cycle was lower than in the standard simulation (Table 3). The number of prey patches decreased to below one at regular intervals. Prey extinction would have been the result, were it not that prey dispersing from predator patches cause the prey population to resurge when the number of dispersing predators became low. Paradoxically, the predator patches served as a refuge to the prey and became a source of dispersing prey that allowed the prey population to increase when the risk of predator invasion becomes sufficiently low again. Figure 5B proves that when prey patches are extinct, there are several predator patches that still have a long time to go before the prey is eliminated, so that these patches are a source of dispersing prey.

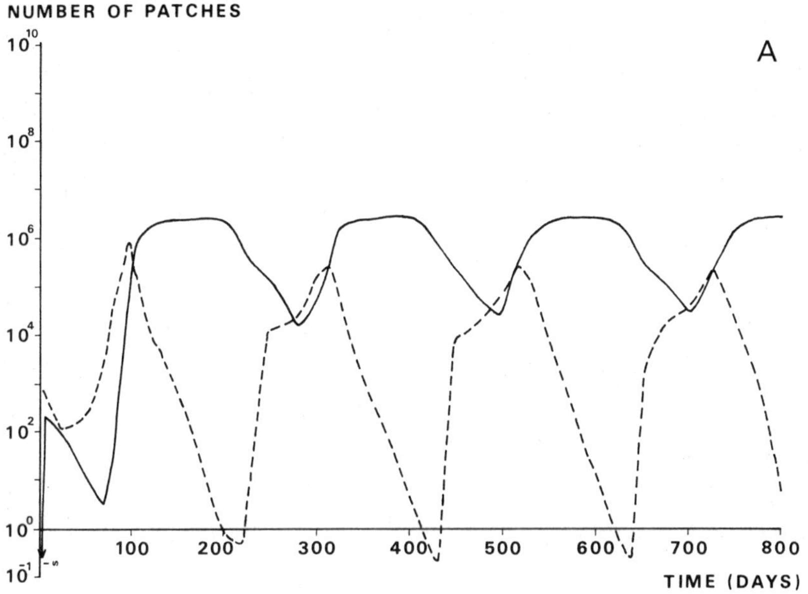

Figure 5A Regional dynamics of the number of prey patches (dashed line) and the number of predator patches (drawn line), when colonizations of host plants by spider mites run the risk of instantaneous discovery and invasion by dispersing predators (i.e. no temporary refuge for prey patches after colonization).

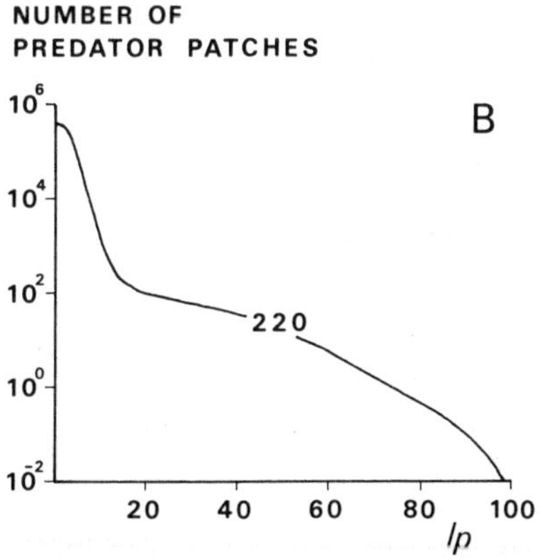

Figure 5B Frequency distribution of predator patches in relation to their future life span (l_p) on day 220, i.e. when the prey patches have gone almost extinct and most of the prey population is to be found in predator patches.

Another type of prey escape may arise if the predators are not capable of completely eradicating the prey in the patches. It is worth considering this case, because the standard model rigorously assumes local prey elimination and because it has never unambiguously been demonstrated that predatory mites continue to search until they have killed the last of the Mohicans. The simulation was done by extending the model with a statement that transforms each predator patch back to a young prey patch at the end of the predator-prey interaction period. The simulation results are presented in Figure 6 and these reveal that patch frequencies rapidly approached a level that seemed to be constant because of optic illusion. Actually, the number of patches keep increasing at a constant rate, but this process is slow and therefore not apparent from the logarithmic plot. The slow rate results from the extremely short life time of both prey and predator patches (1 and 2 days respectively), which causes the number of dispersing prey to be low. Ultimately, the entire area of host-plants would be covered by prey and predator patches if the predators were not capable of exterminating their prey locally. Clearly, this result has no intuitive appeal to any experienced field acarologist. Therefore, we suggest that the local extermination of prey is likely to occur frequently under field conditions.

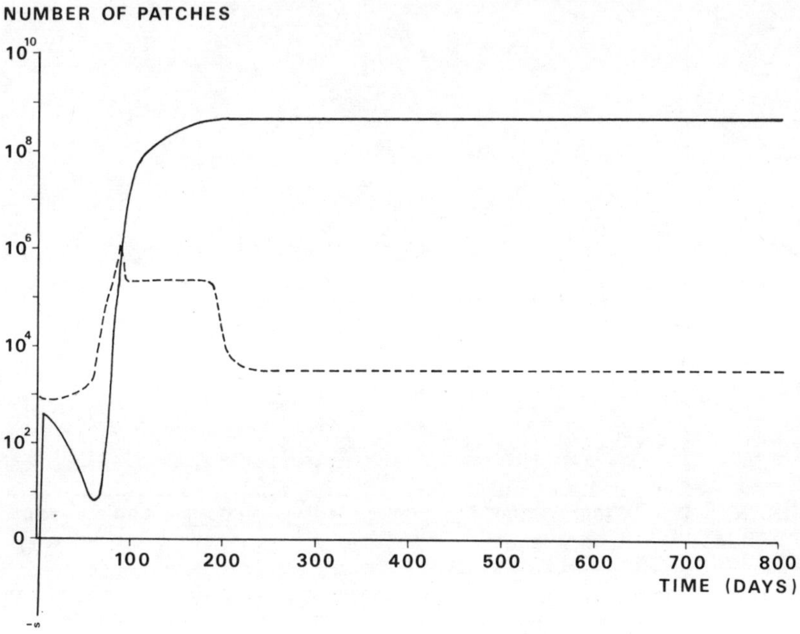

Figure 6 Regional dynamics of the number of prey patches (dashed line) and the number of predator patches (drawn line) for the case where the predators are made unable to eliminate their prey in the invaded prey patches (= predator patches); after the interaction period has elapsed and the predators have dispersed, a small amount of prey is left unpredated.

Risk to prey patches

The risk of a prey patch being discovered by predators depends on the properties of the prey patch, as well as on the abundance and the properties of the dispersing predators. The properties of the prey patch are of concern here. The relevant properties are the age-dependent patch size and the associated odour plume area within which the predator is assumed to find its way to the upwind prey-odour source. Excluding the prey-odour plume from the risk calculations had only minor effects on the position of the limit cycle (Table 4).

Specifications of the simulations		Log (no. prey patches)		Log (No. predator patches)	
In words	In formulas	Lowest value	Highest value	Lowest value	Highest value
Standard conditions	$w_s(t) =$ $\exp(0.171t)$ $a_0(t) =$ $\exp(0.19t - 11.8)$	3.92	5.76	3.91	7.06
odour plume absent	$a_0(t) = 0$	3.97	5.87	4.00	7.17
threshold odour concentration decreased	$a_0(t) =$ $\exp(0.1t - 2.5)$	3.30	5.37	3.29	6.67
prey patch area fixed to size at age of 2 days	$w_s(t) = w_s(2)$	3.96	5.97	3.77	7.27
prey patch area and plume area fixed to size at age $t = 2$	$w_s(t) = w_s(2)$ $a_0(t) = a_0(2)$	4.45	6.55	4.38	7.85

Table 4

Dynamics of the number of patches, as affected by the area covered by prey patches and associated odour plumes and the resulting influence on the risks to prey patches.*

An arbitrary decrease in the threshold odour concentration above which the predator's searching reponse is effective, resulted in larger plume areas, and, hence, in increased risk to the prey patch. As expected, the simulated fluctuations fell, but in view of the substantial change of the plume area the effect is still small compared with the standard simulation. Only slightly larger effects were obtained in simulations where the prey patch area was kept constant at a size equal to a prey patch of two days old. However, when both the odour plume and the prey patch were fixed to their size at $t = 2$ days, the level of the fluctuations increased more than expected from fixing one of the two risk components. Nevertheless, the prey population did not cover more than 5% of the area of the host plants, so that predator control proves to be effective despite the risks to the prey patches being minimized. Hence, as our investigations concerned biologically extreme cases of risks to prey patches, it is clear that the increase in the risk to the prey-patch concomitant with the age of the patch, is hardly relevant to the gross dynamic behaviour of the predator-prey system. But the level of the oscillations is affected by age-dependent risks to prey patches.

Search, invasion and interception of predators

Apart from properties of the prey the risk of prey patch discovery is also determined by the quantity of dispersing predators and their searching capacity. If a predator invades a prey patch or a predator patch, she does not continue to disperse, but tends to stay in the patch as long as prey is available. Since the chance of invasion and the subsequent period of arrestment increases with the age of the prey-patch, the outcome may be the aggregation of dispersing predators in large patches with large odour plumes. Hassell & May (1974) emphasized that if the prey distribution is sufficiently clumped and the predators aggregate in patches of high prey density, small prey patches run less risk of discovery and are more or less free to grow and expand until they are large. The partial refuge that arises from the aggregative response of the predators, can lead to stabilization of the otherwise unstable parasite-host model of Nicholson-Bailey (Hassell & May, 1974). Predatory mites, such as *P. persimilis,* do show an aggregative response to some extent, as they are able to find large patches from a distance by means of chemotaxis and anemotaxis, and as they tend to stay longer when more prey is available in the patch. However, according to simulations, in which the number of dispersing predators were prevented from decreasing as a consequence of predator interceptions in prey

* patch and plume areas are only subject to modifications inside the DO-loop that computes risks to prey patches, and, thus, not in the DO-loop, that computes the interception of dispersing predators in predator patches and prey patches.

and predator patches, the numerical consequences were negligible (Table 5).

Specifications of the simulations		Log (No. prey patches)		Log (No. predator patches)	
In words	In formulas	Lowest value	Highest value	Lowest value	Highest value
Standard conditions	$I = 1 - exp(-i.\Delta\tau)$ $s = 13$ m/day m (Figure 2C)	3.92	5.76	3.91	7.06
100% mortality after 2-day period of dispersal	$m(3) = 1/\Delta\tau$ if $\tau > 100$ days, else m according to Figure 2C	4.49	5.92	4.22	7.22
no interception of predators in prey and predator patches	$I = 0$	3.91	5.78	3.92	7.06
no locomotory search outside plume and patch area	$s = 0$	5.37	7.16	5.53	7.44

Table 5

Dynamics of the number of patches as affected by the survival, interception and searching behaviour of dispersing predators.

Hence, the aggregative response of the predatory mites is not an important stabilizing mechanism in interactions between phytoseiid predators and their tetranychid prey.

Outside the patch and plume area the predators search randomly by passive aerial dispersal and by locomotion. Eliminating locomotory search from the model did not affect the limit cycle character of the population dynamics, but it did give rise to a tenfold increase in the patch frequencies (Table 5). Apparently, this mode of search imposes high risks to the prey patches. As will be shown later, its effect can be compared with reducing to one-tenth the production of dispersing predators within the predator patches. Thus, aerial dispersal and locomotory search significantly contribute to invasions into predator and prey patches, but the number of predators that find these patches and stay is small compared with the number of predators that continue to disperse and search. Hence, interception is of little importance to decreases in the number of dispersing predators.

When the dispersing predators incur high mortality as a consequence of starvation, the risks to prey patches will decrease allowing the prey population to reach higher numbers, but this will be counteracted by an increase in the number of dispersing predators in response to the larger amount of prey available. When the simulations allowed the predators to disperse during a few days only (instead of 3 weeks), patch frequencies converged to a limit cycle at a level that was four times higher than the standard simulation (Table 5). Simulations assuming increased survival were not tried out, as this scenario is biologically improbable. Thus, within the extremes of low and high mortality during predator dispersal regional dynamics appear to be of the limit-cycle type. For this reason, it is unlikely that the relation between regional prey-patch density and the risk of mortality incurred by dispersing predators are of crucial importance to the system's stability. Clearly, this conclusion also applies to the relation between prey-patch density and the predator's travelling time between patches.

Delayed production of dispersing predators

An increase in the number of prey is not immediately followed by an increase in the number of dispersing predators. This is because the predators are assumed not to disperse from a patch until their prey is eliminated. This delayed numerical response of the dispersers was expected to have a destabilizing effect. Indeed, simulation experiments proved that shortening of the interaction period and making dispersal continuous instead of discrete-delayed tends to decrease the amplitude of the oscillations, especially by damping the peak values (Table 6).

Specifications of the simulations		Log (No. prey patches)		Log (No. predator patches)	
In words	In formulas	Lowest value	Highest value	Lowest value	Highest value
Standard conditions	$l_p(t_p)$ (Figure 2A) $n_p(t_p))$ (Figure 2B)	3.92	5.76	3.91	7.06
length of interaction period decreased to one-tenth	max $(1, l_p / 10)$	4.02	5.43	3.86	5.04
uniform take-off of predators during interaction period		4.22	5.05	4.72	5.78
production of predators per patch decreased to one-tenth	$n_p(t_p) / 10$	4.41	6.24	4.38	7.54
production of predators per patch tripled	$n_p(t_p).3$	3.63	5.52	3.69	6.83
within-patch production of predators not influenced by maximum prey patch size	$\log(n_p) \approx \exp(0.073(t_p - 12))$ for $t_p \leqslant 100$	prey extinction		predator extinction	

Table 6

Dynamics of the number of patches, as affected by the period of predator-prey interaction $(l_p)^*$ and by the number of predators dispersing from predator patches (n_p).

A shorter interaction period also leads to shorter cycle periods and a much larger series of cycles in the standard simulation period of 800 days. This suggests that the predator-prey system would have been damped monotonically, if the numerical increase of the dispersing predators were not delayed.

Drastic changes in within-patch production of dispersing predators are biologically relevant, because some species of phytoseiid predators produce more eggs per unit food than others(see e.g. Sabelis, 1981). As mentioned earlier, reducing the production to one-tenth leads to a fourfold increase in the patch frequencies and, of course, increasing the production leads to a decrease (Table 6). However, despite the drastic changes the dynamic behaviour of the patch frequencies is still of the limit cycle type.

Finally, it is important to note that after prey patches become older than ca. 40 days, the predators that will disperse after prey elimination, become fewer and fewer (Figure 2A). This is because the exhaustion of the host plant sets limits to prey population growth and hence also to the population growth of the predators. We fixed the exhaustion phase in a prey patch at an arbitrarily chosen age of 100 days. In a predator patch exhaustion is delayed because predation retards prey population growth and hence also the rate of the prey's exploitation of the host plant. Because

* Changes in the interaction period l_p are brought about only with respect to the production of dispersing predators $(N_p(l))$ and not to any other state variable labelled with l(*i.e.*$N_{s,p}(l)$, $D_{s,p}(l)$, $A_{s,p}(l)$ and $A_{o,p}(l)$). Hence, prey dispersal from predator patches and interception of dispersing predators in predator patches is accounted for in the same way as in the standard simulation.

the moment of hostplant exhaustion was chosen arbitrarily, the consequence of changing this choice should be considered. Figure 2A shows that predator production initially increases drastically with the age of the prey patch at first predator invasion. If hostplant exhaustion occurred much later, extrapolation in the Figure shows that predator production could easily surpass the value of 10^{10} dispersing mites per patch. Recall from the simulation experiments with varying initial values (Table 1) that such high numbers of predators (more than one predator per m^2!) cause the prey population to become scarce and the predator population to subsequently die out because of starvation. In fact, this result was again confirmed by the simulation in which hostplant exhaustion is delayed. It should therefore be recognized that although the area of the hostplants available for colonization is not a limiting factor, the maximum size of the prey patches can be of crucial importance to the dynamic behaviour of the predator-prey system in that it prevents the predators from becoming too numerous. This result may have an important bearing on the biological control of spider mites, because large-scale crops contain large areas of hostplants which may cause first the prey and then the predators to reach numbers that are too high for the predator-prey system to persist. In that case, the reintroduction of predators will be necessary whenever the pest population resurges.

Sensitivity to sets of modifications

So far, the predator-prey system could only be destabilized by introducing large numbers of dispersing predators at the start of the simulation (Table 1), or by increasing the production of predators per patch (e.g. by assuming predator production per patch keeps increasing with the age of the prey patch at first-predator invasion; compare with Figure 2 and see Table 6). The destabilizing effect is caused by the high risks incurred by prey patches to be invaded soon after they are founded. This decreases the production of dispersing prey causing a decline of the number of prey patches and, moreover, it increases the synchrony of local predator-prey cycles, which in turn causes synchronous dispersal of predators. Clearly, this destabilizes the predator-prey interaction. However, all other single parameter modifications were not sufficient to undermine stability. Apparently, there is a complex set of factors that contribute to regional stability of the populations. To find out which of these factors are of crucial importance, combinations of parameter modifications have been tried out, whereby two classes of modifications were to be distinguised:

(1) those contributing to the synchrony of local predator-prey cycles (synchronized predator invasions into prey patches, fixed interaction period, increased production of dispersing predators),

(2) those eliminating (predator/prey) mortality factors that are positively related to prey density at a local and/or regional scale, and that have been shown to be important in stabilizing simple predator-prey models under certain conditions (Murdoch & Oaten, 1975; Oaten, 1977; Hassell & May, 1974; Hassell, 1978).

Following this classification, simulation experiments were carried out, the results of which are summarized in Table 7 by presenting a qualitative description of the simulated dynamical behaviour of the regional populations.

Three major conclusions can be drawn from the results. Firstly, the population fluctuations do not converge to a limit cycle, if the predator-prey interaction period is fixed to a large value (e.g. l_p = 50 days or larger; independent of t_p) and the number of predators dispersing after this period is taken not too low (e.g. n_p = 8000; independent of t_p). An example is shown in Figure 7; the populations fluctuate vigorously, seemingly with increasing amplitude (as far as observed in the simulated time span of 800 days !), and at times prey and predator patches became very scarce, so that their populations would have crashed in a finite space where persistence would be a matter of chance. As shown in Table 7, all other major combinations of l_p and n_p (i.e. short-high, long-low, short-low etc.) failed to destabilize the population fluctuations. Apparently, the delayed numerical response of the predators becomes destabilizing only when both the delay is long and the number of predators dispersing from the patch after this delay is taken not too low.

The second conclusion is that synchronized predator invasion (i.e. predator patches can only start at preselected moments) destabilizes the fluctuations only in combination with a fixed length of the interaction period following predator invasion (Table 7). Without the latter fixation of l_p there appears to be sufficient variation in l_p (in line with variation in t_p, the age of the prey patch at predator invasion !) to prevent the local cycles from becoming synchronized. This clearly demonstrates the stabilizing effect of endogenous variation in the way local predator-prey interactions proceed in the course of time. Paradoxically, delayed dispersal of predators from predator patches tends to exert a destabilizing influence on the regional population dynamics, whereas variation in this delay contributes to the asynchrony of local cycles and thus also to the stability of the interaction.

The third conclusion is that the stability of the limit cycle does not hinge upon mortality factors that are related to local and/or regional prey density (Table 7) and that have been shown to be more or less powerful in stabilizing Lotka-Volterra or Nicholson-Bailey models (Hassell & May, 1974; Murdoch & Oaten, 1975; Oaten, 1977; Hassell, 1978).

Table 7

Summary of qualitative aspects of regional dynamics, when combinations of parameter modification and structural changes are brought about in the model. Class 1 refers to modifications that contribute to the synchrony of local predator-prey cycles. These are specified in the left column. Class 2 refers to modifications that eliminate density related mortality factors. These are: (1) absence of prey refuges, (2) fixed risk to prey patches, of being invaded by predators (fixed patch and plume size), (3) no aggregative response of the predators, (4) fixed travelling time of dispersing predators (or high mortality among the dispersers).

The following symbols are used to describe regional dynamics:

L = population fluctuations converge to a limit cycle;

D = population fluctuations do not converge to a limit cycle, but diverge, become chaotic or get extinct.

Class 1 modifications		Simulations without class 2 modifications	Simulations with all class 2 modifications
Fixed period of predator-prey interaction	$l_p = 6$	L	L
	$l_p = 60$	L	L
Fixed number of predators produced per predator patch	$n_p = 8$	L	L
	$n_p = 8000$	L	L
Fixed interaction period and fixed number of predators dispersing after elapse of the interaction period	$l_p = 6;\ n_p = 8$	L	L
	$l_p = 6;\ n_p = 8000$	L	L
	$l_p = 60;\ n_p = 8$	L	L
	$l_p = 60;\ n_p = 8000$	D	D
Periodically synchronized invasions of predators in prey patches		L	L
Periodically synchronized predator invasions and fixed periods of predator-prey interaction		D	D

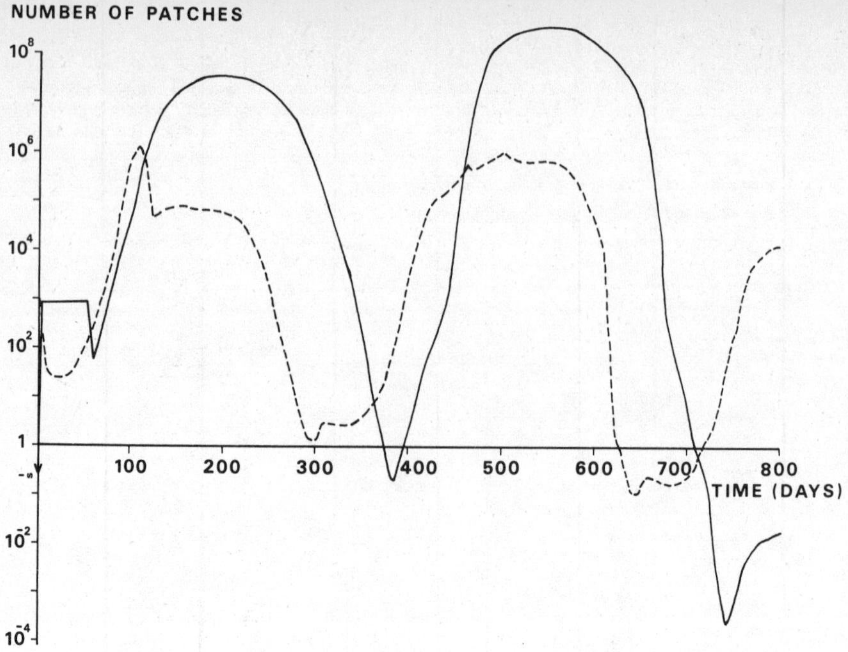

Figure 7 Regional dynamics of the number of prey patches (dashed line) and the number of predator patches (drawn line), when the interaction period was fixed to 60 days and the number of predators dispersing from the patch at the end of the interaction period equals 8000., i.e. l_p is long and n_p is high, both being independent of prey-patch age at first predator invasion (t_p).

Discussion

Using the simulation model discussed above we tried to assess the relative importance of different mechanisms that according to general theory could enhance the persistence and stability of populations of the predatory mite, *P. persimilis,* and of the two-spotted spider mite, *T. urticae.* Because our aim was to test these theories with this particular scenario in mind, the model was provided with most of our biological knowledge to date. Because of this emphasis on 'reality' we were irrevocably committed to a simulation approach and had to seek answers to our questions by testing the response of the model to structural changes and to parameter modifications. Below, we will discuss theories on persistence and stability before continuing with a review of predictions of the model that are amenable to validation under natural and agricultural conditions. Finally, some of these predictions are validated by comparison with published results of population experiments

Persistence and stability

Maynard Smith (1974) was the first to point out that persistence is promoted by processes that keep local predator-prey cycles out of phase relative to each other. He argues that continuous, unbiased population exchange would tend to synchronize the local cycles, and, once synchronized, the net flow of exchange becomes zero. This results in local cycles that can be considered as isolated, and hence, become extinct when prey is eliminated. Zeigler (1977) substantiated Maynard Smith's claim that discrete instead of continuous population exchange promotes persistence.

In our model asynchrony of local cycles is promoted in various ways, the most important being:

(1) differences in the interaction period depending on prey-patch age at first predator invasion (t_p) and

(2) differences in the moment of predator invasion resulting from chance and from the risk of prey patch discovery which depends on prey-patch age.

For these reasons it is unlikely that the prey will be eliminated in all patches simultaneously. The asynchrony can only be disturbed, when the dispersing predators become too numerous, which would cause the risk to prey patches of any age to approach unity. Then, sooner or later the interaction periods in all patches become short and, thus, of equal duration, which increases the chance of synchronous prey elimination in all patches followed by the extinction of prey and predator. Because the asynchrony of the local cycles is promoted by variation in duration and initiation of within-patch cycles, it is not surprising that our attempts to destabilize the limit-cycle dynamics were successful only if this variation was eliminated.

The simulations almost invariably showed limit-cycle dynamics. Sensitivity analysis showed that the amplitude and period of the oscillations decreased if within-patch interaction periods were shortened, or if continuous, uniform dispersal of predators from predator patches was allowed instead of discrete delayed dispersal. In the ideal case without delays in dispersal, deterministic stability would probably be the case, just as Gurney & Nisbet (1978) obtained by modelling Huffaker's experiments without these time delays. Moreover, limit cycle dynamics were found, whenever delays in predator dispersal were included in the model (Zeigler, 1977; Takafuji et al., 1983). Therefore, the delay in predator dispersal since predator invasion in a prey patch can be considered as a destabilizing force similar to the effect of discrete generations in parasite-host models of the Nicholson-Bailey type (Hassell, 1978). It is, however, interesting to note that variation in these time delays is in turn a factor that strongly contributes to regional stability. As argued before, such variation is inherent to the interaction between two-spotted spider mites and predatory mites.

What is the role of the hostplant in the persistence of the tritrophic system? Zeigler (1977) derived a Lotka-Volterra equation with prey self-limitation inspired by Huffaker's experiments. He states that in his simulation model self-limitation of the prey arises from the reduction of space available for colonization when the frequencies of prey and of predator patches are high. However, in our simulations the leaf area available for colonization remained close to the maximum attainable value throughout the simulations. It is therefore unlikely that stability of the limit cycles is caused by prey self-limitation *sensu* Zeigler (1977). In our model, prey limitation within the patch caused by hostplant exhaustion is the main factor preventing the predators from becoming too numerous. In theory this could also have been brought about by the age-related increase in risks to prey patches, because a steep increase may prevent the prey patches from becoming old. However, our estimates of these risks did not prove to be capable of limiting prey-patch age sufficiently. Hence, prey limitation resulting from hostplant exhaustion is essential if the predator-prey system we modelled is to remain stable.

A class of generally acknowledged stabilizing factors are prey refuges or escapes. Huffaker (1958) emphasized the importance of the prey having better capacities to disperse to new areas than their predators. Moreover, several simulation studies (Maynard Smith, 1974; Zeigler, 1977; Crowley, 1981) revealed that multi-cell models tend to persist at high prey mixing rates and intermediate or low predator mixing rates. Using our model, simulations generally resulted in a larger number of dispersing predators than the number of dispersing prey (Figure 4C), because predators have a higher tendency to disperse when food becomes scarce, and because dispersing predators survive much longer. Moreover, as predators are more mobile and search more effectively for food than their prey, it follows that in our model the capacity of the predators to reach new areas is larger than that of the prey. Despite the fact that this difference does not meet the conditions emphasized by the studies mentioned above, persistence was apparent in our model. Hence, differences in the ability to reach new areas are not relevant to understanding the persistence of the predator-prey system modelled by us. The same applies to the role of temporary refuges in the initial phase of hostplant colonizations. Simulations without this type of prey refuge demonstrated that limit cycle behaviour of the predator patches was maintained though the prey patches became almost extinct at regular time intervals. This dynamic behaviour arises because prey dispersal from predator patches ensures the resurgence of prey patches when the number of dispersing predators decreases. Paradoxically, the predator patches appear to be the most important refuges to the prey, because predators need some time before they have eliminated their prey in the patch.

Though persistence in large environments is still guaranteed in absence of temporary refuges, such refuges are certainly relevant in smaller environments in that these increase the lowest level attained by the cycling populations, thereby reducing the role of chance effects. This argument applies of course to all cases, where the lowest level of the cycles was increased. This was the case, for example, when the rate of prey dispersal was attributed with a higher value than actually measured. This is not an unrealistic possibility, because phoretic dispersal of spider mites may occur in addition to aerial dispersal and fugitive responses of the prey to predator attack have been reported (Bernstein, 1983). Moreover, Huffaker's manipulations of prey and predator dispersal rates had an effect on population persistence in a small but complex environment, that can be explained in the way proposed above.

Model predictions

The simulations enable regional dynamics to be calculated from within-patch dynamics and between-patch dispersal. The predictions of regional dynamic behaviour can be compared with population fluctuations in the field. If different concepts of the underlying biological reality give rise to distinct types of regional dynamics, comparison with field observations could lead to inferences about these concepts. However, it is often the case that models predict the same type of dynamics from different mechanisms. Nevertheless, comparisons between measured and simulated fluctuations are worthwile, because they show which theories succeed and which fail to provide an explanation for the observed phenomena. The most pertinent model predictions that are amenable for testing in the field, are summarized below:

(1) Population fluctuations of spider mites and predatory mites converge to a limit cycle only if local prey growth

is sufficiently limited so as to prevent overproduction of predators (and consequently synchronization of local cycles). Then, the populations cycle at such a low level that most of the hostplant area remains free of spider mites.

(2) If hostplant exhaustion does not occur until after a long period of within-patch prey growth, prey and/or predator populations become extinct in the whole region due to excessive numbers of dispersing predators produced in predator patches

(3) Temporary refuges of young spider-mite patches are not essential to the existence of stable limit cycles in large enough environments. If these refuges are absent, situations will occur where all patches with prey are invaded by predators and subsequent resurgence of the non-invaded prey patches is caused by prey dispersal from predator-invaded patches.

(4) Refuges and escape responses of the prey are essential, however, in small but complex environments to increase the lowest level attained by the cycling populations. In this way the role of chance effects at low population densities can be diminished.

(5) If regional populations of prey and predator are close to being deterministically stable, this may indicate that the period between first predator invasion and predator dispersal is very short either because predators disperse during the predator-prey interaction period (instead of staying until prey extermination) or because the interaction period is forced to be short (e.g. when the maximum host plant area alotted to each prey patch is very small). In the special case that the predators do not eliminate their prey in the invaded prey patches, prey and predator patches tend to be extremely ephemeral, but deterministic stability does not arise, because the number of prey and predator patches keep increasing (though very slowly) until all host plant area is colonized.

(6) Increasing the synchrony and length of local cycles will tend to destabilize the predator-prey system even on a regional scale.

Is there any experimental evidence?

Huffaker's laboratory work provided one of the most thought provoking experiments in population ecology. He found self-perpetuating cycles of predator and prey in spatially complex environments. Increasing spatial heterogeneity enhanced the persistence of the populations (Huffaker et al., 1963). Three major features of these experiments seem to be in agreement with the model predictions:

(1) Overall population numbers in the environment do not converge to an equilibrium value, but oscillate with a more or less constant period and amplitude.

(2) In the small but complex environment used in Huffaker's 1958 study facilitation of prey dispersal relative to predator dispersal enhanced the persistence of the populations.

(3) An increase in the amount of food available per prey patch resulted in a generation of too many predators at times of high prey density (in relation to the complexity of the places to be searched). Thus, the areas were then searched too well causing synchronization of local cycles and, therefore, regional extinction (Huffaker et al., 1963).

The latter result may have an important bearing on biological control of spider mites in agricultural crops. The current practice of growing crops in monocultures will not favour the persistence of predator-prey systems such as that studied here. However, temporary control of spider mites may well be achieved by adequately timed releases of mass-reared predatory mites.

Of course, it is dangerous to draw general conclusions from artificial laboratory experiments. There are many reports of population fluctuations of spider mites and predatory mites in the greenhouse. These interactions resulted either in elimination of the prey and subsequently of the predator (Chant, 1961; Bravenboer & Dosse, 1962; Laing & Huffaker, 1969; Takafuji, 1977; Takafuji et al., 1981; Sabelis, in press) or in wide fluctuations of increasing amplitude (Burnett, 1979; Nachman, 1981) or in selfperpetuating oscillations of varying amplitude (Hamai & Huffaker, 1978). Interpretation of these results is very difficult because (1) the spatial scale may be too small to reveal overall population fluctuations predicted by the model, (2) the experimental period is usually not much longer than the period of the local cycles, (3) the spatial heterogeneity resulting from plant and crop structure is generally left undefined, and (4) predators cannot disperse aerially as a consequence of the low rate of air movement in greenhouses. Except for the latter these difficulties also apply to field data. Future long-term experiments should elucidate the relation between spatial scale, spatial heterogeneity and the overall population fluctuations. A prerequisite to gain more insight in these relations is to assess the relevant components of spatial heterogeneity and to relate that to the dispersal capacity of the mites. Only in this way, we may be able to test whether the overall population fluctuations are stable limit cycles.

Acknowledgements:

We thank Odo Diekmann and Hans Metz for stimulating discussions and comments on a preliminary version of the manuscript. For valuable literature references the first author is grateful to Hans-Jürgen Rennau and Heinrich Kaiser. The comments of Piet den Boer on the prefinal version of the manuscript were appreciated. The drawings were made by Rijndert de Fluiter.

REFERENCES

D.E. AYLOR, J.-Y. PARLANGE & J. GRANETT, (1976). *Turbulent dispersion of disparlure in the forest and male gypsy moth response.* Environmental Entomology 5: 1026-1032.

P.J. DEN BOER, (1968). *Spreading of risk and stabilization of animal numbers.* Acta Biotheoretica 18: 165-194.

R.L. BRANDENBURG & G.G. KENNEDY, (1982). *Intercrop relationships and spider-mite dispersal in a corn/peanut agro-ecosystem.* Entomologia experimentalis et applicata 32: 269-276.

C. BERNSTEIN, (1984). *Prey and predator emigration responses in the acarine system Tetranychus urticae-Phytoseiulus persimilis.* Oecologia 61: 134-142.

W.H. BOSSERT, & E.O. WILSON, (1963). *The analysis of olfactory communication among animals.* Journal of theoretical Biology 5: 443-469.

L. BRAVENBOER & G. DOSSE, (1962). *Phytoseiulus riegeli Dosse als prädator einiger schadmilben aus der Tetranychus urticae- gruppe.* Entomologia experimentalis et applicata 5: 291-304.

T. BURNETT, (1979). *An acarine predator-prey population infesting roses.* Researches on Population Ecology 20: 227-234.

D.A. CHANT, (1961) *An experiment in biological control of Tetranychus telarius (L.) (Acarina: Tetranychidae) in a greenhouse using the predacious mite Phytoseiulus persimilis Athias-Henriot (Phytoseiidae).* The Canadian Entomologist 93: 437-443.

T.J.D. COATES, (1974) *The influence of some natural enemies and pesticides on various populations of Tetranychus cinnabarinus (Boisduval), T. lombardini Baker & Pritchard and T. ludeni Zacher (Acari: Tetranychidae) with aspects of their biologies.* Entomology Memoir, Department of agricultural technical services, Republic of South Africa 42: 1-40.

P.H. CROWLEY, (1977). *Spatially distributed stochasticity and the constancy of ecosystems.* Bulletin of Mathematical Biology 39: 157-166.

P.H. CROWLEY, (1978). *Effective size and the persistence of ecosystems.* Oecologia 35: 185-195.

P.H. CROWLEY, (1981). *Dispersal and the stability of predator-prey interactions.* The American Naturalist 118: 673-701.

J.S. ELKINTON, R.T. CARDÉ & C.J. MASON (1984). *Evaluation of time-average dispersion models for estimating pheromone concentration in a deciduous forest.* Journal of chemical Ecology 10: 1081-1108.

E.S. EVELEIGH & D.A. CHANT (1982). *Experimental studies on acarine predator-prey interactions: the distributions of search effort and the functional and numerical responses of predators in a patchy environment (Acarina: Phytoseiidae).* Canadian Journal of Zoology 60: 2979-2991.

Y. FARES, P.J.H. SHARPE & C.E. MAGNUSON (1980). *Pheromone dispersion in forests.* Journal of theoretical Biology 84: 335-359.

K. FUJITA, T. INOUE & A. TAKAFUJI (1979). *Systems analysis of an acarine predator-prey system.* Researches on Population Ecology 21: 105-119.

W.S. GURNEY & R.M. NISBET, (1978). *Predator-prey fluctuations in patchy environments.* Journal of Animal Ecology 47: 85-102.

J. HAMAI & C.B. HUFFAKER (1978). *Potential of predation by Metaseiulus occidentalis in compensating for increased, nutritionally induced, power of increase of Tetranychus urticae.* Entomophaga 23: 225-237.

M.P. HASSELL, (1978). *The dynamics of arthropod predator-prey systems.* Monographs in Population Biology 13, Princeton University Press.

M.P. HASSELL & R.M. MAY, (1974). *Aggregation of predators and insect parasites and its effect on stability.* Journal of Animal Ecology 43: 567-594.

A. HASTINGS, *Spatial heterogeneity and the stability of predator-prey systems.* Theoretical Population Biology 12: 37-48.

A. HASTINGS, (1979). *Spatial heterogeneity and the stability of predator-prey systems:* Population cycles. In: *Applied Nonlinear Analysis,* Academic Press, New York.

R. HILBORN, (1975). *The effect of spatial heterogeneity on the persistence of predator-prey interactions.* Theoretical Population Biology 8: 346-355.

M.A. HOY, (1982). *Aerial dispersal and field efficacy of a genetically improved strain of the spider-mite predator Metaseiulus occidentalis.* Entomologia experimentalis et applicata 32: 205-212.

C.B. HUFFAKER, (1958). *Experimental studies on predation: dispersion factors and predator-prey oscillations.* Hilgardia 27: 343-383.

C.B. HUFFAKER, K.P. SHEA & S.G. HERMAN, (1963). *Experimental studies on predation: complex dispersion and levels of food in an acarine predator-prey interaction.* Hilgardia 34: 305-330.

D.T. JOHNSON & B.A. CROFT (1981). *Dispersal of Amblyseius fallacis (Acarina: Phytoseiidae) in an apple ecosystem.* Environmental Entomology 10: 313-319.

J.E. LAING & C.B. HUFFAKER, (1969). *Comparative studies of predation by Phytoseiulus persimilis Athias-Henriot and Metaseiulus occidentalis (Acarina: Phytoseiidae) on populations of Tetranychus urticae Koch (Acarina: Tetranychidae).* Researches on Population Ecology 11: 105-126.

S.A. LEVIN, (1974). *Dispersion and population interactions.* The American Naturalist 108: 207-228.

S.A. LEVIN, *Population dynamic models in heterogeneous environments,* Annual Review of Ecology and Systematics 7: 287-310.

J. MAYNARD SMITH, (1974). *Models in Ecology,* Cambridge Univiversity Press, New York.

G. NACHMAN, (1981). *Temporal and spatial dynamics of an acarine predator-prey system.* Journal of Animal Ecology 50: 435-351.

W.W. MURDOCH & A. OATEN, (1975). *Predation and population stability.* Advances in Ecological Research 9: 2-131.

A. OATEN, (1977). *Transit time and density-dependent predation on a patchily distributed prey.* American Naturalist 111: 1061-1075.

J. REDDINGIUS, (1971). *Gambling for existence: A discussion of some theoretical problems in animal population ecology.* In: Acta Theoretica Climum, added to Acta Biotheoretica 20, Leiden: Brill, 208 pp.

J. REDDINGIUS & P.J. DEN BOER, (1970). *Simulation experiments illustrating stabilization of animal numbers by spreading of risk.* Oecologia 5: 240-248.

C.J.J. RICHTER, (1971). *Some aspects of aerial dispersal in different populations of wolf spiders, with particular reference to Pardosa amentata (Araneae, Lycosidae).* Miscellaneous papers 8: 77-88 (Landbouwhogeschool, Wageningen).

M.W. SABELIS, (1981). *Biological control of two-spotted spider mites using phytoseiid predators. Part 1: Modelling the predator-prey interaction at the individual level.* Agricultural Research Reports 910, pp. 242, Pudoc, Wageningen, The Netherlands.

M.W. SABELIS, (in press.) *Biological control of two-spotted spider mites using phytoseiid predators. Part 2: Modelling the predator-prey interaction at the population level.* Agricultural Research Reports, Pudoc, Wageningen.

M.W. SABELIS, B.P. AFMAN, & A. GROENEVELD, (1984a). *Factors initiating or suppressing aerial dispersal of the predatory mite, Phytoseiulus persimilis.* Proceedings of the Entomological Congress, August 1984, Hamburg.

M.W. SABELIS, B.P. AFMAN & P.J. SLIM, (1984b) *Location of distant spider-mite colonies by Phytoseiulus persimilis A.-H. (Acarina: Phytoseiidae): Localization and extraction of a kairomone.* Proceedings of the sixth International Congress of Acarology, Edinburgh, U.K., September 1982.

M.W. SABELIS, F. ALEBEEK, A. VAN BAL, J. BILSEN, T. VAN HEIJNINGEN, P. VAN KAIZER, G. KRAMER, H. SNELLEN, R. VEENEBOS & J. VOGELEZANG, (1983). *Experimental validation of a simulation model of the interaction between Phytoseiulus persimilis and Tetranychus urticae on cucumber.* OILB-bulletin SROP/WPRS 6(3): 207-229.

M.W. SABELIS & H.E. VAN DE BAAN, (1983). *Location of distant spider-mite colonies by phytoseiid predators: Demonstration of specific kairomones emitted by Tetranychus urticae and Panonychus ulmi.* Entomologia experimentalis et applicata 33: 303-314.

A. TAKAFUJI, (1977). *The effect of the rate of successful dispersal of a phytoseiid mite, Phytoseiulus persimilis Athias-Henriot (Acarina: Phytoseiidae) on the persistence in the interactive system between the predator and its prey.* Researches on Population Ecology 18: 210-222.

A. TAKAFUJI, T. INOUE & K. FUJITA, (1981). *Analysis of an acarine predator-prey system in glasshouse.* First Japan/USA Symposium on Integrated Pest Management, Tsukuba (Japan), September 29-30, 1981: 144-153.

A. TAKAFUJI, Y. TSUDA & T. MIKI, (1983). *System behaviour in predator-prey interaction, with special reference to acarine predator-prey systems.* Researches on Population Ecology, suppl. 3: 75-92.

D.B. TURNER, (1969). *Workbook of atmospheric dispersion estimates.* U.S. Department of Health, Education and Welfare.

J.M. VANDERMEER, (1973). *On the regional stabilization of locally unstable predator-prey relationships.* Journal of Theoretical Biology 41: 161-170.

B.P. ZEIGLER, (1977). *Persistence and patchiness of predator-prey systems induced by discrete event population exchange mechanisms.* Journal of Theoretical Biology 67: 683-713.

Appendix

The area covered by the prey odour plume is determined from the Gaussian plume model (Bossert & Wilson, 1963; Fares et al., 1980). For a ground level source the odour concentration at ground level (x,y plane) can be described by the relationship:

$$C(x,y) = \frac{Q(1 + \alpha)}{2\pi\sigma_y(x)\sigma_z(x)u} \exp(-\tfrac{1}{2}(y / \sigma_y(x))^2)$$

C = odour concentration at ground level (odour units / m^3)

Q = effective average rate of odour emission (odour units / s)

α = reflection coefficient

σ_y = lateral diffusivity of the odour concentration (m)

σ_z = vertical diffusivity of the odour concentration (m)

u = wind speed (m / s)

The odour unit is defined as the characteristic unit of odour produced by one spider-mite female. Hence, Q is equal to the number of spider-mite females in the patch. The odour concentration C can be calculated when taking the following values for the parameters:

$\alpha = 1$ (i.e. no absorption of molecules at ground level)

$u = 0.5$ m / s

$$\left.\begin{aligned} \sigma_y &= 4(x / 100)^{1.4} \\ \sigma_z &= 6.86(x / 100)^{0.42} \end{aligned}\right\} \text{(assuming neutral to slightly stable air; Pasquill-class D and E; Fares et al., 1980)}$$

The Gaussian plume model is a point source model which can be corrected for source area (= patch area) according to Turner (1969):

$$\sigma_y = 4(x / 100)^{1.4} + \varnothing_s / 4.3$$

$\varnothing_s$ = diameter of the (circular) source (m)

For the predatory mite to respond to prey odour the concentration should exceed ca. 10^4 odour units/m^3, as can be derived from the data presented by Sabelis & Van de Baan (1983). Hence, the area in which the odour concentration C is above this threshold concentration can be calculated from the Gaussian plume model. The part of the plume with the effective concentrations can extend very far away from the source. However, the maximum distance of prey patch detection by predatory mites is not known as yet. We provisionally estimate the maximum distance that a predatory mite can cover in its life time to be ca. 50 m. For this reason, the area covered by the odour plume is calculated up to a distance of 50 m from the leeward side of the source.

It should be recognized that the above calculations provide no more than a guess of the effective plume area. On the one hand it may be smaller in reality because the maximum distance of prey patch detection is chosen too large. On the other hand it may be larger because the Gaussian plume model calculates the time average concentration at a certain downwind point in space, whereas the peak concentrations rather than the average is the concentration relevant to eliciting a response of the searching animal. Field experiments with gypsy moths indicated that peak concentrations can be 10 to 25 times greater than predicted by the time-average Gaussian plume model (Aylor et al., 1976; Elkinton et al., 1984).

II. Field populations

The main reason to study laboratory populations is that they may serve as simplified models of field populations. The *i*-machinery is by definition a property of individuals , and as such independent of whether we deal with these individuals in the lab or in the field. What is different, however, is the structure of the environment. In the field this in general is much less homogeneous, both in space and in time. As our *i*-models are necessarily simplified and the quality of a simplification is dependent both on the range and the temporal dynamics of the inputs encountered by the individuals, the transplantation of lab based models to the field is not straightforward. Moreover, often the species we are interested in are difficult to keep in the laboratory other than for short parts of their life cycle. As a result structured models for field populations often have population data as their main empirical basis, even if they are theoretically based on our preconceptions about the adequate state representation of individuals. The resulting necessity of deducing details of the supposedly underlying *i*-model from population data leads inexorably to the so-called inverse problem: the deduction of (some of) the assumptions of a model from its predictions.

In this section on field populations we have collected three papers which all have a statistical component: how does one use observed characteristics of populations to calculate relevant population parameters. Here "relevant" ultimately refers to some goal one has in mind like exploiting the population as a food source or using it as a pollution indicator.

The first paper by Daan on fisheries biology starts with the classic theme of yield optimisation, stressing age and size as relevant *i*-variables. It then goes on to describe the technique of Virtual Population Analysis as a useful means for estimating the parameters recruitment and fisheries mortality from a set of fishery records stratified by age and year. However, man is not the only predator involved and it may be expected that better results can be obtained if we were to take account of the predation mortalities as influenced by the, partially observed, changes in predator numbers. Predation being strongly size related, we clearly need adequate size stratified data on predator preferences. The paper finishes by describing recent efforts to obtain such data.

The second paper by van Straalen gives an insightful review of the possible statistical procedures for estimating demographic, and in particular stage structure, process parameters from field data.

The final paper by Aldenberg considers another theme broached by Daan, that of biological production. Its main thrust is conceptual: what is the meaning of the various terms, and how do they relate to our view of the processes occurring on the individual level.

AGE STRUCTURED MODELS FOR EXPLOITED FISH POPULATIONS

Niels Daan
Netherlands Institute for Fishery Investigations

I. INTRODUCTION.

One important area of both the development of population dynamic theory and its direct application is covered by the analysis of interactions between exploited fish stocks and their fisheries (BEVERTON and HOLT, 1957; RICKER, 1975). Fishing must still be considered as a primitive human activity, a relic from the early cultures of mankind, when man could only act as competitor of other predators in securing its protein demands. Even in the twentieth century no effective control can be exercized over the various steps in the production processes of this food resource and fisheries are still primarily characterized by competition, both among the parties participating in the industry and between the fishery and natural predators.

Any fisherman out at sea has just one goal: to catch as much fish as possible in the short run, with little care about the future. However, a fish can only be caught once and not only is the adagium of the washer applicable here - where the hogs are many the wash is poor -, but also the long term production capacity of the stocks is affected by the short term catch. In any industry it is common use to invest money in proportion to past profits and fisheries are no exception: as long as fishing is profitable new and more powerful ships are added to the fleet. However, in this case a clear danger becomes imminent, because the entry of a larger fishing effort will not only affect the catch rates of the individual vessels, but at some stage even the total catch may decrease as a result of the reduced productivity of the stocks.

This problem clearly rises beyond the vision of the individual fisherman in trying to detect what type of investment scheme is right or wrong for him in the long run and administrators dealing with industrialized fisheries have soon realized that there is a strong need for management actions at the national or, if stocks are exploited by more than one country, at the international level in order to safeguard a long term profitable fishery.

The formulation of a fishery policy (for a comprehensive discussion see ROTHSCHILD, 1983) obviously contains important socio-economic aspects but the crux of the matter is a biological optimization problem: how much effort is required and how can this effort best be used in order to obtain an optimum, sustainable, yield.

This chapter describes,in the light of the main theme of this book, the essentials of the existing theory on exploited fish population dynamics, much of which refers back to BEVERTON and HOLT (1957) : the significance of structured population models in both simulation and estimation procedures. Rather than a presentation of new ideas it is intended as a short review of the present state of art in fisheries biology for scientists, who are not familiar with the extensive activities in the field.

II. THE BASIC MODEL.

Any biologist will be aware of the vast number of factors affecting a natural population and its production. However, in order not to perish in a model of unmanageable complexity by incorporating all factors, that one could possibly think of, it would seem wise to keep things as simple as possible in the beginning and to elaborate only on particular aspects, if this can not be avoided.

The fishery will be more interested in maximizing the yield in weight than the number of fish caught and therefore the various parameters affecting changes in population size are first considered in weight units rather than in numbers (fig 1).

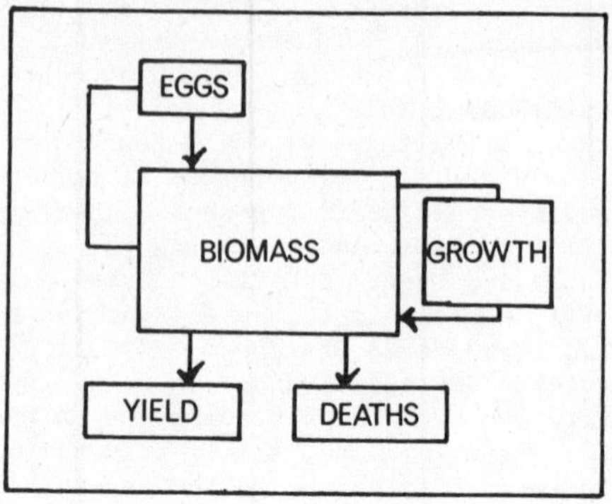

Fig. 1. Schematic representation of the factors affecting biomass of exploited fish populations.

The biomass of the population will be negatively influenced by mortality due to the fishery and due to natural causes and positively by growth in weight of the individuals surviving and by propagation. In mathematical terms the change in biomass (B) from time 0 to time t can be described by the equation due to RUSSELL (1931):

$$B_t = B_0 + G + E - Y - D \tag{1}$$

where G represents the growth, E the egg production, Y the yield and D the weight of all natural deaths between 0 and t.

Growth and egg production will be functions of the biomass and possibly of the amount of food available, but for convenience we will consider the latter quantity to be constant. Natural mortality will be determined mainly by external factors (number of predators etc), but rather than introducing guesswork about intricate relations a constant rate of natural mortality is assumed; i.e. each unit of biomass has a similar and constant chance of dying from natural causes, independent of the size of the biomass. The yield is clearly related to the chance of a unit of biomass being caught and therefore is a function of the mortality rate generated by the fleet (fishing effort).

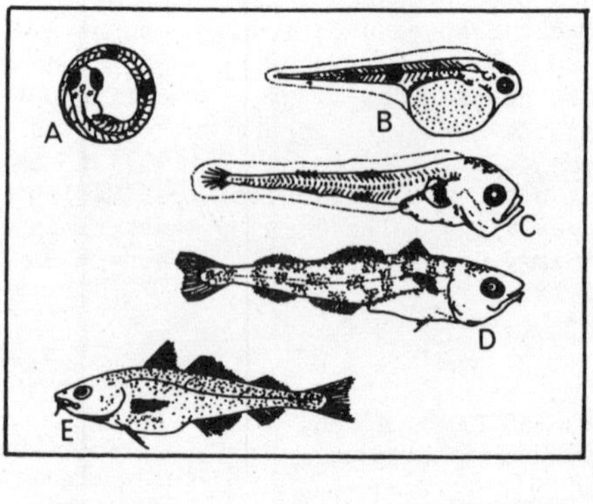

Fig. 2. Development of cod from egg to adult phase:
A - egg (.13 cm)
B - yolk sack larva (.4 cm)
C - feeding larva (.8 cm)
D - metamorphosed juvenile (2 cm)
E - adult (up to 145 cm)

If it were possible to express growth and egg production as a function of biomass then by simple simulation the equilibrium yield could be estimated for any value of the fishing mortality rate. However, in describing G and E as functions of B problems are raised, because fish stock biomasses are not composed of homogeneous elements: there may be juvenile fish, which do not participate in the spawning activities and which are characterized by a different rate of growth than the mature fish. Thus a biomass consisting mainly of juvenile fish will exhibit a very different growth and reproduction potential from an identical biomass consisting mainly of mature individuals and the biomass by itself cannot be used to estimate these quantities without additional information on the structure of the population.

To illustrate this inhomogeneity, fig 2 presents the major features and sizes of various life stages of North Sea cod. Anything between an egg of 1.3 mm diameter to an adult cod of over 1 m must be considered as part of the population biomass and it appears unavoidable to extend the model in the sense that the population must be split in units, which mutually exhibit a larger measure of conformity.

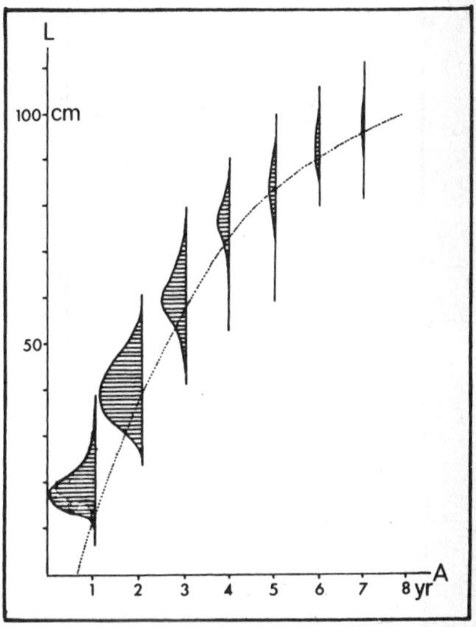

Fig. 3. Estimated mean length (L) at age (A) of North Sea cod with indication of size frequency distributions. Line represents fitted Von Bertalanffy growth curve.

Fig 3 presents the estimated mean length at age of the cod (in practice length is a more efficient measure of size than weight, particularly on board of a rolling vessel; since growth in fish is generally isometric weights can be easily calculated by multiplying length to the power three by a species specific constant) and the estimated growth curves from these, indicating considerable changes in the absolute growth in weight during each year of life. Similarly, egg production is heavily dependent on the weight of an individual (fig 4) and thus on age, whereas length-maturity ogives (fig 5) can be used to estimate the average size and age at first maturity.

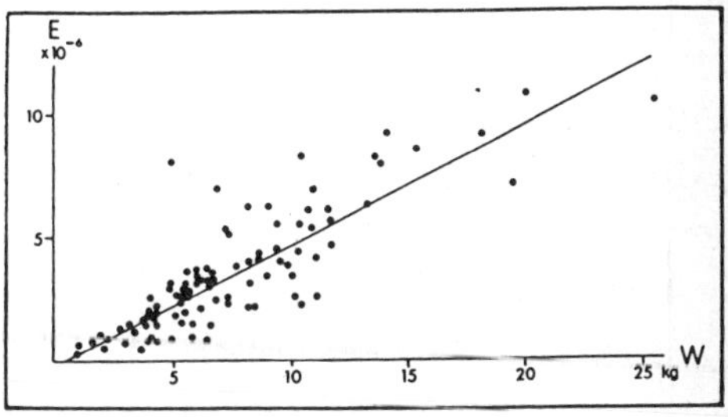

Fig. 4. Fecundity (E: Nr of eggs produced per season) weight (W) relationship of North Sea cod. Points refer to individual observations. Line represents fitted reationship according to equation [4] (Redrawn after OOSTHUIZEN and DAAN, 1974).

On the basis of these patterns it would seem appropriate to structure the population in age groups or cohorts, by which we assume that all relevant parameters within age groups are constant. The extension naturally requires that account has to be taken of the numbers of fish within each age group.

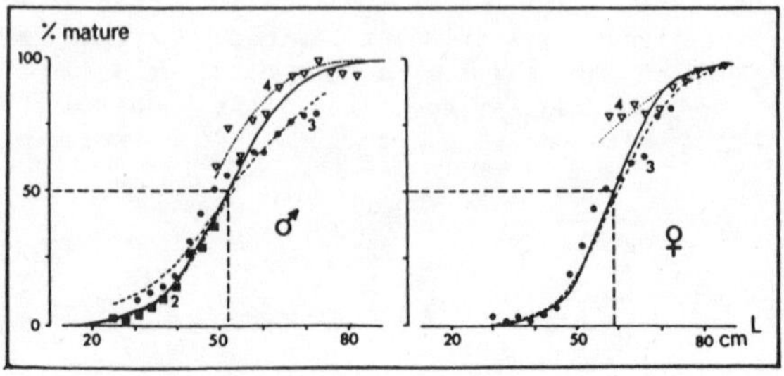

Fig. 5. Maturity length ogives for North Sea cod by sexes and age groups. Squares, dots and triangles represent 2, 3 and 4 year olds respectively. The drawn line applies to all age groups combined and the length at which 50% of the population is mature is indicated by the broken line.

III. THE AGE STRUCTURED MODEL.

Using the index a to indicate age groups, the biomass is by definition:

$$B = \sum_{a=0}^{\lambda} N_a . w_a \qquad [2]$$

where N represents the number in the population, w the average weight of an individual and λ indicates the highest age group considered. In practice λ can be taken as a plus group, because at old age growth tends to become greatly reduced.

For growth of fishes numerous models have been described, but one commonly used in fisheries literature stems from VON BERTALANFFY (1957):

$$w_a = w_\infty .(1 - \text{EXP}(-K.(a - A_0))^3 \qquad [3]$$

where w_∞ is the asymptotic weight reached at old age, K is a growth constant describing the relative speed at which a fish grows to its asymptotic weight and A represents the theoretical age at which w=0 (A_0 can be considered to be a correction factor for the fact that the estimated curve does not necessarily pass through the origin; in fact deviations from zero indicate that growth in the early stage of life may not be described efficiently by the same model parameters applying to later life, as one might expect from animals, which go trough a larval phase and metamorphosis). For an extensive discussion on the Von Bertalanffy growth curve and its derivation the reader is referred to BEVERTON and HOLT (1957).

For the egg production in relation to size also various models have been proposed, but for the present purpose the empirical model found to apply to cod (OOSTHUIZEN and DAAN, 1974) will suffice:

$$\left. \begin{array}{ll} E_a = \varepsilon .(w_a - w_0) & \text{for } a \geq A_m \\ E_a = 0 & \text{for } a < A_m \end{array} \right\} \qquad [4]$$

where ε represents an egg constant, w_0 represents the intersection of the linear relation between E and w and A_m is the age at first maturity.

The age distribution within the population is determined by the total mortality in the preceding time period. If time steps of 1 year are taken corresponding to the structure of the population in age groups and assuming that mortality rates are additive rather then compensatory, we may write:

$$N_{a+1} = N_a \,.\mathrm{EXP}(-(F_a + M)) \qquad [5]$$

where F and M are the coefficients of fishery and natural mortality respectively, both of which are assumed to operate at a constant rate during each year of life and have a dimension of 1/year. Since M has been assumed constant over the total life span, the index a has been dropped.

In respect of the fishing mortality, variation with age is generally allowed, which is particularly useful because the age at first capture (A_ρ) and thus the range of age groups affected by the fishery will depend on mesh sizes in use. Also some age groups may be more vulnerable to the fishery because of concentration effects of the fishery.

One more problem remains to be solved. In structuring the population in age groups, the first group (0-group) still contains all stages between eggs, larvae and juvenile fish. It would seem highly unlikely that the processes governing mortality and growth during this phase are comparable to the older age groups and in fact there rarely is a clear relationship between the number of eggs produced in a particular year and the numbers surviving at one year old, whereas from 1 year old onwards the relative strength of individual year classes (the cohort born in a particular year) is reasonably constant. Apparently only by the end of the first year of life the year class strength has been fixed. Since the assumption of constant natural mortality is obviously not justified for the 0-group, this phase is generally not considered and an age at recruitment to the fishable population is defined (A_R) and only the number of recruits passing this age (the recruitment R) is considered in the model.

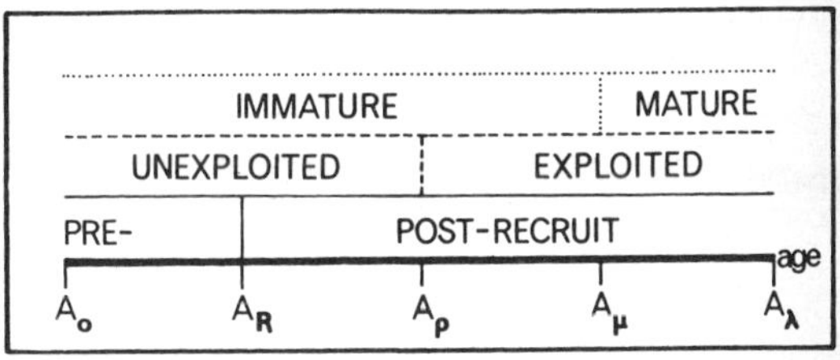

Fig. 6. Schematic representation of the various life phases for exploited fish population models.

In fig 6 a schematic representation of the various life phases is given, which can be superimposed on the age structured population.

The omission of the 0-group results in another problem, because the number of eggs was a straight forward function of the spawning stock biomass but there are considerable uncertainties about the stock/recruitment relationship. RICKER (1954) and BEVERTON and HOLT (1957) have proposed various models which have been recently brought together by SHEPHERD (1982) in one unifying model, but despite a vast literature on this subject (PARRISH ed., 1973) the true nature of stock/recruitment relationship remains generally unresolved. Therefore, the general consensus in fish stock assessment is not to mingle a relatively reliable model with a highly doubtful model and to keep these two aspects separate. In fact this means that as input for the age structured model constant recruitment is used. This is a less severe assumption than one might expect, because for management strategies we are primarily interested in the equilibrium yield for average recruitment as depending on the fishery mortality and if the average yield per recruit and the average recruitment are known the total yield can still be predicted.

IV. THE YIELD PER RECRUIT MODEL.

If the submodels [2] to [5] are incorporated in [1] in connection with the adjustments discussed in the last paragraph, the yield per recruit can be easily simulated for various assumptions about the absolute level of fishing mortality and the exploitation pattern. This latter term may need some explanation.

The fishery mortality coefficient is essentially directly proportional to the fishing effort, because, if the mortality coefficient generated by one unit of effort is represented by F', then the total fishery mortality coefficient F generated by X units equals $X.F'$. However, the fishery mortality coefficient generated by the same unit of effort on different age groups may differ due to age specific differences in catchability. The exploitation pattern is defined as the relative F over age array taking into account these variations in catchability.

In practice it turns out not to be necessary to go through lengthy simulations because solving the differential equations underlying the model leads to a direct solution for the yield per recruit. An extensive description can be found in BEVERTON and HOLT (1957) and the ultimate equation for the yield per recruit, assuming a constant F over age pattern, may suffice here:

$$Y/R = F.w_{\infty}.EXP(-M(A_{\rho} - A_{R})) \sum_{n=0}^{3} \frac{\Omega_{n}.EXP(-nK(A_{\rho} - A_{0}))}{F + M + nK} (1 - EXP(-(F + M + nK)(A_{\lambda} - A_{\rho}))) \quad [6]$$

where $\Omega_n = 1, -3, 3, -1$ for $n = 0, 1, 2, 3$.

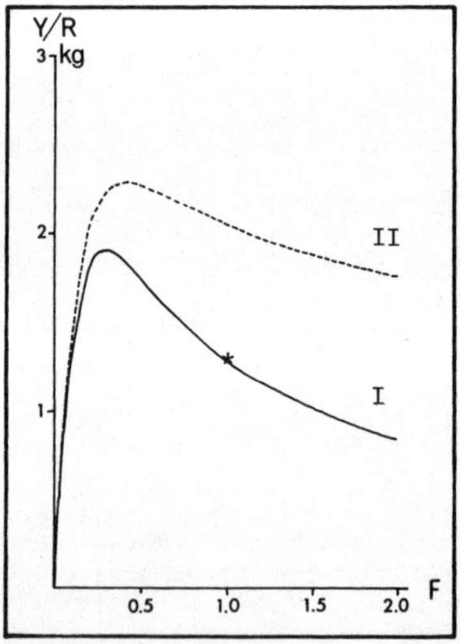

Fig. 7. Yield per recruit curves for North Sea cod for two ages of first capture corresponding to different mesh sizes.
I: A = 2; 80 mm mesh;
II: A = 3; 110 mm mesh;
(The present level of Y/R is indicated on the appropriate curve by an asterix).

In fig 7 results of application of this model to North Sea cod are presented for two different mesh sizes and with an indication of the present level of fishery mortality. It can be easily seen that a considerably higher yield per recruit might be obtained either by reducing fishing mortality (= effort) or by an increase in mesh size.

The dome shaped curves are apparently characterized by a single maximum and these relative Maximum Sustainable Yields per recruit (MSY/R) present convenient reference points for biological advice on fish stock management. Also the term growth overfishing can now be seen to have a clear biological meaning as a level of fishing mortality rate beyond the level required to obtain the MSY/R.

It should be observed that both MSY/R and growth overfishing are related to a particular exploitation pattern: if the mesh size is changed or if the fishery selects different primary fishing grounds, which could also be expected to affect the exploitation pattern, the yield per recruit shifts to a new curve with a different maximum. Thus the same amount of fishing effort may result in considerable growth overfishing or not at all depending on the exploitation pattern. The yield per recruit curves represent equilibrium situations and thus a situation of growth overfishing may safely be maintained without dangers for the stock.

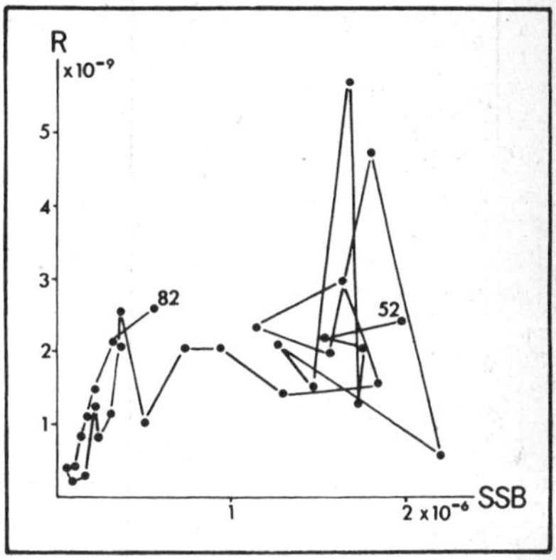

Fig. 8. Plot of stock recruitment observations on North Sea herring. The fishery was closed in 1977, when recruitment had dropped to extremely low values at decimated spawning stock biomass levels (R: number of 1 year old recruits; SSB: spawning stock biomass in tonnes; first and last year of the time series are indicated).

Recruitment overfishing is quite a different matter. This term is reserved for situations, where there is clear evidence that recruitment has become reduced due to a decreasing spawning stock biomass. This type of overfishing may eventually result in a total collapse of the stock, as has been observed in North Sea herring (fig 8), because reduced recruitment leads to a further reduction in spawning stock biomass in a vicious spiral. This situation therefore calls for strong management actions and the closure of the North Sea herring fisheries in the nineteenseventies may serve as an example. In contrast economic considerations may well request that a situation of strong growth overfishing is maintained over prolonged periods of time.

An important aspect of the yield per recruit model is that the various parameters have a clear biological meaning and can be estimated outside the model. Also the underlying assumptions might be tested, because they are clearly identified. If information is available on the amount of error, with which the various parameters are estimated, even confidence limits around the estimated yield per recruit curves might be estimated. However, a real disadvantage is that the directivity of the fishing effort is to a large extent beyond the sphere of influence of management and therefore the exploitation pattern may alter from one year to another uncontrolled. Another problem is that the yield per recruit is very sensitive to the value of M and any uncertainty in this respect is carried over to the management objectives (SPARRE, 1979). Actually, a famous epitaph for the MSY concept has been written by LARKIN (1977) several years ago.

V. VIRTUAL POPULATION ANALYSIS.

In order to produce yield per recruit curves one can rely on direct biological information on growth and natural mortality, but an important next step is of course to estimate the present level of fishing mortality in a stock and the actual exploitation pattern in order to serve as the reference point for management action.

There are quite a number of methods to estimate fishing mortality rates based on catch per unit of effort (CPUE) data, tagging experiments etc., which all have severe limitations. Formerly the CPUE approach has been widely used in stock assessment, but in recent years the method has been left, mainly because it requires some continuity in the type of vessels employed and with the technological development in present days fisheries standardization of effort data has become a major problem in itself. It has been replaced by the Virtual Population Analysis (VPA) and we will concentrate on this method here. The name goes back to FRY (1957), but later amendments by GULLAND (1965) have made the approach particularly suitable for fish stock assessment.

Mortalities are more effectively estimated from numbers than from biomasses and therefore in this context the weight aspects need not be considered. The essential requirement of the method is that reliable estimates of the total annual catch in numbers by age groups are available for all countries exploiting a stock and for prolonged periods of time.

Since the catch in numbers and the number of fish dying within one age group bear the same proportion as F and M, the catch in numbers (C) of age group a can be expressed in terms of the number present at the beginning and at the end of the year, which leads to:

$$C_a = \frac{F_a}{F_a + M} \cdot N_a \, (1 - \mathrm{EXP}(-(F_a + M)) \qquad [7]$$

If C_a and M are known this gives one equation with two unknowns, N_a and F_a. Following this cohort further during its life we get for the next year:

$$C_{a+1} = \frac{F_{a+1}}{F_{a+1} + M} \cdot N_a \cdot \mathrm{EXP}(-(F_a + M)) \cdot (1 - \mathrm{EXP}(-(F_{a+1} + M)) \qquad [8]$$

That is to say, one unknown has been added, F_{a+1}. If data for this same cohort while passing through the fishery are available over n consecutive years, we thus have at our disposal n equations with n+1 unknowns. The trick is now that an arbitrary value is assumed for the fishing mortality rate in the last year for which data are available for this cohort, so that the system can be solved for all F´s and N´s.

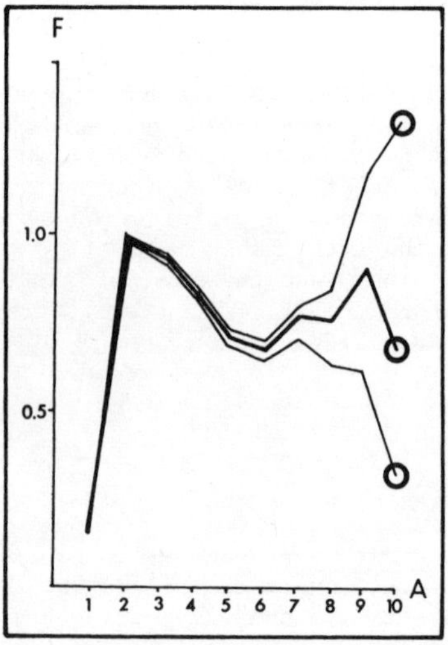

Fig. 9. Estimated fishery mortality rates by age group for one cohort of North Sea cod for three different terminal F-values.

This process can be repeated for different 'terminal F- values', but apparently the effect of the primary assumption is confined to a rather limited number of the older age groups only (fig 9) since the values for the younger age groups turn out to be hardly affected. Moreover, since one might safely assume that the F's on the older age groups should be reasonably constant, it seems likely that the most appropriate terminal F-value is given by the one resulting in a constant F-pattern for these ages.

This process of convergence of estimated F- (and similarly N-) values becomes plausible if one considers that an absolute error is introduced in the value of the terminal F-value and thus the corresponding N-value. As one proceeds to the younger age groups in the calculations, this error becomes relatively smaller, because the accurately known catches in numbers become an increasingly important component of the stock estimate, as is apparent from writing for the recruitment:

$$R = N_1 = N_n + (N_{n-1} - N_n) + \dots + (N_2 - N_1) = N_n + \sum_{a=1}^{n} \frac{F_a + M}{F_a} \cdot C_a \qquad [9]$$

It has been shown, however, that the speed of convergence will be determined by the absolute value of F: if fishing mortality is small then convergence will be slow and in practice the VPA is only applicable to rather heavily exploited fish stocks. Also, again the estimates may be easily biased by uncertainties in the value of M. For a discussion on the various sources of error see BISHOP (1959) and POPE (1972).

Table I presents an example of the results of applying VPA to North Sea cod data. Clear trends in the average level of fishing mortality and also in the exploitation pattern can be observed. It should be noted that for the most recent three or four years the estimated values are to a large extent depending on the terminal F-values and will not be reliable, because these cohorts have not yet passed completely through the fishery.

VI. FURTHER EXTENSION OF THE VPA.

In recent years various amendments and extensions of the VPA have been proposed, among which statistical methods to estimate the total array of F- values by means of least squares approximations (Separable VPA, POPE and SHEPHERD, 1982) and incorporation of migration parameters in a multistock version (ALDENBERG, 1975). Also VPA's on size structured rather than age structured populations have been developed (JONES, 1974; POPE, 1980). In all versions the major problem remains that more parameters have to be estimated than there are equations and some or another simplifying assumption about the exploitation pattern being constant or fishing mortality following a linear trend over time have to be made, which often has to be rejected on the basis of the results obtained. Another problem is that the reliability of the catch data in the various regions of the array may vary due to variations in sampling quality, which reduces the applicability of least squares approximations. The present trend is to use additional CPUE data to solve the shortage of data problem either in connection with the traditional VPA (SHEPHERD and POPE, 1983) or with a statistical version (GUDMUNDSSEN et al, 1982).

Even if ultimately the estimation problems would be solved and if the VPA were perfected, there remain some unsatisfactory underlying assumptions. The single species VPA essentially assumes that the interaction between a fish stock and its fishery can be isolated from the rest of the ecosystem: natural mortality is inherently difficult to estimate and therefore it is assumed constant. If one would wish to relieve this assumption, one immediately runs into the barriers thrown up by the complexity of the ecosystem. Nowadays, anyone seems to be perfectly aware of what is meant by 'ecosystem' in the same way as we know what is meant by the 'hereafter', but it appears almost as difficult to present a useful description. In

A. TOTAL INTERNATIONAL CATCH IN NUMBERS OF NORTH SEA COD BY AGE GROUP AND YEAR (1000 FISH).

YEAR	1972	1973	1974	1975	1976	1977	1978	1979	1980	1981
AGE 1	3608	24408	14877	30305	6238	60267	28358	36314	55522	22804
2	178956	29728	55431	48051	93083	48281	156890	86741	94284	167372
3	47005	52475	10716	18232	17584	23082	14231	39700	29942	24188
4	5460	13567	14869	4220	6608	4307	8469	3596	9702	6807
5	2608	2058	4392	6484	1589	2190	2884	3061	1523	3440
6	3104	1095	920	1732	2439	675	961	660	1037	686
7	1628	1043	417	377	770	926	371	342	384	488
8	603	471	373	149	98	306	364	113	159	124
9	380	69	318	180	49	223	131	127	69	59
10	110	58	75	80	49	20	32	34	46	28

B. ESTIMATED FISHERY MORTALITY COEFFICIENTS FROM VPA (TERMINAL F-VALUES IN BRACKETS).

YEAR	1972	1973	1974	1975	1976	1977	1978	1979	1980	1981
AGE 1	0.05	0.19	0.13	0.16	0.07	0.20	0.15	0.19	0.16	(0.15)
2	0.97	0.77	0.87	0.84	1.03	0.99	1.12	0.93	1.04	(1.00)
3	0.95	0.90	0.71	0.81	0.89	0.80	0.94	1.03	1.03	(0.85)
4	0.68	0.82	0.70	0.69	0.80	0.56	0.79	0.65	0.77	(0.70)
5	0.73	0.60	0.70	0.77	0.61	0.68	0.94	0.75	0.65	(0.70)
6	0.84	0.79	0.59	0.68	0.77	0.57	0.74	0.58	0.63	(0.70)
7	0.79	0.77	0.82	0.51	0.75	0.77	0.73	0.66	0.82	(0.70)
8	1.08	0.56	0.70	0.81	0.24	0.78	0.81	0.52	0.75	(0.70)
9	1.30	0.32	0.97	0.92	0.70	1.35	0.95	0.76	0.70	(0.70)
10	(0.70)	(0.70)	(0.70)	(0.70)	(0.70)	(0.70)	(0.70)	(0.70)	(0.70)	(0.70)

C. ESTIMATED STOCK SIZE FROM VPA ('000 FISH).

YEAR	1972	1973	1974	1975	1976	1977	1978	1979	1980	1981
AGE 1	77810	153957	128542	225015	108858	373995	221679	233441	412145	180180
2	312110	60499	104071	92012	156924	83497	251937	155943	158424	287413
3	83235	96429	22974	35839	32521	45795	25445	67104	50499	45964
4	12079	26340	32237	9242	13090	10971	16914	8173	19673	14748
5	5509	5012	9474	13115	3798	4826	5127	6296	3478	7453
6	5966	2183	2263	3835	4955	1688	1995	1633	2424	1486
7	3233	2118	811	1030	1593	1881	778	776	746	1057
8	989	1195	804	292	505	617	714	306	329	269
9	564	274	557	325	106	325	232	260	149	128
10	238	126	162	173	106	43	69	74	100	61

TABLE I
Estimated total international catches of cod in numbers by age group by year (A) and the estimated fishing mortalities (B) and stock sizes in numbers (C) by applying VPA.

some way an adequate starting point must have been materialized before even more complex mathematical relations are invented.

Undoubtedly, the North Sea ecosystem model developed by ANDERSEN and URSIN (1977) has been a major step forward towards this goal. Although some of the underlying assumptions have frequently been attacked and the applicability of this simulation model in practical stock assessment has been disputed because of the large number of internally estimated parameter values, some of its principles have recently been combined with the traditional VPA in an attempt to create a simultaneous multispecies VPA, in which at least predation among components of the species complex included is taken into account (HELGASON and GISLASON, 1979; POPE, 1979; SPARRE, 1980).

One of the major causes of natural deaths will be predation, because even if starvation or physical deterioration due to infections might occur, fish suffering from these effects would probably become a victim of predators before actually dying from the direct cause. Predators may be manifold, but in these models predation is restricted to what has been so strikingly depicted in Breughel´s allegory ´Large fish eat small fish´, which increasingly appears on the cover of ecological studies related to fisheries (DAAN, 1975; MERCER, 1982; PITCHER and HART, 1982).

In the North Sea all predatory fish species of any numerical significance as well as the vast majority of the fish species serving as possible prey are heavily exploited. This suggests that when the biomass of predatory fish populations is regulated by effective management measures, unwillingly the predation pressure on other, possibly and equally commercially important, species might be changed and consequently their yields. Such predation should of course become apparent in the value of a specific predation mortality rate and therefore [5] is replaced by

$$N_{a+1} = N_a . EXP(-(F_a + P_a + M)) \qquad [10]$$

where P represents the predation mortality rate exercized by all components included in the analysis and M again is a constant representing all remaining sources of natural mortality.

Introducing the notation i for a prey age group and j for a predator age group, the predation coefficient caused by j on i (P_{ij}) equals the fraction of the biomass of i consumed by j:

$$P_{ij} = \frac{\frac{dI_j}{dt} . N_j . \frac{\phi_{ij}}{\phi_j}}{N_i . w_i} \qquad [11]$$

where dI /dt represents the rate of food intake, ϕ_{ij} that part of the biomass of i, that is available as food for j and ϕ_j the total available food for j. The ratio of the latter two factors in fact equals the fraction observed in the food.

Summing the P ´s over j gives the total predation coefficient on i:

$$P_i = \sum_j P_{ij} \qquad [12]$$

That is to say, the predation coefficient is determined by the food intake per predator, their number, the fraction of their food consisting of that particular prey and by the biomass of the prey. If estimates of food intake were available and also of the fraction that each prey consists of the total food of each predator in each year, than the total set of equations could be solved without having to make more assumptions about the terminal F-values than the traditional VPA´s for all species together would require. In this connection it is important to realize that

´big fish eat small fish´, because since the system works gradually backwards in time at each stage the number of fish in the older (=larger) fish has been estimated before.

The rate of food intake of a fish can be relatively easily be expressed in terms of the size of the fish (ANDERSEN and URSIN, 1977):

$$dI/dt = f.h.w^{\alpha} \qquad [13]$$

where f is defined as the feeding level, which may vary according to total food abundance ($0 < f \leq 1$), h is a species specific anabolic constant and α is an almost universally applicable constant, which has some bearing on the physiology of the food intake process.

In contrast the food composition might change with the relative availability of various food resources and continued large scale sampling of stomach contents over prolonged time periods would put very heavy demands on the available research effort. However, food composition has something to do with preference and if the appropriate preference functions were known, food composition might be predicted. A comprehensive preference model has been developed by URSIN (1973). His starting point is that in weighing the contributions of various prey to the total food not the absolute biomasses should be taken, but they should be weighted according to the vulnerability index (v) of prey i to predator j. Thus

$$\phi_{ij} = v_{ij}.N_i .w_i \qquad (0 < v_{ij} \leq 1) \qquad [14]$$

and

$$\phi_i = \sum_i v_{ij}.N_i .w_i \qquad [15]$$

This vulverability will depend on ecological characteristics of the prey, where one could think of the measure of overlap in distribution of predator and prey populations, but moreover on the relative size of prey to predator. In practice it would seem appropriate to distinguish between these two factors:

$$v_{ij} = e_{ij}. s_{ij} \qquad [16]$$

where e is called the ecological vulnerability and s the size suitability index, because the size suitability can be easily approached by direct research. URSIN (1973), on the basis of prey size distributions in relation to predator sizes proposed, that the process of size preference could be imagined to operate in such a way, that the predator evaluates the size of a possible prey in proportion to its own size. Such evaluation appears to be symmetrical on a percentage scale so that a prey ´half as big´ as the one, which could be considered as ´just fine´, has the same chance of being eaten as a prey ´twice that size. Such a process could be written formally as a log-normal distribution function of size:

$$s_{ij} = \mathrm{EXP}(-(\ln (w_j/w_i) - \eta_j)^2 / 2 \sigma_j^2) \qquad [17]$$

where η_j represents the value of $\ln(w_j/w_i)$ which is defined as ´just fine´ and σ_j^2 represents the variance of the distribution.

If we might assume that preference is independent of relative prey abundances, the sampling problem of predator stomachs would be reduced to determining the ecological vulnerability index and size preference function. Such an assumption has often been criticized, but, because we are considering the average preference of total North Sea fish populations with inherently large regional variations in prey spectra, it can be argued that, even if at the individual level preference may be strongly influenced by absolute prey abundance, at the population level such dependence might be largely cancelled out.

VII. THE STOMACH SAMPLING PROJECT 1981.

When the multispecies VPA was developed in the late nineteen seventies as a
manageable assessment tool based on the ideas underlying the North Sea ecosystem
model of ANDERSEN and URSIN (1977), its direct application in practical stock
assessment was hampered by a great lack of essential information. Earlier studies
on food of commercial fishes had not provided adequate information, because, except
in a few cases (e.g., DAAN, 1973), neither predator size nor prey sizes had been
quantified. Moreover, tuning of predation mortalities to actually observed
predation levels in one particular reference year to the estimated stock sizes
requires a much higher sampling intensity than had been reached in those earlier
studies. The model requirements having been clearly identified (ANONYMUS, 1980),
the International Council for the Exploration of the Sea, which channels all fish
stock assessment work in the Northeast Atlantic, set up a sampling scheme for
stomachs of the five main North Sea predator fishes in 1981 in order to obtain a set
of data of the reliability required (ANONYMUS, 1982). To this end during each
quarter of 1981 international sampling surveys were carried out covering the entire
North Sea and a total of over 50 000 fish stomachs have been collected and analysed
(for a review of the project see DAAN, 1983). All predators and all prey were
classified according to predefined size groups, which by means of appropriate
age/size keys could be translated to the age structure of the VPA model. The work
is still in progress and a coherent multispecies VPA has not yet been carried out.
However, a few particularly relevant results may be presented here.

Table II provides the estimated consumption in numbers of some important prey
species by age group by the North Sea cod stock in comparison with estimated stock
sizes of the various prey species estimated by traditional VPA's, assuming a
constant natural mortality of 0.2. The important outcome of this exercise is that
the estimated number 1 year olds consumed for all three species are larger than the
VPA's had estimated to have been present in the sea. This can only be indicative of
a largely underestimated natural mortality in this age group and already at this
stage the exercise has thus presented a type of indirect demonstration that at least
one major assumption of the traditional VPA is evidently incorrect.

Age	C O D N	P	P/N	H A D D O C K N	P	P/N	W H I T I N G N	P	P/N
0	?	3318	?	2278	5486	2.41	1604	2510	1.56
1	131	155	1.18	341	636	1.87	498	681	1.37
2	313	18	.06	1018	212	.21	893	266	.30
3	47	1	.02	255	10	.04	465	56	.12
4+	28	–	–	40	1	.03	197	13	.07

TABLE II. Estimated consumption in numbers by age group of cod, haddock and whiting
by the North Sea cod stock in 1981 (P) in comparison with estimated numbers of these
prey in the sea at the beginning of 1981 (N) from traditional single species
assessment (N and P in 1000 000 fish).

Another interesting point is given by the size preference analysis of North Sea cod.
The prey size distribution in predator stomachs can not be directly interpreted in
terms of preference, because it is a function of both the preference and of the
relative abundance of the various prey size classes in the sea. On the basis of a
model developed by ANDERSEN (1982) to analyse the preference function discussed
earlier from stomach content data, in which the size distribution in the sea is
assumed to reflect an exponential distribution (as the limiting case of a normal
distribution), DEKKER (1983) has analysed the cod data and in fig 10 two examples

are given of the observed prey size distributions in the stomachs and the estimated preference functions for cod eating fish and for cod eating shrimps. The density of the scattered points represent a qualitative indication of the scatter of observations randomized within the various predator and prey size classes. In cod eating fish there is reasonable agreement between size in the stomachs and preferred size over the total range of predators, although clearly fish are more important prey for larger cod. In the case of shrimps the situation is rather different in sofar that the size distribution in the stomachs does hardly follow the preference function. Realizing that shrimps do not grow beyond a particular size, however, this simply indicates shrimp are not suitable as food for large cod. Indeed, although now and again a shrimp may show up in a large cod the scatter of points indicates that there is a rapid decline in the contribution of this prey to the total food, when a cod grows.

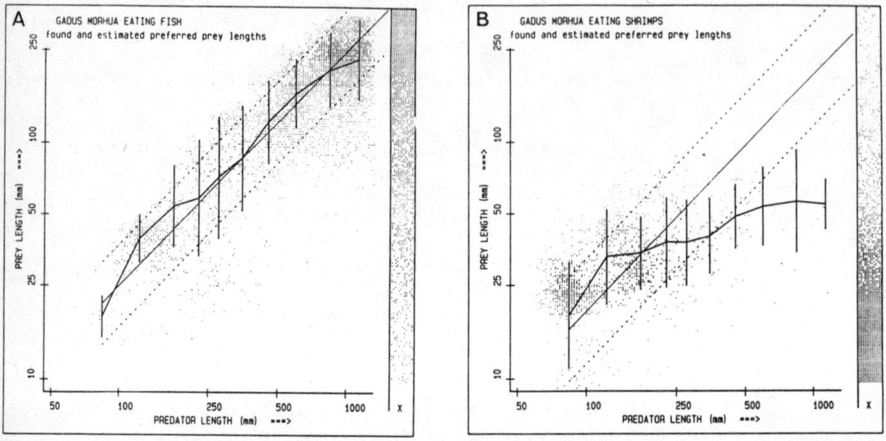

Fig. 10. Observed mean prey length (heavy line) plus/minus standard deviations (vertical lines) in stomachs against predator length and the estimated preference function (thin line) with standard deviations (thin dotted lines). The scattered points represent relative density of observations, randomized within predator and prey categories. The X-column represents the estimated frequency distribution of the available prey in the sea.
A: Fish prey.
B: Shrimps.

If the two prey categories are compared the estimated preference functions appear to be rather similar, although the shrimp preference lies at a slightly lower relative level than the fish. A difficulty with this analysis, however, is that the size distributions have been investigated in terms of length classes rather than weight classes, but the average ´condition coefficients´ to transform lengths to weights may differ considerably between prey categories. Since such estimates are available for a few species only, this correction cannot yet be made and therefore the hypothesis of a constant size suitability among various prey cannot be readily tested.

Ideally such calculations should be made for individual fish species, but as it happens in that case the assumption of a negative exponential abundance function may break down, because within individual prey species complex distribution functions originate in real data due to variations in year class strength and seasonality in the abundance of specific size classes. Apparently, only when many species are grouped, the assumption holds as indicated by research vessel catches given in POPE and KNIGHTS (1982) and the model can be applied effectively.

Nevertheless, considerable confidence has been gained from this study in the applicability of the preference model to real data and the incorporation in the MSVPA appears to have some bearing on what is happening in the ecosystem. For a further discussion on this aspect the reader is referred to the original paper by DEKKER (1983).

VIII. CONCLUSION.

The multicohort composition of commercial fish stocks appears to demand age structured population models and various processes of growth, maturity, egg production and predation can be adequately described as functions of age, although in sampling schemes size in terms of length generally serves as an intermediate parameter and transformation of length classes to ages is performed by collecting appropriate age/length keys. Models applied in fisheries research range from simulation models to find objectives for fish stock management to estimation models directed at describing historical performance of single species and, more recently, multispecies stocks in relation to exploitation by the fisheries. Particularly the latter approach is still fully in development, but as shown by some results, the theoretical formulation of even quite intricate models have stimulated strongly directed research projects considerably. It can only be expected that a strong feed back between theoretical modelling studies and field studies will allow further refinement in the formulation of age structured models of exploited fish populations.

IX. REFERENCES.

ALDENBERG, T., 1975 - Virtual Population Analysis and migration: a theoretical treatment. ICES C.M. 1975/F:32 (mimeo).

ANDERSEN, K.P., 1982 - An interpretation of the stomach contents of fish in relation to prey abundance. Dana, 2 : 1-50.

ANDERSEN, K.P., and E.URSIN, 1977 - A multispecies extension to the Beverton and Holt theory of fishing, with accounts of phosphorous circulation and primary production. Meddr Danm. Fisk.- og Havunders., N.S. 7 : 319-435.

ANONYMUS, 1980 - Report of the ad hoc Working Group on Multispecies Assessment Model Testing. ICES C.M. 1980/G:2 (mimeo).

ANONYMUS, 1982 - Report of the meeting of the coordinators of the North Sea Stomach Sampling Project 1981. ICES C.M. 1982/G:33 (mimeo).

BEVERTON, R.H., and S.J.HOLT, 1957 - On the dynamics of exploited fish populations. Fishery Invest., Lond. (2), 19:1-533.

BISHOP, Y.M.M., 1959 - Errors in estimates of mortality obtained from virtual populations. J.Fish.Res.Board Can., 16 : 73-90.

DAAN, N., 1973 - A quantitative analysis of the food intake of North Sea cod, Gadus morhua. Neth.J.Sea Res., 6 (4): 479-517.

DAAN, N., 1975 - Oecologische gevolgen van de visserij op Noordzee-kabeljauw. (Thesis) E.J.Brill, Leiden.

DAAN, N., 1983 - The ICES Stomach Sampling Project 1981: aims, outline and some results. NAFO SCR Doc. 83/IX/93 (mimeo).

DEKKER, W., 1983 - An application of the Andersen Consumption model in estimating prey size preference of North Sea cod. ICES C.M. 1983/G:63 (mimeo).

FRY, F.E.J., 1957 - Assessment of mortalities by use of the virtual population. Paper p-15 contributed to the Lisbon meeting sponsored by I.C.E.S, F.A.O. and I.C.N.A.F. (mimeo).

GUDMUNDSSEN, G., T.HELGASON and S.A.SCHOPKA, 1982 - Statistical estimation of fishing effort and mortality by gear and season for the Icelandic cod fishery in the period 1972-79. ICES C.M. 1982/G:29 (mimeo).

GULLAND, J.A., 1965 - Estimation of mortality rates. Annex to Arctic fisheries

working group report. ICES C.M. 1965/No 3 (mimeo).
HELGASON, T., 7 H.GISLASON, 1979 - VPA-analysis with species interaction due to predation. ICES C.M. 1979/G:52 (mimeo).
JONES,R., 1974 - Assessing the long-term effects of changes in fishing effort and mesh size from length composition data. ICES C.M. 1974/F:33 (mimeo).
LARKIN, P.A., 1977 - An Epitaph for the Concept of Maximum Sustained Yield. Trans.Amer.Fish.Soc., 107 : 1-11.
MERCER, M.C. [ed], 1982 - Multispecies approaches to fisheries management advice. Can. Spec. Publ. Fish. Aquat. Sci., 59: 169 pp.
OOSTHUIZEN, E., and N.DAAN, 1974 - Egg fecundity and maturity of North Sea cod, Gadus morhua., Neth.J.Sea Res., 8 (4) : 378-397.
PARRISH, B. [ed.], 1973 - Fish Stocks and Recruitment, Proceedings of a Symposium held in Arhus, 7-10 July 1970. Rapp.P.-v.Reun.Cons.int.Explor. Mer, 164 : ?? pp.
PITCHER, T.J., and P.J.B.HART, 1982 - Fisheries Ecology. Croom Helm Ltd. London.
POPE, J.G., 1972 - An investigation of the accuracy of virtual population analysis. ICNAF Research Bulletin, 9 : 65-74.
POPE, J.G., 1979 - A modified Cohort Analysis in which constant natural mortality is replaced by estimates of predation levels. ICES C.M. 1979/H:16 (mimeo).
POPE, J.G., 1980 - Phalanx analysis: an extension of Jones´ length cohort analysis to multispecies chort analysis. ICES C.M. 1980/G:19 (mimeo).
POPE, J.G., and B.J.KNIGHTS, 1982 - Comparison of length distributions of combined catches of all demersal fishes in surveys in the North Sea and at Faroe Bank. p. 116-118. in: M.C.MERCER [ed]. Multispecies approaches to Fisheries management advice. Can. Spec. Publ. Fish. Aquat. Sci., 59.
POPE, J.G., and J.G.SHEPHERD, 1982 - A simple method for the consistent interpretation of catch-at-age data. J. Cons. int. Explor. Mer, 40 : 176-184.
RICKER, W.E., 1954 - Stock and recruitment. J.Fish.Res.Board Can., 11 : 559-623.
RICKER, W.E., 1975 - Computation and interpretation of biological statistics of fish populations. Bull.Fish.Res.Board Can. 191 : 382 pp.
ROTHSCHILD, B.[ed.], 1983 - Global Fisheries. Perspectives for the 1980´s. Springer-verlag New York Inc.
RUSSELL, E.J., 1931 - Some theoretical consideration on the overfishing problem. J.Cons.perm.int.Explor.Mer, 6 : 3-20.
SHEPHERD, J.G., 1982 - A versatile new stock-recruitment relationship for fisheries, and the construction of sustainable yield curves. J. Cons. int. Explor. Mer, 40 (1) : 67-75.
SHEPHERD, J.G., and J.G.POPE, 1983 -
SPARRE, P., 1979 - Some remarks on the application of yield/recruit curves in estimation of maximum sustainable yield. ICES C.M. 1979/G:41 (mimeo).
SPARRE, P., 1980 - A goal function of fisheries (Legion analysis). ICES C.M. 1980/G:40 (mimeo).
URSIN, E., 1973 - On the prey size preference of cod and dab. Meddr Danm. Fisk.- og Havunders., N.S. 7 : 85-98.
VON BERTALANFFY, L., 1957 - Quantitative laws in metabolism and growth. Q.Rev.Biol., 32 : 217-31.

THE "INVERSE PROBLEM" IN DEMOGRAPHIC ANALYSIS OF STAGE-STRUCTURED POPULATIONS

Nico M. van Straalen
Biologisch Laboratorium
Vrije Universiteit
de Boelelaan 1087
Postbus 7161
1007 MC Amsterdam
The Netherlands

ABSTRACT

The reconstruction of demographic processes (such as mortality, birth, and change of stage) from observations on a population's stage-structure is treated here in the context of models where stage is a continuous, monotonic function of age. It is argued that the amount of perturbation or fluctuation exhibited in a population's stage-structure is critical in solving this "inverse problem". If the population is stable, the stage-age relationship cannot be reconstructed from observations on the stage-structure only; this relation must then be known in order to perform a demographic analysis. If it is known, stage-structures can be transformed into age-structures, and the inverse problem can be solved using methods developed for age-structured populations. On the other hand, if the population exhibits fluctuations in stage-structure, these can be used to estimate, in principle, the stage-age relationship. This argument is illustrated by means of an example using length-distributions of Collembola populations with distinct generations. The discrete nature of stage-structure observations (counts) necessitates the use of discrete multivariate statistical analysis. Here, maximum likelihood estimation of parameters is employed using the multiple Poisson distribution as an example.

INTRODUCTION

The object of demographic analysis is the relationship between the state of an age-structured population and the relevant processes passing in that population. Yntema (1977) summarizes this point as follows. If the population's state on time t is denoted by L(t) and the events (processes) occurring in a time interval h, following t, by A(t,t+h), then a model of demographic analysis takes the form

$$A(t,t+h) = \phi\ (L(t)). \tag{1}$$

When the function ϕ is completely specified, the model can be used in two opposite ways:

(i) given the events A as observations, the model can be used to predict the states of the population as a function of time

(ii) given the states L as observations, the model can be used to deduce the events A from them.

The first alternative is usually employed in the study of human populations, and the techniques involved are well developed (Pressat, 1972; Keyfitz, 1968). The second alternative will be the object of the present paper, and will be referred to as the "inverse problem" (a term brought to my attention by the editors of this volume).

In nonhuman populations, there is a very simple reason for concentrating on the second alternative: in many animal and plant populations the population processes are not accessible to direct observation. Whereas in most human populations a registration service keeps statistics of demographic events occurring to the individuals, in animal or plant populations such registration can only be applied in those cases where individuals can be marked, or otherwise identified individually. In many animal populations, particularly so in insects, this is virtually impossible, at least under natural conditions. This situation has led research workers to the study of populations under controlled experimental conditions, allowing direct observation of demographic events. This is exemplified by the approach outlined in Streifer (1974): the demographics of individuals are recorded in the laboratory, taken up in birth and death submodels, which are then combined in an overall model which can be used to predict the state of the population under controlled or natural conditions. This approach essentially follows alternative (i), and it has been succesfull for organisms which can be cultured in the laboratory under conditions closely resembling the natural situation. There are, however, many cases where it is extremely difficult to mimick the field conditions, especially when death from extrinsic causes (predation) is a major mortality factor. One would certainly prefer, wherever possible, to derive the demographic properties of a species from changes in state occurring under natural conditions. Various methods have been developed to this end, in the fields of fisheries, entomology, and mammalogy. The theory of structured populations, as outlined in this volume, makes it possible to tackle this "inverse problem" in a systematic way.

First of all, it is obvious that the "inverse problem" poses more difficulties than the "straight way" (alternative(i)). Usually, there are several different processes (at least both birth and death), which jointly determine the state of the population. From this statement it is clear that there may be situations where the inverse problem cannot be solved, not even in principle, whereas the "straight way" will never pose a problem, at least not in principle. This relates to the "uniqueness problem", as treated, in the context of animal behaviour, by Metz (1981).

The present paper attempts to summarize the present situation with respect to the inverse problem in time-invariant stage-structured populations. At first the problem of identification is discussed in general terms; thereafter the statistical problems of estimation are treated and an example is given.

POPULATIONS WITH "WARPED AGE" STRUCTURE

Consider the case where individuals of a population can be classified according to some continuous variable which is a monotonic function of age. The theory of this "stage-structured" populations is relatively well developed (VanSickle, 1977a; Aldenberg, 1979; Gurney et al., 1983). Because of the monotonic relationship, "stage" is merely "warped age" (called "p-age" by VanSickle, 1977a) and it is not taken as a variable separate from age, as in the original model of Sinko & Streifer (1967).

The use of the term "stage" in this paper differs somewhat from the conventional entomological terminology. In entomology, "stage" usually refers to one of the principle divisions of an arthropod life-cycle, such as the egg, larval and adult stages (see Fink, 1983). In this paper, however, "stage" does not refer to an interval of development, but to a point of development, proceeding with a certain rate. Also, "stage" is used here to characterize the developmental state of any organism (where applicable), not just insects. It is, however, not necessary for the stage-variable to be directly observable over its whole range. The theory of stage-structured populations also applies to situations where only definite levels of the stage-variable can be recognized and where, between these levels, individuals develop without apparent external change (e.g. instars in insects). As long as the transitions from one interval to the next can be seen as marks in an underlying continuous maturation process, the theory can be used to predict the number of animals present at some given time in an interval. In fact, this situation is not very different from the case

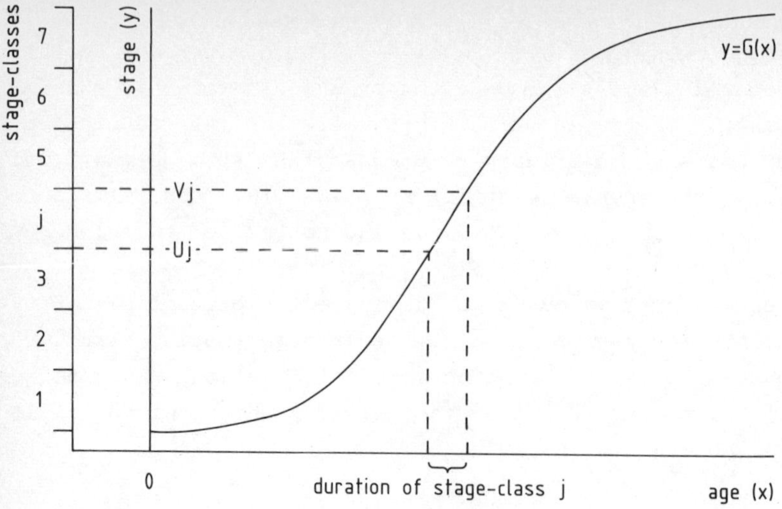

Fig. 1. Example of a monotonic stage-age relationship (G). x = age. y = stage. u_j and v_j = values of the stage-variable which serve as limits for stage-class j. If the stage-variable can be observed continuously, u_j and v_j can take any convenient values on the scale of measurement used by the observer. If the stage-variable can be recognized only at definite levels, u_j and v_j must be given arbitrary values, most logically $u_j = j-1$ and $v_j = j$. In both cases, however, the continuous time theory can be applied.

where the stage-variable can be observed continuously (e.g. body-size in fish). In the latter case also, the observer must create intervals before he can count the number of individuals present in a stage-class. The only difference is that in this situation, depending on the largeness of the sample, the intervals can be made conveniently small, while in the case of discrete external development, the class-limits are set by nature. These arguments are illustrated in Fig. 1. In short, the necessarily discrete measurement of population structure does not necessarily lead to a discrete time model of the Leslie type, as was argued by, e.g. Michod & Anderson (1980).

There are many examples of practically convenient variables of the "warped age" type, such as body-size in fish, instar number in insects, stem-width in trees, or the state of organs subjected to "wear and tear". Southwood (1978) discusses some of these examples with reference to insects. Some stage-variables, however, may be unsuitable in that their relationship with age is too indeterminate, as was shown for stem width in trees by White (1980).

Now the problem of "age-grouping", as discussed, e.g. by Southwood (1978) can be seen to consist of two parts:

(i) to find a suitable (easily measurable) variable satisfying the assumption of monotonicity,

(ii) to establish its precise relationship with age, e.g. by means of an algebraic equation.

In this paper I will consider (ii) as a part of the inverse problem, thereby reducing the problem of age-grouping to (i).

Having assumed that (i) has been solved, we proceed by considering only time-invariant stage-structured populations. This poses serious restrictions, but, to my knowledge, it is the only situation where the inverse problem has been solved at all. Existing methods allowing time-dependence (e.g. Hiby and Mullen, 1980) in fact assume that one of the component processes can be observed directly. Where time-dependence is a consequence of an external factor, such as temperature, this can be eliminated (provided that certain requirements on the similarity of temperature responses are met), by transformation of the time-scale (van Straalen, 1983b). Some cases of density-dependence can be eliminated in a similar fashion (Diekmann et al., 1983).

The model, corresponding to equation (1), which pertains to the time-invariant situation takes the form

$$N_{ij} = \int_{u_j}^{v_j} \frac{1}{g(G^{-1}(y))} F(t_i - G^{-1}(y)) \, S(y) \, dy, \tag{2}$$

where N_{ij} = number of animals in stage-class j at time t_i

$F(t)$ = a function giving the number of animals entering the population (with age zero) per time unit at time t

$G(x)$ = a function giving the stage y of an animal at age x

$g(G^{-1}(y))$ = the derivative of G(x) as a function of y

$S(y)$ = the stage-specific survival function

u_j, v_j = lower and upper limits of stage-class j.

The formulation in terms of N_{ij}, which is obtained by integrating the continuous distribution function over the interval (u_j, v_j), acknowledges the discrete nature of stage-structure observations and allows a precise alignment with statistical methodology discussed later.

It can be seen that (2) corresponds to the general model (1) in the following way: the stage-structure represents the state of the population, and as processes causing changes in state we have: time-dependent recruitment (birth), stage-dependent mortality (survival). and stage-dependent change of stage (growth). So the inverse problem amounts to estimating three independent processes from changes in stage-structure.

The use of a recruitment function F in (2) corresponds to the situation in fisheries and entomology, where a young individual, after being released by its parents, may pass through some intervals of stage before it is considered as a member of the population; for example, eggs or young larvae may be difficult to sample in practice and therefore may be left out of consideration. In mammalian demography, however, it is more convenient to replace the recruitment function by a stage-specific fertility function. This then leads to the stage-specific analog of the familiar Lotka renewal equation, which must be solved before an explicit expression for the stage-structure can be given. Most often, one only considers the asymptotic properties of the model, which lead to a stable stage-structure (N_{ij}^*), given by

$$N_{ij}^* = f_o \, e^{rti} \int_{u_j}^{v_j} \frac{1}{g(G^{-1}(y))} \, e^{-rG^{-1}(y)} \, S(y) dy \qquad (3)$$

where f_0 = recruitment rate at time zero.
The coefficient r (called the intrinsic rate of increase) depends on the stage-specific fertility and survival functions according to the stage-specific analog of the Euler-equation, as given by VanSickle (1977a):

$$1 = \int_{y_o}^{y_m} e^{-rG^{-1}(y)} \, S(y) \, B(y) \, dy \qquad (4)$$

where y_o = stage at recruitment (birth)
y_m = stage at maximal age
$B(y)$ = the stage-specific fertility function giving the number of offspring produced per head *per stage unit* at stage y (this differs from the fertility function used by VanSickle (1977a), in that he considers the offspring produced per *age* (time) unit at stage y).

In this case we have a somewhat more complicated formulation of the inverse problem: the stage-structure (in stable situation) is to be explained in terms of fertility, survival and change of stage. Whereas equation (2) only considers the total birth rate resulting from the joint fertility of all stages, equations (3)-(4) include information on the stage-dependence of fertility. It seems to me that there are many situations where a "lumped fertility" model will suffice; however, as was argued by Birley (1978), it will be essential to include the stage-dependency in those cases where environmental factors affect the fertility of some stages more than others.

The equations (2) - (4) will be the basis for the arguments presented in the following section.

THE PROBLEM OF IDENTIFICATION

In this section the question raised in the introduction will be treated, viz. which general conditions permit a solution of the inverse problem in stage-structured populations. Some pertinent literature is cited to reinforce the arguments.

a. Cohorts

If the population consists of a single cohort, all demographic events can be observed directly. Since a cohort has no ordinary stage-structure, this represents a degenerate case of the theory. A natural population rarely consists of a single cohort for some time, but synthetic cohorts can be created by marking a group of equally aged individuals and studying their subsequent behaviour. In fact, the creation of a cohort means that individuals can be observed directly, and the inverse problem is not solved, but circumvented.

b. Distinct generations

A population can often be divided into several groups, where each group contains individuals of different stages, which can nevertheless be distinguished from their parents and their offspring. This is a frequently occurring case in univoltine and bivoltine insects. A variety of methods have been developed for this situation (see Southwood, 1978 for a review). In van Straalen (1982) some of these methods are placed in the perspective of the theory on structured populations.

Mathematically speaking, under this heading we consider cases where the recruitment function F is such that for some t_1 and some $h>0$ we have $F(t) = 0$ for $t<t_1$ and $t>t_1+h$. Consequently, the integral

$$\int_{t_1}^{t_1+h} F(t)\, dt \qquad (5)$$

can be interpreted as the total number of recruits pertaining to the generation.

This case, like the preceding one, seems to present no a priori problems to the solution of the inverse problem. The point is that one can make use of the time-delay in the recruitment function (see equation (2)) to estimate both the stage-age-relation (the function G), and the survival function S. The amount of precision reached, i.e. the tolerable degree of parameterization, will of course very much depend on the data at hand, especially on the length of the recruitment period, compared to the mean age at death (i.e. the departure from a cohort). An example of this case will be given below.

Birley (1978) demonstrated that, by using the Lotka renewal equation, it is also possible to estimate relevant aspects of the fertility function, provided that recruitment can be observed directly (which is a case we do not consider here).

c. Fluctuating populations

Under this heading we will consider populations where the recruitment function F takes any arbitrary fluctuating form, without marked beginning or end. This type of situation is common in animals with short life-cycles, such as rotifers, cladocerans and copepods. Methods developed for these cases usually assume a stable stage-structure (Edmondson, 1968; Seitz, 1979), or exponential increase or decrease for each stage-class separately (Lynch, 1982). The emphasis in these methods is on estimating birth and mortality rates, with some information on fertility (such as the "egg ratio", or the age distribution of eggs) as given (Johnsen, 1983). Fransz (1980) has devised a method which allows both mortality rates and development rates to be estimated from changes in stage-structure of copepod populations.

It seems to me that by using the theory outlined in this volume more information can be extracted from field data than has been possible by current methods; especially the assumption of steady state where populations are clearly fluctuating seems rather restrictive. The methods developed nevertheless suggest that, in principle, the inverse problem can be solved by utilizing the changes in density of successive stage-classes (cf. Fransz, 1980).

All methods cited above do not include assumptions on the mechanisms that cause populations to fluctuate, such as predator-prey relationships, competition, or seasonally varying abiotic factors. Nisbet & Gurney (1983) have recently developed stage-structure models including some form of competition, thereby enabling an explanation of population fluctuations in terms of an underlying mechanism. The inverse problem however, will be much more complicated in this case.

d. Populations with stable stage-structure

Although a stable stage-structure rarely occurs in natural populations (except for organisms with very short life-cycles, such as bacteria or blood cells), many published methods pertaining to the inverse problem make this assumption, for ease of analysis. Here two statements concerning the stable situation are relevant:

(i) knowing the stage-age-relationship (the growth function G) reduces the inverse problem to the classical analysis of age-structured populations

(ii) the stable situation makes it impossible to derive the stage-age-relation from the stage-structure only.

The first statement is obvious from the starting point that stage is "warped age". In transforming the stage-structure, one must, however, take into account the rate of change of stage with age (VanSickle, 1977a; van Straalen, 1982).

The second statement is obvious from inspection of equations (2) and (3). In equation (3) the function G appears in the exponent under the integral sign. Consequently, the integral contains a product of two independent functions of y, and it will be impossible to estimate both G and S from N^*_{ij}: many combinations of G and S can be made to match the observations. In equation (2) the situation is different; G appears in the argument of the recruitment function F; the time-delay between different stage-classes depends on G and this can be used to estimate G from N_{ij}.

In accordance with statement (ii), published methods directed at the analysis of stable stage-structures, assume that the stage-age relationship is known (e.g. VanSickle, 1977b).

It can be seen from (3) that in a stable population each stage-class will increase exponentially with rate r, so in principle two observations at different times are sufficient to estimate r. The function S can be estimated by considering the interrelationships of demographic functions in stable populations, as discussed by Keyfitz (1968, chapter 7). One of the cases he studied is that the age-structure in combination with an estimate of r determines the life-table or survival function (but not the fertility function). Michod and Anderson (1980) have argued that both the survival function and the rate of increase can be estimated from the age-structure (even from a single observation in time), if the fertility function is given and one point of the survival function.

Conclusion on the problem of identification

The above given arguments suggest that the study of changing stage-structured populations is more profitable, with respect to the inverse problem, than the study of stable populations. Changes in stage-structure allow the stage-age relationship to be estimated from field data, thereby considerably simplifying the problem of "age-grouping". For animals where the stage-age relationship cannot be easily quantified otherwise, this may be a fruitfull approach. For instance, in carabid beetles, the amount of wear of the mandibles is used as a crude indicator of age (Houston, 1981). Such stage-variables will probably increase more or less monotonically with age, but it will be difficult to establish the stage-age relationshop otherwise than from changes in stage-structure (if no other indicator of age is specified). Also, one might want to derive size-age relationships under field conditions, as a check on growth curves established in the laboratory. The approach outlined above aims to maximize the information that can be derived from population data under field conditions.

Von Foerster (1959), in his discussion on the kinetics of cell populations, arrived at a conclusion similar to what has been stated in this paper: only by introducing perturbations in age-structure can the underlying dynamics be revealed. Fortunately, these perturbations often occur naturally, e.g. as a result of an environmentally induced stop on reproduction for part of the year, which automatically leads to fluctuations.

THE PROBLEM OF ESTIMATION

In almost all practical applications of the inverse problem, stage-structure observations are based on samples taken from a natural population. This implies that the observations may be subject to considerable random error, which may drastically limit the amount of information extractable from field data. Furthermore, in many field studies not the population itself is sampled, but the data consist of counts on habitat units, taken at random from the space available to the population. The variation in the observed counts, which is reflected in a sampling distribution to be adopted, relates thus to the spatial pattern of the population.

The estimation problem can be attacked along the following lines:

(i) The functions appearing in the pertinent model (equations 2-4) are specified up to a limited number of unknown constants (parameters).

(ii) Estimates for parameters are obtained by matching the expected counts N_{ij} with the observed counts (henceforth to be denoted by f_{ij}). In this paper the principle of maximum likelihood is used.

(iii) Some statistic is computed that measures the correspondence of expectations and observations. This statistic can either be used to select between alternative models, or to test the goodness of fit.

Concerning parameterization (step (i)) some general comments can be given. Preferably one should choose parameters such that they have a clear biological interpretation. In parameterizing the survival function, for example, several different models have been proposed. Among these, one would prefer models that are based on a physiological interpretation of mortality, like the Gompertz survival function is based on the chance of failure of a system with several vital components (Abernethy, 1979). For many different animals a U-type mortality pattern has been found (Caughley, 1966; Itô, 1980): the chance of dying first decreases when an animal grows up, then increases when getting older. The same argument of interpretation applies to the parameterization of the recruitment and growth (stage-age) functions.

When the principle of maximum likelihood is adopted, one needs to specify a statistical distribution which describes the process of sampling and the variation of the observed counts. The discrete nature of stage-structure observations leads us to consider the use of discrete sampling distributions. One of the most simple discrete distributions is the Poisson distribution. Denote the observations by f_{ij} (the number of animals observed at time t_i in stage-class j), then, under the assumption that the observed counts have independent Poisson distributions with the expected counts N_{ij} as their means, the loglikelihood of the observations is

$$\log L = \sum_{i,j} \left[f_{ij} \ln N_{ij} - N_{ij} - \ln \left\{ f_{ij}! \right\} \right].$$

In this expression, N_{ij} depends on a set of parameters, according to the parameterization step discussed above. Maximum likelihood estimates for these parameters can now be obtained by finding the maximum of logL, regarded as a function of the set of parameters. In many cases no explicit expression for maximum likelihood estimates can be obtained and the problem must be solved numerically. The numerical procedure can be simplified if one of the parameters acts as a multiplicative constant on the expected counts, i.e. if, N_{ij} can be written as $N_{ij} = \alpha n_{ij}$, where α is one of the parameters and n_{ij} depends on the remaining parameters. It can be shown, by elaborating the partial derivative of the loglikelihood with respect to α, that a maximum likelihood estimate for α can be computed directly from maximum likelihood estimates for the other parameters (van Straalen, 1982). The dimension of the numerical maximizing process is thereby reduced by one.

If more sophisticated sampling distributions are adopted, then this accordingly augments the numerical problem of maximizing the loglikelihood but does not alter the general line of the approach given above.

EXAMPLE

Collembola (springtails) are primitive insects occurring in large numbers in forest and grassland soils. They are ametabolous, i.e. they pass through various instars without changing their general appearance, even after they have attained sexual maturity. Their body-length continuously increases during life, and may be treated as a stage-variable in the terminology of this paper.

The population dynamics of several Collembola species in the temperate zone is characterized by the appearance of two generations per year. Using an intensive sampling program, these generations can be separated from each other. At each time, however, a generation consists of a variety of stage-classes, because the period of recruitment to the generation is rather protracted.

A study of two Collembola species, occurring in a coniferous forest in the Netherlands, provided data on changes in length-distributions over one and a half year (van Straalen, 1983a). The counts on two successive generations, of two species, were analysed by applying a stage-structure model. The formulation of the model was based on the following assumptions:

(i) The rate of recruitment to the generation is a Gaussian function of time.

(ii) The per capita age-specific mortality rate is constant.

(iii) The length-age relationship is one of the following functions: (a) linear, (b) power, (c) logistic, (d) Gompertz, (e) Von Bertalanffy.

These assumptions specify the model up to six (or five) parameters: 1) Total number of recruits (area under the Gaussian recruitment function), 2) Time of peak recruitment (mean of the Gaussian function), 3) Width of the recruitment period (standard deviation of recruitment function), 4) Mortality rate, 5) Maximal length (scale parameter of the growth function), 6) Maximum growth rate (shape parameter of the growth function). In the case of growth functions (a) and (b) the number of parameters is five, and 5) and 6) are replaced by a single growth parameter (the growth rate). The length at recruitment, which serves as an initial condition in the growth function, was not taken as a parameter but as a known constant.

The estimation of parameters was based on the principle of maximum likelihood and the multiple Poisson distribution was used as a sampling distribution. Details of the model and of numerical methods can be found in van Straalen (1982, 1983a). Models including various growth curves (see above) were fitted to the data of a generation. As a measure of the goodness-of-fit of a model, the likelihood ratio statistic (X^2), used in the analysis of contingency tables (Bishop et al., 1977), was computed. This statistic was not used to perform a formal test of the model, but to select the best among various alternate models.

Table 1 presents the estimates for one of the generations of the species *Orchesella cincta*. Judged on the value of X^2, it is shown that models using an S-shaped stage-age relationship (logistic, Gompertz) provide a better fit than models using a linear or convex function. This is of course partly a consequence of the greater number of parameters contained in the logistic and Gompertz functions. The model including Von Bertalanffy growth was rejected because it did not lead to a maximum in the likelihood function. The relevance of the reconstruction of the stage-age relationship in this manner may further be illustrated by the following results:

1) For another species (*Tomocerus minor*) the model including the Gompertz growth function provided the best fit. The difference between the species on this point relates to the greater reproductive effort of *Orchesella cincta*.

2) For both species, the model with linear growth provided the best fit in the generations that overwinter. This relates to the deferred reproductive effort in these generations, as a consequence of winter conditions.

It can also be seen from Table 1 that the mean age at death ($\pm$ 4 weeks, being the inverse of the mortality rate) is in the same order of magnitude as the length of the recruitment period ($\pm$ 5 weeks, being the double of the standard deviation of recruitment). This demonstrates that there is considerable overlap of stages; the generation develops rather asynchronously and cannot be analysed as a cohort. If the mean length of the generation is plotted against time, then this does not produce the growth curve. It is essential here that the stage-age relationship is estimated in conjunction with recruitment and mortality.

Table 1. Maximum likelihood estimates for demographic parameters using models with different stage-age relationships. Data for *Orchesella cincta*, spring generation 1979. The time-scale has been transformed so that the rate of moulting is constant and adjusted to 15°C.

		stage-age relation			
parameter	units	linear	power	logistic	Gompertz
mortality rate	1/weeks	0.268	0.343	0.269	0.277
time of maximum recruitment	weeks after 1 Jan 1979	7.65	8.50	7.67	7.78
standard dev. recruitment	weeks	2.64	2.65	2.65	2.64
maximal growth rate	mm/week	0.195	0.565*	0.230	0.219
maximal size	mm	-	-	3.35	3.85
total number of recruits	nrs/0.261 m^2	1965	2506	1977	2031
goodness of fit (X^2)	-	1910.34	2420.76	1528.17	1675.64

* This refers to the slope in a log-log plot of stage against age

This example may serve as an illustration of the statement given in the previous section, namely that for distinct generations, the stage-age relationship can indeed be reconstructed, at least as a parameterized equation, from the stage-structure, even when there is considerable overlap of stages within a generation.

Acknowledgements

I am indebted to Dr. J.A.J. Metz and an anonymous referee for criticism on the first draft of this manuscript, to Mr. I. Berzenczei for drawing the figure, and to Miss D. Hoonhout and Miss S. Richter for typing the manuscript.

REFERENCES

Abernethy, J.D. (1979). The exponential increase in mortality rate with age attributed to the wearing-out of biological components. J. theor. Biol. 80: 333-354

Aldenberg, T. (1979). The calculation of production, reproduction and total growth in the autonomous VanSickle equation. Hydrobiol. Bull. 13: 3-12

Birley, M.H. (1978). The estimation of the net reproductive rate (R_o) of multivoltine pest populations from census data. J. Anim. Ecol. 47: 689-696.

Bishop, Y.M.M., Fienberg, S.E. and Holland, P.W. (1977). Discrete multivariate analysis. MIT Press, Cambridge.

Diekmann, O., Lauwerier, H.A., Aldenberg, T. and Metz, J.A.J. (1983). Growth, fission and the stable size distribution. J. Math. Biology 18: 135-148
Edmondson, W.T. (1968). A graphical model for evaluating the use of the egg ratio for measuring birth and death rates. Oecologia (Berl.) 1: 1-37
Fink, T.J. (1983). A further note on the use of the terms instar, stadium, and stage. Ann. Entomol. Soc. Am. 76: 316-318
Fransz, H.G. (1980). Computation of secondary production of *Calanus finmarchicus* using a multiple regression method. Proc. final ICES/ JONSIS workshop on JONSDAP (1976). ICES C.M. 1980/C: 3: 99-107
Gurney, W.S.C., Nisbet, R. M. and Lawton, L.H. (1983). The systematic formulation of tractable single-species population models incorporating age structure. J. Anim. Ecol. 52: 479-495
Hiby, A.R. and Mullen, A.J. (1980). Simultaneous determination of fluctuating age structure and mortality from field data. Theor. Pop. Biol. 18: 192-203
Houston, W.W.K. (1981). The life cycles and age of *Carabus glabratus* Paykull and *C. problematicus* Herbst (Col.: Carabidae) on moorland in northern England. Ecol. Entomol. 6: 263-271
Johnsen, G. (1983). Egg age distribution, the direct way to cladoceran birth rates. Oecologia (Berl.) 60: 234-236
Keyfitz, N. (1968). Introduction to the mathematics of population. Addison - Wesley Publishing Company, Reading
Lynch, M. (1982). How well does the Edmondson - Paloheimo model approximate instantaneous birth rates? Ecology 63: 12-18
Metz, J.A.J. (1981). Mathematical representations of the dynamics of animal behaviour: an expository survey. The Mathematical Centre, Amsterdam.
Michod, R.E. and Anderson, W.W. (1980). On calculating demographic parameters from age frequency data. Ecology 61: 265-269
Nisbet, R.M. and Gurney, W.S.C. (1983). "Stage structure" models of uniform larval competition. Proc. Res. Symp. - Autumn Course on Mathematical Ecology, Trieste, 1982. (S.A. Levin, ed.), Springer Verlag, Berlin
Pressat, R. (1972). Demographic analysis. Methods, Results, Applications. Aldine - Atherton, Chicago
Sinko, J.W. and Streifer, W. (1967). A new model for age-size structure of a population. Ecology 48: 910-918
Seitz, A. (1979). On the calculation of birth rates and death rates in fluctuating populations with continuous recruitment. Oecologia (Berl.) 41: 343-359
Southwood, T.R.E. (1978). Ecological methods. With particular reference to insect populations. 2nd ed. Chapman and Hall, London
Straalen van, N.M. (1982). Demographic analysis of arthropod populations using a continuous stage-variable. J. Anim. Ecol. 51: 769-783
Straalen van, N.M. (1983a). Recruitment, body-growth and mortality in populations of forest floor Collembola. In: Vergelijkende demografie van springstaarten. Ph.D. Thesis Vrije Universiteit, Amsterdam
Straalen van, N.M. (1983b). Physiological time and time-invariance. J. theor. Biol. 104: 349-357
Streifer, W. (1974). Realistic models in population ecology. In: A. Macfadyen (ed.) Adv. Ecol. Res. 8: 199-266
VanSickle, J. (1977a). Analysis of a distributed - parameter population model based on physiological age. J. theor. Biol. 64: 571-586
VanSickle, J. (1977b). Mortality from size distributions. The application of a conservation law. Oecologia (Berl.) 27: 311-318

Von Foerster, H. (1959). Some remarks on changing populations. In: F. Stohlman (ed.) The kinetics of cellular proliferation. Grune and Stratton, New York

White, J. (1980). Demographic factors in populations of plants. In: O.T. Solbrig (ed.) Demography and evolution of plant populations. Blackwell, Oxford

Yntema, L. (1977). Inleiding tot de demometrie. Van Loghum Slaterus, Deventer (In Dutch)

Structured Population Models and Methods of Calculating Secondary Production

Tom Aldenberg

RIVM (Nat. Inst. Publ. Health & Env. Hyg.)
p.o. box 1, 3720 BA Bilthoven
The Netherlands

ABSTRACT

There are many different methods of estimating secondary production from growth-survivorship data pertaining to cohorts of equally aged animals. This paper, which is theoretical in nature, tries to explore why some of the methods are basically inconsistent and to investigate whether the growth-survivorship diagram contains sufficient information to unambiguously calculate production. It turns out that it does not. The type of reproduction (parent/offspring or fission) does matter and related to this the way the data have been collected and processed. Different models lead to different methods being valid. Methods developed for cohorts can be generalized to methods applicable to number-per-weight distributions at one moment of time. These distributions can be described by the partial differential equations of structured population dynamics. With these analogous methods for distributed populations the rate of production of the total population is obtained at once. The so-called size-frequency method turns out to be a special case. However, the fundamental difficulties encountered with cohorts transfer to the distribution methods too.

Contents

1. Introduction.
2. Defining production on the individual, cohort and population level.
3. The growth-survivorship curve may not suffice to calculate production.
4. Some recent methods of calculating production compared to the first ever applied.
5. A paradox implicit in cohort models employing a continuous growth-curve.
6. Some continuously structured population models.
7. The rate of production of continuously structured populations.
8. Appendix: The rationale behind estimating production.
9. Acknowledgements.
10. References.

1. Introduction

The principal aim of this paper is to extend the methodology of calculating biological secondary productivity, as developed for cohorts (i.e. generations separated in time), into the realm of equations describing continuously structured populations, which form the main theme of these Lecture Notes. (A general introduction into the rationale behind estimating biological production is given in the Appendix.)

Things have turned out not to be that simple, though, as is not unusual, but not because the partial differential equations of continuously structured population dynamics tend to be more difficult than the ordinary differential equations describing a particular cohort: it has turned out that the theory of secondary productivity itself, as developed for cohorts, is a mess. The uninitiated meets a plethora of different 'methods', many still in arithmetic form, most of them sharing the property of doing something with the so-called growth-survivorship curve, a plot of the surviving number of individuals of a cohort against their mean individual weight. When turned into calculus* , some methods can be proved to be equivalent, others may lead to conflicting results. The field of calculating productivity 'needs review and translation into genuine mathematics' (Edmondson 1974), indeed. Gillespie and Benke's (1979) inspiring paper, reviewing four cohort methods, does employ calculus and I owe much to their treatment. Unfortunately, they fail to notice that their model, which has a distinguished history, can show most clearly what paradoxes may be arrived at in this field (Section 5). Their analysis has led to the erroneous statement: 'There is only one simple method of calculating production' (Downing and Rigler 1984).

Previous attempts to study productivity in relation to continuously structured population models have focussed on one equation, a version of the McKendrick (1926) equation† , where linear chronological age is replaced by curvilinear physiological age (e.g. individual weight) with the important proviso that both ages can be invertibly mapped onto each other. VanSickle (1975) is the first author who has done this: he assumes however that no reproduction takes place at the moment of consideration. Independently, I derived identical expressions (Aldenberg 1978, 1979), and showed that reproduction can be fully taken into account. Van Straalen (1985) studies the same model again, but now using it as a framework for comparing several secondary production calculation methods in current use. He is the first author to show that some celebrated methods of calculating production, when evaluated with respect to a *model,* may be in conflict and therefore cannot be consistent. Much of the spirit of this paper is derived from Van Straalen's.

In this paper, I will show that methods which are faulty for one model, may be exactly the right ones for other models. This is not only true for different equations of continuously structured population dynamics, but can be shown for different cohort models as well, which are much easier to study. In fact, the approach followed by Boysen Jensen (1919) can be used to explain most clearly, why it is impossible to decide between production calculating methods without exactly knowing how the data were gathered and processed and, therefore, what model is appropriate (Section 4). Apparently, more than one model may fit the growth-survivorship data.

Transitions in the growth-survivorship diagram may be due to a multitude of different processes which take place on the individual level: growth, respiration, reproduction, migration, being swallowed by a predator, and so on (Section 2). Some subsets of this set of processes allow production to be calculated from the growth-survivorship curve only. Other combinations of processes make the calculation of production from growth-survivorship data infeasible, without further information on, or assumptions about, what processes are going on at each moment, at what rate, and even, on where the biomass goes, a point raised previously by Chapman (1978a, b). Unless one is satisfied with crude estimates, this is disappointing, since the literature seems to suggest that growth-survivorship data are sufficient to calculate production, and this would be true for a diverse range of different types of organisms.

Continuously structured population models explicitly deal with a specific type of reproduction: egg-laying, cell-fission, etc. (Section 6). Assuming a limited number of processes to take place, we can show that the theoretical correctness of a production calculation method depends on the model (Section 7). The model itself will not be unique, either, since the way the data are gathered and handled may matter. In the continuous reproduction case, production calculation 'methods' can be devised that are analogous to some cohort methods. Now integrals are involved with regard to the distribution of the number of individuals over individual weight at a particular moment of time (here called a 'snapshot' distribution), instead of with regard to a cohort which develops through time. Again, these snapshot distributions do not constitute sufficient knowledge to calculate production. Experiments have to be done to estimate the individual growth-rate. The so-called size-frequency method, which has caused considerable debate in the biological literature, can be regarded as a special case of an integral derived in Section 7. This method involves a

* That much of production biology is in its pre-calculus days can be illustrated with Downing and Rigler's (1984) valuable handbook on secondary productivity, summarizing a zillion references, but not containing one single derivative.

† On many occasions indicated as the Von Foerster (1959) equation.

hidden assumption about the growth-rate.

All this seems to boil down to the following observation. We have applied production calculation methods to models belonging to the field of population dynamics, and also derived new expressions. From the biological literature one may have gotten the impression that there is a theory of estimating (secondary) production in addition to the theory of population dynamics. There may only be one left in the (near) future.

2. Defining production on the individual, cohort and population level

Production of, say, a fish population can be defined as the increase or decrease in biomass of the population during some time period plus the biomass of the fish that died during that time period (e.g. Bagenal 1978, p.294). Denoting weight produced during Δt as ΔP, denoting the change in biomass as ΔB and the biomass eliminated through mortality or other causes as ΔE, we have:

$$\Delta P = \Delta B + \Delta E ,$$

or, for infinitesimal time-spans:

$$dP = dB + dE .$$

Here we adopt the notation of LeBlond and Parsons (1977) and Gillespie and Benke (1979) who denote the rate of production as dP / dt. (Petrusewicz and Macfadyen (1970), for instance, denote this rate as P.) So, here P, ΔP and dP all have the same dimension as B, i.e. weight, and $\Delta P / \Delta t$ or dP / dt are expressed in weight per unit of time.

To motivate this definition, one may go back to Thienemann (1931) who pointed to the possibility of a sustained yield ($\Delta E > 0$) from a fish pond, without major change in the 'standing stock', i.e. $\Delta B = 0$. From the economical standpoint, this yield is 'produced' by the farmer, and some areas or ponds may be more 'productive' than others. From the biological standpoint it is better to attribute the act of producing to the fish, which makes also sense in non-exploited situations. Production, then, measures how much one species or trophic level contributes to other species or trophic levels.

Another way of viewing the above relationship is:

$$\frac{dB}{dt} = \frac{dP}{dt} - \frac{dE}{dt}$$

which may be perceived as the rough outlines of a total population model with the single statevariable B.

A population that reproduces discontinuously may consist of one or more cohorts (groups of individuals of equal age). A cohort can be characterized by $N(t)$, the number of individuals surviving as a function of time, and $w(t)$ the (mean) individual weight at t. The curve that results from plotting w at the abscissa and N on the ordinate is called the growth-survivorship curve and visualizes the wax and wane of a cohort. Mathematically, this curve is parameterized by time.

The cohort comes into existence with the appearance of N_0 individuals at some point in time, which have mean individual weight w_0. It ceases to exist when $N(t)$ becomes zero for some t, which may be infinite in some models, but not in practice of course. Several cohorts may be simultaneously present in a population.

The biomass of a cohort is defined as:

$$B(t) = w(t)N(t) ,$$

although on many occasions B and N constitute the raw data and w is calculated from the data as B / N. The differential of the biomass is:

$$dB = d[wN] = wdN + Ndw .$$

Assume the cohorts to be labelled with integer j (compare wines, yearclass numbering, etc.). The characteristics of the population are nothing more than the sum of the characteristics of the cohorts that constitute the population:

$$B = \sum_j B_j , \qquad P = \sum_j P_j ,$$

etc. One also has:

$$dB_j = w_j dN_j + N_j dw_j$$

for each j.

It is tempting to associate $N_j\, dw_j$ with dP_j and $w_j\, dN_j$ with dE_j, as is done by Gillespie and Benke (1979), but it is advantageous to investigate whether the production definition given above will lead to that. We therefore have to go to the individual level.

Several processes may affect the (mean) individual weight of the members of a cohort: consumption of food (C), loss of faeces (F), excretion of dissolved substances (U), respiration (R), which is the utilization of biomass for energetic purposes, and the release of reproductive materials, like eggs, gametes, neonates (G). This is the standard list (cf. Edmondson 1974).

I am going to apply the production definition to each term separately. In order to do so, the concept of elimination of biomass has to be sharpened somewhat. Elimination of biomass is taken to be all those losses of biomass from the population that lead to biomass or tissue becoming available, whether alive or dead, to other species or trophic levels, or to the environment (cf. Petrusewicz and Macfadyen 1970, p.20). The processes F, U, R and G all cost biomass or body-weight, but none of them has to be counted as elimination. For example, respired biomass is not available, as biomass, to other species. It may recycle dissolved carbon, however. Reproductive products are available to other species, but, when alive, still belong to the population. With regard to the definition of production, these processes consequently account for *negative* productivities. (The reader may not appreciate associating reproduction with negative production, at this stage, but hold on.)

The (mean) individual growth-rate is a net balance of the processes:

$$\frac{dw}{dt} = C - F - U - R - G \,.$$

Consumption leads to an increase in biomass, when acting alone; it does not involve elimination either, and therefore counts as a positive contribution to production. The production associated with the individual growth-rate is simply the sum of the productivities of the processes. So it is justified, at this stage, to regard individual growth and individual production as synonymous.

Suppose $\Delta N_j = 0$ for some time period. We then clearly have:

$$\Delta P_j = N_j \, \Delta w_j \,,$$

where Δw_j may be negative, zero or positive. These three possibilities are depicted in Fig. 1. The weight produced corresponds to the shaded areas. For $\Delta w_j < 0$ this area is taken to be negative.

Returning to the deduction that the term $-G$, loss of body-weight through reproduction, has to be counted as negative production, we observe that the question of whether or not to include $-G$ into the differential equation for (mean) individual weight and how to account for it with respect to calculating production is the crucial point that causes the debate and confusion about methods, definitions, etc. Edmondson (1974) and Crisp (1971) consider $-G$ to be a component of the growth-rate. Mann (1969) and Petrusewicz and Macfadyen (1970) leave the term out, while Ricker (1978a) employs two production definitions, one incorporating $-G$ and one leaving it out.

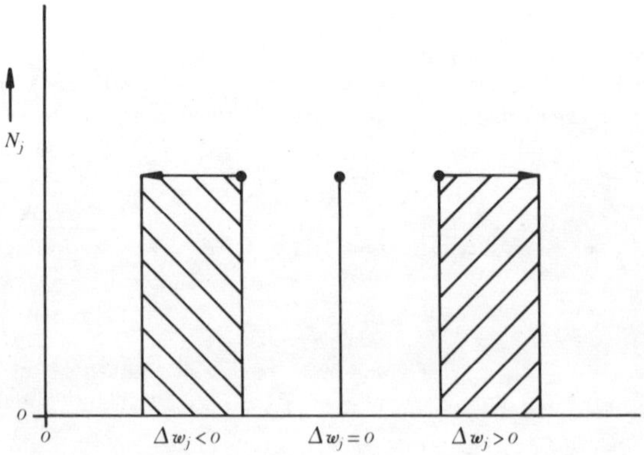

Fig. 1. Three qualitative possibilities for the change in mean individual weight drawn in the growth-survivorship diagram. $\Delta N_j = 0$. Weight produced (ΔP_j) corresponds to the hatched areas. For $\Delta w_j < 0$, ΔP_j is negative.

Transitions in the growth-survivorship diagram that affect N_j can be treated similarly. Assume $\Delta w_j = 0$ and consider the appearance of neonates, when a new cohort comes into existence (Fig. 2). The increase in number from zero until N_{j0} (the initial number of individuals of cohort j) corresponds to an increase in biomass of cohort 'j', while no elimination is involved. So this transition is a positive contribution to production. Since the production of the population is the sum of the production terms of all cohorts, the negative production of the cohorts that are reproducing is counterbalanced by the positive production associated with the appearance of new cohorts. Individual reproduction, then, has no effect on the population production whatsoever, while it is, and should be incorporated in the (mean) net individual growth-rate of cohort members.

To complete the list of transitions in the growth-survivorship diagram, we will consider elimination of biomass due to mortality, predation, etc. Suppose $\Delta w_j = 0$ and $\Delta N_j < 0$. Let number decline because of predation. The decrease of biomass equals the biomass that benefits the predator, which is counted as elimination, so $-\Delta B = \Delta E$ (Fig. 3). The weight produced is zero, then: $\Delta P = 0$. The reader may be familiar with the 'removal summation method', where the eliminated biomass is used as an estimate of production. We observe that the process of elimination of biomass itself has no productivity associated with it, according to the definition.

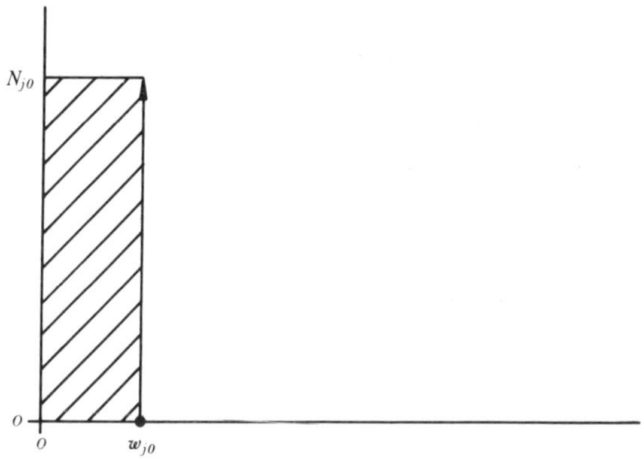

Fig. 2: The appearance of a new cohort depicted in the growth-survivorship diagram.
N_{j0}: initial number of individuals of cohort 'j'
w_{j0}: initial (mean) individual weight of these individuals.

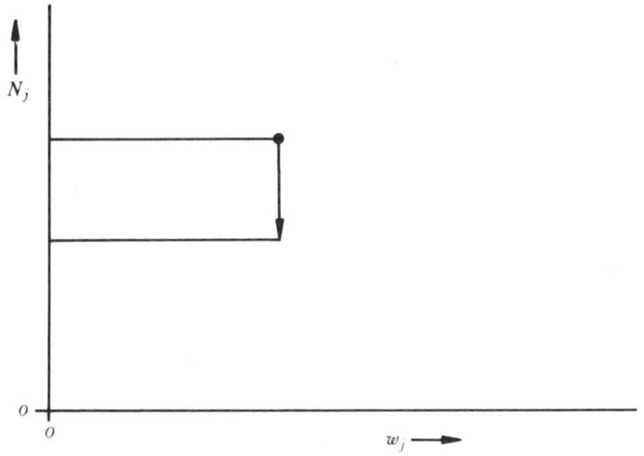

Fig. 3: The loss of individuals, for example due to predators, depicted in the growth-survivorship diagram; productivity of this transition is zero.

In Section 4, we will observe that Boysen Jensen (1919) in his pioneering calculations did add the biomass of a new cohort ('upgrowth') to the growth-increment of the parental cohort to estimate annual production, which exactly conforms to the treatment above. Although the 'Boysen Jensen method' is still alive and is incorporated into the so-called 'summation methods', his adding of the 'upgrowth' is neglected in later treatments (cf. Winberg 1971). The definition of production adopted by the International Biological Programme (see Appendix) illustrates this: 'Production is the total of all growth-increments of all individuals alive at the beginning of the time interval whether or not surviving to the end of the interval, plus the growth-increment of those born during the interval' (Winberg 1971, Edmondson 1974). This definition only stresses growth-increments summed over all existing individuals and therefore underestimates production as defined in this paper. The appearance of neonates is apparently not counted. The IBP definition is said to be due to Thienemann (1931), but his formulation is in much vaguer terms. It is interesting to note that the definition of production employed in this paper may also be derived from Thienemann (1931). We will see later on that the total growth-increment suffices to calculate production when it is calculated from growth-rates corrected for the reproductive material released up to that moment. In fission models growth-increment suffices too. We will come back to that also.

3. The growth-survivorship curve may not suffice to calculate production

The standard processes that determine the increase or decrease of body weight (C,F,U,R,G in Section 2) showed a remarkable uniformity: no elimination is involved, where as before elimination is to be understood as the 'yield' to other species or the environment. (In the literature the term yield is reserved for biomass used by man.) It therefore does not matter whether we apply the production definition to each process separately (provided we know their intensity) or to the net result observable through the change in body weight. When consumption and respiration balance, while the other processes do not take place, production is nil, no matter how large consumption and respiration are. (Incidentally, we observe here that production has little or no relationship to the turnover or flow of matter through the species considered.) Unfortunately, there exist other processes on the individual level that destroy this uniformity or additivity. For example, a predator or parasite may not swallow the whole prey, but eat only part of it. Think of blood- or sap-sucking creatures, plaice feeding on polychaete tails and so on. If the prey doesn't die, we have a loss of body weight which is accompanied by a non-zero elimination. For this process acting alone, productivity is not negative, but zero. If this 'incomplete predation' or 'nibbling' cannot be excluded, one cannot calculate productivity from the net change in body weight. The same is true for exuviation which stands for a collection of processes: loss of skin, carapaces, or feathers during molting, loss of placenta in mammals. These tissues become available to the environment and represent a (potential) elimination quantity.

Lactation can be treated as reproduction when the milk goes the natural way. In the exploited situation, elimination or yield is involved.

In retrospect, one may re-evaluate the standard process of faeces production. We did not attach a positive elimination to it, but the stuff is food for bacteria and fungi. We could better define production, then, as the change in biomass plus the biomass lost because of whatever process is involved. This gives consumption or gross production and amounts to the real turnover of material. Consumption cannot be calculated from the growth-survivorship curve, however.

Migratory movements constitute another problem. A cohort is nearly always studied through samples in a well-defined area. Then we talk about representatives of the cohort with respect to that area only. Immigration, for example of neonates, counts as productivity when we apply the definition again. They sure have been produced somewhere; but it makes our per-square-meter-values biased. Incidentally, immigration is one way of solving a paradox implicit in many cohort models (Section 5).

Another trouble-making process is colony-formation or cell-fusion. For example, algal cells may form colonies by sharing one mucous hull: number declines, mean individual weight increases, while total biomass is conserved and no other species benefits. Productivity is zero, of course. How can we tell the difference from number losses due to predation accompanied by individual growth amounting to a positive productivity, with only the growth-survivorship curve at hand? The same difficulty is encountered with cell-fission. We will study cell-fission in Sections 6 and 7.

To summarize, we observe that identical transitions in the (w,N) plane may be due to different processes, yielding sometimes different productivities. We clearly have to know what species is under consideration, what processes are causing certain transitions in N and w, at what rates, and we may even need to know where the biomass goes. But that is as much as saying that, generally speaking, the growth-survivorship curve, alone, is of limited use for calculating production.

4. Some recent methods of calculating production compared to the first ever applied

Boysen Jensen (1919) calculated the annual production of some marine benthic species fed upon by the commercially fished plaice. His method contains the essential elements fundamental to later developments with regard to calculating production; it only involves arithmetic and may be used to easily clarify some of the points I want to make.

N_0 denotes the number of individuals per square meter recently hatched in spring. N_1 individuals still remain next spring. There may also be N_1' newly hatched individuals, then. The corresponding biomasses (total weight in grams per square meter) are B_0, B_1 and B_1' leading to average weights per individual (grams): w_0, w_1 and w_1' respectively. The two data points, (w_0,N_0) and (w_1,N_1), referring to the parental cohort only, are plotted into the (w,N) plane and are interpolated with a straight line giving a crude growth-survivorship curve (Fig. 4).

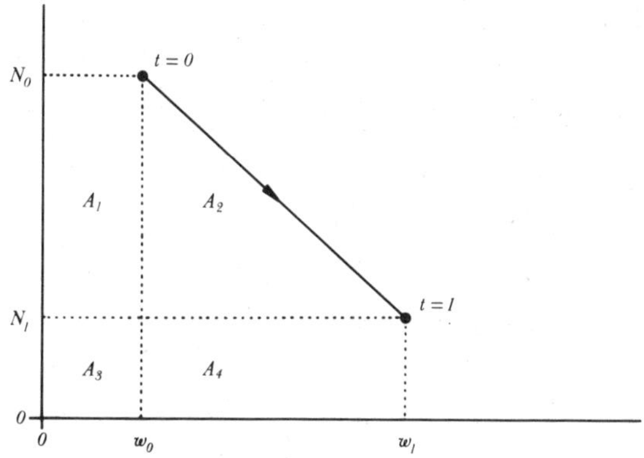

Fig. 4: Boysen Jensen/Blegvad diagram displaying two samples of a cohort at $t = 0$ and $t = 1$, interpolated with a straight line. Projection lines (dotted) cut off areas $A_1, \cdots, A_4$. (see text). Arrow indicates the course of time. $A_1 + A_2$ is the yield (elimination) to other species. $A_2 + A_4$ is the (net) growth-increment of this cohort.

The areas $A_1, \cdots, A_4$ relate to the data as follows:

$$A_1 = w_0(N_0 - N_1)\,, \qquad A_3 = w_0 N_1\,,$$
$$A_2 = (w_1 - w_0)(N_0 - N_1)/2\,, \qquad A_4 = (w_1 - w_0)N_1\,.$$

The biomasses correspond to rectangles:

$$B_0 = w_0 N_0 = A_1 + A_3\,, \qquad B_1 = w_1 N_1 = A_3 + A_4\,.$$

The change in biomass over the year is:

$$\Delta B = B_1 - B_0 = A_4 - A_1\,.$$

To estimate the eliminated biomass, let $N_e = N_0 - N_1$ be the number of individuals eliminated and let $w_e = (w_0 + w_1)/2$ be the mean weight of eliminated individuals. The elimination or yield to other species over the year then is:

$$\Delta E = N_e w_e = (N_0 - N_1)(w_0 + w_1)/2 = A_1 + A_2\,.$$

The annual production, i.e. the weight produced during the year by this parental cohort, according to the definition is:

$$\Delta P = \Delta B + \Delta E = A_4 - A_1 + A_1 + A_2 = A_2 + A_4 = (w_1 - w_0)(N_0 + N_1)/2\,.$$

This result can also be interpreted as the growth- increment of individuals 'whether surviving or not':

$$A_2 = N_e(w_e - w_0)\,, \qquad A_4 = N_1(w_1 - w_0)$$

(Winberg 1971, Winberg et al. 1971). The way Boysen Jensen proceeds fits entirely into our treatment here. The second measurement showed the presence of a new cohort, presumably 'born' to the old one (it could have been migration as well). Its appearance can be depicted as the 'growth-survivorship curve' of Fig. 2. The change in biomass of that cohort is:

$$\Delta B' = w_1' N_1' - 0 = w_1' N_1'\,.$$

Let us call this A_5. The yield to other species is naught:

$$\Delta E' = 0$$

and production becomes:

$$\Delta P' = \Delta B' + \Delta E' = w_1' N_1' .$$

Annual production of the population is the sum of the productivities of the two cohorts, the population being studied from $t = 0$ up to and including $t = 1$:

$$\Delta P_{pop} = \Delta P + \Delta P' = A_2 + A_4 + A_5 .$$

So, Boysen Jensen correctly adds what he calls 'upgrowth' to the growth-increment of the parental generation.

Next, several recent methods can be illustrated in the framework of Boysen Jensen's considerations. These are: the 'removal summation method', the 'increment summation method' and the 'Allen curve integral method'. The above estimation of the elimination, $A_1 + A_2$, is the rudimentary form of the 'removal summation method' (Waters 1977). When successive data sampled from the growth-survivorship curve are available, one may estimate the elimination in the Boysen Jensen way for each time-interval t_i, t_{i+1} as:

$$(N(t_i) - N(t_{i+1}))(w(t_i) + w(t_{i+1}))/2 .$$

The word 'summation' refers to the addition of these estimates over all such time intervals. If N and w are thought to change continuously and N decreases monotonically this summation converges to the integral:

$$-\int_{N_0}^{N(t)} w dN = \int_{N(t)}^{N_0} w dN .$$

If this integral is applied to the Boysen Jensen 'curve' (i.e. straight line), of course the same ΔE results.

Alternatively, the removal summation method can be considered to be a straightforward, not necessarily the best, numerical approximation method of calculating the integral, using straightline segments (Gillespie and Benke 1979).

The area $A_2 + A_4$, the growth-increment of the parental cohort, is a rudimentary form of the 'Allen curve integral method' (Allen 1951). Applied iteratively to successive data it is called the 'increment summation method' (Waters 1977). For each interval $[t_i, t_{i+1}]$ one calculates:

$$(w(t_{i+1}) - w(t_i))(N(t_i) + N(t_{i+1}))/2$$

and these are likewise summed. The Allen integral is the limiting integral:

$$\int_{w_0}^{w(t)} N dw .$$

Both the removal summation method and the increment summation method or the Allen integral are said to give an estimate of the production of a cohort over some period of time (Mann 1969, Waters 1977, Gillespie and Benke 1979). In Section 5 it will be shown by way of an example model that the results need not be identical.

The Allen integral is used by many authors as an estimate of production. Its time-parameterized form is more general:

$$\int_0^t N \frac{dw}{dt} dt$$

in that it is able to cope with non-monotone w, but that is almost never considered. The latter form is identical to Pechen's graphical method (Winberg 1971, p. 102, the first variant). When summed over all cohorts, this growth-increment of all individuals is a direct implementation of Winberg's definition of production, adopted by the IBP (cf. Section 2).

The Boysen Jensen approach can be used to illustrate one of the major points of this paper: the fact that the correctness of the method of calculating production may depend on the data-handling and/or the model assumed to be valid. Suppose that the second sample was taken just *before* the release of reproductive material, instead of just after that moment. Suppose further that reproductive growth had started only when N_1 individuals were left and that no individuals with eggs had been consumed. The Boysen Jensen diagram then would look like the one in Fig. 5. There

$$\tilde{w}_1 = B_1' / N_1$$

is the extra reproductive weight attained and soon to be released. By applying our production definition as before we get:

$$\Delta B = (w_1 + \tilde{w}_1)N_1 - B_0 = B_1 - B_0 + B_1' = A_4 - A_1 + A_5 ,$$

$$\Delta E = A_1 + A_2,$$

as before, so:

$$\Delta P = \Delta B + \Delta E = A_2 + A_4 + A_5 .$$

That is, we obtain the correct result for the production of the population at once. Note that $A_2 + A_4 + A_5$ can be seen as the growth-increment of the parental generation until just before the time of reproductive loss. So growth-increment is an ambiguous term. When growth-increment includes the formation of reproductive material and either the release has not taken place or the release is not counted, growth-increment suffices to calculate production.

This way of calculating growth-increment is equivalent to Greze's graphical method (Winberg 1971, p. 107, the second variant), where one talks about a growth-increment calculated from an average body-weight curve including the weight of all eggs laid up to that moment.

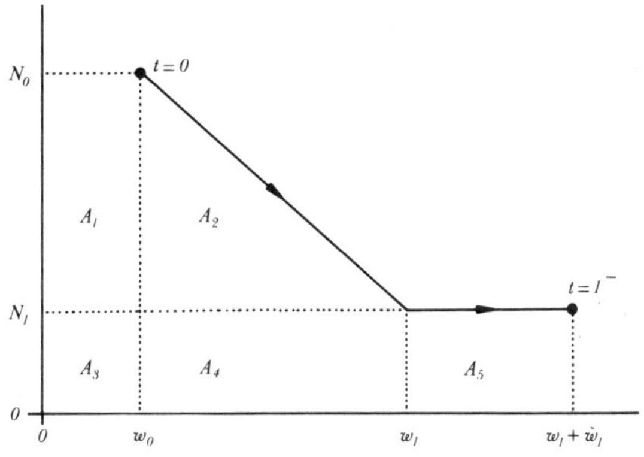

Fig. 5: Growth-increment ($A_2 + A_4 + A_5$) of the parental cohort including the formation of reproductive material (A_5); The second sample is taken just before the release of reproductive material.

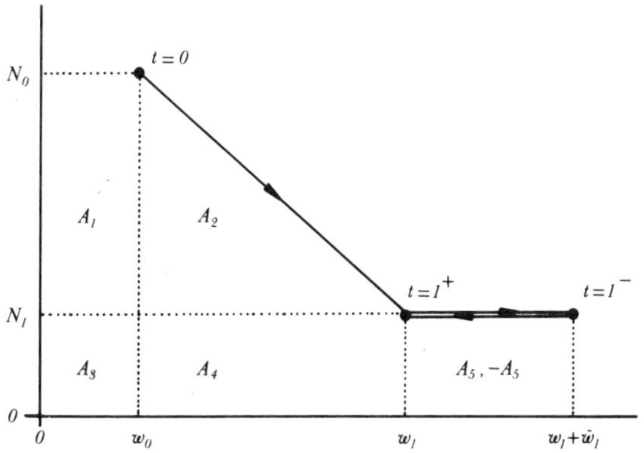

Fig. 6: Growth-increment $A_2 + A_4 + A_5 - A_5$ including formation and release of reproductive material.

Fig. 6 indicates the situation where a sample is taken just before and just after the release of reproductive material. The successive growth-increments become:

$$(A_2 + A_4) + A_5 - A_5 = A_2 + A_4$$

again.
Here $-A_5$ accounts for the negative productivity due to the release of reproductive material, compensated by A_5 due to the appearance of a new cohort. So Boysen Jensen was entirely right in adding his 'upgrowth' afterwards, most probably having missed reproductive growth at the moment of the second sample.

In later treatments (Allen 1951, Winberg 1971, Waters 1977, Downing and Rigler 1984) the viewpoint that the growth-increment of existing individuals is sufficient dominates, which leads to underestimates, unless reproductive growth is accounted for by way of sampling or during the data-handling. Then growth-increment gives the right answer. We observe here that the sampling, the data-handling, the model and ultimately the choice of method to calculate production are intertwined.

5. A paradox implicit in cohort models employing a continuous growth-curve

In a population closed with respect to migration, a continuous growth-curve cannot be consistent with periodic reproduction, sharply peaked in time. The sudden appearance of neonates forming a new cohort must be accompanied by a negative jump in mean individual weight in one or more older cohorts. Most cohort models do employ a continuous growth-curve, however. By way of an example model, I am going to show how one arrives at a paradox when the growth-curve is thought to be continuous, while on the other hand it is assumed that the population is closed.

The following cohort model is the first explicitly mathematical approach in estimating production (Clarke et al. 1946, Ricker 1946, Allen 1949):

$$w(t) = w_0 e^{kt} ,$$

$$N(t) = N_0 e^{-mt} ,$$

with $w(t)$ mean individual weight as a function of time and $N(t)$ surviving number of individuals as a function of time. This model is studied by Neess and Dugdale (1959), LeBlond and Parsons (1977) and Gillespie and Benke (1979). Clearly $w(t)$ is a continuous function of time.

The biomass $B(t) = w(t)N(t)$ is given by:

$$B(t) = B_0 e^{(k-m)t}$$

and satisfies

$$\frac{dB}{dt} = (k-m)B .$$

Assuming the decrease in number to benefit other species, mB is the rate of elimination:

$$\frac{dE}{dt} = mB .$$

Applying the definition, production per unit of time, i.e. the rate of production, becomes:

$$\frac{dP}{dt} = \frac{dB}{dt} + \frac{dE}{dt} = (k-m)B + mB = kB .$$

Weight produced by the cohort up to t follows from integrating the rate of production:

$$P(t) = \int_0^t kBdt = \frac{k}{k-m} \int_0^t \frac{dB}{dt}\, dt = \frac{k}{k-m}(B(t) - B_0)$$

(Allen 1949, Ricker 1946, later re-invented by LeBlond and Parsons 1977, see Ricker 1978). Note that by the fundamental theorem of calculus, weight produced over some time interval $[t_1, t_2]$ would be $P(t_2) - P(t_1)$; one assumes $P(0) = 0$.

In the next interlude, some related methods are touched upon.

The average biomass up to t for this model is:

$$\overline{B}[0,t] = \frac{1}{t} \int_0^t B(\tau)d\tau = \frac{B(t) - B_0}{t(k-m)} .$$

It follows that:

$$P(t) = tk\overline{B}$$

(Clarke et al. 1946). The so-called 'Ricker's instantaneous growth-rate method' is based on this relationship (Chapman 1978, Waters 1977). This method is generally applied to intervals between successive samples of the growth-survivorship curve, just as the summation methods (Section 4). We have

$$P(t_{i+1}) - P(t_i) = (t_{i+1} - t_i)k_i\overline{B}[t_i, t_{i+1}]$$

where $k_i(t_{i+1} - t_i)$ can be estimated from $\log w(t_{i+1}) - \log w(t_i)$. Note that k may differ from interval to interval. $\overline{B}[t_i, t_{i+1}]$ is usually incorrectly estimated as $(B(t_i) + B(t_{i+1}))/2$ (cf. Waters and Crawford 1973). Then productivities for all such intervals are summed. For $t_{i+1}\downarrow t_i$ the method converges to the Allen integral, because the infinitesimal instantaneous growth-rate element is:

$$dP = dt \cdot k(t) \cdot B(t) = \frac{dw}{w(t)}B(t) = N(t)dw(t) = Ndw$$

where k may depend on the infinitesimal time-interval. Allen (1949, 1951) may have derived his later called Allen curve integral in this way. So, if we explicitly calculate the Allen curve integral from the growth-survivorship curve corresponding to this model, we should arrive at the above result. The growth-survivorship curve can be obtained by eliminating t:

$$\frac{N}{N_0} = \left(\frac{w}{w_0}\right)^{-m/k}.$$

The Allen integral up to t is:

$$P(t) = N_0(w_0)^{\nu}(w(t)^{1-\nu} - w_0^{1-\nu})/(1-\nu)$$

with $\nu = m/k$ (Neess and Dugdale 1959), which is easily shown to be equal to $k(B(t) - B_0)/(k-m)$ (Gillespie and Benke 1979). So $P(t)$ would also be found with the instantaneous equivalent of the increment summation method. End of interlude.

Similar calculations give the eliminated biomass up to t:

$$E(t) = \frac{m}{k-m}(B(t) - B_0).$$

It is easy to show that this value would be found by the instantaneous equivalent of the removal summation method.

If k and m are constants, only $k<m$ makes sense. Otherwise, there would be biomass for ever or an ever increasing amount of it. It turns out that the Allen curve integral converges only for $k<m$ (Van Straalen 1984). (Neess and Dugdale (1959) draw only curves for $k = m, 2m, \cdots, 8m$ which violate this condition; their midge larvae fortunately become pupae after finite time.)

With $B(t)\rightarrow 0$ for $t\rightarrow\infty$, total cohort production and elimination become:

$$P(\infty) = kB_0/(m-k) \quad \text{(increment sum)}, \quad E(\infty) = mB_0/(m-k) \quad \text{(removed sum)}.$$

Clearly, we have:

$$P(\infty) + B_0 = E(\infty),$$

i.e.:

$$P(\infty) < E(\infty).$$

So one can eat more from this cohort than it produces. Since nothing prevents us from assuming all cohorts to be identical, assuming the population to be in stationary state, we arrive at the paradox that one can eat more from this population than it produces, every year.

A more precise picture of what happens at the population level can be obtained by actually calculating the population biomass. Let j be the year of birth of a cohort 'j' and label each quantity correspondingly:

$$B_j(t) = w_j(t)N_j(t)$$

(cf. Section 2). For notational purposes it is convenient to shift time in such a way that the birth of a cohort is always at integral t-values. From now on, j denotes the largest integer less than or equal to t, the moment of consideration, and define $f = t - j$ as the fraction of the year gone by: $0 \leq f < 1$. Cohort 'j' is the youngest cohort, then, cohort '$j-1$' is one year old (or $1+f$ year old to be precise). Since the life-span of each cohort is infinite, always an infinity of cohorts is present in the population.

Stationarity is guaranteed if k and m are independent of time and $N_l(l) = N_0$, $w_l(l) = w_0$ for $l = \cdots, j-1, j, j+1, \cdots$.

An r-year old cohort, at t, has biomass:

$$B_{j-r}(t) = B_0 e^{(k-m)(r+f)} = B_0 \eta^{(r+f)}$$

with $\eta = e^{(k-m)}$. Population biomass satisfies:

$$B(t) = \sum_{r=0}^{\infty} B_{j-r}(t) = B_0 \eta^f (1+\eta+\eta^2+...) .$$

Since $k<m$, this geometric series can be summed to give:

$$B(t) = B_0 \eta^f / (1-\eta) .$$

For $f = 0$ the population biomass is $B_0/(1-\eta)$, while for $f \uparrow 1$, just before '$j+1$' will start, population biomass approaches $B_0 \eta / (1-\eta)$, which is B_0 less than at $f = 0$. So population biomass has a discontinuity each time a new cohort starts: it oscillates in a steady saw-toothed manner.

During the year $(0 \leq f < 1)$ population biomass satisfies:

$$\frac{dB}{dt} = (k-m)B$$

with kB the rate of production of the population and mB the rate of elimination from the population, which is larger since $k<m$. At $f \uparrow 1$ the deficit is B_0, which is balanced by an injection of neonates amounting B_0 in weight. So, *during* the year one can eat more from this population than it produces. (Incidentally, there is no time-independent ordinary differential equation that can match the behaviour of $B(t)$.)

I can think of a realistic situation in which this phenomenon may occur. In our treatment (Section 2), the appearance of neonates in the cohort would be counted as production, whether caused by birth or by immigration. During reproductive loss biomass should be conserved. If the neonates were descendants from the population under study, there should not be a jump in the population biomass when a new cohort starts. However, one or more cohorts should show a jump in average individual weight to account for the neonates, but they can't: they obey a continuous growth curve. So the neonates apparently come from somewhere else. The production refers to growth in the area under study, but the neonates are introduced, i.e. produced elsewhere causing elimination to be larger than the production (growth-increment) on the spot itself, which was to be proven.

In Allen's (1971) list of growth models the paradox would not arise, since in each model, except that of exponential growth, initial weight at birth is zero. Initial cohort biomass is zero, too, and increment sum and removed sum are equal even for continuous growth-curves.

This section is another illustration of the fact that it is impossible to decide between methods that differ in the way they deal with the growth-survivorship curve, when we don't know what processes are going on and how the data were gathered.

6. Some continuously structured population models

When reproduction continues over extended periods of time, a continuum of cohorts arises. At any moment of time, a continuum of individual weights or ages may therefore be present in the population, while in the sharply peaked periodical reproduction case only a finite, or at most a countably infinite, number of individual weights or ages is present at any moment of time. These continua naturally lead to descriptions in terms of density functions, i.e. number-per-age distribution functions, or number-per-weight distribution functions, that satisfy certain partial differential equations.

Let $n(t,w)$ be the number of individuals of a population, distributed over weight, at time t. That is:

$$\int_{w_1}^{w_2} n(t,w)dw$$

is the number of individuals between weights w_1 and w_2 at t. Clearly n has dimension: number . weight^{-1}. These individuals are 'passing' w with 'speed':

$$\frac{dw}{dt} = g(w) .$$

For the time being, we only consider growth-rate g to be independent of time and strictly positive, leading to strictly monotone w ('physiological age'). Heuristically speaking, $g(w)n(t,w)$, with dimension: number . time^{-1}, is the number of individuals passing point (t,w) per unit of time, i.e. the 'flux' of individuals at that point. If w_0 is the weight at birth,

$$R(t) = g(w_0)n(t,w_0)$$

is the birth-rate or rate of recruitment.*

Suppose we conceptually follow a particular group of individuals born at the same time through the time-weight continuum, and suppose there is no mortality. Then the number of individuals passing at $(0,w_0)$ still passes at $(t,w(t))$ for any $t>0$, i.e.:

$$g(w(t))n(t,w(t)) = \text{constant.}$$

Differentiation with respect to t yields:

$$\frac{\partial g}{\partial w} gn + g\left[\frac{\partial n}{\partial t} + \frac{\partial n}{\partial w} g\right] = 0 ,$$

or, more compactly, since $g \neq 0$ (which is true whenever $g(w_0) \neq 0$):

$$\frac{\partial n}{\partial t} + \frac{\partial}{\partial w}[gn] = 0 .$$

In case of mortality, we may assume the number of passing individuals to decrease exponentially in time, i.e.:

$$\frac{d}{dt}[gn] = -\mu[gn]$$

with μ a rate-constant, where μ may depend on w.† This leads to:

$$\frac{\partial n}{\partial t} + \frac{\partial}{\partial w}[gn] = -\mu n .$$

This equation is studied by VanSickle (1975, 1977).

We observe that $[gn]$ plays the role N, i.e. number of individuals in a cohort, played in the separate cohorts case: it is conserved when there is no mortality and decreases accordingly when there is. This analogy is important, when studying production in this framework (Section 7). (Unfortunately, the analogy seems limited to the case where w is one-dimensional and g does not depend on time explicitly.)

In the above partial differential equations, reproduction is generally accounted for with the aid of an accompanying boundary condition, for example:

$$R(t) = g(w_0)n(t,w_0) = \int_{w_0}^{\infty} b(w)n(t,w)dw ,$$

where b is a weight-dependent birth-rate coefficient with dimension: time^{-1}. No jumps in individual weight are allowed in this formulation.

In fission models the rate at which one or more parts split off is explicitly accounted for in the partial differential equation itself.

Suppose $b(w,u)$ is the rate at which w-organisms split into pieces u and $w-u$, where u is the smaller part: $u \leqslant w-u$, that is $u \leqslant w/2$. Then, one would write:

$$\frac{\partial n}{\partial t} + \frac{\partial}{\partial w}[gn] + \mu n = -\int_0^{w/2} b(w,u)n(t,w)du \tag{1}$$

$$+ \int_0^{w} b(w+u,u)n(t,w+u)du \tag{2}$$

---

* More usual notation is $B(t)$, but we already used that for cohort biomass.

† Both g and μ may also depend on time. A more rigorous derivation of the equation is needed when g is time dependent (part A, chapter III).

$$+\int_w^{\infty} b(w+u,w)n(t,w+u)du \qquad (3)$$

The three terms on the right-hand side of the equation are to be interpreted respectively as:

(1) the rate of disappearing of w-organisms that split

(2) the 'birth' of w-organisms that happen to be the bigger 'daughter'-organisms

(3) the 'birth' of w-organisms that happen to be the smaller 'daughters'.

A special case of this equation arises, when

$$b(w,u) = \delta(u-\alpha w)\beta(w)$$

where δ is Dirac's delta function, and $\alpha \leqslant \frac{1}{2}$. Now, the u-part is a constant fraction α of the parental weight w. Using the more common properties of the delta function, as well as the rule:

$$\delta(ax-b) = \frac{1}{a}\,\delta(x-\frac{b}{a})$$

(Courant and Hilbert 1962, p.790), the three terms on the right-hand side become:

$$-\beta(w)n(t,w) \qquad (1)$$

$$\int_0^w \delta(u(1-\alpha)-\alpha w)\beta(w+u)n(t,w+u)du = \frac{1}{1-\alpha}\beta(\frac{w}{1-\alpha})n(t,\frac{w}{1-\alpha}) \qquad (2)$$

$$\frac{1}{\alpha}\,\beta(\frac{w}{\alpha})n(t,\frac{w}{\alpha})\,, \qquad (3)$$

analogously. This equation is the fission model due to Sinko and Streifer (1971). Note that for $\alpha = 1/3$ the two source terms, (2) and (3), become:

$$\frac{3}{2}\,\beta(\frac{3}{2}w)n(t,\frac{3}{2}w) + 3\beta(3w)n(t,3w)\,,$$

while for $\alpha = \frac{1}{2}$, i.e. fission into two exactly equal parts, both proportional to the parental weight, leads to one source term:

$$4\beta(2w)n(t,2w)\,.$$

This equation is studied by Diekmann et al. (1983) and elsewhere in these Lecture Notes (Part A, section I.4 and chapter II).

Yet another equation results when the smaller u-part, split off from a w-organism, always has weight w_0, irrespective of the parental weight. A true egg. We have:

$$b(w,u) = \delta(u-w_0)\beta(w)\,.$$

We assume $w_0 < w/2$, or, more precisely, $\beta = 0$ for $w \leqslant 2w_0$. The three integrals in the partial differential equation reduce to:

$$-\beta(w)n(t,w) \qquad (1)$$

$$\int_0^w \delta(u-w_0)\beta(w+u)n(t,w+u)du = \beta(w+w_0)n(t,w+w_0) \qquad (2)$$

$$\delta(w-w_0)\int_w^{\infty}\beta(w+u)n(t,w+u)du \qquad (3)$$

The latter term is a point-source term in the partial differential equation. In the autonomous case, $g(w_0)>0$ implies that w_0 is the minimum individual weight present, which means that the source term can be omitted from the partial differential equation and becomes an appropriate boundary condition (see also Part A, chapter III). The resulting partial differential equation now becomes:

$$\begin{cases} \dfrac{\partial n}{\partial t} + \dfrac{\partial}{\partial w}[gn] + \mu n = -\beta(w)n(t,w) + \beta(w+w_0)n(t,w+w_0) \\ R(t) = g(w_0)n(t,w_0) = \displaystyle\int_{2w_0}^{w_\infty} \beta(w)n(t,w)dw \end{cases}$$

7. The rate of production of continuously structured populations

In the continuous reproduction case, studying 'snapshots' of the number-per-weight distribution of the whole population is as natural as it is to study cohorts when reproduction is discontinuous. There is no need to try to study cohorts when reproduction is continuous. Moreover, in non-stationary situations, quantities calculated from cohorts are difficult to interpret, while quantities calculated from a 'cross-section' of the population are valid for that moment and informative about the state of the population.

Given the knowledge of what processes are going on, expressions analogous to those dealing with the growth-survivorship curve, can be derived for the rate of production of the population based on the number-over-weight distribution and the individual growth-rate. The latter may be difficult to obtain, however, and one usually does experiments in the lab to estimate this rate.

Suppose the autonomous equation studied by VanSickle (1977) holds:

$$\frac{\partial n}{\partial t} + \frac{\partial}{\partial w}[gn] = -\mu n$$

Multiplying by w and integrating with respect to w, we find:

$$-\int_{w_0}^{w_\infty} w\frac{\partial}{\partial w}[gn]dw = \int_{w_0}^{w_\infty} w\frac{\partial n}{\partial t}dw + \int_{w_0}^{w_\infty} w\mu n dw = \frac{dB}{dt} + \frac{dE}{dt},$$

where $B(t) = \int_{w_0}^{w_\infty} wn(t,w)dw$ is the population biomass. So the term on the left-hand side of the equation equals the rate of change of population biomass plus the rate of biomass elimination through mortality. So:

$$\frac{dP}{dt} = -\int_{w_0}^{w_\infty} w\frac{\partial}{\partial w}[gn]dw = -\int_{w_0}^{w_\infty} wd[g(w)n(t,w)]$$

is the rate of production of the population. (I prefer dP/dt, with regard to the foregoing sections, to my $P(t)$ published previously (Aldenberg 1978, 1979).) Remembering the analogy between $[gn]$ and N, surviving number in a cohort, we recognize this expression as the continuous analogue of the removal summation method (Van Straalen 1984). But it differs from the cohort removed sum in that it equals an instantaneous rate, expressed in weight per unit of time, instead of weight removed over some period of time (dimension weight). Moreover, the formula refers to a snapshot of the number over weight distribution and not to a cohort developing through time.

The latter formula relates to a method which has caused considerable debate in the literature: the 'size-frequency method' due to Hynes and Coleman (1968), as corrected by Hamilton (1969). In the size-frequency method one obtains a snapshot of the number-per-*length* distribution of a population, say $\tilde{n}(t,l)$. Let $h = \frac{dl}{dt}$ be the speed with which individuals travel through length-classes. The rate of production becomes:

$$-\int_{l_0}^{l_\infty} w(l)d[h\tilde{n}(t,l)]$$

Note that $h\tilde{n}$ has dimension: number.time^{-1}. Essential to the size-frequency method, as adapted by Hamilton, is that h is supposed to be a constant, that is to say that an equal amount of time is spent in each length-class. Then h is independent of l and one obtains

$$\frac{dP}{dt} = -h\int_{l_0}^{l_\infty} w(l)d[\tilde{n}(t,l)]$$

which, with $w(l) = l^3$, is exactly the continuous equivalent of the method proposed by Hamilton (1969). With h a constant, individual weight is supposed to increase with the cube of time. Thus, the size-frequency method assumes a particular growth-rate:

$$g(w) = 3hw^{2/3}$$

So while normally the growth-rate has to be known to use the integral expression for the rate of production, the size-frequency method involves a hidden assumption about that growth-rate.

One may further integrate the rate of production formula for the population by parts (VanSickle 1975):

$$\frac{dP}{dt} = -\int_{w_0}^{w_\infty} wd[gn] = w_0 g(w_0)n(t,w_0) + \int_{w_0}^{w_\infty} gndw$$

(one assumes $w_\infty g(w_\infty)n(t,w_\infty) = 0$)

$$= w_0 R(t) + \int_{w_0}^{w_\infty} g(w)n(t,w)dw \ ,$$

i.e. reproduction expressed in weight plus a term analogous to the Allen curve integral, a similar analogy as above. As pointed out by Van Straalen (1984), this 'Allen integral' underestimates the rate of production.

On the other hand, think of g being the net result of growth including the formation of reproductive material minus the rate of release of reproductive material:

$$\frac{dw}{dt} = g(w) = g_1(w) - g_2(w) \ ,$$

where $g_2(w) = w_0 b(w)$, and b is a birth-rate coefficient, dimension: time^{-1}. Assume the usual renewal boundary condition to hold:

$$R(t) = \int_{w_0}^{w_\infty} b(w)n(t,w)dw \ .$$

When substituted in the latter expression for $\frac{dP}{dt}$, we have:

$$\frac{dP}{dt} = \int_{w_0}^{w_\infty} w_0 bndw + \int_{w_0}^{w_\infty} gndw = \int_{w_0}^{w_\infty} [g_2 + g]ndw = \int_{w_0}^{w_\infty} g_1 ndw$$

(Aldenberg 1978). This is the total growth-increment of the population, neglecting the individual release of reproductive material, and known in the literature as Greze's graphical method for calculating production (the second variant) (Winberg 1971, p.107). In this case, the individual growth-increment is calculated from an averaged body-weight curve including the weight of all eggs laid up to that moment, i.e.:

$$\int_0^t (g + g_2)dt$$

which is differentiated (graphically) to yield: $g + g_2$. The discussion parallells the one in Section 3. We observe again that the validity of the method depends on the way the data are gathered.

Similar manipulations may be applied to other equations of continuously structured population dynamics. Let's take the egg-laying equation, in which egg-laying is considered as a true fission process:

$$\frac{\partial n}{\partial t} + \frac{\partial}{\partial w}[gn] + \mu n = -\beta n + \beta(w+w_0)n(t,w+w_0)$$

The rate of population production now is:

$$\frac{dP}{dt} = -\int_{w_0}^{w_\infty} w\frac{\partial}{\partial w}[gn]dw - \int_{w_0}^{w_\infty} w\beta ndw + \int_{w_0}^{w_\infty} w\beta(w+w_0)n(t,w+w_0)dw \ .$$

The second term on the right-hand side can be written as:

$$-\int_{2w_0}^{w_\infty} w\beta ndw \ ,$$

since $\beta = 0$ for $w \leqslant 2w_0$. The third term equals:

$$\int_{2w_0}^{w_\infty} (w-w_0)\beta(w)n(t,w)dw \ .$$

So, second and third term added yield:

$$-w_0 \int_{2w_0}^{w_\infty} \beta ndw = -w_0 R(t) \ .$$

The first term on the right-hand side becomes, with integration by parts:

$$w_0 g(w_0)n(t,w_0) + \int_{w_0}^{w_\infty} gndw \ ,$$

so the total rate of production becomes:

$$\frac{dP}{dt} = \int_{w_0}^{w_\infty} gndw \ .$$

Conclusion: in a fission model the Allen integral analogue or total growth-increment analogue suffices.

An alternative formulation with respect to this model can be:

$$\frac{dP}{dt} = w_0R(t) + \int_{w_0}^{w_\infty} (g-w_0\beta)ndw \ ,$$

so with an appropriately adjusted g one could model this population in the VanSickle way, which, once more, indicates that one should attune the measurements, the model and the method to calculate production to each other.

Appendix: The rationale behind estimating production

The objective of this appendix is to give some background information, especially for the non-biologist reader, on what the estimation of biological productivity is all about. It is strongly biased towards aquatic ecology.

The concept of biological production or productivity relates to the flow of biomass and energy through various ecological entities. Many productivity studies derive their motivation from Lindeman, who, in the footsteps of G.E. Hutchinson, developed the 'trophic dynamic viewpoint of ecology' (cf. Cook 1977). In this viewpoint, the set of species of an ecosystem is partitioned into subsets called trophic levels, which are more or less homogeneous with respect to feeding habits. The first trophic level is formed by the primary producers, algae and weeds in the aquatic environment, which convert sunlight and inorganic substances into biomass. Secondary producers or herbivores constitute the second trophic level and feed on primary produced material or on detritus, i.e. dead and decaying organic matter. Primary producers feed themselves (autotrophs), secondary producers feed on others and therefore belong to the much wider class of heterotrophs. Predators or carnivores, heterotrophic as well, constitute the third trophic level. They feed on secondary producers and on each other. Decomposers, i.e. bacteria and fungi, are also heterotrophic organisms and convert detritus back into inorganic substances which can be re-used. They are not given a cardinal trophic number usually - one could put them in level zero or infinity -, but their role is of immense importance for the functioning of an ecosystem.

Excluding decomposers, one may, in first approximation, say that trophic level '$j+1$' feeds on trophic level 'j' , $j=1, 2, \cdots$. Such a relationship is also called a food-chain. (More commonly known are food-chains of individual species, which may contain more than three links.) Lindeman and Hutchinson called the rate of contribution of energy from a lower level the productivity of the level under consideration. Given, first of all, that level '1' is the only one to bring in energy, second, that energy is conserved, third, that energy is dissipated at each level, it follows that productivity must decrease going up the trophic 'ladder'. Elton (1927) observed that animals at the base of a food-chain are relatively abundant, while those at the end are relatively few in number with a gradual transition between the extremes. So this 'Eltonian Pyramid of Numbers' seemed to hold for productivity as well.

Lindeman's main interest was in lake succession. The lake is not considered to be a casual aggregate of water, sediment, fish, plankton, etc., but is perceived as an organismic whole, coming of age through various stages - indeed attaining 'senescence', i.e. a stage dominated by littoral and terrestrial vegetation, after which a completely terrestrial succession follows. This way of thinking typically illustrates the ecological viewpoint.

The Hutchinson/Lindeman model of the trophic-dynamic concept, in their notation, reads (Lindeman 1942):

$$\frac{d\Lambda_j}{dt} = \lambda_j + \lambda_j' \ ,$$

where Λ_j is the energy content of trophic level 'j'; λ_j is the productivity of level 'j' and λ_j' denotes the *sum* of two rates: the rate of energy dissipated at level 'j' plus the rate of energy content passed on to the next trophic level, '$j+1$' ; λ_j' is negative.

Then Lindeman estimates λ's (in calories . $cm^{-2}.year^{-1}$) for a few lakes, as well as the 'progressive efficiencies', $100\lambda_j / \lambda_{j-1}$, which are also due to Hutchinson.

Lindeman's discussion of the role played by the productivities and the efficiencies - the latter seemed to increase going up the trophic ladder - in the succession of an ecosystem was mere speculation. But the idea of possibly being able to understand the functioning of an ecosystem in this way, was so inspiring that a vast amount of subsequent, mostly experimental, research, during the fifties, sixties and even up to now, may be interpreted as a further filling-in of this scheme. These investigations comprise the measurement of both λ's and Λ's, e.g. caloric content of various organisms and tissues, rates of production and respiration in terms of calories, amount of oxygen produced or consumed, of radio-active carbon taken up, etc. (cf. Wiegert 1976).

Useful, as this scheme may be, having set the stage for decades, the above differential equations do not form a complete set in the dynamical sense: the λ's weren't expressed in the Λ's, as the state space approach requires.

From 1920 onwards, population dynamics based on the state space approach, developed (e.g. Lotka 1956, Scudo and Ziegler 1978). It encompassed the manipulation of speculative equations, practical problems in epidemics and methods inherited from actuarial practice. Gradually developing over the decades, population dynamics matured into what is nowadays called ecological modeling. A typical ecological model consists of coupled, on many occasions time-forced, non-linear ordinary differential equations, describing the dynamics of species or trophic levels expressed in biomass units, grams of dry weight, say, or equivalent amounts of inorganic nutrients (carbon, nitrogen, phosphorus), or pigments (chlorophyll). Energy, originally the common currency in the trophic-dynamic concept can only be found back in physical terms: light intensity, heat balances to predict water movements, etc.

Productivity, however, did remain important, as is exemplified by the world-wide coordinated effort under the name of International Biological Programme, which started in the midsixties and lasted until the midseventies. The IBP adopted the study of productivity as one of its main themes. This has led to a further development of the methodology of estimating production of species and trophic levels, cf. the IBP handbooks: Edmondson and Winberg (1971), Ricker/Bagenal (1968/1978), Downing and Rigler (1984). The present paper is strongly related to this approach. The IBP generated several integrated ecosystem modeling studies, especially in the U.S. 'biome' groups: tundra, coniferous forest, deciduous forest, etc. (Patten 1975).

Succession does not seem to be the motivating problem so much in present-day ecological modeling or in the practical methodology of estimating production. The problem seems either forgotten or collectively avoided as being insufficiently understood.

Although many questions regarding structure, functioning and stability of ecosystems are still open, or haven't even been posed yet, ecological modeling is generally considered very promising, acting as a framework against which new measurements or experiments are planned. Trophic levels and individual species are normally modeled in terms of total biomass, which leads to ordinary differential equations. Individual species may have a very complicated life-history, though. Present-day ecological models generally do not deal with such life-history data; that would lead to hundreds or thousands of statevariables and massive data requirements. On the other hand, structured population models, to which this book is devoted, are not sufficiently formulated and studied to be able to couple them, routinely, and calculate the behaviour of the system. The role of estimating productivity of species from detailed life-history data (cohorts, age- and weight-distributions, etc.), therefore, can be seen as the estimation of the main positive term in the ordinary differential equation describing the rate of change of biomass of that species. The model should correctly describe the role played by the species in the trophic food-web, which may include man, as in fisheries for example.

Production-biologists may regard this statement as an underrating of their work. They don't work for the estimation of a rate-constant to help calibrate a model. This may be true to the extent that on many occasions nobody makes such a model for the area in hand. Then, the estimation of production is only done to compare it to similar estimates elsewhere, as a goal in itself.

The theory of secondary productivity, to which this article is devoted, is not strictly limited to estimating the productivity of secondary producers only. Secondary production has become a general term for the production of fish, all kinds of zooplankton, insect larvae, bivalves and other zoobenthos (animals living on the bottom of a sea or lake), etc. The equations presented possess an even more general character, e.g. the fission models, which may hold for certain primary producers or decomposers. A general production biology may serve as a link between structured population models and ecosystem models.

Acknowledgements

I would like to thank Odo Diekmann for his encouragement, critical comments with respect to earlier versions and patience, Nico van Straalen for his empathetic and critical comments (moreover: without his paper on the subject this one would not have been written), Leo Boekhoff for correcting the English of a previous version, Theo Belterman for discussing the biological basis of the production definition employed, which led to major revisions and Mrs. G. Verloop for the excellent typescript.

References

T. ALDENBERG (1978). *Theoretical considerations regarding the calculation of production in the case of continuous reproduction.* Hydrobiol. Bull., 12, 119-126.

T. ALDENBERG (1979). *The calculation of production, reproduction and total growth in the autonomous VanSickle equation.* Hydrobiol. Bull., 13, 3-12.

K.R. ALLEN (1949). *Some aspects of the production and cropping of fresh waters.* N.Z. Science Congress. 1947, Trans. R. Soc. N.Z., 77, 222-228.

K.R. ALLEN (1951). *The Horokiwi stream. A study of a trout population.* New Zealand Mar. Dept. Fish. Bull., 10.

K.R. ALLEN (1971). *Relation between production and biomass.* J. Fish. Res. B. Canada, 28, 1573-1581.

T. BAGENAL (ed.) (1978). 3rd ed. of: W.E. Ricker (ed.) (1968) op. cit.

P. BOYSEN JENSEN (1919). *Valuation of the Limfjord I:* Studies on the Fish-Food in the Limfjord 1909-1917, its quantity, variation and annual production. Rep. Dan. Biol. Station, 26.

D.W. CHAPMAN (1978a). *Production in Fish populations.* In: Ecology of freshwater fish production, S.D. Gerking (ed.). Blackwell, p. 5-25. (Originally: (1967). The biological basis of freshwater fish production, S.D. Gerking (ed.).)

D.W. CHAPMAN (1978b). *Production.* In: T. Bagenal (1978) op.cit., p. 202-217.

G.L. CLARKE, W.T. EDMONDSON, W.E. RICKER (1946). *Mathematical formulation of biological productivity.* Ecol. Monogr., 16, 336-337.

R.E. COOK (1977). *Raymond Lindeman and the Trophic-Dynamic Concept in Ecology.* Science, 198, 22-26.

R. COURANT and D. HILBERT (1962). *Methods of mathematical physics.* II. Partial differential equations. Wiley.

D.J. CRISP (1971). *Energy flow measurements.* In: Methods for the Study of Marine Benthos. N.A. Holme, A.D. McIntyre (eds.) IBP Handbook 16, Blackwell, 197-279.

O. DIEKMANN, H.A. LAUWERIER, T. ALDENBERG and J.A.J. METZ (1983). *Growth, fission and the stable size distribution.* J. Math. Biol., 18, 135-148.

J.A. DOWNING and F.H. RIGLER (eds.) (1984). 2nd ed. of: W.T. Edmondson and G.G. Winberg (eds.) (1971) op. cit.

W.T. EDMONDSON (1974). *Secondary production.* Mitt. Internat. Verein. Limnol., 20, 229-271.

W.T. EDMONDSON and G.G. WINBERG (eds.) (1971). *A manual on methods for the assessment of secondary productivity in fresh waters.* IBP Handbook 17, Blackwell.

C. ELTON (1927). *Animal Ecology.* Sidgwick & Jackson.

D.M. GILLESPIE and A.C. BENKE (1977). *Methods of calculating cohort production from field data - some relationships.* Limnol. Ocean., 12, 171-176.

A.L. HAMILTON (1969). *On estimating annual production.* Limnol. Ocean., 14, 771-782.

H.B.N. HYNES and M.J. COLEMAN (1968). *A simple method of assessing the annual production of stream benthos.* Limnol. Ocean., 13, 569-573.

P.H. LEBLOND, and T.R. PARSONS (1977). *A simplified expression for calculating cohort production.* Limnol. Ocean., 22, 156-157.

R.L. LINDEMAN (1942). *The trophic-dynamic aspect of ecology.* Ecol., 23, 399-418.

A.J. LOTKA (1956). *Elements of mathematical biology.* Dover. (origin.: 1924).

K.H. MANN (1969). *The dynamics of aquatic ecosystems.* In: Advances in Ecological Research, J.B. Cragg (ed.). Academic Press, 1-81.

A.G. MCKENDRICK (1926). *Applications of mathematics to medical problems.* Proc. Edinb. Math. Soc., 44, 98-130.

J. NEESS and R.C. DUGDALE (1959). *Computation of production for populations of aquatic midge larvae.* Ecol., 40, 425-430.

B.C. PATTEN (1975). *Systems Analysis and Simulation in Ecology,* vol. III, Academic Press.

K. PETRUSEWICZ and A. MACFADYEN (1970). *Productivity of Terrestrial Animals,* IBP Handbook 13, Blackwell.

W.E. RICKER (1946). *Production and utilization of fish populations.* Ecol. Monogr., 16, 375-391.

W.E. RICKER (1968). *Methods for assessment of fish production in fresh waters.* IBP Handbook 3, Blackwell.

W.E. RICKER (1978a). *Introduction.* In: T. Bagenal (ed.) op. cit., 1-6.

W.E. RICKER (1978b) *On computing production.* Limnol. Ocean., 23, 379-380.

F.M. SCUDO and J.R. ZIEGLER (1978) *The golden age of theoretical ecology:* 1923-1940. Springer.

J.W. SINKO and W. STREIFER (1971). *A model for populations reproducing by fission.* Ecology, 52, 330-335.

A. THIENEMANN (1931). *Der Produktionsbegriff in der Biologie.* Arch. Hydrob., 22, 616-622.

J. VANSICKLE (1975). *The theory of physiological age distributions in biological populations.* Ph. D. Thesis, Michigan State University, East Lansing, MI.

J. VANSICKLE (1977). *Analysis of a distributed-parameter population model based on physiological age.* J. theor. Biol., 64, 571-586.

N.M. VAN STRAALEN (1985). *Production and Biomass Turnover in Stationary Stage-structured Populations.* J. theor. Biol., 113, 331-352.

H. VON FOERSTER (1959). *Some remarks on changing populations.* In: The kinetics of cellular proliferation, F. Stohlmann (ed.). Grune & Stratton, p.382-407.

T.F. WATERS (1977) *Secondary production in inland waters.* In: A. MacFadyen (ed.). Advances in ecological research, 10, 91-164, Acad. Press.

T.F. WATERS and G.W. CRAWFORD (1973). *Annual production of a stream mayfly population: a comparison of methods.* Limnol. Ocean., 18, 286-296.

R.G. WIEGERT (1976). *Ecological energetics.* Benchmark papers in ecology, 4, Dowden, Hutchinson & Ross.

G.G. WINBERG (1971) *Methods for the estimation of production of aquatic animals.* Academic Press.

G.G. WINBERG, K. PATALKAS, J.C. WRIGHT, A. HILLBRICHT-ILKOWSKA, W.E. COOPER and K.H. MANN (1971). *Methods for calculating productivity.* In: W.T. Edmondson, G.G. Winberg (eds.) (1971) op. cit., p.296-317.

III. Cell populations

In the case of cell populations we seldomly can do any direct measurements on the level of the individuals. So in cell biology the inverse problem is paramount: we have to infer the dynamical properties of the individuals from population observations. The paper by Voorn & Koch exemplifies this (also compare part A section I.4.4). In it a very general procedure is described for connecting various size related statistics on the *p*- and *i*-levels.

Many populations have a distinct spatial character. This is especially manifest for cells occurring in tissues: there our main interest is not so much in the number of cells as well as in their spatial relations. The paper by Raats reviews the formalism for dealing with this issue.

CHARACTERIZATION OF THE STABLE SIZE DISTRIBUTION OF CULTURED CELLS BY MOMENTS

Wim J. Voorn* and Arthur L. Koch**

* Department of Electron Microscopy and Molecular Cytology, University of Amsterdam. Present address: Department of Medical Microbiology, University of Amsterdam, Meibergdreef 15, 1105 AZ Amsterdam, The Netherlands.

** Department of Biology, Indiana University, Bloomington, Indiana 47405, U.S.A.

1. INTRODUCTION

In recent years the size distribution of cells taken from asynchronous balanced growing cultures, alternatively referred to as steady-state cultures, have become of importance in bacteriology and cell biology. This is because such distributions contain information about the kinetics of cell growth. In some cases, notably when the cells are small and must be fixed for electron microscopy, there is no true alternative for the study of growth kinetics. Besides, the constancy of distributions of cell properties with time in steady-state cultures may serve to define the physiology of a particular cell strain in a particular environment because of the reproducibility in the laboratory.

Cell size is believed to play an important role in the division cycle of several bacteria (c.f. Koch and Schaechter, 1962; Donachie, 1968; Marr et al., 1969; Pritchard et al., 1969; Koch and Higgins, 1982). Therefore there is much interest in the size growth rate during the cell cycle and in the distribution of size at important cell cycle events such as cell division and initiation of DNA replication in steady-state cultures. The problem is how this growth rate and these distributions are related to the observable size distribution in the culture. In Section 2 we explore Collins and Richmond's (1962) solution to this problem in the case of no cell loss. This brings about a phenomenological description of a steady-state cell growth model which is discussed in Section 3. In Section 4 the use of moments as a basis for easy estimating and testing such models is proposed. In Section 5 the Collins and Richmond equation is generalized to the case where cell loss is possible and a formula is given for the mean value of an arbitrary real function of cell size (Theorem 1). In Section 6 this formula is applied to the moments of cell size in the case of no cell loss. In the further specialized case of exponential growth of cell size these moments prove to be proportional to those of size at cell division so that the latter can be estimated very easily. In Section 7 slightly more complicated expressions are found when growth rate is an arbitrary linear function of cell size. In Section 8 this "general exponential" growth mode and so-called "bilinear" growth are used in studying the sensitivities of the mean and coefficients of variation, skewness and kurtosis of the size distribution of all cells to deviations from exponential growth, to the size distribution at cell division and to inequality of newborn sister cells.

2. THE STEADY-STATE SIZE DISTRIBUTION

In this paper we mainly consider steady-state cultures in which each cell divides into two daughter cells and in which cell death is negligible. These conditions are satisfied to a good approximation in some laboratory experiments. For such cultures Collins and Richmond (1962) derived the equation,

$$v(x) = \frac{\mu}{\lambda(x)} \int_a^x \{2\psi(y)-\phi(y)-\lambda(y)\}dy \qquad (a<x<b), \tag{1}$$

based upon the net increase per unit of time of the number of cells of size less than x by birth, division and growth. For any x in the interval (a,b) of possible cell sizes the mean size growth rate of all cells of size x, denoted by $v(x)$, is expressed by (1) in terms of the specific population growth rate, μ, and the probability densities of size of all cells or so-called "extant" cells, $\lambda(x)$, of newborn or "baby" cells, $\psi(x)$, and of dividing or "mother" cells, $\phi(x)$. A cell is called a mother cell if it will divide within Δt time units where Δt is arbitrarily small and has the same value for all cells. Similarly a cell is called a baby if its age is less than Δt. These hypothetical cells are introduced to avoid the concept of conditional probability on a zero measure subpopulation.

It is easy to solve (1) for $\lambda(x)$ (e.g. Painter and Marr, 1968), giving

$$\lambda(x) = \frac{\mu}{v(x)} e^{-\mu t(x)} \int_a^x e^{\mu t(y)} \{2\psi(y)-\phi(y)\}dy, \tag{2}$$

where $t(x)$ is defined such that

$$t(x_2) - t(x_1) = \int_{x_1}^{x_2} \frac{dx}{v(x)} \qquad \text{if} \quad x_1 < x_2, \tag{3}$$

that is, $t(x_2) - t(x_1)$ is the time it takes for a cell to grow with the mean growth rate $v(x)$ from size x_1 to size x_2.

To avoid analytical complications let us assume that the interval (a,b) of possible cell sizes is finite and that $v(x)$ has a positive lower bound. Then it follows from (2) that μ satisfies the relation

$$\int_a^b e^{\mu t(y)} \phi(y)dy = 2 \int_a^b e^{\mu t(y)} \psi(y)dy, \tag{4}$$

since $t(x)$ is bounded on (a,b) and (1) implies the vanishing of $v(x)\lambda(x)$ if x tends to b.

We note that (1)-(4) do not assume that the sizes of two newborn sister cells add up to the size of their mother so that "size" admits very different interpretations and need not even be positive. For example, "size" may be interpreted as cell age or minus remaining life length (Voorn, 1983).

When "size" has one of its usual meanings, e.g. volume or mass, ϕ and ψ are related. If each mother produces two equally sized daughter cells, the case of symmetric or equal division, this relation is given by

$$\psi(x) = 2\phi(2x). \tag{5}$$

When division is not symmetric it is sometimes assumed (Koch and Schaechter, 1962; Harvey et al., 1967) that the ratio of daughter size to mother size in a randomly sampled mother (with one of the two daughters randomly chosen to calculate this ratio) has a probability density function K which is independent of mother size. For enteric bacteria there is some experimental support for the assumption (Errington et al., 1965). By elementary probability theory it implies the relationship (Painter and Marr, 1968)

$$\psi(x) = \int_0^1 \frac{1}{p}\,\phi(x/p)K(p)dp, \tag{6}$$

of which (5) is the limiting case when the variance of K tends to zero.

In several estimations of K from electron micrographs of deeply constricted bacteria cells division was almost precisely even and was observed to have small values of its CV (coefficient of variation, defined as the standard deviation divided by the mean) of 2-9% (Harvey et al., 1967; Koppes et al., 1978). By definition, K is symmetric about 0.5, and it is reported to be approximately normal or Gaussian (Marr et al., 1966; Koppes et al., 1978; Trueba, 1982). So K may be considered as known in some cases. In view of relations (1)-(6) it will be clear that each of the functions λ, v, ϕ can then be calculated from the other two. This has frequently been applied to calculate v (e.g. Harvey et al., 1967; Marr et al., 1969; Koppes et al., 1978) and sometimes to calculate λ to test a hypothetical v, or to calculate ϕ (Harvey et al., 1967).

3. NOTE ON MODEL DESCRIPTION

Collins and Richmond's description of the steady-state size distribution in terms of v, ϕ and ψ is a purely phenomenological one and v, ϕ and ψ, as defined in the previous section, do exist in any real steady-state cell culture.

Estimating ϕ seems a little bit unnatural and not quite to the point if we are in search for a model of individual cell growth and division which would produce, in the long run, the observed steady-state density function λ. However, it should be noted that such a model is not uniquely determined by (v,ϕ,ψ) or (v,ϕ,K), not even when division is symmetric and v(x), as the expected growth rate, applies to each individua cell of size x, whereas there exists only one density function ϕ of size at division in the steady-state for a given triple λ, v and K. The special choice of model defined by a deterministic growth rate function v, valid for any cell, a probability $\mu\,\phi(x)/\lambda(x)$ of dividing per unit of time for any cell of size x and a probability density function K of the ratio of daughter size to mother size for any dividing cell will, in general except in some very special cases, generate λ. But this choice may be far from reality since real cultures may have oscillating mean growth and division rates of equally sized cells, due to dependence on age, before a steady-state is reached.

In conclusion we can say that ϕ and the concept of a size determined division probability are equally appropriate in the phenomenological description of a steady-state size distribution and that the division probability concept is also useful in defining simple models that generate this distribution. Of course, analogous remarks can be made on cell death or other causes of cell loss.

4. MOMENTS

Let θ_1 and θ_2 denote the mean and CV of ϕ. Let λ and ϕ be related by (6) and let θ_3 denote the CV of K. Koch and Schaechter (1962) found ϕ to be approximately normal, in which case ϕ is determined by θ_1 and θ_2. Assuming a known function for v and, neglecting thin tails, normality for ϕ and also for K (see Section 2), λ is completely determined by the parameters θ_1, θ_2 and θ_3 according to relations (2)-(6). A parameter of v, describing a family of growth patterns, may serve as a fourth parameter of λ.

The practical problem of estimating some or all of these parameters or testing hypotheses about them on the basis of an observed λ can be solved by the well-known statistical method of maximum likelihood (though simultaneous estimation of four parameters may not be very precise). However, it seems convenient to have also a method based on some simple statistics to do this job in a relatively easy way because maximum likelihood requires an iterative computing program, the convergence of which depends on initial parameter estimates. To this end the method of moments, using the lower moments of the observed λ (the k-th moment being the mean value of the k-th power of the variate considered), seems to be a good choice, since Koch (1966), comparing exponential ($v(x)=\mu x$) and linear ($v(x)$=constant) growth, already noted that the CV of λ is very sensitive to θ_2 and that the skewness coefficient γ_1 (the standardized third moment) strongly depends on the acceleration of growth rate during the cell cycle whereas the CV is relatively insensitive to this acceleration.

Besides, moments are in frequent use to characterize statistical distributions and for many well-known distributions, e.g. the normal distribution, the usual parameters themselves are moments or very simple functions of them. All the probability distributions of the Pearson system, which includes the normal, the beta, the gamma and many others, are determined by their first four moments and their parameters can be expressed in these moments. The Pearson system is often used to approximate a distribution of which only the first four moments are known since, in general, these determine to a high degree any unimodal distribution. Most likely this is also the case with stable cell size distributions, but, rather than to relate the first four moments of λ to the parameters of a Pearson distribution it seems more significant, from a biological viewpoint, to relate them to v, ϕ and K.

Therefore we will explore here the relations between the moments of λ and the parameters of parametrized hypothetical functions v, ϕ and K which determine λ. In particular we will try to express the lower moments of ϕ in those of λ when v and K are known or assumed. It will turn out that, without restricting ϕ to some type of distribution, this is possible in the important case of exponential size growth and, in good approximation also, for a more general exponential growth pattern with v(x) an arbitrary linear function of cell size x. The latter model, which includes both exponential and linear growth as special cases, permits the demonstration of the above mentioned findings of Koch (1966) in more detail.

5. GENERALLY VALID EQUATIONS

Let us consider the slightly more general situation where cell loss by death or other causes must be taken into account and let δ be the steady-state probability density function of size of dying cells, defined analogously to mother cells. If N is the total number of cells at a given point of time, the population has a net increase of μN cells per time unit. So there are constants f_m and f_d such that there are $f_m \mu N$ divisions and $f_d \mu N$ deaths per time unit, with

$$f_m - f_d = 1. \tag{7}$$

The Collins-Richmond equation (1) is now easily extended to this more general situation We get

$$v(x) = \frac{\mu}{\lambda(x)} \int_a^x \{2f_m\psi(y) - f_m\phi(y) - f_d\delta(y) - \lambda(y)\}dy. \tag{8}$$

The proof is completely analogous to that of (1), which is the special case $f_m=1$ and $f_d=0$.

We can use (8) to compute the expectation Eg(X) of an arbitrary function g of cell size in the population defined by

$$Eg(X) = \int_a^b g(x)\lambda(x)dx, \tag{9}$$

where the capital X denotes the random variable or "variate" cell size of a randomly sampled cell from the population. The sizes of randomly sampled mothers, babies and dying cells will be denoted by M, B and D respectively, and expectations calculated on these special samples by E_m, E_b and E_d. Using these notations it is possible to derive from (8) the following theorem on $Eg(X)$:

Theorem 1. For any real function g

$$Eg(X) = f_m E_m R(M) + f_d E_d R(D) - 2f_m E_b R(B), \tag{10}$$

where R is a function defined by

$$R(x) = e^{\mu t(x)} H(x), \tag{11}$$

where t satisfies (3) and H is a function such that

$$H(x_2) - H(x_1) = \int_{x_1}^{x_2} g(x) \frac{\mu}{v(x)} e^{-\mu t(x)} dx \qquad \text{if } x_1 < x_2. \tag{12}$$

Theorem 1 is proved by partial integration of

$$\int_a^b \{\frac{d}{dx} R(x)\} \int_a^x \{2f_m \psi(y) - f_m \phi(y) - f_d \delta(y) - \lambda(y)\} dy\, dx$$

(see Voorn (1983) for the details in the case $f_d = 0$).

Choosing $g(x)=0$ and $H(x)=1$ for all x we find from (10)-(12) a condition on μ, expressed by

Theorem 2. The specific population growth rate μ satisfies

$$f_m E_m e^{\mu t(M)} + f_d E_d e^{\mu t(D)} = 2f_m E_b e^{\mu t(B)}. \tag{13}$$

Equation (4) is the special case $f_d=0$.

We note that, like equations (1)-(4), Theorems 1 and 2 do not require any relation between ϕ and ψ, so that "size" admits a wide range of interpretations, e.g. cell age.

Theorems 1 and 2 may be used to calculate $Eg(X)$ from (v,ϕ,ψ,δ,f_d) without first evaluating λ.

Reversely, in estimating parameters of ψ, ϕ and δ from an observed λ it may be interesting to know whether there exists a function g corresponding to a given function R according to (11) and (12), e.g. the function $R(x) = x^k$ $(k=1,2,\ldots)$, when the unknown parameters of these probability density functions are simple functions of their moments. We may then use the equation

$$g(x) = \frac{v(x)}{\mu} \frac{dR(x)}{dx} - R(x), \tag{14}$$

which clearly is satisfied for any pair (g,R) related by (11)-(12) when R is differentiable at x.

6. MOMENTS OF THE SIZE DISTRIBUTION AT CELL DIVISION

Returning to the case of no cell loss ($f_d=0$) and to cell division in two unequally sized daughters according to (6), let us denote the ratio of daughter size to mother size in a randomly sampled mother cell (with only one of the two daughters randomly selected to get a mother-daughter pair) by P.

Let v be an arbitrary growth rate function. Using (14) to find functions g for $R(x)=x \ln x$ and $R(x)=x^k$, and the obvious relation

$$\mu = Ev(X)/EX, \tag{15}$$

we get from (10), after some algebra and using the independence of M and P in mothers,

$$-2E_m(P \ln P)E_m M = EX + \frac{EX}{Ev(X)} E(v(X)\ln X) - E(X \ln X) \tag{16}$$

and, for all real k,

$$(1-2E_m P^k)E_m M^k = k \frac{EX}{Ev(X)} E(v(X)X^{k-1}) - EX^k. \tag{17}$$

Equations (16) and (17) permit direct calculation of the moments of ϕ from λ, v and K.

When v(x) is an r degree polynomial in x, the first k moments of ϕ can be calculated from the mean values of $X^i \ln X$ $(i=0,1,\ldots,r)$ and X^i $(i=0,1,\ldots,r+k-1)$ for any set of values of the polynomial coefficients.

In the case of exponential size growth, $v(x)=\mu x$, the right hand sides of (16) and (17) reduce to functions of EX and EX^k respectively, that is, to the moments of λ for $k=1,2,\ldots$. This is an important result because this growth pattern admits simple interpretations and has often been hypothesized (e.g. Koch and Schaechter, 1962; Koppes et al., 1978) or found to be approximately true in the study of bacterial cell growth (e.g. Trueba, 1981; Kubitschek and Woldringh, 1983). Equations (16) and (17) then simplify to

$$-2E_m(P \ln P)E_m M = EX \tag{18}$$

and

$$(1-2E_m P^k)E_m M^k = (k-1)EX^k \qquad \text{(all real k)}, \tag{19}$$

so that the k-th moments of ϕ and λ are proportional to each other with proportionality constants determined by K, the probability density function of P. In the equal division case P takes the value 0.5 with probability 1. Denoting the k-th moment of λ by ν_k and that of ϕ by $\nu_{k\phi}$ $(k=1,2,\ldots)$, we get in this case

$$\nu_1 = \ln 2 \cdot \nu_{1\phi} \quad \text{and} \quad \nu_k = \frac{1-2^{1-k}}{k-1} \nu_{k\phi} \quad (k=2,3,\ldots). \tag{20}$$

7. GENERAL EXPONENTIAL GROWTH

Now let us consider the case that v(x) is a linear function of x,

$$v(x) = c(x-s) \qquad (v(x)>0), \tag{21}$$

which includes linear growth ($s\to-\infty$ with $c>0$ and cs=constant) and exponential growth ($s=0$, $c>0$) as special cases. We shall call this type of growth "general exponential". It was used earlier by Koppes and Nanninga (1980) interpreting x-s as that part of cell size which contributes to growth. Define a growth rate parameter ω by

$$\omega = s/(\nu_1-s) \qquad (s\neq\nu_1), \tag{22}$$

then, for general exponential growth, (16) and (17) imply

$$\nu_{1\phi} = \frac{\nu_1}{-2E_m(P\ \ln P)}\{1+\omega\nu_1^{-1}E((X-\nu_1)\ln X)\} \tag{23}$$

and

$$\nu_{k\phi} = \frac{(k-1)\nu_k}{1-2E_mP^k}\{1+\omega\frac{k}{k-1}(1-\nu_1\nu_{k-1}/\nu_k)\} \qquad (k=2,3,\ldots) \tag{24}$$

these relations display the relative differences, $\{\nu_{1\phi}(\omega)-\nu_{1\phi}(0)\}/\nu_{1\phi}(0)$ (k=1,2,...), of the moments of ϕ with the exponential case ($\omega=0$). When λ and K are kept fixed, these are proportional to ω with positive proportionality constants since $\nu_k>\nu_1\nu_{k-1}$ (k=2,3,...) and $E((X-\nu_1)\ln X)>0$. The latter inequality follows from the convexity of the curve $y=(x-\nu_1)\ln x$ and Jensen's inequality.

In order to find an approximation of $\nu_{1\phi}$ ($=E_mM=\theta_1$) in terms of the mean and the coefficients of variation, skewness and kurtosis of λ, let us denote the standard deviation of λ, $\{E(X-\nu_1)^2\}^{\frac{1}{2}}$, by σ, its coefficient of variation, σ/ν_1, by q, and its coefficients of skewness, $E(X-\nu_1)^3/\sigma^3$, and of kurtosis, $E(X-\nu_1)^4/\sigma^4$, by the usual symbols γ_1 and β_2 respectively. With respect to ϕ we may use similarly σ_ϕ, q_ϕ ($=\theta_2$), $\gamma_{1\phi}$ and $\beta_{2\phi}$. Further let us assume that all cells are smaller than $2\nu_1$, which seems to be very common or at least approximately so in real bacterial steady-state cultures (c.f. Powell, 1964, in his defense of the assumption that the largest baby is smaller than the smallest mother). Then, expanding $(x-\nu_1)\ln x$ in a power series in $x-\nu_1$ on the interval $(0, 2\nu_1)$ and using simple properties of expectations, we derive from (23)

$$\theta_1 = \frac{\nu_1}{-2E_m(P\ \ln P)}\{1+\omega(q^2-\tfrac{1}{2}\gamma_1q^3+\tfrac{1}{3}\beta_2q^4-\ldots)\}. \tag{25}$$

Since q seldom exceeds 0.3, neglecting powers of q beyond the fourth gives a good approximation to θ_1 in many applications. We may use this to calculate θ_1 and θ_2 ($=q_\phi$) approximately from (25) and (24) when K, ω and the first four moments of λ (which have simple relations to ν_1, q, γ_1, β_2) are given. For example, let $\omega=-1$ (linear growth) and let K be approximately normal, such that its first eight moments equal those of a normal distribution, with CV θ_3=0.05. In this case we may use the equations (see Voorn, 1983)

$$-2E_m(P \ln P) = \ln 2 - (\tfrac{1}{2}\theta_3^2 + \tfrac{1}{4}\theta_3^4 + \tfrac{1}{2}\theta_3^6 + \frac{15}{8}\theta_3^8 + \ldots) \tag{26}$$

and

$$1 - 2E_m P^2 = (1-\theta_3^2)/2, \tag{27}$$

where the terms in (26) which are not explicitly written can be neglected in all practical applications. Applying (24)-(27) to ν_1=1.456287, q=0.223379, γ_1=0.358988 and β_2=2.43162, corresponding to θ_1=2.0, θ_2=0.1 and θ_3=0.05 with ϕ and K normal (see Voorn, 1983), we get the approximations $\theta_1 \approx 1.9997$ and $\theta_2 \approx 0.1014$.

8. SENSITIVITIES OF ν_1, q, γ_1 AND β_2 TO K, ϕ AND THE GROWTH RATE FUNCTION

The general exponential growth model is also useful in studying the sensitivity of λ in terms of ν_1, q, γ_1 and β_2 to K, ϕ and the acceleration of growth rate during the cell cycle.

E.g., from (25) and (24) (for k=2) it is possible to derive, after some manipulation of power series in q (see Voorn, 1983), the equation

$$q^2 = \{\frac{1-\theta_3^2}{8(E_m P \ln P)^2}(1+\theta_2^2) - 1\} - \omega\{\gamma_1 q^3 - (2+\tfrac{2}{3}\beta_2+\omega)q^4 + \ldots\}. \tag{28}$$

The vanishing of the q^0, q and q^2 terms in $\{\gamma_1 q^3 - (2+\frac{2}{3}\beta_2+\omega)q^4+\ldots\}$ in the right hand side of (28) explains the relative insensitivity of q to the acceleration of growth rate (indicated here by ω) found by Koch (1966) in comparing linear (ω=-1) and exponential (ω=0) growth. Equation (28) demonstrates this to hold within the whole general exponential growth family.

In order to get an impression of many kinds of sensitivities we made some calculations of ν_1, q, γ_1 and β_2 for different growth rate patterns and for different values of θ_2 (=q_ϕ), $\gamma_{1\phi}$ and $\beta_{2\phi}$. Fig. 1 summarizes the results in a matrix of smaller figures. Only a few of the great number of possible combinations of variations are considered and equal division is assumed in all but the first column, which shows the sensitivities to θ_2 in the exponential growth case (ω=0) for the values 0 and 0.05 of θ_3 with K normal. In the remaining columns only two values of θ_2, 0.1 and 0.15, are considered. The second and third columns show the sensitivities to $\gamma_{1\phi}$ and $\beta_{2\phi}$, again for exponential size growth. The fourth and fifth columns show the sensitivities to variation in the growth rate parameter for general exponential and bilinear growth respectively, with ϕ normal. The latter growth pattern assumes a constant growth rate until the cell reaches a critical size c at which the rate is doubled. For the calculation of ν_1, q, γ_1 and β_2 for this and the general exponential growth pattern equations (2) and (4) were used.

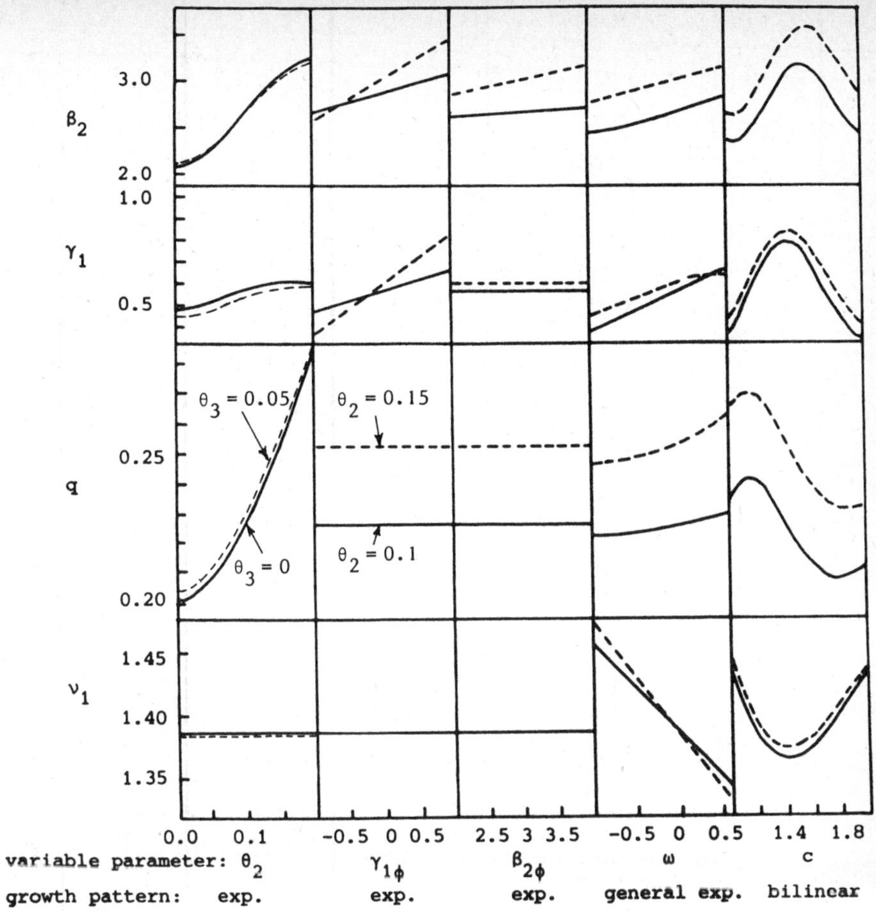

Fig. 1. Sensitivities of the mean, ν_1, and the coefficients of variation, q, skewness, γ_1, and kurtosis, β_2, of the steady-state size distribution to θ_2, θ_3, $\gamma_{1\phi}$, $\beta_{2\phi}$ and to the growth rate parameters ω and c of the general exponential and the bilinear growth patterns respectively (see text). θ_1, θ_2, θ_3 denote the mean and CV of the distribution ϕ of mother size and the CV of the distribution K of the ratio of daughter size to mother size in dividing cells respectively; they serve as parameters of the size distribution if the growth rate function v (see text) and the types of ϕ and K are fixed.
Column 2 to 5: ——— : $\theta_2 = 0.1$; ------- : $\theta_2 = 0.15$.
All columns assume equal division, except the first where ——— : $\theta_3 = 0$, -------: $\theta_3 = 0.05$ with normal K.
Column 3 assumes that $\gamma_{1\phi} = 0$, columns 4 and 5 assume that ϕ is normal.
In all columns the mean birth size of babies is taken as the unit measure of size (i.e. $\theta_1 = 2$) and the scaling in the vertical direction is such that, when measured in the figure, the standard errors of the sample values of ν_1, q, γ_1 and β_2 are equal in the first column at $\theta_2 = 0.1$.

The figure clearly shows the rather great sensitivity of γ_1 to the growth rate function which was already noted by Koch (1966), comparing linear and exponential growth. It is also clear, however, that the relative insensitivity of q to the growth rate function within the general exponential growth family cannot be extended to bilinear growth. Consequently, knowledge of the growth pattern remains important for estimating the CV of ϕ from that of λ. But the growth pattern of bacterial cells is difficult to determine (see Burdett and Kirkwood, 1983) and it is not to be expected that the mere use of moment statistics, though attractive because of simplicity, will better this situation substantially.

LITERATURE CITED

Burdett, I.D.J. and T.B.L. Kirkwood. 1983. How does a bacterium grow during its cycle? J. Theor. Biol. 103: 11-20.

Collins, J.F. and M.H. Richmond. 1962. Rate of growth of Bacillus cereus between divisions. J. Gen. Microbiol. 28: 15-88.

Donachie, W.D. 1968. Relationship between cell size and time of initiation of DNA replication. Nature (London) 219: 1077-1079.

Errington, F.P., E.O. Powell and N. Thompson. 1965. Growth characteristics of some Gram-negative bacteria. J. Gen. Microbiol. 39: 109-123.

Harvey, R.J., A.G. Marr and P.R. Painter. 1967. Kinetics of growth of individual cells of Escherichia coli and Azotobacter agilis. J. Bacteriol. 93: 605-617.

Koch, A.L. and M. Schaechter. 1962. A model for the statistics of the cell division process. J. Gen. Microbiol. 29: 435-454.

Koch, A.L. 1966. Distribution of cell size in growing cultures of bacteria and the applicability of the Collins-Richmond principle. J. Gen. Microbiol. 45: 409-417.

Koch, A.L. and M.L. Higgins. 1982. Cell cycle dynamics inferred from the static properties of cells in balanced growth. J. Gen. Microbiol. 128: 2877-2892.

Koppes, L.J.H., C.L. Woldringh and N. Nanninga. 1978. Size variations and correlation of different cell cycle events in slow-growing Escherichia coli. J. Bacteriol. 134: 423-433.

Koppes, L.J.H. and N. Nanninga. 1980. Positive correlation between size at initiation of chromosome replication in Escherichia coli and size at initiation of cell constriction. J. Bacteriol. 143: 89-99.

Kubitschek, H.E. and C.L. Woldringh. 1983. Cell elongation and division probability during the Escherichia coli growth cycle. J. Bacteriol. 153: 1379-1387.

Marr, A.G., R.J. Harvey and W.C. Trentini. 1966. Growth and division of Escherichia coli. J. Bacteriol. 91: 2388-2389.

Marr, A.G., P.R. Painter and E.H. Nilson. 1969. Growth and division of individual bacteria. Symp. Soc. Gen. Microbiol. 19: 237-259.

Painter, P.R., and A.G. Marr. 1968. Mathematics of microbial populations. Annu. Rev. Microbiol. 22: 519-548.

Powell, E.O. 1964. A note on Koch and Schaechter's hypothesis about growth and fission of bacteria. J. Gen. Microbiol. 37: 231-249.

Pritchard, R.H., P.T. Barth and J. Collins. 1969. Microbial growth. Symp. Soc. Gen. Microbiol. 19: 263-297.

Trueba, F.J. 1981. A morphometric analysis of Escherichia coli and other rod-shaped bacteria. Thesis. University of Amsterdam, Amsterdam.

Trueba, F.J. 1982. On the precision and accuracy achieved by Escherichia coli cells at fission about their middle. Arch. Microbiol. 131: 55-59.

Voorn, W.J. 1983. Statistics of cell size in the steady-state with applications to Escherichia coli. Thesis, University of Amsterdam, Amsterdam.

THE KINEMATICS OF GROWING TISSUES

P.A.C. Raats
Institute for Soil Fertility
Haren (Gr.), The Netherlands

Growth and division of cells are determined on the one hand by genetic properties and on the other hand by the environment in which the growth occurs. Observations related to growth of stems, leaves and roots were reported already in the 18th century. Since 1950 there is also a lot of interest in detailed models for growth of tissues. Most theories proposed thus far are rather special. The purpose of this paper is to formulate a general framework which can accomodate all the earlier work. This helps making the widely spread literature more accessible and removing some of the conceptual confusion.

It is shown that the kinematics of continuous media can be used to describe the growth of tissues. Not only the comparison of tissues at different moments, but also the gradual change of tissues can be dealt with. Cell division is treated on the basis of balance equations for various categories of cells.

DEFORMATION AND MOTION OF TISSUES

Two approaches can be used to describe motion (Truesdell and Toupin, 1960; Lin and Segel, 1974; Segel, 1977; Gurtin, 1981):

1) The spatial approach describes what happens in the course of time t at specific places $\underset{\sim}{x}$;

2) The material approach describes what happens in the course of time t to specific pieces of tissue $\underset{\sim}{X}$.

Specifically applied markings or recognizable, lasting anatomical features can be used as labels of pieces of tissue. Hales (1727, Fig. 1) used red lead mixed with oil to mark stems and leaves. Erickson (1976; see also references given in this comprehensive review) photographed a root of *Zea mays* brushed with a suspension of lampblack and a leaf of *Xanthium* with a distinctive venation pattern.

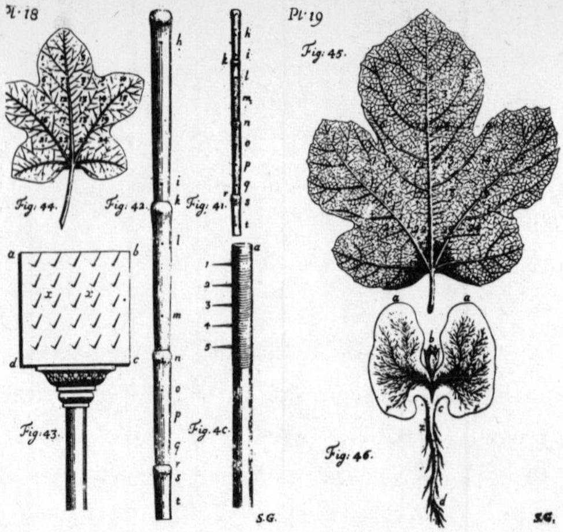

Figure 1. Growth of labeled pieces of tissue: Fig. 40-42. Growth of a stem; Fig. 43-45. Growth of a leaf; Fig. 46 shows a sketch of a seedling (from Hales, 1727).

In theoretical developments it is convenient to use as labels $\underset{\sim}{X}$ for pieces of tissue their locations at some reference time.

The material description gives for any piece of tissue $\underset{\sim}{X}$ the places $\underset{\sim}{x}$ occupied in the course of time t:

$$\underset{\sim}{x} = \underset{\sim}{x}\,[X, t]. \tag{1}$$

Differentiation of this functional relationship gives the two key concepts for describing deformation and motion:

1) The deformation gradient tensor $\underset{\sim}{F}$ is defined by

$$\underset{\sim}{F} \equiv \partial\underset{\sim}{x}/\partial\underset{\sim}{X}\Big|_t . \tag{2}$$

2) The velocity vector $\underset{\sim}{v}$ is defined by

$$\underset{\sim}{v} \equiv \partial\underset{\sim}{x}/\partial t\Big|_{\underset{\sim}{X}} . \tag{3}$$

If, from any configuration actually occupied, the reference configuration can be reached by a continuous motion, then $J \equiv \det \underset{\sim}{F} > 0$ and then the polar decomposition theorem gives two unique, multiplicative decompositions of $\underset{\sim}{F}$:

$$\underset{\sim}{F} = \underset{\sim}{R}\,\underset{\sim}{U} = \underset{\sim}{V}\,\underset{\sim}{R}, \tag{4}$$

where $\underset{\sim}{R}$ is the rotation tensor and $\underset{\sim}{U}$ and $\underset{\sim}{V}$ are the right and left hand stretch tensors. The tensor $\underset{\sim}{R}$ is orthogonal:

$$\underset{\sim}{R}\,\underset{\sim}{R}^{T} = \underset{\sim}{I}, \tag{5}$$

where the superscript T denotes the transpose and $\underset{\sim}{I}$ is the identity tensor. The tensors $\underset{\sim}{U}$ and $\underset{\sim}{V}$ are positive-definite symmetric:

$$\underset{\sim}{U} = \underset{\sim}{U}^{T}, \qquad \underset{\sim}{V} = \underset{\sim}{V}^{T}. \tag{6}$$

The geometric interpretation of the two expressions for $\underset{\sim}{F}$ given by (4) is very simple: the deformation corresponding locally to $\underset{\sim}{F}$ may be obtained by effecting pure stretches along three suitable, mutually orthogonal directions, followed by a rigid rotation of those directions, or by performing the same rotation first and then effecting the same stretches along the appropriate directions. Numerous other measures of deformation can be derived from $\underset{\sim}{F}$ (Truesdell and Toupin, 1960).

The velocity $\underset{\sim}{v}$ defined by (3) can be used to relate the spatial time derivative $\partial./\partial t|_{\underset{\sim}{x}}$ and the material time derivative $\partial./\partial t|_{X}$:

$$\partial\phi/\partial t|_{X} = \partial\phi/\partial t|_{\underset{\sim}{x}} + \underset{\sim}{v}.\partial\phi/\partial\underset{\sim}{x}, \tag{7}$$

where ϕ may be any variable in the tissue varying continuously with position and time, e.g. the cell density, the phosphorus content, the DNA content. This relationship will be used later to transform the spatial forms of the balances of cell density to material forms.

Applying (7) to the velocity $\underset{\sim}{v}$ gives an expression for the accelaration $\underset{\sim}{a} = \partial\underset{\sim}{v}/\partial t|_{\underset{\sim}{X}}$:

$$\underset{\sim}{a} = \partial\underset{\sim}{v}/\partial t|_{\underset{\sim}{x}} + \underset{\sim}{v}.\partial\underset{\sim}{v}/\partial\underset{\sim}{x}, \tag{8}$$

where $\partial\underset{\sim}{v}/\partial\underset{\sim}{x}$ is the velocity gradient tensor. The tensor $\partial\underset{\sim}{v}/\partial\underset{\sim}{x}$ can be additively decomposed in a symmetric part $\frac{1}{2}(\partial\underset{\sim}{v}/\partial\underset{\sim}{x} + \partial\underset{\sim}{v}/\partial\underset{\sim}{x}^{T})$ and a skew-symmetric part $\frac{1}{2}(\partial\underset{\sim}{v}/\partial\underset{\sim}{x} - \partial\underset{\sim}{v}/\partial\underset{\sim}{x}^{T})$:

$$\partial\underset{\sim}{v}/\partial\underset{\sim}{x} = \tfrac{1}{2}(\partial\underset{\sim}{v}/\partial\underset{\sim}{x} + \partial\underset{\sim}{v}/\partial\underset{\sim}{x}^{T}) + \tfrac{1}{2}(\partial\underset{\sim}{v}/\partial\underset{\sim}{x} - \partial\underset{\sim}{v}/\partial\underset{\sim}{x}^{T}). \tag{9}$$

Again the geometric interpretation is very simple: the symmetric part describes the rate of stretch, the skew-symmetric part describes the rate of rotation.

The deformation gradient tensor $\underset{\sim}{F}$ compares the current configuration of a piece of tissue relative to its neighbours with the configuration at some reference time. The velocity gradient tensor $\partial\underset{\sim}{v}/\partial\underset{\sim}{x}$ compares the current velocities of neighbouring pieces of tissue. Differentiating (2) with respect to t for fixed X or (3) with respect to X for fixed t, it can be shown that the tensors $\underset{\sim}{F}$ and $\partial\underset{\sim}{v}/\partial\underset{\sim}{x}$ are related by

$$\partial\underset{\sim}{F}/\partial t\Big|_{\underset{\sim}{X}} = (\partial\underset{\sim}{v}/\partial\underset{\sim}{x})\underset{\sim}{F}. \tag{10}$$

In principle, this *material, tensorial* partial differential equation enables one to calculate $\underset{\sim}{F}$ from the time course of $\partial\underset{\sim}{v}/\partial\underset{\sim}{x}$. Here attention is restricted to a *material, scalar* partial differential equation for $J = \det \underset{\sim}{F}$ implied in (10):

$$\partial J/\partial t\Big|_{X} = (\nabla.\underset{\sim}{v})J. \tag{11}$$

Integration of (11) from some past time $t_\circ$ to the current time t gives

$$J = J_\circ \exp \int_{t_\circ}^{t} (\nabla.\underset{\sim}{v})dt. \tag{12}$$

Equation (12) describes the volumetric growth of a piece of tissue. The determinant J of the deformation gradient tensor is a measure of the volume of a piece of tissue. One- and two-dimensional analogs of (12), giving expressions for, respectively, the length and the area of a piece of tissue, abound in the literature on plant morphology.

BALANCE EQUATIONS FOR CELL DENSITY

A description of growth should consider not only the distribution of tissue in space, but also the formation and fate of cells. Viewed spatially, the cell density at a point in space changes due to movement of cells to and away from the point and due to division of cells at the point. Viewed materially, the cell density of a piece of tissue changes due to expansion or contraction of the piece of tissue and due to division of cells in the piece of tissue. What happens to a piece of tissue is inherently of more interest than what happens at a point in space. A point of space is generally occupied by different pieces of tissue in the course of time. In the following, differential balance equations according to both viewpoints will be formulated for the fractional cell density based on size. Integration of these equations over all sizes will give differential balance equations for the total cell density. By integrations of the differential balance equations over an entire tissue, integral balance equations will be derived. The latter balance equations can be compared with the balance equation for a population subject to fission reproduction derived by Bell and Anderson (1967).

Differential balance equations taking into account size structure

For size σ, let δ be the number of cells per unit size per unit volume such that

$$\int_0^\infty \delta\sigma \, d\sigma = 1 - f, \tag{13}$$

where f is the volume fraction not occupied by cells. For a fixed place $\underset{\sim}{x}$, the balance equation for cells of size σ can be written as:

$$\frac{\partial}{\partial t} \delta[t, \underset{\sim}{x}, \sigma]\Big|_{\underset{\sim}{x}} + \frac{\partial}{\partial\sigma} g[t, \underset{\sim}{x}, \sigma] \delta [t, \underset{\sim}{x}, \sigma] + \nabla\cdot(\underset{\sim}{v}[t, \underset{\sim}{x}] \delta [t, \underset{\sim}{x}, \sigma] =$$

$$\lambda_\delta [t, \underset{\sim}{x}, \sigma] \delta [t, \underset{\sim}{x}, \sigma], \tag{14}$$

where g is the growth rate and λ_δ is the net source strength per cell per unit time. The growth rate is defined by (cf. definition of $\underset{\sim}{v}$ by (3)):

$$g[t, \underset{\sim}{x}, \sigma] \equiv \partial\sigma/\partial t\Big|_X . \tag{15}$$

If a cell shrinks, then its growth rate is negative. The product $\underset{\sim}{v}\delta$ is the flux of cells per unit area per unit time. Similarly, the product $g\delta$ can be interpreted as the flux of cells past the size σ. The growth rate λ_δ is defined by

$$\lambda_\delta [t, \underset{\sim}{x}, \sigma] = -b[t, \underset{\sim}{x}, \sigma] + 4b[t, \underset{\sim}{x}, 2\sigma] \frac{\delta[t, \underset{\sim}{x}, 2\sigma]}{\delta[t, \underset{\sim}{x}, \sigma]}, \tag{16}$$

where b is the fission rate. The factor 4 in the last term of (16) is easily understood if one realizes that division of a cell in the range 2σ to $2\sigma + 2\Delta\sigma$ produces 2 cells in the range σ to $\sigma + \Delta\sigma$.

The spatial balance equation can be transformed to a material balance equation by expanding the divergence of the flux $\underset{\sim}{v}\delta$ and using (7) and (11);

$$\frac{\partial}{\partial t} J[t, X]\, \delta\, [t, X, \sigma]\Big|_{\tilde{X}} + \frac{\partial}{\partial \sigma} g[t, X, \sigma]\, J[t, X]\, \delta\, [t, X, \sigma] =$$

$$\lambda_\delta [t, X, \sigma]\, J[t, X]\, \delta\, [t, X, \sigma]. \qquad (17)$$

At time t and for a piece of tissue X, the product $J\delta$ can be interpreted as the number of cells per unit size per unit volume of the piece of tissue in the reference configuration. Equation (17) can also be written in the form

$$\frac{1}{\delta}\frac{\partial \delta}{\partial t}\Big|_X - \frac{1}{\delta}\frac{\partial}{\partial t} g\delta - \lambda_\delta = -\frac{1}{J}\frac{\partial J}{\partial t}\Big|_{\tilde{X}} . \qquad (18)$$

Equation (13) is a constraint to be satisfied by solutions of equation (14), and hence (17) and (18). For any piece of tissue the expansion, of which J is a measure, will adjust itself in such a manner that the distribution of δ satisfies (13). The resulting growth may show the consequences: wrinkles on leaves, cracks in fruits, etc.

Differential balance equation disregarding size structure

Integration of (14) over all sizes gives the spatial differential balance equation for cells of all sizes:

$$\frac{\partial n[t, \tilde{x}]}{\partial t} + \nabla.(v[t, \tilde{x}]\, n[t, \tilde{x}]) = \lambda_n [t, \tilde{x}]\, n[t, \tilde{x}], \qquad (19)$$

where n is the total number of cells per unit volume defined by

$$n \equiv \int_0^\infty \delta d\sigma, \qquad (20)$$

and λ_n is the number of divisions per cell per unit time defined by

$$\lambda_n \equiv \frac{\int_0^\infty \lambda_\delta \delta\, d\sigma}{n} . \qquad (21)$$

Expansion of the second term in (19) and using (7) and (11) gives the material differential balance equation for cells of all sizes

$$\left.\frac{\partial J[t, X]\, n[t, X]}{\partial t}\right|_X = \lambda_n[t, X]\, J[t, X]\, n[t, X] \tag{22}$$

At time t and for a piece of tissue X, the product Jn can be interpreted as the total number of cells per unit volume of the piece of tissue in the reference configuration. Equation (22) also follows from integrating (17) over all sizes.

Bodily integrated balance equations

Integration of either (14) over a region V chosen large enough so that the flux $\underset{\sim}{v}\delta$ is zero at the boundary or (17) over a set of pieces of tissue comprising an entire tissue E gives

$$\frac{\partial}{\partial t} \Delta[t,\sigma] + \frac{\partial}{\partial \sigma} \bar{g}[t, \sigma]\, \Delta[t, \sigma] = \lambda_\Delta[t, \sigma]\, \Delta[t, \sigma], \tag{23}$$

where the total number of cells Δ per unit size is defined by

$$\Delta \equiv \int_V \delta \, d\underset{\sim}{x} = \int_E J\delta \, d X, \tag{24}$$

$\bar{g}$ is the average growth rate defined by

$$\bar{g} \equiv \frac{\int_V g\, \delta\, d\underset{\sim}{x}}{\Delta} = \frac{\int_E g\, J\delta d X}{\Delta}, \tag{25}$$

and λ_Δ is the average net source strength per cell per unit time defined by

$$\lambda_\Delta \equiv \frac{\int_V \lambda_\delta \delta d\underset{\sim}{x}}{\Delta} = \frac{\int_E \lambda_\delta J\delta dX}{\Delta}. \tag{26}$$

Clearly, substitution of (16) into (26) allows one to write the average net source strength λ_Δ as the sum of two averages.

Integration of either (19) over a region V chosen large enough so that the flux $\underset{\sim}{v}n$ is zero at the boundary or (23) over a set of pieces of tissue comprising an entire tissue E gives

$$\frac{\partial}{\partial t} N[t] = \lambda_N[t] \, N[t], \tag{27}$$

where the total number of cells N is defined by

$$N \equiv \int_V n \, d\underset{\sim}{x} = \int_E J n \, d X, \tag{28}$$

$$= \int_0^\infty \Delta d\sigma, \tag{29}$$

$$= \int_0^\infty \int_V \delta \, d \underset{\sim}{x} \, d \sigma = \int_0^\infty \int_E J \delta \, d X d \sigma, \tag{30}$$

and the average net source strength per unit time is defined by

$$\lambda_N \equiv \frac{\int_V \lambda_n n d\underset{\sim}{x}}{N} = \frac{\int_E J\lambda_n n dX}{N}, \tag{31}$$

$$= \frac{\int_0^\infty \bar{\lambda}_\Delta \Delta d\sigma}{N}, \tag{32}$$

$$= \frac{\int_0^\infty \int_V \lambda_\delta \delta d\underset{\sim}{x} d\sigma}{N} = \frac{\int_0^\infty \int_E J\lambda_\delta \delta dX d\sigma}{N}. \tag{33}$$

Equations (23) and (27) are the bodily integrated balance equations. Size structure is taken into account in (23) and is disregarded in (27). Naturally, (27) also follows from integrating (23) over all sizes.

The bodily balance equation (23) for the number of cells Δ in the entire tissue E within the spatial region V is of the same form as the balance equation for a population subject to fission reproduction derived by Bell and Anderson (1967; see also Diekmann, 1983; and Diekmann, *et al.*, 1983). It is interesting to note that, except for the presence of X, the material balance equation (17) for a piece of tissue X is also of this form, with the roles of Δ in (23) and $J\delta$ in (17) being analoguous. For a piece of tissue X, the product $J\delta$ at time t can be interpreted as the number of cells per unit size per unit volume of the piece of tissue in the reference configuration.

Temporal integrated balance equations

Integration of the bodily integrated balance equation (27) for the entire tissue E gives

$$N = N_\circ \exp \int_{t_\circ}^{t} \lambda_N \, dt. \qquad (34)$$

Of course, the innocent looking equation (34) involves the complicated average net source strength per unit time, λ_N.

Integration of the material differential balance equation (22) for cells of all sizes gives:

$$Jn = J_\circ \, n_\circ \exp \int_{t_\circ}^{t} \lambda_n dt, \qquad (35)$$

or

$$n/n_\circ = (J/J_\circ)^{-1} \exp \int_{t_\circ}^{t} \lambda_n dt. \qquad (36)$$

According to (35) the ratio of the cell densities at two instants is the product of two factors, one due to volume change and one due to cell division.

One - dimensional, steady growth

It should be realized that (22), and hence (35) and (36), applies to some piece of tissue, only very exceptionally at some point in space. Integration of (19), the spatial counterpart of (22), is generally more difficult. An important exception is 1-dimensional, steady growth, for example of a plant root.

Let x be a spatial coordinate coinciding with the axis of a root and attached to the mature part of the root. With respect to the x-coordinate, let v be the velocity of any piece of tissue and let w be the velocity of the root tip. Then v - w is the speed of a piece of tissue relative to the root tip. Let y be a spatial coordinate coinciding with the axis of a root with origin at the root tip at all times. Then y is related to t, w, and x by

$$y = wt - x. \tag{37}$$

Adding $w\partial n/\partial x$ to both sides of the 1-dimensional form of (19) gives

$$\partial n/\partial t\big|_x + w\partial n/\partial x = -\partial\{n(v - w)\}/\partial x + \lambda_n n. \tag{38}$$

Introducing (37) into (38) gives

$$\partial n/\partial t\big|_y = \partial\{n(v - w)\}/\partial y + \lambda_n n. \tag{39}$$

For growth which is *steady in the y-frame*, (39) reduces to

$$\partial\{n(w - v)\}/\partial y = -\lambda_n n, \tag{40}$$

with λ_n and n being merely functions of y. Integration of (40) from y = 0 til y = L gives

$$n(v - w) = \int_0^L \lambda_n\, n\, dy. \tag{41}$$

In (41) the left hand side can be interpreted as the flux of cells relative to the y-frame at y = L and the right hand side can be interpreted as the integral fission rate between y = 0 and y = L (cf. Erickson, 1976, p. 419). If n and (v - w) are measured as functions of y then λ_n can be calculated from equation (40).

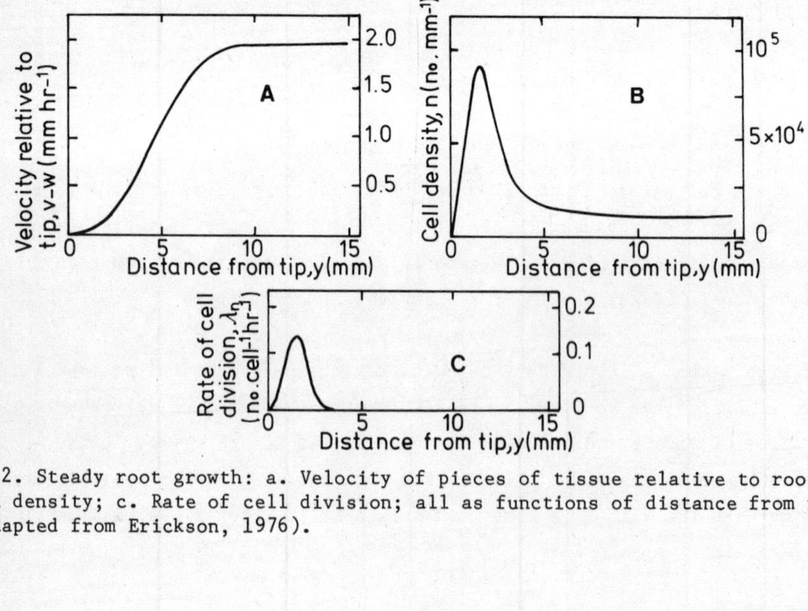

Figure 2. Steady root growth: a. Velocity of pieces of tissue relative to root tip; b. Cell density; c. Rate of cell division; all as functions of distance from root tip (adapted from Erickson, 1976).

Figure 2 shows an example. Note that at large distance from the root tip $\lambda_n \rightarrow 0$, and (v - w) and n approach constant values

CONCLUDING REMARKS

The theory of growth and division of cells in tissues presented in this paper is based on experience with kinematics of continuous media and dynamics of structured populations. The resulting framework can serve as a basis for synthesis of concepts and interpretation of observations. The framework still needs 1) to be extended to include the possibility of accretionary growth and 2) to be made more specific by formulating constitutive assumptions, for example for $g|t, \underset{\sim}{x}, \sigma|$ and $\lambda_\delta|t, \underset{\sim}{x}, \sigma|$ defined by equations (15) and (16).

ACKNOWLEDGEMENT

I wish to thank Dr. O. Diekmann of the Centre for Mathematics and Computer Science at Amsterdam for his constructive comments.

REFERENCES

Bell, G.I. and E.C. Anderson, 1967. Cell growth and division. I. A mathematical model with applications to cell volume distributions in mammalian suspension cultures. Biophys. J. 7: 329-351.

Diekmann, O., 1983. The dynamics of structured populations: some examples. Preprint TW 241/83, Mathematical Centre, Amsterdam, 12 pp.

Diekmann, O., H.J.A.M. Heymans and H.R. Thieme, 1983. On the stability of the cell size distribution. Preprint TW 242/83, Mathematical Centre, Amsterdam, 32 pp.

Erickson, R.O., 1976. Modeling of plant growth. Annu. Rev. Plant Physiol. 27: 207-234.

Gurtin, M.E., 1981. An introduction to continuum mechanics. Academic Press, New York, 265 pp.

Hales, S., 1727. Vegetable Staticks. Republished in History of Science Library: Primary Sources. MacDonald, London, 216 pp.

Lin, C.C. and L.A. Segel, 1974. Mathematics applied to deterministic problems in the natural sciences. Macmillan, New York, 604 pp.

Segel, L.A., 1977. Mathematics applied to continuum mechanics. Macmillan, New York, 590 pp.

Truesdell, C. and R.A. Tonpin, 1960. The classical field theories. Handbuch der Physik III/1: 226-793. Springer Verlag, Berlin.

IV. Numerical approaches.

To arrive at a workable modelling methodology the theoretical approaches have to be complemented by efficient numerical techniques for exploring the models so conceived. Thus far the problem of numerically calculating solutions of structured population equations has received little attention except for the simplest case of pure age dependence. The two papers in this section both deal with models in which the *i*-state variable under consideration is physiological age (compare part A, remark IV.1.2.3).

The author of the first paper, Goudriaan, is a member of a very active group at the Agricultural University Wageningen, which succesfully intertwines experimentation and computer simulation. (Sabelis, the (co)author of three papers in section I, also started his career in Wageningen). Goudriaan's "boxcar trains" should be seen in this light. They were originally conceived as a means for representing experimentally observed developmental successions in the simulation language CSMP. Their basis is the physical analogy of solute flow through a sequence of well stirred compartments (fixed boxcar train) and the Jacob's ladder (escalator boxcar train). The main emphasis of the paper is on the interplay between modelling methodology and numerical processing.

Gurney, Nisbet and Blythe approach the same problem in a more mathematical spirit. Their aim is the construction of classes of relatively tractable parameter sparse models, which yet are capable of incorporating an amount of biological detail sufficient for concrete applications. To this end they propose dividing up a life cycle into a finite succession of developmental stages which may be considered homogeneous in the sense of all individuals in one stage having the same death rate and the same developmental rate. A stage is left when the developmental index crosses a threshold which is fixed in advance or "determined stochastically at the moment of entering a stage" (compare part A I.4.1 and III.3.2.1, for a more detailed exposition concerning this type of modelling assumption). The use of a linear chain trick (see part A IV.5.2) then allows the reduction of the resulting *p*-equations to an equivalent set of (delay)-differential equations. Generally for models of this type it is not unduly complicated to study equilibria and their linearized stability. And, more importantly, for numerical work one can fall back on methods developed for ordinary differential equations.

Both Goudriaan's and Gurney's, Nisbet's and Blythe's formulations have in common that the coefficient of variation of the stage duration under different constant conditions is the same. Moreover, the fixed boxcar train corresponds to a special sort of linear chain. Goudriaan's "mixed" boxcar train, however, does not have a counterpart in the Gurney, Nisbet & Blythe formalism, as can be seen from the fact that stage duration necessarily has a maximum set by the "escalator" component.

The methodologies described in the two papers in this section clearly form a significant addition to the toolkit available for modelling real populations. Yet we wish to end with making a general plea for the development of efficient numerical techniques for dealing with larger classes of structured population models!

BOXCARTRAIN METHODS FOR MODELLING OF AGEING, DEVELOPMENT, DELAYS AND DISPERSION

J.Goudriaan
Department of Theoretical Production Ecology
Agricultural University
Bornsesteeg 65
6708 PD Wageningen
The Netherlands

1. Introduction

1.1. Development

Development means a coherent and irreversible change of a number of properties of a being (usually living) when it gets older. Basically the pattern of change is repeated in individuals of the same species, so that it is possible to establish a series of stages that every individual goes through.
Usually a reproductive phase of life succeeds a juvenile phase, without reproductive activity. In the last phase of life, when senescence occurs, the probability to die will strongly increase. For interaction with other species it is important to know how predation activity or also vulnerability for predation develop with age.
In population dynamics not just one individual is studied, and a means to describe the age or development composition of a whole population is needed. When the average composition is always the same, an average population characteristic may suffice. But seasonal changes often trigger the onset of population growth, so that the age distribution is changing all the time. Then, also population characteristics as relative reproductive rate change with the season.

1.2. Ageing and a development scale

Age dependent characteristics, such as reproductive activity or relative death rate, are in fact not a function of age but of internal development. Still age is often used as an indicator of the development because development and age are highly correlated. Especially in warm-blooded animals the internal environment is well stabilised so that external conditions have little influence on the rate of ageing. In cold-blooded animals and in plants, development is poorly related with age, because the rate of development is all but constant. Under influence of the environmental conditions the rate of development may be altered. The way in which environmental conditions influence the rate of development,is an important subject of experimental research. Ideally one would hope for this research to lead to the construction of a scale of development which is uniquely and linearly related to the integral of the developmental effect of the environmental conditions. Important developmental transitions, such as onset of flowering, are then not necessarily equally spaced on the developmental scale.

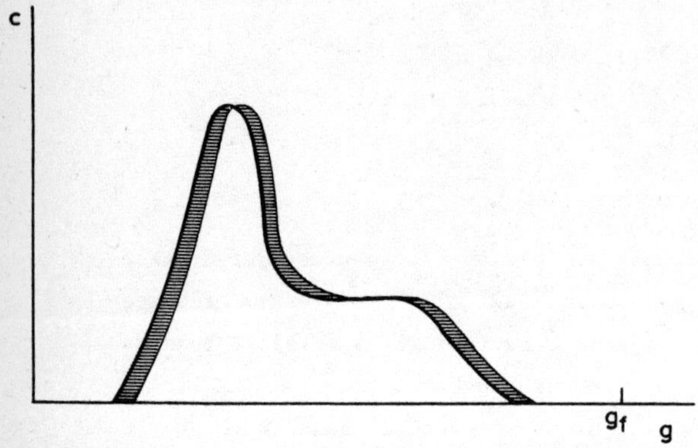

Fig. 1 A distribution of population density c with stage of development g

For most plant species the basis for the development scale is the temperature sum, defined as the integral of the excess of the daily mean temperature above a threshold temperature. In this simple situation the rate of development is proportional to temperature, but the different stages may well be separated by different increments in the temperature sum.

1.3. A mathematical formulation of development.

On a linear scale of development, the rate of development v is constant for constant environmental conditions. Rate of ageing is essentially dimensionless, because age is expressed in time, but rate of development has the unit development (g) per time (t).
To describe the age (or development) distribution of the population a density function c (Fig. 1) is needed. This density will also be called concentration because it is a measure of the degree of concentration of individuals at a certain age. Its unit is number of individuals per development unit (g^{-1}). It is clear

that the numerical value of the concentration depends on the arbitrary choice of this unit. The size of the total developing population is given by

$$H = \int_{0}^{g_f} c\,(g)\,dg \qquad (1)$$

where g_f is the upper limit of the particular development bracket, e.g. the moment of molting, the onset of fruiting etc. Development means that the value of g of each individual increases with a rate v. Therefore development can be visualised as a shift to right in Fig. 1. Mathematically this can be expressed as

$$\frac{\partial c}{\partial t} = - \frac{\partial vc}{\partial g} = - v \frac{\partial c}{\partial g} \qquad (2)$$

where the latter equality is based on the assumption that v is independent of g. This assumption can always be realized by a linearization of the developmental scale versus some accumulated external factor, and if necessary by a breaking up of this scale in pieces that are internally homogeneous in environmental response.

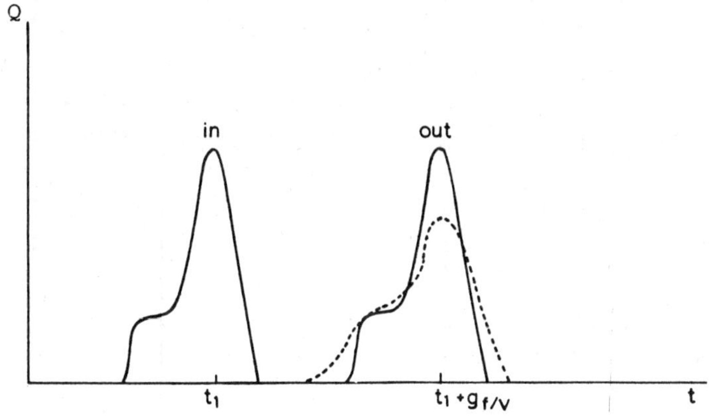

Fig. 2 Outflow shows a delay , and may be dispersed (dashed line)

1.4. Delay and dispersion

In the simplest situation there is no death, and only when an individual has reached the final development stage g_f a transition to another phase occurs. This will happen some time after the entrance of this individual into the development process. When the development rate is constant, there is a constant delay and the shape of the outflow curve is exactly equal to the shape of the inflow curve. In this simple situation the total delay period is by definition equal to the final development state g_f divided by the development rate v so that:

$$Q_{out}(t) = Q_{in}(t - g_f / v) \qquad (3)$$

where Q_{out} is outflow and Q_{in} is inflow.

When the inflow Q_{in} shows a peak at time t_1, the outflow Q_{out} should show exactly the same peak at time $t_1 + g_f / v$. (Fig. 2). Of course, in reality some dispersion may occur during development up to stage g_f.
Then a time course of Q_o may be observed as given by the dashed line in Fig 2. In this situation, Eqn 2 is not adequate and must be extended to allow for dispersion, as described in the next pages.

At this point the concept of dispersion deserves some attention. Dispersion is vaguely described as moving away from a densely populated centre to underpopulated areas. This means that dispersion tends to level out peaks and dips in the distribution of the population. Usually the process of dispersion occurs in true space, but here it occurs in the degree of development. The quantitative description of dispersion is often based on the analogy with physical diffusion processes, because diffusion also causes a levelling out of peaks and dips. Because of its simplicity of formulation we shall also use this analogy, but still a warning is necessary. In physical diffusion the driving force for the levelling process is formed by a gradient in concentration. In dispersion of development the driving force is the inherent variability of the rate of development. An important consequence of this difference is that backward movement is impossible, whereas in a physical diffusion process a strong concentration gradient may drive the net flow opposite to the general mass flow.

First the situation must be considered when there is no dispersion at all. The problem we face in numerical modelling is that it is very hard to avoid an artificial dispersion, that is caused by the model structure itself. The reasons for such an artificial dispersion will be discussed in the next pages.

2. Some numerical methods

2.1. Discretization in time

To solve numerically the differential equations for the rate of change of development and of concentration a discretization of time is necessary.
Errors introduced by this discretization depend on the time resolution of the integration interval, and also on the chosen integration method (Goudriaan 2.3. In: Penning de Vries and Van Laar, 1982). Simulation languages such as CSMP provide convenient software for the organization of the time axis. Ideally the time axis should be discretized to such a high degree of resolution that further refinement does not improve the results. With CSMP or a similar simulation language this ideal situation can be very closely approximated, except in extremely large or complicated simulation models.
Given these software tools, also the representation of the inflow curve (Fig. 2) does not present particular problems. When its shape must be externally provided, that means as an independent function of time, sufficient accuracy can be obtained by reading a table of points on the input curve and using an appropriate interpolation method. When on the other hand the inflow must be generated in the model itself, it is the time resolution of the integration method that determines the accuracy, and in this respect there is no difference with the simulation of any other flow or state variable in the model.

2.2. Fixed boxcar train, the escalator boxcar train, and the fractional boxcar train

Next, the development axis of the population is discretized into a number of classes, equal in width. It is clear that the resolution increases with the number of classes that we choose, but so does the computational effort.
After the discretization into N classes we can rewrite the integral expression (Eqn 1) into a summation:

$$H = \sum_{i=1}^{N} c_i \gamma \qquad (4)$$

where γ is the width of the developmental classes expressed in developmental units, and c_i is the average concentration in each class i.
The distribution of the population with respect to development is assumed to be known at time zero. For further handling of the classes a crucial choice must be

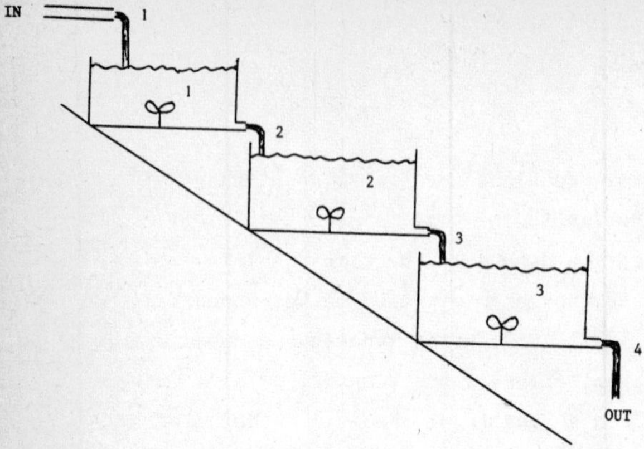

Fig. 3 A cascade of water tanks as a physical model of the fixed boxcartrain

made. It is possible to fix the class boundaries with respect to development, but it is also possible to follow the population classes in their development and let the class boundaries move with the same development rate. The two methods will be termed the fixed boxcar train and the escalator boxcar train, respectively. Which one we choose is largely determined by the amount of dispersion that we want to simulate.

In the escalator boxcar train the boundaries move with the same development rate as the individuals. Therefore there is no exchange of individuals across the boundaries and dispersion does not occur.

In the fixed boxcar train a continuous through flow of individuals occurs across the fixed boundaries. This flow does not only mean development, but also implies dispersion, because it establishes forward exchange between boxes.

In mathematical terms the fixed boxcar train describes the population distribution in Eulerian coordinates, and the escalator boxcar train in Lagrangian coordinates. The fractional boxcar train is a hybridization of these two methods. It offers the possibility to modify the dispersion during simulation relative to the delay.

3. Fixed boxcartrain

3.1. Its delay

A cascade of water tanks (Fig.3) is a model of the fixed boxcar train. The flow out of a box is assumed to be proportional to its contents H and inversely proportional to a time constant τ :

$$Q_{i+1} = H_i / \tau \tag{5}$$

where i denotes the number of the box, and Q_i the flow from box i-1 to box i. The average residence time $\bar{\tau}$ in box i can be found as the difference between the average time of outflow t_{i+1} and the average time of inflow t_i:

$$\bar{\tau} = \frac{\int_0^{t_f} t\, Q_{i+1}\, dt}{\int_0^{t_f} Q_{i+1}\, dt} - \frac{\int_0^{t_f} t\, Q_i dt}{\int_0^{t_f} Q_i\, dt} \tag{6}$$

To evaluate this expression it is necessary to assume that the box starts empty, and is practically empty again at time t_f. This means that the inflow shows a flush, and that t_f is chosen sufficiently long after the inflow has returned to zero, so that the outflow has become zero as well (Fig. 1). Under this condition the integrals of inflow and outflow are the same so that Eqn. (6) can be written as

$$\bar{\tau} = \frac{\int_0^{t_f} t\, (Q_{i+1} - Q_i) dt}{\int_0^{t_f} Q_{i+1}\, dt} \tag{7}$$

Because $Q_{i+1} - Q_i$ is equal to $-\frac{dH_i}{dt}$, this expression becomes

$$\bar{\tau} = \frac{-\int_0^{t_f} t\, dH_i}{\int_0^{t_f} Q_{i+1}\, dt} \tag{8}$$

Integration by parts of $\int_0^{t_f} t\, dH_i$, and using that $t\, H_i$ equals zero at both time zero and time t_i gives

$$\bar{\tau} = \frac{\int_o^{t_f} H_i\, dt}{\int_o^{t_f} Q_{i+1}\, dt} \qquad (9)$$

Now the relationship between outflow and contents (Eqn 5) must be used, which says that H_i is always τ times as large as Q_{i+1}. This relation also holds for their integrals so that indeed $\bar{\tau}$ is equal to τ.
It should be noted that in this derivation about the average residence time no assumption was made about the shape of the inflow. Therefore it is valid for each box in the cascade, even though the shape of the inflow curve may be altered on its way through the cascade. Because the boxes are connected in series, and all material must pass through all boxes the important conclusion can be drawn that:

- The total mean delay in the boxcar train is the sum of the delays in the individual boxes.

3.2. Its dispersion

The variance of the time of outflow from a box will usually be larger than that of the time of inflow. The difference is the dispersion added by the residence in the box. The definition of the dispersion σ_t^2 is

$$\sigma_t^2 = \frac{\int_o^{t_f} (t - t_{i+1})^2\, Q_{i+1}\, dt}{\int_o^{t_f} Q_{i+1}\, dt} - \frac{\int_o^{t_f} (t - t_i)^2\, Q_i\, dt}{\int_o^{t_f} Q_i\, dt} \qquad (10)$$

Using the same technique as in the previous paragraph it can be shown that

$$\sigma^2_t = \tau^2 \qquad (11)$$

Because each box in the cascade will add this amount of dispersion, whatever its position in the cascade, the conclusion can be drawn that:

- The total dispersion by a boxcar train is the sum of the dispersions by the individual boxes.

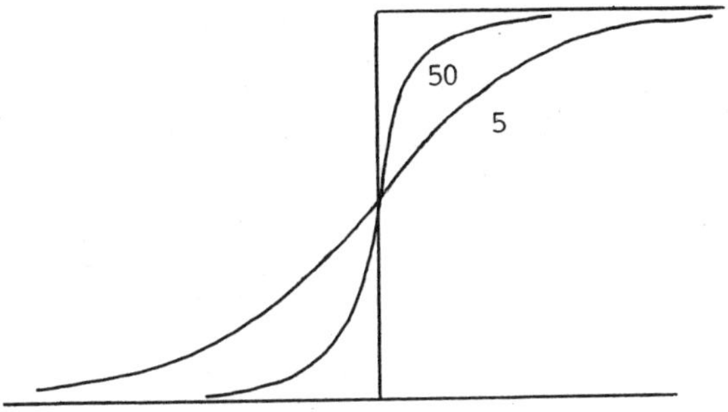

Fig 4. Higher order delays show less dispersion

3.3. The overall behaviour

It follows from the preceding paragraphs that the total delay time T_N and the total variance of the time of outflow σ^2_N are given by:

$$T_N = N\tau \tag{12}$$

$$\sigma^2_N = N\,\tau^2 \tag{13}$$

Consequently, the coefficient of variation of the outflow is equal to:

$$\frac{\sigma_N}{T_N} = \frac{1}{\sqrt{N}} \tag{14}$$

These simple relationships show that the relative dispersion, caused by the fixed boxcar train, decreases as the number of boxes in the boxcar train increases. This type of chained exponential delays is also often termed an Nth- order delay, where N indicates the number of boxes. The dynamic response of the outflow to a step wise change in the inflow is given in Fig. 4, for different values of N (Ferrari, 1978). An analytical expression for the contents of each box in the cascade was given by Goudriaan (1973).

3.4 Variable rate of development

When the rate of development v is variable with time, Eqn 5 must be replaced by the following equation:

$$Q_{i+1} = v\, c_i \tag{15}$$

To obtain the differential equation for H_i the concentration c_i must be expressed in H_i and in the developmental width γ of box i (see also Eqn 4):

$$c_i = H_i/\gamma \tag{16}$$

These two equations can be combined to

$$\frac{Q_{i+1}}{v} = \frac{H_i}{\gamma} \tag{17}$$

The rate of increase of the physiological time g is equal to the rate of development v:

$$dg = v\, dt \tag{18}$$

Using Eqn 17 and 18, the derivation of the average residence time and of the dispersion in paragraph 3.1 and 3.2 can be entirely written in terms of development instead of time. The coefficient γ replaces the time constant τ, and so the duration of residence in a box expressed in physiological time units is also γ. The physiological time g itself is replaced by $\int_0^t v\, dt'$. The physiological age of the individuals in the boxcartrain is given by the index number of the box, multiplied with γ.

4. The escalator boxcar train

4.1. Its functioning

The discretization of the population into N boxes of width γ is not different from the fixed boxcar train. But because the boundaries between the boxes now also "age", they move together with the population and the flow across the boundaries is zero. The following example can clarify the principle of the escalator boxcar train. In an imaginary school children are admitted on their 6th birthday and they leave the school on their 12th birthday. The school year starts on September 1. At that moment the school has 6 grades (or classes or boxes). In the first class are only children that were born between August 31 6 years ago and September 1 7 years ago, in the second class born between August 31 7 years ago and September 1 8 years ago etc. No children repeat a class. At the beginning of the school year all children in the first class are six years old, and all children in the sixth class are eleven years old. During the school year all children celebrate their birthday, but only those of the last class must leave the school when they become 12 years old. The children in the other classes stay in their class. The sixth class will gradually loose all its pupils, but for the school as a whole this loss is compensated by admission of children who become six years old. To accommodate these children a zeroth class is established. At September 1 it is still empty, but it will gradually receive newcomers and by the next first of September this class will be termed the first class. Then also the number of the other classes is increased by unity. From then on these children will stay together, untill they leave the sixth class due to their age.

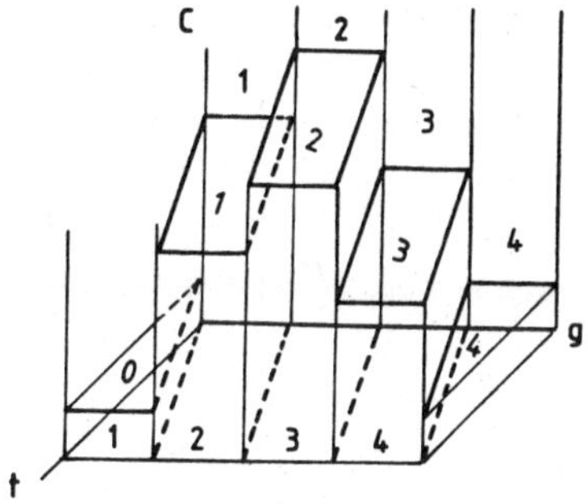

Fig. 5 Shift of the classes over the development scale with time in the escalator boxcartrain

4.2. The mathematical formulation

The only two boundaries that are fixed are the beginning and the end of the development scale. Therefore the inflow of the boxcar train is not zero, and neither is the outflow. The inflow is collected in an additional zeroth box:

$$\frac{dH_o}{dt} = Q_{in} \tag{19}$$

The outflow is calculated with Eqn 15 and removed from the last box:

$$Q_{out} = v\ c_N \tag{20}$$

To illustrate the computation of c_N reference is made to Fig. 5. Because the boundaries between the boxes "age" themselves, the development width covered by the last box continuously decreases with increasing physiological age, until it completely vanishes. Before this moment the concentration can be simply calculated as

$$c_N = H_N \ / \ (\gamma - g') \tag{21}$$

where g' stands for the physiological age elapsed since the present last reached its position of "last box".
The ratio c_N is constant.
This can be seen by combining Eqns 20 and 21 to

$$\frac{dH_N}{dt} = - \frac{v\ H_N}{\gamma - g'} \tag{22}$$

or

$$\frac{dH_N}{H_N} = \frac{-v\ dt}{\gamma - g'} \tag{23}$$

Because vdt is the same as dg' also:

$$\frac{dH_N}{H_N} = \frac{d(\gamma - g')}{\gamma - g'} \tag{24}$$

so that H_N and γ-g' have the same relative rate of change. Therefore their ratio, which is c_N, is constant, until g' reaches the value of γ. At that moment the last box vanishes entirely, and the preceding box takes its function. Physically nothing happens to the intermediate boxes and they can smoothly move along. However, we must close the zeroth box and create a new box of newcomers in the population. To prevent an unwieldy growth of the number of boxes it is convenient to perform an act of renumbering right at this moment (Fig.5). The last box H_N will receive the contents of the preceding box and by working backwards to the zeroth box which will start with zero, the whole boxcar train has undergone a complete shift. Also the value of g', the development elapsed since the last shift, must be reset and be decreased with a quantity γ.

4.3. Delay in the escalator boxcar train

The total delay is the sum of the delays in the individual boxes.
The delay in the zeroth box, and also in the last box (with number N) is different from the delays in those numbered between 1 through N-1. In all these boxes the residence time τ is the same and equal to γ/v when v is constant. Their combined delay is equal to $(N-1)\tau$. The total delay in the fixed boxcartrain is $N\tau$, so that the delay of the zeroth and the last box together must be equal to τ, to obtain the same total delay.
For the last box the inflow is pulsed, but the outflow is continuous, and constant during the period τ between two shifts. Therefore the mean residence time in the last box is the arithmetic mean of 0 and τ, and equal to $\frac{1}{2}\tau$.
In the zeroth box the situation is the opposite; the inflow is continuous and the outflow is pulsed with intervals τ. Only for a constant rate of inflow is the mean residence time in the zeroth box now also equal to $\frac{1}{2}\tau$. For a growing rate of inflow there will be a bias towards a shorter mean residence time, and vice versa. The first approximation of the bias can be found by assuming a linear increase of inflow and neglecting the higher order terms. When the inflow Q at time zero is denoted by Q(o), and its rate of increase by Q'(o) the mean residence time in the box is equal to

$$\tfrac{1}{2}\tau \left(1 - \frac{\tau}{6} (Q'(o)/Q(o)\right)$$

This expression can be found by evaluation of

$$\tau - \int_o^t t\,Q\,dt / \int_o^t Q\,dt$$

When τ is small enough this bias can be neglected. Then both the zeroth and the last box each cause a mean delay of $\frac{1}{2}\tau$, making up for the difference that remained between the total delay in the fixed boxcartrain and in the central boxes in the escalator boxcartrain. The methods therefore only differ in their influence on the dispersion.

4.4. Dispersion in the escalator boxcartrain

No dispersion occurs during the movement from box 1 to box N-1, because their exchange is zero. Within a box, however, the exchange is perfect. During a development cycle γ the inflow is collected in box zero, and whatever its variation, it is levelled out. How much dispersion is added by this process? The answer is, unfortunately, that it depends on the shape of the inflow. The stronger the variation of inflow within the time period of one development cycle, the higher the apparent dispersion. The strongest possible concentration of inflow is in the form of a single pulse. The outflow from the last box will occur some time later, and cover a time span τ, equal to γ/v. The dispersion around the average time of outflow t_d can be derived by

$$\sigma_t^2 = \frac{1}{H_o} \int_{t_d-0.5}^{t_d+0.5} (t-t_d)^2 \; Q_{out} \; dt \qquad (25)$$

Because during the time span τ an amount H_o must flow out, Q_{out} will be equal to H_o/τ and H_o cancels.
Substitution, and solving the integral gives $\sigma_t^2 = (1/12)\tau^2$. But in this equation the outflow is centred symmetrically around t_d. Dependent on when the pulse arrived in the cyclic development stage, the lower and upper boundaries may vary between $t_d - \tau$, t_d and t_d, $t_d + \tau$ resp.
Under the assumption of random arrival the dispersion is doubled so that:

$$\sigma_t^2 = \tau^2/6 \qquad (26)$$

and the coefficient of variation (c.v.)

$$\frac{\sigma_t}{N\,\tau} = \sqrt{6/N} \qquad (27)$$

This equation permits to choose the right number of boxes when the observed c.v. is small. For instance, when it is 5%, 8 boxes are required. An entirely dispersion free boxcar train does not exist.

5. The fractional boxcartrain

5.1. Its functioning

With the method of the fixed boxes it is possible to influence the dispersion by the choice of the number of boxes.

This number determines the coefficient of variation, and fixes it at $1/\sqrt{N}$. Once it is chosen, it cannot be changed during the simulation. But in several experimental data sets there is evidence that the delay and the dispersion are not equally influenced by for instance temperature, so that the coefficient of variation also varies. To allow for this change during the simulation a more flexible method than that of the fixed boxcar train is needed. Such a flexible method can be obtained by a hybridization of the methods of the fixed and of the escalator boxcar train. This method will be termed the fractional boxcar train, because it is based on a fractional repeated shift.

In the escalator boxcar train a cyclic renumbering occurs, but this can also be considered as a complete shift to the next box. In the fractional boxcar train not the complete contents is shifted, but only a fraction F of it. To compensate for the smaller amount, it must occur more frequently. Whereas in the escalator boxcar train the renumbering (or shift) occurs upon completion of the development cycle γ, in this method the fractional shift occurs upon completing a fraction F only of the development cycle γ.

The fraction F ranges between 0 and 1. The imaginary escalator school in the preceding chapter may be turned into a fractional shift school by a quarterly promotion system. Each quarter of the year one quarter of the box is moved to the next one (F = 0.25). The choice of the children is not based on age, but determined by lottery. If this system is always maintained, once can imagine that the children in the sixth grade show a wide variety of ages. This variety is an expression of dispersion. By choosing F equal to unity the escalator boxcar train can be retained with annual promotion of the whole box. On the other hand, F can also be chosen very small. If F is chosen at 1/365, every day a fraction 1/365 of each box (selected by lottery) is transferred to the next one. Now effectively the fixed boxcar train is obtained with a time step of integration of one day.

5.2. Its mathematical function

Because the movement through the boxes is pulsewise, the differential equations must be replaced by difference equations. In paragraph 4.2. the cyclic development g', elapsed since the last shift, was introduced. In the escalator boxcar train g' triggers the renumbering when it exceeds γ. Here, in the fractional boxcar train, the trigger level is set at $F\gamma$. When this level is exceeded the fractional shift occurs and g' is decreased with the quantity $F\gamma$. Also the contents of box i is decreased:

$$H_{i,j} = H_{i,j-1} - F\,H_{i,j-1} \tag{28}$$

where j counts the number of shifts since the start. Here, just like for Equation 6, it is assumed that the preceding box is kept zero. Then $H_{i,j}$ will be given by

$$H_{i,j} = H_{i,o} (1-F)^{j} \qquad (29)$$

A special situation occurs in the zeroth box. The contents of this box are entirely transferred to the first one, so that upon the shift $H_{o,j} = 0$.
The last box will receive the pulsewise transfer from the previous one, but it will not loose pulsewise. It will only release its contents gradually according to Eqn 20.

5.3. Its delay

The first fractional shift does not occur at time zero, but only when g' equals $F\gamma$. When the development speed v is constant, this happens at time $F\gamma/v$, or at time $F\tau$. The expression for the average residence time $\bar{\tau}$ is:

$$\bar{\tau} = \frac{1}{H_o} \sum_{j=1}^{\infty} \underbrace{j\,F\tau}_{\text{time}} \; \underbrace{H_o(1-F)^{j-1} F}_{\text{quantity transferred}} \qquad (30)$$

Evaluation of this expression gives

$$\bar{\tau} = \tau \qquad (31)$$

This result shows that the delay per box is independent of the value of F. Also the total delay of the boxcar train is independent of F, and equal to $N\tau$.

5.4. Its dispersion

The dispersion can be evaluated by

$$\sigma_t^2 = \frac{1}{H_o} \sum_{j=1}^{\infty} \underbrace{(j\,F\tau - \tau)^2}_{\text{deviation}} \; \underbrace{H_o(1-F)^{j-1}F}_{\text{quantity transferred}} \qquad (32)$$

which gives

$$\sigma_t^2 = \tau^2 (1-F) \tag{33}$$

This result shows that the dispersion is linearly related to the value of the fraction F. This dispersion occurs in each box so that the total dispersion of the whole boxcar train is

$$\sigma_t^2 = N \tau^2 (1-F) \tag{34}$$

A complication is offered by the dispersion upon inflow, precisely as in the escalator boxcar train (Chapter 4.4). The width of the zeroth box is $F\tau$ at most. Using the same proportionality as in Chapter 4.4, we must add $F^2 \tau^2/6$ to the result of Eqn 34

$$\sigma_t^2 = \tau^2(N(1-F) + F^2/6) \tag{35}$$

6. The escalator boxcar train, applied to a demographic problem

For demography the best method is the escalator boxcar train, because age which is used as a characteristic, does not disperse. To illustrate its use, the same example will be given for the growth of the Dutch population as by De Wit and Goudriaan (1978). For clarity only the female part of the population is simulated, the male part being taken for granted.

The age dependence of relative death rate and of relative birth rate is given in Fig. 6. The corresponding fraction of survival FS is found by simulating a single cohort from birth onwards. Mathematically the relative death rate RDR and the fraction survival FS are related by:

$$RDR = -\frac{d(FS)}{dA}/(FS)$$

where A stands for age.

The listing of the CSMP-program used is given below. First data are supplied about the initial age distribution of the population . This is done by a TABLE specifying the contents of the 20 5-year classes of the population array W. Then two FUNCTIONS with a list of coordinate points of the relationship between relative death and

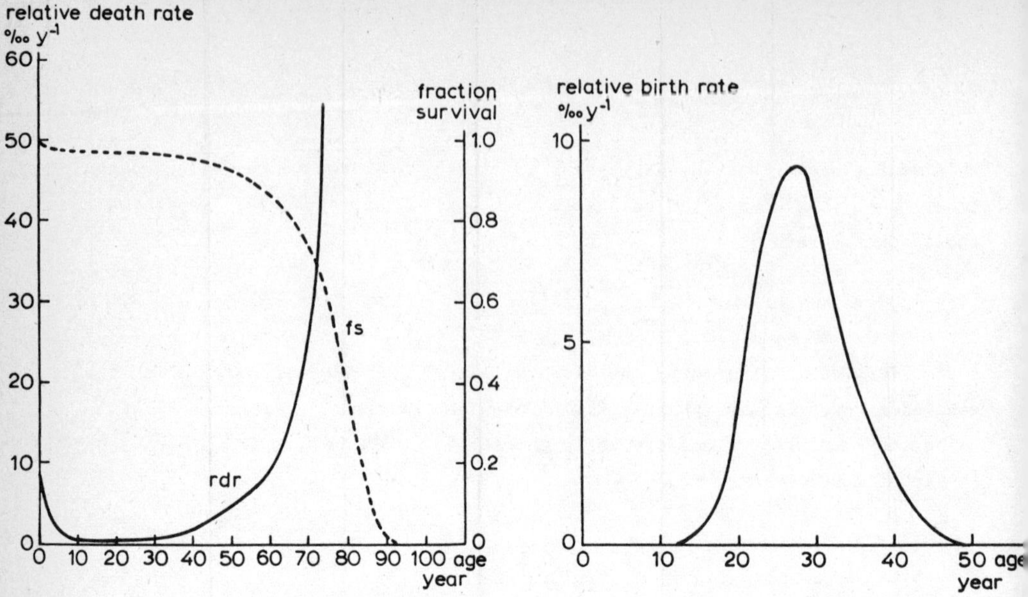

Fig. 6 The age dependence of relative death rate rdr and relative birth rate. The fraction survival fs (dashed line) is a function of rdr.

birth rates and age are supplied. In the INITIAL segment some computations are done for the discretization of age and development scale, necessary before the actual simulation in the DYNAMIC segment. The simulation itself requires computation of the rates of change, and of course the integration of these rates, which is done by the INTGRL statement. Also whole population totals are computed by summation over all age classes in a regular FORTRAN DO-loop. The data supplied for the FUNCTIONs are read by an AFGEN statement.
The shift as explained by the end of chapter 4.2 is performed in subroutine, called from the main program upon g' reaching γ.
This method results in a smooth time course of total population size and of birth rate, even on a time scale of one year, which is much smaller than the width of the boxes. In the method of De Wit and Goudriaan (1978) there was always the danger of a sawtooth behaviour of total population size, if one wanted a finer time resolution than the class width. Another disadvantage was that the age boundaries were defined in a much more complicated way, especially for the so-called pre-class. Here the role of this pre-class has been taken over by the zero class.This class has a clear meaning with well defined age boundaries.

```
TITLE   GROWTH OF A HUMAN POPULATION
FIXED N,I
*  INITIAL CONTENTS OF AGECLASSES OF 5 YEARS WIDE, IN THOUSANDS
TABLE WI(1-20)=582.,587.,553.,543.,554.,420.,380.,381.,378.,376., ...
      330.,323.,298.,226.,150., 70., 25., 13., 0.
*  AGE DEPENDENCE OF RELATIVE DEATH RATE, IN PROMILLE PER YEAR (FIG 6A):
FUNCTION WRDRT=0.,10.,2.5,4.,5.,1.8,7.5,0.8,10.,0.5,15.,0.3,20.,0.3,...
      30.,0.6,40.,1.6,50.,4.9,60.,8.5,65.,14.,70.,25.,75.,55., ...
      82.5,180.,87.5,380.,92.5,760.,97.5,900.,105.,900.
*  AGE DEPENDENCE OF RELATIVE BIRTH RATE; PER YEAR (FIG 6B):
FUNCTION RBRT=0.,0.,12.5,0.,17.5,0.02,22.5,.137,25.,.166,27.5,.188,...
      30.,.166,32.5,.113,37.5,.055,42.5,.016,47.5,.002,50.,0.,100.,0.
*  RATIO OF YOUNG BORN BOYS TO GIRLS
PARAM SEXR=1.048

INITIAL
*  100 YEARS OF AGE IS COVERED IN 20 CLASSES:
PARAM N=20,AGETOT=100.
*  RESIDENCE TIME IN ONE AGE CLASS:
      TC=AGETOT/N
*  DEVELOPMENT RATE; AND DEVELOPMENT WIDTH:
      DEVR=1.0/AGETOT
      GAMMA=TC*DEVR
*  THE FRACTION OF GIRLS IN THE YOUNG BORNS:
      FRGIRL=1.0/(1.0+SEXR)
NOSORT
*  CONVERSION OF THE INITIAL THOUSANDS TO INDIVIDUALS:
      DO 20 I=1,N
20    WI(I)=WI(I)*1000.

DYNAMIC
NOSORT

*  INTEGRAL OF THE DEVELOPMENT RATE
   GACC=INTGRL(0.,DEVR)
*  WHEN GACC EXCEEDS GAMMA, THE SHIFT IS APPLIED:
      IF(GACC.GE.GAMMA) CALL SHIFT(N,GAMMA,GACC,W,WO)
```

```
*  SUMMATION OF TOTAL BIRTH RATE AND OF TOTAL FEMALE POPULATION:
      TBR=0.
      TW=0.
      DO 100 I=1,N
      FLI=I
*  THE AGE AT THE CENTRE OF EACH CLASS:
      AGE=TC*(FLI-0.5) + GACC*AGETOT
      TBR=TBR + W(I)*AFGEN(RBRT,AGE)
*  THE DEATH RATE OF EACH CLASS:
      WR(I)= -W(I)*0.001*AFGEN(WRDRT,AGE)
100   CONTINUE

*  THE RATE OF BIRTH OF GIRLS:
      WBR=TBR*FRGIRL

*  THE ZERO CLASS RECEIVES THE BIRTH RATE, BUT ALSO DEATH
*  OCCURS AT AN AGE OF HALF ITS CURRENT WIDTH:
      WRO=WBR - WO*0.001*AFGEN(WRDRT,0.5*GACC*AGETOT)

*  INTEGRATIONS
      WO = INTGRL(0.,WRO)
      W  = INTGRL(WI,WR,20)

*     TIME IS EXPRESSED IN YEARS
TIMER FINTIM=50.,DELT=0.5,PRDEL=5.
*  TOTAL FEMALE POPULATION TW AND TOTAL BIRTH RATE TBR ARE PRINTED:
PRINT TW,TBR
METHOD RECT
END
STOP
      SUBROUTINE SHIFT(N,GAMMA,GACC,H,HO)
      DIMENSION H(N)
      DO 300 I=N,2,-1
300   H(I)=H(I-1)
      H(1)=HO
      HO  =0.
      GACC=GACC - GAMMA
      RETURN
      END
ENDJOB
```

Literature

Ferrari, Th.J., 1978. Elements of system-dynamics simulation; a textbook with exercises. Simulation Monographs, Pudoc, Wageningen, 89 pp.

Goudriaan, J., 1973. Dispersion in simulation models of population growth and salt movement in the soil. Neth. J. agric. Sci. 21: 269-281.

Penning de Vries, F.W.T. and H.H. van Laar, 1982. Simulation of plant growth and crop production. Simulation Monographs, Pudoc, Wageningen, 308 pp.

Wit, C.T. de and J. Goudriaan, 1978. Simulation of ecological processes. Revised version. Simulation Monographs, Pudoc, Wageningen, 175 pp.

THE SYSTEMATIC FORMULATION OF MODELS OF STAGE-STRUCTURED POPULATIONS

W.S.C. Gurney, R.M. Nisbet and S.P. Blythe
University of Strathclyde, Glasgow, G4 ONG

1. INTRODUCTION

It is widely believed that variations in the proportion of a given population in a particular physiological state (as determined by age, size, morphology or whatever) can be vital in determining its dynamic behaviour. The formalism for describing such variations is well known (see for example McKendrick 1926, von Foerster 1959, Sinko and Streifer 1967) and has been widely studied in the biomathematical literature. However, it has not proved widely acceptable as a tool for practical population modelling both because of its notorious technical intractability and because of its pronounced tendency to produce models whose parameter richness makes them very hard to test experimentally.

In response to the evident need for models which are both tractable and parameter sparse a number of workers have formulated *ad hoc* models of particular population systems in which age structure is represented via some form of time delay. Despite their popularity such models are frequently fatally flawed because their lack of a rigorous conceptual underpinning makes unambiguous interpretation impossible. Perhaps the most notorious example of this genre is the "time delayed logistic" (Hutchinson 1948, May 1974) which signally fails to distinguish between births, which clearly affect the size of the reproductively active population after some kind of maturation period, and deaths, which self-evidently act immediately, but the literature is overflowing with similar (if less blatant) examples. Moreover although it is perfectly possible (given sufficient care) to formulate an ad hoc age-structure model which at least obeys the elementary constraints of proper book-keeping, this is by no means the case when population structure is determined by some variable such as mass or physiological age which does not change at a constant rate, since this introduces book-keeping subtleties of considerable complexity.

Recently, however, a number of workers (Tognetti 1975, Vansickle 1977, Hastings and Wollkind 1982, Hastings, 1983, Gurney, Nisbet and Lawton 1983, Nisbet and Gurney 1983) have shown that with the aid of suitable "modelling approximations" it is possible to produce rigorous models of populations structured in a variety of different ways which are tractable, parameter sparse and readily interpretable. Examination of these apparently disparate formulations makes it clear that they have a number of essential features in common. In this paper we therefore discuss a unifying formalism which achieves a high degree of generality while incorporating ab initio the essential simplifications which facilitate the formulation of parameter sparse models.

2. THE DYNAMICS OF A STAGE-STRUCTURED POPULATION

The key to the formulation of tractable structured population models is the recognition that individuals of many species have life-histories composed of a sequence of "stages" within which their characteristics are broadly similar to those of other individuals in the same stage and markedly different from those of individuals in other stages. In insects such stages are particularly easy to recognise, being separated by short, clearly identifiable, events such as moults or pupation, but most species show discernible morphological, behavioural or biochemical changes which define similar "natural" stage boundaries, perhaps the most universal being the onset of reproductive activity.

We shall consider populations closed to immigration and emigration, so that individuals can be recruited into the ith life-history stage only from other stages within the population and can be lost from it only by death or maturation into another stage. Thus if we denote the sub-population of stage i at time t by $N_i(t)$, and define $R_i(t)$, $D_i(t)$ and $M_i(t)$ to represent the instantaneous total rates of recruitment into, death within, and maturation out of that stage, then the instantaneous rate of change of $N_i(t)$ must be given by the simple balance equation

$$\dot{N}_i(t) = R_i(t) - D_i(t) - M_i(t). \tag{1}$$

We now adopt the convention that if the life-history comprises a total of S stages then the stage index i shall run from 1 to S with the "newborn" stage being the first in the sequence. Hence if, at time t, the sub-population of the jth stage produces offspring at a rate $B_j(t)$ then the instantaneous rate of recruitment to the first stage must be

$$R_1(t) = \sum_{j=1}^{S} B_j(t). \tag{2}$$

Recruitment to stages other than the first comes only from immigration or maturation from other stages. Many species have simple linear life-histories in which individuals maturing out of stage i enter stage i+1 or die in the attempt. In such cases if the probability of a successful transition from stage i-1 to stage i is $\gamma_{i-1}(t)$, then the instantaneous total recruitment rate to stage i at time t is

$$R_i(t) = \gamma_{i-1}(t)\, M_{i-1}(t). \qquad i \geqslant 2 \tag{3}$$

There are, however, species whose life histories are somewhat more complex than that implied by equation (3). For example, to avoid adult emergence under unfavourable climatic conditions many damselfly species can switch into a "holding" or "overwintering" state (Corbett 1962, Norling 1981). To extend our formalism to cover such branched life-histories it is only necessary to define $\Phi_{ij}(t)$ as the probability that an individual maturing out of stage i at time t makes a successful transition to stage j, so that

$$R_i(t) = \sum_{j=1}^{S} \Phi_{ji}(t)\ M_j(t), \tag{4a}$$

where clearly the transition probability Φ_{ij} must obey the elementary restriction that

$$\sum_{j=1}^{S} \Phi_{ij}(t) \leqslant 1. \tag{4b}$$

In order to turn this purely formal structure into a useful modelling tool we need to establish systematic relationships between the "macroscopic" quantities, $N_i(t)$, $R_i(t)$, $D_i(t)$, $M_i(t)$ and $B_i(t)$, which characterize the overall behaviour of each stage. Such relationships clearly reflect the way in which individuals develop within the stage and must therefore be obtained by a consideration of internal stage dynamics. The key to an economical internal stage description lies in extending the concept of physiological age used by a number of previous workers (Gilbert and Gutierez 1973, Vansickle 1977, Taylor 1981, van Straalen 1983) and defining, for each stage, a single stage development index q_i, whose value is a unique measure of the state of development of a given individual within that stage.

We differ from previous workers in our use of this concept in several important respects. Firstly we regard the development index as describing the progress of an individual through a given stage rather than through its entire life-history, thus leaving ourselves free to adopt the most appropriate definition of q_i for each stage. For example in an insect species we might choose a suitably nomalized time-temperature product for the egg stage, normalized body size for the larvae, and some suitable measure of senescence for the adults. Secondly we differ from van Straalen (1983) in particular, in believing that the stage development index should not have the dimension of time. It is the fact that physiological age does not change with time at a constant rate which introduces many of the subtlties which bedevil ad hoc models, and it is thus helpful to be able to distinguish readily between time and developmental state; an aim which may be assissted by making the development index dimensionless.

The last group of modelling assumptions needed to complete our conceptual framework concern the characteristics of the individual within a given stage. We observed earlier that in general individuals within a stage naturally tend to have rather similar characteristic parameters, a tendency which it is generally possible to accentuate by skilful choice of stage definition. We now extrapolate from this observed similarity to the modelling approximation that individuals within a given stage are in fact identical in three important respects:

1. they were all recruited into the stage with the same initial development index,
2. at time t they all have the same development rate $q_i = g_i(t)$,
3. at time t they all have the same per capita death rate $\delta_i(t)$.

In fact we can gain considerably in convenience without any further loss of generality if we assume that each stage development index is defined so that the new recruits arrive with $q_i = 0$.

In addition to those vital rates which we assume identical for all individuals in a given stage we also assume that the values of any other vital rates which may vary from individual to individual are systematically related to the development indices of the individuals concerned. In particular we describe the production of offspring by individuals in stage i by a per capita fecundity $\beta_i(q_i,t)$ defined so that

$$\beta_i(q_i,t)dt = \text{Probability that an individual with development index } q_i \text{ produces an offspring during } t \to t + dt \tag{5}$$

and we also follow Diekmann, Lauwerier, Aldenberg and Metz (1983) in describing the process of maturation out of stage i by an index-dependent maturation probability $m_i(q_i)$ defined so that

$$m_i(q_i)dq_i = \text{Probability that an individual with development index } q_i \text{ matures out of stage i before reaching index } q_i + dq_i. \tag{6}$$

This latter quantity is related to $\Psi_i(q_i)$ the distribution of development indices at maturation which would be observed (after correction for mortality) for a single cohort passing through stage i, by

$$\Psi_i(q_i) = m_i(q_i) \exp \{- \int_o^{q_i} m(x)dx\}. \tag{7}$$

3. THE INTERNAL DYNAMICS OF A STAGE

We now consider in detail the internal dynamics of a stage and seek to establish connections between the rates of recruitment, maturation, death and reproduction which characterize its behaviour. In this section we are considering a given stage in isolation from the remainder of the life-history, and thus for economy of presentation we shall drop all stage identifying suffixes.

3.1 The Stage Development-index Distribution

The modelling assumptions discussed in section 2 together imply that the characteristics of an individual within the stage depend wholly on its development index q. A complete description of the internal state of the stage at time t is thus embodied in the stage development index distribution f(q,t) defined so that

$$f(q,t)dq = \text{Number of individuals in the stage at time t with development indices in the range q to q + dq.} \tag{8}$$

and related to the instantaneous stage population by

$$N(t) = \int_o^\infty f(q,t)dq. \tag{9}$$

We now recall our fundamental modelling approximation that all individuals in the stage at time t have the same development rate g(t) and the same per capita death rate $\delta(t)$. We also note that our description of maturation (equation (6)) implies that the per capita maturation rate for individuals with development index q at time t is g(t)m(q), and hence that the total loss rate of such individuals by death and maturation is $\delta(t) + g(t)m(q)$. It follows (see for example Sinko and Streiffer 1967) that the stage development index distribution f(q,t) must obey the continuity equation

$$\frac{\partial f(q,t)}{\partial t} + g(t)\frac{\partial f(q,t)}{\partial q} + [\delta(t)+g(t)m(q)]f(q,t) = 0 \qquad (10)$$

which must (see for example Vansickle 1977, Nisbet and Gurney 1983) be solved subject to the boundary conditions

$$f(0,t) = R(t)/g(t) \quad t \geq 0 \quad ; \quad f(q,0) = f_o(q) \quad q \geq 0. \qquad (11)$$

The solution of equation (10) is facilitated by first defining four auxiliary quantities:

$$S_D(t_1,t_2) \equiv \exp\{-\int_{t_1}^{t_2} \delta(x)dx\} \qquad (12)$$

which represents the probability that an individual who does not mature out of the stage between times t_1 and t_2, also does not die during that interval;

$$S_M(q_1,q_2) \equiv \exp\{-\int_{q_1}^{q_2} m(x)dx\} \qquad (13)$$

which represents the probability that an individual who does not die while making the transition from development index q_1 to index q_2 also fails to mature out of the stage during that time;

$$q_M(t) \equiv \int_o^t g(x)dx \qquad (14)$$

which represents the total development increment achieved by an individual who remains in the stage between time zero and time t; and $t_R(q,t)$ which is defined as the solution of

$$\int_{t_R(q,t)}^{t} g(x)dx = q \qquad (15)$$

and thus represents the time at which an individual with development index q at time t was recruited into the stage. It now follows from our assumption that all individuals in the stage at a given time develop at the same rate, that any member of the initial population still present in the stage at time t must have a development index in excess of $q_M(t)$, while by the same token all individuals recruited into the stage after time zero must, at time t, have development indices less than $q_M(t)$. The quantity $q_M(t)$ thus forms a (moving) boundary between two domains of solution. The part of the stage population with development indices below the boundary is composed solely of survivors of cohorts recruited into the stage after t = 0, so that in this region ($q \leqslant q_M(t)$) the distribution function f(q,t) is related to the rate of recruitment at time $t_R(q,t)$ by

$$f(q,t) = \frac{R(t_R(q,t))}{g(t_R(q,t))} S_D(t_R(q,t),t)\ S_M(0,q) \qquad q \leqslant q_M(t). \qquad (16)$$

By constrast the part of the stage population with development indices above the boundary is composed solely of survivors of the original population, so that in this region the distribution function is related to the initial distribution by

$$f(q,t) = F_o(q-q_M(t))\ S_D(0,t)\ S_M(0,q) \qquad q \geqslant q_M(t) \qquad (17)$$

where we have, for future convenience, re-expressed the raw initial development index distribution $f_o(q)$ in terms of a modified initial distribution $F_o(q)$ defined by

$$F_o(q) \equiv f_o(q)/S_M(0,q). \qquad (18)$$

3.2 Death and Maturation

Integrating the continuity equation (10) over the range q = 0 to ∞ now confirms that the overall balance equation for the stage population is

$$\dot{N}(t) = R(t) - M(t) - D(t) \tag{19}$$

where, as we should expect, the overall death rate D(t) is

$$D(t) = \delta(t)N(t) \tag{20}$$

and where we can identify the maturation rate at time t, M(t), as

$$M(t) = g(t) \int_0^\infty f(q,t)m(q)dq. \tag{21}$$

Substituting the formal solutions, equations (16) and (17), for f(q,t) and then recognizing $m(q)S_M(0,q)$ as the mortality corrected cohort maturation distribution $\Psi(q)$, enables us to restate equation (21) as the combination of an initial condition term and a recruitment history integral back to t = 0, thus

$$M(t) = g(t)P^+(t)O_M(q_M(t)) + g(t) \int_0^t \Psi(\Delta q(x,t))S_D(x,t)R(x)dx. \tag{22}$$

The quantity $O_M(q_M)$ in the term related to the initial conditions is an overlap integral between the modified initial distribution $F_o(q)$ and the cohort maturation distribution $\Psi(q)$ displaced by the growth increment q_M thus:

$$O_M(q_M) \equiv \int_0^\infty F_o(q)\Psi(q+q_M)dq, \tag{23}$$

$\Delta q(t_1,t_2)$ is the development increment achieved by an individual present in the population between times t_1 and t_2:

$$\Delta q(t_1,t_2) \equiv \int_{t_1}^{t_2} g(x)dx, \tag{24}$$

and for convenience we have defined $P^+(t)$ to represent the overall survival (against death alone) of the initial population:

$$P^+(t) \equiv S_D(0,t). \tag{25}$$

Equation (22) can be dramatically simplified under two special sets of circumstances. The first of these is where the cohort maturation distribution $\Psi(q)$ is a member, say the Lth, of some set of Q+1 functions $X_i(q)$ $\{i = 0..Q\}$ all of which have the property that

$$\frac{dX_i(q)}{dq} = \sum_{j=0}^{Q} \alpha_{ij} X_j(q). \tag{26}$$

In this case we use the "linear chain trick" which dates back to Fargue (1973), was developed by MacDonald (1978) in another context and was applied to age dependent maturation by Blythe, Nisbet and Gurney (1984a). We define a set of Q+1 auxiliary variables

$$A_i(t) = \int_0^t R(x)S_D(x,t)\ X_i(\Delta q(x,t))dx \tag{27}$$

whose dynamics can be shown by routine manipulation to be described by a set of Q+1 ordinary differential equations

$$\dot{A}_i(t) = X_i(0)R(t) - \delta(t)A_i(t) + g(t) \sum_{j=0}^{Q} \alpha_{ij}A_j(t). \tag{28}$$

solved subject to the initial condition

$$A_i(0) = 0, \tag{29}$$

and then use the solution of these equations to calculate the maturation rate from

$$M(t) = g(t)\{S_D(0,t)0_M(q_M(t)) + A_L(t)\}. \tag{30}$$

The second class of circumstances in which we can simplify equation (22) is where no individual can mature with a development index less than some critical value $q = q^*$, but any individual who survives to this point matures out of the stage immediately (Nisbet and Gurney 1983). In this case the mortality corrected cohort maturation distribution $\Psi(q)$ is a right-sided delta function at $q = q^*$ and the maturation rate at time t is given by

$$M(t) = \begin{cases} g(t)\ P^{+}(t)f_o(q^{*}-q_M(t)) & q_M(t) \leq q^{*} \\ g(t)\ P(t)R(T_R(t))/g(T_R(t)) & q_M(t) > q^{*} \end{cases} \qquad (31)$$

where for convenience we have defined

$$T_R(t) \equiv \begin{cases} 0 & q_M(t) \leq q^{*} \\ t_R(q^{*},t) & q_M(t) > q^{*} \end{cases}, \qquad (32)$$

and

$$P(t) \equiv S_D(T_R(t),t). \qquad (33)$$

3.3 Reproduction

If the per capita fecundity of an individual with development index q at time t is $\beta(q,t)$ then it is clear that the total rate at which the stage population produces offspring must be

$$B(t) = \int_0^\infty \beta(q,t)f(q,t)dq. \qquad (34)$$

If we now make the additional assumption that the time and development dependence of β act multiplicatively so that

$$\beta(q,t) = \beta_t(t)\beta_q(q). \qquad (35)$$

then equation (34) is structurally identical to equation (21) and we can proceed by analogy with section 3.2.

We begin by defining a "cohort fecundity distribution"

$$\theta(q) \equiv \beta_q(q)S_M(0,q) \qquad (36)$$

which, if the "fecundity multiplier" $\beta_t(t)$ is appropriately defined, represents the relative fecundity distribution which would be observed after correction for mortality only, as a single cohort passed through the stage under constant conditions. It is then possible to re-express equation (34) as the combination of a term describing the (decaying) effects of the initial population and a recruitment history integral

$$B(t) = \beta_t(t)S_D(0,t)O_B(q_M(t)) + \beta_t(t)\int_0^t \Theta(\Delta q(x,t))S_D(x,t)R(x)dx \tag{37}$$

where the overlap integral in the initial condition term is here defined as

$$O_B(q_M) = \int_0^\infty F_o(q)\Theta(q+q_M)dq. \tag{38}$$

As in the previous section, the history integral in equation (37) can be calculated from a set of ordinary differential equations if $\Theta(q)$ is a member (say the Lth) of a set of Q+1 functions $Y_i(q)$ $\{i = 0..Q\}$ each with the property that

$$\frac{dY_i(q)}{dq} = \sum_{j=0}^{Q} \nu_{ij}Y_j(q) \qquad \text{for all } q. \tag{39}$$

As before we use the linear chain trick (applied to age dependent fecundity in a maturation free stage by Blythe Nisbet and Gurney 1984b). We define a set of Q+1 auxiliary variables

$$C_i(t) \equiv \int_0^t R(x)S_D(x,t)\ Y_i(\Delta q(x,t))dx \tag{40}$$

which, by analogy with equations (27) and (28) can be seen to obey the differential equation

$$\dot{C}_i(t) = Y_i(0)R(t) - \delta(t)C_i(t) + g(t)\sum_{j=0}^{Q} \nu_{ij}C_j(t) \tag{41}$$

solved subject to

$$C_i(0) = 0. \tag{42}$$

The overall reproduction rate B(t) is then calculated from

$$B(t) = \beta_t(t)P^+(t)O_B(q_M(t)) + \beta_t(t)C_L(t). \tag{43}$$

In the case of a fixed end-point stage where

$$S_M(0,q) = 1 \qquad q \leqslant q^* \tag{44}$$

$$f(q,t) = 0 \qquad q > q^* \tag{45}$$

we can restate B(t) as

$$B(t) = \beta_t(t)\, P^+(t)O_F(q_M(t)) + \beta_t(t) \int_{T_R(t)}^{t} \beta_q(\Delta q(x,t))S_D(x,t)R(x)dx \tag{46}$$

where we have here defined the initial condition overlap integral as

$$O_F(q_M) = \begin{cases} 0 & q_M(t) > q^* \\ \int_{q_M(t)}^{q^*} f_o(q-q_M(t))\beta_q(q)dq & q_M(t) \leqslant q^* \end{cases} \tag{47}$$

In this case too we can use the linear chain trick provided that $\beta_q(q)$ is a member of a set of Q+1 functions $Z_i(q)$ {i = 0..Q} each with the property that

$$\frac{dZ_i(q)}{dq} = \sum_{j=0}^{Q} \eta_{ij}\, Z_j(q) \qquad q \leqslant q^*. \tag{48}$$

Here we define a set of Q+1 auxiliary variables

$$H_i(t) = \int_{T_R(t)}^{t} Z_i(\Delta q(x,t))S_D(x,t)R(x)dx. \tag{49}$$

which can, by routine manipulation be shown to obey the ordinary differential equation

$$\dot{H}_i(t) = Z_i(0)R(t) - Z_i(q^*)M_R(t) - \delta(t)H_i(t) + g(t)\sum_{j=0}^{Q} \eta_{ij}H_j(t) \tag{50}$$

where we define

$$M_R(t) = \begin{cases} 0 & q_M(t) < q^* \\ M(t) & q_M(t) \geq q^*. \end{cases} \tag{51}$$

These equations are solved subject to the initial condition

$$H_i(0) = 0 \tag{52}$$

and then B(t) is obtained from

$$B(t) = \beta_t(t)P^+(t)\Theta_F(q_M(t)) + \beta_t(t)H_L(t). \tag{53}$$

4. PRACTICAL STAGE-STRUCTURE MODELLING

4.1 Summary of the Formalism

Formulating a stage-structured population model involves approximating the life-history of the species concerned as a sequence of stages within each of which individuals are as near identical as can be arranged. For each stage an appropriate stage development index q_i is defined, whose rate of change (the development rate $g_i(t)$) describes the rate at which individuals progress through the stage. The pattern of reproduction and outward maturation for each stage is characterized by time-independent fecundity and maturation distributions ($\Theta_i(q_i)$ and $\Psi_i(q_i)$ respectively) which would be measured by observing the mortality corrected behaviour of a single cohort passing through the stage under constant conditions. The time (or density) dependence of reproduction and maturation are assumed to act multiplicatively and are respectively described by a fecundity multiplier $\beta_{ti}(t)$ and a maturation success probability matrix $\Phi_{ij}(t)$, which also specifies the (possibly time dependent) pattern of stage linkage.

The further modelling approximations that all individuals in a given stage at a given time were recruited with $q_i = 0$ and have identical per capita death and development rates ($\delta_i(t)$ and $g_i(t)$ respectively) then enable us to show that the population dynamics are described by a set of sub-population balance equations

$$\dot{N}_i(t) = R_i(t) - \delta_i(t)N_i(t) - M_i(t) \tag{54}$$

where the recruitment rate to the newborn (i=1) stage is related to the total offspring production rate of each stage $B_i(t)$, by

$$R_1(t) = \sum_{j=1}^{S} B_j(t) \tag{55}$$

and the recruitment rate to all other stages is related to the rates of maturation $M_j(t)$ out of the appropriate previous stage (or stages) by

$$R_i(t) = \sum_{j=1}^{S} \Phi_{ji}(t)M_j(t) \qquad i \geq 2. \tag{56}$$

The maturation and offspring production rates for each stage are then given by the combination of a term describing the decaying effect of the initial population with a recruitment history integral

$$M_i(t) = g_i(t)P_i^+(t)O_{Mi}(q_{Mi}(t)) + g_i(t) \int_o^t R_i(x)S_{Di}(x,t)\Psi_i\{\int_x^t g_i(y)dy\}dx \tag{57}$$

$$B_i(t) = \beta_{ti}(t)P_i^+(t)O_{Bi}(q_{Mi}(t)) + \beta_{ti}(t) \int_o^t R_i(x)S_{Di}(x,t)\Theta_i\{ \int_x^t g_i(y)dy\}dx \tag{58}$$

where the maturation independent survival $S_{Di}(x,t)$ is given by

$$S_{Di}(x,t) \equiv \exp\{-\int_x^t \delta_i(y)dy\}, \tag{59}$$

the boundary index value $q_{Mi}(t)$ is given by

$$q_{Mi}(t) \equiv \int_o^t g_i(y)dy \tag{60}$$

and for convenience we have defined an "initial condition survival"

$$P_i^+(t) \equiv S_{Di}(0,t) \tag{61}$$

and two overlap integrals

$$O_M(q') \equiv \int_o^\infty F_o(q)\Psi(q+q')dq \quad ; \quad O_B(q') \equiv \int_o^\infty F_o(q)\Theta(q+q')dq \tag{62}$$

4.2 Numerical integration of Stage-Structure models

Specific population models are erected on this framework by choosing appropriate functional forms for the cohort fecundity and maturation distributions $\Theta_i(q_i)$ and $\Psi_i(q_i)$, the per capita growth and death rates $g_i(t)$ and $\delta_i(t)$, the fecundity multipliers $\beta_{ti}(t)$ and the stage linkage matrix $\Phi_{ij}(t)$. Population regulation is of course provided by making some of the time-dependent factors explicitly dependent upon some function of the stage populations. It is normally possible to obtain stationary states and local stability properties for such models without undue analytic labour, but non-linear properties such as limit cycle frequencies are almost always more conveniently obtained from numerical solutions. A predictor-correction algorithm is well suited to this purpose provided that a historical record of stage recruitment and development rates is maintained (usually in a ring buffer for the sake of efficiency).

Evaluation of the initial condition terms is extremely simple provided the functions modelling the modified initial distributions F_{oi} and the cohort maturation and fecundity distributions Ψ_i and Θ_i are chosen so that the overlap integrals O_{Mi} and O_{Bi} can be evaluated analytically. We then observe that $q_{Mi}(t)$ and $P^+{}_i(t)$ obey the simple ordinary differential equations

$$\dot{q}_{Mi}(t) = g_i(t) \qquad ; \qquad q_{Mi}(0) = 0 \tag{63}$$

$$\dot{P}_i^+(t) = -P_i\,(t)\delta_i(t) \qquad ; \qquad P_i^+(0) = 1 \tag{64}$$

whose solutions (analytic or numerical) can be plugged directly into the initial condition terms in (57) and (58).

In principle the history integrals in equations (57) and (58) can be evaluated directly from the historical record, but this is clearly rather time consuming and so it pays to simplify wherever possible. The simplest case is that of a stage in which per capita fecundity is index independent so that the cohort fecundity distribution is

$$\theta_i(q_i) = \beta_{qi}^0 \exp\{- \int_o^{q_i} m_i(x)dx\} \tag{65}$$

and hence, as we should expect

$$B_i(t) = \beta_{qi}^0\ \beta_{ti}(t)N_i(t). \tag{66}$$

This result is so simple that it implies that if adult fecundity varies rather slowly with development index it is well worth dividing the "proper" adult stage into a series of dummy stages within which constant per capita fecundity is a viable approximation.

More generally we have shown in sections 3.2 and 3.3 that if Ψ_i or Θ_i are members of a set of functions all of whose first differentials can be expressed as linear combinations of members of the set, then the appropriate history integrals can be evaluated as the solution of a set of ordinary differential equations ((28) or (41)). Since this qualifying condition applies to numerous functions including polynomials, sinusoids, gamma distributions and products of these things it is not unduly restricting, but it is possible for the number of auxiliary equations to become rather large. In general one finds that if the data can be fitted with a low order polynomial or gamma function (i.e. if the quantity concerned varies fairly smoothly) then few auxiliary equations are needed and great gains can be made, but if very sharp variations are to be modelled very large numbers of auxiliary variables are needed and it may be preferable to go back to direct evaluation of the history integrals, or reformulate the model with different stage boundaries.

Considerable simplifications are also possible in the case of a fixed end-point stage, in which all maturation occurs at a fixed q (say q^*_i), in which case $\Psi_i(q_i)$ is a delta function at q^*_i and it can be shown that

$$M_i(t) = \begin{cases} g_i(t)P_i^+(t)f_{oi}(q_i^*-q_{Mi}(t)) & q_{Mi}(t) \leq q_i^* \quad (67) \\ R_i(T_{Ri}(t))P_i(t)g_i(t)/g_i(T_{Ri}(t)), & q_{Mi}(t) > q_i^* \quad (68) \end{cases}$$

where the auxiliary variables $T_{Ri}(t)$, the recruitment time for individuals maturing at time t, and $P_i(t)$, the through stage survival for such individuals can be shown to obey the ordinary differential equations

$$\frac{dT_{Ri}}{dt} = \begin{cases} 0 & q_{Mi}(t) \leq q_i^* \\ g_i(t)/g_i(T_{Ri}(t)). & q_{Mi}(t) > q_i^* \end{cases} \qquad (69)$$

$$\frac{dP_i}{dt} = P_i(t) \, \{\delta_i(T_{Ri}(t)) \frac{dT_{Ri}}{dt} - \delta_i(t)\}. \qquad (70)$$

which must be solved subject to

$$T_{Ri}(0) = 0 \quad ; \quad P_i(0) = 1. \qquad (71)$$

Additional savings in numerical work can also be made by noting that $q_{Mi}(t)$ increases monotonically with time (provided $g_i(t) \geq 0$) so that once $q_{Mi}(t)$ has passed q^*_i neither it nor the accompanying auxiliary variable P^+_i play any further role in the calculation.

We must however enter a cautionary note that this "fixed end point" variant of the formalism suffers from a technical difficulty in circumstances where R_i remains finite when g_i goes to zero; a condition which might plausibly arise if adverse conditions bore less heavily upon a given stage then upon its successor in the life history. In this case all individuals recruited while the growth rate is zero will eventually mature out of the stage at exactly the same time, thus producing a singularity in the maturation rate. Although relatively innocuous, as singularities go, since it still corresponds to the maturation of a finite number of individuals, it still produces very serious numerical problems. This problem does not occur in the general, distributed maturation, version of the formalism which is consequently to be preferred to the fixed endpoint variant under circumstances where development rates can become very small without an accompanying cessation of recruitment.

4.3 Discussion

We have shown that with the aid of some intuitively reasonable modelling approximations, populations of organisms with a stage-structured life-history can be represented by parameter-sparse models which are both analytically tractable and computationally convenient. These techniques were developed in direct response to the need to formulate falsifiable models of specific experimental and field systems, and have already been successfully used to model laboratory populations of the sheep blowfly *Lucilia cuprina* (Gurney et al 1983) and of the stored product moth *Plodia interpunctella* (Gurney et al 1983, Gurney and Nisbet 1984). Work is now underway to extend our repertoire to field populations of damselflies, daphnia, red-scale, marine copepods and/or fish larvae growing in large sea enclosures, and to more complex laboratory cultures in which a stored product beetle of the *Callosobruchus* family interacts with a parasitic wasp.

The techniques are most readily applicable to species such as insects whose life-history stages are clearly defined, but even where stage boundaries are blurred (or inconveniently numerous, as with both marine copepods and damselflies) it usually proves possible to define arbitrary pseudo-stages which yield an acceptably accurate description. A number of technical problems remain unsolved - notably the question of index dependent death rates - but only one entire class of organisms remains resistant to the stage structure approach. Organisms which reproduce by fission (algae, bacteria etc.) cannot sensibly be regarded as producing identical offspring, since it appears to be central to the dynamics of such populations that the properties of offspring correlate strongly with those of their parents. We have thus been unable to formulate a useful stage-structured cell-cycle model. However with this one (admittedly important) group of exceptions the technique has proved widely useful, particularly because the structured approach to model formulation which it encourages is itself a valuable discipline.

Acknowledgment. We thank Horst Thieme for helping to elucidate some of the technical problems associated with zero development rates.

References

Blythe, S.P. Nisbet, R.M. and Gurney, W.S.C. (1984a), "The Dynamics of Population Models with Distributed Maturation Period". Theor. Pop. Biol. - in press.

Blythe,S.P. Nisbet, R.M. and Gurney, W.S.C. (1984b), "Stage Structure Models with Age and Density-Dependent Fecundity". In preparation.

Corbet, P.S. (1962), "Dragonflies", Witherby, London.

Diekmann, O, Lauwerier, H., Aldenberg, T. and Metz, J. (1983), "Growth, Fission and the Stable Size Distribution", J. Math. Biol. 18, 135-148.

Fargue, D. (1973) "Reductibilite des Systems Hereditaires", Comptes Rendue Acad. Sci. Paris Serie B, 277, 471-473.

von Foerster, H., (1959), "Some Remarks on Changing Populations" in "The Kinetics of Cellular Proliferation", (Ed. F. Stohlman, Jr.) Grune & Stratton, New York.

Gilbert N. and Gutierez, A.P. (1973). "A Plant -Aphid- Parasite Relationship", J. Anim. Ecol. 42, 323.

Gurney, W.S.C. Nisbet, R.M. and Lawton, J.H. (1983), "The Systematic Formulation of Tractable Single Species Population Models Incorporating Age-structure", J. Anim. Ecol. 52, 479-495.

Gurney, W.S.C. and Nisbet, R.M. (1984) "Generation Separation, Fluctuation Periodicity and the Expression of Larval Competition". Submitted to American Naturalist.

Hastings, A. and Wollkind, D, (1982), "Age Structure in Predator-Prey Systems: I A General Model and a Specific Example", Theor. Pop. Biol. 21: 44-56.

Hastings, A., (1983), "Age-dependent Predation is Not a Simple Process. I continuous time models", Theor. Pop. Biol. 23, 347-362.

Hutchinson, G.E., (1948), "Circular Causal Systems in Ecology", Annals N.Y. Academy of Sciences, 50: 221-246.

May, R.M.(1974), "Stability and Complexity in Model Ecosystems", Princeton U.P.

McKendrick, A.G. (1926), "Applications of Mathematics to Medical Problems", Proc. Edinburgh Math. Soc. 44: 98-130.

MacDonald, N. (1978), "Time Delays in Biological Models", Springer-Verlag Lecture, Notes in Biomathematics, Vol. 27.

Nisbet, R.M. and Gurney, W.S.C., (1983), "The Systematic Formulation of Population Models for Insects with Dynamically Varying Instar Duration", Theor. Pop. Biol. 23, 114-135.

Norling, U. (1981), "Evolutionary Traits in Life Cycles and Gill Development in Odonata", Thesis - University of Lund.

Vansickle, J. (1977), "Analysis of a Distributed Parameter Population Model Based on Physiological Age", J. Theor. Biol. 64: 571-586.

van Straalen, (1983) "Physiological Time and Time-invariance" J. Theor. Biol. 104, 349-357.

Sinko, J. and Streifer, W. (1967), "A New Model for Age-size Structure of a Population" Ecology, 48, 6, 910-918.

Taylor, (1981), "Ecology and Evolution of Physiological Time in Insects", Am. Nat. 117: 1-23.

Tognetti, K. (1975), "The Two Stage Integral Population Model", Math. Biosciences, 24, 61-70.

V. Analytical approaches and a novel type of *i*-state.

Population dynamical models generally are constructed either to mimic the behaviour of some specific population or to illustrate some conceptual issue. Models of the first type are either completely specified or at worst belong to a low dimensional parametric family, allowing their properties to be studied numerically. The quantitative matching of observations and model predictions also instills an implicit trust in the accuracy of the numerical technique. In the second case numerical methods may be used in the construction of examples or counter examples, but we have to ascertain the essential correctness of the numerics in an independent manner. However, analytical methods are to be preferred as only these allow us to infer in a general fashion the potential consequences of large classes of mechanisms.

The *i*-variables occurring in the models encountered till now, such as age or size, were fairly concrete. This is less the case for the *i*-state variable which figures in the only paper in this section: "social rank". Here rank is considered a unique attribute of an individual, higher rank conferring higher "fitness". Size of territory provides one concrete example. It turns out that, given some plausible assumptions on the way individuals jump ranks, a rank structure bestows a strong type of stability upon the associated population dynamics.

A differential-integral equation modelling the dynamics of populations with a rank structure

Horst R. Thieme

Sonderforschungsbereich 123, Universität Heidelberg *

Introduction

This paper is intended as a mathematical contribution to the ecological question whether and to what extent population structures can stabilize the growth of populations. We concentrate upon structures which result in a stratification of the population according to ranks such that individuals with higher ranks do better (i.e. produce more offspring in their lives) than individuals with lower ranks. Examples are provided by the territorial or hierarchical organisation of populations and rank can be interpreted as the quality of the territory or the status in a dominance hierarchy (see, e.g., [5], II,12,13).

A mathematical model cannot prove, of course, that such population structures which originate from social behaviour actually regulate populations; our model will show however, that they potentially do so (this can already be seen without any mathematics) and that they are potentially very effective in so far as they drive the population into stable states (See [5], p.24, for the distinction between potential and operational factors in population regulation, further see the section on territories and population regulation, p.274).

To make the stabilizing potential of rank structures as clear as possible the model is chosen so that the instantaneous individual birth and death rates depend on the instantaneous rank of the individual, but not on the population density (or on something else like resources, predators or infectious agents the abundance of which might depend on the present or past population density). It is only the transition from lower to higher ranks where density-dependence comes in. Individuals are born at a low rank and then work up the rank hierarchy. We assume that it is the more difficult to get to a higher rank the more it is already occupied by other individuals. We do not model the rise in rank as proceeding smoothly but instead as occurring in jumps. Thus we obtain a differential-integral equation (differentiation in time and integration over ranks) instead of the partial differential equations arising in the models which are discussed in chapter 3 of this book. For simplicity we confine our consideration to a time-homogeneous model, i.e. the individual birth, death and rank transition rates are constant in time; but generalization to a periodic or completely irregular time-dependence seems possible. We show that the population either dies out or tends to a globally asymptotically stable non-zero equilibrium state. Which of these alternatives actually occurs depends on whether the reproductive potential of the population and the permeability of the rank structure at low population densities are high enough to prevent extinction. See section 1 for the model and a precise statement and a discussion of the main results. Proving the existence and uniqueness of solutions is interesting in so far as we neither assume a Lipschitz condition nor the equation exhibits obvious compactness properties. See section 2. Whether the population dies out or not will eventually depend on whether the zero solution is stable or not. See section 3 and section 7. Section 4 shows that there exist non-zero equilibria, if the zero equilibrium looses its local asymptotic stability. Since our assumptions make the equation quasi-monotone, global asymptotic stability reduces to uniqueness of non-zero steady states. See section 5. (Further see [1] for a general treatment of the asymptotic behaviour of quasi-monotone differential equations). Proving this uniqueness is the main mathematical point of this paper (see section 6). For the steady state equation does not permit an immediate application of the usual techniques, e.g. of Krasnosel'skii's sub- or superlinearity methods [2]. It is only after a non-obvious transformation that a modification of these techniques will work in producing additional assumptions which, astonishing enough, make biological sense, namely that more highly ranking individuals produce more offspring in their lives and the chance of getting a certain higher rank increases with the height of rank already held. Whether these assumptions are also necessary, remains an open problem, however.

---

* This paper presents work supported by the Deutsche Forschungsgemeinschaft.

1. The model. main results. discussion

It should be stated first that our model does not claim to describe any really existing population. It is a completely theoretical model caricaturing some mechanisms operating in territorially or hierarchially organised populations. See, e.g. [5], sections 12,13.

We assume that the population is stratified according to a rank structure ranging from a minimum rank 0 up to a maximum rank 1. The individuals of the population can occupy any rank in the interval [0,1] with their instantaneous birth and mortality rates b and m depending on their instantaneous ranks. Individuals with rank x can jump to a higher rank y at a rate g depending on x,y and the number of individuals occupying rank y. Hereby the upgrading of one individual is not linked with the degrading of another. Actually degradation is assumed to be such a seldom event that it can be ignored. Newborns have a lower rank than their parents, at least some of them must start from the lowest rank 0.

To make our model precise let $u(t,x)$ denote the "number" of individuals with rank x at time t. As usually $u(t,x)$ has to be considered a density in x, i.e. only $\int_{x_1}^{x_2} u(t,x)dx$, the number of individuals ranking from x_1 to x_2 at time t, has a proper biological meaning. The following equation links the three processes on the level of the individual - birth, death and rank transition - to the development of the population

$$\dot{u}(t,x) = \int_x^1 u(t,y)b(y,x)dy - m(x)u(t,x) + \int_0^x u(t,y)\tilde{g}(y,x,t)dy - \int_x^1 u(t,x)\tilde{g}(x,y,t)dy. \tag{1.1}$$

The temporal change $\dot{u}(t,x) = d/dt\, u(t,x)$ of the number of individuals with rank x at time t results from four events: the birth of individuals with rank x from parents with rank y higher than x; $b(y,x)$ denotes the related birth rate. The death of individuals with rank x with $m(x)$ denoting the rank-dependent mortality rate. The immigration to rank x from lower ranks and the emigration from x to higher ranks with $\tilde{g}(y,x,t)$ indicating the rate at which an individual jumps from rank y to rank x at time t.

Since we want to explore the regulatory power of rank structures we assume that the birth and mortality rates b and m do not depend on the population density, i.e. on u. We do, however, assume density-dependence for $\tilde{g}(y,x,t)$ in the following way:

$$\tilde{g}(y,x,t) = g(y,x,u(t,x)) \tag{1.2}$$

with $g(y,x,u)$ being a monotone non-increasing function of u. This means that jumping from rank y to a higher rank $x>y$ at time t is impeded by the individuals holding rank x at that time who do not want to share their relative privileged position with others. The resistance is the stronger the more rank x is already occupied. The implicit assumption that individuals with ranks different from x do not interfere fits better with a territorial than hierarchical organisation of the population. For the attempt to settle in a territory leads to confrontation with the holders of the same territories whereas jumping from a lower to a higher rank in a dominance hierarchy may be affected by individuals ranking in between.

To complete the model we prescribe initial values at time $t=0$, namely

$$u(0,x) = u_0(x), \quad 0 \leqslant x \leqslant 1 \tag{2}$$

with a given non-negative continuous function u_0.

Let us combine the assumptions made so far with some technical conditions.

ASSUMPTIONS 1.1.

a) $b(y,x), m(x), g(x,y,u)$ are non-negative, jointly continuous functions of $0 \leqslant x \leqslant y \leqslant 1, u \geqslant 0$.

b) $m(x) > 0$ for $0 \leqslant x \leqslant 1$.

c) $g(x,y,u)$ is a monotone non-increasing function of u for $x \leqslant y$.

The assumptions 1.1 are supposed to hold throughout this paper without further mentioning. They are sufficient to guarantee the existence and uniqueness of non-negative solutions to (1) for $t > 0$, if non-negative initial values are prescribed at $t = 0$.

THEOREM 1.2 *Let $u_0(x)$ be a continuous non-negative function of $0 \leqslant x \leqslant 1$. Then there exists a continuous non-negative solution $u(t,x)$ to (1),(2) for all $t \geqslant 0, 0 \leqslant x \leqslant 1$.*

Since non-negative solutions are of biological interest only and $g(x,y,u)$ only makes sense for $u \geqslant 0$ we agree on a 'solution to (1)' being automatically non-negative and, for simplicity, continuous in (t,x).

Though we have not assumed a Lipschitz condition for $g(x,y,u)$ in u, we obtain not only existence but also uniqueness of solutions. Even existence seems to be a problem because no compactness is visible at the first glance. But the decrease of g in u helps to establish compactness properties. It also provides a comparison principle which settles the uniqueness problem. The comparison principle also relies on the following observation: We obtain an equation for the growth of the total population by integrating (1.1) over x from 0 to 1. The terms containing $\tilde{g}$ drop out (as they should do because rank transition does not change the total size of the population) and, after a change of the order of integration and of variables, we obtain

$$\frac{d}{dt}\int_0^1 u(t,x)dx = \int_0^1 u(t,x)(\int_0^x b(x,z)dz)dx - \int_0^1 m(x)u(t,x)dx. \tag{3}$$

(3) does not only provide the last step in proving the comparison principle (see section 2) but also gives some first intuitive insight how population regulation by internal structures works. (3) looks very linear such that one might not understand at the first glance why regulation takes place at all. To see this more clearly we set $N(t) = \int_0^1 u(t,x)dx$ and $\tilde{u}(t,x) = u(t,x)/N(t)$. $N(t)$ indicates the total size of the population while $\tilde{u}(t,x)$ indicates the rank distribution. Now (3) takes the form

$$\dot{N}(t) = (B(t)-M(t))N(t) \tag{4}$$

with

$$B(t) = \int_0^1 \tilde{u}(t,x)(\int_0^x b(x,z)dz)dx \qquad M(t) = \int_0^1 m(x)\tilde{u}(t,x)dx. \tag{5}$$

$B(t)$ and $M(t)$ are the average birth and mortality rates of the total population at time t (recall that b and m are individual rates). $B(t)$ and $M(t)$ change if and only if the rank distribution of the population changes. So population increase changes the rank distribution in such a way that $B(t)/M(t)$ decreases until $B(t) \leqslant M(t)$. The mechanism changing the rank distribution is provided by the transition from lower to higher ranks which is strongly impeded at high population densities so that individuals accumulate at low ranks undergoing high mortality and low reproduction and the population stops increasing.

In order to understand how rank structures can not only limit populations, but even drive them to stable states we must leave intuition and go back to mathematics. Let us first make precise our intuitive ideas about what happens on the level of the individual.

ASSUMPTIONS 1.3.

a) There exists some $\epsilon>0$ such that $g(x,y,u)$ strictly decreases in $u \geqslant 0$ if $0 \leqslant x \leqslant y \leqslant 1$, $|x-y|<\epsilon$.

b) $g(x,y,u)\downarrow 0$ for $u\uparrow \infty, 0 \leqslant x \leqslant y \leqslant 1$.

c) $b(y,0)>0$ for some $0<y<1$.

d) $\int_x^y b(y,z)dz \,/\, m(y)$ is a monotone non-decreasing function of $y \in [x,1]$ for $0 \leqslant x \leqslant 1$.

e) $g(x,y,u)/m(x)$ is a monotone non-decreasing function of $x \in [0,y]$ for $0 \leqslant y \leqslant 1, u \geqslant 0$.

These assumptions are supposed to hold from now on throughout this paper with the exception of section 2. In their full strength, however, we only use them in the sections 6 and 7. Since the sections 3,4,5 contain results of own interest, we mention there how the assumptions 1.3 can be weakened. But next let us explain their meaning:

The assumptions a) and b) sharpen assumption 1.1c) that individuals impede lower ranking ones getting at their rank; if a rank is overcrowded it is impossible to get there.

Assumption c) guarantees that at least some newborns have the minimum rank 0.

To interpret d) and e) we recall that $1/m(y)$ is the average life expectancy of an individual staying at rank y from birth to death. So $\int_x^y b(y,z)dz \,/\, m(y)$ is the expected number of offspring of rank between x and y produced from birth to death by an individual staying at rank y throughout its life.

Assumption d) states that higher ranking individuals do not only produce more offspring on the whole in their lives but also more higher ranking offspring than lower ranking individuals do. Assumption e) means that the chance of an individual born at rank x to jump directly to rank $y>x$ before it dies increases with x. Assumption d) makes good biological sense, actually it is the justification of speaking about high and low ranks (see [5],p.287). Assumption

e) seems reasonable for dominance hierarchies. In territorially organised populations individuals with poor or even no territories (the so-called 'floaters') should be very eager to settle in a good territory because they have nothing to loose, whereas individuals living in a relatively good territory might only try to improve if the success is guaranteed. On the other hand mortality is much higher in the poor than in the rich territories, but it is questionable whether this can compensate for the first effect such that assumption e) holds. See [5],II.12.

After these explanations let us write down in what respect rank structures stabilize populations.

THEOREM 1.4 *Let the assumptions 1.1 and 1.3 be satisfied. Then the following alternative holds. Either*

a) *all solutions* $u(t,x)$ *of (1) converge to zero for* $t\to\infty$ *uniformly in* $0\leqslant x\leqslant 1$, *or*

b) *there exists a uniquely determined equilibrium solution* $\mathcal{U}$ *to (1),* $\mathcal{U}\not\equiv 0$, *such that*

$$u(t,x)\to\mathcal{U}(x),\quad t\to\infty,$$

uniformly in $0\leqslant x\leqslant 1$, *for any solution* u *of (1),* $u\not\equiv 0$.

So either all populations of the species under consideration (provided they live under indentical conditions) die out or they are all attracted by a uniquely determined equilibrium state. It is of interest, of course, to elucidate the conditions under which the first or the second alternative occurs. Mathematically this will amount to whether the zero equilibrium solution has linearized local asymptotic stability or not, or equivalently whether the spectral radius of a linear positive operator associated with the linearization of (1) around zero does not or does exceed one. Here the theory of compact strongly positive operators on Banach lattices (see, e.g.[4]) will be useful. See section 3. Unfortunately there is no biologically meaningful way of expressing this *threshold condition* in terms of b,m,g. A rough qualitative feeling of what the threshold condition means biologically can be obtained by considering (1) as a two parameter problem. To this end we set

$$b(y,x) = \beta b_0(y,x) \qquad g(x,y,u) = \gamma g_0(x,y,u) \tag{6}$$

and normalize b_0 and g_0 such that

$$\int_0^1 b_0(1,z)dz \;/\, m(1) = 1 \tag{7}$$

$$\int_0^1 g(0,y,0)dy \;/\, m(0) = 1. \tag{8}$$

So β indicates the expected number of offspring which an individual can produce living at the highest rank, i.e. under optimum reproductive conditions, throughout its life. γ is a measure for the permeability of the rank structure. So we call β the reproductive potential and γ the permeability of the population. One could also introduce a mortality parameter but this one can be absorbed into β and γ.

We obtain the following result.

Theorem 1.5 *Let the assumptions 1.1 and 1.3 be satisfied with b,g being replaced by* b_0, g_0. *Then there exists a continuous strictly decreasing function* $\Phi(\gamma),\gamma>0$, *with* $\lim_{\gamma\to 0}\Phi(\gamma) = \infty$ *and* $\lim_{\gamma\to\infty}\Phi(\gamma) = 1$ *such that the following holds:*

a) *If* $0\leqslant\beta\leqslant\Phi(\gamma)$, *then all solutions* $u(t,x)$ *of (1) converge to zero for* $t\to\infty$ *uniformly in* $0\leqslant x\leqslant 1$.

b) *If* $\beta>\Phi(\gamma)$, *then there exists a uniquely determined equilibrium solution* $\mathcal{U}$ *of (1),* $\mathcal{U}\not\equiv 0$, *such that*

$$u(t,x)\to\mathcal{U}(x) \quad \text{for} \quad t\to\infty$$

uniformly in $0\leqslant x\leqslant 1$ *for any solution* u *of (1),* $u\not\equiv 0$.

In particular the population dies out if $\beta\leqslant 1$, respectless of the values of γ. This is intuitively clear because $\beta\leqslant 1$ means that even individuals living at the highest rank, i.e. under optimal reproductive conditions, throughout their life have at most one descendant on the average. If $\beta>1$ the population stabilizes at a non-trivial equilibrium, if the permeability γ is large enough, i.e. if, at low population densities, the individuals can get quickly enough to the highest rank. In general a low permeability γ requires a large reproductive potential β and vice versa in order to prevent the population from extinction.

The last two theorems show that at least in principle, rank structures can regulate a population living under temporally constant conditions to a stable equilibrium state. Exterior regulatory factors like resources, predators, infectious diseases etc. usually operate with a time delay and so have the tendency to drive the population into oscillations. At the bottom of large oscillations the population runs the risk to go extinct by random effects. An interior regulation of the population by a social structure can avoid or at least decrease this risk if it is not delayed itself and if it limits the growth of the population before exterior regulatory factors come into operation. The first condition is implicitely contained in our model by setting $\tilde{g}(x,y,t) = g(x,y,u(t,y))$ which involves that jumping to rank y at time t is impeded by the individuals staying at y at that very time t independently of how many individuals were there before and how many resources are available. The second condition is contained in letting the birth and death rates b and m neither directly nor indirectly depend on the density u of the population. In nature both conditions are probably not satisfied completely but our results indicate the potential of social structures to buffer the fluctuations caused by exterior regulatory factors. (See GURNEY and NISBET [6], e.g., who, in a quite different model, study the stabilizing effects of social hierarchies in predator populations in which higher ranking individuals have a higher feeding success than lower ranking ones.) So our mathematical results confirm the intuitive understanding of biologists concerning the regulatory role of social structures (see [5],p.59,p.274).

2. Existence and uniqueness of solutions

In order to prove theorem 1.2 concerning the existence and uniqueness of solutions to (1) we write (1) in the form

$$\dot{u}(t,x) = f(x,u(t,x),u(t,\cdot)) \tag{20}$$

with f being given by the right hand side of (1). f is a map from $[0,1]\times[0,\infty)\times X_+$ to $\mathbb{R}$. $X = C[0,1]$ is the Banach lattice of continuous real-(or complex-) valued functions on $[0,1]$ and $X_+ = C_+[0,1]$ the cone of non-negative functions. In particular X is ordered by the pointwise ordering.

f has the following properties by the assumptions 1.1.

LEMMA 2.1.

a) *f is continuous.*

b) $f(x,0,0) = 0$.

c) $f(x,0,v) \geq 0$ for $v \in X_+$.

d) *$f(x,u,v)$ non-increases in $u \in [0,\infty)$ and non-decreases in $v \in X_+$.*

e) *$f(x,u,v)$ is continuous in x uniformly for u in bounded subsets of $[0,\infty)$ and v in bounded subsets of X_+.*

f) *$f(x,u,v) \leq c\|v\|$ with $\|\cdot\|$ denoting the sup-norm on X.*

The crux of the proof is this: Since we do not assume that $g(x,y,u)$ satisfies some Lipschitz condition in u, $f(x,u,v)$ is Lipschitz continuous neither in u nor in v. This excludes the standard application of Banach's fixed point theorem. Further the form of (20) does not suggest an obvious transformation to a fixed point equation with a compact (or condensing) operator and therefore an immediate application of Schauder's or Tychnoff's fixed point theorem is not possible. An appropriate combination of Banach's fixed point theorem, compactness arguments and the monotonicity properties of f, however, will settle the problem. We start with an a-priori estimate of solutions to (20) which will also be helpful for studying the asymptotic behaviour of solutions in theorem 1.4 (see lemma 5.3). Let u be a solution of (20). Then, since $f(x,u,v)$ decreases in u,

$$\frac{d}{dt}|u(t,x_1)-u(t,x_2)| = sign(u(t,x_1)-u(t,x_2))\frac{d}{dt}(u(t,x_1)-u(t,x_2)) \tag{21}$$
$$\leq |f(x_1,u(t,x_1),u(t,\cdot))-f(x_2,u(t,x_1),u(t,\cdot))| - |f(x_2,u(t,x_1),u(t,\cdot))-f(x_2,u(t,x_2),u(t,\cdot))|$$

Here the derivative at the left hand side of (21) has to be considered the generalized derivative of an absolutely continuous function. Integrating (21) with respect to t yields

$$|u(t,x_1)-u(t,x_2)| \leq |u(0,x_1)-u(0,x_2)| + \int_0^t |f(x_1,u(s,x_1),u(s,\cdot))-f(x_2,u(s,x_1),u(s,\cdot))|\,ds$$

and shows that the continuity of u in x only depends on the continuity of $u(0,\cdot)$ and $f(\cdot,u,v)$.

In a next step we approximate g by g_ϵ ,$\epsilon>0$,

$$g_\epsilon(y,x,u) = \frac{1}{\epsilon}\int_0^\epsilon g(y,x,u+v)dv,$$

and consider the approximating equations

$$\dot{u}_\epsilon(t,x) = f_\epsilon(x,u_\epsilon(t,x),u_\epsilon(t,\cdot)) \qquad u_\epsilon(0,x) = u_0(x). \tag{23}$$

Here f_ϵ is given by the right hand side of (1) with g being replaced by g_ϵ . Note that lemma 2.1 remains valid for f_ϵ with e) and f) holding uniformly in $0\leq\epsilon\leq1$. The reason for this approximation is the Lipschitz continuity of $f_\epsilon(x,u,v)$ in u and v which allows to find non-negative continuous solutions u_ϵ of (23) on $[0,\infty)\times[0,1]$ by a standard application of Banach's fixed point principle. Since lemma 2.1 f) holds for f_ϵ uniformly in $0\leq\epsilon\leq1$, the family $\{u_\epsilon;0\leq\epsilon\leq1\}$ is equibounded on $[0,t]\times[0,1]$ for any $t>0$. As (21) and (22) also hold for f_ϵ instead of f and lemma 2.1 e) holds for f_ϵ uniformly in $0\leq\epsilon\leq1$, we realize that $\{u_\epsilon;0\leq\epsilon\leq1\}$ is equicontinuous on $[0,t]\times[0,1]$ for any $t>0$. Thus, by Ascoli's theorem (see [3],IX,§ 4,e.g.), for any $t>0$ we find a subsequence of $\{u_\epsilon\}$, converging for $\epsilon\downarrow0$ uniformly on $[0,t]\times[0,1]$. By a diagonalization procedure we find a continuous non-negative function u and a subsequence of $\{u_\epsilon\}$ converging towards u for $\epsilon\downarrow0$ uniformly on any compact subset of $[0,\infty)\times[0,1]$. Integrating (23) and taking the limit for $\epsilon\downarrow0$ shows that u is a solution of (20). This completes the existence proof.

In order to show the uniqueness of solutions to (1) we derive a comparison principle which will also be useful for studying the asymptotic behaviour. Recall

$$\frac{d^+}{dt}u(t,x) = \limsup_{h\downarrow0}\frac{1}{h}(u(t+h,x)-u(t,x))$$

$$\frac{d_+}{dt}u(t,x) = \liminf_{h\downarrow0}\frac{1}{h}(u(t+h,x)-u(t,x))$$

and the following definition of lower and upper solutions of (1).

DEFINITION 2.2. *A continuous non-negative function u on* $[0,\infty)\times[0,1]$ *is called a lower solution of (1) iff*

$$\frac{d^+}{dt}u(t,x) \leq f(x,u(t,x),u(t,\cdot)).$$

It is called an upper solution of (1) iff

$$\frac{d_+}{dt}u(t,x) \geq f(x,u(t,x),u(t,\cdot)).$$

Here again f is the right hand side of (1).

We can now formulate the following comparison principle which immediatly implies the uniqueness of solutions to (1).

PROPOSITION 2.3. *Let* u_1,u_2 *be a lower and an upper solution of (1) respectively,* $u_1(0,x)\leq u_2(0,x)$ *for* $0\leq x\leq1$. *Then* $u_1(t,x)\leq u_2(t,x)$ *for* $t\geq0$, $0\leq x\leq1$.

PROOF. Let us first consider the case that u_1 is not only a lower solution, but a solution of (1). Set $\tilde{u}(t,x) = min(u_1(t,x),u_2(t,x))$. Since $f(x,u,v)$ increases as v increases, $\tilde{u}$ is an upper solution, too. Set $u = u_1-\tilde{u}$. Integration over x from 0 to 1 yields that

$$\int_0^1\frac{d^+}{dt}u(t,x)dx\leq\int_0^1 u(t,x)(\int_0^x b(x,z)dz)dx - \int_0^1 m(x)u(t,x)dx.$$

(Recall the derivation of (3).). It follows from the definition of lower and upper solutions that Fatou's lemma can be applied. Since $u\geq0$ we find some $c>0$ such that

$$(\frac{d^+}{dt}-c)\int_0^1 u(t,x)dx\leq0.$$

As $u(0,\cdot) = 0, u(t,\cdot) = 0$ for $t \geq 0$. Thus $u_1 = min(u_1, u_2) \leq u_2$. The case that u_2 is a solution is treated analogously by regarding $\underset{\sim}{u} = max(u_1, u_2)$ and $u = \underset{\sim}{u} - u_2$. In the case that neither u_1 nor u_2 are solutions we take a solution u of (1) with $u(0,\cdot) = u_1(0,\cdot)$. Then, by the preceding considerations, $u_1 \leq u \leq u_2$.

Imposing a Lipschitz condition on $g(x,y,u)$ in u would allow to prove theorem 1.2 by Banach's fixed point theorem and to show existence and uniqueness at one blow. One could even dispense with the monotonicity assumption 1.1 c) in such an approach. We took the other route for several reasons. The monotonicity asumption 1.1c) (decrease of $g(x,y,u)$ in u) is biologically interpretable whereas Lipschitz continuity is a technical condition which might be an unnecessary restriction. Further the proof by Banach's fixed theorem alone is standard and it was a nice mathematical exercise to do it without. Finally the a-priori estimate (21) and the comparison principle in proposition 2.3 will also be needed for the proof of theorem 1.4. (21) provides uniform instead of pointwise convergence of $u(t,x)$ for $t \to \infty$. See lemma 5.3.

3. The linearization around zero

Usually the local asymptotic stability of a trivial equilibrium solution is of some importance for the asymptotic behaviour of a dynamical system often deciding whether something interesting happens or not. In our case this amounts to the question whether the linearization of (1) around zero has solutions $u(t,x) = e^{\lambda t} U(x)$ with the real part of λ exceeding 0 or not. We will pose this problem in a different way, namely whether the spectral radius of the following operator K exceeds 1 or not:

$$(KU)(x) = \frac{\int_x^1 U(y) b(y,x) dy + \int_0^x U(y) g(y,x,0) dy}{m(x) + \int_x^1 g(x,y,0) dy} \tag{30}$$

$U \in C[0,1], 0 \leq x \leq 1$. The Arzela-Ascoli theorem tells that K is a compact linear operator on the Banach space $X = C[0,1]$. Moreover it is a positive operator on the Banach lattice X mapping the cone $X_+ = C_+[0,1]$ of non-negative functions into itself. The following notations will be useful for $u,v \in X$:

$$\begin{aligned} &u \leq v \quad \textit{iff} \quad u(x) \leq v(x), \quad 0 \leq x \leq 1, \\ &u < v \quad \textit{iff} \quad u \leq v, u(x) \neq v(x) \text{ for } \textit{some } x, \\ &u << v \quad \textit{iff} \quad u(x) < v(x), \quad 0 \leq x \leq 1. \end{aligned} \tag{31}$$

In the next lemma 3.1 we draw some easy consequences from the assumptions 1.3 a),c),d). We emphazise that the results of this section hold if we take the results of lemma 3.1 as assumptions instead of the assumptions 1.3.

LEMMA 3.1

a) There exists some $\epsilon > 0$ such that $g(y,x,0) > 0$ if $0 \leq y \leq x \leq 1, |y-x| < \epsilon$.

b) *For any* $0 \leq x < 1$ *there exist* z,y *such that* $0 \leq z \leq x < y \leq 1$ *and* $b(y,z) > 0$.

PROOF

a) follows from assumption 1.3a)

b) Let $0 \leq x < 1$. If $x = 0$, there is $y > 0$ such that $b(y,0) > 1$ by assumption 1.3c). If $x > 0$ and

$$\int_0^x b(x,z) dz = 0 \quad \textit{then} \quad \int_0^{\tilde{x}} b(\tilde{x},z) dz = 0$$

for $0 \leq \tilde{x} \leq x$, by assumption 1.3d). As b is continuous, $b(\tilde{x},0) = 0$ for $0 \leq \tilde{x} \leq x$. By assumption 1.3c) there is some $y > 0$ such that $b(y,0) > 0$. Necessarily $y > x$. If

$$\int_0^x b(x,z) dz > 0,$$

then $b(x,z) > 0$ for some $z < x$ and, since b is continuous, $b(y,z) > 0$ for some y,z with $z < x < y$.

The results of lemma 3.1 imply strong positivity properties for the operator K.

PROPOSITION 3.2. *For any $u\in X, u>0$ there exists $j\in\mathbb{N}$ such that $K^j u>>0$.*

In order to show this proposition we first draw an easy consequence from lemma 3.1a).

LEMMA 3.3 *There exists $j\in\mathbb{N}$ such that $(K^j u)(y)>0$ for all $y\in[x, 1]$, if $u\in X_+$, $0\leqslant x\leqslant 1$ and $u(x)>0$.*

PROOF OF PROPOSITION 3.2: Let $u\in X, u>0$. Further let x_0 be the greatest lower bound of those $0\leqslant x\leqslant 1$ such that $(K^j u)(y)>0$ for all $y\in(x, 1]$ with some j possibly depending on x, but not on y. $x_0<1$ by lemma 3.3. By lemma 3.1b) we find z,y such that $0\leqslant z\leqslant x_0<y\leqslant 1$ and $b(y,z)>0$. By definition of x_0, $(K^j u)(y)>0$ for some $j\in\mathbb{N}$. Hence $(K^{j+1}u)(z)>0$. Thus $(K^i u)(x)>0$ for all $x\geqslant x_0$, with some $i>j$ not depending on x, by lemma 3.3. In particular $x_0 = 0$ because otherwise we get a contradiction to the definition of x_0 from the continuity of $K^i u$.

By proposition 3.2 every $u\in X, u>0$ is mapped into the open interior of X_+ by some power of K. In particular K is irreducible and has a non-zero spectral radius. So theorem 5.2 and proposition 5.6 in [4],V., hold:

PROPOSITION 3.4. *Let r denote the spectral radius of K. Then the following holds:*

a) $r>0$.

b) *r is a simple eigenvalue of K with an eigenvector $\tilde{U}>>0$.*

c) *r is the only eigenvalue with an eigenvector in X_+.*

d) *r is strictly larger than the moduli of all other eigenvalues.*

e) *r has a strictly positive eigenfunctional $U'\in X'$, i.e. $U'u>0$ for $u>0$.*

It follows easily from proposition 3.4 that the linearization of (1) around zero has a solution $u(t,x) = e^{\lambda t}U(x)$ with Re $\lambda>0$ iff $r>1$. Further we can draw the first information on the relation between the spectral radius r of K and the rates b,m,g.

PROPOSITION 3.5. *If $\int_0^y b(y,x)dx \,/\, m(y)\leqslant 1$ for all $0\leqslant y\leqslant 1$, then $r\leqslant 1$.*

PROOF. Let $r\geqslant 1$. By proposition 3.4 we find $\tilde{U}\in X, \tilde{U}>>0$ such that

$$rm(x)\tilde{U}(x)\leqslant \int_x^1 \tilde{U}(y)b(y,x)dy \;+\; \int_0^x \tilde{U}(y)g(y,x,0)dy \;-\int_x^1 \tilde{U}(x)g(x,y,0)dy.$$

Integrating from 0 to 1 yields

$$r\int_0^1 m(x)\tilde{U}(x)dx\leqslant \int_0^1\Big(\int_x^1 \tilde{U}(y)b(y,x)dy\Big)dx.$$

Changing the order of integration in the second integral and using the assumption of this proposition yields $r\leqslant 1$. So $r\geqslant 1$ is only possible if $r=1$. This proves the proposition.

4. Existence of non-trivial equilibrium solutions

Obviously $u\equiv 0$ is an equilibrium solution of (1), i.e. a solution not depending on t. We are interested in whether there exist equilibrium solutions $U>0$ because these are the candidates for being non-trivial asymptotic states of the population. So we look for solutions U of

$$f(x,U(x),U) = 0, \quad 0\leqslant x\leqslant 1 \tag{40}$$

with $U\in X = C[0,1]$, $U>0$, and f being given by the right hand side of (1.1) and by (1.2). Our procedure is similar to that in section 2. Now it is easier, however, to consider (40) in the form

$$m(x)U(x) = \tilde{f}(x, U(x), U), \quad 0 \leqslant x \leqslant 1 \tag{41}$$

with $\tilde{f}$ having the properties in lemma 2.1. We want to transform (41) into an operator equation with a monotone increasing operator. To this end we fix $V \in X_+$ and consider the equation

$$m(x)U(x) = \tilde{f}(x, U(x), V), \quad 0 \leqslant x \leqslant 1. \tag{42}$$

The properties of $\tilde{f}$ (see lemma 2.1) imply that, for any $0 \leqslant x \leqslant 1$, there is a unique $U(x)$ solving (42) and that $U(x)$ depends on V in a monotone non-decreasing way. Similarly as in (21) we derive an a-priori estimate for solutions of (42). Let

$$m(x_i)U_i(x_i) = \tilde{f}(x_i, U_i(x_i), V_i), \tag{43}$$

$i = 1,2$, $0 \leqslant x_i \leqslant 1$, $V_i \in X_+$. Then, since $\tilde{f}(x,u,V)$ non-increases in u,

$$|U_1(x_1) - U_2(x_2)|$$

$$\leqslant \left|\frac{1}{m(x_1)}\tilde{f}(x_1, U_1(x_1), V_1) - \frac{1}{m(x_2)}\tilde{f}(x_2, U_1(x_1), V_1)\right| + \frac{1}{m(x_2)}|\tilde{f}(x_2, U_2(x_2), V_1) - \tilde{f}(x_2, U_2(x_2), V_2)| \tag{44}$$

(43), (44) implies that the solution $U(x)$ to (42) is continuous in x. So, by defining

$$\Psi(V) = U \tag{45}$$

with U being the solution of (42), we obtain a monotone increasing operator Ψ on X_+, i.e. $V_1 \leqslant V_2$ implies $\Psi(V_1) \leqslant \Psi(V_2)$. Fixed points of Ψ are solutions of (41), i.e. equilibrium solutions to (1). (43), (44) imply that Ψ is a compact continuous operator on X_+. See lemma 2.1e) and apply the Arzela-Ascoli-theorem. Moreover $m(x)U(x) \leqslant \tilde{f}(x, U(x), U)$, $0 \leqslant x \leqslant 1$, implies $U \leqslant \Psi(U)$ and the same implication holds for $\geqslant$.

It is convenient to make the following

Definition 4.1. $U \in X_+$ is called a *lower equilibrium solution* of (1) (or a lower solution of (40) or of (41)) iff

$$m(x)U(x) \leqslant \tilde{f}(x, U(x), U) \quad ,0 \leqslant x \leqslant 1. \tag{46}$$

It is called an *upper solution* iff (46) holds with $\geqslant$ instead of $\leqslant$.

The following result is now almost obvious.

Proposition 4.2. *Let $U_0, V_0 \in X_+$ be a lower and an upper equilibrium solution of (1), $U_0 \leqslant V_0$. Then there exist equilibrium solutions U, V of (1) such that $U_0 \leqslant U \leqslant V \leqslant V_0$ and $U \leqslant W \leqslant V$ for any other equilibrium solution W of (1) with $U_0 \leqslant W \leqslant V_0$.*

Proof. Actually the definitions $U_{n+1} = \Psi(U_n)$, $V_{n+1} = \Psi(V_n)$ provide monotone sequences $U_0 \leqslant U_1 \leqslant \cdots \leqslant U_n \leqslant V_n \leqslant \cdots V_1 \leqslant V_0$ with $U_n \leqslant W \leqslant V_n$ for any n and any equilibrium solution W of (1) with $U_0 \leqslant W \leqslant V_0$. $U = \lim_{n\to\infty} U_n$ and $V = \lim_{n\to\infty} V_n$ satisfy the statement of the proposition.

In order to find non-trivial equilibrium solutions of (1) we look for lower and upper solutions of (46) with $U > 0$. It is relatively easy to find upper solutions and the proof is left to the reader. Only the assumptions 1.1 and 1.3.b) are needed.

Proposition 4.3. $\tilde{U}(x) = ce^{-\mu x}$, $0 \leqslant x \leqslant 1$, *is an upper equilibrium solution of (1), i.e. an upper solution of (46), if μ, c are sufficiently large.*

Finding a lower equilibrium solution is closely related to the question whether the spectral radius r of the operator K in (30) exceeds 1. If $r > 1$ choose an eigenvector $\underset{\sim_0}{U} >> 0$ of K belonging to the eigenvalue r and make the 'Ansatz'

$$\underset{\sim}{U} = \epsilon \underset{\sim_0}{U}$$

with $\epsilon>0$ being small. Since $r>1$ and $\underset{\sim}{U}_0>>0$, $\underset{\sim}{U}$ is a lower solution of (46) if ϵ is small enough. This result also holds if we take lemma 3.1 as an assumption instead of assumption 1.3.

PROPOSITION 4.4. *Let the spectral radius r of K in (30) satisfy $r>1$. Then there exists $\underset{\sim}{U}_0\in X$, $\underset{\sim}{U}_0>>0$ such that $\underset{\sim}{U}=\epsilon\underset{\sim}{U}_0$ is a lower equilibrium solution of (1) if $\epsilon>0$ is small enough.*

Let us combine the propositions 4.2, 4.3 and 4.4 to the following result on the existence of non-trivial equilibrium solutions of (1). It also holds if we take lemma 3.1 and assumption 1.3b) as assumptions instead of all the assumptions 1.3.

THEOREM 4.5 *Let the spectral radius r of K in (30) satisfy $r>1$. Then there exists an equilibrium solution U of (1), $U>>0$.*

It will turn out in section 6 that under the assumptions 1.1 and 1.3 $r>1$ is also necessary for the existence of non-trivial equilibria.

5. Preliminary results on the asymptotic behaviour of solutions

In this section we want to state some results which one can obtain taking lemma 3.1 and assumption 1.3b) as assumptions instead of all the assumptions 1.3. We start with a boundedness result for solutions to (1) which follows from proposition 4.3 and the comparison principle in proposition 2.3.

PROPOSITION 5.1 *Any solution u of (1) is bounded on* $[0,\infty)\times[0,1]$.

The next result gives the first rough idea how the existence of non-trivial equilibrium solutions is related to the asymptotic behaviour of solutions of (1).

PROPOSITION 5.2 *Let u be a solution of (1). Then there exists an equilibrium solution U of (1) such that*

$$\limsup_{t\to\infty} u(t,x)\leqslant U(x),\ 0\leqslant x\leqslant 1.$$

If $U\equiv 0$ is the only equilibrium solution to (1), then $u(t,x)\to 0$ for $t\to\infty$ even uniformly in $0\leqslant x\leqslant 1$.

PROOF. By proposition 5.1 u is bounded. We define $\bar{u}(x)=\limsup_{t\to\infty} u(t,x)$ and claim that

$$f(x,\bar{u}(x),\bar{u})\geqslant 0,\quad 0\leqslant x\leqslant 1, \tag{50}$$

with f being given by the right hand side of (1.1) and by (1.2), i.e. $\bar{u}$ is a lower equilibrium solution of (1). To see (50) fix x and choose $t_n\to\infty$ $(n\to\infty)$ such that $u(t_n,x)\to\bar{u}(x)$ and $\liminf_{n\to\infty}\dot{u}(t_n,x)\geqslant 0$. Then

$$0\leqslant\limsup_{n\to\infty} f(x,,u(t_n,x),u(t_n,\cdot)).$$

Applying Fatou's lemma to the integral terms in f yields (50). Proposition 4.2 and 4.3 now imply the existence of an equilibrium solution $U\geqslant\bar{u}$. If 0 is the only equilibrium solution, $\bar{u}\equiv 0$. The uniform convergence of $u(t,x)$ towards 0 for $t\to\infty$ now follows from the subsequent

LEMMA 5.3. *Let u be a bounded solution of (1) such that $u(t,x)$ converges towards a function $U(x)$ for $t\to\infty$ pointwise in $0\leqslant x\leqslant 1$. Then U is continuous and the convergence is uniform in x.*

PROOF. We recall that $f(x,u,v)=\tilde{f}(x,u,v)-m(x)u$ with $\tilde{f}$ also having the properties in lemma 2.1. We first prove that U is continuous. Specializing (21) yields

$$\frac{d}{dt}|u(t,x_1)-u(t,x_2)|\leqslant|\tilde{f}(x_1,u(t,x_1),u(t,\cdot))-\tilde{f}(x_2,u(t,x_1),u(t,\cdot))|\ -m(x)|u(t,x_1)-u(t,x_2)|. \tag{51}$$

Remember that $\tilde{f}(x,u,v)$ non-increases in u. Since u is bounded, this inequality implies that u is continuous in x uniformly for all $t\geqslant 0$. See lemma 1.2e) which also holds for $\tilde{f}$. Now $U(x)=\lim_{t\to\infty} u(t,x)$ is continuous.

Let us now suppose that the convergence is not uniform. Then we find $\epsilon>0$, $t_n\to\infty$ $(n\to\infty)$ and $0\leqslant x_n\leqslant 1$ such that

$$|u(t_n,x_n)-U(x_n)|\geqslant\epsilon \text{ for } n\in\mathbb{N}. \tag{52}$$

Choosing a subsequence we can assume that $x_n\to x$ for $n\to\infty$. Since the continuity of $u(t,x)$ in x is uniform for $t\geqslant 0$ we obtain $|u(t_n,x_n)-u(t_n,x)|\to 0$ for $n\to\infty$, hence $|u(t_n,x_n)-U(x)|\to 0$ for $n\to\infty$. The continuity of U yields a contradiction to (52).

We continue our discussion with a positivity result for solutions of (1). After the proof of proposition 3.2 it is not difficult to prove the forthcoming result and the proof is left to the reader. Note that the form of (1) implies that $u(t,x)>0$ involves $u(s,x)>0$ for $s\geqslant t$.

PROPOSITION 5.4. *Let u be a solution of (1),* $u(0,\cdot)>0$. *Then* $u(t,\cdot)>>0$ *for all* $t>0$.

In a next step we establish a further relation between the asymptotic behaviour of solutions of (1) and the existence of non-trivial equilibrium solutions, provided that the spectral radius of K in (30) strictly exceeds 1.

PROPOSITION 5.5. *Let the spectral radius r of the operator K in (30) satisfy* $r>1$. *Further let u be a solution of (1),* $u(0,\cdot)>0$ *Then there exists an equilibrium solution* $U>>0$ *of (1) with*

$$\liminf_{t\to 0} u(t,x)\geqslant U(x), \quad 0\leqslant x\leqslant 1.$$

In particular the population does not die out.

PROOF. By proposition 5.4, $u(1,\cdot)>>0$. By proposition 4.4 we find a strict lower equilibrium solution $\underset{\sim}{U}$ of (1) with $0<<\underset{\sim}{U}<<u(1,\cdot)$. The comparison principle in proposition 2.3 implies that $\underset{\sim}{U}<<u(t,\cdot)$ for all $t>1$. Let $\underline{u}(x)=\liminf_{t\to\infty}u(t,x)$. As in the proof of proposition 5.2 we find that $\underline{u}$ is an upper equilibrium solution of (1). Proposition 4.2 implies the existence of an equilibrium U with $\tilde{U}\leqslant U\leqslant\underline{u}$.

Combining proposition 5.2 and 5.5 and applying lemma 5.3 yields

THEOREM 5.6. *Let the spectral radius r of the operator K in (30) satisfy* $r>1$. *Further let u be a solution to (1),* $u(0,\cdot)>0$. *Then there exist equilibrium solutions* $U_2\geqslant U_1>>0$ *of (1) such that*

$$U_1(x)\leqslant\liminf_{t\to\infty}u(t,x)\leqslant\limsup_{t\to\infty}u(t,x)\leqslant U_2(x)$$

for $0\leqslant x\leqslant 1$. *If there is only one equilibrium solution* $U>0$ *to (1), then* $u(t,x)\to U(x)$ *for* $t\to\infty$, *uniformly in* $0\leqslant x\leqslant 1$.

So, in view of the global stability claimed in theorem 1.4, proposition 5.2 and 5.6 leave us with the task of finding conditions for non-existence and for uniqueness of non-trivial equilibrium solutions of (1). We mention the work by HIRSCH [1] on quasi monotone differential equations or, more general, on strongly monotone semiflows which indicate what kind of results one can still obtain if uniqueness of non-trivial equilibrium solutions cannot be proved.

6. Uniqueness and non-existence of non-trivial equilibrium solutions

The question adressed to in the heading is of interest for the qualitative discussion of any dynamical system. But, as we have seen in proposition 5.2 and 5.6, it has particularly far-reaching consequences for our model (as for many other quasi-monotone differential equations). Let us state the main result of this section. Now all the assumptions 1.1 and 1.3 are supposed to hold.

THEOREM 6.1. *Let K be the operator in (30) and r the spectral radius of K. Then the following holds*

a) *If* $r\leqslant 1$, *then* $U\equiv 0$ *is the only equilibrium solution of (1).*

b) *If $r>1$, then there exists a unique equilibrium solution $U>0$ of (1). Further $U>>0$.*

Proposition 5.2 and 5.6 now immediately imply theorem 1.4.

We start the proof of theorem 5.1 with specializing proposition 5.4 to equilbrium solutions.

LEMMA 6.2. *Let U be an equilibrium solution of (1). If $U>0$, then $U>>0$.*

We now give a rather detailed

PROOF OF THEOREM 6.1b),

because similar arguments will be used in the proof of part a) and in the proofs in chapter 7 which will only be sketched.

Let U_1, U_2 be two different equilibrium solutions of (1), $U_i>0$. By proposition 4.2, 4.3, and 4.4 and by lemma 6.2 we can assume that

$$0 \ll U_1 \leqslant U_2 \tag{60}$$

and then show $U_1 = U_2$. We suppose

$$U_1 < U_2. \tag{61}$$

We now use the following

LEMMA 6.3. *Let $0 \leqslant U_1 < U_2$ be two equilibrium solutions of (1). Then $U_1 \ll U_2$.*

Its proof is postponed in order not to interrupt the flow of arguments. In order to derive a contradiction we use a technique presumably going back to fKrasnosel'skii [2], 2.3.3, 6.1.3. We define

$$V_j(z) = \int_z^1 m(x)U_j(x)dx, \quad j = 1,2, \tag{62}$$

and

$$\xi = inf\left\{\frac{V_1(z)}{V_2(z)}\ ; \quad 0 \leqslant z < 1\right\}. \tag{63}$$

Take into account that

$$\frac{V_1(z)}{V_2(z)} \rightarrow \frac{U_1(1)}{U_2(1)} \quad \text{for } x \rightarrow 1. \tag{64}$$

The following three cases are possible:

case 1: $\xi = \dfrac{U_1(1)}{U_2(1)}$

case 2: $\xi = \dfrac{V_1(0)}{V_2(0)}$ and $V_1(z) > \xi V_2(z)$ for all $0 < z < 1$

case 3: $\xi = \dfrac{V_1(z)}{V_2(z)}$ for some $0 < z < 1$.

We show that each case leads to a contradiction.

case 1: $\xi = \dfrac{U_1(1)}{U_2(1)}$

By (1),

$$m(1)U_j(1) = \int_0^1 U_j(y)g(y,1,U_j(1))dy, \quad j = 1,2. \tag{65}$$

Since $U_1(1) < U_2(1)$ and $g(y,1,u)$ strictly decreases in u for some y by assumption 1.3a), we obtain that

$$m(1)U_1(1) > \int_0^1 U_1(y)g(y,1,U_2(1))dy.$$

Using that $U_j(y) = -\frac{1}{m(y)}\frac{d}{dy}V_j(y)$ by (62) and integrating by parts yields

$$m(1)U_1(1) > \int_0^1 V_1(y)d_y(g(y,1,U_2(1))/m(y)) + V_1(0)g(0,1,U_2(1)).$$

Note that, by assumption 1.3e, the Stieltjes integral makes sense and the following inequality holds

$$m(1)U_1(1) > \xi\int_0^1 V_2(y)d_y(g(y,1,U(1))/m(y)) + \xi V_2(0)g(0,1,U_2(1))$$

Integrating by parts again and using (65) for $j = 2$ yields

$$m(1)U_1(1) > \xi m(1)U_2(1),$$

a contradiction.

case 2: $\xi = \frac{V_1(0)}{V_2(0)}$ and $V_1(z) > \xi V_2(z)$ for all $0 < z < 1$.

Integrating (1) over x from 0 to 1 yields

$$V_j(0) = \int_0^1 U_j(y)(\int_0^y b(y,x)dx)dy = \int_0^1 V_j(y)d_y(\int_0^y b(y,x)dx / m(y)).$$

Note that, by assumption 1.3d) the Stieltjes integral makes sense. It follows from assumption 1.3c) that

$$V_1(0) > \xi\int_0^1 V_2(y)d_y(\int_0^y b(y,x)dx / m(y)) = \xi V_2(0),$$

a contradiction.

case 3: $\xi = \frac{V_1(z)}{V_2(z)}$ for some $0<z<1$.

Integrating (1) over x from z to 1 yields

$$V_j(z) = \int_z^1 U_j(y)(\int_z^y b(y,x)dx)dy + \int_0^z U_j(y)(\int_z^1 g(y,x,U_j(x))dx)dy. \tag{66}$$

As $U_1 \ll U_2$ and $g(y,x,u)$ strictly decreases in u, if $|x-y|$ is small, by assumption 1.3a) we obtain that

$$V_1(z) > \int_z^1 U_1(y)(\int_z^y b(y,x)dx)dy + \int_0^z U_1(y)(\int_z^1 g(y,x,U_2(x))dx)dy$$

Using $U_j(y) = -\frac{d}{dy}V_j(y)/m(y)$ and integrating by parts yields

$$V_1(z) > \int_z^1 V_1(y)d_y(\int_z^y b(y,x)dx / m(y)) + \int_1^z V_1(y)d_y(\int_z^1 g(y,x,U_2(x))dx / m(y))$$

$$+ V_1(0)\int_z^1 g(0,x,U_2(x))dx / m(0) - V_1(z)\int_z^1 g(z,,x,U_2(x))dx / m(z). := G_1(V_1)(z).$$

Note that, by assumption 1.3d),e) the Stieltjes integrals make sense and that

$$G(V_1) \geqslant G(\xi V_2) = \xi G(V_2).$$

Integrating by parts again and using (66) for $j=2$ now yields

$$V_1(z) > \xi V_2(z),$$

a contradiction.

This completes the proof of theorem 6.1b) up to the postponed

PROOF OF LEMMA 6.3.
As in section 2 and 4 we write (1) for U_i in the form

$$0 = f(x, U_i(x), U_i), \quad 0 \leqslant x \leqslant 1$$

with f having the properties in lemma 2.1. By assumption 1.3a) we have the following property in addition: If $U_2(x) > U_1(x)$, then $f(y, U_1(y), U_1) < f(y, U_1(y), U_2)$ for $x \leqslant y \leqslant x + \epsilon$. In particular $U_1(y) < U_2(y)$ for $x \leqslant y \leqslant x+\epsilon$ because $f(y,u,v)$ non-increases in u. $\epsilon>0$ is independent of x. A similar argumentation derives the following from assumptions 1.3c) and d) (which imply lemma 3.1b)). If $U_2(y) > U_1(y)$ for $y > x$, then there exists $z \leqslant x$ such that $U_2(z) > U_1(z)$. Combining these two arguments (see the proof of proposition 3.2) yields the lemma.

The *Proof of theorem 6.1a)* is quite similar to the proof of part b) and so we give only some hints. We suppose that there is an equilibrium solution $U_2 > 0$ of (1). By lemma 6.2, $U_2 >> 0$. On the other hand, by proposition 3.4, there exist $0 < r \leqslant 1$, $U_1 > 0$ such that

$$rU_1 = KU_1,$$

with the operator K in (30), which implies that

$$m(x)U_1(x) \geqslant \int_x^1 U_1(y)b(y,x)dy + \int_0^x U_1(y)g(y,x,0)dy - \int_x^1 U_1(x)g(x,y,0)dy.$$

Proceeding as in the proof of part b) we obtain a contradiction.

7. The Threshold Condition

As we have seen from proposition 5.2, theorem 5.6 and theorem 6.1 the populations either die out or tend to a non-zero equilibrium state in dependence on the threshold condition whether the spectral radius r of the operator K in (30) satisfies $r \leqslant 1$ or $r > 1$. Apart from special cases which are much too far from realism even conceptually the spectral radius of K cannot be expressed explicitly in terms of g,b and m. One could derive estimates, of course, which do not provide qualitative insight, however. Any quantiative information one might obtain would be a fake for a completely theoretical model like ours.

In order to obtain rough qualitative insight at least we study the dependence of the spectral radius of K on b,g,m as a two parameter problem via (6). One could introduce a third parameter which can be absorbed into the other two, however. So we consider operators $K_{\gamma,\beta}$ with $\gamma,\beta>0$ and $K_{\gamma,\beta}$ being defined by (30) and (6). Theorem 1.5 follows from proposition 5.2, theorem 5.6 and theorem 6.1 and the subsequent

LEMMA 7.1. *Let the assumptions 1.1 and 1.3 be satisfied with b and g being replaced by b_0 and g_0. Let $r_{\gamma,\beta}$ denote the spectral radius of the operator $K_{\gamma,\beta}$. Then the following holds:*

a) *$r_{\gamma,\beta}$ depends jointly continuously on $\gamma,\beta>0$.*
b) *$r_{\gamma,\beta} \leqslant 1$ for all $\gamma>0$, $\beta \leqslant 1$.*
c) *For every $\gamma>0$ there exists $\beta>0$ with $r_{\gamma,\beta}>1$.*
d) *$r_{\gamma,\beta} \to 0$ for $\gamma \downarrow 0$, $\beta>0$.*
e) *If $\beta>1$, $r_{\gamma,\beta}>1$ for sufficiently large γ.*
f) *$r_{\gamma,\beta} > r_{\gamma,\beta'}$ if $\beta>\beta'$.*
g) *If $r_{\gamma,\beta} \geqslant 1$, then $r_{\gamma',\beta}>1$ for $\gamma'>\gamma$. If $r_{\gamma,\beta} \leqslant 1$, then $r_{\gamma',\beta}<1$ for $\gamma'<\gamma$.*

a),b),c) imply the existence of a function Φ such that $r_{\gamma,\beta}=1$ for $\beta=\Phi(\gamma),\gamma>0$, via the intermediate value theorem, f) and g) imply the strict decrease and, together with a), the continuity of Φ; b),c),d),e) imply the asymptotic behaviour for $\gamma\to 0$ and $\gamma\to\infty$.

Proof. a) follows from proposition 3.4 and the continuous dependence of the compact operators $K_{\gamma,\beta}$ on $\gamma,\beta>0$ in the uniform operator topology.

b) follows from proposition 3.5 and the normalization (7).

c) The proof proceeds with the same technique as the proof of theorem 6.1: Suppose that there is some $\gamma>0$ such that $r_{\gamma\beta}\leqslant 1$ for all $\beta>0$. Fix γ. By proposition 3.5 there exist $U_{1,\beta}\gg 0$ such that

$$m(x)U_{1,\beta}(x)\geqslant \beta\int_x^1 U_{1,\beta}(y)b_0(y,x)dy+\gamma\int_0^x U_{1,\beta}(y)g_0(y,x)dy-\gamma\int_x^1 U_{1,\beta}(x)g_0(x,y,0)dy$$

A contradiction follows in the same way as in the proof of theorem 6.1 if, for large β, we can find a continuous function $U_2>0$ with the following properties (see the cases 1,2,3 in the proof of theorem 6.1b):

$$m(1)U_2(1)<\gamma\int_0^1 U_2(y)g_0(y,1,0)dy\,, \tag{71}$$

$$\int_z^1 m(x)U_2(x)dx<\beta\int_z^1 U_2(y)\Big(\int_z^y b_0(y,x)dx\Big)dy+\gamma\int_0^z U_2(y)\Big(\int_z^1 g_0(y,x,0)dx\Big)dy \tag{72}$$

for $0\leqslant z<1$.

We make the 'Ansatz'

$$m(x)U_2(x)=(1-x)^n$$

with $n\in\mathbb{N}$. Then (71) is obviously satisfied by assumption 1.3a). By assumption 1.3c),d) we choose $0<\delta<1$ such that

$$\int_z^y b_0(y,x)dx>0 \quad \text{for} \quad z\leqslant\delta, y>z, 1-\delta\leqslant y\leqslant 1. \tag{73}$$

In order to satisfy (72) for $\delta\leqslant z\leqslant 1$ we choose n such that

$$\frac{2}{n+1}\leqslant\frac{1}{1-z}\int_z^1\Big(\int_0^z g(y,x,0)\,/\,m(y)dy\Big)dx.$$

This is possible because the right hand side is positive for $0<z<1$ by assumption 1.3a) and converges to

$$\int_0^1 g(y,1,0)\,/\,m(y)dy>0 \quad \text{for} \quad z\to 1.$$

Note that the choice of n is independent of β. By (73) we can now choose β so large that (72) holds for $0\leqslant z\leqslant\delta$.

d) Fix $\beta>0$. It is sufficient to find $U_\gamma\gg 0$ with $K_{\gamma,\beta}\,U_\gamma\leqslant\delta(\gamma)U_\gamma$ and $\delta(\gamma)\to 0$ for $\gamma\to 0$. For then $K^n_{\gamma,\beta}U_\gamma\leqslant\delta(\gamma)^n U_\gamma$ and $\|K^n_{\gamma\beta}\|\leqslant c\,\delta(\gamma)^n$ for all n with some $c>0$ implying $r_{\gamma\beta}\leqslant\delta(\gamma)\to 0$ for $n\to\infty$. We make the 'Ansatz' $U_\gamma(x)=e^{-\mu x}$. By (30),

$$(K_{\gamma\beta}U_\gamma)(x)\,/\,U_\gamma(x)\leqslant\Big(\int_x^1 b(y,x)e^{\mu(x-y)}dy+\gamma\int_0^x e^{\mu(x-y)}g_0(y,x,0)dy\Big)\,/\,m(x).$$

By choosing μ large the first term on the right hand side of the inequality gets as small as we want and then by choosing γ small (in dependence of μ) the second term gets arbitrarily small.

e) Fix β. As in the proof of c) it is sufficient to find a continuous function $U_2\gg 0$ such that (71),(72) hold for large $\gamma>0$. We choose a continuous function U_2 with the following properties:

$$m(x)U_2(x)=\frac{1}{n^2},\quad 0\leqslant x\leqslant 1-\frac{1}{n}-\frac{1}{n^2}\,,$$

$$m(x)U_2(x)=1,\quad 1-\frac{1}{n}\leqslant x\leqslant 1,$$

$$\frac{1}{n^2}\leqslant m(x)U_2(x)\leqslant 1,\quad 1-\frac{1}{n}-\frac{1}{n^2}\leqslant x\leqslant 1-\frac{1}{n}\,.$$

Let $0 \leqslant z \leqslant 1/n$. Then

$$\int_z^1 m(x)U_2(x)dx \leqslant \frac{1}{n}(1+\frac{2}{n})$$

and

$$\beta \int_z^1 U(y)(\int_z^y b_0(y,x)dx)dy \geqslant \beta \int_{1-\frac{1}{n}}^1 (\int_z^y b_0(y,x)dx)dy.$$

Thus (72) holds for $0 \leqslant z \leqslant 1/n$ with n being large by the normalization (7) and $\beta > 1$ independently of the size of γ. In order to satisfy (71) we choose γ so large that

$$1 < \gamma(\frac{1}{n})^2 \int_0^1 g_0(y, 1,0) / m(y)dy$$

(72) is satisfied for $1/n \leqslant z < 1$, if γ is chosen so large that

$$1 < \gamma(\frac{1}{n})^2 \frac{1}{1-z} \int_z^1 (\int_0^z g_0(y,x, 0) / m(y)dy)dx.$$

See assumption 1.3a). Note that, for $z \to 1$, the last condition reduces to the last but one condition.

f) follows by a standard argument in positive operator theory because the operators $K_{\gamma,\beta}$ depend on β in a strictly monotone increasing way: from proposition 3.4 we obtain that

$$r_{\gamma\beta_i} U_i = K_{\gamma\beta_i} U_i, \quad i = 1,2,$$

with $U_i >> 0$, $\beta_2 > \beta_1$. Obviously $K_{\gamma\beta_2} U_{\beta_1} > K_{\gamma\beta_1} U_{\beta_1}$. Choosing a strictly positive eigenfunctional

$$r_{\gamma\beta_2} U_2' = K_{\gamma\beta_2}' U_2'$$

yields

$$r_{\gamma\beta_2} = \frac{U_2'(K_{\gamma\beta_2} U_1)}{U_2'(U_1)} > \frac{U_2'(K_{\gamma\beta_1} U_1)}{U_2'(U_1)} = r_{\gamma\beta_1}.$$

g) Unfortunately the operators $K_{\gamma\beta}$ do not depend on γ in a monotone way such that we cannot proceed as in the proof of f) but again have to employ the technique of the proof of theorem 6.1b). Let us suppose that $\gamma' > \gamma$ and $r_{\gamma',\beta} \leqslant 1 \leqslant r_{\gamma,\beta}$. Then proposition 3.4 provides $U_1, U_2 >> 0$ such that

$$m(x)U_1(x) \geqslant \int_x^1 U_1(y)b(y,x)dy + \gamma' \int_0^x U_1(y)g_0(y,x, 0)dy - \gamma' \int_x^1 U_1(x)g_0(x,y, 0)dy$$

$$m(x)U_2(x) \leqslant \int_x^1 U_2(y)b(y,x)dy + \gamma \int_0^x U_2(y)g_0(y,x, 0)dy - \gamma \int_x^1 U_2(x)g_0(x,y, 0)dy .$$

A contradiction is now obtained by proceeding in the same way as in the proof of theorem 6.1b). The second part of g) is proved analogously.

References

M.W. Hirsch, *Stability and convergence in stongly monotone flows,* Preprint PAM-245. Berkely.

M.A. Krasnosel'skii, *Positive solutions of operator equations,* Groningen: Noordhoff 1964.

S. Lang, *Analysis II,* Reading-Menlo Park-London-Don Mills: Addison Wesley 1969.

H.H. Schaefer, *Banach lattices and positive operators,* Berlin-Heidelsberg-New York: Springer 1974.

E.O. Wilson, *Sociobiology. The new synthesis,* Cambridge, Massachusetts, and London, England: The Belknapp Press of Harvard University Press 1975.

W.S.G. Gurney, R.M. Nisbet : *Ecological stability and social hierarchy.* Theor. Pop. Biol. 16 (1979), 48-80.

Journal of Mathematical Biology

ISSN 0303-6812 Title No. 285

Editorial Board: K. P. Hadeler, Tübingen; S. A. Levin, Ithaca (Managing Editors); H. T. Banks, Providence; J. D. Cowan, Chicago; J. Gani, Santa Barbara; F. C. Hoppensteadt, Salt Lake City; D. Ludwig, Vancouver; J. D. Murray, Oxford; T. Nagylaki, Chicago; L. A. Segel, Rehovot

For mathematicians and biologists working in a wide spectrum of fields, the **Journal of Mathematical Biology** publishes:

- papers in which mathematics is used for a better understanding of biological phenomena
- mathematical papers inspired by biological research, and
- papers which yield new experimental data bearing on mathematical models.

Contributions also discuss related areas of medicine, chemistry, and physics.

Fields of interest: Mathematics, genetics, demography, ecology, neurobiology, epidemiology, morphogenesis, cell biology, and other branches of biology.

Abstracted/Indexed in: Biosis, Current Contents, Excerpta Medica, Inspec, Math. Reviews, Medlars, Physics Briefs, SCI Abstracts, Technical Information Center/Energinfo, Zentralblatt für Mathematik.

For subscription information and sample copies, contact Springer-Verlag, Dept. ZSW, Heidelberger Platz 3, D-1000 Berlin 33, W. Germany

Springer-Verlag
Berlin Heidelberg New York
London Paris Tokyo

Springer

Bio-mathematics

Managing Editor: **S. A. Levin**

Editorial Board: **M. Arbib, H. J. Bremermann, J. Cowan, W. M. Hirsch, J. Karlin, J. Keller, K. Krickeberg, R. C. Lewontin, R. M. May, J. D. Murray, A. Perelson, T. Poggio, L. A. Segel**

Volume 17

Mathematical Ecology

An Introduction

Editors: **Th. G. Hallam, S. A. Levin**

1986. Approx. 87 figures. Approx. 495 pages
ISBN 3-540-13631-2

Contents: Introduction. - Physiological and Behavioral Ecology. - Population Ecology. - Communities and Ecosystems. - Applied Mathematical Ecology. - Subject Index.

Volume 16

Complexity, Language, and Life: Mathematical Approaches

Editors: **J. L. Casti, A. Karlqvist**

1986. XIII, 281 pages. ISBN 3-540-16180-5

Contents: Allowing, forbidding, but nor requiring: a mathematic for human world. - A theory of stars in complex systems. - Pictures as complex systems. - A survey of replicator equations. - Darwinian evolution in ecosystems: a survey of some ideas and difficulties together with some possible solutions. - On system complexity: identification, measurement, and management. - On information and complexity. - Organs and tools; a common theory of morphogenesis. - The language of life. - Universal principles of measurement and language functions in evolving systems.

Volume 15
D. L. DeAngelis, W. Post, C. C. Travis

Positive Feedback in Natural Systems

1986. 90 figures. Approx. 305 pages. ISBN 3-540-15942-8

Contents: Introduction. - The Mathematics of Positive Feedback. - Physical Systems. - Evolutionary Processes. - Organisms Physiology and Behavior. - Resource Utilization by Organisms. - Social Behavior. - Mutualistic and Competitive Systems. - Age-Structured Populations. - Spatially Heterogeneous Systems: Islands and Patchy Regions. - Spatially Heterogeneous Ecosystems; Pattern Formation. - Disease and Pest Outbreaks. - The Ecosystem and Succession. - References. - Appendices A to H.

Springer-Verlag
Berlin Heidelberg New York
London Paris Tokyo

Springer

Lecture Notes in Biomathematics

Vol. 58: C. A. Macken, A. S. Perelson, Branching Processes Applied to Cell Surface Aggregation Phenomena. VIII, 122 pages. 1985.

Vol. 59: I. Nåsell, Hybrid Models of Tropical Infections. VI, 206 pages. 1985.

Vol. 60: Population Genetics in Forestry. Proceedings, 1984. Edited by H.-R. Gregorius. VI, 287 pages 1985.

Vol. 61: Resource Management. Proceedings, 1984. Edited by M. Mangel. V, 138 pages 1985.

Vol. 62: B. J. West, An Essay on the Importance of Being Nonlinear. VIII, 204 pages. 1985.

Vol. 63: Carme Torras i Genís, Temporal-Pattern Learning in Neural Models. VII, 227 pages, 1985.

Vol. 64: Peripheral Auditory Mechanisms. Proceedings, 1985. Edited by J.B. Allen, J.L. Hall, A. Hubbard, S. Neely and A. Tubis. VII, 400 pages. 1986.

Vol. 65: Immunology and Epidemiology. Proceedings, 1985. Edited by G.W. Hoffmann and T. Hraba. VIII, 242 pages. 1986.

Vol. 66: Nonlinear Oscillations in Biology and Chemistry. Proceedings, 1985. Edited by H.G. Othmer. VI, 289 pages. 1986.

Vol. 67: P.H. Todd, Intrinsic Geometry of Biological Surface Growth. IV, 128 pages. 1986.

Vol. 68: The Dynamics of Physiologically Structured Populations. Edited by J.A.J. Metz and O. Diekmann. XII, 511 pages. 1986.

This series reports new developments in biomathematics research and teaching – quickly, informally and at a high level. The type of material considered for publication includes:

1. Original papers and monographs
2. Lectures on a new field or presentations of new angles in a classical field
3. Seminar work-outs
4. Reports of meetings, provided they are
   a) of exceptional interest and
   b) devoted to a single topic.

Texts which are out of print but still in demand may also be considered if they fall within these categories.

The timeliness of a manuscript is more important than its form, which may be unfinished or tentative. Thus, in some instances, proofs may be merely outlined and results presented which have been or will later be published elsewhere. If possible, a subject index should be included. Publication of Lecture Notes is intended as a service to the international scientific community, in that a commercial publisher, Springer-Verlag, can offer a wide distribution of documents which would otherwise have a restricted readership. Once published and copyrighted, they can be documented in the scientific literature.

Manuscripts

Manuscripts should be no less than 100 and preferably no more than 500 pages in length.
They are reproduced by a photographic process and therefore must be typed with extreme care. Symbols not on the typewriter should be inserted by hand in indelible black ink. Corrections to the typescript should be made by pasting in the new text or painting out errors with white correction fluid. The typescript is reduced slightly in size during reproduction; best results will not be obtained unless on each page a typing area of 18 x 26.5 cm (7 x 10½ inches) is respected. On request, the publisher can supply paper with the typing area outlined. Move detailed typing instructions are also available on request.

Manuscripts generated by a word-processor or computerized typesetting are in principle acceptable. However if the quality of this output differs significantly from that of a standard typewriter, then authors should contact Springer-Verlag at an early stage.

Authors of monographs receive 50 free copies; editors of proceedings receive 75 free copies: all authors are free to use the material in other publications.

Manuscripts should be sent to Prof. Simon Levin, Section of Ecology and Systematics, 345 Corson Hall, Cornell University, Ithaca, NY 14853-0239, USA; or directly to Springer-Verlag Heidelberg.

Springer-Verlag, Heidelberger Platz 3, D-1000 Berlin 33
Springer-Verlag, Tiergartenstraße 17, D-6900 Heidelberg 1
Springer-Verlag, 175 Fifth Avenue, New York, NY 10010/USA
Springer-Verlag, 37-3, Hongo 3-chome, Bunkyo-ku, Tokyo 113, Japan

---

ISBN 3-540-16786-2
ISBN 0-387-16786-2

RANDALL LIBRARY-UNCW
3 0490 0036882 0